MW01258988

ONLY
ASST $4.99 ZZNR FROM 712 9/02
$4.99
SORTED @ 7 9992 Bargain
7053824 R 475 92602 # ZZN" NR C
712-04D 662865 0000499REMAINDER
243 0

The Queen of Mathematics

A Historically Motivated Guide to Number Theory

The Queen of Mathematics

A Historically Motivated Guide to Number Theory

Jay R. Goldman
School of Mathematics
University of Minnesota
Minneapolis, Minnesota

A K Peters
Wellesley, Massachusetts

Editorial, Sales, and Customer Service Office

A K Peters, Ltd.
289 Linden Street
Wellesley, MA 02181

Author's address:
Professor Jay R. Goldman
School of Mathematics
University of Minnesota
Minneapolis, MN 55454
goldman@math.umn.edu

The photographs on the cover depict, from left to right: Leonhard Euler, Carl Friedrich Gauss, and Pierre de Fermat (front cover); Hermann Minkowski (spine); David Hilbert, Lejeune Dirichlet, and Richard Dedekind (back cover).

Library of Congress Cataloging-in-Publication Data

Goldman, Jay R.
The queen of mathematics: a historically motivated guide to number theory / Jay R. Goldman
p. cm.
Includes bibliographical references and index.
ISBN: 1-56881-006-7
1. Number theory. I. Title
QA241.G6295 1998 98-20017
512'.7--dc20 CIP

Printed in the United States of America
02 01 00 99 98 10 9 8 7 6 5 4 3 2 1

Dedicated to the memory of my mother and father,

Frieda and Joe Goldman

Preface

When I was in high school, I read Eric Temple Bell's *Men of Mathematics* which started my continuing fascination with the lives of mathematicians and the history of mathematics. Understanding the evolution of ideas leading to many modern mathematical definitions and theorems gives great insight into their significance and even greater insight into the goals and open questions of mathematical theories.

Number theory, which Gauss called the "Queen of Mathematics", is particularly susceptible to historical understanding. Starting in the fall of 1986, I combined my interests in number theory and history by teaching a one-year course in number theory from a historical perspective. The course used historical highlights to show the development of various topics, and I included some discussion about the lives of famous number theorists. My lecture notes for the course turned into a more formal set of notes, which turned into this book. Both the spirit and organization of material in the book remain those of my original lectures.

The book is not self-contained. I don't hesitate to state theorems without proof. This enables me to present a broad perspective and introduce important concepts, without the details needed by a specialist in the area. Sometimes such results are used to prove other theorems because the arguments involve new ideas.

The first part of the book (chapters 1–6) is devoted to the work of Fermat, Euler, Lagrange and Legendre. Together with Gauss they were the major players in founding modern number theory, and almost everything that is of current interest (quadratic forms, reciprocity laws, distribution of prime numbers, equations over finite fields, algebraic number theory, rational points on curves, ...) can be traced back to their ideas. Many of their discoveries are illustrated by example, leaving the systematic developments for later in the book. The only subject I cover in some depth is the theory of continued fractions, a topic with roots in ancient Greece, unknown to most mathematicians, yet very much in tune with the current interest in algorithmic methods.

Next we come to Gauss and his *Disquisitiones Arithmeticae*, certainly one of the most important mathematics books ever written. The *Disquisitiones*, published in 1801, is the first presentation of number theory in a rigorous and modern style. Gauss treated the divisibility theory of integers, introducing the idea of congruence and, even more importantly, he introduced the congruence notation. He reworked known developments, solved old problems such as finding the first proofs of the law of quadratic reciprocity, and added new ideas such as genus and composition for quadratic forms. He solved the ancient problem of finding criteria for the constructibility of regular polygons by straightedge and compass. This led to the theory of cyclotomy (circle division), a new branch of number theory.

In the second part of the book, I give a systematic introduction to the elements of number theory, following Gauss's organization in the *Disquisitiones*, but not including all of his details. After an overview of Gauss's work and comments on his life (chapter 7), there is an English translation of chapter 1 of the *Disquisitiones*, which introduces congruences together with their basic properties and some applications (chapter 8). Gauss wrote this chapter so clearly and in such a modern style that it could replace the first chapter of most elementary books on the subject. Then I use the unifying algebraic ideas of group and ring to place the concepts in a modern context. After all, the theory of congruences was one of the original motivations for introducing these algebraic ideas. Next comes the further development of congruence theory including two proofs of the law of quadratic reciprocity (chapters 9–11), an introduction to the arithmetic theory of binary quadratic forms (chapter 12), and some geometric theory of forms (some of which postdates Gauss — chapter 13). Finally I present an introduction to cyclotomy (chapter 14), including a third proof of quadratic reciprocity and the outline of a fourth.

These fourteen chapters provide a somewhat sophisticated historically oriented introduction to elementary number theory.

The founding fathers of number theory (Fermat, Euler, Lagrange, Legendre and Gauss) were not just number theorists; they made contributions to all the mathematics of their time. After Gauss, the bulk of mathematical knowledge grew at a fast pace and more mathematicians concentrated in just a few areas, although there was not as much specialization as we see today. Number theory has become so vast that it is impossible to present even a basic introduction to all its aspects in a single volume. So, for the final part of the book, I chose a few themes; quadratic algebraic number theory as a prototype for general algebraic number theory (chapters 15–17), arithmetic on curves (chapters 18–20), geometry of numbers (chapter 22), p-adic numbers and valuations (chapters 23), and the interconnected topics of irrational and transcendental numbers and Diophantine approximation (chapter 21). These themes are fairly independent of each other and can be read in any order. I sometimes repeat definitions to increase the independence. However, the topics all assume an understanding of chapters 1–14.

Chapter 21 is my favorite; it exemplifies what I am trying to achieve with the book. The history of the subject is not complicated, allowing an interweaving with the mathematics which provides a good introduction and an overview of the important results and open problems. The subject is one of the most difficult in mathematics with a small number of very deep results and hard but not very abstract proofs (many of which can be understood knowing only calculus). There are many easily stated open problems. For example, we prove that at least one of $e + \pi$ and $e\pi$ must be irrational. A proof that they are both irrational would be very nice; of course everyone believes it, but no proof exists.

My original course was designed for graduate students and advanced undergraduates who wanted a broad introduction to number theory rather than a specialized course. I also tried to appeal to students planning further study in number theory and algebraic geometry by providing background, examples, and motivation for the abstract developments. This is the audience I had in mind as I wrote the book. I hope that it will also attract mathematicians and mathematically educated laymen intrigued by today's exciting developments in number theory.

The prerequisites for a fairly complete understanding of the book are an undergraduate course in abstract algebra (the elements of groups, rings and ideals, fields and vector spaces), calculus of one and several variables and a little advanced calculus. In sections 19.7 and 19.8, some knowledge of complex analysis would be useful, but the basic ideas can be understood without this. For those who have not studied abstract algebra but who have that elusive prerequisite "some mathematical sophistication", chapters 1–15, 18, 21, 22 and parts of chapter 23 can be read with "judicious skipping". I also recommend Harold Davenport's *The Higher Arithmetic* as a companion volume for the first 12 chapters of this book. It is an inexpensive gem and should be on the bookshelf of anyone interested in number theory.

Number theory is in a golden age of development. I hope this book conveys some of the flavor of its many accomplishments.

Jay R. Goldman
Minneapolis, MN

Personal Thanks

Writing this book has been a big project and there are many people to thank.

How do I thank James Vaughn? Without his encouragement and the support of the Vaughn Foundation Fund, I would not have learned enough to even think about writing this book, nor would I have had the time to begin the project.

My publisher Klaus Peters deeply believes in the type of mathematical exposition with historical perspective that I have tried to achieve. When he saw the first version of a few chapters, he immediately offered me a contract. Over the many years of writing, Klaus has put up with my delays and kept me at my desk by always insisting that I had a worthy goal. All of the people at A K Peters, Ltd. have been helpful in every way possible.

Barry Cipra and the late Garrett Birkhoff read an early version of my notes. Barry's detailed comments and Garrett's remarks on organization, goals and incorporating the history were of great help in turning the notes into a book.

Debbie Borkovitz read what I thought was basically a final version of the manuscript. Her corrections, incisive questions, and invaluable suggestions led me to rewrite almost every chapter. I hope she approves of the final product.

George Brauer saved me a great amount of time by providing English translations of Gauss's two memoirs on biquadratic reciprocity and other mathematical works in German.

I thank Larry Smith, Steve Sperber, Mike Rosen, and Bill Messing for many corrections in the manuscript and for valuable mathematical conversations over the years.

Victor Grambsch and his staff at Electric Images Inc. have done a nice job redrawing over a hundred crude diagrams.

Sarah Jaffe has transformed a manuscript written in MacWrite II into the final TEXdocument ready for filming. What I told her would be a four- or five-month task has, due to my delays, taken over two years. Her patience and fine work is really appreciated.

Finally I thank my wife, Anne. She has read many parts of the manuscript and tried very hard to improve my writing style. Over the last three years her repeated question "How much more needs to be done?", and her response to my answer "You said that last week", helped overcome my procrastination and shamed me into finishing the book in finite time.

Acknowledgements

I would like to thank the following authors and publishers for permission to quote selected portions from the works listed below.

C. Reid, *Hilbert*, copyright 1970, Springer-Verlag Inc., New York, Heidelberg, Berlin; 1996, Springer-Verlag, New York, Inc. (Copernicus Imprint). All rights reserved. Reprinted with the permission of the publisher and the author.

H. Edwards, *Fermat's Last Theorem*, copyright 1977 by Harold M. Edwards. Published by Springer-Verlag, New York, Inc. 1977. All rights reserved. Reprinted with the permission of the publisher and the author.

Dictionary of Scientific Biography, Vol. V, copyright 1972 by the American Council of Learned Societies. Published by Charles Scribner's Sons (NY). All rights reserved. Reprinted with the permission of the American Council of Learned Societies.

Dictionary of Scientific Biography, Vol. VIII, copyright 1973 by the American Council of Learned Societies. Published by Charles Scribner's Sons (NY). All rights reserved. Reprinted with the permission of the American Council of Learned Societies.

G. Polya, *Mathematics and Plausible Reasoning*, Vol. 1, Induction and Analogy in Mathematics, copyright 1990, Princeton University Press. All rights reserved. Reprinted with the permission of the publisher.

F. Klein, *Development of Mathematics in the 19th Century* (English Translation), copyright 1979 by Robert Hermann, Math-Sci Press. All rights reserved. Reprinted with the permission of the publisher.

Notation and Numbering

Z denotes the set of integers.
Q denotes the set of rational numbers.
R denotes the set of real numbers.
C denotes the set of complex numbers.

Equation (3.6.7) refers to chapter 3, section 6, equation 7.
Equation (6.7) refers to section 6, equation 7 in the chapter where the reference appears.
Equation (7) refers to equation 7 in the section where the reference appears.

Section 9.5 refers to section 5 of chapter 9.
Section 5 refers to section 5 in the chapter where the reference appears.

The diagrams are numbered sequentially within each chapter.

Contents

PART 1: FROM FERMAT TO LEGENDRE

Chapter 1. The Founding Fathers

Chapter 2. Fermat

Chapter 3. Euler

Chapter 4. From Euler to Lagrange; The Theory of Continued Fractions

Chapter 5. Lagrange

Chapter 6. Legendre

PART 2: GAUSS AND THE *DISQUISITIONES ARITHMETICAE*

Chapter 7. Gauss

Chapter 8. Theory of Congruence 1

Chapter 9. Theory of Congruences 2

Chapter 10. Primitive Roots and Power Residues

Chapter 11. Congruences of the Second Degree

Chapter 12. Binary Quadratic Forms 1: Arithmetic Theory

Chapter 13. Binary Quadratic Forms 2: Geometric Theory

Chapter 14. Cyclotomy

PART 3: ALGEBRAIC NUMBER THEORY

Chapter 15. Algebraic Number Theory 1: The Gaussian Integers and Biquadratic Reciprocity

PART 4: ARITHMETIC ON CURVES

Chapter 18. Arithmetic on Curves 1: Rational Points and Plane Algebraic Curves

Chapter 19. Arithmetic on Curves 2: Rational Points and Elliptic Curves

Chapter 20. Arithmetic on Curves 3: The Twentieth Century

Chapter 1

The Founding Fathers

1. The Beginnings

Number Theory, or the Higher Arithmetic, is the study of those properties of integers and rational numbers which go beyond the ordinary manipulations of everyday arithmetic.[1]

The founding fathers of modern number theory are Fermat, Euler, Lagrange, Legendre and Gauss. From about 1630, when Fermat began to work in number theory, to Gauss' publication of the *Disquisitiones Arithmeticae* in 1801, each of these mathematical giants built upon on the work of his predecessors and made the primary contribution to the number theory of his times.

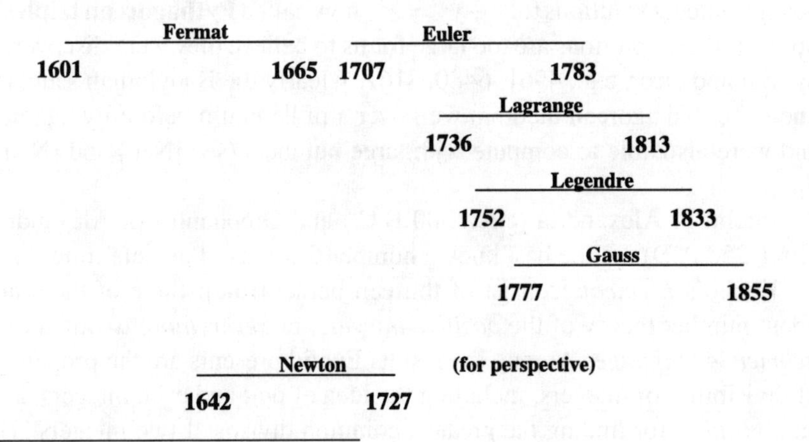

[1] See the Appendix to this chapter for a review of some basic notions and notation.

These men developed the foundations of number theory, discovered and proved many of the fundamental theorems and raised many of the questions that still guide our thinking today. In figure 1, we show the span of their lives and their potential for interacting with each other. After Gauss, mathematical knowledge increased at such a rapid pace that mathematicians had to become specialists and no single figure could dominate the field. The first half of this book is primarily devoted to explaining their work, which comprises an introduction to the fundamentals of number theory.

Of course, Fermat did not begin his studies in a vacuum. Throughout history, almost every major civilization has been fascinated by properties of the integers and has produced number theorists. In ancient and medieval times these were usually geometers (mathematicians) or more generally scholars, calender calculators, astronomers, astrologers, priests or magicians.

Chinese myths talk about an interest in numbers as far back as 3000 B.C., but there are no documents from this time. The oldest known mathematical documents are the Golenischev papyrus and the Rhind papyrus, both from Egypt (12th Dynasty - circa 2000 B.C.). The latter papyrus hangs in the British Museum and the former is in Moscow. They show how the Egyptians solved geometric problems and how they used fractions to solve arithmetic problems (see [New, J] and [Rob - Shu]).

The oldest number theoretical record we have is a tablet from Babylonia (circa 1900-1600 B.C.), a table of right triangles with integer sides, i.e., positive integer solutions to $x^2+y^2 = z^2$ (now called **Pythagorean triples**). Some of these solutions are too large for us to believe they were discovered by trial and error, e.g., 4961, 6480, 8161. Clearly the Babylonian scholars knew the Pythagorean theorem well over a millennium before Pythagoras and were also able to compute with large numbers (see [Neu] and [Neu - Sac]).

Euclid of Alexandria (circa 300 B.C.) and Diophantus of Alexandria (circa 250 A.D) are the best known number theorists of ancient times.

Euclid's *Elements* consist of thirteen books [Euc]; three of them are about number theory of the *positive integers*, but *everything is stated in a geometric language*. Among the results Euclid presents are the properties of divisibility of integers, including the idea of odd and even integers, and an algorithm for finding the greatest common divisor of two integers. He derives formulas for the sum of a finite geometric progression and for all Pythagorean triples. He introduces the notion of a prime number and shows that if a prime divides a product of two integers, then it must divide at least one of them. Euclid's proof that there are infinitely many primes is very brief (if there were only a finite number of primes, say $p_1, p_2, \ldots, p_k$, then $(p_1 p_2 \ldots p_k) + 1$ is either a prime or has a prime factor which cannot be any of the p_i, contradicting our assumption). We will discuss most of these results in more detail later in the chapter.

Diophantus' *Arithmetica* is the first known book devoted exclusively to number theory. The parts of the book that have survived to modern times ([Hea], [Ses]) contain about 150 problems worked out in terms of specific numerical examples, although in some cases the author seems to have known general methods. Although Diophantus is primarily interested in solving equations in positive rational numbers, equations with solutions to be found in integers are now referred to as **Diophantine equations**. The study of rational solutions of equations has evolved into the modern theory of arithmetic on curves (chapters 18 - 20).

To complete this quick overview of the early history of number theory and to provide a transition into modern times, I freely translate from the preface to the first edition (1798) of Legendre's *Theory of Numbers* [Leg]. Any unfamiliar terms will be explained in the next few chapters.

> "To judge by the fragments and the results in manuscripts attributed to Euclid, it appears that the ancient philosophers did rather extensive research on the properties of numbers. But they lacked two tools needed to reach the depths of this science: the art of numeration, which serves to express numbers with great ease, and algebra, which generalizes results and can operate equally on the known and the unknown. The invention of both of these arts greatly influenced the progress of the science of numbers. We also see the works of Diophantus of Alexandria, the oldest known writer on algebra, whose work is entirely dedicated to numbers and solves difficult questions with skill and sagacity.
>
> From Diophantus until the time of Viète and Bachet, mathematicians continued to be occupied with numbers but without much success and without appreciably advancing the science. Viète, by adding new degrees of perfection to algebra, solved several difficult problems on numbers. Bachet, in his work entitled "Pleasant and Delectable Problems [actually the title is: Pleasant and delectable problems concerning numbers . . . very useful for all kinds of curious persons who make use of arithmetic], solved the indeterminate [Diophantine] equation of the first degree by a general and very ingenious method. We owe to the same savant [scientist-scholar] an excellent commentary on Diophantus, which was later enriched by Fermat's marginal notes.

Fermat, among the geometers whose work contributed the most to accelerate the discovery of the new calculus, cultivated the science of numbers with great success and opened up new paths. We credit him with a large number of new theorems but he left almost all of them without proof. It was in the spirit of the times to propose problems [challenges] to one another. One often concealed his method in order to save the new triumph both for himself and his nation, for there was great rivalry between French and English geometers. Thus most of Fermat's proofs have been lost and the few that remain are unfortunately not those that we need.

Geometers from Fermat to Euler, devoted entirely to the discoveries or the applications of the new calculus, paid no attention to the theory of numbers. Euler was the first to become attached to this specialty; the numerous memoirs he published on this subject in the Commentaries of the Petersburg Academy and in other works proves how, with all his heart, he made the same progress in the science of numbers for which most of the other parts of mathematics are indebted to him. It is also believed that Euler had a particular liking for this type of research and that he devoted himself to it with a kind of passion, as he succeeded in almost all that concerned him. Be that as it may, his clever research led to the proof of two of the principal theorems of Fermat, viz.,

i) if a is a prime number and x an arbitrary number, not divisible by a, $x^{a-1} - 1$ is always divisible by a,

ii) every prime number of the form $4n + 1$ is the sum of two squares.

A multitude of other important discoveries can be found in Euler's memoirs. One finds there the theory of divisors of the quantity $a^n \pm b^n$, the treatise "Partitions of a Number"..., the use of irrational or imaginary factors in the solution of indeterminate equations; the solution of the general indeterminate equation of the 2nd degree by assuming one knows a special solution; the demonstration of many theorems on powers of numbers and particularly the negative propositions stated by Fermat that the sum or difference of two cubes cannot be a cube, and that the sum or difference of two fourth powers cannot be a square. Finally we find in the same writings a large number of indeterminate [Diophantine] questions solved by very ingenious analytical tricks.

Euler was for a long time the only Geometer concerned with the theory of numbers. Finally Lagrange entered the same arena and his first steps were signaled by successes of a most sublime kind, equal to those he had already achieved in his research. A method for solving indeterminate equations of the second degree in rational numbers and, what is more difficult, a method for finding integer solutions was the 'trial coup' of this illustrious savant. Shortly afterwards, he applied continued fractions to this branch of analysis. He gave the first proof that the continued fraction of a root of a rational equation of the second degree must be periodic, and he then concluded that Fermat's problem concerning the equation $x^2 - Ay^2 = 1$ is always solvable; a proposition which had not been rigorously established until then, although several geometers had given methods for the solution of this equation.

This same scholar, in his later research which appears in the Memoirs of the Berlin Academy, gave the first proof that every integer is the sum of four squares ..."

Three important references on the general history of number theory are the books by Weil [Wei] and Dickson [Dic], and the long essay by the Ellisons [Ell, W - Ell, F].

2. Fermat's Mathematical Background

In order to understand Fermat's achievements, it is important to have some idea of his mathematical background. He studied the Greek works of Euclid, Diophantus and Archimedes and he also knew the basic properties of the integers and rational numbers (positive, negative, and zero), the Arabic notation we currently use for integers and its computational algorithms, and the algebra developed during the Renaissance.

Fermat knew the standard form of the principle of mathematical induction (n to $n+1$). He also made use, in what he called the "method of infinite descent" (sec. 2.2), of the well ordering of the integers, namely, every set of positive integers contains a smallest element. Well ordering is equivalent to induction (exercise).

Fermat was certainly aware of the unique factorization of integers into primes, often referred to as the Fundamental Theorem of Arithmetic. Unique factorization seems never to have been explicitly stated or proved before Gauss, but was certainly used by all the founding fathers. This is rather puzzling unless it was regarded as so obvious that there was no need to state it. Legendre, prior to Gauss, explicitly states the existence of a prime factorization, but not uniqueness. Of course uniqueness follow immediately from Euclid's result that if a prime divides the product of two numbers, then it divides one of the factors. Perhaps the solution to the puzzle is that Gauss was the first to realize what is required for a rigorous exposition of number theory.

For a more detailed discussion of Fermat's background see Weil [Wei].

Now we present some topics from the early years of number theory which were part of Fermat's toolkit. One of the fundamental number theoretic results in Euclid's *Elements* is the **division algorithm**:

If a and b are integers, then there exist integers numbers q and r such that $a = bq + r$ where $0 \leq r < |b|$.

This is just another way of saying that either a is a multiple of b or it lies between two multiples of b. Assume $a, b > 0$, and let qb be the largest multiple of b which is less than or equal to a. Then setting $r = a - qb$, we have $qb \leq a = qb + r \leq (q+1)b$ or $0 \leq r \leq b$ (fig. 1). The case where a or b is negative is essentially the same.

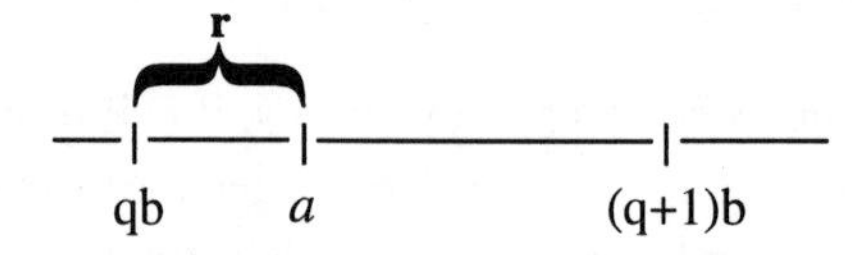

Figure 1

Recall that Euclid dealt only with positive integers. However, once the negative integers and zero were established during the middle ages and the Renaissance, most basic results were extended in a straightforward way.

To prove the existence of the greatest common divisor (and provide a method for computing it), we follow Euclid in repeated applications of the division algorithm. This is the **Euclidean algorithm**:

Given the integers a and b, form the sequence of equations

$$
\begin{aligned}
a &= q_1 b + r_1 \\
b &= q_2 r_1 + r_2 \\
r_1 &= q_3 r_2 + r_3 \\
&\vdots \\
r_i &= q_{i+2} r_{i+1} + r_{i+2} \\
&\vdots
\end{aligned}
$$

where the q_i's and the r_i's are defined by the division algorithm. Since $b > r_1 > r_2 > \cdots > 0$ the process must stop, say with

$$\begin{aligned} r_n &= r_{n+1}q_{n+2} + r_{n+2} \\ r_{n+1} &= r_{n+2}q_{n+3} \qquad (r_{n+3} = 0). \end{aligned}$$

We prove that $r_{n+2} = \gcd(a, b)$ by showing that r_{n+2} satisfies the two conditions for the greatest common divisor.

First we prove that r_{n+2} divides a and b. By the last equation $r_{n+2}|r_{n+1}$. By the next to the last equation, since r_{n+2} divides itself and r_{n+1}, it also divides r_n. Repeating this reasoning and working up the sequence of equations, we see that $r_{n+2}|a$ and $r_{n+2}|b$.

Now suppose that some integer e divides a and b. We must prove that $e|r_{n+2}$. By the first equation, since $e|a$ and $e|b$, we have $e|r_1$. By the second equation, since $e|b$ and $e|r_1$, we have $e|r_2$. Repeating this reasoning and working down the sequence of equations, we see, by the next to the last equation that $e|r_2$. Hence $r_{n+2} = \gcd(a, b)$.

Starting with the Euclidean algorithm, the following exercise leads to a proof of unique factorization.

Exercise: i) Beginning with $r_{n+2} = r_n - r_{n+1}q_{n+2}$ in the Euclidean algorithm and substituting each equation in the one above, show that there exist integers k, m such that $\gcd(a, b) = ka + mb$. In particular if a and b are relatively prime then $ka + mb = 1$. From this result and a little extra reasoning, it follows that

ii) the Diophantine equation $ax + by = 1$ has a solution if and only if $(a, b) = 1$.

ii) Prove that if p is a prime and $p|ab$, then $p|a$ or $p|b$.

Now prove the Fundamental Theorem of Arithmetic, i.e.,

iii) *Existence*: prove that any positive integer can be factored into primes.

iv) *Uniqueness*: prove that any two factorizations of a positive integer into primes are the same after rearrangement of the order of the primes.

Part ii) of the last exercise also leads to the solution of the general two variable linear Diophantine equation.

Exercise: Prove that the Diophantine equation $ax+by=c$ has a solution if and only if $\gcd(a,b)|c$. If $d=(a,b)$ and if x_0, y_0 is one solution to our equation, then all the solutions are given by

$$x = x_0 + k\left(\frac{b}{d}\right), y = y_0 + k\left(-\frac{a}{d}\right), \quad k = 0, \pm 1, \pm 2, \dots .$$

3. Pythagorean Triples

Recall that the problem of finding all Pythagorean triples, which we have traced back to the Babylonians, is to find all right triangles with integer sides, i.e., all solutions in *positive* integers to $x^2+y^2=z^2$. Since this problem is one of the oldest in number theory, the motivation for Fermat's most famous 'claimed result' (sec. 2.3), and the precursor of a large part of modern number theory, we shall now present a complete solution. Although we present the solution in modern algebraic notation, it is essentially the solution that Euclid presented in geometric language [Euc, Book 10, lemma 1 to prop. 29] .

Now assume the positive integers x, y, z satisfy $x^2+y^2=z^2$. If $d|x, y, z$, then we can divide both sides of the equation by d^2. Hence we may assume $\gcd(x,y,z)=1$. Such a solution is called a **primitive** solution. Conversely, if (x,y,z) is a solution, then so is (dx,dy,dz); thus if we know the primitive solutions, we know all solutions. So now we assume *our solution* (x,y,z) *is primitive*, and find all such primitive triples.

1) $(x,y)=(y,z)=(x,z)=1$.

Proof: If a prime p divides x and y, then it divides x^2 and y^2 and hence z, contradicting primitivity. Similarly for the other pairs.

2) z is odd and x, y have opposite parity.

Proof: i) x, y, z all even contradicts primitiveness.
x, y, z odd $\Longrightarrow x^2, y^2, z^2$ odd $\Longrightarrow$ odd + odd = odd, a contradiction.
If two are even then the third is even, a contradiction. Thus two of x, y, z are odd and one is even.

ii) Suppose z is even and x, y are odd, say $z=2n$, $x=2k+1$, and $y=2r+1$. Then $x^2=4(k^2+k)+1$, $y^2=4(r^2+r)+1$, $z^2=4n^2$

and, from $x^2 + y^2 = z^2$, we have $4n^2 = 4(k^2 + k + r^2 + r) + 2$ or $4(n^2 - k^2 - k - r^2 - r) = 2$, a contradiction. *Therefore z is odd* and exactly one of x or y is even.

Note that I am deliberately avoiding the notion of 'congruence', introduced by Gauss in 1801, which would simplify our reasoning, and I will do so until we discuss Gauss' work. I'm doing it the way Fermat would have done it.

3) Assume x is even, y odd, and z odd. Then $x^2 = z^2 - y^2 = (z-y)(z+y)$ and $z - y$, $z + y$ are even. Thus x, $z - y$, and $z + y$ are all even.

Let

$$x = 2u, \quad z + y = 2v, \quad z - y = 2w;$$

then

$$(2u)^2 = (2v)(2w) \text{ or } u^2 = vw.$$

Note that $(v, w) = 1$ since if a prime $p|v, w$, then $p|v + w = z$ and $p|v - w = y$; hence $p|x$, which, by primitivity, implies $p = 1$.

Main Step. By unique factorization,

$$u^2 = vw, \quad u, v > 0, \quad (v, w) = 1 \implies v, w \text{ are squares}$$

i.e.,

$$v = p^2, \quad w = q^2, \text{ for some positive integers } p \text{ and } q.$$

Thus we have

$$z = v + w = p^2 + q^2, \quad y = v - w = p^2 - q^2$$

and since $x^2 = z^2 - y^2$, we have $x = 2pq$.

Moreover, $p > q$ since $y > 0$, and $(p, q) = 1$ since $d|p, q$ implies $d|z, y$, which, by property 1) above, means $d = 1$. The numbers p and q have opposite parity for otherwise y and z are both even.

To summarize:

If the positive integers x, y, z are a primitive solution of $x^2 + y^2 = z^2$, with x even, then y and z are odd and there are relatively prime numbers p, q of opposite parity, such that $p > q$, $(p, q) = 1$, and

$$x = 2pq, \quad y = p^2 - q^2, \quad z = p^2 + q^2.$$

If y is even, then x and z are odd, and we get the same formulas with x and y interchanged

The converse is clear since

$$(2pq)^2 + (p^2 - q^2)^2 = (p^2 + q^2)^2.$$

Thus, allowing for the possibility of interchanging x and y, all positive solutions to $x^2 + y^2 = z^2$, primitive or not, are given by

$$x = d(2pq), \quad y = d(p^2 - q^2), \quad z = d(p^2 + q^2)$$

where d is any positive integer.

Note that since changing the signs of x, y or z yields integer solutions of $x^2 + y^2 = z^2$, all integer solutions can be included in our formulas by dropping the requirements that $p > q$ and $d > 0$.

The following exercise presents a different derivation of our formula for Pythagorean triples using methods of analytic geometry.

Exercise: Dividing by z^2 we see that integer solutions of $x^2 + y^2 = z^2$ correspond to rational solutions of $\left(\frac{x}{z}\right)^2 + \left(\frac{y}{z}\right)^2 = 1$ or $X^2 + Y^2 = 1$, where$X = \frac{x}{y}$ and $Y = \frac{y}{z}$. Hence the rational solutions correspond to *rational points* (rational coordinates) on the unit circle $X^2 + Y^2 = 1$. Conversely, every rational point on the circle, after clearing fractions, corresponds to an integer solution of $x^2 + y^2 = z^2$.

If L is the line through $(-1, 0)$ intersecting the unit circle in the point (x', y'), with slope t, then its equation is $y = t(x + 1)$ (fig. 2). Show that x', y' are rational functions of t and thus we have a parametrization of the circle with t as parameter.

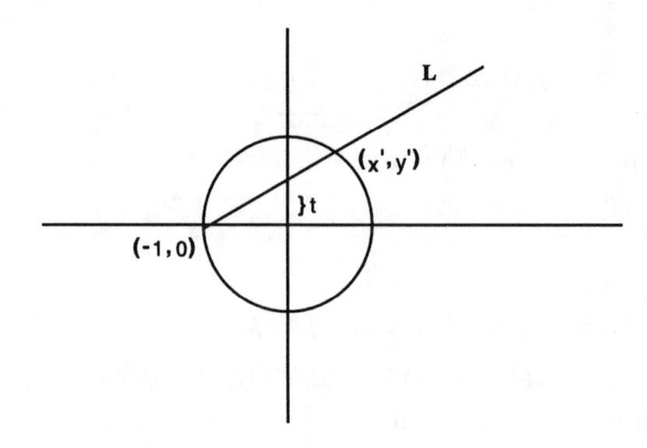

Figure 2

Prove: t is rational $\Longleftrightarrow$ (x', y') is a *rational point.*

Hence we have a formula for rational points on the circle. Use the formula for rational points to derive our formula for Pythagorean triples.

What can one say for conics other than the circle (assuming the conic has at least one rational point) ? This example will be the basis for our later theme of rational points on curves (chapters 18 - 20).

Appendix: Properties of the Integers

In this chapter, it is assumed that the reader is familiar with the basic properties of the integers as presented informally by Davenport [Dav, chap. 1, sec. 1 - 3]. We review a few thing in order to fix our notation and terminology.

If the integer a divides the integer b, i.e., $b = ac$ for some integer c, we write $\mathbf{a|b}$, while $\mathbf{a \nmid b}$ means a does not divide b. We also write $\mathbf{x|x_1, x_2, \ldots, x_n}$ to mean that x divides each of the integers $x_1, x_2, \ldots, x_n$.

Given the integers a and b, a positive integer d is called a *greatest common divisor* of a and b if

i) $d|a, d|b$,

ii) if $e|a$ and $e|b$, then $e|d$.

A greatest common divisor of a and b exists and is unique. The existence is shown in section 2. The uniqueness follows from property ii), which says that d is the *largest* common divisor of a and b. Hence we can talk about **the greatest common divisor of a and b**, which will be denoted by **gcd(a, b)** or, when no confusion can arise, just by **(a,b)**. If $\gcd(a, b) = 1$, we say that a and b are **relatively prime**. Similarly, there exists a unique *greatest common divisor of a finite set of integers* $\{a_1, a_2, \ldots, a_n\}$, which is denoted it by $\gcd(a_1, a_2, \ldots, a_n)$ or just $(a_1, a_2, \ldots, a_n)$. A positive integer p is a **prime number** if it has no divisors other than 1 and p.

Chapter 2

Fermat

1. Fermat (1601-1665)

Pierre Fermat's work in analytic geometry, calculus, algebra, optics and probability theory made him one of the most famous mathematicians of his day. Although he produced brilliant work and we now see him as the first of the founding fathers of modern number theory, number theory in Fermat's time was poorly studied and not appreciated by such other leading lights as Descartes and Pascal.

Fermat was born in a small town near Toulouse, in the south of France. He spent his life among the prosperous middle class, raising a family of five. He received a law degree and became a 'councilor' in the High Court of Toulouse, which entitled him to be called Pierre *de* Fermat. Over the years, he advanced to higher posts within this system and was still practicing the law when he died.

It is hard to find any outward excitement or adventure in Fermat's life. Although he mastered many languages and was consulted as an expert on Greek manuscripts, which he also collected, his true passion and creativity was reserved for his mathematics. Fermat became interested in mathematics in his late 20's, but most of his important work began in his 30's. With the exception of an essay on the rectification of curves, published anonymously as an appendix to a friends book, *Fermat never published a single work in number theory.* He circulated manuscripts, corresponded with other mathematicians, and challenged them to prove theorems he already knew (recall Legendre's discussion of challenges in section 1.1). In number theory we have *almost* none of his proofs, but enough hints and statements to be assured that he had proofs of most of his results.

Fermat's most consistent correspondents were Father Marin Mersenne and Pierre de Carcavi. Neither of these men were creative mathematicians but, with great enthusiasm for mathematics, they corresponded with many of the leading scientists of the day and served as clearinghouses for the spread of new ideas. Both were certainly champions of Fermat's work.

Our knowledge of Fermat's number theoretic work comes from the notes he made in the margins of his copy of Bachet's translation of Diophantus and from his correspondence [Fer]. Heath [Hea] contains an English translation of Diophantus and of many of Fermat's notes and some of his letters. For a comprehensive discussion of Fermat's number theory and its influence, see [Wei 1] and [Edw 1] and for his other mathematics and personal life see [Bel 1], [Bel 2] and [Mah].

We now discuss most of Fermat's important results. Almost all of Fermat's results later lead to major new developments.

2. Infinite Descent

In the margin of his copy of Bachet's translation of *Diophantus*, Fermat states and proves the proposition that *the area of a right triangle whose sides have rational length cannot be the square of a rational number*. This is Fermat's only known proof of a result in number theory that was written out in detail and not just hinted at.

First Fermat says (translation by [Hea, pg. 293]):

> This proposition, which is my own discovery, I have at length succeeded in proving, though not without much labor and hard thinking. I give the proof here, as this method will enable extraordinary developments to be made in the theory of numbers.

Then he shows that the existence of such a triangle is equivalent to a statement about squares of positive integers and, from this, arrives at a contradiction. He continues:

> Thus, if there exist two squares [of relatively prime positive integers] such that their sum and difference are both squares, there will also exist two other integer squares [of relatively prime positive integers] which have the same property but have a smaller sum. By the same reasoning we find a sum still smaller than the last found, and we can go on *ad infinitum* finding integer square numbers smaller and smaller which have the same property. This is, however, impossible because *there cannot be an infinite series of numbers [positive integers] smaller than any given*

> *integer we please.* – The margin is too small to enable me to give the proof completely and with all detail.

(See [Edw 1] for details of this proof).

The last statement in italics is what Fermat called his method of infinite descent. In his correspondence, he claimed to have proved almost all of his results in number theory by this method. We state it formally in two ways.

THE METHOD OF INFINITE DESCENT: If the assumption that a positive integer satisfies a set of properties implies that a smaller positive integer satisfies these properties, then no positive integer satisfies these properties. Equivalently, there cannot exist an infinite decreasing sequence, $n_1 > n_2 > \cdots$, of positive integers.

The proof is immediate. By the well ordering of the positive integers (sec 1.2), the set of positive integers satisfying the given properties has a smallest member which contradicts the assumption that there is an even smaller one.

Why the big fuss about such a simple statement, which is just a minor variation of induction and well ordering? Fermat's genius was to realize that this reformulation provided him with a powerful method for proving deep results about the integers.

We will use this method in a moment.

3. Fermat's Last Theorem

In another annotation of his copy of *Diophantus*, next to Diophantus' problem "to divide a given square into two squares" (Pythagorean triples - sec. 1.3), Fermat wrote his now famous and most often quoted note (translation by [Hea, pp. 144-145]):

> "On the other hand, it is impossible to separate a cube into two cubes, or a biquadrate [fourth power] into two biquadrates, or generally any power except a square into two powers with the same exponent. I have discovered a truly marvelous proof of this which however the margin is not large enough to contain."

What Fermat is stating in symbols is that, for any integer $n > 2$, the equation

$$x^n + y^n = z^n$$

has no solutions in positive integers. This is equivalent to saying that there are no solutions in integers except for the trivial ones where one of the variables is zero. This conjecture, which was one of the most famous unsolved problems in mathematics for about 350 years, is generally referred to as **Fermat's Last Theorem**. Attempts to prove it helped motivate the development of a great deal of important mathematics, such as algebraic number theory (chap. 3, 15-17) and arithmetic on curves (chap. 18–20). [Edw 1] is devoted to the last theorem and its influence in the development of algebraic number theory.

In 1983, Gerd Faltings [Fal] proved the Mordell conjecture, a very general theorem on rational solutions to polynomial equations in two variables, which immediately implies that, for each $n > 3$, $x^n + y^n = z^n$ has at most a finite number of primitive solutions. We will discuss rational solutions in Chapters 18–20.

Finally in 1995, after about 350 years since Fermat's claim, Andrew Wiles (with the assistance of Richard Taylor) proved Fermat's Last Theorem ([Wile], [Wile-Tay]). The proof, which draws on many areas of mathematics, but is primarily based on the theory of arithmetic on curves (Chap. 18–20), is beyond the scope of this book. The recent book by A. van der Poorten [Poo 2] is the best place to get an introduction to the ideas behind Wiles proof. It is written at the level of the latter half of this book. It is reasonable to regard Wiles proof as the high point of 20th century number theory.

Here and in some other parts of the book, we will discuss special cases of Fermat's Last Theorem.

Note that the conjecture for arbitrary n reduces to the cases $n = 4$ and $n = p$, a prime, since if d divides n, then

$$x^n + y^n = \left(x^{\frac{n}{d}}\right)^d + \left(y^{\frac{n}{d}}\right)^d .$$

It is generally believed that Fermat did not have a proof of the general theorem. In his correspondence he only mentions the cases $n = 3, 4$. When we get to Euler, we shall discuss the case $n = 3$ and why Fermat may have thought he had a proof for the general case. It seems certain that Fermat proved the case $n = 4$, since it follows from his proof of the theorem on rational right triangles (see sec. 2 and [Edw 1]). We present a more direct proof.

Since $x^4 + y^4 = z^4$ implies $x^4 + y^4 = (z^2)^2$, the case $n = 4$ is a corollary of the following more general theorem.

Theorem: $x^4 + y^4 = z^2$ *has no solution in positive integers.*

Proof: The main idea is to use our formula for Pythagorean triples (sec. 1.2) and milk it dry.

We use the method of infinite descent. Assume that x, y, z is a positive integer solution. We shall show that this implies that there is another positive integer solution X, Y, Z with $Z^2 < z^2$, and thus, by infinite descent, there are no solutions.

A With the same reasoning used for Pythagorean triples we see that we can assume that $\gcd(x, y, z) = 1$ and any pair of the numbers are relatively prime (these are called the **primitive** solutions).

B If $x^4+y^4 = z^2$, then $(x^2)^2+(y^2)^2 = z^2$ and (x^2, y^2, z) is a Pythagorean triple. By A), we can assume it is primitive. Hence (if necessary after interchanging x and y) there exist p, q such that $p > q > 0$, $(p, q) = 1$, where p and q have opposite parity and

(i) $x^2 = 2pq$,

(ii) $y^2 = p^2 - q^2$,

(iii) $z = p^2 + q^2$.

Since we have $y^2+q^2 = p^2$ and since $(p, q) = 1$ implies $\gcd(p, q, y) = 1$, (y, q, p) is a primitive Pythagorean triple.

By our general results on Pythagorean triples, p must be odd; therefore q is even (opposite parity) and y is odd. Hence there exist positive integers a, b of opposite parity such that $(a, b) = 1, a > b > 0$, and

$$q = 2ab\ , \quad y = a^2 - b^2\ , \quad p = a^2 + b^2\ .$$

By (i) we have

$$x^2 = 2pq = 4ab(a^2 + b^2)\ . \tag{1}$$

We prove that a, b and a^2+b^2 are squares. First, by (1), $ab(a^2+b^2) = (\frac{x}{2})^2$ is a square (recall that x is even). If a prime p divides ab then, since $(a, b) = 1$, p divides a or b, but not both; say p divides a. Then p cannot divide $a^2 + b^2$; hence $(ab, a^2 + b^2) = 1$. Since $(a, b) = 1$ and the product of the relatively prime integers ab and $a^2 + b^2$ is a square, each of them is is a square.

Let $a = X^2$, $b = Y^2$ and $Z^2 = a^2 + b^2$. Then $X^4 + Y^4 = Z^2$. Thus we have shown that if there exist positive integers x, y, z such that $x^4 + y^4 = z^2$, then there exists a new triple (X, Y, Z) of positive integers such that $X^4 + Y^4 = Z^2$. But we have

$$Z^2 = X^4 + Y^4 = a^2 + b^2 = p < p^2 + q^2 = z < z^2 \; ;$$

hence our infinite descent is started and we have a contradiction.

4. Pell's Equation

Desiring to stimulate interest in number theory, especially among the English, Fermat issued the following problem, which he claimed to have solved, as a challenge to mathematicians (translation by [Hea, pp. 286-287]):

> " There is hardly anyone who proposes purely arithmetical questions, hardly anyone who understands them. Is this due to the fact that up to now arithmetic has been treated geometrically rather than arithmetically? This has indeed been the case both in ancient and modern works; even Diophantus is an instance. For, although he has freed himself from geometry a little more than others have in that he confines his analysis to the consideration of rational numbers, yet even there geometry is not entirely absent, as is sufficiently proved by the Zetetica of Vieta, where the method of Diophantus is extended to continuous magnitudes and therefore to geometry.
>
> Now arithmetic has, so to speak, a special domain of its own, the theory of integral numbers. This was only lightly touched upon by Euclid in his Elements, and was not sufficiently studied by those who followed him (unless perchance, it is contained in those books of Diophantus of which the ravages of time have robbed us); arithmeticians have therefore now to develop it or restore it.
>
> To arithmeticians therefore, by way of lighting up the road to be followed, I propose the following theorem to be proved or problem to be solved. If they succeed in discovering the proof or solution, they will admit that questions of this kind are not inferior to the more celebrated questions in geometry in respect of beauty, difficulty or method of proof.
>
> *Given any number whatever which is not a square, there are also given an infinite number of squares such that, if the square is multiplied into the given number and unity is added to the product, the result is a square.*"

Fermat is claiming that if d is not the square of an integer, then

$$dy^2 + 1 = x^2$$

has infinitely many integer solutions. He continues:

> " Example. Let 3, which is not a square, be the given number; when it is multiplied into the square 1, and 1 is added to the product, the result is 4, being a square.
>
> The same 3 multiplied by the square 16 gives a product which, if increased by 1, becomes 49, a square.
>
> And an infinite number of squares besides 1 and 16 can be found which have the same property.
>
> But I ask for a general rule of solution when any number not a square is given.
>
> For example, let it be required to find a square such that, if the product of the square and the number 149, or 109, or 433 etc. be increased by 1, the result is a square."

The English mathematicians Brouncker and Wallis took up the challenge. For some reason, they had only seen Fermat's statement of the problem, and not his preamble. Hence they interpreted the problem as finding *rational* solutions and they gave the following easy method. Let

$$x = 1 + \left(\frac{m}{n}\right) y \, .$$

Then

$$dy^2 + 1 = x^2 = 1 + \frac{2m}{n} y + \frac{m^2}{n^2} y^2 \, .$$

Solving for y and then x we get

$$y = \frac{2mn}{dn^2 - m^2} \, , \qquad x = \frac{dn^2 + m^2}{dn^2 - m^2} \, .$$

When Brouncker and Wallis heard about the "integer" requirement they complained that Fermat had changed the problem and a solution in integers was of no value. Fermat replied that it was ridiculous to think he would propose such a trivial problem "since any novice can obtain a solution in fractions" [Wei 1, pg. 92]. A few months later he received a complete solution in integers from them. Fermat later pointed out, in a letter to Huygens, that the English had given a method but no proof that a solution always exists. He said he had proved the existence by infinite descent. This is an example of applying infinite descent to prove a positive result.

Euler later misnamed the equation Pell's equation (Pell was English but had nothing to do with the equation). About 100 years after Fermat, Lagrange published a proof of the existence of solutions. We present this later as an application of continued fractions (sec. 4.9). In modern terms, the solutions determine the 'units of real quadratic fields' (sec. 17.11).

Interest in special cases of Pell's equation have been traced back as far as Archimedes. Important contributions were made by Indian mathematicians. Brahmagupta (approx. 600 A.D.) gave the general solution of the linear Diophantine equation and treated special cases of the Diophantine equations $x^2 - Ay^2 = 1$, where A is a positive integer, and Bhaskara (12th century) knew general rules for finding solutions of the latter equation. For more details about the history and methods of solution, see the books by Edwards [Edw 1], Heath [Hea] and Weil [Wei 1], a paper by Weil [Wei 2], and the collected works of Fermat [Fer].

5. $y^3 = x^2 + k$.

Fermat stated that

$x = 5, y = 3$ is the only solution of $y^3 = x^2 + 2$, and

$x = 11, y = 5$ and $x = 2, y = 2$ are the only solutions of $y^3 = x^2 + 4$.

He gave no idea about where the results come from. Later we will present Euler's attack on the first equation.

The study of rational solutions of $y^3 = x^2 + k$ plays a central role in number theory up to the present day (see [Mor 1] and [Mor 2]). Fermat's interest in rational solutions to equations is discussed in section 19.9.

6. Sums of Squares

From Diophantus and Bachet, Fermat knew of the conjecture that every positive integer is the sum of at most four squares, i.e., for all positive integers m

$$m = x^2 + y^2 + z^2 + u^2$$

has a solution in non-negative integers.

Writing to Pascal in 1654 (see[Fer] and [Ell, W - Ell, F]), Fermat stated he had proved a more general theorem which included the four square conjecture and that in order to prove the theorem it was necessary to:

(a) prove that every prime of the form $4n + 1$ is the sum of two squares,

(b) develop a general method for finding these two squares for any prime $p = 4n + 1$,

(c) prove that every prime of the form $3n + 1$ can be written as $x^2 + 3y^2$,

(d) prove that every prime of the form $8n + 1$ or $8n + 3$ can be written as $x^2 + 2y^2$, and

(e) prove there does not exist a right triangle with rational sides whose area is the square of a rational number.

Fermat never wrote down his proof of the four square theorem. Euler was unable to prove it. The theorem was finally proved by Lagrange in 1770, and the proof was later simplified by Euler.

EXERCISE: Prove that a prime p of the form $4n + 3$ cannot be written as the sum of two squares. Hint: look at the parity of the two squares.

Results (a), (b) and (c) were later proved by Euler. In fact, (a), the representation of numbers as a sum of two squares, plays a central role in later developments and we shall give several proofs. Fermat claimed to have proved the theorem by infinite descent and he further claimed that he could prove each number had only one such representation [Wei 1, chap. II, sec. VIII, IX]. Result (c) played a major role in Euler's proof of FLT for $n = 3$ [Edw 1, chap. 2]. As discussed in section 2, result (e) was proved by Fermat and, in fact, is the only proof of his in number theory that was written out in full.

The above results lead naturally to the more general problems: for a given positive integer n, which positive integers m can be represented in the form

$$m = x^2 + ny^2,$$

for x, y integers, in how many ways can this be done, and how do we find the representations? A very large part of Fermat's work in number theory is devoted to these questions.

One approach to this problem is to study primes of the form $x^2 + ny^2$ and generalize these results to composite integers by using the identity

$$(a^2 + nb^2)(c^2 + nd^2) = (ac - nbd)^2 + n(ad + bc)^2.$$

Fermat's results are the beginning of the theory of representations of numbers by quadratic forms, one of the central themes in the development of number theory. We will eventually prove many of his results. His genius is evident from results like the following. Fermat conjectured that

if primes p_1 and p_2 are both of the form $4n + 3$ and both have 3 or 7 as their last digit then

$$p_1 p_2 = x^2 + 5y^2 \text{ is solvable.}$$

This is a important result of a new order of difficulty compared to his earlier results and was first proved by Lagrange [Lag, Vol. 3, pp. 788 - 789]. Results of this type only became natural in the context of Gauss' theory of composition of quadratic forms (see sec. 13.9 and [Edw 1, sec. 1.7, 8.6]).

Exercise: Prove that a prime of the form $4n + 3$ has 3 or 7 as its last digit if and only if it is of the form $20k + 3$ or $20k + 7$.

Exercise: Prove that a prime of the form $4n + 3$ cannot be represented in the form $x^2 + 5y^2$.

Primes of the Form $x^2 + ny^2$ by David Cox [Cox] explores the role of the studies of theses primes in the development of some very deep areas of modern number theory. It is much more advanced than our treatment.

7. Perfect Numbers and Fermat's Little Theorem

A positive integer n is called a perfect number if it equals the sum of all its *proper divisors* (all the numbers which divide n except n itself). Thus, for example, $6 = 1 + 2 + 3$ and $28 = 1 + 2 + 4 + 7 + 14$ are perfect numbers. Euclid proved that if $1 + 2 + 2^2 \cdots + 2^{n-1} = 2^{n-1}$ is prime then $2^{n-1}(2^n - 1)$ is perfect (exercise). In the 18th century Euler showed that every even perfect number is necessarily of this form. Probably the oldest unsolved problem in number theory (and possibly in mathematics) is the question:

Are there any odd perfect numbers ?

Euler's theorem leads to the question: for which n is $2^n - 1$ prime? Primes of the form $2^n - 1$ are called *Mersenne* primes after Father Marin Mersenne, a correspondent of Fermat. In studying this question Fermat was led to a theorem which is one of the most important for the later development of number theory, viz.,

Fermat's Little Theorem:

(i) *If p is a prime and a is an integer, then $p|a^p - a$, or equivalently,*

(ii) *if p is a prime and a is an integer such that $p \nmid a$, then $p|a^{p-1} - 1$.*

The first published proofs were by Euler who gave two proofs. We present his first proof, which depends on the following

Lemma: *For a, b integers and any prime p,*

$$p|(a+b)^p - a^p - b^p \ , \quad \textit{i.e.,} \quad (a+b)^p - a^p - b^p = mp \ ,$$

for some integer m.

Proof: We know that the binomial coefficient

$$\binom{p}{k} = \frac{p(p-1)\cdots(p-k+1)}{k!}$$

is an integer for $0 < k < p$ since it counts the number of subsets of size k of a set of p elements. The factor p occurs in the numerator and not the denominator. Hence, by unique factorization, we have

$$p \left| \binom{p}{k} \right.$$

for $0 < k < p$. The rest follows by the binomial theorem.

proof of (i). The case $a < 0$ follows immediately from the case $a \geq 0$, so we prove the latter. For any fixed prime p, we proceed by induction on a.

If $a = 0$ then $p|0$.

If the theorem is true for $a = k$ then for $a = k + 1$ we have, by the lemma,

$$(k+1)^p = k^p + mp + 1$$

or

$$(k+1)^p - (k+1) = (k^p - k) + mp\ .$$

But, by the induction hypothesis, $p|k^p - k$ and thus

$$p|(k+1)^p - (k+1)$$

and we are done.

Euler's second proof is essentially a proof of Lagrange's Theorem, which says that for finite groups, the order of a subgroup divides the order of the group (see sec 8.2 and the English translation of Euler's paper [Eul 1]).

8. Fermat's Error

There is one mistake among Fermat's claims. He claimed he could prove that

$$F_k = 2^{2^k} + 1 \quad \text{is prime for all } k \geq 0\ ,$$

and thus he had a formula which takes only prime values. Note that $2^n + 1$ can only be a prime when $n = 2^k$ for some k (proof: if n has an odd prime factor q, then $n = qm$ and $2^n + 1 = (2^m + 1)(2^{m(q-1)} - 2^{m(q-2)} + \cdots + 1)$).

Euler disproved Fermat's claim when he showed that $641 | 2^{32} + 1$. Nevertheless those values of F_k which are prime, the *Fermat primes*, played a major role in Gauss's later theory of polygon constructibility by ruler and compass (chapter 14).

F_1 through F_4 are primes and F_5 through F_9 have been factored. In fact, F_9 has 155 digits and was completely factored in 1990, as a result of a huge effort using over 1000 computers (see [Cip] and [Cam]).

Chapter 3

Euler

1601 Fermat 1665	1642 Newton 1727	1707 Euler 1783

1. Euler

Although Newton bridges the mathematical gap between Fermat and Euler, he was not particularly interested in number theory and Euler was Fermat's true successor.

Leonard Euler was phenomenal. He was certainly the greatest mathematician of the 18th century and perhaps its leading scientific figure. The 18th century has often been referred to as the age of Euler, on both scientific and mathematical grounds. He contributed to every area of the pure and applied mathematics of his time, laid the foundations of mathematical physics, and studied many problems of technology. The most prolific mathematician of all time, his published collected works total over 70 large volumes, yet they do not contain all of his work. This unusual productivity continued through the last twelve years of his life when he was totally blind. Only four volumes of Euler's collected works are devoted to number theory; nevertheless he would be a major figure in the history of mathematics even if these has been his only contributions.

Euler was born and educated in Switzerland. He received his beginning general and mathematical education from his father, a Protestant minister who had attended Jacques Bernoulli's mathematical lectures at the University of Basel. After spending several years studying a difficult algebra book by C. Rudolph, he studied mathematics privately with Johann Burkhardt, an amateur mathematician.

Euler entered the University of Basel at the age of thirteen. Johann Bernoulli I (Jacques' brother) was the professor of mathematics, but he only gave public lectures on elementary topics. Although he did not have the time to give private lessons to Euler, he advised him to study difficult mathematical works and let him visit on Saturday afternoons, when he would explain any mathematical points Euler did not understand. This was an extraordinary opportunity for a young mathematical genius, and Euler regarded it as the "best method to succeed in mathematics" [You]. Euler began his first research when he was eighteen.

Nicolas II and Daniel Bernoulli (Johann's sons) went to Russia's newly formed St. Petersburg Academy in 1725. Shortly after, they arranged a position for Euler and he arrived in 1727. Euler spent the next 14 years in St. Petersburg, but in 1741, Frederick the Great of Prussia recruited him for the Berlin Academy. After 25 years in Berlin, Catherine the Great brought Euler back to St. Petersburg where he spent the remaining 17 years of his life. We will say more about the role of the academies when we discuss Lagrange's life (sec. 5.1).

Euler's interest in number theory seems to have been stimulated by Christian Goldbach, an amateur mathematician fascinated with number theory as well as a man with broad intellectual interests. He is remembered in mathematics for his still unsolved conjecture, now known as *Goldbach's Conjecture*, that every even integer greater than two is the sum of two primes. Euler and Goldbach carried on an extensive correspondence for several decades, with Euler often discussing his newest discoveries in number theory.

In 1729, in response to Euler's first letter to him, Goldbach mentions Fermat's conjecture that all numbers of the form 2^{2^n} are primes. Euler responds to this with scepticism and not much interest, but only five months later he writes Goldbach that he is reading Fermat's works. He has been bitten by the number theory bug, retaining an intense interest in number theory for the rest of his life. As André Weil points out in a fascinating article on the history of number theory [Wei 3], Euler begins his research only with the mysterious statements by Fermat about what he claimed to have proven. Euler has to reconstruct everything including what we now put in an elementary textbook on number theory. He has no way of knowing which of Fermat's statements are relatively easy and which are very deep.

In addition to reconstructing the elementary results and proving (or in one case disproving) most of Fermat's statements, Euler made major new contributions. His research touched every part of number theory: the theory of divisibility including the multiplicative group modulo n, a generalization of Fermat's little theorem (sec. 2.7), and the law of quadratic reciprocity, sums of squares and the more general representation of integers by quadratic forms, Diophantine equations, continued fractions and Pell's equation, elliptic integrals, zeta functions and Bernoulli numbers, complex integers and algebraic number theory, partitions of a number and formal power series, and the theory of prime numbers. We will explain most of these topics later in the book. Weil [Wei 1] devotes over a hundred pages to the historical development of Euler's number theoretic work.

In addition to the books and papers already mentioned ([You], [Wei 3], [Wei 1]), the sources for this discussion of Euler's life and work, I recommend C. Truesdell's very beautiful and scholarly biography "Leonard Euler, Supreme Geometer" [Tru]. E.T. Bell [Bel 1] discussed the large and talented Bernoulli family.

Euler's textbooks on algebra and calculus are still the prototypes of todays texts, and superior to many of them. The recently reprinted English translation of Euler's *Elements of Algebra* [Eul 2] (to be referred to simply as Euler's *Algebra*) includes a substantial part of the Truesdell biography. Part 1 of the *Algebra* is a systematic introduction, starting with the most elementary principles and ending with the solutions of equations of the third and fourth degrees, including prototypes of many of the "word problems" that appear in present day high school texts. Part 2 is the first treatise since Diophantus on Diophantine equations. The book is a delight to read and still very informative. Everything is explained in a clear and well motivated way; the expressions "it is easy to see" and "obviously" do not seem to be part of Euler's vocabulary.

Euler's *Introduction in Analysin Infinitorum* (Analysis of the Infinite), recently translated into English [Eul 3], contains a mixture of precalculus topics, infinite series, continued fractions and geometric problems, and is the perfect place to experience Euler's enthusiasm and great skill in formal manipulation. I also recommend the English translations of some Euler papers in the Smith [Smi, D] and Struik [Str] sourcebooks, an Euler memoir in [Pol, Vol. 1, chap. 6] and the special Euler issue of Mathematics Magazine (November, 1983 - Vol. 56, No. 5), which contains seven articles describing his work and its influence.

Euler's enthusiasm for any interesting question about the integers led to an enormous variety of contributions as described earlier. We present a few of them here, and discuss the rest throughout this book.

It would be natural to begin our discussion with Euler's results in divisibility, however, because Gauss' introduction of the congruence notation leads to a great simplification of this theory, we will present it when we discuss Gauss' work (chap. 8 - 11).

2. Partitions of a Number

In 1740, the Berlin mathematician Naudé wrote to Euler to ask in how many ways a given positive integer n can be written as the sum of r distinct positive integers. Euler quickly solved the problem and within a few months sent a memoir on the subject to the Petersburg Academy. In this rather innocuous way, he started a new area of number theory which he continued to think about for many years. He proved many beautiful theorems and saw connections between his results and other parts of number theory.

First Euler introduced the concept of a **partition** of a positive number n into r **parts** as a sequence, $n_1 \leq n_2 \leq \cdots \leq n_r$, of positive integers such that $n = n_1 + n_2 + \cdots + n_r$, where the n_i are the parts. For example, the partitions of 4 are

$$1+1+1+1\,,\ \ 1+1+2\,,\ \ 1+3\,,\ \ 2+2\,,\ \ 4\,.$$

Let **p(n)** denote the number of partitions of n into any number of parts, where we take $p(0) = 1$. To study the sequence $\{p(n)\}$, Euler introduced its generating function $\sum p(n)x^n$. This is probably the first time that the generating function of a sequence $a_0, a_1, \cdots$, i.e., the power series $\sum a_n x^n$, was used as a systematic tool to study the properties of the sequence.

Euler showed that

$$1 + p(1)x + p(2)x^2 + \cdots = \frac{1}{1-x} \cdot \frac{1}{1-x^2} \cdot \frac{1}{1-x^3} \cdots \tag{1}$$

To see this, we expand the right hand side into a product of geometric series:

$$\frac{1}{1-x} \cdot \frac{1}{1-x^2} \cdot \frac{1}{1-x^3} \cdots = \left(1 + x^{1\cdot 1} + x^{2\cdot 1} + x^{3\cdot 1} + \cdots\right) \cdot \\ \left(1 + x^{1\cdot 2} + x^{2\cdot 2} + x^{3\cdot 2} + \cdots\right)\left(1 + x^{1\cdot 3} + x^{2\cdot 3} + \cdots\right) \cdots \tag{2}$$

Recall that a term from an infinite product is defined to be a product of monomials, one from each factor, such that all but a finite number are 1. Thus, e.g.,

$$x^{2\cdot 1} x^{1\cdot 3} x^{4\cdot 7} = x^{33}$$

is a term of the product in (1) and contributes a 1 to the coefficient of x^{33} on the left hand side. At the same time, the exponent $2 \cdot 1 + 1 \cdot 3 + 4 \cdot 7 = 33$ can be thought of as representing the partition

$$1 + 1 + 3 + 7 + 7 + 7 + 7$$

of 33. Thus each term of the product determines a partition of its exponent; the term from the first factor gives the number of 1's in the partition, the term from the second factor gives the number of 2's and so on. Note that the 1 chosen from the other factors is just x^0, indicating that the partition has no parts from that factor. So, in our example, the partition has no parts equal to $2, 4, 5, 6, 8, 9, 10, \cdots$. Conversely, each partition of an exponent determines a term of the product. This one to one correspondence between partitions and terms of the infinite product shows that the infinite product is identical with the left hand side of (1).

Note that we did not worry about convergence (neither did Euler). Our manipulations can be justified by an algebra of formal power series and infinite products without any regard for convergence (see [Wil], [Van-Wil] and[Niv]).

If we want to count partitions of a number with restrictions on the parts, then we can often derive the generating function by dropping those terms that violate the restrictions from the factors of (2). For example, the number of partitions of n into distinct parts is the coefficient of x^n in the series for

$$(1 + x)(1 + x^2)(1 + x^3)(1 + x^4) \cdots, \tag{3}$$

since we have dropped the terms corresponding to repeated parts. This, of course is one answer to Naudé's question to Euler. Likewise, dropping the factors in (2) corresponding to even parts, we see that

$$\frac{1}{1-x} \cdot \frac{1}{1-x^3} \cdot \frac{1}{1-x^5} \cdot \frac{1}{1-x^7} \cdots \tag{4}$$

is the generating function for the number of partitions of n into odd parts.

Multiplying (3) by

$$\frac{(1 + x)(1 + x^2)(1 + x^3) \cdots}{(1 + x)(1 + x^2)(1 + x^3) \cdots} = 1$$

and rearranging terms in the numerator, we have

$$(1+x)(1+x^2)(1+x^3)\cdots$$

$$=\frac{(1+x)(1-x)}{(1-x)}\frac{(1+x^2)(1-x^2)}{(1-x^2)}\frac{(1+x^3)(1-x^3)}{(1-x^3)}\frac{(1+x^4)(1-x^4)}{(1-x^4)}\cdots$$

$$=\frac{(1-x^2)(1-x^4)(1-x^6)(1-x^8)\cdots}{(1-x)(1-x^2)(1-x^3)(1-x^4)(1-x^5)\cdots},$$

which, by cancelling common terms in the numerator and the denominator,

$$=\frac{1}{(1-x)(1-x^3)(1-x^5)}\cdots,$$

the right hand side of (4). Therefore *the number of partitions of n into distinct parts equals the number of partitions of n into odd parts.* This illustrates the power of generating functions.

Euler found the theory of partitions a perfect subject to exercising his great skill in formal manipulation, and he proved many important identities. In the 19th century, C.G.J. Jacobi had some interest in partitions and J.J. Sylvester introduced combinatorial ideas (see their collected works). In the present century, Hardy, Littlewood and Ramanujan revived interest in partitions with their work on the asymptotic behavior of p(n), and partitions are currently of intense interest to number theorists and combinatorialists.

Euler's *Analysis of the Infinite* [Eul 3, Vol. 1, ch. 16] provides a very readable introduction to partitions and Hardy and Wright [Har - Wri] present a more modern introduction. The *Theory of Partitions* [And 1] by George Andrews is the most thorough book on the subject, and his introductory book on number theory [And 2] gives a good idea of how one of the world's leading experts thinks about partitions. Stanton and White [Sta - Whi] discuss recent combinatorial advances.

Recently Steven Weinberg, a Nobel prize winner in physics, needed information about $p(n)$ to calculate the number of states that would arise in a vibrating string of a given mass [Wein]. Partition questions also arise in problems in statistical mechanics [Bax].

Beyond partitions, Euler saw an even wider use for power series in number theory. In a letter to Goldbach, he noted that the coefficient a_n in the series

$$\sum a_n x^n = \left(\sum x^{n^2}\right)^4$$

is the number of ways to represent n as a sum of four integer squares. Hence Fermat's claim that every positive integer is the sum of four squares would be true if one could prove that $a_n > 0$, for all n. In the 1800's, Jacobi used this representation when he applied the theory of elliptic functions to prove the four square theorem.

3. The Beginning of Analytic Number Theory; Prime Numbers, Zeta Functions, Bernoulli Numbers

From the time of Fermat to that of Euler, there was no progress and basically no interest in number theory. This period was primarily devoted to the development of analysis. Euler not only revived number theory but applied analytic methods to these studies. Leibniz, Euler and some of the Bernoulli's (Jacques, Johann I, Nicolas II, and Daniel) were particularly interested in summing infinite series since they were fascinated by results such as Leibniz's, viz.,

$$\frac{\pi}{4} = 1 - \frac{1}{3} + \frac{1}{5} - \frac{1}{7} + \cdots.$$

Jacques Bernoulli had tried, without success, to find the sum of the reciprocals of the squares

$$1 + \frac{1}{4} + \frac{1}{9} + \frac{1}{16} + \frac{1}{25} + \cdots.$$

Euler succeeded where Bernoulli had failed and, in fact, went much further. We present some of this work as explained by George Polya [Pol, Vol. 1, pp. 17–22], as it beautifully illustrates Euler's very daring use of analogy for finding new results.

> "He found various expressions for the desired sum (definite integrals, other series), none of which satisfied him. He used one of these expressions to compute the sum numerically to seven places (1.644934). Yet this is only an approximate value and his goal was to find the exact value. He discovered it, eventually. Analogy led him to an extremely daring conjecture.
>
> (1) We begin by reviewing a few elementary algebraic facts essential to Euler's discovery. If an equation of degree n
>
> $$a_0 + a_1x + a_2x^2 + \cdots + a_nx^n = 0$$

has different roots

$$\alpha_0, \alpha_1, \ldots, \alpha_n$$

the polynomial on its left hand side can be represented as a product of linear factors

$$a_0 + a_1 x + a_2 x^2 + \cdots + a_n x^n = a_n (x - \alpha_0)(x - \alpha_1) \cdots (x - \alpha_n) .$$

By comparing the terms with the same power of x on both sides of this identity, we derive the well known relations between the roots and the coefficients of an equation, the simplest of which is

$$a_{n-1} = -a_n(\alpha_1 + \alpha_2 + \cdots + \alpha_n) ;$$

we find this by comparing the terms with x^{n-1}.

There is another way of presenting the decomposition in linear factors. If none of the roots $\alpha_0, \alpha_1, \ldots, \alpha_n$ is equal to 0, or (which is the same) if α_0 is different from zero, we have also

$$a_0 + a_1 x + a_2 x^2 + \cdots + a_n x^n$$
$$a = a_0 \left(1 - \frac{x}{\alpha_0}\right)\left(1 - \frac{x}{\alpha_1}\right) \cdots \left(1 - \frac{x}{\alpha_n}\right)$$

and

$$a_1 = -a_0 \left(\frac{1}{\alpha_1} + \frac{1}{\alpha_2} + \cdots + \frac{1}{\alpha_n}\right) .$$

There is still another variant. Suppose that the equation is of degree $2n$, has the form

$$b_0 - b_1 x^2 + b_2 x^4 - \cdots + (-1)^n b_n x^{2n} = 0$$

and $2n$ different roots

$$\beta_1,\ -\beta_1,\ \beta_2,\ -\beta_2,\ \cdots \beta_n,\ -\beta_n .$$

Then

$$b_0 - b_1 x^2 + b_2 x^4 - \cdots + (-1)^n b_n x^{2n}$$
$$= b_0 \left(1 - \frac{x^2}{\beta_0}\right)\left(1 - \frac{x^2}{\beta_1}\right) \cdots \left(1 - \frac{x^2}{\beta_n}\right)$$

and

$$b_1 = -b_0 \left(\frac{1}{\beta_1^2} + \frac{1}{\beta_2^2} + \cdots + \frac{1}{\beta_n^2}\right) .$$

(2) Euler considers the equation

$$\sin x = 0$$

or

$$\frac{x}{1} - \frac{x^3}{1 \cdot 2 \cdot 3} + \frac{x^5}{1 \cdot 2 \cdot 3 \cdot 4 \cdot 5} - \frac{x^7}{1 \cdot 2 \cdot 3 \cdot 4 \cdot 5 \cdot 6 \cdot 7} + \cdots = 0 .$$

The left hand side has an infinity of terms, is of "infinite degree." Therefore, it is no wonder, says Euler, that there is an infinity of roots

$$0,\ \pi,\ -\pi,\ 2\pi,\ -2\pi,\ 3\pi,\ -3\pi,\ \cdots .$$

Euler discards the root 0. He divides the left hand side of the equation by x, the linear factor corresponding to the root 0, and obtains so the equation

$$1-\frac{x^2}{1\cdot2\cdot3}+\frac{x^4}{1\cdot2\cdot3\cdot4\cdot5}-\frac{x^6}{1\cdot2\cdot3\cdot4\cdot5\cdot6\cdot7}+\cdots=0$$

with the roots

$$0,\ \pi,\ -\pi,\ 2\pi,\ -2\pi,\ 3\pi,\ -3\pi,\ \cdots .$$

We have seen an analogous situation before, under (1), as we discussed the last variant of the decomposition in linear factors. Euler concludes by analogy, that

$$\begin{aligned}\frac{\sin x}{x} &= 1-\frac{x^2}{1\cdot2\cdot3}+\frac{x^4}{1\cdot2\cdot3\cdot4\cdot5}-\frac{x^6}{1\cdot2\cdot3\cdot4\cdot5\cdot6\cdot7}+\cdots\\ &= \left(1-\frac{x^2}{\pi^2}\right)\left(1-\frac{x^2}{4\pi^2}\right)\left(1-\frac{x^2}{9\pi^2}\right)\cdots,\\ \frac{1}{2\cdot3} &= \frac{1}{\pi^2}+\frac{1}{4\pi^2}+\frac{1}{9\pi^2}+\cdots,\\ 1+\frac{1}{4}+\frac{1}{9}+\frac{1}{16}+\cdots &= \frac{\pi^2}{6}.\end{aligned}$$

This is the series that withstood the efforts of Jacques Bernoulli - but it was a daring conclusion.

(3) Euler knew very well that his conclusion was daring. "The method was new and never used yet for such a purpose," he wrote ten years later. He saw some objections himself and many objections were raised by his mathematical friends when they recovered from their first admiring surprise.

Yet Euler had his reasons to trust his discovery. First of all, the numerical value for the sum of the series which he has computed before, agreed to the last place with $\frac{\pi^2}{6}$. Comparing further coefficients for his expression of $\sin x$ as a product, he found the sum of other remarkable series, as that of the reciprocals of the fourth powers,

$$1+\frac{1}{16}+\frac{1}{81}+\frac{1}{256}+\frac{1}{625}+\cdots=\frac{\pi^4}{90}.$$

Again he examined the numerical value and again he found agreement.

(4) Euler also tested his method on other examples. Doing so he succeeded in rederiving the sum $\frac{\pi^2}{6}$ for Jacques Bernoulli's series by various

modifications of his first approach. He succeeded also in rediscovering by his method the sum of an important series due to Leibniz.

Let us discuss the last point. Let us consider, following Euler, the equation

$$1 - \sin x = 0 .$$

It has the roots

$$\frac{\pi}{2}, \frac{-3\pi}{2}, \frac{5\pi}{2}, \frac{-7\pi}{2}, \frac{9\pi}{2}, \frac{-11\pi}{2}, \dots .$$

Each of these roots is, however, a double root. (The curve $y = \sin x$ does not intersect the line $y = 1$ at these abscissas, but is tangent to it. The derivative of the left hand side vanishes for the same values of x, but not the second derivative.) Therefore, the equation

$$1 - \frac{x}{1} + \frac{x^3}{1 \cdot 2 \cdot 3} - \frac{x^5}{1 \cdot 2 \cdot 3 \cdot 4 \cdot 5} + \cdots = 0$$

has the roots

$$\frac{\pi}{2}, \frac{\pi}{2}, \frac{-3\pi}{2}, \frac{-3\pi}{2}, \frac{5\pi}{2}, \frac{5\pi}{2}, \frac{-7\pi}{2}, \frac{-7\pi}{2}, \dots$$

and Euler's analogical conclusion leads to the decomposition in linear factors

$$\begin{aligned} 1 - \sin x &= 1 - \frac{x}{1} + \frac{x^3}{1 \cdot 2 \cdot 3} - \frac{x^5}{1 \cdot 2 \cdot 3 \cdot 4 \cdot 5} + \cdots \\ &= \left(1 - \frac{2x}{\pi}\right)^2 \left(1 + \frac{2x}{3\pi}\right)^2 \left(1 - \frac{2x}{5\pi}\right)^2 \left(1 + \frac{2x}{7\pi}\right)^2 \cdots . \end{aligned}$$

Comparing the coefficients of x on both sides, we obtain

$$-1 = -\frac{4}{\pi} + \frac{4}{3\pi} - \frac{4}{5\pi} + \frac{4}{7\pi} - \cdots ,$$

$$\frac{\pi}{4} = 1 - \frac{1}{3} + \frac{1}{5} - \frac{1}{7} + \frac{1}{9} - \frac{1}{11} + \cdots .$$

This is Leibniz's celebrated series; Euler's daring procedure led us to a known result. "For our method" says Euler, "which may appear to some as not reliable enough, a great confirmation comes here to light. Therefore, we should not doubt at all of the other things which are derived by the same method."

(5) Yet Euler kept on doubting. He continued the numerical verification described above under (3), examined more series and more decimal places, and found agreement in all cases examined. He tried other approaches, too, and, finally, he succeeded in verifying not only numerically, but exactly, the value $\frac{\pi^2}{6}$ for Jacques Bernoulli's series. He found a new proof. This proof, although hidden and ingenious, was

based on more usual considerations and was accepted as completely rigorous. Thus, the most conspicuous consequence of Euler's discovery was satisfactorily verified.

These arguments, it seems, convinced Euler that his result was correct[3]"

---

[3] Much later, almost ten years after his first discovery, Euler returned to the subject, answered the objections, completed to some extent his original heuristic approach, and gave a new, essentially different proof. See L. Euler, *Opera Omnia*, ser. 1, vol. 14, p. 73–86, 138–155, 177–186, and also p. 156–176, containing a note by Paul Stackel on the history of the problem."

Euler did not stop here but went on to consider the sums of the reciprocals of higher powers. In 1736, he discovered one of his most beautiful results:

$$\frac{1}{1^{2k}} + \frac{1}{2^{2k}} + \frac{1}{3^{2k}} + \cdots + = \frac{2^{2k-1}\pi^{2k}\,|B_{2k}|}{(2k)!}.$$

The B_{2k} are the Bernoulli numbers defined by

$$\frac{x}{e^x - 1} = B_0 + \frac{B_1 x}{1!} + \frac{B_2 x^2}{2!} + \cdots.$$

Even today, almost nothing is known about the sums of reciprocals of odd powers, beyond Apéry's recent proof that $\sum \frac{1}{n^3}$ is irrational (see chapter 21 and [Poo]).

Exercise: Prove that $B_{2k+1} = 0$, for $k > 0$. Hint: Differentiate $B(x) = \frac{x}{e^x-1}$ twice and set $x = 0$ to find B_1. Let $f(x) = B(x) - B_1 x$ and show that $f(x) - f(-x) = 0$.

The Bernoulli numbers were introduced by Jacques Bernoulli in his *Ars Conjectandi* (1713) to evaluate $1^k + 2^k + \cdots + (n-1)^k$. He showed that

$$1^k + 2^k + \cdots + (n-1)^k = \frac{1}{k+1}\left((n+B)^{k+1} - B^{k+1}\right),$$

where, after expanding the right hand side, we identify B^m with the Bernoulli number B_m. Thus, e.g., when $k = 1$, we have

$$\begin{aligned}1^1 + 2^1 + \cdots + (n-1)^1 &= \frac{1}{2}(n^2 + 2B^1 n) \\ &= \frac{1}{2}(n^2 + 2B_1 n) \\ &= \frac{1}{2}\left(n^2 + 2\left(\frac{-1}{2}\right)n\right) \qquad \left\{B_1 = -\frac{1}{2}\right\} \\ &= \frac{n(n-1)}{2}.\end{aligned}$$

See [Ire - Ros], [Sch - Opo] and [Kli] for more details.

After discovering his result on $\sum \frac{1}{n^{2k}}$, Euler then began the study of the **zeta function,**

$$\zeta(s) = \sum_n \frac{1}{n^s},$$

as a function of s, *for real* s.

By the comparison test for series and improper integrals (from elementary calculus), we have, for $s > 1$,

$$\frac{1}{s-1} = \int_1^\infty \frac{1}{x^s} dx \le \zeta(s) \le 1 + \int_1^\infty \frac{1}{x^s} dx = 1 + \frac{1}{s-1}. \qquad (1)$$

Thus $\zeta(s)$ is convergent and well defined for $s > 1$. Since $\frac{1}{n^s}$ is continuous and since for any $\epsilon > 0$ and all $s > 1 + \epsilon$ we have

$$\sum \frac{1}{n^s} < \sum \frac{1}{n^{1+\epsilon}},$$

$\sum \frac{1}{n^s}$ converges uniformly for $s > 1 + \epsilon$, and $\zeta(s)$ is a continuous function of s. Euler did not know about uniform continuity and just assumed that $\zeta(s)$ was continuous. By (1), as $\mathbf{s} \to \mathbf{1}^+ (s \to 1, s > 1)$, $\zeta(s)(s-1) \to 1$, and we see that $\lim_{s \to 1^+} \zeta(s) = \infty$, i.e., $\zeta(s)$ has a pole of order one at $s = 1$.

Euler looked at the infinite product

$$\prod_{p \text{ prime}} \frac{1}{1 - \frac{1}{p^s}} = \prod_p \left(1 + \frac{1}{p^s} + \frac{1}{p^{2s}} + \frac{1}{p^{3s}} + \cdots\right).$$

As with our reasoning about infinite products for partitions, a term of the product will be of the form

$$\left(\frac{1}{p_1^{a_1} p_2^{a_2} \cdots p_k^{a_k}}\right)^s = \frac{1}{n^s},$$

where $n = p_1^{a_1} p_2^{a_2} \cdots p_k^{a_k}$. By unique factorization every n appears once and only once.

Thus

$$\zeta(s) = \sum \frac{1}{n^s} = \prod_p \frac{1}{1 - \frac{1}{p^s}},$$

which is known as the **Euler product formula**. This is essentially an analytic way of stating unique factorization.

Corollary: *The number of primes is infinite.*

Proof: If there are only a finite number of primes, then, as $s \to 1^+$, $\prod_p \frac{1}{1-\frac{1}{p^s}}$ must converge to a finite value. But this contradicts $\zeta(s) \to \infty$.

This may seem a rather difficult way of proving a theorem that Euclid had already proved by very elementary reasoning, but in the 19th century Dirichlet vastly extended the ideas used here and they form the basis for the most important applications of analysis in modern number theory.

Theorem: (Euler): $\displaystyle\sum_{p \text{ a prime}} \frac{1}{p}$ *diverges.*

Proof:

$$\lim_{s \to 1^+} \zeta(s) = \infty \implies \lim_{s \to 1^+} (\log \zeta(s)) = \infty .$$

But

$$\log \zeta(s) = \log \left(\prod_p \frac{1}{1 - \frac{1}{p^s}}\right) = \sum_p \log\left(\frac{1}{1 - p^{-s}}\right),$$

which, by Taylor series expansion,

$$= \sum_p \sum_{n=1}^{\infty} \frac{p^{-ns}}{n}$$

$$= \sum_p \frac{1}{p^s} + \sum_p \sum_{n=2}^{\infty} \frac{p^{-ns}}{n} .$$

If the latter term is bounded, independent of s for $s > 1$, then, since $\lim_{s\to 1^+} \zeta(s) = \infty$, we must have

$$\lim_{s\to 1^+} \sum \frac{1}{p^s} = \infty .$$

Since

$$\sum_p \sum_2^\infty \frac{1}{np^{ns}} < \sum_p \sum_2^\infty \frac{1}{p^{ns}} = \sum_p \frac{1}{p^{2s}(1-p^{-s})} = \sum_p \frac{1}{(p^s(p^s-1))}$$
$$< \sum_p \frac{1}{p(p-1)} < \sum_2^\infty \frac{1}{n(n-1)} < \sum \frac{1}{(n-1)^2} < \infty ,$$

we are done.

Euler's proof of the theorem was based directly on the infinite product $\prod_p \left(1 - \frac{1}{p}\right)$ [Edw 2, pg. 1]. Note that $\sum \frac{1}{n^2} < \infty$; thus our result implies that the primes are 'denser' than the squares.

In the 1930's Paul Erdos conjectured that if $\{a_n\}$ is a sequence of positive integers, such that $\sum \frac{1}{a_n} = \infty$, then the sequence contains arbitrarily long arithmetic progressions. This conjecture, which is still unsolved, has motivated the development of deep connections between number theory and ergodic theory [Fur].

In 1859, Riemann published his fundamental paper on the zeta function and the distribution of prime numbers, where, for the first time, $\zeta(s)$ was regarded as a function of a complex variable. This paper implicitly includes Riemann's conjecture on the location of the complex zeros of the zeta function, which is probably the most famous unsolved problem in mathematics. There is a functional equation for the zeta function which is usually credited to Riemann. However, at least as a formal identity, it appeared in Euler's work some hundred years earlier with remarks as to its potential importance. Edwards [Edw 2] gives a thorough historically oriented presentation of the theory of the zeta function and Weil [Wei 1] provides a detailed history of the topics in this section and their connection with other work of Euler on infinite series.

4. Arithmetic Functions

Euler introduced the following functions and studied their properties:

$$\begin{aligned}\varphi(\mathbf{n}) &= \text{the number of positive integers } k \text{ such that} \\ &\qquad 1 \le k \le n \quad \text{and} \quad (k, n) = 1 \, . \\ \mathbf{d}(\mathbf{n}) &= \text{number of positive divisors of } n. \\ \sigma(\mathbf{n}) &= \text{sum of the positive divisors of } n.\end{aligned}$$

$\varphi(n)$ is known as the **Euler phi function**. (Exercise - prove that $\sum_{d|n} \varphi(d) = n$.)

These functions are all multiplicative, i.e.,

$$(m, n) = 1 \implies f(mn) = f(m) f(n) \, ,$$

and this result is one way to explore their properties. We do not prove this here (see sec. 9.2 for a proof for $\varphi(n)$ and [Ada -Gol] for the others).

Assuming the multiplicative property, we derive a formula for $\varphi(n)$. If $n = p_1^{a_1} \cdots p_m^{a_m}$, where the p_i are distinct primes, then, by the multiplicative property,

$$\varphi(n) = \varphi\left(p_1^{a_1}\right) \cdots \varphi(p_1^{a_1}) \, ,$$

and our problem is reduced to calculating $\varphi(p^k)$for all k and all primes p. The only numbers from the set $\{1, 2, \ldots, p^k - 1, p^k\}$ which are not relatively prime to p^k are those divisible by p. These are the p^{k-1} number $ps, s = 1, 2, \ldots, p^{k-1}$. Therefore

$$\varphi(p^k) = p^k - p^{k-1} = p^k\left(1 - \frac{1}{p}\right)$$

and

$$\begin{aligned}\varphi(n) &= p_1^{a_1} \cdots p_m^{a_m}\left(1 - \frac{1}{p_1}\right) \cdots \left(1 - \frac{1}{p_m}\right) \\ &= n\left(1 - \frac{1}{p_1}\right) \cdots \left(1 - \frac{1}{p_m}\right) .\end{aligned}$$

Euler generalized Fermat's little theorem (sec. 2.7) and proved that if $(x, m) = 1$, then m divides $x^{\varphi(m)} - 1$ (sec. 9.3).

With the same notation as above and the assumption of multiplicativity, we see that

$$d(n) = (a_1 + 1)(a_2 + 1) \cdots (a_m + 1) \, ,$$

since, for the prime factor p_i of a divisor of n, we may choose any of the powers p_i^k with $0 \leq k \leq a_i$. It is a simple exercise to derive a formula for $\sigma(n)$.

Euler's memoir on $\sigma(n)$ [Pol, Vol. 1, pp. 91–98] contains many beautiful results relating to the primes and partitions. It is a perfect illustration of Euler's methods, viz., experimenting, guessing results, verifying their plausibility, and his conviction of their correctness even when he cannot prove it.

The following equalities [Zag, sec. 1.2]

$$\sum \frac{\sigma(n)}{n^s} = \zeta(s)\zeta(s-1)\,,$$

$$\sum \frac{\varphi(n)}{n^s} = \frac{\zeta(s-1)}{\zeta(s)}\,,$$

$$\sum \frac{d(n)}{n^s} = (\zeta(s))^2\,,$$

relate the zeta function to arithmetic functions. I don't know if Euler knew these relations, but they allow the use of analytic methods to study these arithmetic functions. For more on arithmetic functions, see Hardy and Wright [Har - Wri].

5. The Beginning of Algebraic Number Theory

The first glimmerings of extending the integers (and rationals) to a larger system of number theoretic interest appear in Euler's work on the solutions of Diophantine equations.. This marks the early history of algebraic number theory. Euler presented many of his results in chapters XI, XII and XV of the second part of his *Algebra*. We discuss some of his solutions.

In section 188, Euler discusses a technique for finding integer solutions of equations of the form

$$z^3 = ax^2 + by^2\,, \tag{1}$$

where $a, b \in \mathbf{Z}, a > 0$. In section 193 he specializes these techniques, setting $a = 1, b = 2$ and $y = \pm 1$, to solve $z^3 = x^2 + 2$. We use the latter equation to illustrate his method.

First Euler factors the equation over the complex numbers; so

$$z^3 = x^2 + 2 = (x + \sqrt{-2})(x - \sqrt{-2})\,. \tag{2}$$

Then he takes the very bold step of using arithmetic properties of the set of complex numbers,

$$\mathbf{Z}[\sqrt{-2}] = \{a + b\sqrt{-2} | a, b \in \mathbf{Z}\} ,$$

as the main tool for solving the equation. (It is important to realize that most mathematicians of that time were hesitant to use complex numbers for anything.)

If A, B, C are integers, A and B relatively prime, such that $AB = C^3$, then A and B are the cubes of integers. Euler assumes that the same should be true for numbers in $\mathbf{Z}[\sqrt{-2}]$, i.e., if $\alpha, \beta, \gamma \in \mathbf{Z}[\sqrt{-2}]$, with α and β relatively prime (no common factors in $\mathbf{Z}[\sqrt{-2}]$), and $\alpha\beta = \gamma^3$, then α and β are the cubes of numbers in $\mathbf{Z}[\sqrt{-2}]$. He also assumes that if u and v are relatively prime integers, then $u + v\sqrt{-2}$ and $u - v\sqrt{-2}$ are relatively prime in $\mathbf{Z}[\sqrt{-2}]$.

Let x, z be integer solutions of our equation (2). Then we see that $x + \sqrt{-2}$ and $x - \sqrt{-2}$ are relatively prime in $\mathbf{Z}[\sqrt{-2}]$, and $x + \sqrt{-2}$ is the cube of a number in $\mathbf{Z}[\sqrt{-2}]$, i.e.,

$$x + \sqrt{-2} = \left(a + b\sqrt{-2}\right)^3 , \tag{3}$$

for some $a, b \in \mathbf{Z}$. By taking complex conjugates, we have $x - \sqrt{-2} = \left(a - b\sqrt{-2}\right)^3$; therefore

$$\begin{aligned} z^3 &= \left(a + b\sqrt{-2}\right)^3 \left(a - b\sqrt{-2}\right)^3 \\ &= (a^2 + 2b^2)^3 \end{aligned}$$

and

$$z = a^2 + 2b^2 .$$

Expanding the right hand side of (3) and equating the imaginary parts, we have

$$1 = b(3a^2 - 2b^2) ,$$

which easily implies $b = 1, a = \pm 1, z = 3, x = \pm 5$. Euler says, without proof: "it follows that there is no square, except 25, which has the property required." For a further discussion of role of equation of the form $x^2 = z^3 + k$, with regard to both integer and rational solutions, in the development of number theory, see [Mor 1].

In the very next section (194), Euler applies a slight generalization of the same ideas to $z^3 = 5x^2 + 7$. First he considers the factorization of the more general equation

$$z^3 = 5x^2 + 7y^2 = \left(x\sqrt{5} + y\sqrt{-7}\right)\left(x\sqrt{5} - y\sqrt{-7}\right) .$$

He assumes that $x\sqrt{5} + y\sqrt{-7}$ is the cube of a number of the form $p\sqrt{5} + q\sqrt{-7}$, $p, q \in \mathbf{Z}$, i.e., $x\sqrt{5} + y\sqrt{-7} = \left(p\sqrt{5} + q\sqrt{-7}\right)^3$, then he expands the latter equation and gets

$$x = 5p^3 - 21pq^2 , \quad y = 15p^2q - 7q^3 .$$

To quote Euler [Eul 2]:

> "... so that in our example y being $= \pm 1$, we have $15p^2q - 7q^3 = q(15p^2 - 7q^2) = \pm 1$; therefore q must be a divisor of 1; that is to say $q = \pm 1$; consequently , we shall have $15p^2 - 7 = \pm 1$; from which, in both cases, we get irrational values for p: but from which we must not, however, conclude that the question is impossible, since p and q might be such fractions, that $y = 1$, and that x would become an integer; and this is what really happens; for if $p = \frac{1}{2}$, and $q = \frac{1}{2}$, we find that $y = 1$, and $x = 2$; but there are no other fractions which render the solution possible."

Euler explicitly states, without proof, that these methods yield all solutions of $z^3 = x^2 + 2$ and $z^3 = 5x^2 + 7$. He goes on to discuss the solutions of equations of the form $z^3 = ax^2 + by^2$, when $b < 0$, together with an unsatisfactory explanation of why his methods work when $b > 0$ and not when $b < 0$. In fact, Euler could not have determined the range of validity of his methods with the tools at his command. A more complete treatment requires a deeper knowledge of factorization in algebraic number fields, a subject which is still of great research interest and to which we will return later (chap. 15–17).

Euler never justified his assumptions about the arithmetic properties of complex numbers. His reasoning by analogy with the properties of the integers is reminiscent of his use of analogy is discovering that $\sum \frac{1}{n^2} = \frac{\pi^2}{6}$.

The existence of gaps in Euler's proofs is not the important point here. The *important point* is the boldness of Euler's idea of using complex numbers to derive information about the integers.

EXERCISE: Mimic Euler's loose reasoning to find our formula for Pythagorean triples (sec. 1.3), i.e., find all integer solutions of $z^2 = x^2 + y^2$.

Later, in chapter XV of the *Algebra*, Euler presents a proof of Fermat's last theorem for $n = 3$, i.e. he purports to prove that there are no non trivial integer solutions of $z^3 = x^3 + y^3$. He uses an infinite descent argument and reasoning with numbers in $\mathbf{Z}[\sqrt{-3}] = \{a+b\sqrt{-2} | a, b \in \mathbf{Z}\}$. As in the earlier examples, there are gaps and, in this case, errors in his reasoning. Edwards [Edw 1, sec. 2.3, 2.5] gives a complete analysis of Euler's arguments and shows how to correct and complete the proof using techniques that Euler knew. For a somewhat different proof in the more modern context of factorization in integral domains, see [Gol, L].

Euler was very proud of his methods. In a letter to Lagrange, he compliments him for his use of them. However, Gauss, in his studies of quadratic forms, cyclotomy and reciprocity laws, was the first to fully realize the need for rigor and the great potential of introducing complex numbers in number theory. The systematic development of algebraic number theory began with Gauss and his successors.

Chapter 4

From Euler to Lagrange; The Theory of Continued Fractions

1. Introduction

We now come to continued fractions, our first theme to be treated in some depth. Continued fractions form a natural transition between Euler and Joseph Louis Lagrange (1736–1813), the next great number theorist after Euler. We leave his biographical details until later. Euler's introduction to the subject in his *Analysis of the Infinite* [Eul 3], which emphasizes computation and the connection with infinite series, is the first modern introduction to appear in a text, and Lagrange's appendix to Euler's *Algebra* [Eul 2] is the first systematic introduction to continued fractions in number theory. I recommend both of these beautifully written expositions. In the last two hundred years of number theory the basic theory of continued fractions has not changed significantly. Some good references include [Dav], [Har-Wri], [Lan 1], [Chr], [Per] and [Hur-Krit]. Jones and Thron [Jon-Thr] is the most up to date exposition of the analytic theory of continued fractions, a related subject we do not discuss.

As we shall see, continued fraction's are particularly useful in number theory because they often provide explicit constructions for solving problems, whereas other methods may only give existence proofs.

2. The Basic Notions: Finite and Infinite Continued Fractions

A (finite or infinite) expression of the form

$$\mathbf{a}_0 + \cfrac{\mathbf{b}_1}{\mathbf{a}_1 + \cfrac{\mathbf{b}_2}{\mathbf{a}_2 + \cfrac{\mathbf{b}_3}{\mathbf{a}_3 + \ddots}}}$$

where the a_i and b_i are real or complex numbers, is called a **continued fraction.** Here we will only consider the case where the a_i *and* b_i *are integers.* More general expressions where the a_i and b_i are polynomials are treated in the analytic theory.

For space and typographical reasons, we usually will use the notation

$$\left[\mathbf{a}_0; \frac{\mathbf{b}_1}{\mathbf{a}_1}, \frac{\mathbf{b}_2}{\mathbf{a}_2}, \dots\right],$$

to represent the continued fraction. However, I strongly recommend that when reading this chapter for the first time, you write out all continued fractions. The notation

$$a_0 + \frac{b_1}{a_1+}\frac{b_2}{a_2+}\cdots$$

is also often used in the literature, but we avoid it here.

The $\mathbf{n}^{\mathbf{th}}$ **convergent** of the continued fraction is the finite continued fraction

$$\left[a_0; \frac{b_1}{a_1}, \dots, \frac{b_n}{a_n}\right] = a_0 + \cfrac{b_1}{a_1 + \cfrac{b_2}{a_2 + \cfrac{b_3}{a_3 + \ddots + \cfrac{a_n}{b_n}}}}.$$

We are primarily concerned with the case where $b_1 = b_2 = \cdots = 1$ *and the* a_i *are integers,* $a_i > 0$ *for* $i > 0$, i.e.

$$\left[a_0;\ \frac{1}{a_1}, \frac{1}{a_2}, \ldots\right] = a_0 + \cfrac{1}{a_1 + \cfrac{1}{a_2 + \cfrac{1}{a_3 + \ddots}}}$$

These are **regular** or **simple** continued fractions. The a_i are the **partial quotients** of the continued fraction.

Unless otherwise indicated in what follows, all continued fractions are assumed to be regular.

At first glance, continued fractions are strange looking expressions, but they do arise naturally. One way to obtain a finite continued fractions is by rewriting the sequence of divisions in the Euclidean algorithm (sec. 1.2). For example, applying the algorithm to find gcd(60, 22), which is 2, we have

$$\begin{aligned} 60 &= \underline{2} \times 22 + 16 \\ 22 &= \underline{1} \times 16 + 6 \\ 16 &= \underline{2} \times 6 + 4 \\ 6 &= \underline{1} \times 4 + 2 \\ 4 &= \underline{2} \times 2\,. \end{aligned}$$

We rewrite these equations as

$$\begin{aligned} \frac{60}{22} &= 2 + \frac{16}{22} = \underline{2} + \frac{1}{\frac{22}{16}} \\ \frac{22}{16} &= 1 + \frac{6}{16} = \underline{1} + \frac{1}{\frac{16}{6}} \\ \frac{16}{6} &= 2 + \frac{4}{6} = \underline{2} + \frac{1}{\frac{6}{4}} \\ \frac{6}{4} &= \underline{1} + \frac{1}{2}\,. \end{aligned} \tag{1}$$

Substituting the last equation in the third, the third in the second and the second in the first, we have

$$\frac{60}{22} = \left[2; \frac{1}{1}, \frac{1}{2}, \frac{1}{2}\right] = 2 + \cfrac{1}{1 + \cfrac{1}{2 + \cfrac{1}{1 + \cfrac{1}{2}}}}$$

where the underlined numbers in (1) are the partial quotients.

If we start with the continued fraction and combine the terms, beginning with the last, we obtain $1 + \frac{1}{2} = \frac{3}{2}$, $2 + \frac{1}{\frac{3}{2}} = \frac{8}{3}$, $1 + \frac{1}{\frac{8}{3}} = \frac{11}{8}$, and $2 + \frac{1}{\frac{11}{8}} = \frac{30}{11} = \frac{60}{22}$. Note that at each step, the corresponding convergent is in lowest terms, i.e., the greatest common divisor of the numerator and denominator are relatively prime. Thus the continued fraction does not explicitly give the greatest common divisor of 60 and 22, but it encodes the information indirectly by yielding the reduced fraction $\frac{30}{11}$. In section 5, we will prove that, as in our example, the convergents are always in lowest terms.

Similarly, applying the Euclidean algorithm to find gcd(-18, 7), we have

$$\begin{aligned} -18 &= -\underline{3} \times 7 + 3 \\ 7 &= \underline{2} \times 3 + 1 \\ 3 &= \underline{3} \times 1 \end{aligned}$$

and therefore $\frac{-18}{7} = \left[-3; \frac{1}{2}, \frac{1}{3}\right]$.

One can form a continued fraction of any rational number $\frac{a}{b}$, just as in our examples, by applying the Euclidean algorithm to find $\gcd(a, b)$. If $\frac{a}{b}$ is positive, then all the steps in the Euclidean algorithm yield positive partial quotients. If $\frac{a}{b}$ is negative, then the first partial quotient is negative and all subsequent ones are positive. In either case the continued fraction is regular. Thus we have

Theorem: *Every rational number equals a finite regular continued fraction. Conversely, every finite regular continued fraction equals a rational number.*

We will discuss the uniqueness of the continued fraction in section 5.

Exercise: Compute the continued fractions of $\frac{23}{2}$, $\frac{23}{3}$, $\frac{23}{4}$, $\frac{23}{5}$, $-\frac{3}{5}$ and $\frac{6}{33}$. Verify that the convergents are always in lowest terms.

Since finite continued fractions equal rational numbers, it seems that if there is any natural way to associate irrational numbers with continued fractions, the continued fractions should be infinite.

One way an infinite continued fraction can arise is from playing with special quadratic equations. Consider, e.g., $x^2 - 2x - 1 = 0$. If α is a root, then $\alpha^2 = 2\alpha + 1$ or $\alpha = 2 + \frac{1}{\alpha}$. Iterating the last equation, we have

$$\alpha = 2 + \cfrac{1}{\alpha} = 2 + \cfrac{1}{2 + \cfrac{1}{\alpha}} = 2 + \cfrac{1}{2 + \cfrac{1}{2 + \cfrac{1}{\alpha}}} = \cdots$$

This leads us to consider the infinite continued fraction $\left[2; \frac{1}{2}, \frac{1}{2}, \cdots\right]$. Later we will prove that the sequence $2, \left[2; \frac{1}{2}\right], \left[2; \frac{1}{2}, \frac{1}{2}\right], \left[2; \frac{1}{2}, \frac{1}{2}, \frac{1}{2}\right], \cdots$ converges to $1 + \sqrt{2}$, the positive root of our equation. We will explore the relationship between roots of quadratic equations and infinite continued fraction in section 8. Unfortunately this example gives no hint as to whether every irrational number can be represented by a continued fraction. For that we return to the relationship between rational numbers and finite continued fractions.

Recall that the **greatest integer function** $[\alpha]$ is defined, for α a real number, by

$$[\alpha] = \text{ the unique integer } k \text{ such that } k \leq a < k + 1\,.$$

For example, $[2] = 2$, $[2.55] = 2$, $[-2] = -2$, $[-2.55] = -3$ and $\left[\sqrt{11}\right] = 3$.

We now consider another way of viewing the set of equations (1). The first equation $\frac{60}{22} = 2 + \frac{1}{\frac{22}{16}}$ can be rewritten as

$$\frac{60}{22} = \left[\frac{60}{22}\right] + \frac{1}{\alpha_1}\,, \quad \text{where } \alpha_1 = \frac{22}{16} > 1,$$

and the second, third and fourth equations as

$$\alpha_1 = [\alpha_1] + \frac{1}{\alpha_2}\ \text{,where } \alpha_2 = \frac{16}{6} > 1\ ,$$
$$\alpha_2 = [\alpha_2] + \frac{1}{\alpha_3}\ \text{,where } \alpha_3 = \frac{6}{4} > 1\ ,$$
$$\alpha_3 = [\alpha_3] + \frac{1}{\alpha_4}\ \text{,where } \alpha_4 = 2 \textit{ is an integer.}$$

This format generalizes immediately to any real number α. We follow Lagrange, but with a more systematic notation (Euler, Lagrange and Gauss avoided the use of subscripts. Why? Was it a printing problem?).

THE CONTINUED FRACTION ALGORITHM

A) If α is an integer, set $a_0 = \alpha$ and stop.

If α is not an integer, let

$$\alpha = a_0 + \frac{1}{\alpha_1},\quad \text{where } a_0 = [\alpha] \text{ and } \alpha_1 = \frac{1}{\alpha - a_0} > 1;\ \text{so}$$
$$\alpha = \left[a_0; \frac{1}{\alpha_1}\right].$$

B) If α_1 is an integer, set $a_1 = \alpha_1$ and stop.

If α_1 is not an integer, let

$$\alpha_1 = a_1 + \frac{1}{\alpha_2},\quad \text{where } a_1 = [\alpha_1] \text{ and } \alpha_2 = \frac{1}{\alpha_1 - a_1} > 1;\ \text{so}$$
$$\alpha = \left[a_0; \frac{1}{a_1}, \frac{1}{\alpha_2}\right].$$

$$\vdots$$

C) If α_n is an integer, set $a_n = \alpha_n$ and stop.

If α_n is not an integer, let

$$\alpha_n = a_n + \frac{1}{\alpha_{n+1}}, \text{ where } a_n = [\alpha_n] \text{ and } \alpha_{n+1} = \frac{1}{\alpha_n - a_n} > 1; \text{ so}$$

$$\alpha = \left[a_0; \frac{1}{a_1}, \frac{1}{a_2}, \ldots, \frac{1}{a_n}, \frac{1}{\alpha_{n+1}} \right].$$

$$\vdots$$

Then $\left[a_0; \frac{1}{a_1}, \frac{1}{a_2}, \ldots, \frac{1}{a_n}, \cdots \right]$, where the sequence of a_i's may be finite or infinite, is the **continued fraction of** α, and we write

$$\alpha = \left[a_0; \frac{1}{a_1}, \frac{1}{a_2}, \ldots, \frac{1}{a_n}, \cdots \right].$$

Of course, the use of the equal sign will only make sense if the nth convergent

$$\left[a_0; \frac{1}{a_1}, \ldots, \frac{1}{a_{n-1}}, \frac{1}{a_n} \right] \longrightarrow \alpha, \quad \text{as } n \longrightarrow \infty$$

(this means that the last convergent equals α when the sequence is finite). We will prove this in section 6.

If α is rational, this algorithm is just a reformulation of our use of the Euclidean algorithm and we get the same finite continued fraction.

If α is irrational, then $\alpha_1 = \frac{1}{\alpha - a_0}$ is irrational (a_0 is an integer). Similarly, if α_n is irrational, then $\alpha_{n+1} = \frac{1}{\alpha_n - a_n}$ is irrational. Hence, by induction, all the α_i are irrational and we are led to an infinite continued fraction.

Exercise: Prove that $a_i > 0$ for $i > 0$.

α_i is called the ith **partial remainder** or **complete quotient** of the continued fraction.

Example: Let $\alpha = \sqrt{2}$. Then

$$a_0 = [\alpha] = 1 \Longrightarrow \sqrt{2} = 1 + \frac{1}{\alpha_1}, \text{ where } \alpha_1 = \frac{1}{\sqrt{2} - 1} = \sqrt{2} + 1.$$

$$a_1 = [\alpha_1] = 2 \Longrightarrow \sqrt{2} + 1 = 2 + \frac{1}{\alpha_2} \text{ where } \alpha_2 = \frac{1}{\sqrt{2} - 1} = \sqrt{2} + 1.$$

So $\alpha_1 = \alpha_2$, our process repeats, i.e., $\alpha_1 = \alpha_2 = \alpha_3 = \cdots$, and $\sqrt{2} = \left[1; \frac{1}{2}, \frac{1}{2}, \cdots\right]$.

Adding one to both sides of the equation, we see that $1 + \sqrt{2} = [2; \frac{1}{2}, \frac{1}{2}, \cdots]$, which is consistent with our earlier discussion of the equation $x^2 - 2x - 1 = 0$.

Exercise: Show that

$$\sqrt{3} = \left[1; \frac{1}{1}, \frac{1}{2}, \frac{1}{1}, \frac{1}{2}, \cdots\right] = \left[1; \overline{\frac{1}{1}, \frac{1}{2}}\right] \quad \text{and}$$

$$-1 - \sqrt{3} = \left[-3; \frac{1}{3}, \frac{1}{1}, \frac{1}{2}, \overline{\frac{1}{2}, \frac{1}{1}}\right]$$

where the overlined sequence is repeated periodically. The periodicity in these examples is not an accident (see sec. 8).

Example: One can use the decimal expansion of a number to get terms of its continued fraction (see [Eul 3, pg. 326] and [Eul 2, chap 1, sec. 8 of the appendix]). One such result is

$$\pi = \left[3; \frac{1}{7}, \frac{1}{15}, \frac{1}{1}, \frac{1}{292}, \frac{1}{1}, \frac{1}{1}, \cdots\right],$$

where there is no known pattern for the denominators. The reader is invited to verify this result.

I want to stress that the continued fraction of a real number is a *natural* or *intrinsic* object since it does not depend on any special notation for the number. On the other hand, the decimal expansion of a number is not intrinsic since it depends on the arbitrary choice of the base 10.

3. The Early History

The theory of continued fractions was developed by many mathematicians, some of them unaware of each others work. Euler was the first to use general algebraic formulas in the theory (for example, in his *Analysis of the Infinite* [Eul 3]). As mentioned earlier, Lagrange's appendix to Euler's *Algebra* is the first systematic introduction to continued fractions in number theory. In section 2 of his exposition, he presents the history, prior to Euler, as he knew it:

"Lord Brouncker, I believe, was the first who thought of continued fractions. We know that the continued fraction, which he devised to express the ratio of the circumscribed square to the area of the circle [namely $\frac{4}{\pi}$] was this:

$$1 + \cfrac{1}{2 + \cfrac{9}{2 + \cfrac{25}{2+ \ddots}}}$$

but we are ignorant of the means which led him to it. We only find in the *Arithmetica Infinitorum* some researches on this subject, in which Wallis demonstrates, in an indirect, although ingenious manner, the identity of Brouncker's expression to his, which is $\frac{3\times3\times5\times5\times7\times7\cdots}{2\times4\times4\times6\times6\times\cdots}$. He there also gives the general method of reducing all sorts of continued fractions to vulgar [regular] fractions; but it does not appear that either of those great mathematicians knew the principal properties and singular advantages of continued fraction; and we shall afterwards see, that the discovery of them is chiefly due to Huygens."

Lagrange's version of the history is far from complete or correct. For example, he seems to be unaware of the sixteenth century contributions of Bombelli and Cataldi (see [Smi]), the real creators of continued fractions in their modern form.

The history of continued fractions has always seemed a bit fuzzy to me. However, the recently published "History of Continued Fractions and Padé Approximants" by Claude Brezinski [Bre] is a successful first attempt to present a history of the modern theory. Weil [Wei 1] puts many of these discoveries in the context of the general development of number theory. For the evolution of ideas prior to the modern development, such as the role of the Euclidean algorithm, I recommend the fascinating historical essay by Fowler [Fow]. Fowler is primarily concerned with a new interpretation of large parts of Greek mathematics. A key thesis is that the concept of a continued fraction, in various geometric guises, was central in Greek mathematical thought.

4. The Algebra of Finite Continued Fractions

We now study some algebraic properties of $\left[a_0; \frac{1}{a_1}, \ldots, \frac{1}{a_n}\right]$, where the a_i are independent variables. These results will be immediately applied to obtain arithmetic properties of continued fractions.

Summing a continued fraction: Computing the first three convergents, we have

$$\left[a_0; \frac{1}{a_1}\right] = a_0 + \frac{1}{a_1} = \frac{a_0 a_1 + 1}{a_1},$$

$$\left[a_0; \frac{1}{a_1}, \frac{1}{a_2}\right] = a_0 + \frac{1}{\frac{a_1+1}{a_2}} = a_0 + \frac{1}{\frac{a_1 a_2 + 1}{a_2}} = a_0 + \frac{a_2}{a_1 a_2 + 1}$$

$$= \frac{a_0 a_1 a_2 + a_0 + a_2}{a_1 a_2 + 1},$$

$$\left[a_0; \frac{1}{a_1}, \frac{1}{a_2}, \frac{1}{a_3}\right] = a_0 + \frac{1}{\left[a_1; \frac{1}{a_2}, \frac{1}{a_3}\right]},$$

which by the previous calculation

$$= a_0 + \frac{a_2 a_3 + 1}{a_1 a_2 a_3 + a_1 + a_3}$$

$$= \frac{a_0 a_1 a_2 a_3 + a_0 a_1 + a_0 a_3 + a_2 a_3 + 1}{a_1 a_2 a_3 + a_1 + a_3}.$$

Since

$$\left[a_0; \frac{1}{a_1}, \ldots, \frac{1}{a_n}\right] = a_0 + \frac{1}{\left[a_1; \frac{1}{a_2}, \ldots, \frac{1}{a_n}\right]},$$

it follows by induction that $\left[a_0; \frac{1}{a_1}, \ldots, \frac{1}{a_n}\right]$ is a rational function of a_0, a_1, ..., a_n, i.e., a quotient of polynomials. Let $\langle \mathbf{a_0}, \mathbf{a_1}, \ldots, \mathbf{a_n} \rangle$ denote the numerator of this rational function; $\langle a_0, a_1, \ldots, a_n \rangle$ is a polynomial in the a_i. For example,

$$\langle a_0 \rangle = a_0,\ \langle a_0, a_1 \rangle = a_0 a_1 + 1,\ \langle a_0, a_1, a_2 \rangle = a_0 a_1 a_2 + a_0 + a_2\ ,$$
$$\text{and } \langle a_0, a_1, a_2, a_3 \rangle = a_0 a_1 a_2 a_3 + a_0 a_1 + a_0 a_3 + a_2 a_3 + 1\ .$$

Again, since $\left[a_0; \frac{1}{a_1}, \ldots, \frac{1}{a_n}\right] = a_0 + \frac{1}{\left[a_1; \frac{1}{a_2}, \ldots, \frac{1}{a_n}\right]}$, the denominator of $\left[a_0; \frac{1}{a_1}, \ldots, \frac{1}{a_n}\right]$ is just the numerator of $\left[a_1; \frac{1}{a_2}, \ldots, \frac{1}{a_n}\right]$. Therefore

$$\left[a_0; \frac{1}{a_1}, \ldots, \frac{1}{a_n}\right] = \frac{\langle a_0, a_1, \ldots, a_n\rangle}{\langle a_1, a_2, ..., a_n\rangle}. \tag{1}$$

Analyzing our earlier calculations, as Euler must have done, it is straightforward to conjecture, and an easy exercise to prove by induction, that

$$\langle a_0, a_1, \cdots a_n\rangle = a_0\langle a_1, \ldots, a_n\rangle + \langle a_2, a_3, \ldots, a_n\rangle\,, \quad \text{for } n > 1\,. \tag{2}$$

where we set $\langle \text{empty}\rangle = 1$. For example, $\langle a_0, a_1\rangle = a_0\langle a_1\rangle + 1 = a_0a_1 + 1$. Although we haven't proved that cancellation is not possible in the right hand side of (1), it is not both for polynomials and integers. We have no need of the former and later we will prove the latter.

Euler also gave a direct method for calculating $\langle a_0, a_1, \ldots, a_n\rangle$.

Euler's Rule: *$\langle a_0, a_1, \ldots, a_n\rangle$ is a polynomial obtained by adding the following products of variables:*

(i) the product $a_0a_1 \cdots a_n$,

(ii) every product that can be obtained from $a_0a_1 \cdots a_n$ by omitting any pair of consecutive variables a_ia_{i+1},

(iii) every product that can be obtained from $a_0a_1 \cdots a_n$ by omitting any two non overlapping pairs of consecutive variables a_ia_{i+1} and $a_j\, a_{j+1}$, where $j > i + 1$, and so on.

If n is even, we stop after omitting any $\frac{n}{2}$ pairs. If n is odd, we stop after omitting $\frac{n+1}{2}$ pairs with the usual convention that the final product, the empty product, is 1.

Proof: (exercise - induction using the recursion (2))

It is easy to see that our earlier calculations of $\langle a_0\rangle$, $\langle a_0, a_1\rangle$, $\langle a_0, a_1, a_2\rangle$ follow Euler's rule.

Since Euler's rule is symmetric in the variables, we have

$$\langle a_0, a_1, \ldots, a_n\rangle = \langle a_n, a_{n-1}, \ldots, a_0\rangle\,. \tag{3}$$

Substituting the sequence of symbols $a_n, a_{n-1}, \ldots, a_0$ in equation (2) yields

$$\langle a_n, a_{n-1}, \ldots, a_0 \rangle = a_n \langle a_{n-1}, \ldots, a_0 \rangle + \langle a_{n-2}, \ldots, a_0 \rangle \ ,$$

and, by (3), we have

$$\langle a_0, a_1, \ldots, a_n \rangle = a_n \langle a_0, a_1, \ldots, a_{n-1} \rangle + \langle a_0, a_1, \ldots, a_{n-2} \rangle \ . \qquad (4)$$

Let $\mathbf{p_m} = \langle a_0, a_1, \ldots, a_m \rangle$ and $\mathbf{q_m} = \langle a_1, \ldots, a_m \rangle$, with $\mathbf{p_0} = a_0$ and $\mathbf{q_0} = 1$; so

$$\frac{p_m}{q_m} = \left[a_0; \frac{1}{a_1}, \ldots, \frac{1}{a_m} \right] .$$

Then, by (4), we have

Theorem: *For $m > 0$, p_m and q_m satisfy the* **basic recursions**

$$\mathbf{p_m = a_m p_{m-1} + p_{m-2}} \qquad (5a)$$

$$\mathbf{q_m = a_m q_{m-1} + q_{m-2}} \ , \qquad (5b)$$

where we set $\mathbf{p_{-1}} = 1$, $\mathbf{q_{-1}} = 0$.

These are probably the most important relations for deriving properties of continued fractions. In particular, we have

$$\frac{p_m}{q_m} = \frac{a_m p_{m-1} + p_{m-2}}{a_m q_{m-1} + q_{m-2}} \ . \qquad (6)$$

Finally, there is a basic relation between successive convergents.

Theorem:

$$p_m q_{m-1} - q_m p_{m-1} = (-1)^{m-1} \ . \qquad (7)$$

Proof: We proceed by induction on m. Let $\Delta_m = p_m q_{m-1} - q_m p_{m-1}$. If $m = 1$, then $\Delta_1 = (a_0 a_1 + 1) \cdot 1 - a_0 a_1 = 1$.
Assume the theorem is true for $m = n$, i.e., $\Delta_n = (-1)^{n-1}$. Then

$$\begin{aligned}
\Delta_{n+1} &= p_{n+1} q_n - q_{n+1} p_n \\
&= (a_{n+1} p_n + p_{n-1})\, q_n - (a_{n+1} q_n + q_{n-1})\, p_n, \quad \text{by (5),} \\
&= -\,(p_n q_{n-1} - q_n p_{n-1}) \\
&= -(-1)^{n-1}, \quad \text{by the induction hypothesis,} \\
&= (-1)^n.
\end{aligned}$$

Dividing equation (7) by $q_{m-1}q_m$, we obtain

Corollary:

$$\frac{p_m}{q_m} - \frac{p_{m-1}}{q_{m-1}} = \frac{(-1)^{m-1}}{q_{m-1}q_m} . \tag{8}$$

5. The Arithmetic of Finite Continued Fractions

We first apply the results of the last section to obtain arithmetic properties of finite continued fractions. If a rational number $\frac{a}{b} = \left[a_0; \frac{1}{a_1}, \ldots, \frac{1}{a_n}\right]$, *where the a_i are positive integers for $i > 0$*, then, in the notation of section 4, its convergents are

$$\frac{p_m}{q_m} = \left[a_0; \frac{1}{a_1}, \ldots, \frac{1}{a_m}\right], \quad m \leq n ,$$

where $\frac{p_0}{q_0} = \frac{a_0}{1}$ and $\frac{p_n}{q_n} = \frac{a}{b}$. Since $q_{-1} = 0, q_0 = 1, q_1 = a_0$, and $a_i > 0$ for $i > 0$, we have, by equation (4.5b),

$$q_n = a_n q_{n-1} + q_{n-2} \geq q_{n-1} + q_{n-2} > q_{n-1} + 1$$

for $n \geq 2$, i.e.,

Proposition: $1 = q_0 \leq q_1 < q_2 < \cdots$.

From equation (4.7), we see that if $d|p_m$ and $d|q_m$, then $d|(-1)^{m-1}$; hence

Proposition: *$gcd(p_m, q_m) = 1$, i.e., $\frac{p_m}{q_m}$ is in lowest terms.*

How are the convergents related to $\frac{a}{b}$?

Theorem: $\frac{p_0}{q_0} < \frac{p_2}{q_2} < \cdots < \frac{p_n}{q_n} = \frac{a}{b} < \cdots < \frac{p_3}{q_3} < \frac{p_1}{q_1}$.

Proof: By equation (4.8),

$$\frac{p_m}{q_m} - \frac{p_m - 1}{q_{m-1}} = \frac{(-1)^{m-1}}{q_{m-1}q_m} .$$

Therefore the difference is positive if m is odd, and negative if m is even. Since $\{q_i\}$ is an increasing sequence, each difference is smaller in absolute value than the preceding one; therefore

$$\frac{p_0}{q_0} < \frac{p_1}{q_1},$$
$$\frac{p_0}{q_0} < \frac{p_2}{q_2} < \frac{p_1}{q_1},$$
$$\frac{p_2}{q_2} < \frac{p_3}{q_3} < \frac{p_1}{q_1}, \quad \text{and so on,}$$

where $\frac{a}{b}$ is the last convergent.

The last theorem only gives us information about the relative positions of the convergents. As regards distance from $\frac{a}{b}$, we have

Theorem: *Each convergent is closer in absolute value to $\frac{a}{b}$ than the preceding convergent.*

Proof: See [Har-Wri].

This theorem shows how to approximate $\frac{a}{b}$ by other fractions with a smaller denominator (we shall see that these are best approximations, in a well defined sense). This motivated Christian Huygens (1629 - 1695), a mathematician - scientist interested in the construction of planetariums, to study continued fractions. Since the ratios of the periods of the planets are represented by fractions with very large denominators, this requires the construction of gears with a large number of teeth, which is quite difficult. Approximating these fractions by fractions with small denominators allows the construction of gears with fewer teeth and still yields a good representation of planetary motion.

Returning to mathematics, we now reformulate our application of the Euclidean algorithm to solve linear Diophantine equations (section 1. 2).

Example: *The Diophantine equation $ax - by = 1$.*

Recall that if there is a solution, then, since $(a, b)|a$ and $(a, b)|b$, we must have $(a, b) = 1$. Now assume that $(a, b) = 1$.

Suppose that $\frac{a}{b} = \left[a_0; \frac{1}{a_1}, \ldots, \frac{1}{a_n}\right]$. Then, since $(a, b) = 1$ and the convergents are always in lowest terms, $\frac{p_n}{q_n} = \frac{a}{b}$. Setting $m = n$ in equation (4.7), we have

$$aq_{n-1} - bp_{n-1} = (-1)^{n-1};$$

so if n is odd then $x = q_{n-1}$, $y = p_{n-1}$ is a solution of $ax - by = 1$.

If n is even and $a_n > 1$, we can change the continued fraction by setting

$$a_n = a_{n-1} + \frac{1}{1}$$

and proceeding as in the odd case, i.e., use the next to last convergent in the new continued fraction to find a solution. If $a_n = 1$, n even, then this method still works if we adopt the conventions $\frac{1}{0} = \infty$, $\frac{1}{\infty} = 0$, and $k + \infty = \infty$, for all integers k (see [Dav] for a different formulation of this case).

We have seen that every rational number $\frac{a}{b}$ has a finite continued fraction expansion (sec. 2). Is there another finite expansion of $\frac{a}{b}$ as a continued fraction? Not really, as the following proposition shows.

Proposition: *(Uniqueness) If* $\frac{a}{b} = \left[a_0; \frac{1}{a_1}, \ldots, \frac{1}{a_n}\right] = \left[b_0; \frac{1}{b_1}, \ldots, \frac{1}{b_m}\right]$ *and if* $a_n, b_m > 1$, *then* $m = n$ *and* $a_i = b_i$ *for all* i.

Proof: First we show that $\left[0; \frac{1}{a_k}, \ldots, \frac{1}{a_n}\right] < 1$, for all $k \leq n$. Since $a_i > 0$, for $i > 0$, and $a_n > 1$,

$$\frac{1}{a_n} = \left[0; \frac{1}{a_n}\right] < 1 \implies a_{n-1} + \frac{1}{a_n} = \left[a_{n-1}; \frac{1}{a_n}\right] > 1$$
$$\implies \left[0; \frac{1}{a_{n-1}}, \frac{1}{a_n}\right] < 1 \implies \cdots \implies \left[0; \frac{1}{a_1}, \ldots, \frac{1}{a_n}\right] < 1 .$$

Similarly $\left[0; \frac{1}{b_k}, \ldots, \frac{1}{b_m}\right] < 1$, for all $k \leq n$.

Therefore, from

$$a_0 + \left[0; \frac{1}{a_1}, \ldots, \frac{1}{a_n}\right] = \left[a_0; \frac{1}{a_1}, \ldots, \frac{1}{a_n}\right]$$
$$= \left[b_0; \frac{1}{b_1}, \ldots, \frac{1}{b_m}\right] = b_0 + \left[0; \frac{1}{b_1}, \ldots, \frac{1}{b_m}\right],$$

we see that $a_0 = b_0 = \left[\frac{a}{b}\right]$. Subtracting $\left[\frac{a}{b}\right]$ from both sides and inverting yields

$$\left[a_1; \frac{1}{a_2}, \ldots, \frac{1}{a_n}\right] = \left[b_1; \frac{1}{b_2}, \ldots, \frac{1}{b_m}\right] .$$

As before, we have $a_1 = b_1$. Similarly $a_2 = b_2$ and so on.

If $n < m$, then $a_n = [a_n] = \left[b_n; \frac{1}{b_{n+1}}, \ldots, \frac{1}{b_m}\right] = b_n + \left[0; \frac{1}{b_{n+1}}, \ldots, \frac{1}{b_m}\right]$. But $0 < \left[0; \frac{1}{b_{n+1}}, \ldots, \frac{1}{b_m}\right] < 1$, and a_n and b_n are integers, a contradiction. Therefore $n = m$ and $a_i = b_i$ for all i.

Note that if we do not require that both $a_n, b_m > 1$ then the proposition is not true since if $a_n > 1$, $a_n = (a_n - 1) + \frac{1}{1}$. Thus e.g., $\left[1; \frac{1}{2}, \frac{1}{3}\right] = \left[1; \frac{1}{2}, \frac{1}{3-1}, \frac{1}{1}\right]$. Hence every rational number, except 1, has a unique finite continued fraction expansion with $a_n = 1$ and one with $a_n > 1$. In section 6, we will prove that an infinite continued fraction cannot equal a rational number.

6. Infinite Continued Fractions

We have already described the continued fraction algorithm (sec. 2), which associates an infinite continued fraction with an irrational number,

$$\alpha \longleftrightarrow \left[a_0; \frac{1}{a_1}, \frac{1}{a_2}, \cdots\right].$$

Our first goal is to prove that $\frac{p_n}{q_n} \longrightarrow \alpha$.

At the nth step of the algorithm, we have

$$\alpha = \left[a_0; \frac{1}{a_1}, \ldots, \frac{1}{a_n}, \frac{1}{\alpha_n + 1}\right],$$

where $\alpha_{n+1} > 1$ is irrational. Since we put no restriction on the value of the variables in equation (4.1), we have

$$\alpha = \frac{\langle a_0, a_1, \ldots, a_n, \alpha_{n+1}\rangle}{\langle a_1, a_2, \ldots, a_n, \alpha_{n+1}\rangle}.$$

Applying equation (4.6), we have

Proposition:

$$\alpha = \frac{\alpha_{n+1} p_n + p_{n-1}}{\alpha_{n+1} q_n + q_{n-1}}. \tag{1}$$

Theorem:

(i) $\alpha - \frac{p_n}{q_n} = \pm\frac{1}{q_n}\left(a_{n+1} q_n + q_{n-1}\right),$

(ii) $\left|\alpha - \frac{p_n}{q_n}\right| < \frac{1}{q_n q_{n+1}}$.

Proof: (i) By (1),

$$\alpha - \frac{p_n}{q_n} = \frac{\alpha_{n+1} p_n + p_{n-1}}{\alpha_{n+1} q_n + q_{n-1}} - \frac{p_n}{q_n}$$
$$= \frac{p_{n-1} q_n - q_{n-1} p_n}{q_n \left(\alpha_{n+1} q_n + q_{n-1}\right)},$$

which, by equation (4.7),

$$= \pm \frac{1}{q_n} \left(\alpha_{n+1} q_n + q_{n-1}\right).$$

(ii) Since $a_{n+1} = [\alpha_{n+1}] < \alpha_{n+1}$ it follows that $a_{n+1} q_n + q_{n-1} < \alpha_{n+1} q_n + q_{n-1} < q_{n+1}$ (by equation (4.5b)) and our inequality follows from part (i).

Since $\left|a - \frac{p_n}{q_n}\right| < \frac{1}{q_n q_{n+1}}$ and the $\{q_i\}$ are an increasing sequence of integers, we see that

Corollary:

$$\lim_{n \to \infty} \frac{p_n}{q_n} = \alpha .$$

We say that an infinite continued fraction $\left[a_0; \frac{1}{a_1}, \frac{1}{a_2}, \cdots\right]$ **converges to** α **or is equal to** α if $\lim_{n\to\infty} \frac{p_n}{q_n} = \alpha$.

Our next goal is to prove that, just as finite continued fractions correspond to rational numbers, infinite continued fractions correspond in a unique way to irrational numbers.

Does every infinite continued fraction converge to a real number? The answer is yes.

Proposition: *Given any sequence of integers,* $a_0, a_1 > 0, a_2 > 0, a_3 > 0, \ldots$, *the continued fraction* $\left[a_0; \frac{1}{a_1}, \frac{1}{a_2}, \cdots\right]$ *converges to a real number.*

Proof: From section 5, we know that the convergents satisfy

$$\frac{p_0}{q_0} < \frac{p_2}{q_2} < \frac{p_4}{q_4} < \cdots < \frac{p_5}{q_5} < \frac{p_3}{q_3} < \frac{p_1}{q_1} .$$

Recall that any bounded increasing (resp. decreasing) sequence of real numbers which is bounded from above (resp. below) has a limit, namely, its least upper bound (resp. greatest lower bound). Hence the lower increasing sequence $\left\{\frac{p_{2n}}{q_{2n}}\right\}$, which is bounded from above by $\frac{p_1}{q_1}$, has a limit r_1 and the upper decreasing sequence $\left\{\frac{p_{2n+1}}{q_{2n+1}}\right\}$, which is bounded from below by $\frac{p_0}{q_0}$, has a limit r_2.

By equation (4.6), the difference of the sequences $\frac{p_n}{q_n} - \frac{p_{n-1}}{q_{n-1}} = \frac{(-1)^{n-1}}{q_{n-1}q_n}$. Since $\lim_{i\to\infty} q_i = \infty$, this difference approaches zero as $n \to \infty$. Therefore $r_1 = r_2$ and $\lim_{n\to\infty} \frac{p_n}{q_n} = r_1$.

Theorem: *(Uniqueness) If two infinite continued fractions are equal, i.e., if they both converge to the same real number α, then they are identical.*

Proof: Let $\alpha = \left[a_0; \frac{1}{a_1}, \frac{1}{a_2}, \cdots\right]$ where $a_i > 0$ for $i > 0$. By the last proposition, $\left[a_n; \frac{1}{a_{n+1}}, \frac{1}{a_{n+2}}, \cdots\right]$ converges to a real number α_n. Since $a_i > 0$ for all $i > 0$, $\alpha_n > a_n > 1$. Now

$$\alpha = \lim_{N\to\infty}\left[a_0; \frac{1}{a_1}, \ldots, \frac{1}{a_N}\right] = a_0 + \frac{1}{\lim_{N\to\infty}\left[a_1; \frac{1}{a_2}, \ldots, \frac{1}{a_N}\right]}$$

$$= a_0 + \frac{1}{\alpha_1}.$$

Since $\alpha_1 > 1$, $a_0 = [\alpha]$. Similarly, $\alpha_1 = \lim_{N\to\infty}\left[a_1; \frac{1}{a_2}, \ldots, \frac{1}{a_N}\right] = a_1 + \frac{1}{\lim_{N\to\infty}\left[a_2; \frac{1}{a_3}, \ldots, \frac{1}{a_N}\right]} = a_1 + \frac{1}{\alpha_2}$. Since $\alpha_2 > 1$, $\alpha_1 = [\alpha_1]$. Continuing, we see that in general $\alpha_n = a_n + \frac{1}{\alpha_{n+1}}$, where $\alpha_{n+1} > 1$, and hence $a_n = [\alpha_n]$. It is an easy exercise to show that $\alpha = \left[a_0; \frac{1}{a_1}, \ldots, \frac{1}{a_n}, \frac{1}{\alpha_{n+1}}\right]$, i.e., these are the same a_i used at the beginning of the section.

Similarly, if $\alpha = \left[b_0; \frac{1}{b_1}, \frac{1}{b_2}, \cdots\right]$, where $b_i > 0$ for $i > 0$, and $\left[b_n; \frac{1}{b_{n+1}}, \frac{1}{n_{n+2}}, \cdots\right]$ converges to β_n, then $b_0 = [\alpha]$, $\beta_n = b_n + \frac{1}{\beta_{n+1}}$, and $b_n = [\beta_n]$.

Now we prove, by induction, that $\alpha_i = \beta_i$ and $a_i = b_i$, for all i. First $a_0 = [\alpha] = b_0$. From this and $a_0 + \frac{1}{\alpha_1} = \alpha = b_0 + \frac{1}{\beta_1}$, we see that $\alpha_1 = \beta_1$ and hence $a_1 = [\alpha_1] = [\beta_1] = b_1$. If $\alpha_n = \beta_n$ and $a_n = b_n$, then since $a_n + \frac{1}{\alpha_{n+1}} = \alpha_n = \beta_n = b_n + \frac{1}{\beta_{n+1}}$, we have $\alpha_{n+1} = \beta_{n+1}$ and $a_{n+1} = [\alpha_{n+1}] = [\beta_{n+1}] = b_{n+1}$, and we are done.

To summarize: the infinite continued fraction associated with an irrational number α by the continued fraction algorithm equals α, and it is the unique continued fraction equal to α. Moreover, we know that every infinite continued fraction converges to a real number. To complete the correspondence between infinite continued fractions and irrational numbers, we prove

Proposition: *If* $\alpha = \left[a_0; \frac{1}{a_1}, \frac{1}{a_2}, \cdots\right]$, *where* $a_i > 0$ *for* $i > 0$, *then* α *is irrational.*

Proof: The a_i's are uniquely determined by the continued fraction algorithm. But, if α is rational, we know that the algorithm terminates after a finite number of steps yielding a finite continued fraction, a contradiction.

It's worth mentioning a property of continued fractions that plays a key role in the study of *indefinite binary quadratic forms* (chapter 13). Two irrational numbers, α and β, are **equivalent** if there exist integers a, b, c, d, with $ad - bc = \pm 1$, such that

$$\beta = \frac{a\alpha + b}{c\alpha + d}.$$

Theorem: *(Serret) Two irrational numbers* $\alpha = \left[a_0; \frac{1}{a_1}, \frac{1}{a_2}, \cdots\right]$ *and* $\beta = \left[b_0; \frac{1}{b_1}, \frac{1}{b_2}, \cdots\right]$ *are equivalent if and only if there exist positive integers* m, n *such that the partial remainders* $\alpha_{n+1} = \beta_{m+1}$, *or equivalently* $a_{n+k} = b_{m+k}$ *for all* $k > 0$ *(they have equal tails).*

See [Har-Wri] or [Lan 1] for a proof.

7. Diophantine Approximation and Geometry

Diophantine approximation refers to the approximation of irrational numbers by rational numbers, where the strength of approximation is measured as a function of the denominator of the approximating fraction. The subject started as an application of continued fractions and we present a short introduction. Several theorems are stated without proof ([Har-Wri] and [Lan 1] are good references for the proofs as is Lagrange [Eul 2, appendix] for some of the results). In chapter 21, we present a more thorough historically oriented development of Diophantine approximation and its connection to transcendental numbers.

In the last section, we proved that the convergents of the continued fraction of an irrational number α satisfy

$$\left|\alpha - \frac{p_n}{q_n}\right| < \frac{1}{q_n q_{n+1}} .$$

Example: From the continued fraction for π (sec. 2), we have $\frac{p_1}{q_1} = \frac{22}{7}$ and $\frac{p_2}{q_2} = \frac{333}{106}$. Thus $\left|\pi - \frac{22}{7}\right| < \frac{1}{7 \cdot 106} = \frac{1}{742}$, which explains why $\frac{22}{7}$ is so often used as an approximation to π.

Since the q_i form an increasing sequence, $\left|\alpha - \frac{p_n}{q_n}\right| < \frac{1}{q_n q_n}$. Thus, changing notations, we have proved:

Theorem: *(Lagrange) If $\frac{x}{y}$ is a convergent of the continued fraction of α, then*

$$\left|\alpha - \frac{x}{y}\right| < \frac{1}{y^2} .$$

Hence this last inequality, with x and y now regard as variables, is satisfied by infinitely many rational numbers $\frac{x}{y}$.

The next two theorems show that using only some of the convergents yields infinitely many stronger approximations.

Theorem: *Given any two successive convergents to a, one of them satisfies*

$$\left|\alpha - \frac{x}{y}\right| < \frac{1}{2y^2} .$$

Theorem: *(Hurwitz,1891) Given any three successive convergents to α, one of them satisfies*

$$\left|\alpha - \frac{x}{y}\right| < \frac{1}{\sqrt{5}y^2} .$$

These results cannot be strengthened in the sense that if $\alpha = \frac{\sqrt{5}-1}{2} = [0; \frac{1}{1}, \frac{1}{1}, \frac{1}{1}, \cdots]$, then for any, $k > \sqrt{5}$, $\left|\alpha - \frac{x}{y}\right| < \frac{1}{ky^2}$ has only finitely many solutions.

To show how good the convergents are as approximations, we have:

Theorem: *If*

$$\left|\alpha - \frac{x}{y}\right| < \frac{1}{2y^2},$$

then $\frac{x}{y} = \frac{p_n}{q_n}$ *for some n.*

Approximation results are a tool for deriving results on the minimum value of a binary quadratic form $ax^2 + bxy + cy^2$, for integer values of the variables. The connection arises from the factorization

$$\begin{aligned} ax^2 + bxy + cy^2 &= a\,y^2\left(\left(\frac{x}{y}\right)^2 + \left(\frac{b}{a}\right)\left(\frac{x}{y}\right) + \frac{c}{a}\right) \\ &= a\,y^2\left(\frac{x}{y} - \alpha_1\right)\left(\frac{x}{y} - \alpha_2\right), \end{aligned}$$

where α_1 and α_2 are the roots of $z^2 + \frac{b}{a}z + \frac{c}{a} = 0$.

Lagrange introduced the quantity $|q\alpha - p|$ as opposed to $|\alpha - \frac{p}{q}|$ to measure approximation. We call $\frac{p}{q}$ a **best approximation** to α if

$$0 < q' \le q, \frac{p}{q} \ne \frac{p'}{q'} \implies |q\alpha - p| < |q'\alpha - p'|.$$

The measure $|q\alpha - p|$ will be seen in a more natural geometric setting in a moment. One of the basic theorems of the subject characterizes best approximations.

Theorem: *The set of convergents of the (finite or infinite) continued fractions for a real number α, except possibly for $\frac{p_0}{q_0}$, is the set of best approximations to α.*

The relation between the measures $|q\alpha - p|$ and $\left|\alpha - \frac{p}{q}\right|$ is given by

Proposition: *If the convergent $\frac{p}{q}$ is a best approximation to α and if $n > 1$, $0 < q' \le q$, $\frac{p}{q} \ne \frac{p'}{q'}$, then*

$$\left|\alpha - \frac{p}{q}\right| < \left|\alpha - \frac{p'}{q'}\right| .$$

Proof:

$$|q\alpha - p| < |q'\alpha - p'| \implies \left|\alpha - \frac{p}{q}\right| < \left|\frac{q'\alpha}{q} - \frac{p'}{q}\right| = \frac{q'}{q}\left|a - \frac{p'}{q'}\right|$$
$$\le \left|a - \frac{p'}{q'}\right| , \quad \text{since } q' \le q .$$

These measures also have a geometric meaning using **integer points** in the plane (points with integer coordinates). We represent the fraction $\frac{p}{q}$ by the integer point (q, p) in the plane and draw the lines $L : y = \alpha x$ and $L' : y = \frac{p}{q}x$ (figure 1). Then

$$|q\alpha - p| = \text{the vertical distance from } (q, p) \text{ to the line } L,$$
$$\left|\alpha - \frac{p}{q}\right| = \text{the difference between the slopes of } L \text{ and } L'.$$

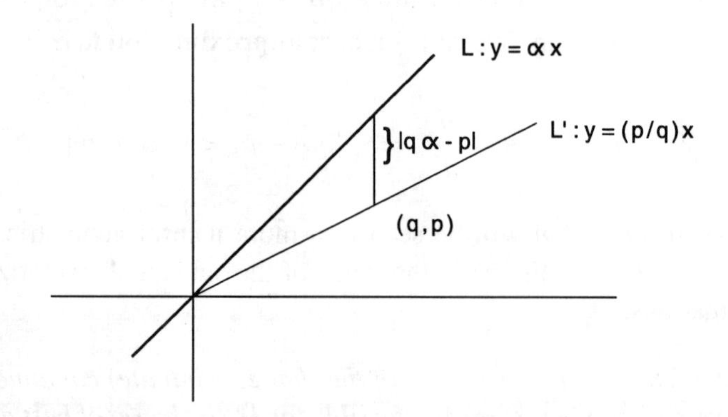

Figure 1

Since distance is a "stronger measure" than slope, the idea behind the proof of the corollary becomes more intuitive.

This idea of representing the convergents by integer points in the plane is very powerful. Felix Klein [Kle 1, pp. 42-44] used it to develop the theory of continued fractions as a theory of vectors in the plane. The idea also appears in earlier works of H.J.S. Smith [Smi, H 1] and Poincare [Poi].

If $\left\{\frac{p_n}{q_n}\right\}$ are the convergents of α, then the basic recursions (4.5) translate to the vector equation

$$(q_n, p_n) = a_n\,(q_{n-1}, p_{n-1}) + (q_{n-2}, p_n - 2)\ ,$$

Klein saw how each point was constructed geometrically from the previous two. The points in the sequence $(q_0, p_0), (q_1, p_1), \cdots$ alternately lie below and above the line $y = \alpha x$ (corresponding to the fact that the convergents $\frac{p_n}{q_n}$ are alternately less that and greater than α), and the distance from these points to the line approaches zero.

A quick and illuminating illustration of some of these ideas is in Davenport [Dav, pp. 111-113]. A complete and elementary development of the geometric theory is given by Stark [Sta], including a very beautiful geometric proof that every best approximation is a convergent. In many respects, this two dimensional picture provides a more natural theory than the original one dimensional development. [Min 1], [Hur 1] and [Hum] present alternate geometric formulations of both regular and other types of continued fractions.

All our work on continued fractions can be reformulated in terms of 2×2 matrices. This is most important for exploring the relationship between continued fractions and quadratic forms (chapter 13) and for generalizing continued fractions. It is also very natural.

For $n \geq 0$, let

$$A_n = \begin{pmatrix} 0 & 1 \\ 1 & a_n \end{pmatrix} \quad S_n = \begin{pmatrix} q_{n-1} & q_n \\ p_{n-1} & p_n \end{pmatrix} \quad S_0 = \begin{pmatrix} 0 & 1 \\ 1 & a_0 \end{pmatrix} = A_0$$

Then our basic recursions (4.5) say

$$S_n = S_{n-1}A_n\ , \quad \text{for } n > 0\ ,$$

and iterating

$$S_n = A_0 A_1 \cdots A_n\ .$$

Much of what we did can be treated more elegantly in this notation. For example, since $\det A_i = -1$, we see that $\det S_n = p_n q_{n-1} - p_{n-1} q_n =$

$(-1)^{n+1} = (-1)^{n-1}$. The fact that S_n is **unimodular** (a matrix with determinant ± 1) is very important in studying the action of these matrices on integer points in the plane, which ties in with our earlier geometric discussion. For example, unimodularity of S_n tells us that the triangle with vertices $(0, 0)$, (p_{n-1}, q_{n-1}), (p_n, q_n) always has area $\frac{1}{2}$. See [Sta] for more details.

8. Quadratic Irrationalities

One of the earliest and most successful applications of continued fractions is the characterization of quadratic irrationalities. A **quadratic irrationality**, or **quadratic number,** α is the root of an irreducible quadratic equation with integer coefficients, i.e., a number of the form $r + s\sqrt{t}$, where r and s are rational, and t is a square free integer. In this section, we assume $t > 0$, i.e., α is a real number.

Recall that $\sqrt{2} = \left[1; \overline{\frac{1}{2}}\right]$ and $\sqrt{3} = \left[1; \overline{\frac{1}{1}, \frac{1}{2}}\right]$, *where the overlined sequence is repeated periodically*, e.g., $\sqrt{3} = \left[1; \frac{1}{1} \cdot \frac{1}{2}, \frac{1}{1}, \frac{1}{2}, \cdots\right]$. Some other examples are

$$\frac{24 - \sqrt{15}}{17} = \left[1; \frac{1}{5}, \overline{\frac{1}{2}, \frac{1}{3}}\right], \quad \sqrt{28} = \left[5; \overline{\frac{1}{3}, \frac{1}{2}, \frac{1}{3}, \frac{1}{10}}\right]$$

In general, we say that a continued fraction $\left[a_0; \frac{1}{a_1}, \frac{1}{a_2}, \cdots\right]$ is **periodic** if there exist positive integers n, k such that the continued fraction is of the form

$$\left[a_0; \frac{1}{a_1}, \ldots, \frac{1}{a_{n-1}}, \overline{\frac{1}{a_n}, \cdots \frac{1}{a_{n+k}}}\right].$$

This periodic phenomena characterizes quadratic irrationalities.

Lagrange's Theorem: *α is a quadratic number $\Longleftrightarrow$ its continued fraction is periodic.*

Proof: $\Longleftarrow$ (the easy part)

Suppose that the continued fraction of α is periodic, as defined above. Then

$$\begin{aligned}\alpha_n &= \left[a_n; \frac{1}{a_{n+1}}, \frac{1}{a_{n+2}}, \cdots\right] \\ &= \left[a_{n+k+1}; \frac{1}{a_{n+k+2}}, \cdots\right] \\ &= \alpha_{n+k+1}\,.\end{aligned}$$

But, by equation (6.1),

$$\begin{aligned}\alpha &= \frac{\alpha_n p_{n-1} + p_{n-2}}{\alpha_n q_{n-1} + q_{n-2}} \qquad (1)\\ &= \frac{\alpha_{n+k+1} p_{n+k} + p_{n+k-1}}{\alpha_{n+k+1} q_{n+k} + q_{n+k-1}} \\ &= \frac{\alpha_n p_{n+k} + p_{n+k-1}}{a_n q_{n+k} + q_{n+k-1}} \quad (\text{because } \alpha_{n+k+1} = \alpha_n).\end{aligned}$$

After clearing fractions in the equation

$$\frac{\alpha_n p_{n-1} + p_{n-2}}{\alpha_n q_{n-1} + q_{n-2}} = \frac{\alpha_n p_{n+k} + p_{n+k-1}}{\alpha_n q_{n+k} + q_{n+k-1}} \qquad (2),$$

we get a quadratic equation for α_n which is not identically zero (just check the coefficient of α_n). α_n is irrational since the continued fraction for α_n is infinite. Hence α_n is a quadratic number. Thus by (1), α is a quadratic number.

$\Longrightarrow$):

We give Lagrange's proof with some simplification (see [Str]). The main idea is to prove that two of the α_i are equal, and thus the sequences of partial quotients following them are the same.

Let $\alpha = \left[a_0; \frac{1}{a_1}, \frac{1}{a_2}, \cdots\right]$. Suppose that α is a root of

$$A_0 x^2 + B_0 x + C_0 = 0\,.$$

Then Lagrange sets $\alpha = a_0 + \frac{1}{\alpha_1}$, where $a_0 = [\alpha]$, and, by substituting in the equation, derives an equation for α_1, say

$$A_1 x^2 + B_1 x + C_1 = 0\,.$$

Again setting $\alpha_1 = a_1 + \frac{1}{\alpha_2}$, $a_1 = [\alpha_1]$, we get an equation for α_2,

$$A_2 x^2 + B_2 x + C_2 = 0 \, .$$

Continuing in this way we derive a sequence of equations

$$A_n x^2 + B_n x + C_n = 0 \, , \quad n = 0, 1, 2, \ldots \, ,$$

with *integer coefficients* and α_n as one of the roots. An induction proof with a straightforward computation shows that the discriminant, $D_n = B_n^2 - 4A_n C_n$, is a constant independent of n; call it D. Since α is real, $D > 0$.

We now show that the coefficients of all the equations are bounded by a single number which is independent of n. Thus there are only a finite number of distinct equations in the sequence, and thus only a finite number of roots of these equations. It follows that the sequence $\alpha_1, \alpha_2, \cdots$ can only take a finite number of values. Thus, two of the α_i must be equal and the continued fraction is periodic.

To prove the boundedness, we substitute

$$\alpha = \frac{\alpha_n p_{n-1} + p_{n-2}}{\alpha_n q_{n-1} + q_{n-2}}$$

into

$$A_0 \alpha^2 + B_0 \alpha + C_0 = 0$$

to get the equation

$$A_n \alpha_n^2 + B_n \alpha_n + C_n = 0 \, ,$$

where

$$\begin{aligned} A_n &= A_0 p_{n-1}^2 + B_0 p_{n-1} q_{k-1} + C_0 q_{n-1}^2 \, , \\ B_n &= \text{mess} \, , \\ C_n &= A_0 p_{n-2}^2 + B_0 p_{n-2} q_{k-2} + C_0 q_{n-2}^2 \, . \end{aligned}$$

Let

$$f(x) = A_0 x^2 + B_0 x + C_0 \, ;$$

then

$$A_n = q_{n-1}^2 f\left(\frac{p_{n-1}}{q_{n-1}}\right) , \quad C_n = q_{n-2}^2 f\left(\frac{p_{n-2}}{q_{n-2}}\right) .$$

By the Taylor expansion of $f(x)$ about α,

$$\begin{aligned} A_n &= q_{n-1}^2 f\left(\frac{p_{n-1}}{q_{n-1}}\right) \\ &= q_{n-1}^2 \left\{ f(\alpha) + f'(\alpha)\left(\frac{p_{n-1}}{q_{n-1}} - \alpha\right) + f''\frac{\alpha}{2}\left(\frac{p_{n-1}}{q_{n-1}} - \alpha\right)^2 \right\}. \end{aligned}$$

The higher derivatives are zero since f is quadratic. But we have assumed $f(\alpha) = 0$ and we know that $\left|\alpha - \frac{p_{n-1}}{q_{n-1}}\right| < \frac{1}{q_{n-1}^2}$. Hence

$$|A_n| \leq |f'(a)| + \frac{|f''(\alpha)|}{2q_{n-1}^2}.$$

Similarly,

$$|C_n| \leq |f'(\alpha)| + |f''(\alpha)|/2q_{n-2}^2 .$$

Since $q_i \to \infty$ as $i \to \infty$, we see that $|A_n|$ and $|B_n|$ are bounded by some constant L. But

$$B_n^2 = D + 4A_nC_n ,$$

and thus

$$|B_n^2| < D + 4L^2 \quad \text{(recall that } D > 0 \text{ since a is real)}$$

or

$$|B_n| < \sqrt{D + 4L2} .$$

Therefore there is a constant M such that

$$|A_n| < M, \ |B_n| < M, \ |C_n| < M, \ \text{for all } n .$$

Hence there are only finitely many quadratic equations in our sequence and we are done.

E. Galois gave a different proof of the last theorem which can be used to characterize those continued fractions which are **purely periodic**, i.e., those of the form $\left[\overline{a_0, \frac{1}{a_1}, \cdots \frac{1}{a_n}}\right] = \left[a_0; \frac{1}{a_1}, \frac{1}{a_2}, \ldots, \frac{1}{a_n}, \frac{1}{a_0}, \ldots, \frac{1}{a_n}, \frac{1}{a_0}, \ldots\right]$. We state the general result (see [Dav]).

Theorem: *α has a purely periodic continued fraction if and only if α is a quadratic number > 1 whose* **conjugate** *(the other root of the irreducible equation for α) lies between -1 and 0. In this case, its conjugate is $\frac{-1}{\beta}$, where β is derived from α by reversing its period.*

From this it quickly follows that

Corollary: *$\sqrt{d}$, d not a perfect square, has a "symmetric" continued fraction of the form*

$$\sqrt{d} = \left[a_0; \overline{\frac{1}{a_1}, \frac{1}{a_2}, \cdots \frac{1}{a_2}, \frac{1}{a_1}, \frac{1}{2a_0}}\right],$$

i.e., the period, without its last term $\left(\frac{1}{2a_0}\right)$, is symmetric about its center.

Example: $\sqrt{7} = \left[2; \overline{\frac{1}{1}, \frac{1}{1}, \frac{1}{1}, \frac{1}{4}}\right]$ and $\sqrt{31} = \left[5; \overline{\frac{1}{1}, \frac{1}{1}, \frac{1}{3}, \frac{1}{5}, \frac{1}{3}, \frac{1}{1}, \frac{1}{1}, \frac{1}{10}}\right]$.

See [Davenport] for a table of $\sqrt{d}$, $d = 1, 2, \ldots, 50$.

Very little is known about the continued fractions of non-quadratic irrationals.

Example: (Lambert and Euler)

$$\text{(i)} \qquad \frac{e-1}{e+1} = \left[0; \frac{1}{2}, \frac{1}{6}, \frac{1}{10}, \frac{1}{14}, \cdots\right].$$

More generally

$$\text{(ii)} \qquad \frac{e^{\frac{2}{k}}-1}{e^{\frac{2}{k}}+1} = \left[0; \frac{1}{k}, \frac{1}{3k}, \frac{1}{5k}, \cdots\right].$$

Note that in these examples the partial quotients form an arithmetic progression. Using (i), Euler showed that

$$e = \left[2; \frac{1}{1}, \frac{1}{2}, \frac{1}{1}, \frac{1}{1}, \frac{1}{4}, \frac{1}{1}, \frac{1}{1}, \frac{1}{6}, \cdots\right],$$

where the partial quotients are obtained from an arithmetic progression by inserting two 1's between every pair of terms and putting 2; $\frac{1}{1}$ in front (see [Lan 1, sec. V.2]).

Hurwitz [Hur 2] studied continued fractions whose quotients are obtained from arithmetic progression by inserting 1's. Very little is known, but the subject is tantalizing.

9. Pell's Equation

We discussed the history of Pell's equation in chapter 2. Now we solve it. This is a good example of the constructive methods afforded by the use of continued fractions.

Suppose x, y is a solution in *positive integers* of

$$x^2 - dy^2 = 1\,, \quad d > 0 \text{ not a perfect square.}$$

Then $\frac{x^2}{y^2} - d = \frac{1}{y^2}$ and

$$\frac{x^2}{y^2} = d + \frac{1}{y^2} \implies \frac{x}{y} > \sqrt{d} \implies \frac{x}{y} + \sqrt{d} > 2\sqrt{d}\,.$$

But

$$x^2 - dy^2 = 1 \implies \left(x - y\sqrt{d}\right)\left(x + y\sqrt{d}\right) = 1$$
$$\implies \left|\frac{x}{y} - \sqrt{d}\right| = \frac{1}{y^2\left|\frac{x}{y} + \sqrt{d}\right|} < \frac{1}{2y^2\sqrt{d}} < \frac{1}{2y^2}\,.$$

We saw in section 7 that $\left|\frac{x}{y} - \sqrt{d}\right| < \frac{1}{2}y^2$ implies that $\frac{x}{y}$ *is a convergent of the continued fraction for* $\sqrt{d}$.

Now that we see that every solution determines a convergent, we must decide which convergents solve Pell's equation. By the corollary in section 8 (ignoring the symmetry), we know that

$$\sqrt{d} = \left[a_0; \frac{1}{a_1}, \frac{1}{a_2}, \cdots \frac{1}{a_n}, \frac{1}{2a_0}\right]. \tag{1}$$

Looking at the two convergents before the term $2a_0$, viz., $\frac{p_{n-1}}{q_{n-1}}$ and $\frac{p_n}{q_n}$, we see that

$$\sqrt{d} = \frac{\alpha_{n+1}p_n + p_{n-1}}{\alpha_{n+1}q_n + q_{n-1}}\,. \tag{2}$$

But

$$\alpha_{n+1} = \left[2a_0; \frac{1}{a_1}, \cdots\right] = a_0 + \left[a_0; \frac{1}{a_1}, \cdots\right]$$
$$= a_0 + \sqrt{d}\,.$$

Substituting $\alpha_{n+1} = a_0 + \sqrt{d}$ in (2) and equating coefficients of 1 and $\sqrt{d}$ yields

$$p_{n-1} = dq_n - a_0 p_n ,$$
$$q_{n-1} = p_n - a_0 q_n .$$

Substituting in $p_n q_{n-1} - q_n p_{n-1} = (-1)^{n-1}$, we have

$$p_n^2 - dq_n^2 = (-1)^{n-1} .$$

Hence if n *is odd*, $x = p_n, y = q_n$ is a solution of Pell's equation. (The question of characterizing those d for which n is odd is an unsolved problem.)

If n *is even*, we use the next period of (1) with the same reasoning. Our p_i and q_i are now chosen at the $(2n+1)^{\text{st}}$ position and

$$p_{2n+1}^2 - dq_{2n+1}^2 = (-1)^{2n} = 1 .$$

As we look at successive periods, we can derive infinitely many solutions. These are *all* the positive solutions of Pell's equation [Sch-Opo]. It is easy to derive the non-positive solutions since, if we change the sign of x or y, we still have a solution.

There is yet more structure in the set of solutions to Pell's equation. Noting that $x^2 - dy^2 = (x + y\sqrt{d})(x - y\sqrt{d})$, the following can be proved:

(i) The first solution we derived, call it (x_0, y_0), is the smallest positive solution, in the sense of minimizing $x + y\sqrt{d}$. We call (x_0, y_0) the *fundamental solution.*

(ii) There is a solution (x_n, y_n), for any integer n (negative values allowed), defined by

$$x_n + y_n\sqrt{d} = \left(x_0 + y_0\sqrt{d}\right)^n ,$$

since

$$\begin{aligned} x_n^2 - dy_n^2 &= (x_n + y_n\sqrt{d})(x_n - y_n\sqrt{d}) \\ &= \left(x_0 + y_0\sqrt{d}\right)^n \left(x_0 - y_0\sqrt{d}\right)^n \\ &= \left[\left(x_0 + y_0\sqrt{d}\right)^n \left(x_0 + y_0\sqrt{d}\right)\right]^n \\ &= \left(x_0^2 - dy_0^2\right)^n = 1 . \end{aligned}$$

Note that we assumed $x + y\sqrt{d} = \left(a + b\sqrt{d}\right)^n \Longrightarrow x - y\sqrt{d} = \left(a - b\sqrt{d}\right)^n$ (exercise).

(iii) Every solution Pell's equation is of the form given in (ii).

More generally, it is also true that every solution (p, q) of

$$x^2 - dy^2 = \pm M \,, \quad 0 < M < \sqrt{d} \,,$$

corresponds to a convergent $\frac{p}{q}$ of the continued fraction of $\sqrt{d}$.

See [Hua], [Sch-Opo] and [Chr] for proofs of these statements. Chrystal has the most thorough treatment in English of many aspects of continued fractions.

Pell's equation is central in the study of quadratic fields. It also plays a role in the proof, that there is no algorithm for deciding if any Diophantine equation in any number of variables is solvable [Dav-Mat-Rob]

10. Generalizations

In number theoretic studies, it is often useful to consider a wider class of continued fractions than the regular ones. A **unitary continued fraction** is a continued fractions $\left[a_0; \frac{b_1}{a_1}, \frac{b_2}{a_2}, \dots\right]$ with $|b_i| = 1$, for all i. Thus when all $b_i = 1$, we have the regular continued fractions, and when all $b_i = -1$, we have the **negative continued fractions**. The negative continued fractions, which are defined by replacing the function $[\alpha]$ in the continued fraction algorithm by the function $[\alpha] + 1$, are quite useful in the reduction theory of indefinite binary quadratic forms (see [Zagier] and chap. 13).

The relation between various classes of unitary continued fractions can be seen by introducing the intermediate convergents. By equation (6) of section 4, the convergents of the regular continued fraction of α satisfy the recursion $\frac{p_n}{q_n} = \frac{a_n p_{n-1} + p_{n-2}}{a_n q_{n-1} + q_{n-2}}$. The **intermediate convergents** between $\frac{p_{n-1}}{q_{n-1}}$ and $\frac{p_n}{q_n}$ are the fractions

$$\frac{p_{i,s}}{q_{i,s}} = \frac{s p_{n-1} + p_{n-2}}{s q_{n-1} + q_{n-2}} \,, \qquad s = 1, 2, \dots, a_n - 1 \,.$$

The intermediate convergents are useful in questions of Diophantine approximations (see [Lan 1]). The sequence $\frac{p_0}{q_0}, \frac{p_{0,1}}{q_{0,1}}, \frac{p_{0,2}}{q_{0,2}}, \dots, \frac{p_1}{q_1}, \frac{p_{1,1}}{q_{1,1}}, \frac{p_{1,2}}{q_{1,2}},$

$\ldots, \frac{p_2}{q_2}, \cdots$ of regular and intermediate convergents of the regular continued fraction of a is called the **Hurwitz sequence** of α.

For any α, one can form a unitary continued fraction $\left[a_0; \frac{b_1}{a_1}, \frac{b_2}{a_2}, \cdots\right]$, with $\frac{b_i}{a_i} = \frac{1}{1}$ or $-\frac{1}{2}$, the **full continued fraction** of α, which converges to α. The sequence of convergents of the full continued fraction is the Hurwitz sequence of α. Moreover, the convergents of any unitary continued fraction which converges to α are a subsequence of the Hurwitz sequence of α. For example, the convergents of the negative continued fraction of α are exactly those fractions in the Hurwitz sequence that are greater than α. See Goldman [Gol, J 1] for the details of this theory.

In addition to their importance in number theory, unitary continued fractions have, in recent years, proved to be useful in such diverse areas as ergodic theory ([Ser, C] and [Moe]) and knot theory ([Con] and [Gol-Kau 1, 2]).

It seems that even though continued fractions are intrinsic descriptions of real numbers, they are not natural objects for describing arithmetic operations. Given regular continued fractions for α and β, there are some algorithms for finding the continued fractions of $k\alpha$, $k \in \mathbf{Z}$, $\alpha + \beta$, and $\alpha\beta$, but there is very little understanding. See Fowler [Fow] for a description of Gosper's algorithm and Raney [Ran] and Hall [Hal] for other algorithms.

As mentioned earlier, almost nothing is known about the regular continued fraction of non quadratic irrational numbers. For example, are the partial quotients in the continued fraction for $\sqrt[3]{2}$ bounded? Many leading mathematicians of the 19th and early 20th centuries, including Jacobi, Perron, Hermite and Minkowski, were interested in generalizing continued fractions so as to find characterizations for *algebraic numbers of degree n* (roots of irreducible polynomial equations of degree n with integer coefficients). Many believe that a good generalization will involve sequences of finite sets of points in $\mathbf{R^n}$ or sequences of $n \times n$ matrices (analogous to the sequence of convergents represented by integer points). Among the proposed algorithms are:

(i) the Jacobi-Perron algorithm, which generalizes the Euclidean algorithm, and

(ii) the Minkowski algorithm which starts with the idea of best approximation.

Fowler [Fow] discusses some of these generalizations and provides references to both early and recent work.

When generalizing continued fractions, one would like to retain appropriate generalizations of the key properties, viz.,

periodicity for algebraic numbers,

good Diophantine approximation, and

an efficient method for computing the proposed objects.

It is certainly not *a priori* clear that that all of these properties can be achieved in one generalized algorithm.

Chapter 5

Lagrange

1707	Euler	1783	
	1736	Lagrange	1813

1. Lagrange and His Work

Joseph Louis Lagrange (1736 - 1813) lived through some very turbulent times, including the French Revolution. Following the biography by Jean Itard [Ita 1], we divide Lagrange's life into three periods:

> 1736-1766-Turin-born there-left at age 30 and never returned,
>
> 1766-1787-Berlin Academy under Frederick the Great,
>
> 1787-1813-French Academy, Senator and Count under Napoleon, Ecole Polytechnic.

Lagrange did his principal research during the first two periods and wrote major memoirs summarizing his results during the third.

When Lagrange was eighteen he began to correspond with Euler, who immediately recognized a mathematician of the first rank and strongly encouraged him. At this time he began creating the Calculus of Variations, one of his greatest achievements.

At the age of thirty, he replaced Euler at Frederick the Great's Berlin Academy (Euler went back to Russia), strongly supported by Euler and D'Alembert. Frederick was very happy to have Lagrange. He disliked Euler who wouldn't pay attention to the pomp and etiquette of the court, whereas Lagrange fit in very well, always avoided politics and intrigue.

A very modest and self-critical man, Lagrange was always saying that he wished he hadn't published some of his papers. He led as quite a life as he could and paid almost no attention to his surroundings, even on trips. A very systematic person, he did his research from 6-12 in the evening, every day.

The Academies were government institutes under the patronage of rulers. They hired creative scholars in the arts and sciences. Almost no research was done in the universities. Leaders, such as Catherine of Russia, Frederick the Great, and Napoleon considered it of singular importance to have the leading scientists of the time under their direct patronage.

Lagrange was very cautious. He realized that when Frederick died Berlin would not be a good place for him (this was *10 years* before Frederick's death in 1786), so he began to look around for a new position. In fact just before Frederick's death the French finance minister sent an agent, Count Mirabeau, to Berlin to report on the political situation. Mirabeau, who became a friend of Lagrange, sent a letter back to France pointing out that many charlatans and idiots were being supported in luxury by the French government and that, for a not too large sum, Lagrange, the greatest mathematician of his time, could be brought to serve the great glory of France. It is really astonishing that politicians of that time were so interested in the sciences.

Due to Mirabeau and D'Alembert, Lagrange came to the French academy after Frederick's death. During the French revolution the academy was suppressed and all foreigners ordered arrested. Lavoisier intervened on Lagrange's behalf and saved him. Under Napoleon, Lagrange was made a senator, a count of the empire, and a professor at the Ecole Polytechnic (a school founded by Napoleon to train military officers and, even today, the principal training ground for the technical and business leaders of France).

In addition to Itard's biography, mentioned earlier, George Sarton wrote about Lagrange's personality [Sar].

Lagrange made major contributions to almost every part of mathematics. He invented the calculus of variations and worked on celestial mechanics (his famous memoir "Mechanique Analytique"), analysis (including elliptic integrals), algebra (where his work set the stage for Galois's discoveries), and of course number theory.

Lagrange's number theoretical work was done primarily during the period 1767-1778. He was the first to publish a proof of the existence of solutions of Pell's equation and to later recast it in the language of continued fractions (sec. 2.4, 4.9). As already discussed in the last chapter, Lagrange's appendix to Euler's *Algebra* was the first systematic presentation of continued fractions. In sections 4.8, we proved Lagrange's theorem characterizing the continued fractions of quadratic irrationalities.

Lagrange published the first proof of the *four square theorem* (which Fermat claimed to have proved - sec. 2.6), namely, that every positive integer is the sum of four integer squares. Euler was very pleased when Lagrange extended his use of irrational and complex numbers to solve Diophantine problems. Among Lagrange's contributions to divisibility problems is a proof that if $f(x)$ is a polynomial of degree n and p is a prime, then there are at most n numbers k, $-\frac{p}{2} < k < \frac{p}{2}$, such that p divides $f(k)$. To Euler and Lagrange, this was a very difficult problem. However we shall see that with Gauss's congruence notation it becomes quite easy.

Certainly one of Lagrange's great contributions was the creation of a theory of the representations of integers by binary quadratic forms. For now, we present a short introduction. In chapters 12 and 13, we will present a more systematic and extensive treatment, including many of Gauss's simplifications and new contributions.

2. Quadratic Forms

The basic question in the theory of binary quadratic forms is given such a form $ax^2 + bxy + cy^2$, where $a, b, c \in \mathbf{Z}$, which integers are **represented** by this form, i.e., for which integers n are there integers x, y such that

$$n = ax^2 + bxy + cy^2 \ ?$$

In addition, we would like to know how many such solutions exist and algorithms for finding them.

As an example, consider the form $f(x, y) = 2x^2 + 6xy + 5y^2$. Note that

$$2x^2 + 6xy + 5y^2 = (x + y)^2 + (x + 2y)^2 \ .$$

Let

$$X = x + y \ , \quad Y = x + 2y \ . \tag{1}$$

Under this substitution our form becomes

$$g(X, Y) = X^2 + Y^2 \ .$$

Solving for x, y we have

$$x = 2X - Y\,, \quad y = -X + Y. \tag{2}$$

Clearly equations (1) and (2) define a 1-1 correspondence between the *integer* pairs (x, y) and (X, Y), so the forms f and g represent the same integers. $X^2 + Y^2$ is intuitively the "simpler" of the two forms and it is reasonable to think that its representation problem would be easier to solve. (Recall Fermat's claim (sec. 2.6) concerning which primes are the sums of two squares.)

More generally, given

$$f(x, y) = ax^2 + bxy + cy^2\,,$$

we want to find other quadratic forms which represent the same integers. Consider the linear transformation T, given by

$$\begin{aligned} x &= AX + BY \\ y &= CX + DY\,, \end{aligned} \tag{3}$$

with determinant $\Delta = AD - BC$, where for the moment we take the coefficients to be real numbers. Substitute (3) into $f(x, y)$ to get a new form

$$g(X, Y) = a'X^2 + b'XY + c'Y^2\,.$$

As in our example, we would like T to define a bijection between the integer pairs (x, y) and (X, Y), so that f and g will represent the same integers.

Assume $\Delta = 0$. Then, since $Y\Delta = Ay - Cx = 0$, (X, Y) cannot take all integer values as x and y run through the integers; thus T is not a bijection.

Therefore we assume $\Delta \neq 0$, and solve (1) for X and Y to obtain T^{-1} given by

$$\begin{aligned} X &= \frac{D}{\Delta}x - \frac{B}{\Delta}y \\ Y &= \frac{-C}{\Delta}x + \frac{a}{\Delta}y\,. \end{aligned} \tag{4}$$

Equations (3) and (4) define a bijection between all pairs (x, y) and (X, Y), for real values of the variables. To decide when this is also a bijection between integer pairs, *we assume it is true* and analyze the problem to force out the appropriate conditions. Setting $(X, Y) = (1, 0)$ and $(0, 1)$, we see by (3) that $A, B, C, D \in \mathbf{Z}$. Similarly, setting $(x, y) = (1, 0)$ and $(0, 1)$, we see by (4) that $\frac{A}{\Delta}, \frac{B}{\Delta}, \frac{C}{\Delta}, \frac{D}{\Delta} \in \mathbf{Z}$. Therefore

$$\left(\frac{A}{\Delta}\right)\left(\frac{D}{\Delta}\right) - \left(\frac{B}{\Delta}\right)\left(\frac{D}{\Delta}\right) = \frac{AD - BC}{\Delta^2} = \frac{\Delta}{\Delta^2} = \frac{1}{\Delta} \in \mathbf{Z}\,;$$

hence $\Delta = \pm 1$.

Conversely, when $A, B, C, D \in \mathbf{Z}$ and $\Delta = \pm 1$, T defines a bijection between integer pairs.

This leads to one of Lagrange's major contributions to the theory of quadratic forms, the concept of equivalence. Assume $f(x, y)$ and $g(X, Y)$ are related by (3), with the conditions $A, B, C, D \in \mathbf{Z}$ and $\Delta = \pm 1$. Lagrange, who was always reluctant to introduce new terms, called f and g "forms which can be transformed into one another". Later Gauss, who was a master at introducing suggestive terminology, called them **equivalent forms**. Equivalence of forms is an equivalence relation (exercise). Lagrange, and later Gauss, studied the equivalence classes of forms. For example, they looked for a "simple" form, or set of forms, in each equivalence class to try to simplify the representation problem. We return to these matters in chapter 12.

Chapter 6

Legendre

1707	Euler	1783
1736	Lagrange	1813
1752	Legendre	1833
1777	Gauss	1855

1. Legendre

Adrien-Marie Legendre (1752-1833) spent his life in France as one of the leading mathematicians in Europe, but not the equal of Euler, Lagrange, Laplace or Gauss, whose lives overlapped his.

His favorite areas of research were celestial mechanics, elliptic integrals and number theory. His name is associated with a variety of topics including Legendre polynomials, the method of least squares and the Legendre symbol in number theory. Legendre's texts and treatises were very influential, and his geometry text dominated the teaching of geometry in Europe and America throughout the 19th century.

Many of his results were found independently by Gauss and this led to some serious priority disputes. To quote Jean Itard's biography of Legendre [Ita 2]:

> "Gauss considered that a theorem was his if he gave the first rigorous proof of it. Legendre, twenty five years his senior, had a much broader and hazier sense of rigor. For Legendre, a belated disciple of Euler, an argument that was merely plausible often took the place of a proof. Consequently all discussion of priority between the two resembled a dialog of the deaf."

In 1798, Legendre published his *Theorie des Nombres* [Leg 1], the first systematic book devoted exclusively to number theory. This book had a strong influence on the mathematicians of the time and was revised several times. But even in the last edition (1830), he did not adopt many of the superior methods developed by Gauss.

We shall only discuss some of his more important work in number theory.

2. Rational Points on Conics

In an exercise in section 1.2, we introduced the relationship between Pythagorean triples and rational points on the unit circle, and we invited the reader to explore the question of rational points on other conics. In other words, we asked for rational solutions to second degree polynomial equations in two variables. Legendre found an algorithm to decide if such an equation has at least one rational solution. We shall return to this question when we discuss arithmetic on curves (chap. 18) and later we will see that Legendre's algorithm is most naturally stated in the language of p-adic numbers (chap. 23).

3. Distribution of Prime Numbers

Legendre made several conjectures about primes, which stimulated much subsequent research up to the present day.

Conjecture: If $\pi(x)$ denotes the number of primes $< x$, then there exist constants A, B such that

$$\pi(x) \sim \frac{x}{A \log x + B} .$$

($f(x) \sim g(x)$, read "$f(x)$ is asymptotic to $g(x)$", means that the limit of $\frac{f(x)}{g(x)}$ is 1 as x approaches infinity.) About a hundred years later it was proved that $\pi(x) \sim \frac{x}{\log x}$.

Conjecture: If $(k, m) = 1$ then the arithmetic progression $k, k+m, k+2m, \ldots$ contains infinitely many primes. This conjecture was proved by Dirichlet in 1837, and it marked the real beginning of the use of analytic methods in number theory.

Conjecture: For all $n > 1$ there exists a prime between n^2 and $(n+1)^2$. This is still unknown.

4. Quadratic Residues and Quadratic Reciprocity

Fermat's little theorem arises from asking which primes divide numbers of the form $a^n - 1$. Euler became very interested in the companion question as to which primes divide $a^n + 1$. This leads to the study of quadratic residues [Wei 1].

Let a be an integer and p a prime not dividing a. We say that a is a **quadratic residue** of p if there exists an integer x such that p divides $x^2 - a$. Otherwise a is a **quadratic nonresidue** of p. Quadratic residues yield a strong necessary condition for a prime p to be of the form $x^2 - ay^2$.

Theorem: *Let p be a prime and a an integer not divisible by p. If*

$$p = x^2 - ay^2$$

for some integers x, y, then a is a quadratic residue of p.

Proof: By a simple divisibility argument, we have $(p, y) = 1$ (exercise). Hence there exist integers k, m such that $kp + my = 1$. Substituting $my = 1 - kp$ in

$$pm^2 = x^2m^2 - ay^2m^2 = (xm)^2 - a(ym)^2 ,$$

we have

$$pm^2 = (xm)^2 - a(1 - kp)^2$$

or

$$(xm)^2 - a = p(m^2 - 2ka + k^2 pa) .$$

Thus p divides $(xm)^2 - a$ and a is a quadratic residue of p.

Legendre found it very useful to introduce the symbol $\left(\frac{a}{p}\right)$, now called the **Legendre symbol**, as follows: if p is an odd prime not dividing the integer a, then

$$\left(\frac{a}{p}\right) = 1 , \quad \text{if } a \text{ is a quadratic residue of } p ,$$

$$\left(\frac{a}{p}\right) = -1 , \quad \text{otherwise} .$$

Euler made an extensive study of quadratic residues and among his results are:

Theorem: (Euler's Criterion) *Let p be an odd prime not dividing the integer a. Then*

$$p \mid a^{\frac{p-1}{2}} - \left(\frac{a}{p}\right) .$$

Euler did not have the notation of the Legendre symbol and thus his statements were less compact.

Corollary: $\left(\frac{-1}{p}\right) = (-1)^{\frac{p-1}{2}}$.

We shall prove the theorem in section 9.3. The corollary is an easy exercise.

In the course of his research on the representation of p by $x^2 \pm qy^2$, where p and q are distinct odd primes, Legendre found that he had to consider $\left(\frac{q}{p}\right)$ as well as $\left(\frac{p}{q}\right)$. These studies led him to formulate his celebrated **law of quadratic reciprocity** [Leg 1]:

> " For any (distinct odd) prime numbers m and n, if they are not both of the form $4x + 3$ one will always have $\left(\frac{n}{m}\right) = \left(\frac{m}{n}\right)$, and if they are both of the form $4x + 3$, one will have $\left(\frac{n}{m}\right) = -\left(\frac{m}{m}\right)$. These two general cases are combined in the formula
>
> $$\left(\frac{n}{m}\right) = (-1)^{\left(\frac{m-1}{2}\right)\left(\frac{n-1}{2}\right)} \left(\frac{m}{n}\right) ."$$

In present day books this is more commonly written as

$$\left(\frac{m}{n}\right)\left(\frac{n}{m}\right) = (-1)^{\frac{(m-1)(n-1)}{4}} .$$

Legendre gave a "proof" of the law based on the assumption that if m is a prime of the form $4x + 1$ then there exists a prime of the form $4x + 3$ such that $\left(\frac{m}{n}\right) = -1$. He knew that this followed from his conjecture on the distribution of primes in an arithmetic progression (sec. 3), but he was never able to prove either of these results (see [Sch - Opo] and [Wei 1] for more details).

Unknown to Legendre or later to Gauss, who discovered it independently, Euler had also conjectured the law of quadratic reciprocity in a slightly different but equivalent form, but he was unable to prove it. It appeared in a paper published in 1772 and in his Opuscula Analytica of 1783 [Stru]. In fact, in 1742, when he was close to having discovered the full law, he wrote to Clairaut: "I feel sure that I am far from having exhausted this topic, but that innumerably many splendid properties of numbers remain still to be discovered" [Wei 1, pg. 187].

The law of quadratic reciprocity has been an important theme in the development of number theory. Later, we will present a more systematic formulation, equivalent forms and several proofs, and we will see how its generalizations were a key stimulus in the creation of algebraic number theory.

Chapter 7

Gauss

1. Gauss and His Work

Generally regarded, with Archimedes and Newton, as one of the three greatest mathematicians of all time, Carl Friedrich Gauss's work spans the full range of mathematical science as well as physics, astronomy and engineering.

We present two accounts of Gauss's early life, the first from May's biography [May] and the second from Felix Klein, *Development of Mathematics in the 19th Century* [Kle 2]. I highly recommended the biography of Gauss by Walter Kaufmann-Buhler [Kau], especially for its insights into the European culture of Gauss's time.

> "GAUSS, CARL FRIEDRICH (b. Brunswick, Germany, 30 April 1777; d. Gottingen, Germany, 23 February 1855), mathematical sciences.
>
> The life of Gauss was very simple in external form. During an austere childhood in a poor and unlettered family he showed extraordinary precocity. Beginning when he was fourteen, a stipend from the duke of Brunswick permitted him to concentrate on intellectual interests for sixteen years. Before the age of twenty-five he was famous as a mathematician and astronomer. At thirty he went to Gottingen as director of the observatory. There he worked for forty-seven years, seldom leaving the city except on scientific business, until his death at almost seventy-eight.
>
> In marked contrast to this external simplicity, Gauss's personal life was complicated and tragic. He suffered from the political turmoil and financial insecurity associated with the French Revolution, the Napoleonic period, and the democratic revolutions in Germany. He found no mathematical collaborators and worked alone most of his life. An unsympathetic father, the early death of his first wife, the poor health of his

second wife, and unsatisfactory relations with his sons denied him a family sanctuary until late in life.

In this difficult context Gauss maintained an amazingly rich scientific activity. An early passion for numbers and calculations extended first to the theory of numbers and then to algebra, analysis, geometry, probability, and the theory of errors. Concurrently he carried on intensive empirical and theoretical research in many branches of science, including observational astronomy, celestial mechanics, surveying, geodesy, capillarity, geomagnetism, electromagnetism, mechanics, optics, the design of scientific equipment, and actuarial science. His publications, voluminous correspondence, notes, and manuscripts show him to have been one of the greatest scientific virtuosos of all time.

Early Years. Gauss was born into a family of town workers striving on the hard road from peasant to lower middle-class status. His mother, a highly intelligent but only semiliterate daughter of a peasant stonemason, worked as a maid before becoming the second wife of Gauss's father, a gardener, laborer at various trades, foreman ("master of waterworks"), assistant to a merchant, and treasurer of a small insurance fund. The only relative known to have even modest intellectual gifts was the mother's brother, a master weaver. Gauss described his father as worthy of esteem" but "domineering, uncouth, and unrefined." His mother kept her cheerful disposition in spite of an unhappy marriage, was always her only son's devoted support, and died at ninety-seven, after living in his house for twenty-two years.

Without the help or knowledge of others, Gauss learned to calculate before he could talk. At the age of three, according to a well-authenticated story, he corrected an error in his father's wage calculations. He taught himself to read and must have continued arithmetical experimentation intensively, because in his first arithmetic class at the age of eight he astonished his teacher by instantly solving a busy-work problem: to find the sum of the first hundred integers. Fortunately, his father did not see the possibility of commercially exploiting the calculating prodigy, and his teacher had the insight to supply the boy with books and to encourage his continued intellectual development.

During his eleventh year, Gauss studied with Martin Bartels, then an assistant in the school and later a teacher of Lobachevsky at Kazan. The father was persuaded to allow Carl Friedrich to enter the Gymnasium in 1788 and to study after school instead of spinning to help support the family. At the Gymnasium, Gauss made very rapid progress in all subjects, especially classics and mathematics, largely on his own. E. A. W. Zimmermann, then professor at the local Collegium Carolinum and later privy councillor to the duke of Brunswick, offered friendship, encouragement, and good offices at court. In 1792 Duke Carl Wilhelm Ferdinand began the stipend that made Gauss independent.

When Gauss entered the Brunswick Collegium Carolinum in 1792, he possessed a scientific and classical education far beyond that usual for his age at the time. He was familiar with elementary geometry, algebra, and analysis (often having discovered important theorems before reaching them in his studies), but in addition he possessed a wealth of arithmetical information and many number-theoretic insights. Extensive calculations and observation of the results, often recorded in tables, had led him to an intimate acquaintance with individual numbers and to generalizations that he used to extend his calculating ability. Already his lifelong heuristic pattern had been set: extensive empirical investigation leading to conjectures and new insights that guided further experiment and observation. By such means he had already independently discovered Bode's law of planetary distances, the binomial theorem for rational exponents, and the arithmetic-geometric mean.

During his three years at the Collegium, Gauss continued his empirical arithmetic, on one occasion finding a square root in two different ways to fifty decimal places by ingenious expansions and interpolations. He formulated the principle of least squares, apparently while adjusting unequal approximations and searching for regularity in the distribution of prime numbers. Before entering the University of Gottingen in 1975 he had rediscovered the law of quadratic reciprocity (conjectured by Lagrange in 1785), related the arithmetic-geometric mean to infinite series expansions, conjectured the prime number theorem (first proved by J. Hadamard in 1896), and found some results that would hold if "Euclidean geometry were not the true one."

In Brunswick, Gauss had read Newton's Principia and Bernoulli's Ars Conjectandi, but most mathematical classics were unavailable. At Gottingen, he devoured masterworks and back files of journals, often finding that his own discoveries were not new. Attracted more by the brilliant classicist G. Heyne than by the mediocre mathematician A. G. Kastner, Gauss planned to be a philologist. But in 1796 came a dramatic discovery that marked him as a mathematician. As a by-product of a systematic investigation of the cyclotomic equation (whose solution has the geometric counterpart of dividing a circle into equal arcs), Gauss obtained conditions for the constructibility by ruler and compass of regular polygons and was able to announce that the regular 17-gon was constructible by ruler and compasses, the first advance in this matter in two millennia.

The logical component of Gauss's method matured at Gottingen. His heroes were Archimedes and Newton. But Gauss adopted the spirit of Greek rigor (insistence on precise definition, explicit assumption, and complete proof) without the classical geometric form. He thought numerically and algebraically, after the manner of Euler, and personified the extension of Euclidean rigor to analysis. By his twentieth year, Gauss was driving ahead with incredible speed according to the pattern he was to continue in many contexts-massive empirical investigations in close interaction with intensive meditation and rigorous theory construction.

During the five years from 1796 to 1800, mathematical ideas came so fast that Gauss could hardly write them down. In reviewing one of his seven proofs of the law of quadratic reciprocity in the *Gottingische gelehrie Anzeigen* for March 1817, he wrote autobiographically:

"It is characteristic of higher arithmetic that many of its most beautiful theorems can be discovered by induction with the greatest of ease but have proofs that lie anywhere but near at hand and are often found only after many fruitless investigations with the aid of deep analysis and lucky combinations. This significant phenomenon arises from the wonderful concatenation of different teachings of this branch of mathematics, and from this it often happens that many theorems, whose proof for years was sought in vain, are later proved in many different ways. As soon as a new result is discovered by induction, one must consider as the first requirement the finding of a proof by any possible means. But after such good fortune, one must not in higher arithmetic consider the investigation closed or view the search for other proofs as a superfluous luxury. For sometimes one does not at first come upon the most beautiful and simplest proof, and then it is just the insight into the wonderful concatenation of truth in higher arithmetic that is the chief attraction for study and often leads to the discovery of new truths. For these reasons the finding of new proofs for known truths is often at least as important as the discovery itself [Werke, II, 159-160]."

In the next account, Klein breaks Gauss's work in pure mathematics into two periods, the latter ending in 1801. This is not to say that Gauss did not pursue research in pure mathematics after 1801. However, except for number theory, much of this work arose from more applied considerations. For instance his differential geometry was motivated by his field work in geodesy.

"First, the prehistoric period, as I would like to name the time up to the start of the diary.

A natural interest, I might even say a certain children curiosity, first led the boy to mathematical questions, independently of any outside influence. Indeed, it was simply the art of calculating with numbers that first attracted him. He calculated continually, with overpowering industry and untiring perseverance. By this incessant exercise in manipulating numbers (for example, calculating decimals to an unbelievable number of places) he acquired not only the astounding virtuosity in computational technique that marked him throughout his life, but also an immense memory stock of definite numerical values, and thereby an appreciation and overview of the realm of numbers such as probably no one, before or after him, has possessed. Aside from arithmetic he was occupied with numerical operations on infinite series. From his activity with numbers, and thus in an inductive, "experimental" way, he arrived quite early at a knowledge of their general relations and laws. This method of work has already been mentioned in connection with the theorems aureum. It was not so rare in the 18th century–for example, with Euler–but stands in sharp contrast to the normal practice of today's mathematicians.

One of the earliest topics to arouse Gauss's appetite for discovery is the so-called *arithmetic-geometric mean.* To unite, as it were, the advantages of the two means $m' = \frac{m+n}{2}$ and $n' = \sqrt{mn}$ by a mixture of both, he continued their method of construction:

$$m'' = \frac{m' + n'}{2}, \quad n'' = \sqrt{m'n'}\,.$$

He remarked–of course computing with definite numbers, for example, $m = 1, n = \sqrt{2}$, that this procedure converges to a value he determined to many decimal places. At this time, of course, Gauss did not suspect the importance this fact would some day have in the theory of elliptic functions. Here we encounter a strange, certainly not accidental phenomenon. All these early intellectual games, devised solely for his own pleasure, were first steps towards a great goal that became conscious only later. It is part of the anticipatory wisdom of genius to place the pick-ax precisely on the rock vein where the gold mine lies concealed, and to do this even in the half-playful first testings of its powers, unconscious of its deeper meaning.

We now come to the year 1795, of which we have more detailed evidence. According to Gauss's own statement, it was then that he discovered the method of least squares. Then, still before his Goettingen period, a passionate interest in the integers seized him, even more tenaciously than before, as is vividly evidenced by the preface to the Disquisitiones Arithmeticae. Unacquainted with the literature, he had to create everything for himself. Here again it was the untiring calculator who blazed the way into the unknown. Gauss set out huge tables: of prime numbers, of quadratic residues and non-residues, and of the fraction $\frac{1}{p}$ for $p = 1$ to $p = 1000$ with their decimal expansions carried out to a complete period, and therefore sometimes to several hundred places! With this last table Gauss tried to determine the dependence of the period on the denominator p. What researcher of today would be likely to enter upon this strange path in search of a new theorem?

But for Gauss it was precisely this path, followed with such unheard of energy – he himself maintained that he differed from other men only in his diligence – that led to his goal. Thus, like Euler before him, he discovered the law of quadratic reciprocity, the *theorema aureum*, in a numerical - inductive way.

In the autumn of 1795 he moved to Göttingen, where he must have devoured the works of Euler and Lagrange, presented to him for the first time. And on March 30, 1796 he underwent his conversion on the road to Damascus. With this began the

Second period, marked by the regular use of his diary, 1796 - 1801. For a long time Gauss had been busy with grouping the roots of unity $x^n = 1$ on the basis of his theory of "primitive roots". Then suddenly one morning, still in bed, he saw clearly that the construction of the 17-gon follows from his theory. As already mentioned, this discovery marked a turning point in Gauss's life. He decided to devote himself entirely to mathematics, not philology. On this date begins the diary, the most interesting document we have on Gauss's development. Here we do not find the inaccessible, isolated, cautious man; here we see Gauss as he lived and experienced his great discoveries. He expressed his joy and satisfaction in the most lively way; he congratulated himself and broke out in enthusiastic exclamations. We see the proud series of great discoveries in arithmetic, algebra and analysis parade before us (though not, of course, all of them) and live through the genesis of the *Disquisitiones Arithmeticae*. Among these traces of the burgeoning of a mighty genius one finds, touchingly, little miniatures of school exercises, which even a Gauss was not spared. Here we find a record of conscientious exercises in differentiation; and just before a section on the division of the lemniscate, there are totally banal integral substitutions, such as every student must practice.

This intensive period led to the publication of the *Disquisitiones Arithmeticae* in 1801. Klein comments:

"In the *Disquisitiones Arithmeticae* Gauss created modern number theory in its true sense and fixed its whole subsequent development. Our amazement at this achievement increases when we consider that Gauss created this whole world of thought purely out of himself and by himself, without any outside stimulus. Historical research shows, as we shall see, that Gauss had already made most of his discoveries before he became acquainted with the relevant literature in Goettingen. These were the works of Euler, Lagrange and Legendre, in which he took a passionate interest because of his own creations. Aside from this reading and his rare visits to his colleague Kaestner, Gauss was not influence at

Goettingen by anything but the bidding of his own inexorable urge to create."

Klein gives a selection of characteristic remarks from the diary (see [Gra] for an English translation of the diary).

"I would now like to give a small selection of especially characteristic remarks from the diary. Their numbering follows that of Volume 10.1 of the *Werke*.

1. March 30, 1796: The geometric division of the circle into 17 parts.

2. April 8, 1796: The first exact proof of the theorema aureum. This extremely long proof, containing eight different special cases, is still highly noteworthy because of the intrepid consistency of its execution. Kronecker calls it a "test of strength for Gauss's genius". With

51. January 7, 1797 begins his work on the lemniscate, and in the note

60. March 19, 1797, he has found the reason–the "*cur*"–for the appearance of the exponent n^2 in the equation of the division of the lemniscate. That is, he has recognized, with the use of the complex domain, the double periodicity of the lemniscate integral

$$\int \frac{dx}{\sqrt{1-x^4}} = \int \frac{dx}{(1-x)(1+x)(1-ix)(1+ix)} .$$

In the note

80. In October, we find the proof of the fundamental theorem of algebra that was to earn him his doctorate in 1799. But then with

98. May 30, 1799, he arrived at a highly significant result: he found the relation between the arithmetic-geometric mean and the length of the lemniscate, and again in a purely computational way, using the value of $\frac{1}{[M(1,\sqrt{2})]}$ worked out to eleven decimal places. Although he still did not recognize the relation clearly, he did appreciate the importance of this discovery, that it "opened up a whole new field of analysis". From then on the development of the area of elliptic functions went forward rapidly. At first he was still occupied with the "lemniscate" function, i.e., with the special case of a square period parallelogram. But the notes

105-109. May 6 - June 3, 1800, record the discovery of the general doubly-periodic functions. The square was replaced by a general parallelogram. With this the full theory of elliptic and modular functions was created, and with one stroke it was developed beyond even Abel and Jacobi.

With his increasing astronomical activity, this great period of discovery drew to a close. However, I should just mention

144. October 23, 1813, where the true theory of biquadratic residues is given, along with the introduction of the numbers $a + bi$ into number

theory. Apparently this discovery filled Gauss with a special joy. For he added a remark that this solution, which he had sought in vain for seven years, had been granted him at the same time as the birth of a son.

I do not want to leave the summary of this unique document without adding a few remarks of a general nature.

Perhaps there are people who wonder that Gauss spent so much energy on problems that were already settled, that he had to overcome again, without any guidance or assistance, all those difficulties whose defeat was already part of the common knowledge of the science. Against this opinion I would most emphatically urge the blessings of independent discovery. From just this example we can learn the pedagogical truth that the successful development of an individual has much less to do with the acquisition of knowledge than with the development of abilities. The obstinacy with which Gauss followed a path once chosen, the youthful impetuosity with which he regularly and recklessly took the steepest way towards his goal–these hard tests strengthened his powers and made him capable of striding recklessly over all obstacles, even when they had already been removed by earlier investigations."

We will discuss most of the results that Klein selected from Gauss's diary - the law of quadratic reciprocity in chapter 11, circle division in chapter 14, biquadratic residues in chapter 15, and elliptic functions in chapter 19. See Cox [Cox 2] for the role played by the arithmetic-geometric mean in Gauss work on elliptic functions.

None of Gauss's work on elliptic functions was published during his lifetime. However from his notes we know that he had anticipated much of the most important work of Abel and Jacobi on this subject. Gauss had the very annoying habit of writing to his friends to claim priority for his unpublished results after others had rediscovered and published them. Although, as we know from his unpublished notes, Gauss's claims were true, they did not endear him to those he had anticipated. For example, a letter to his friend Wolfgang Boylai claimed priority for Boylai's son Johann's work on non-Euclidean geometry. Despite Gauss's "assurance of my special esteem" for the younger Boylai, the lack of any public recognition from Gauss discouraged Johann from pursuing any further work in mathematics (see [Hal, T]).

There are some instances where Gauss acted to help younger mathematicians. He admired the work of Eisenstein and Dirichlet, and tried to advance their careers. See D. E. Rowe [Row] for a discussion of Gauss's relations with Dirichlet and some of their correspondence.

In addition to the biographies already mentions, Waldo Dunnnington's [Dun] contains much information on Gauss's personal and professional life.

2. An Overview of the Disquisitiones Arithmeticae

A short discussion and table of contents of the *Disquisitiones Arithmeticae*, to be referred to hereafter as the *Disquisitiones* will give us some perspective. The *Disquisitiones* is now available in English translation [Gau 1] and is still well worth serious study. .

Gauss's loyalty and appreciation to the Duke of Brunswick, who provided the funds for his education, is evident in the dedication to the *Disquisitiones* which reads like a prayer. However, this must be taken as a sincere expression of Gauss's feelings since the Duke's patronage made it possible for Gauss to have an education and career.

In the introduction to the *Disquisitiones*, Gauss discusses his discoveries and his desire to present a complete exposition of major parts of the higher arithmetic, including results of earlier scholars, many of which he had rediscovered.

The *Disquisitiones* is divided into three parts (by me, not Gauss), of which the second takes up more than 60% of the book. Gauss divided the *Disquisitiones* into seven sections (chapters) which comprise a total of 366 articles.

Congruence

I. Congruent Numbers in General
II. Congruences of the First Degree
III. Residues of Powers
IV. Congruences of the Second Degree

Quadratic Forms

V. Forms and Indeterminate Equations of the Second Degree
VI. Various Applications of the Preceding Discussion

Cyclotomy

VII. Equations Defining Sections of the Circle

This is just the table of contents and does not even begin to exhibit the fantastic richness of the 366 articles. We devote the next seven chapters to a discussion of the *Disquisitiones*.

Chapter 8

Theory of Congruence 1

1. Section I of the Disquisitiones

We begin a systematic introduction to the theory of congruence by reproducing the first section of the *Disquisitiones* (translation from [Gau 1]). This illustrates Gauss's careful attention to rigor and complete presentation of even the most elementary concepts, as well as the revolutionary introduction of the congruence notation. Particularly interesting is the fact that in many current elementary textbooks the introduction to congruence is only a minor variation of this section.

Together with Lagrange's work on algebraic equations, Gauss's study of congruence can be regarded as the beginning of modern algebra.

"Section I

CONGRUENT NUMBERS IN GENERAL

Congruent numbers, moduli, residues, and nonresidues

1. If a number a divides the difference of the numbers b and c, b and c are said to be *congruent relative to* a; if not, b and c are *noncongruent*. The number a is called the *modulus*. If the numbers b and c are congruent, each of them is called a *residue* of the other. If they are noncongruent they are called *nonresidues*.

The numbers involved must be positive or negative integers,[1] not fractions. For example, -9 and $+16$ are congruent relative to 5; -7 is a residue of $+15$ relative to 11, but a nonresidue relative to 5; -7 is a residue of $+15$ relative to 11, but a nonresidue relative to 3.

---

[1] The modulus must obviously be taken *absolutely*, i.e. without sign.

Since every number divides zero, it follows that we can regard any number as congruent to itself relative to any modulus.

2. Given a, all its residues modulo m are contained in the formula $a + km$ where k is an arbitrary integer. The easier propositions that we state below follow at once from this, but with equal ease they can be proved directly.

Henceforth we shall designate congruence by the symbol $\equiv$, joining to it in parentheses the modulus when it is necessary to do so; e.g. $-7 \equiv 15 (\text{mod. } 11)$, $-16 \equiv 9 (\text{mod. } 5)$.[2]

3. Theorem. *Let m successive integers $a, a+1, a+2, \cdots a+m-1$ and another integer A be given; then one, and only one, of these integers will be congruent to A relative to m.*

If $\frac{a-A}{m}$ is an integer then $a \equiv A$; if it is a fraction, let k be the next larger integer (or if it is negative, the next smaller integer not regarding sign). $A + km$ will fall between a and $a + m$ and will be the desired number. Evidently all the quotients $\frac{a-A}{m}, \frac{a+1-A}{m}, \frac{a+2-A}{m}$, etc. lie between $k-1$ and $k+1$, so only one of them can be an integer.

Least residues

4. Each number therefore will have a residue in the series $0, 1, 2, \cdots m-1$ and in the series $0, -1, -2, \cdots -(m-1)$. We will call these the *least residues*, and it is obvious that unless 0 is a residue, they always occur in pairs, one *positive* and one *negative*. If they are unequal in magnitude one will be $< \frac{m}{2}$; otherwise each will $= \frac{m}{2}$ disregarding sign. Thus each number has a residue which is not larger than half the modulus. It will be called the *absolutely least* residue. Relative to the modulus 7, $+5$ is its own least positive residue; -2 is the least negative residue and the *absolutely least* residue.

For example, relative to the modulus 5, -13 has 2 as least positive residue. It is also the absolutely least residue, whereas -3 is the least negative residue.

Elementary propositions regarding congruences

5. Having established these concepts, let us collect the properties of congruences that are immediately obvious.

[2] We have adopted this symbol because of the analogy between equality and congruence. For the same reason Legendre, in the treatise which we shall often have occasion to cite, used the same sign for equality and congruence. To avoid ambiguity we have made a distinction.

Numbers that are congruent relative to a composite modulus are also congruent relative to any divisor of the modulus.

If several numbers are congruent to the same number relative to the same modulus, they are congruent to one another (relative to the same modulus).

This identity of moduli is to be understood also in what follows.

Congruent numbers have the same least residues; noncongruent numbers have different least residues.

6. *Given the numbers A, B, C, etc. and other numbers a, b, c, etc. that are congruent to them relative to some modulus, i.e.. $A \equiv a$, $B \equiv b$, etc., then $A + B + C +$ etc. $\equiv a + b + c +$ etc.*

If $A \equiv a$, $B \equiv b$, then $A - B \equiv a - b$.

7. *If $A \equiv a$, then also $kA \equiv ka$.*

If k is a positive number, then this is only a particular case of the preceding article (art. 6) letting $A = B = C$ etc., $a = b = c$ etc. If k is negative, $-k$ will be positive. Thus $-kA \equiv -ka$ and so $kA \equiv ka$.

If $A \equiv a$, $B \equiv b$ then $AB \equiv ab$, because $AB \equiv Ab \equiv ba$.

8. *Given any numbers whatsoever A, B, C, etc. and other numbers a, b, c, etc. which are congruent to them, i.e. $A \equiv a$, $B \equiv b$, etc., the products of each will be congruent; i.e. ABC etc. $\equiv abc$ etc.*

From the preceding article $AB \equiv ab$ and for the same reason $ABC \equiv abc$ and any number of factors may be adjoined.

If all the numbers A, B, C, etc. and the corresponding a, b, c, etc. are taken equal, then one gets the following theorem: *If $A \equiv a$ and k is a positive integer, $A^k \equiv a^k$.*

9. *Let X be an algebraic function of the indeterminate x of the form*

$$Ax^a + Bx^b + Cx^c + \text{etc.}$$

where A, B, C, etc. are any integers; a, b, c, etc. are nonnegative integers. Then if x is given values which are congruent relative to some modulus, the resulting values of the function X will also be congruent.

Let f, g be congruent values of x. Then from the preceding article $f^a \equiv g^a$ and $Af^b \equiv Ag^a$, and in the same way $Bf^b \equiv Bg^b$ etc. Thus

$$Af^a + Bf^b + Cf^c + \text{etc.} \equiv Ag^a + Bg^b + Cg^c + \text{etc.} \qquad \text{Q.E.D.}$$

It is easy to understand how this theorem can be extended to functions of several indeterminates.

10. Thus, if all integers are substituted consecutively for x, and the corresponding values of the function X are reduced to least residues, they will form a sequence in which after an interval of m terms (m being the modulus) the same terms will recur; that is, the sequence will be formed by a *period* of m terms repeated infinitely often. For example, let $X = x^3 - 8x + 6$ and $m = 5$; then for $x = 0, 1, 2, 3$, etc. the values of X produce these least positive residues: $1, 4, 3, 4, 3, 1, 4$, etc. where the first five numbers $1, 4, 3, 4, 3$ are repeated infinitely often; and if the sequence is continued in a contrary sense, that is, if one gives negative values to x, the same period appears with the order of the term inverted. From this it follows that no terms other than those that make up this period can occur in the whole sequence.

11. In this example X cannot become $\equiv 0$, nor $\equiv 2 \pmod{5}$ and still less can it $= 0$ or $= 2$. Thus the equations $x^3 - 8x + 6 = 0$ and $x^3 - 8x + 4 = 0$ cannot be solved in integers and consequently, as we know, not by rational numbers. More generally, suppose X is a function in unknown x of the form

$$x^n + Ax^{n-1} + Bx^{n-2} + \text{etc.} + N$$

where A, B, C, etc. are integers, n a positive integer (it is known that all algebraic equations can be reduced to this form). It is evident that in the equation $X = 0$ there exist no rational root if for some modulus the congruence $X \equiv 0$ cannot be satisfied. But this criterion will be discussed more fully in Section VIII.[3] From this example some small idea of the usefulness of these investigations can be gained.

Certain applications

12. Many things that are customarily taught in treatises on arithmetic depend on theorems expounded in this section, e.g. rules for deciding whether given numbers are divisible by 9, 11, or any other number. *Relative to the modulus 9* all powers of the number 10

[3] Gauss planned eight sections for the *Disquisitiones* and had essentially worked out the eighth section treating of congruences of higher degrees. However, he decided to publish only seven sections in order not to increase the cost of printing the book. See the Author's Preface.

are congruent to unity; hence if a number is of the form $a+10b+100c+$ etc. it will have, relative to the modulus 9, the same least residue as $a+b+c+$ etc. Thus it is clear that if the single digits of a number expressed in decimal notation are added without regard to position in the number, this sum and the given number will have the same least residue; and thus the latter can be divided by 9 if the former can and vice versa. The same is true of the divisor 3. And since *relative to the modulus 11*, $100 \equiv 1$, in general $10^{2k} \equiv 1$, $10^{2k}+1 \equiv 10 \equiv -1$, and a number of the form $a+10b+100c+$ etc. will have the same least residue relativeto the modulus 11 as $a-b+c$ etc. From this the well-known rule is derived immediately. And from the same principle we can easily deduce all similar rules.

From the preceding argument we can also discover the underlying principles governing the rules that are ordinarily used to verify arithmetic operations. Specifically, if from given numbers others are to be derived by addition, subtraction, multiplication, or by raising to powers, we substitute least residues in place of the given numbers relative to an arbitrary modulus (usually we use 9 or 11 because in our decimal system the residues are easily found, as we have just seen). The resulting numbers must be congruent to those deduced from the given numbers, otherwise there is a defect in the calculation.

But since these and similar results are well known, it would be superfluous to dwell on them."

2. Residue Classes

We now discuss and supplement Gauss's chapter, bringing in the unifying algebraic concepts of group, ring and field. The *Disquisitiones* provided one of the basic sets of examples for the later development of the theory of finite abelian groups.

In articles 1 and 2, Gauss introduces the formal definition of congruence, viz., integers a, b are **congruent** modulo the integer n, written $\mathbf{a} \equiv \mathbf{b}(\mathbf{mod}\ \mathbf{n})$ or $\mathbf{a} \equiv \mathbf{b}(\mathbf{n})$, if n divides $a-b$. This is equivalent to saying that *a and b leave the same remainder when divided by n*.

It is difficult to stress sufficiently the revolutionary impact on number theory of this fruitful notation, which Gauss says he adopted because of the analogy between congruence and equality. The congruence notation unifies and clarifies a large body of earlier work and raises a huge new field of study, i.e., to what extent are the results of number theory and algebra true when we replace the = symbol by $\equiv \pmod{n}$. This analogy is one of the unifying threads in our study of congruence.

In a footnote to article 2, Gauss credits Legendre with using the = sign for congruence. However a quick perusal of Legendre's, *Theorie des Nombres*, shows that Legendre uses this idea in a casual way in the proof of Fermat's theorem and does not realize its significance. The real credit for this advance belongs to Gauss.

Congruence is an *equivalence relation* on the integers **Z**. The reflexivity, $a \equiv a \pmod{n}$, is obvious. The symmetry, $a \equiv b \pmod{n} \Longrightarrow b \equiv a \pmod{n}$, and transitivity, $a \equiv b \pmod{n}, b \equiv c \pmod{n} \Longrightarrow a \equiv c \pmod{n}$, whose proofs are also immediate, are essentially combined in the second property of article 5. The equivalence class containing a, which we denote by $\bar{\mathbf{a}}$, is just

$$\bar{\mathbf{a}} = \{\mathbf{a} + \mathbf{kn} | \mathbf{k} \in \mathbf{Z}\} .$$

Thus $\equiv \pmod{n}$ partitions the integers into a finite number of disjoint arithmetic progressions

$$\cdots a - 2n,\ a - n,\ a,\ a + n,\ a + 2n, \ldots,$$

infinite in both directions. Note that the modulus n does not appear in the symbol $\bar{a}$ since it is almost always clear from context. When we need to stress the modulus we will use the symbol **a(mod n)** instead of $\bar{a}$. The equivalence classes modulo n are generally called **residue classes modulo n** or **congruence classes modulo n.**

Gauss never used the terminology of residue classes; he just refers to all numbers congruent to a given number or residue. This is particularly interesting since his use of equivalence classes and operations on them in his theory of composition of forms (sec. 13.9) is probably the first known use of this abstract concept.

We see, from the division algorithm, that *there are exactly n distinct residue classes modulo n.*

The simple, but essential, property that we use to translate between congruence of integers and equality of residue classes is

$$\mathbf{a} \equiv \mathbf{b} \pmod{\mathbf{n}} \iff \bar{\mathbf{a}} = \bar{\mathbf{b}} .$$

A *system of distinct representatives* of the equivalence classes, i.e., a set of numbers such that each class contains exactly one number of the set, is called a **complete system of residues modulo n.** In article 3, Gauss shows that any n consecutive integers form a complete system of residues modulo n. In fact, any n incongruent integers are a complete system of residues, since they lie in different residue classes and there are only n such classes. The most commonly used systems are

i) the *least non-negative residues modulo* n : $\{0, 1, \ldots, n-1\}$,
ii) the *least positive residues modulo* n : $\{1, 2, \ldots, n\}$,
iii) the *absolute least residues modulo* n : $\{-\frac{n-1}{2}, \ldots -1, 0, 1, \ldots, \frac{n-1}{2}\}$ for n odd and $\{-\frac{n}{2}+1, \ldots, -1, 0, 1, \ldots, \frac{n}{2}-1, \frac{n}{2}\}$ or $\{-\frac{n}{2}, \ldots, -1, 0, 1, \ldots, \frac{n}{2}-1\}$ for n even.

Unless otherwise stated, we shall use the least non-negative residues.

3. Congruences and Algebraic Structures

$\mathbf{Z}$ is an abelian group under addition, $\overline{0} = \{kn | k \in \mathbf{Z}\} = n\mathbf{Z}$ is a subgroup, and the residue classes $\overline{a} = a + n\mathbf{Z}$ are the cosets of $\overline{0}$. Hence the quotient group $(\mathbf{Z}/n\mathbf{Z})^+$, the set of residue classes, with addition defined by

$$\overline{a} + \overline{b} = \overline{a+b}\,,$$

is a finite abelian group of order n. Gauss's theorem in article 6, $A \equiv a(\operatorname{mod} n)$ and $B \equiv b(\operatorname{mod} n) \Longrightarrow A + B \equiv a + b(\operatorname{mod} n)$, proves that this operation is well defined.

The results of article 7 show that multiplication of residue classes defined by $\overline{a} \cdot \overline{b} = \overline{ab}$ is well defined. With these two operations, the residue classes form a ring. In fact, since $\mathbf{Z}$ is a ring and $n\mathbf{Z}$ is an *ideal*, this ring is the quotient or residue class ring $\mathbf{Z}/n\mathbf{Z}$.

Unfortunately our rings are not in general fields or even integral domains since, if n is *composite* (not a prime), they have divisors of zero. E.g., we have $4 \cdot 2 \equiv 0(\operatorname{mod} 8)$ and thus, in $\mathbf{Z}/8\mathbf{Z}$, $\overline{4} \cdot \overline{2} = \overline{0}$.

The following properties are very useful for the study of $\mathbf{Z}/n\mathbf{Z}$.

Exercises:

i) $a \equiv b(\operatorname{mod} n),\ m|n \Longrightarrow a \equiv b(\operatorname{mod} m)$,
ii) $(a, n) = 1$ and $ax \equiv ay(\operatorname{mod} n) \Longrightarrow x \equiv y(\operatorname{mod} n)$; in particular, if $n = p$, a prime, then $a \not\equiv 0(\operatorname{mod} p)$, $ax \equiv ay(\operatorname{mod} p) \Longrightarrow x \equiv y(\operatorname{mod} p)$,
iii) $(a, n) = d,\ ax \equiv ay(\operatorname{mod} n) \Longrightarrow x \equiv y(\operatorname{mod} n/d)$.

Theorem: *$ax \equiv 1 \pmod{n}$ (or equivalently $\overline{a}x = \overline{1}$) has a solution $\Longleftrightarrow$ $(a, n) = 1$.*

Proof:

$$(a, n) = 1 \Longleftrightarrow \text{there exist } u, v \text{ such that } ua + vn = 1$$
$$\Longleftrightarrow au \equiv 1 \pmod{n} \Longleftrightarrow \overline{a}\,\overline{u} = \overline{1}\,.$$

Another rather nice proof of $\Longleftarrow$) follows from exercise (ii); the n numbers, $0, a, 2a, \ldots, (n-1)a$, are all incongruent and therefore form a complete system of residues modulo n. Thus one of them is $\equiv 1 \pmod{n}$.

If $ab \equiv 1 \pmod{n}$, we call b an **inverse** of a modulo n. It follows from exercise (ii) that any two inverses of a are congruent modulo n and thus $\overline{b}$ is the *unique solution* of $\overline{a}x = \overline{1}$. We use the symbol a^{-1} to denote an inverse of a modulo n, understanding that it is only determined modulo n. If a has an inverse modulo n, then $\overline{a^{-1}} = \overline{a}^{-1}$ is the unique multiplicative inverse of $\overline{a}$ in $\mathbf{Z}/n\mathbf{Z}$.

Of course, inverses may not exist. For example, 2 has no inverse modulo 4 and 3 has none modulo 6. However, if $n = p$, a prime, and $a \not\equiv 0 \pmod{p}$, then $(a, p) = 1$ and, by the theorem, we have

Corollary: *If $a \not\equiv 0 \pmod{p}$, then a has an inverse modulo p.*

Reformulating this in terms of algebraic structures, we have

Corollary: *For p a prime, the non zero residue classes modulo p have multiplicative inverses and thus form a group. Hence $\mathbf{Z}/p\mathbf{Z}$ is a field with p elements. The order of the multiplicative group of the field is $p-1$. (Recall from the theory of fields that, up to isomorphism, a finite field is uniquely characterized by its size.)*

From now on, we shall also use the notation $\mathbf{F_p} = \mathbf{Z}/p\mathbf{Z}$ for the finite field with p elements, $\mathbf{F_p^+} = (\mathbf{Z}/p\mathbf{Z})^+$ for its additive group, and $\mathbf{F_p^\times} = (\mathbf{Z}/p\mathbf{Z})^\times$ for its multiplicative group. Results from the theories of groups, rings and fields now yield number theoretic results.

Example: (Fermat's Little Theorem) - With congruence notation, the two forms of the theorem (sec. 2.7) become

1) $a \not\equiv 0 \pmod{p} \implies a^{p-1} \equiv 1 \pmod{p}$
   (equivalently $\overline{a} \neq \overline{0} \implies \overline{a}^{p-1} = \overline{1}$),

2) for all a, $a^p \equiv a (\bmod p)$
(equivalently $\overline{a}^p = \overline{a}$).

To prove 1), let k be the order of the element $\overline{a}$ in the group $F_p^\times$, i.e., the smallest positive integer m such that $\overline{a}^m = \overline{1}$. Then k is the order of the subgroup $\{\overline{a}\}$ generated by $\overline{a}$. But, by Lagrange's theorem in group theory, the order of the subgroup divides the order of the group. Hence $k | p - 1$, i.e., $p - 1 = kt$ for some t, and we have

$$\overline{a}^{p-1} = \overline{a}^{kt} = \left(\overline{a}^k\right)^t = \overline{1}^t = \overline{1} .$$

Multiplying both sides by a, we have form 2). From a structural point of view, this is the most natural proof of the theorem.

It is interesting to note that Lagrange's theorem came not from his number theoretic work but from his studies of the role of permutation groups in the theory of equations (see [Edw 2]).

4. Applications

In article 9, Gauss proves that if $f(x) = a_0 + a_1 x + \cdots + a_m x^m$ is a polynomial with integer coefficients, then

$$x_1 \equiv x_2 (\bmod n) \implies f(x_1) \equiv f(x_2) (\bmod n) .$$

Combined with the fact that

$$a = b \implies a \equiv b (\bmod n), \quad \textit{for all } n ,$$

this provides a powerful method for finding necessary conditions for the solution of Diophantine equations and for showing the nonexistence of such solutions. Gauss illustrates these methods in article 11. These ideas were commonly used before Gauss, but the use of congruence notation yields tremendous conceptual and computational advantages.

Example: Suppose there exist integers x, y such that $x^3 - 4y^2 = 2$. Then $x^3 - 4y^2 \equiv 2 (\bmod 4)$ or, since $4 \equiv 0 (\bmod 4)$, $x^3 \equiv 2 (\bmod 4)$. But for the complete system of residues $x \equiv 0, 1, 2, 3 (\bmod 4)$, x^3 can only take the values 0, 1 and 3 modulo 4, a contradiction; thus the equation has no integer solutions.

Example: (Sums of two squares) Let $p \equiv 3 \pmod 4$ be a prime and assume that $p = x^2 + y^2$ has integer solutions. A square can only take the values 0 and 1 modulo 4 and thus $x^2 + y^2$ must be 0, 1 or $2 \pmod 4$, a contradiction. *Therefore, a prime $p \equiv 3 \pmod 4$ cannot be written as the sum of two squares.*

Congruence arguments can also be used to derive positive results.

Proposition: *If (x, y, z) is a solution of*

$$x^3 + y^3 = z^3 ,$$

then at least one of x, y, z is divisible by 3.

Proof: We work modulo 3 and 9. Since $x^3 + y^3 = z^3$,

$$x^3 + y^3 \equiv z^3 \pmod 9 \qquad (1)$$

and

$$x^3 + y^3 \equiv z^3 \pmod 3 . \qquad (2)$$

By Fermat's little theorem, $w^3 \equiv w \pmod 3$ and thus, by (2),

$$x + y \equiv z \pmod 3 \qquad (3)$$

$$\implies z = x + y + 3k , \quad \text{for some } k.$$

Substituting in (1), and reducing modulo 9, we have

$$x^3 + y^3 \equiv (x + y + 3k)^3 \equiv x^3 + y^3 + 3x^2y + 3xy^2 \pmod 9 ,$$

or $3xy(x + y) \equiv 0 \pmod 9$. But

$$\begin{aligned} 3xy(x + y) \equiv 0 \pmod 9 &\implies xy(x + y) \equiv 0 \pmod 3 , \\ &\implies xyz \equiv 0 \pmod 3), \quad \text{by (3)} \\ &\implies 3 | xyz \\ &\implies 3| \text{ at least one of x, y, z.} \end{aligned}$$

A special but very useful result is

Wilson's Theorem: p *a prime* $\iff (p - 1)! \equiv -1 \pmod p$.

We present Gauss' proof of the theorem.

Proof: $\Longrightarrow$) Every number $a, 1 \leq a \leq p-1$, has an inverse a' in this set. Now $a = a' \Longrightarrow aa' = a^2 \equiv 1 (\text{mod } p) \Longrightarrow (a-1)(a+1) \equiv 0 (\text{mod } p)$, and this happens only when $a \equiv 1 (\text{mod } p)$ or $a \equiv -1 \equiv p-1 (\text{mod } p)$.

The other $p-3$ numbers, $2, 3, \ldots, p-2$, can be paired off with their inverses in the set, and their products are $\equiv 1(p)$. Hence

$$2 \times 3 \times \cdots \times p-2 \equiv 1 (\text{mod } p) ,$$

and therefore

$$(p-1)! = 1 \times (2 \times 3 \cdots \times p-2) \times p-1 \equiv p-1 \equiv -1 (\text{mod } p) .$$

The converse is a simple exercise.

An immediate application of Wilson's theorem is

Proposition (Euler): *Let p be an odd prime. Then*

$$x^2 \equiv -1 (\text{mod } p) \text{ has a solution } \Longleftrightarrow p \equiv 1 (\text{mod } 4) .$$

In the language of quadratic residues and the Legendre symbol (sec. 6.4), this can be restated as

$$-1 \text{ is a quadratic residue of } p, \text{ i.e., } \left(\frac{-1}{p}\right) = 1 \Longleftrightarrow p \equiv 1 (\text{mod } 4) .$$

Proof: $\Longrightarrow$) $x^2 \equiv -1 (\text{mod } p) \Longrightarrow x^{p-1} \equiv (x^2)^{\frac{p-1}{2}} \equiv (-1)^{\frac{p-1}{2}}$ $(\text{mod } p)$. By Fermat's little theorem, $x^{p-1} \equiv 1 (\text{mod } p)$; hence

$$1 \equiv (-1)^{\frac{p-1}{2}} (\text{mod } p) \quad \text{and} \quad p \mid 1 - (-1)^{\frac{p-1}{2}} .$$

But $1 - (-1)^{\frac{p-1}{2}} = 0$ or 2. If it is equal to 2, then $p|2$, contradicting our assumption that p is an odd prime. Hence it is 0, i.e., the exponent

$$\frac{p-1}{2} \text{ is even} \Longleftrightarrow \frac{p-1}{2} = 2k , \quad \text{for some } k,$$
$$\Longleftrightarrow p-1 = 4k \Longleftrightarrow p \equiv 1 (\text{mod } 4) .$$

The condition $p \equiv 1 (\text{mod } 4)$ *will arise in many contexts. The reason is almost always that we need* $\frac{p-1}{2}$ *to be even.*

Note that this part of the proof is independent of Wilson's theorem.

$\Longleftarrow$) Since p is assumed odd $p-1$ is even. Therefore

$$(p-1)! = \left(1 \cdot 2 \cdots \frac{p-1}{2}\right) \cdot (p-1)(p-2) \cdots \underbrace{\left(p - \frac{p-1}{2}\right)}_{\left(\frac{p-1}{2}+1\right)}$$

$$\equiv \left(1 \cdot 2 \cdots \frac{p-1}{2}\right) \cdot (-1)(-2) \cdots \left(-\frac{p-1}{2}\right) (\text{mod } p)$$

$$\equiv (-1)^{\frac{p-1}{2}} \left(1 \cdot 2 \cdots \frac{p-1}{2}\right)^2 (\text{mod } p) ,$$

and, since $p \equiv 1 (\text{mod } 4) \implies \dfrac{p-1}{2}$ is even,

$$(p-1)! \equiv \left(1 \cdot 2 \cdots \frac{p-1}{2}\right)^2 (\text{mod } p).$$

Let $x = \left(1 \cdot 2 \cdots \frac{p-1}{2}\right)$. Then, by Wilson's theorem,

$$x^2 = \left(1 \cdot 2 \cdots \frac{p-1}{2}\right)^2 \equiv (p-1)! \equiv -1 (\text{mod } p) ,$$

and we have found a solution to $x^2 \equiv -1 (\text{mod } p)$.

This proposition can be used to prove a special case of Legendre's conjecture (Dirichlet's Theorem) that if $(k, n) = 1$, then the arithmetic progression, $k, k+n, k+2n, \ldots$, contains infinitely many primes; equivalently, there exist infinitely many primes $p \equiv k (\text{mod } n)$.

Proposition: *There exist infinitely many primes* $p \equiv 1 (\text{mod } 4)$.

Proof: Our proof is just an adaptation of Euclid's proof that there are infinitely many primes (sec. 1.2).

Let $p_1, \ldots, p_r$ be the first r primes $\equiv 1 (\text{mod } 4)$. Since 5 is such a prime, this statement is not vacuous. We show that there is another such prime p. Let

$$N = (p_1 p_2 \cdots p_r)^2 + 1 .$$

If p is a prime divisor of N, then $N = (p_1 p_2 \cdots p_r)^2 + 1 \equiv 0(\mathrm{mod}\, p)$, and our last proposition implies that $p \equiv 1(\mathrm{mod}\, 4)$. But none of the $p_i | N$, $p \neq p_i$, and thus p is our required prime.

Exercise: Prove that there are infinitely many primes of the form $p \equiv 3 \equiv -1(\mathrm{mod}\, 4)$ by showing that if $p_1, \ldots, p_r$ are the first r such primes, then some prime divisor of $4(p_1 \cdots p_r) - 1$ must be another such prime.

Similar methods can be used for other special values and the more general case $p \equiv 1(\mathrm{mod}\, n)$ can be proved using properties of the cyclotomic polynomials [Gol, L]. However, the only known proofs of Dirichlet's full theorem require deep analytic methods [Ire - Ros].

5. Linear Congruences

By the definition of congruence, the *solution of* $ax \equiv b(\mathrm{mod}\, n)$ *is equivalent to the solution of the linear Diophantine equation* $ax - ny = b$. We outlined the solution of $ax - ny = b$ in a series of exercises in section 1.2 and again using continued fractions in section 4.5. Now we present it using congruence notation.

Note that when we ask for the number of solutions of a congruence such as $ax \equiv b(\mathrm{mod}\, n)$, we count two solutions which are congruent modulo n as the same and therefore this is the same as asking for the number of solutions in a complete system of residues. Thus we are also counting the residue classes modulo n satisfying the equation $\overline{a}\,\overline{x} = \overline{b} \in \mathbf{Z}/p\mathbf{Z}$. More generally, if $(a, n) = 1$, then, by the theorem of the last section, $ax \equiv b(\mathrm{mod}\, n)$ has the solution $x = a^{-1}b$. The solution is unique modulo n, since $ax \equiv ax'(\mathrm{mod}\, n)$ and $(a, n) = 1$ implies $x \equiv x'(\mathrm{mod}\, n)$.

Example: We solve $3x \equiv 6(\mathrm{mod}\, 18)$. Since $3x \equiv 6(\mathrm{mod}\, 18)$ if and only if $x \equiv 2(\mathrm{mod}\, 6)$, the solutions of the latter congruence, and thus the former, are $2 \pm 6k$, $k = 0, 1, 2, \cdots$. However, there are only three distinct solutions modulo 18, namely, $2, 2+6, 2+2\cdot 6$. This generalizes to the following

Theorem: *Let* $d = (a, n)$. *Then*

$$ax \equiv b(\mathrm{mod}\, n) \text{ has a solution } \Longleftrightarrow d | b.$$

Moreover, if x_0 *is a solution and* $n' = \frac{n}{d}$, *then the congruence has exactly* d *incongruent solutions, given by*

$$x_0,\, x_0 + n',\, x_0 + 2n', \ldots, x_0 + (d-1)n' .$$

Proof: (exercise).

In the next chapter we will solve systems of linear congruences.

Chapter 9

Theory of Congruences 2

1. Introduction

Up to now, we have presented the contents of Section I and part of Section II of the *Disquisitiones* (unique factorization of integers and solutions of linear congruences). In this and the next two chapters, we shall present much of the contents of Sections II, III and IV. We do not follow Gauss's order of development because we want to make 'light use' of the notions of group and ring to provide a certain unity to the ideas. Again, we emphasize that this is not a major deviation from our desire to show the historical evolution of these ideas, since most of the algebraic generalizations were motivated by the number theory and the proofs are straightforward generalizations of the original number theoretic ones.

For an elementary introduction to these ideas, using algebraic concepts in a more systematic way, see Ireland and Rosen [Ire - Ros], and for a more sophisticated and concise treatment, including more advanced results, see Serre [Ser].

2. Reduced Residue Classes

If n is not a prime, then we have seen that the non - zero residue classes modulo n are not a group under multiplication. However, we can single out an important subset of these classes which is a group.

It is easy to prove that if $(x, n) = 1$ and $x \equiv x' \pmod{n}$, then $(x', n) = 1$. Thus, either all or none of the numbers in a residue class are relatively prime to n, so it makes sense to call some equivalence classes relatively prime to n. The class $\overline{a}$ is *relatively prime* to n if $(a, n) = 1$. The residue classes which are relatively prime to n are called the **reduced residue classes modulo n,** and any set of representatives from them is a **reduced residue system modulo n**. For example, 1, 5, 7, 11 is a reduced residue system modulo 12.

Since

$$(a,n) = (b,n) = 1 \Longleftrightarrow (ab,n) = 1 \ ,$$

the reduced residue classes are closed under multiplication. In section 8.5 we proved that $(a,n) = 1$ if and only if there exists an x such that $ax \equiv 1 (\bmod n)$. Therefore reduced classes are the only residue classes with multiplicative inverses. Thus we have proved the following

Theorem: *The reduced residue classes modulo n form a group under multiplication. We denote this group by* $\mathbf{Z}_{\mathbf{n}}^{\mathbf{x}}$. *This group is the largest multiplicative subgroup of the ring* $\mathbf{Z}/n\mathbf{Z}$.

Since the set of integers which are in the interval $[1,n]$ and relatively prime to n are a reduced residue system and the number of these is given by the Euler phi function $\varphi(n)$ (sec. 3.4), we have

Corollary: $|\mathbf{Z}_n^\times| = \varphi(n)$.

An immediate consequence of this is Euler's generalization of Fermat's little theorem.

Theorem (Euler):

$$(a,n) = 1 \implies a^{\varphi(n)} \equiv 1 (\bmod n)$$
$$(\textit{equivalently } \overline{a}^{\varphi(n)} = \overline{1}) \ .$$

Proof: We present two proofs.

1) We essentially mimic the proof of Fermat's little theorem given in section 8.3. If $(a,n) = 1$, then $\overline{a}$ is in $\mathbf{Z}_n^x$ and its order k is the order of the subgroup $\{\overline{a}\}$ it generates by $\overline{a}$. By Lagrange's theorem, k divides $\varphi(n)$, the order of Z_n^x. Hence $\varphi(n) = kt$ for some t,

$$\overline{a}^{\varphi(n)} = \overline{a}^{kt} = (\overline{a}^k)^t = (\overline{1})^t = \overline{1} \ ,$$

and we are done.

2) Our second proof generalizes J. Ivory's proof (1808) of Fermat's theorem.

If $r_1, r_2, \ldots, r_{\varphi(n)}$ is a reduced residue system modulo n, then so is $ar_1, ar_2, \ldots, ar_{\varphi(n)}$ and thus the latter system is, after rearrangement, congruent term by term to the former. Hence $r_1 r_2 \cdots r_{\varphi(n)} \equiv (ar_1)(ar_2) \cdot (ar_{\varphi(n)}) \equiv a^{\varphi(n)} r_1 r_2 \cdots r_{\varphi(n)} (\bmod n)$ and cancellation yields the theorem.

Since $\varphi(p) = p - 1$, Fermat's little theorem is a special case of Euler's theorem.

3. The Structure of $\mathbf{Z}/n\mathbf{Z}$

Now we use some ideas about groups and rings to elucidate the structure of the group $\mathbf{Z}_n^\times$ and the ring $\mathbf{Z}/n\mathbf{Z}$. *All unproven statements are to be regarded as exercises.*

Let R be a commutative ring with identity 1_R. Recall, that if $a, b \in R$, then we say a **divides** b (written $\mathbf{a|b}$), if there exists an element c in R such that $b = ac$. We say that a in R is a **unit** if a divides 1_R. Let $\mathbf{U(R)}$ be the set of units in R.

Proposition: *$U(R)$ is a group under multiplication.*

Our proof that $\mathbf{Z}_n^\times$ is a group can be reformulated as

Proposition: $U(\mathbf{Z}/n\mathbf{Z}) = \mathbf{Z}_n^\times$.

Thus we see that, in a more general setting, the reduced residue classes are natural objects of study.

The key to our structure theorems is the

Chinese Reminder Theorem: *If $m_1, m_2 \cdots m_k$ are relatively prime in pairs and $m = m_1 m_2 \cdots m_k$ and $b_1, \ldots, b_k$ are integers, then the system of congruences*

$$x \equiv b_1 (\bmod m_1), \ldots, x \equiv b_k (\bmod m_k)$$

has a solution which is unique modulo m.

A Chinese mathematician Sun-Tsu knew a special case of this theorem, without our modern notation, during the second century A.D., and the full theorem was first stated and proved by Euler (see [Dic, Vol. 2] for historical details).

Proof:

uniqueness. If $x' \equiv x (\bmod m_1), \ldots, x' \equiv x (\bmod m_k)$, and $(m_i, m_j) = 1, i \neq j$, then

$$m_1, \ldots, m_k | x' - x \implies m_1 \cdots m_k | x' - x \implies x' \equiv x (\bmod m) .$$

existence. We first prove the theorem for $k = 2$. Suppose that

$$x \equiv b_1 (\bmod m_1), x \equiv b_2 (\bmod m_2) .$$

The solutions of the first congruence are

$$x = b_1 + m_1 t, \quad t = 0, \pm 1, \pm 2, \cdots .$$

Such an x will satisfy the second congruence if and only if t satisfies the linear congruence

$$b_1 + m_1 t \equiv b_2 (\bmod m_2) \quad \text{or} \quad m_1 t \equiv b_2 - b_1 (\bmod m_2) .$$

Since $(m_1, m_2) = 1$, there is such a t.

To prove the theorem for arbitrary k, we proceed by induction. For example for $k = 3$, to solve

$$x \equiv b_1 (\bmod m_1), \quad x \equiv b_2 (\bmod m_2), \quad x \equiv b_3 (\bmod m_3) , \tag{1}$$

we apply the case $k = 2$ to find a y satisfying

$$y \equiv b_1 (\bmod m_1), \quad y \equiv b_2 (\bmod m_2) . \tag{2}$$

Again applying the case $k = 2$, we can find an x satisfying

$$x \equiv y (\bmod m_1 m_2), \quad x \equiv b_3 (\bmod m_3) .$$

Then

$$x \equiv y (\bmod m_1 m_2) \implies x \equiv y (\bmod m_1), x \equiv y (\bmod m_2)$$

and, by (2), x is the solution we seek for the system of congruences (1).

The induction step from k to $k + 1$ is immediate (exercise).

A more direct proof of the existence of solutions can be given as follows: Let $n_i = \frac{m}{m_i}$. Then $(m_i, n_i) = 1$ and there exist integers d_i, e_i such that $d_i m_i + e_i n_i = 1$. Let $r_i = e_i n_i$, and note that $r_i \equiv 0 (\bmod m_j)$ for $j \neq i$ and $r_i \equiv 1 (\bmod m_i)$. Then $x = \Sigma b_i r_i$ is a solution.

See Knuth [Knu] for yet another proof and for an application of the theorem to the problem of storing integers in computers.

For the remainder of this section *all rings are assumed to be commutative with identity*. Recall that if $R_1, \ldots, R_n$ are rings, then the *direct sum* is the set

$$R_1 \oplus R_2 \oplus \cdots \oplus R_n = \{(r_1, \ldots, r_n) \,|\, r_i \in R_i\} ,$$

with the operations

$$(r_1, \ldots, r_n) + (s_1, \ldots, s_n) = (r_1 + s_1, \ldots, r_n + s_n) ,$$
$$(r_1, \ldots, r_n) \cdot (s_1, \ldots, s_n) = (r_1 s_1, \ldots, r_n s_n) .$$

If $G_1, \ldots, G_n$ are groups, then the *direct product* is the set

$$G_1 \times G_2 \times \cdots \times G_n = \{(g_1, \ldots, g_n) \,|\, g_i \in G_i\} ,$$

with the operation

$$(g_1, \ldots, g_n) \cdot (h_1, \ldots, h_n) = (g_1 h_1, \ldots, g_n h_n)$$

Proposition: $R_1 \oplus R_2 \oplus \cdots \oplus R_n$ *is a commutative ring with identity.*

Proposition: $G_1 \times G_2 \times \cdots \times G_n$ *is a group.*

Proposition: $U(R_1 \oplus \cdots \oplus R_n) = U(R_1) \times \cdots \times U(R_n)$.

Theorem: *Let* $m = m_1 \cdots m_k$, *where* $(m_i, m_j) = 1$ *for all* $i \neq j$. *Then*

$$\mathbf{Z}/m\mathbf{Z} \cong \mathbf{Z}/m_1\mathbf{Z} \oplus \cdots \oplus \mathbf{Z}/m_k\mathbf{Z} ,$$

where $\cong$ *denotes isomorphism of rings.*

(Hint: Consider the homomorphisms $\pi_i : \mathbf{Z} \to \mathbf{Z}/m_i\mathbf{Z}$, defined by $n \mapsto n (\operatorname{mod} m_i)$, and $f : \mathbf{Z} \to \mathbf{Z}/m_1\mathbf{Z} \oplus \cdots \oplus \mathbf{Z}/m_k\mathbf{Z}$, defined by $n \mapsto (\pi_1(n), \ldots, \pi_k(n))$, and apply the Chinese remainder theorem.)

Corollary: $U(\mathbf{Z}/m\mathbf{Z}) \cong U(\mathbf{Z}/m_1\mathbf{Z}) \times \cdots \times U(\mathbf{Z}/m_k\mathbf{Z})$, *where* $\cong$ *denotes isomorphism of groups.*

Corollary: *If* $m = p_1^{k_1} \cdots p_r^{k_r}$, *where the* p_i *are distinct primes, then*

$$U(\mathbf{Z}/m\mathbf{Z}) \cong U(\mathbf{Z}/p_1^{k_1}\mathbf{Z}) \oplus \cdots \oplus U(\mathbf{Z}/p_r^{k_r}\mathbf{Z}) .$$

Since $|\mathbf{Z}/n\mathbf{Z}| = \prod_i |\mathbf{Z}/p_i^{k_i}\mathbf{Z}|$, we have

Corollary: *If* $m = p_1^{k_1} \cdots p_r^{k_r}$, *where the* p_i *are distinct primes, then*

$$\varphi(m) = \prod \varphi(p_i^{k_i}) .$$

The multiplicativity of the phi function can also be proved directly from the Chinese Remainder Theorem without all this algebraic superstructure (exercise).

4. Polynomial Congruences

If $f(x) = f_0 + f_1 x + \cdots + f_k x^k \in \mathbf{Z}[x]$, then $r \in \mathbf{Z}$ is a *root* or *solution* of the *congruence*

$$f(x) \equiv 0(\operatorname{mod} n) ,$$

if $f(r) \equiv 0(\operatorname{mod} n)$. Equivalently, if $\overline{f}(x) = \Sigma \overline{f_i} x^i \in (\mathbf{Z}/n\mathbf{Z})[x]$, where $\overline{f_i}$ is the residue class of f_i modulo n, and $\overline{f}(\overline{r}) = \overline{0}$, then $\overline{r}$ is a root of the equation $\overline{f}(x) = \overline{0}$.

Given such a polynomial congruence we ask three questions:

i) Does a solution exist?

ii) How many incongruent solutions are there (i.e., how many solutions in $\mathbf{Z}/n\mathbf{Z}$)?

iii) What are the solutions?

Since for any given $f(x)$ and n we need only check a finite number of cases, clearly the questions refer to general results for classes of polynomials or sets of integers. For example, we might ask, given positive integers k and a, for which primes p does $x^k \equiv a(\operatorname{mod} p)$ have a solution. These questions have led to the development of a large part of number theory during the past two centuries and are still of great interest.

We shall reduce these problems to the study of polynomial congruences with prime moduli in two steps.

1) If $n = p_1^{n_1} p_2^{n_2} \cdots p_k^{n_k}$ is the unique decomposition of n into primes, we reduce the study of $f(x) \equiv 0(\operatorname{mod} n)$ to the study of $f(x) \equiv 0(\operatorname{mod} p_i^{n_i})$, for all i.

2) We reduce the study of $f(x) \equiv 0(\operatorname{mod} p^n)$, p prime, to the study of $f(x) \equiv 0(\operatorname{mod} p)$.

STEP 1: Since $(p_i^{n_i}, p_j^{n_j}) = 1$, for all $i \neq j$, then for any x,

$$f(x) \equiv 0 (\bmod n) \iff f(x) \equiv 0 (\bmod p_i^{n_i}), \text{ for all } i. \quad (1)$$

Hence every solution of $f \equiv 0(\bmod n)$ is a solution of $f \equiv 0(\bmod p_i^{n_i})$. Conversely, if x_i is a solution of $f \equiv 0(\bmod p_i^{n_i})$, $i = 1, \ldots, k$, then, by the Chinese remainder theorem, there exists a solution x of the congruences

$$x \equiv x_1 (\bmod p_1^{n_1}), \ldots, x \equiv x_k (\bmod p_k^{n_k}) \quad (2)$$

But (2) implies $f(x) \equiv 0(\bmod p_i^{n_i})$, for all i, and thus, by (1), x is a solution of $f \equiv 0(\bmod n)$.

Now we restrict x and the x_i to least positive residue solutions modulo n and $p_i^{n_i}$ respectively. Then, by the uniqueness part of the Chinese remainder theorem, the correspondence

$$x \longleftrightarrow (x_1, \ldots, x_k)$$

is a bijection. In particular, if there are N_i incongruent solutions of $f \equiv 0(\bmod p_i^{n_i})$, then there are $N = N_1 \cdots N_k$ incongruent solutions of $f \equiv 0(\bmod n)$.

STEP 2: We begin with an example which illustrates the general procedure.

Example: Consider the congruence

$$x^2 \equiv 2(\bmod 7^n), \quad \text{for all } n \geq 1 . \quad (3)$$

If $n = 1$, the two incongruent solutions are $x_0 \equiv 3(\bmod 7)$ and $x_0' \equiv -3(\bmod 7)$.

Now let $n = 2$. If

$$x^2 \equiv 2(\bmod 7^2), \quad (4)$$

then $x^2 \equiv 2(\bmod 7)$ and $x \equiv \pm 3(\bmod 7)$. Suppose $x \equiv 3(\bmod 7)$. Then x must be of the form $3 + 7t$, and, substituting this into (4) to find which values of t yield solutions, we have

$$(3 + 7t)^2 \equiv 2(\bmod 7^2),$$

$$9 + 6 \cdot 7t + 7^2t^2 \equiv 2(\bmod 7^2),$$

$$7 + 6 \cdot 7t \equiv 0(\bmod 7^2),$$

$$1 + 6t \equiv 0(\bmod 7),$$

$$t \equiv 1(\bmod 7),$$

or $t = 1 + 7u$, $u \in \mathbf{Z}$. *Since the steps are all reversible*, $3 + 7(1 + 7u) = 3 + 1 \cdot 7 + 7^2 u$, $u \in \mathbf{Z}$, are all solutions of (4). But these solutions are all congruent modulo 7^2. Therefore, setting $u = 0$, we see that

$$x_1 \equiv 3 + 1 \cdot 7 (\mathrm{mod}\, 7^2)$$

is the unique solution modulo 7^2 of (4) such that $x_1 \equiv 3 (\mathrm{mod}\, 7)$ (similarly $x_1' \equiv -3 + 6 \cdot 7 \equiv 4 + 5 \cdot 7 (\mathrm{mod}\, 7^2)$ is the unique solution of (4) congruent to -3 modulo 7).

Let $n = 3$. If

$$x^2 \equiv 2 (\mathrm{mod}\, 7^3)\,, \tag{5}$$

then $x^2 \equiv 2 (\mathrm{mod}\, 7^2)$, and x is congruent to one of our solutions of (4) modulo 7^2. Suppose $x \equiv x_1 (\mathrm{mod}\, 7^2)$; so x is of the form $x_1 + 7^2 t = 3 + 1 \cdot 7 + 7^2 t$. Substituting the latter expression in (5) to find which t yield solutions of (5), we have

$$\left(3 + 1 \cdot 7 + 7^2 t\right)^2 \equiv 2 (\mathrm{mod}\, 7^3)$$

or $t \equiv 2 (\mathrm{mod}\, 7)$. Therefore $3 + 1 \cdot 7 + 7^2(2 + 7u) = 3 + 1 \cdot 7 + 2 \cdot 7^2 + 7^3 u$, $u \in \mathbf{Z}$, are all solutions of (5). But these solutions are all congruent modulo 7^3. Therefore, setting $u = 0$, we see that

$$x_2 \equiv 3 + 1 \cdot 7 + 2 \cdot 7^2 (\mathrm{mod}\, 7^3)$$

is the unique solution of (5) such that $x_2 \equiv x_1 (\mathrm{mod}\, 7^2)$. (Similarly, if we had chosen $x \equiv x_1' (\mathrm{mod}\, 7^2)$, we would have found the other solution of (5).)

Continuing this procedure, it can be easily proved (exercise) that each solution a_0 of $x^2 \equiv 2 (\mathrm{mod}\, 7)$ determines a unique sequence $a_0, a_1, \ldots$, $0 \leq a_i < 7$, such that

$$a_0 + a_1 7 + a_2 7^2 + \cdots + a_{n-1} 7^{n-1}$$

is a solution of $x^2 \equiv 2 (\mathrm{mod}\, 7^n)$. Since every solution of the latter congruence is congruent modulo 7 to a unique solution of $x^2 \equiv 2 (\mathrm{mod}\, 7)$, these are the only solutions.

The fact that the formal power series

$$a_0 + a_1 7 + \cdots + a_n 7^n + \cdots$$

encodes a solution of $x^2 \equiv 2 (\mathrm{mod}\, 7^n)$, for each n, is one motivation for constructing the so called p-adic numbers, which we will study later (chap. 23).

Now we consider an arbitrary polynomial $f(x)$ and show that the solutions of $f \equiv 0(\text{mod } p^n)$, for all $n > 1$, can be constructed from the solutions of $f(x) \equiv 0(\text{mod } p)$.

For a fixed n, every solution of

$$f(x) \equiv 0 \left(\text{mod } p^{n+1}\right) \tag{6}$$

is a solution of

$$f(x) \equiv 0 \left(\text{mod } p^{n}\right) . \tag{7}$$

We shall determine which solutions of (7) are also solutions of (6). Let x_{n-1} be a solution of (7) and x_n a solution of (6) such that $x_n \equiv x_{n-1}(\text{mod } p^n)$. Then x_n is of the form $x_{n-1} + tp^n$, $t \in \mathbf{Z}$, and we want to find those t which yield solutions of (6), i.e., those t satisfying

$$f\left(x_n\right) = f\left(x_{n-1} + tp^n\right) \equiv 0 \left(\text{mod } p^{n+1}\right) . \tag{8}$$

Recall Taylor's formula, an algebraic identity for polynomials:

$$f(x+h) = f(x) + f^{(1)}(x)h + f^{(2)}(x)\frac{h^2}{2!} + \cdots + f^{(k)}(x)\frac{h^k}{k!} ,$$

where $\deg f = k$. If $f(x)$ has integer coefficients, then for any fixed integer a, $f(a+h)$ has integer coefficients. But the coefficient $\frac{f^{(i)}(a)}{i!}$ of h^i in $f(a+h)$ is uniquely determined by Taylor's formula, so $\frac{f^{(i)}(a)}{i!}$ is an integer, for all i.

Let $x = x_{n-1}$ and h = tp^n in Taylor's formula; then we have

$$\begin{aligned} f\left(x_{n-1} + tp^n\right) =& f\left(x_{n-1}\right) + f^{(1)}\left(x_{n-1}\right) tp^n + f^{(2)}\left(x_{n-1}\right)\frac{(tp^n)^2}{2!} + \\ & \cdots + f^{(k)}\left(x_{n-1}\right)\frac{(tp^n)^k}{k!} . \end{aligned}$$

We note that when $i > 1$, $\left(\frac{f^{(i)}(x_{n-1})}{i!}\right)(tp^n)^i \equiv 0 \left(\text{mod } p^{n+1}\right)$. Therefore

$$f\left(x_{n-1} + tp^n\right) \equiv f\left(x_{n-1}\right) + f^{(1)}\left(x_{n-1}\right) tp^n \left(\text{mod } p^{n+1}\right) .$$

Hence, finding those t which solve equation (8) is equivalent to finding the solution of

$$tp^n f^{(1)}\left(x_{n-1}\right) \equiv -f\left(x_{n-1}\right) \left(\text{mod } p^{n+1}\right) . \tag{9}$$

Since $f(x_{n-1}) \equiv 0 (\bmod\, p^n)$, $\frac{f(x_{n-1})}{p^n}$ is an integer and (9) is equivalent to

$$tf^{(1)}(x_{n-1}) \equiv -\frac{f(x_{n-1})}{p^n} (\bmod\, p) \,, \tag{10}$$

a linear congruence in t. Therefore $x_n = x_{n-1} + tp^n$ will be a solution of (6) if and only if t satisfies (10).

Our theorem on linear congruences (sec. 8.5) now applies. We work out the details.

(i) $f^{(1)}(x_{n-1}) \not\equiv 0 (\bmod\, p)$: In this case, there is a unique solution t' modulo p of (10). The integers in the residue class $t'(\bmod\, p)$ are of the form $t = t' + up$, $u \in \mathbf{Z}$. Hence

$$x_n = x_{n-1} + (t' + up)p^n = x_{n-1} + t'p^n + up^{n+1}$$

are solutions of (6). But these solutions are all congruent modulo p^{n+1}, and therefore there is a unique solution $x_{n-1} + t'p^n$ of (6) satisfying $x_n \equiv x_{n-1} (\bmod\, p^n)$.

(ii) $f^{(1)}(x_{n-1}) \equiv 0 (\bmod\, p)$ and $-\frac{f(x_{n-1})}{p^n} \not\equiv 0 (\bmod\, p)$: Then (10) has no solutions and there are no solutions of (6) satisfying $x_n \equiv x_{n-1} (\bmod\, p^n)$.

(iii) $f^{(1)}(x_{n-1}) \equiv 0 (\bmod\, p)$ and $-\frac{f(x_{n-1})}{p^n} \equiv 0 (\bmod\, p)$: Then $x_n = x_{n-1} + tp^n$ is a solution of (6), for every $t \in \mathbf{Z}$.

We see that we can construct the solutions of $f \equiv 0 (\bmod\, p^{n+1})$ from the solutions of $f \equiv 0 (\bmod\, p^n)$, $n = 1, 2, \ldots$, and therefore for all n we can construct the solutions of $f \equiv 0 (\bmod\, p^n)$ from the solutions of $f \equiv 0 (\bmod\, p)$.

The general problem of solving polynomial congruences in one variable modulo an integer n has been reduced to solving $f(x) \equiv 0 (\bmod\, p)$, for all primes p dividing n. *Similar reduction steps can be used for congruences involving polynomials in several variables.*

Even in the quadratic case, the existence of patterns in the solutions of congruences is a deep question, leading to the law of quadratic reciprocity (sec. 6.2, chap. 11). For higher degrees, there are many unsolved problems. We digress for a moment to discuss some recent work related to step 2 in the reduction process.

Let p be a fixed prime and $f(x_1, x_2, \ldots, x_k)$ a polynomial with integer coefficients. If c_n denotes the number of solutions of $f \equiv 0 (\bmod\, p^n)$, then the *Poincaré series* of f, $P_{f,p}(t)$, is the formal power series

$$P_{f,p}(t) = \sum_{0}^{\infty} c_i t^i ,$$

where $c_0 = 1$. This series was introduced by Borevich and Shafarevich [Bor - Sha, pg. 47] in the 1950's, where they conjectured that $P_{f,p}(t)$ is a rational function of t for all polynomials f. This was proved, in much more generality, by Igusa in 1975 [Igu], using a mixture of analytic and algebraic methods and the resolution of singularities for algebraic varieties. Igusa's proof is not constructive and there is a great deal of current interest in computing Poincaré series and extending the results in various ways (see, e.g., [Den] or [Meu]). In [Gol, J 2] and [Gol, J 3], I presented elementary algebraic-combinatorial techniques for deriving the Poincaré series for strongly nondegenerate forms in k variables and for a large class of two variable polynomials.

5. Polynomial Congruences and Polynomial Functions

Let $\overline{f} = \Sigma \overline{f_i} x^i$ and $\overline{g} = \Sigma \overline{g_i} x^i$ be polynomials in $(\mathbf{Z}/n\mathbf{Z})[x]$, where f_i and g_i are integers and $\overline{f_i}$ and $\overline{g_i}$ are their residue classes modulo n. The equation $\overline{f} = \overline{g}$ can have two possible meanings.

First, it can mean *equality of polynomials*, i.e., the degrees are equal and $\overline{f_i} = \overline{g_i}$, for all i. Second, since any polynomial $\overline{f}(x)$ defines a function

$$\overline{f} : (\mathbf{Z}/n\mathbf{Z}) \longrightarrow (\mathbf{Z}/n\mathbf{Z}) , \quad \text{by}$$
$$\overline{r} \longmapsto \overline{f}(\overline{r}) ,$$

$\overline{f} = \overline{g}$ can mean *equality of functions.*

For polynomials over the rational, real or complex numbers, these two concepts are equivalent, but not for polynomials over finite rings and fields. Clearly, equality of polynomials implies equality of functions, *but the converse is not true*. For example, by Fermat's theorem $x^p - x = \overline{0}$ for all x in $\mathbf{F}_p$, but $x^p - x$ and $\overline{0}$ are not equal as polynomials.

For the moment, $\bar{\mathbf{f}} = \bar{\mathbf{g}}$ will denote equality of functions and $\bar{\mathbf{f}} =_{\mathbf{x}} \bar{\mathbf{g}}$ will denote equality of polynomials. Translating from equality in $\mathbf{Z}/n\mathbf{Z}$ to congruence modulo n, we now make similar distinction for polynomials modulo n.

Let $f(x) = \Sigma f_i x^i$ and $g(x) = \Sigma g_i x^i$ be in $\mathbf{Z}[x]$. Then we say that **f(x) is congruent to g(x) modulo n**, denoted by $\mathbf{f(x) \equiv g(x) (mod\, n)}$, if $f(i) \equiv g(i) (\mathrm{mod}\, n)$ for all integers i. This corresponds to equality of functions.

We say **f(x) is congruent sub x to g(x) modulo n**, denoted by

$$\mathbf{f(x) \equiv_x g(x) (mod\ n)}\ ,$$

if $f_i \equiv g_i (\mathrm{mod}\, n)$ for all i. It follows from the definition that

$$\mathbf{f(x) \equiv_x g(x) (mod\ n)} \iff \mathbf{\bar{f}(x) =_x \bar{g}(x)}\ . \tag{1}$$

Note that if $f(x) \equiv_x g(x) (\mathrm{mod}\, n)$, then $f(x) \equiv 0 (\mathrm{mod}\, n)$ and $g(x) \equiv 0 (\mathrm{mod}\, n)$ have the same solutions. *The distinction between $=$ and $=_x$ and between $\equiv$ and $\equiv_x$ carries over immediately to congruences in several variables.* (The notation $=_x$ and $\equiv_x$ seems to have first been introduced by Adams and Goldstein [Ada - Gol].)

Now we specialize to the case $n = p$, a prime, so $\mathbf{Z}/p\mathbf{Z} = \mathbf{F}_p$ is a field. We assume the basic theory of factorization of polynomials over a field and over the integers (as discussed in any text on modern algebra) and translate some of that theory to $\equiv_x$ by means of (1). The proofs for these more general results are essentially the same as the original proofs for polynomials modulo p.

By the **degree** of $f(x) = \Sigma f_i x^i$ modulo p, denoted by $\mathbf{deg_p}\, f$, we mean the largest integer i such that $f_i \not\equiv 0 (\mathrm{mod}\, p)$.

Proposition: *Let r be an integer and $f(x) \in \mathbf{Z}[x]$. Then*

$$f(r) \equiv 0 (\mathrm{mod}\, p) \iff f(x) \equiv_x (x - r) q(x) (\mathrm{mod}\, p)\ , \textit{for some}$$
$$q(x) \in \mathbf{Z}[x],\ \textit{where } deg\ q(x) \le deg\ f(x) - 1\ .$$

Corollary: *Let $r_1, \ldots, r_t$ be t incongruent roots of $f(x) \equiv 0 (\mathrm{mod}\, p)$. Then there exists a polynomial $q(x) \in \mathbf{Z}[x]$, $deg_p q(x) \le deg_p f(x) - t$, such that*

$$f(x) \equiv_x (x - r_1) \cdots (x - r_t) q(x) (\mathrm{mod}\, p)\ .$$

Lagrange's Theorem (Form 1): *If $f(x) \in \mathbf{Z}[x]$, $f(x) \equiv_x 0 (\mathrm{mod}\, p)$, then the number of incongruent roots of $f(x) \equiv 0 (\mathrm{mod}\, p)$ is $\le deg_p f(x)$.*

This can be reformulated as

Lagrange's Theorem (Form 2): *If the number of incongruent roots of $f(x) \equiv 0 (\mathrm{mod}\, p)$ is greater than $deg\ f$, then $f \equiv_x 0 (\mathrm{mod}\, p)$.*

Note that Lagrange's theorem is not necessarily true modulo n, when n is not prime. For example, $x^2 - 1 \equiv_x 0 \pmod 8$ has the four solutions $1, 3, 5, 7$.

Corollary: *Let $f(x), g(x) \in \mathbf{Z}[x]$, deg $f = \deg g \leq n$, and assume $f \equiv g \pmod p$ has at least $n + 1$ incongruent solutions. The $f \equiv_x g$.*

Example: Since $1, 2, \ldots, p-1$ are roots of $x^{p-1} - 1 \equiv 0 \pmod p$, the first corollary of this section implies that $x^{p-1} - 1 \equiv_x (x-1)(x-2)\cdots(x-(p-1))b \pmod p$, where b is a constant. Comparing the coefficients of x^{p-1} on both sides, we see that $b \equiv 1 \pmod p$ and obtain the

Corollary: $x^{p-1} - 1 \equiv_x (x-1)(x-2)\cdots(x-(p-1)) \pmod p$.

Setting $x = 0$, we have a new proof of

Wilson's Theorem: $(p-1)! \equiv -1 \pmod p$.

Now having made the distinction between $=_x$ and $=$, and $\equiv_x$ and $\equiv$, *we shall use $=$ and $\equiv$ for both concepts* when the meaning of our statements will be clear from context.

6. Congruences in Several Variables; Chevally's Theorem

We present a simple general result conjectured by Emil Artin and proved by Claude Chevally in the 1930's [Che]. I find it very surprising that this was not discovered by Euler, Lagrange or Gauss.

Chevally's Theorem: *Let $f(x_1, .., x_n)$ be a polynomial with integer coefficients, constant term zero, and degree $< n$. Then $f \equiv 0 \pmod p$ has a non-trivial solution for all primes p.*

Example: There exists a non-trivial solution to $x_1^2 + x_2^2 + x_3^2 + x_4^2 \equiv 0 \pmod p$, for all primes p. This is central to a proof of Lagrange's theorem that every positive integer is the sum of four squares which we will present in chapter 22 (also see sec. 2.6 and[Dav]).

The proof of Chevally's theorem requires a preliminary lemma. For clarity, we continue to use the notation $\equiv$ and $\equiv_x$ of the last section.

By Fermat's little theorem $x^p \equiv x \pmod p$ for all x, and thus every polynomial f is $\equiv$ modulo p to a polynomial g, where the exponents in each term are less than p-1 (just reduce each exponent modulo $p-1$). For example $x^7y^5 + x^3y^4z^{18} \equiv x^3y + x^3z^2 \pmod 5$.

Lemma: *If $g(x_1, \ldots, x_n)$ is a polynomial with integer coefficients with the exponents in every term $\leq p-1$ and $g \equiv 0 (\text{mod } p)$, for all integers $x_1, \ldots, x_n$, then*

$$g \equiv_x 0 (\text{mod } p) ,$$

i.e., every coefficient is $\equiv 0 (\text{mod } p)$.

Proof of lemma. We use induction on n, the number of variables.

$n = 1$: This is just the second form of Lagrange's theorem (sec. 5).

$n = 2$: Assume that $g(x_1, x_2) \equiv 0 (\text{mod } p)$ for all x_1, x_2. Grouping terms in $g(x_1, x_2)$ according to powers of x_1, we have

$$g(x_1, x_2) = \sum_i^{p-1} g_i(x_2) x_1^i .$$

We fix x_2 and view $g(x_1, x_2)$ as a function of x_1 which satisfies $g(x_1, x_2) \equiv 0 (\text{mod } p)$ for all x_1. Hence, by the case $n = 1$, $g_i(x_2) \equiv 0 (\text{mod } p)$. But this is true for all p possible values for x_2 and using the case $n = 1$ again, we have $g_i(x) \equiv_x 0 (\text{mod } p)$. But the coefficients of $g_i(x)$ are the coefficients of g, and we are done.

$n = k$: The general case proceeds as for $n = 2$. Write

$$g(x_1, \ldots, x_k) = \Sigma g_i(x_2, \ldots, x_k) x_1^i$$

and use the case $n = 1$ to reduce to $n = k - 1$.

Proof of Chevally's theorem. Suppose there is an $f(x_1, \ldots, x_n)$ such that the theorem is false, i.e., $f(x_1, \ldots, x_n)$ is a polynomial with integer coefficients, constant term zero, and degree $< n$, and for some prime p, $f \not\equiv 0 (\text{mod } p)$ for all non-trivial $(x_1, \ldots, x_n)$. Then we claim that

$$1 - f(x_1, \ldots, x_n)^{p-1} \equiv (1 - x_1^{p-1}) \cdots (1 - x_n^{p-1}) (\text{mod } p) , \quad (1)$$

for all integer values of the variables.

If $(x_1, \ldots, x_n) = (0, \ldots, 0)$, then, since f has constant term zero, both sides of (1) are one. If $(x_1, \ldots, x_n) \neq (0, \ldots, 0)$, then there is some $x_i \neq 0$, so $x_i^{p-1} \equiv 1 (\text{mod } p)$ and the right hand side of (1) is congruent to zero modulo p. By assumption $f(x_1, \ldots, x_n) \equiv 0 (\text{mod } p)$, so $f^{p-1} \equiv 1 (\text{mod } p)$, and the left hand side of (1) is also congruent to zero modulo p.

Now, as discussed before the lemma, f^{p-1} is $\equiv$ to a polynomial g, such that every exponent in any term is $< p-1$. Therefore

$$1 - g(x_1, \dots, x_n) \equiv (1 - x_1^{p-1}) \cdots (1 - x_n^{p-1}) (\bmod p)$$

for all integers $x_1, \dots, x_n$, and, by the lemma,

$$1 - g(x_1, \dots, x_n) \equiv_x (1 - x_1^{p-1}) \cdots (1 - x_n^{p-1}) (\bmod p) . \qquad (2)$$

Since g is a polynomial in n variables with every exponent $< p-1$, the maximum degree of a term on the left hand side of (2) is $< n(p-1)$. On the other hand, $(-x_1^{p-1})(-x_2^{p-1}) \cdots (-x_n^{p-1})$, on the right hand side of (2), has degree $n(p-1)$. This contradicts the $\equiv_x$ in (2) and we are done.

There is a very extensive literature on the problem of counting the number of solutions of congruences and of systems of congruences. This includes the famous Weil conjectures (sec. 20.9). We will discuss the problem of counting solutions of $f(x, y) \equiv 0 (\bmod p)$ when we deal with arithmetic on curves.

7. Solutions of Congruences and Solutions of Equations; The Hasse Principle

We know that if $f(x_1, \dots, x_n) = 0$, for a polynomial f with integer coefficients and some integers $x_1, \dots, x_n$, then $f(x_1, \dots, x_n) \equiv 0 (\bmod m)$, for every integer m. In section 8.4, we saw that this trivial fact could be used to prove that $f = 0$ has no integer solutions by simply giving one value of m for which $f \equiv 0 (\bmod m)$ is not solvable. Moreover, $f \equiv 0 (\bmod m)$ is solvable if and only if $f \equiv 0 (\bmod p^v)$ is solvable for all prime powers p^v dividing m (sec. 4).

It is natural to ask about a converse: if $f \equiv 0 (\bmod p^v)$ is solvable for all primes p and all $v > 0$, and if $f = 0$ has a solution in real numbers, does $f = 0$ have a non-trivial (non-zero) solution in integers, or at least in *rational numbers?* The condition that $f = 0$ has a real solution is included to guarantee the possibility of a solution. For example, $x^2 + y^2 + 1 = 0$ has no real solutions, so it cannot have a rational solution, but $x = y = 1$ is a solution modulo 3.

A solution modulo p^v or a real solution is called a *local solution*, and a rational solution is called a *global solution*. We say the **Hasse principle** holds for a set C of polynomials if the existence of all local solutions of $f = 0$, for any f in C, implies the existence of a global solution for f. The Hasse principle holds for quadratic forms $\Sigma c_{i,j}x_i x_j$ (Hasse - Minkowski theorem). In 1942, Reichardt proved that $x^4 - 17y^4 - 2z^4 = 0$ has all local solutions but no global ones, and later Selmer gave a large number of such examples, including $3x^3 + 4y^3 + 5z^3 = 0$ [Cas 2].

These questions both motivate and are treated most naturally in the language of p-adic numbers, which we return to in chapter 23. An excellent reference is Borevich and Shafarevich [Bor - Sha].

Chapter 10

Primitive Roots and Power Residues

1. Primitive Roots

We study the solutions of the binomial congruences

$$x^n \equiv d (\text{mod } p) \,,$$

where p is a prime. Our long range goal is to understand the detailed structure of the solutions of quadratic congruences.

Our first theorem provides the central tool for our studies:

Theorem: *The group* $(\mathbf{Z}/p\mathbf{Z})^\times$ *of residue classes modulo p under multiplication is cyclic. In congruence language this means that for every prime p, there exists a* **primitive root** *modulo p, i.e., an integer g such that every integer* $n \not\equiv 0 (\text{mod } p)$ *is congruent modulo p to exactly one of the integers* $g, g^2, g^3, \ldots, g^{p-1}$. *Equivalently the sequence* $g, g^2, g^3, \ldots, g^{p-1}$ *can be rearranged so that it is congruent modulo p to the sequence* $1, 2, \ldots, p-1$.

Example: 2 is a primitive root modulo 5, since $2, 2^2, 2^3, 2^4$ are $\equiv$ to 2, 4, 3, 1 (mod 5).

3 is a primitive root modulo 5, since $3, 3^2, 3^3, 3^4$ are $\equiv$ to 3, 4, 2, 1 (mod 5).

2 is a primitive root modulo 19.

Euler introduced the term "primitive root" but his existence proof was unsatisfactory. Legendre gave the first correct proof. Most proofs of the theorem, which is a special case of the fact that any finite subgroup of the multiplicative group of a field is cyclic (see [Art]), use the Euler phi function. However, our proof, due to Legendre, is direct and elegant.

If a is an integer, we let $\mathbf{o}(a)$ denote the order of the subgroup generated by the residue class $\overline{a}$ in the group $(\mathbf{Z}/p\mathbf{Z})^{\times}$, i.e., $o(a)$ is the smallest positive integer k such that $a^k \equiv 1 \pmod{p}$. We call $o(a)$ **the order of a mod p.**

Lemma: *If $a^v \equiv 1 \pmod{p}$, then $o(a)|v$.*

Proof: If the lemma is not true, then $v = o(a)m + r$, where $0 < r < o(a)$. But then $1 \equiv a^v \equiv a^{o(a)m+r} \equiv a^r \pmod{p}$, contradicting the definition of $o(a)$.

Proof of the theorem. The proof follows from two interesting results:

i) if $o(a) = m, o(b) = k$ and $(m, k) = 1$, then $o(ab) = mk$.

ii) the congruence $x^d \equiv 1 \pmod{p}$, where $d|p-1$, has exactly d solutions.

Assuming i) and ii), let $p - 1 = q_1^{s_1} \cdots q_t^{s_t}$, where the q_i's are primes. If there exists an integer a_i of order $q_i^{s_i} \pmod{p}$ for each i, then, by repeated application of (i), $a_1 a_2 \cdots a_t$ has order $\prod_{q_i^{s_i}} = p - 1$ as required.

First we prove that such integers a_i exist. Let q^s be one of the $q_i^{s_i}$. If x satisfies

$$x^{q^s} \equiv 1 \pmod{p}, \tag{1}$$

and its order mod p is not q^s, then $o(x)$ divides q^s and it must be one of $1, q, q^2, ..., q^{s-1}$, all of which divide q^{s-1}. This x satisfies

$$x^{q^{s-1}} \equiv 1 \pmod{p}. \tag{2}$$

By (ii), (1) has exactly q^s solutions and (2) has exactly q^{s-1} solutions. Clearly any solution of (2) is a solution of (1); thus there are $q^s - q^{s-1}$ solutions of (1) which are not solution of (2), i.e., there are $q^s - q^{s-1}$ elements of order q^s mod p. Therefore the a_i's exist.

To complete the proof of the theorem, we prove (i) and (ii).

Proof of (i). Working modulo p, we have

$$o(a) = m, o(b) = k \implies (ab)^{mk} = (am)^k (b^k)^m \equiv 1 \pmod{p},$$

and therefore $o(ab)|mk$. Let $o(ab) = r = m_1 k_1$, where $m_1|m$ and $k_1|k$. We prove that $k_1 = k$ and $m_1 = m$.

Since $(ab)^r = a^{m_1 k_1} b^{m_1 k_1} \equiv 1 \pmod{p}$, we have

$$1 \equiv \left(a^{m_1 k_1} b^{m_1 k_1}\right)^{\frac{m}{m_1}} \equiv (a^m)^{k_1} (b^{k_1})^m \equiv b^{mk_1} \pmod{p};$$

hence $k|mk_1$. But, by assumption, $(k, m) = 1$; so $k|k_1$ and, since $k_1|k$, we see that $k_1 = k$. Similarly, $m_1 = m$ and we are done.

Proof of (ii). By Lagrange's theorem (sec. 9. 5), $x^d - 1 \equiv 0 (\text{mod } p)$ has at most d solutions. We now show it has exactly d solutions.

Since $d|p-1$, we let $p - 1 = de$ and $y = x^d$. Then $x^{p-1} - 1 = y^e - 1 = (y-1)(y^{e-1} + y^{e-2} + \cdots + 1)$ or

$$x^{p-1} - 1 = (x^d - 1) f(x) ,$$

where $\deg f(x) = p - 1 - d$. By Fermat's little theorem, $x^{p-1} - 1 \equiv 0 (\text{mod } p)$ has $p - 1$ distinct solutions and every solution is a solution of

$$x^d - 1 \equiv 0 (\text{mod } p) \quad \text{or} \quad f(x) \equiv 0 (\text{mod } p) .$$

The second congruence has at most $p - 1 - d$ solutions, and therefore the first has *at least d* solutions and thus *exactly d* solutions.

A well known conjecture of E. Artin claims that if an integer $a > 1$ is not a square, then a is a primitive root for infinitely many primes. In the case $a = 10$, this conjecture is due to Gauss and is equivalent to the existence of infinitely many primes p such that the length of the period of the decimal expansion of $\frac{1}{p}$ is p-1(recall that Gauss calculated the periods of $\frac{1}{p}$, for all $p \leq 1000$ - (Klein biography in sec. 7.1)). See the notes to chapter 4 of Ireland and Rosen [Ire - Ros] for further references on this conjecture and Hardy and Wright [Har - Wri] for an introduction to the theory of decimal expansions and their relation to primitive roots.

2. Indices

We can now define an analog of logarithms for the finite field $\mathbf{F_p}$. Recall that the map $x \mapsto e^x$ defines an isomorphism between the additive group of real numbers $\mathbf{R}^+$ and the multiplicative group of positive reals $\mathbf{R}_+^\times$, and the inverse map is given by the logarithm to the base e.

Similarly each primitive root g modulo p defines an isomorphism between the additive group $(\mathbf{Z}/(p-1)\mathbf{Z})^+$ and the multiplicative group $(\mathbf{Z}/p\mathbf{Z})^\times = F_p^\times$, defined by

$$f_g(i (\text{mod } p - 1)) = g^i (\text{mod } p) .$$

Since

$$g^i \equiv g^j (\text{mod } p) \implies g^{i-j} \equiv 1 (\text{mod } p)$$
$$\implies i \equiv j (\text{mod } p-1), \text{ (since } o(g) = p-1 \text{)} ,$$

the map is well defined. Any integer x must satisfy $x \equiv g^i (\text{mod } p)$, for some i; thus $f(i(\text{mod } p-1)) = x(\text{mod } p)$ and the map is surjective. Since

$$f_g(i(\text{mod } p-1) + j(\text{mod } p-1)) = g^{i+j}(\text{mod } p)$$
$$= g^i g^j (\text{mod } p) = g^i (\text{mod } p) \cdot g^j (\text{mod } p) ,$$

f_g is a homomorphism.

The kernel of $f_g$is $\{\overline{0}\}$ since

$$i(\text{mod } p-1) \in \text{Ker } f_g \implies g^i(\text{mod } p) = 1(\text{mod } p)$$
$$\implies o(g) = (p-1)|i$$
$$\implies i \equiv 0(\text{mod } p-1)$$
$$\implies i(\text{mod } p-1) = 0(\text{mod } p-1);$$

hence f_g is injective and f_g is an isomorphism. The inverse map, denoted by $\mathbf{ind_g}$, is the analog of the logarithm.

This isomorphism provides a very powerful tool, since we can reduce multiplicative problems mod p to additive problems mod $p-1$. We translate it into congruence terminology.

If g is a primitive root mod p, the index of x relative to g, denoted by **ind x**, is the *unique* integer i, $1 \le i \le p-1$, such that $x \equiv g^i (\text{mod } p)$. Similarly, in $\mathbf{F}_p^\times$ we say $\text{ind}\,\overline{x} = i$. Properly speaking, we should write $\text{ind}_{g,p}$. However, *in any given argument involving indices, p and g will be fixed*, so that no confusion will arise. From our proof of isomorphism and the properties of isomorphism, we have

Proposition:

(i) $x \equiv y(\text{mod } p) \iff \text{ind } x \equiv \text{ind } y(\text{mod } p-1)$,
(ii) $\text{ind } xy \equiv \text{ind } x + \text{ind } y(\text{mod } p-1)$,
(iii) $\text{ind } a^{-1} \equiv -\text{ind } a(\text{mod } p-1)$
(iv) $\text{ind } x^k \equiv k \text{ ind } x(\text{mod } p-1)$.

Indices are so important that Jacobi published a table for all primes less than 1000. Now we explore the power of this tool.

3. kth Power Residues

We say that $a \not\equiv 0$ is a **kth power residue mod p**, if

$$x^k \equiv a(\text{mod } p) \quad \text{is solvable,}$$

i.e., if a has a kth root modulo p. If the equation is not solvable then a is a **kth power non residue mod p**. This generalizes **quadratic residues**, the special case where $\mathbf{k = 2}$ (sec. 6.4).

By the properties of indices we have

Proposition: *If $a \not\equiv 0(\text{mod } p)$, then*

$x^k \equiv a(\text{mod } p)$ *is solvable* $\iff k \text{ ind } x \equiv \text{ind } a(\text{mod } p{-}1)$ *is solvable.*

Hence, by the theory of linear congruences (sec. 8.5), if $(k, p-1) = 1$, then every a is a kth power residue and $x^k \equiv a(\text{mod } p)$ has a unique solution. However, in many cases this information is insufficient; even in the case $k = 2$, p an odd prime, we have $(2, p-1) = 2$ and we need more information.

Example: $x^3 \equiv a(\text{mod } 19)$.

Fix a primitive root mod 19. The congruence is solvable, i.e., a is a *cubic residue*, if and only if

$$3 \text{ ind } x \equiv \text{ind } a(\text{mod } 18)\ .$$

But this latter congruence is solvable if and only if $(3, 18) = 3$ divides ind a. If $3|\text{ ind } a$, then $\text{ind } a = 3b$ for some b, and our congruence is equivalent to

$$\text{ind } x \equiv b(\text{mod } 6)\ .$$

By the theory of linear congruences, this congruence has 3 solutions modulo 8, namely, $\text{ind } x = b, b+6, b+12$. Since 2 is a primitive root mod 19, a short calculation to determine when 3 divides ind a yields

$$a \text{ is a cubic residue mod } 19 \iff a \equiv 1, 7, 8, 11, 12, \text{ or } 18(\text{mod } 19).$$

Note that *the criteria for a to be a cubic residue is given by congruence conditions.*

The general situation is similar. Let $d = (k, p-1)$. Then, by the theory of linear congruences, $k \text{ ind } x \equiv \text{ind } a(\text{mod } p-1)$ is solvable if and only if $d|\text{ ind } a$, and, in this case, there are d solutions. Hence we have

Theorem: *The kth power residues modulo p are those numbers whose indices are divisible by $d = (k, p-1)$. If*

$$x^k \equiv a (\text{mod } p)$$

is solvable, then there are exactly d solutions. $\frac{p-1}{d}$ of the numbers $1, 2, \ldots,$ $p-1$ *are k^{th} power residues.*

Example: Quadratic Residues.

Let $k = 2$. Then $(2, p-1) = 2$ and $x^2 \equiv a(\text{mod } p)$ is solvable if and only if $2 | \operatorname{ind} a$. Hence

> the quadratic residues are those integers with even index,
>
> the quadratic non residues are those integers with odd index.

Thus there are $\frac{p-1}{2}$ of each of these. $x^2 \equiv a(\text{mod } p)$ has 0 or 2 solutions. If b is a solution, then so is $-b$.

A useful criterion for residues (to be applied later to quadratic and higher power reciprocity laws), which doesn't involve the index in its formulation, and doesn't require finding primitive roots, is

Euler's Criterion: *If p is a prime, $a \in \mathbf{Z}$, $a \not\equiv 0(\text{mod } p)$, then*

$$x^k \equiv a(\text{mod } p) \text{ is solvable} \iff a^{\frac{p-1}{d}} \equiv 1(\text{mod } p),$$

where $d = (k, p-1)$.

Proof: $\Longrightarrow$) If $x^k \equiv a(\text{mod } p)$, then since $d|k$,

$$a^{\frac{p-1}{d}} \equiv (x^k)^{\frac{p-1}{d}} \equiv (x^{p-1})^{\frac{k}{d}} \equiv 1^{\frac{k}{d}} \equiv 1(\text{mod } p).$$

$\Longleftarrow$) Let g be a primitive root mod p and $a' = \operatorname{ind} a$. Then, by our last theorem, we must show $d|a'$. Now

$$1 \equiv a^{\frac{p-1}{d}} \equiv g^{a' \frac{p-1}{d}} (\text{mod } p).$$

But then $o(g) = p-1$ divides $a'\frac{p-1}{d}$, which implies that $\frac{a'}{d}$ is an integer, i.e., $d|a'$.

Example: (Cubic residues) We refine our example of $x^3 \equiv a(\text{mod } 19)$ by giving a criterion for the solvability of $x^3 \equiv a(\text{mod } p)$, for any prime p.

If $p \equiv 0 \pmod 3$, then since p is prime, $p = 3$. Since $1^3 \equiv 1, 2^3 \equiv 2 \pmod 3$, every a is a cubic residue.

If $p \equiv 1 \pmod 3$, $p - 1 \equiv 0 \pmod 3$, then $d = (3, p-1) = 3$ and, by Euler's criterion, a is a cubic residue if and only if $a^{\frac{p-1}{3}} \equiv 1 \pmod p$.

If $p \equiv 2 \pmod 3$, $p - 1 \equiv 1 \pmod 3$, then $d = (3, p-1) = 1$. By our first theorem, a is always a cubic residue.

Returning to the case $p = 19$, we see that since $7 \equiv 1 \pmod 3$ and $7^{\frac{19-1}{3}} = 7^6 = 49^3 \equiv 11^3 = 121 \cdot 11 \equiv 7 \cdot 11 = 77 \equiv 1 \pmod{19}$, 7 is a cubic residue mod 19.

Chapter 11

Congruences of the Second Degree

1. Introduction

In this chapter, we study the general congruence of the second degree,

$$f(x) = ax^2 + bx + c \equiv 0 (\mathrm{mod}\, p) ,$$

where $a \not\equiv 0(\mathrm{mod}\, p)$, culminating in the law of quadratic reciprocity which we discussed briefly in chapter 6. The reciprocity law is certainly the deepest theorem we have so far considered. Recall that the solution of a quadratic congruence modulo any integer n can be reduced to the solutions modulo the prime divisors of n (sec. 9.4).

Unless stated otherwise, *p is always an odd prime.*

To solve the congruence, we proceed as for quadratic equations, i.e., by completing the square. Thus we have

$$\begin{aligned} ax^2 + bx + c &\equiv a(x^2 + a^{-1}bx + a^{-1}c) \\ &\equiv a((x + 2^{-1}a^{-1}b)^2 + ca^{-1} - b^2 4^{-1}a^{-2})(\mathrm{mod}\, p) . \end{aligned}$$

Hence $f(x) \equiv 0(\mathrm{mod}\, p)$ if and only if

$$(x + 2^{-1}a^{-1}b)^2 \equiv b^2 4^{-1}a^{-2} - ca^{-1}(\mathrm{mod}\, p) .$$

If we set $y = x + 2^{-1}a^{-1}b$ and $d = b^2 4^{-1}a^{-2} - ca^{-1}$, then $f(x) \equiv 0(\mathrm{mod}\, p)$ is solvable if and only if

$$y^2 \equiv d(\mathrm{mod}\, p)$$

is solvable, and we have reduced our problem to the study of quadratic residues (sec. 10.3).

2. Elementary Properties of Quadratic Residues

For the moment, let p be a fixed prime. *The terms residue and nonresidue will refer to quadratic residues and nonresidues modulo p.*

(i) We have seen (sec. 10.3) that

the number of residues = the number of nonresidues.

To prove this directly, just note that the numbers $1^2, 2^2, \ldots, \left(\frac{p-1}{2}\right)^2$ are all distinct modulo p and, since $(p-x)^2 \equiv (-x)^2 \pmod{p}$, these are all the quadratic residues.

(ii) Since a is a quadratic residue if and only if the index of a is even, and $\operatorname{ind} ab = \operatorname{ind} a + \operatorname{ind} b$, we have

the product of two residues is a residue,

the product of two nonresidues is a residue, and

the product of a residue and a nonresidue is a nonresidue.

(iii) It was probably (ii) which induced Legendre to introduce his **residue symbol** defined, for $a \not\equiv 0 \pmod{p}$, by

$$\left(\frac{\mathbf{a}}{\mathbf{p}}\right) = 1, \qquad \text{if } a \text{ is a quadratic residue mod } p,$$

$$= -1, \qquad \text{if } a \text{ is a quadratic nonresidue mod } p.$$

From (i)-(iii), we have immediately

1) $\left(\frac{a}{p}\right) = (-1)^{\operatorname{ind} a}$,

2) $\left(\frac{ab}{p}\right) = \left(\frac{a}{p}\right)\left(\frac{b}{p}\right)$,

3) $\left(\frac{1}{p}\right) = 1$, since $1^2 \equiv 1 \pmod{p}$,

4) $\left(\frac{a^2}{p}\right) = 1$, for all a,

5) if $a \equiv b \pmod{p}$, then $\left(\frac{a}{p}\right) = \left(\frac{b}{p}\right)$.

It will be convenient when we discuss algebraic number theory to define $\left(\frac{a}{p}\right) = 0$, if $p|a$. *Note* that when $p|a$, property 1) makes no sense and 4) is false.

Specializing Euler's criterion for k^{th} power residues (sec. 10.3), we prove

Euler's Quadratic Residue Criterion:

$$p \text{ an odd prime, } a \not\equiv 0 (\text{mod } p) \Rightarrow \left(\frac{a}{p}\right) \equiv a^{\frac{p-1}{2}} (\text{mod } p)\ .$$

Proof:

$$a^{p-1} - 1 \equiv 0(\text{mod } p) \Rightarrow \left(a^{\frac{p-1}{2}} - 1\right)\left(a^{\frac{p-1}{2}} + 1\right) \equiv 0(\text{mod } p)$$
$$\Rightarrow a^{\frac{p-1}{2}} \equiv \pm 1(\text{mod } p)\ .$$

But by Euler's k^{th} power residue criterion for $k = 2$,

$$a^{\frac{p-1}{2}} \equiv 1(\text{mod } p) \iff a \text{ is a quadratic residue.}$$

Corollary: *For* $p > 2$, $\left(\frac{-1}{p}\right) = (-1)^{\frac{p-1}{2}}$. *Hence*

$$\left(\frac{-1}{p}\right) = 1, \quad \textit{if } p \equiv 1(\text{mod } 4)\ ,$$
$$= -1, \quad \textit{if } p \equiv 3(\text{mod } 4)\ .$$

Proof: By the theorem, $\left(\frac{-1}{p}\right) \equiv (-1)^{\frac{p-1}{2}} (\text{mod } p)$. Hence, since each side of the congruence equals ± 1 and $p > 2$, $\left(\frac{-1}{p}\right) = (-1)^{\frac{p-1}{2}}$.

Corollary: $\left(\frac{ab}{p}\right) = \left(\frac{a}{p}\right)\left(\frac{b}{p}\right)$

Moreover, since $\left(\frac{-a^2}{p}\right) = \left(\frac{-1}{p}\right)\left(\frac{a^2}{p}\right) = \left(\frac{-1}{p}\right) = (-1)^{\frac{p-1}{2}}$, we have

Corollary: $\left(\frac{-a^2}{p}\right) = 1 \iff p \equiv 1(\text{mod } 4)$.

For example, $19 \equiv -4(\text{mod } 23)$, and since $23 \equiv 3(\text{mod } 4)$, $x^2 \equiv -4 \equiv 19(\text{mod } 23)$ is not solvable.

Remarks:: i) If $a = \pm p_1^{a_1} \dots p_r^{a_r}$, then $\left(\frac{a}{p}\right) = \left(\frac{\pm 1}{p}\right) \prod \left(\frac{p_i}{p}\right)^{a_i}$. Hence to compute $\left(\frac{a}{p}\right)$, it suffices to compute $\left(\frac{q}{p}\right)$, where p and q are distinct primes. Euler's criterion gives a method of computation, but it can be cumbersome and doesn't give any insight or patterns, if any exist. The law of quadratic reciprocity will, to some extent, show us the patterns.

ii) The congruence classes of the quadratic residues modulo p form a subgroup of $F_p^\times$. This subgroup is the kernel of the homomorphism of $F_p^\times$ to $F_p^\times$, defined by $x \mapsto x^{\frac{p-1}{2}}$.

3. Gauss's Lemma

Our first proof of the reciprocity law will be one of Gauss's many proofs, published several years after the *Disquisitiones*. The proof is based on a lemma, discovered by Gauss, which gives yet another method for computing $\left(\frac{a}{p}\right)$, that we will use extensively in the next section.

Gauss's Lemma: *Let p be an odd prime and a an integer such that $a \not\equiv 0 (\mathrm{mod}\ p)$. Form the list*

$$a, 2a, 3a, \ldots, \left(\frac{p-1}{2}\right) a\,, \tag{1}$$

and replace each number by its least absolute residue, i.e., the number congruent to it modulo p which lies between $-\frac{p}{2}$ and $\frac{p}{2}$ (neither of which is an integer), to get the new list

$$b_1, b_2, \ldots, b_{\frac{p-1}{2}}\,. \tag{2}$$

If v is the number of b_i which are negative, then

$$\left(\frac{a}{p}\right) = (-1)^v\,.$$

This lemma may sound very complicated, but it is very useful.

Proof: We first show that the absolute values of the b_i's are distinct.

If $b_i = b_j$, then $ia \equiv ja (\mathrm{mod}\ p)$. Since $(a, p) = 1$, this means that $i \equiv j (\mathrm{mod}\ p)$, contradicting $1 \leq i, j \leq \frac{p-1}{2}$. Similarly, if $b_i = -b_j$, then $ia + ja = (i + j)a \equiv 0 (\mathrm{mod}\ p)$ and $i + j \equiv 0 (\mathrm{mod}\ p)$, contradicting $i + j \leq p - 1$. Hence the absolute values of the b_i's, possibly after rearrangement, consists of the numbers $1, 2, \ldots, \frac{p-1}{2}$. Thus we have

$$a \cdot 2a \cdot 3a \cdots \frac{p-1}{2} a \equiv (\pm 1)(\pm 2) \cdots \left(\pm \frac{p-1}{2}\right) (\mathrm{mod}\ p)\,,$$

and, cancelling $\left(\frac{p-1}{2}\right)!$ from both sides, we have

$$a^{\frac{p-1}{2}} = (-1)^{\upsilon} ,$$

where $\upsilon =$ the number of negative b_i's.

By Euler's criterion, $a^{\frac{p-1}{2}} \equiv \left(\frac{a}{p}\right)$ (mod p) and thus

$$\left(\frac{a}{p}\right) \equiv (-1)^{\upsilon} (\text{mod } p) .$$

Since both sides of the congruence are equal to ± 1 and $p > 2$, they are equal.

4. Computing $\left(\frac{a}{p}\right)$

We compute the Legendre symbol for several values of a to motivate the law of reciprocity, following the exposition of Adams and Goldstein [Ada - Gol].

Key Observation Lemma: In applying Gauss's lemma, the positive integers x congruent to an integer in the interval $\left[-\frac{p-1}{2}, -1\right]$ or, equivalently, $\left[-\frac{p}{2}, -1\right]$ (since $-\frac{p}{2}$ is not in $\mathbf{Z}$), are those x in the intervals

$$\frac{p}{2} < x < p$$
$$\frac{3p}{2} < x < 2p$$
$$\vdots$$
$$\left(s - \frac{1}{2}\right) p < x < sp$$
$$\vdots$$

since $x + p \equiv x (\text{mod } p)$ and it is true for the first interval.

|. . . . .|____|. . . . .|____| .

0 $\quad$ $\mathbf{\frac{p}{2}}$ $\quad$ **p** $\quad$ $\mathbf{\frac{3p}{2}}$ $\quad$ **2p**

Example: $\left(\frac{-1}{p}\right)$, p an odd prime: We form the list

$$1(-1),\ 2(-1), \ldots, \frac{p-1}{2}(-1)\,,$$

and apply Gauss's lemma. All the numbers are between $-\frac{p-1}{2}$ and -1 and they are all integers; thus $v = \frac{p-1}{2}$ and $\left(\frac{-1}{p}\right) = (-1)^{\frac{p-1}{2}}$, a result we derived earlier.

Example: $\left(\frac{2}{p}\right)$: The list required for Gauss's lemma is

$$1 \cdot 2,\ 2 \cdot 2, \ldots, \left(\frac{p-1}{2}\right) \cdot 2.$$

All the numbers lie between 1 and $p-1$. By the Key Observation Lemma, the numbers $2k$ in our list with negative least absolute residue modulo p are those between $\frac{p}{2}$ and p (we don't have to worry about the endpoints, since $2 \nmid p$ - this will be true in all our arguments).

|. . . . |___|

0 $\frac{p}{2}$ p

Thus we count those k in $[1, p-1]$ such that

$$\frac{p}{2} \le 2k \le p \quad \text{or} \quad \frac{p}{4} \le k \le \frac{p}{2}\,. \tag{1}$$

Now let $p = 8m + r, 0 < r < 8$, i.e., find the least positive residue of p mod 8. Since p is odd $r \equiv 1, 3, 5, 7 (\text{mod}\, 8)$. *Why choose 8 and not 2?* Because Euler, Legendre and Gauss found, by experimenting, that the value of $\left(\frac{a}{p}\right)$ will be determined by the residue classes of $p (\text{mod}\, 4a)$ (this is essentially one part of the law of reciprocity). Substituting $p = 8m + r$ in (1), we must count those k such that

$$2m + \frac{r}{4} \le k \le 4m + \frac{r}{2}\,. \tag{2}$$

We only need to check the four possibilities $r = 1, 3, 5, 7$. But *note* that it is only the parity of our count that matters for Gauss's lemma.

To simplify the parity count, here and in all later examples we use the following basic lemma, whose proof is left as an easy exercise.

Let **[x]** be the **greatest integer function**, i.e., the unique integer satisfying $x - 1 < [x] \le x$. For example $\left[\frac{3}{2}\right] = 1$, $[3] = 3$, and $[-4.5] = -5$.

Basic Lemma: *Let $u, w \in \mathbf{R}$, $u \notin \mathbf{Z}$, and $u \le w$. Then*

(i) *the number of integers k satisfying $u \le k \le w$ equals $[w] - [u]$,*

(ii) $n \in \mathbf{Z} \Rightarrow [n + w] = n + [w]$,

(iii) *if $n_1, n_2 \in \mathbf{Z}$ and $n_1 \le n_2$, then the number of integers k such that $2n_1 + u \le k \le 2n_2 + w$ has the same parity as the number of integers k such that $u \le k \le w$, i.e., we can drop the even numbers from a parity count,*

(iv) $[-x] = -[x] - 1$.

Now back to $\left(\frac{2}{p}\right)$. To apply Gauss's lemma, we need the parity of the number v of those k satisfying (2). By the Basic Lemma (iii), the parity of the number v is the same as the parity of the number v' of the k satisfying

$$\frac{r}{4} \le k \le \frac{r}{2} .$$

$r = 1$: $\frac{1}{4} \le k \le \frac{1}{2}$, $v' = 0$, and the parity is even. Hence

$$p \equiv 1 (\mathrm{mod}\, 8) \Rightarrow \left(\frac{2}{p}\right) = 1 .$$

$r = 3$: $\frac{3}{4} \le k \le \frac{3}{2}$, $v' = 1$, and the parity is odd. Hence

$$p \equiv 3 (\mathrm{mod}\, 8) \Rightarrow \left(\frac{2}{p}\right) = -1 .$$

$r = 5$: $\frac{5}{4} \le k \le \frac{5}{2}$, $v' = 1$, and the parity is odd. Hence

$$p \equiv 5 (\mathrm{mod}\, 8) \Rightarrow \left(\frac{2}{p}\right) = -1 .$$

$r = 7$: $\frac{7}{4} \le k \le \frac{7}{2}$, $v' = 2$, and the parity is even. Hence

$$p \equiv 7 (\mathrm{mod}\, 8) \Rightarrow \left(\frac{2}{p}\right) = 1 .$$

We summarize in the following table, where $p \equiv r \pmod 8$:

r	:	1	3	5	7
$\left(\frac{2}{p}\right)$	:	1	-1	-1	1

Thus we have

Proposition: *If p is an odd prime, then*

$$\left(\frac{2}{p}\right) = 1 \quad \textit{if} \quad p \equiv \pm 1 \pmod 8\,,$$

$$-1 \quad \textit{if} \quad p \equiv \pm 3 \pmod 8\,.$$

Corollary: *If p is an odd prime, then* $\left(\frac{2}{p}\right) = (-1)^{\frac{p^2-1}{8}}$. *(Just check all cases.)*

Example: $\left(\frac{3}{p}\right)$, $p \neq 2, 3$: The required list for Gauss' lemma is

$$1 \cdot 3, 2 \cdot 3, \ldots, \left(\frac{p-1}{2}\right) \cdot 3\,.$$

Since $3 \cdot \frac{p-1}{2} < \frac{3p}{2}$, all the numbers lie between 1 and $\frac{3p}{2}$. Considering the intervals of length $\frac{p}{2}$,

$$\begin{array}{cccc} 0 & \frac{p}{2} & p & \frac{3p}{2} \end{array}$$

the numbers on our list in the first interval and those in the third (which are congruent mod p to those in the first) have positive least absolute residue, and those in the second interval have negative least absolute residue. Thus, to apply Gauss's lemma we need the parity of the numbers of the form $3k$ in the second interval, i.e., those such that

$$\frac{p}{2} \leq 3k \leq p \quad \text{or} \quad \frac{p}{6} \leq k \leq \frac{p}{3}\,.$$

Setting $p = 12m + r$, where $r = 1, 5, 7$ or 11 (the other cases do not give primes), we are reduced to counting those k such that

$$2m + \frac{r}{6} \leq k \leq 4m + \frac{r}{3}\,.$$

Applying the basic lemma, we only need the parity of those k for which

$$\frac{r}{6} \leq k \leq \frac{r}{3} .$$

Computing this explicitly, as in the previous example, we have

$r:$	1	5	7	11
$\left(\frac{3}{p}\right):$	1	-1	-1	1

where $p \equiv r \pmod{12}$, and thus we have

Proposition: *If $p \neq 2, 3$ is a prime, then $\left(\frac{3}{p}\right) = 1$ if and only if $p \equiv \pm 1 \pmod{12}$.*

Example: $\left(\frac{5}{p}\right)$, $p \neq 2, 5$: The required list is

$$1 \cdot 5, 2 \cdot 5, \ldots, \left(\frac{p-1}{2}\right) \cdot 5$$

Since $5 \cdot \frac{p-1}{2} < \frac{5p}{2}$, all the numbers lie between 1 and $\frac{5p}{2}$. Considering intervals of length $\frac{p}{2}$,

$$0 \qquad \frac{p}{2} \qquad p \qquad \frac{3p}{2} \qquad 2p \qquad \frac{5p}{2}$$

the numbers on our list in the first, third and fifth intervals (which are congruent mod p to those in the first) have positive least absolute residue and do not contribute to the parity count in Gauss's lemma. Those in the second and fourth intervals have negative least absolute residue. Thus, to apply Gauss's lemma we need the parity of the numbers of the form $5k$ in the second and fourth interval, i.e., those such that

$$\frac{p}{2} \leq 5k \leq p \quad \text{or} \quad \frac{3p}{2} \leq 5k \leq 2p ,$$

or equivalently

$$\frac{p}{10} \leq k \leq \frac{p}{5} \quad \text{or} \quad \frac{3p}{10} \leq k \leq \frac{2p}{5} .$$

Let $p = 20m + r$, where $r = 1, 3, 7, 9, 11, 13, 17$ or 19, and proceed as in the previous example (remember that for each r we must find the parity of the number of integers in both intervals). Then we have the table

r :	1	3	7	9	11	13	17	19
$\left(\frac{5}{p}\right)$:	1	−1	−1	1	1	−1	−1	1

where $p \equiv r \pmod{20}$, and thus

Proposition: *If $p \neq 2, 5$ is a prime, then $\left(\frac{5}{p}\right) = 1$ if and only if $p \equiv \pm 1$ or $\pm 9 \pmod{20}$.*

Exercise: Show that for $\left(\frac{5}{p}\right)$, we can also work modulo 5, viz.,

$$\begin{aligned}\left(\frac{5}{p}\right) &= 1 \qquad \text{if} \quad p \equiv \pm 1 \pmod 5\ , \\ &= -1 \qquad \text{if} \quad p \equiv \pm 2 \pmod 5\ .\end{aligned}$$

Exercise: Determine the value of $\left(\frac{7}{p}\right)$ for $p \neq 2, 7$.

5. Quadratic Reciprocity 1

As mentioned earlier, after computing many special examples, Euler, Legendre and Gauss found two important properties of $\left(\frac{a}{p}\right)$ (for Euler see [Eul 4] with an English translation in [Str] and for Legendre see [Leg]).

1) The value of $\left(\frac{a}{p}\right)$ depends only on the residue class of p modulo a or modulo $4a$, and thus always modulo $4a$. Thus our conditions are always given by *congruences*, i.e.,

$$\left(\frac{a}{p}\right) = 1 \quad \text{if and only if} \quad r \equiv \ldots\ldots \pmod{4a}\ .$$

Hence $\left(\frac{a}{p}\right)$ behaves the same for all integers in an *arithmetic progression* with difference $4a$.

2) The table of values for $\left(\frac{a}{p}\right)$, for fixed a, is symmetric about its center, i.e., $\left(\frac{a}{p}\right)$ takes the same value for the remainders r and $4a - r$.

These results are essentially one form of the law of quadratic reciprocity,

Law of Quadratic Reciprocity:

(Form 1) *Let a be a positive integer, p an odd prime, and* $a \not\equiv 0 \pmod{p}$*; thus* $p \nmid 4a$*. Suppose that* $p = 4am + r$, $0 < r < 4a$*. Then*

1) $\left(\frac{a}{p}\right)$ *depends only on the remainder r and not on p.*

2) *Furthermore,* $\left(\frac{a}{p}\right)$ *has the same value for the remainders r and* $4a - r$*.*

Note that if $p \equiv q \pmod{4a}$, then p and q have the same remainders r, and if $p \equiv -q \pmod{4a}$, one has a remainder r and the other the remainder $4a - r$. Hence a cleaner way of stating the law, under the above assumptions is

(Form 1′) $p \equiv \pm q \pmod{4a} \Rightarrow \left(\frac{a}{p}\right) = \left(\frac{a}{q}\right)$.

Proof: The theorem is trivially true for $a = 1$, so we assume $a > 1$. We proceed just as in our examples. We form the list

$$a,\ 2a,\ 3a, \ldots, \left(\frac{p-1}{2}\right) a\ ,$$

find the parity of the number of integers in the list with negative least absolute residues, and apply Gauss's lemma.

Break up the positive integers into intervals of length $\frac{p}{2}$.

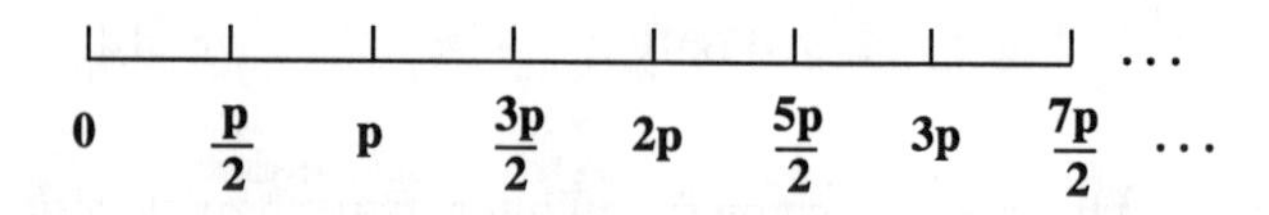

The integers in the 1st, 3rd, 5th, $\cdots$ intervals, i.e., intervals of the form $\left[tp, (t+\frac{1}{2})p\right]$, have positive least absolute residue, and those in in the even intervals, $\left[(t-\frac{1}{2})p, tp\right]$, have negative least absolute residues. Therefore, for the purpose of applying Gauss's lemma, we are only interested in the latter intervals. Hence we want to find the numbers of the form ka,

$1 < k < \frac{p-1}{2}$, satisfying one of the conditions

$$\begin{aligned}\frac{p}{2} &\le ka \le p\,,\\ \frac{3p}{2} &\le ka \le 2p\,,\\ \frac{5p}{2} &\le ka \le 3p\,,\\ &\vdots\end{aligned}$$

or, equivalently,

$$\begin{aligned}\frac{p}{2a} &\le k \le \frac{p}{a}\,,\\ \frac{3p}{2a} &\le k \le \frac{2p}{a}\,,\\ \frac{5p}{2a} &\le k \le \frac{3p}{a}\,,\\ &\vdots\end{aligned} \tag{1}$$

where we must determine the last interval.

To do so, recall that the endpoints cannot be in our list. The intervals are all of the form $\left[(s-\frac{1}{2})\frac{p}{a}, \frac{sp}{a}\right] = \left[(2s-1)\frac{p}{2a}, \frac{sp}{a}\right]$. Since $\frac{p}{2}$ lies between the integers $\frac{p-1}{2} = \frac{p}{2} - \frac{1}{2}$ and $\frac{p}{2} + \frac{1}{2}$, the condition $1 \le k \le \frac{p-1}{2}$ is equivalent to $1 \le k \le \frac{p}{2}$. Thus we don't have to go beyond the interval containing $\frac{p}{2}$, viz.,

$$(2t-1)\frac{p}{2a} \le \frac{p}{2} \le \frac{tp}{a}$$

or

$$\frac{2t-1}{2} \le \frac{a}{2} \le t\,,$$

which implies

$$\frac{a}{2} \le t \le \frac{a}{2} + \frac{1}{2}\,.$$

Therefore we must find the integers k satisfying one of the conditions

$$\frac{p}{2a} \le k \le \frac{p}{a}\,,$$

$$\begin{aligned} \frac{3p}{2a} \le k \le \frac{2p}{a}\,, \\ \frac{5p}{2a} \le k \le \frac{3p}{a}\,, \\ \vdots \\ (2t-1)\frac{p}{2a} \le k \le \frac{tp}{a}\,. \end{aligned} \tag{2}$$

If a is even, set $t = \frac{a}{2}$. Then the last k in our list, $k = \frac{p-1}{2}$ will be in the corresponding interval. Furthermore, since $\frac{tp}{a} = \frac{p}{2}$, the condition $1 \le k \le \frac{p}{2}$ is implied by (2) in this case.

If a is odd, set $t = \frac{a}{2} + \frac{1}{2}$. Then the last interval of (2) is

$$\frac{p}{2} \le k \le \frac{p}{2} + \frac{p}{2a}\,.$$

Since we require $1 \le k \le \frac{p}{2}$, we can drop this interval. Thus we take $\frac{a}{2} - \frac{1}{2}$ instead of $\frac{a}{2} + \frac{1}{2}$ and the condition, $1 \le k \le \frac{p}{2}$, is automatically satisfied. Therefore let

$$u = \begin{cases} \frac{a}{2} & \text{if } a \text{ is even} \\ \frac{a}{2} - \frac{1}{2} & \text{if } a \text{ is odd}\,. \end{cases}$$

To apply Gauss' lemma, we must find the parity of the number of integers satisfying one of the conditions

$$\begin{aligned} \frac{p}{2a} \le k \le \frac{p}{a}\,, \\ 3\frac{p}{2a} \le k \le \frac{2p}{a}\,, \\ 5\frac{p}{2a} \le k \le \frac{3p}{a}\,, \\ \vdots \\ \frac{(2u-1)p}{2a} \le k \le \frac{up}{a}\,. \end{aligned} \tag{3}$$

Now, as in our examples, let $p = 4am + r$, $0 < r < 4a$. Then the conditions (3) become

$$\begin{gathered} 2m + \frac{r}{2a} \le k \le 4m + \frac{r}{a}, \\ \vdots \\ 2(2u-1)m + \frac{(2u-1)r}{2a} \le k \le 4um + \frac{ur}{a}, \end{gathered}$$

and we must determine the parity of the number of k satisfying one of these conditions. By our basic lemma, we can drop the even integers from these inequalities and retain the same parity. Thus, to apply Gauss's lemma, we must find the parity of the number of k satisfying one of the conditions

$$\begin{gathered} \frac{r}{2a} \le k \le \frac{r}{a}, \\ \vdots \\ \frac{(2u-1)r}{2a} \le k \le \frac{ur}{a}. \end{gathered} \tag{4}$$

But this number clearly depends only on r and not on p. Thus, if $p \equiv q \equiv r \pmod{4a}$, then $\left(\frac{a}{p}\right) = \left(\frac{a}{q}\right)$ and we have proved the first part of the law.

To prove the second part, we replace r by $4a - r$ in (4) and show that the parity of the number of k satisfying one of these new conditions is the same as the parity of the number of k satisfying one of the conditions (4). This is left as an exercise.

6. Quadratic Reciprocity 2

We have proved the reciprocity law as stated by Euler. As discussed in section 6.3, Legendre conjectured the law in what has become the more commonly stated form, which also explains the use of the term "reciprocity".

Law of Quadratic Reciprocity:

(Form 2) *If p and q are distinct odd primes, then*

$$\left(\frac{p}{q}\right)\left(\frac{q}{p}\right) = (-1)^{\frac{p-1}{2}\frac{q-1}{2}}$$

(Form 2′) *If at least one of p or q is* $\equiv 1 \pmod 4$, *then* $\left(\frac{p}{q}\right) = \left(\frac{q}{p}\right)$.

If $p \equiv q \equiv 3 \pmod 4$, *then* $\left(\frac{p}{q}\right) = -\left(\frac{q}{p}\right)$.

In section 11.4, we have already proved the

Completion Theorem:

$$\left(\frac{-1}{p}\right) = (-1)^{\frac{p-1}{2}},$$
$$\left(\frac{2}{p}\right) = (-1)^{p^2-1} \quad (= -1 \textit{ if and only if } p \equiv \pm 5 \pmod 8)\ .$$

This is called the completion theorem since, together with form 2, it completes the results needed to compute all Legendre symbols.

Example:

$$\left(\frac{-54}{71}\right) = \left(\frac{(-1)2 \cdot 3^3}{71}\right) = \left(\frac{-1}{71}\right)\left(\frac{2}{71}\right)\left(\frac{3}{71}\right)^3$$
$$= (-1)^{\frac{71-1}{2}}\left(\frac{2}{7}\right)\left(\frac{3}{71}\right), \text{ by the Completion Theorem,}$$
$$= (-1)(1)\left(\frac{3}{71}\right), \text{ by the Completion Theorem,}$$
$$= (-1)\left(\frac{71}{3}\right)(-1)^{\frac{71-1}{2}\cdot\frac{3-1}{2}}, \text{ by quadratic reciprocity,}$$
$$= \left(\frac{71}{3}\right) = \left(\frac{2}{3}\right) = -1.$$

Hence $x^2 \equiv -54 \pmod{71}$ is not solvable.

Now we show that each of forms 1 and 2 can be derived from the other and then we give a direct proof of the second form, again using Gauss's lemma.

Form 1 $\Rightarrow$ *Form 2* : We suppose, with no loss of generality, that $p > q$. There are two cases:

i) $p \equiv q \pmod 4$: Then $p = q + 4a$, for some $a > 0$ and $p \equiv q \pmod{4a}$. Therefore

$$\left(\frac{p}{q}\right) = \left(\frac{q+4a}{q}\right) = \left(\frac{4a}{q}\right) = \left(\frac{4}{q}\right)\left(\frac{a}{q}\right) = \left(\frac{a}{q}\right)$$

and

$$\left(\frac{q}{p}\right) = \left(\frac{p-4a}{p}\right) = \left(\frac{-4a}{p}\right) = \left(\frac{-1}{p}\right)\left(\frac{4}{p}\right)\left(\frac{a}{p}\right)$$
$$= (-1)^{\frac{p-1}{2}}\left(\frac{a}{p}\right);$$

hence

$$\left(\frac{p}{q}\right)\left(\frac{q}{p}\right) = \left(\frac{a}{p}\right)(-1)^{\frac{p-1}{2}}\left(\frac{a}{q}\right).$$

By form 1, $p \equiv q \pmod{4a} \Rightarrow \left(\frac{a}{p}\right) = \left(\frac{a}{q}\right)$, so

$$\left(\frac{p}{q}\right)\left(\frac{q}{p}\right) = (-1)^{\frac{p-1}{2}}.$$

Since $\frac{p-1}{2} = \frac{q-1}{2} + 2a$, $\frac{p-1}{2}$ and $\frac{q-1}{2}$ have the same parity; hence

$$(-1)^{\frac{p-1}{2}} = (-1)^{\frac{p-1}{2}\frac{q-1}{2}}.$$

ii) $p \not\equiv q \pmod 4$: Either $p \equiv 1 \pmod 4$, $q \equiv 3 \pmod 4$ or $p \equiv 3 \pmod 4$, $q \equiv 1 \pmod 4$. In each case $p \equiv -q \pmod 4$, $p = -q + 4a$, for some $a > 0$, and $p \equiv -q \pmod{4a}$. Therefore

$$\left(\frac{p}{q}\right) = \left(\frac{-q+4a}{q}\right) = \left(\frac{4a}{q}\right) = \left(\frac{a}{q}\right)$$

and

$$\left(\frac{q}{p}\right) = \left(\frac{-p+4a}{p}\right) = \left(\frac{4a}{p}\right) = \left(\frac{a}{p}\right).$$

By form 1′ (sec. 5), $p \equiv -q \pmod 4 \Rightarrow \left(\frac{a}{p}\right) = \left(\frac{a}{q}\right)$, so $\left(\frac{p}{q}\right)\left(\frac{q}{p}\right) = 1$. But $\frac{p-1}{2} + \frac{q-1}{2} = 2a - 1$ is odd. Therefore exactly one of $\frac{p-1}{2}$ and $\frac{q-1}{2}$ is odd and

$$(-1)^{\left(\frac{p-1}{2}\right)\left(\frac{q-1}{2}\right)} = 1.$$

Form 2 $\Rightarrow$ *Form 1*: We proceed in several steps. First we prove the theorem when a is a prime.

1) $a = 2$: Our result for $\left(\frac{2}{p}\right)$ in section 4, trivially proves this.

2) $a =$ an odd prime:

i) $p \equiv q (\bmod 4a)$: By form 2, $\left(\frac{a}{p}\right) = \left(\frac{p}{a}\right)(-1)^{\frac{p-1}{2}\frac{a-1}{2}}$. But $p \equiv q (\bmod 4a) \Rightarrow p \equiv q (\bmod a) \Rightarrow \left(\frac{p}{a}\right) = \left(\frac{q}{a}\right)$. Therefore, by form 2,

$$\begin{aligned}\left(\frac{a}{p}\right) &= \left(\frac{p}{a}\right)(-1)^{\frac{p-1}{2}\frac{a-1}{2}} \\ &= \left(\frac{q}{a}\right)(-1)^{\frac{p-1}{2}\frac{a-1}{2}} \\ &= \left(\frac{a}{q}\right)(-1)^{\frac{q-1}{2}\frac{a-1}{2}}(-1)^{\frac{p-1}{2}\frac{a-1}{2}} \\ &= \left(\frac{a}{q}\right)(-1)^{\frac{a-1}{2}\frac{p+q-2}{2}} .\end{aligned}$$

Now $p \equiv q (\bmod 4a) \Rightarrow p \equiv q (\bmod 4) \Rightarrow p + q - 2 \equiv 2q - 2 (\bmod 4)$. But q is odd and thus $q \equiv 1, 3 (\bmod 4)$. Either case yields $p + q - 2 \equiv 0 (\bmod 4)$ and thus $\frac{p+q-2}{2}$ is even and $\left(\frac{a}{p}\right) = \left(\frac{a}{q}\right)$.

ii) $p \equiv -q (\bmod 4a)$: Proceeding as in i), we can derive

$$\left(\frac{a}{p}\right) = \left(\frac{a}{q}\right)(-1)^{\frac{a-1}{2}\frac{p+q}{2}} .$$

But $p \equiv -q (\bmod 4a) \Rightarrow p + q \equiv 0 (\bmod 4a) \Rightarrow p + q \equiv 0 (\bmod 4)$, i.e., $\frac{p+q}{2}$ is even and we are done.

3) Now assume a is not a prime, $a = \prod p_i$. If $p \equiv \pm q (\bmod 4a)$, then $p \equiv \pm q (\bmod 4p_i)$ and, since form 1 is assumed true for primes, we have

$$\left(\frac{p_i}{p}\right) = \left(\frac{p_i}{q}\right) \Rightarrow \prod\left(\frac{p_i}{p}\right) = \prod\left(\frac{p_i}{q}\right) \Rightarrow \left(\frac{a}{p}\right) = \left(\frac{a}{p}\right) .$$

The following direct proof of form 2 of the reciprocity law is due to Eisenstein [Eis]. First we need a lemma, very similar in spirit to Gauss's lemma and based on it.

Lemma: *Let m be an odd integer, p an odd prime, and $(m, p) = 1$. Then*

$$\left(\frac{m}{p}\right) = (-1)^k ,$$

where $k = \sum_{i=1}^{\frac{p-1}{2}} \left[\frac{im}{p}\right]$*, and* $[x]$ *is the greatest integer function.*

Proof: Form the list $m, 2m, \ldots, \left(\frac{p-1}{2}\right) m$. Then either

$$im = \left[\frac{mi}{p}\right] p + r_i , \quad 0 < r_i < \frac{p}{2} ,$$

or (1)

$$im = \left[\frac{mi}{p}\right] p + (p - s_i), \quad 0 < s_i < \frac{p}{2} .$$

The r_i and $-s_i$ are the least absolute residues of the integers in our list and thus

the number of s_i = *the* v *of Gauss's lemma.*

Summing the equations (1) over all r_i, and s_i, we have

$$m \sum_{1}^{\frac{p-1}{2}} i = p \sum \left[\frac{mi}{p}\right] + pv + \sum r_i - \sum s_i . \tag{2}$$

Note that the r_i and s_i consist of all the integers $1, 2, \ldots, \frac{p-1}{2}$, since in the proof of Gauss' lemma we showed that they are distinct. Therefore adding and subtracting $\sum s_i$ from the right hand side of (2), we have

$$m \sum_{1}^{\frac{p-1}{2}} i = p \sum \left[\frac{mi}{p}\right] + pv + \sum_{1}^{\frac{p-1}{2}} i - 2 \sum s_i$$

or

$$(m - 1) \sum i = p \sum \left[\frac{mi}{p}\right] + pv - 2 \sum s_i .$$

But $p \equiv 1 \pmod 2$, $m \equiv 1 \pmod 2$ and $2 \equiv 0 \pmod 2$; so we have

$$v \equiv -\sum_i \left[\frac{mi}{p}\right] \pmod 2 .$$

Recall that $k = \sum \left[\frac{mi}{p}\right]$; thus v and k have the same parity. Therefore, by Gauss's lemma,

$$\left(\frac{m}{p}\right) = (-1)^v = (-1)^k .$$

Eisenstein's proof of form 2 : Let

$$k = \sum_{i=1}^{\frac{q-1}{2}} \left[\frac{pi}{q}\right] \quad \text{and} \quad k' = \sum_{i=1}^{\frac{p-1}{2}} \left[\frac{qi}{p}\right] .$$

Then, by the lemma,

$$\left(\frac{p}{q}\right) = (-1)^k \quad \text{and} \quad \left(\frac{q}{p}\right) = (-1)^{k'} .$$

We shall now show that $k + k' = \frac{p-1}{2}\frac{q-1}{2}$, by counting *integer points* (points with integer coordinates) *inside* the rectangle with vertices $(0, 0)$, $(0, \frac{p}{2})$, $(\frac{q}{2}, 0)$, and $(\frac{q}{2}, \frac{p}{2})$ (see fig. 1).

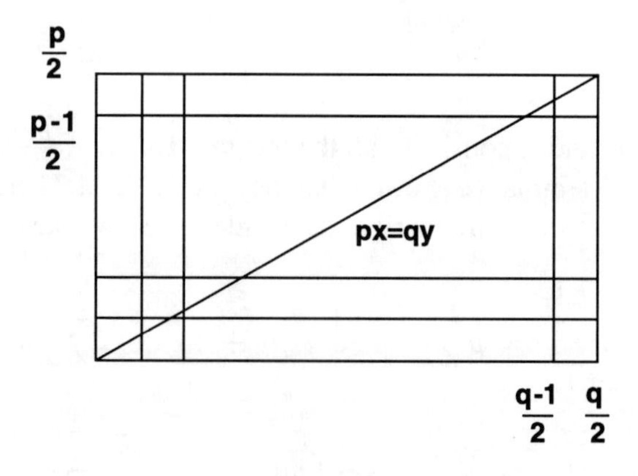

Figure 1

There are $\frac{p-1}{2} \cdot \frac{q-1}{2}$ integer points inside the rectangle.

Since $\left(i, \frac{pi}{q}\right)$ is on the diagonal, there are $\left[\frac{pi}{q}\right]$ integer points on the line $x = i$ and below the diagonal, and thus there are $\sum_1^{\frac{q-1}{2}} \left[\frac{pi}{q}\right] = k$ integer points below the diagonal. Similarly the number of integer points on the line $y = i$, to the left of the diagonal is $\left[\frac{qi}{p}\right]$ and thus there are k' integer points above the diagonal.

Since there are no integer points on the diagonal, $k + k' = \frac{p-1}{2}\frac{q-1}{2}$ and we are done.

The second form of the reciprocity law is quite startling. It gives a totally unobvious relation between the solutions of

$$x^2 \equiv p (\mathrm{mod}\, q) \quad \text{and} \quad x^2 \equiv q (\mathrm{mod}\, p) .$$

André Weil remarked that the fact that there is a law, even though the distribution of primes in any arithmetic progression seems random, is itself amazing [Wei 4, vol. 1, pp. 244 - 255].

7. Some History and Other Proofs

In article 151 of the *Disquisitiones*, Gauss discusses the history of the law of quadratic reciprocity (translation from [Gau 1]).

> "*The work of other mathematicians*
> *concerning these investigations*
>
> 151. The fundamental theorem must certainly be regarded as one of the most elegant of its type. No one has thus far presented it in as simple a form as we have done above. This is even more surprising since Euler already knew other propositions which depend on it and from which it can easily be recovered. He was aware that certain forms existed which contain all the prime divisors of numbers of the form $x^2 - A$, that there were others containing all prime nondivisors of numbers of the same form, and that the two sets were mutually exclusive. And he knew further the method of finding these forms, but all his attempts at demonstration were in vain, and he succeeded only in giving a greater degree of verisimilitude to the truth that he had discovered by induction. In a memoir entitled "*Novae demonstrationes circa divisores numerorum formae* $xx + nyy$" which he read in the St. Petersburg Academy (Nov. 20, 1775) and which was published after his death,[10] he seems to believe that he had fulfilled his resolve. But an error did creep in, for on p. 65

he tacitly presupposed the existence of such forms of the divisors[p] and nondivisors, and from this it was not difficult to discover what the forms should be. But the method he used to prove this supposition does not seem to be suitable. In another paper, "*De criteriis aequationis fxx + gyy = hzz utrumque resolutionem admittat necne,* "[11] Opuscula Analytica, 1, 211 (f, g, h are given, x, y, z are indeterminate) he finds by induction that if the equation is solvable for one value of $h = s$, it will also be solvable for any other value congruent to s relative to the modulus $4fg$ provided it is a prime number. From this proposition the supposition we spoke of can easily be demonstrated. But the demonstration of this theorem also eluded his efforts.[q] This is not remarkable because in our judgment is necessary to start from the fundamental theorem. The truth of the proposition will flow automatically from what we will show in the following section.

After Euler, the renowned Legendre worked zealously on the same problem in his excellent tract, "Recherches d'analyse indétérminée", *Hist. Acad. Paris* , 1785, p. 465 ff. He arrived at a theorem basically the same as the fundamental theorem. He stated that if p, q are two positive prime numbers, the absolute least residues of the powers $p^{\frac{q-1}{2}}, q^{\frac{p-1}{2}}$ relative to the moduli q, p respectively, are either both $+1$ or both -1 when either p or q is of the form $4n + 1$: but when both p and q are of the form $4n + 3$ one residue will be $+1$, the other -1 (p. 516). From this, according to article 106, we derive the fact that the *relation* (taken according to the meaning in article 146) of p to q and of q to p will be the *same* when either p or q is of the form $4n + 1$, *opposite* when both p and q are of the form $4n + 3$. This proposition is contained among the propositions of article 131 and follows also from 1, 3, 9 of article 133: on the other hand the fundamental theorem can be derived from it. Legendre also attempted a demonstration and, since it is extremely ingenious, we will speak of it at some length in the following section. But since he presupposed many things without demonstration (as he himself confesses on p. 520: "*Nous avons supposé seulement* ..."), some of which have not been demonstrated by anyone up till now, and some of which cannot be demonstrated in our judgment without the help of the fundamental theorem itself, the road he has entered upon seems to lead to an impasse, and so our demonstration must regarded as the first."

---

[p] Namely that there do exist numbers r, r', r'', etc., n, n', n'', etc., all distinct and $< 4A$ such that all prime divisors of $x^2 - A$ are contained in one of the forms $4Ak + r, 4Ak + r'$, etc., and all prime nondivisors in one of the forms $4Ak + n, 4Ak + n'$, etc. (k being indeterminate).

[10] Nova acta acad. Petrop., 1 [1783], 178, 47-74.
[11] The original article reads "utrum ea" instead of "utrumque."
[q] As he himself confesses (*Opuscula Analytica*, I, 216): "A demonstration of this most elegant theorem is still sought even though it has been investigated in vain for so long and by so many... And anyone who succeeds in finding such a demonstration must certainly be considered most outstanding." With what ardor this great man searched for a proof of this theorem and of others which are only special cases of the fundamental theorem can be seen in many other places, e.g. *Opuscula Analytica*, I, 268 (Additamentum ad Diss. 8) and 2,275 (Diss. 13) and in many dissertations in *Comm. acad. Petrop.* which we have cited on several occasions. [For Dissertation 8 see p. 62. Dissertation 13 is entitled "De insigni promotione scientiae numerorum" Ed. Note.]

Gauss gave a total of six proofs of quadratic reciprocity, his principal goal was, as he stated it, to find the approach that would allow generalizations to higher powers. We will discuss some of these generalizations later.

The first proof (given in articles 125–135 of the *Disquisitiones*) is a very complicated induction argument. It held very little interest until recently, when the ideas were used for calculations in K-theory [Tat 1].

The second proof (given in article 262 of the *Disquisitiones*) is based on the theory of quadratic forms.

The third and fifth proofs, which are similar, are essentially based on Gauss's lemma.

The fourth and sixth proofs are based on Gauss's theory of the division of the circle, i.e., the theory of complex roots of 1. We will discuss this theory in chapter 14 and present the sixth proof.

Many of these proofs were rediscovered, sometimes in a slightly different form. Cauchy, Jacobi and Eisenstein gave proofs (independently of each other), and argued over priority, without being aware that they were variations on Gauss's sixth proof. For a further discussion of Gauss' and other proofs see H.J.S. Smith [Smi, H 3, art. 18-22] and Ireland and Rosen [Ire - Ros, chap. 6]. Many proofs are just minor variations of each other.

This concludes our discussion of quadratic congruences. As mentioned earlier, Gauss's many proofs of quadratic reciprocity were motivated by his desire to generalize the law to higher powers. The search for higher reciprocity laws led to the development of algebraic number theory (see chapters 15 - 17).

We have now finished the major contents of sections 1-4 of the *Disquisitiones* plus some later developments.

Chapter 12

Binary Quadratic Forms 1: Arithmetic Theory

1. Introduction

Sections V and VI of the *Disquisitiones* present the theory of representations of integers by binary quadratic forms and some applications. We have seen the genesis of this theory in the work of Fermat and Euler and the beginning of a general theory with Lagrange (sec. 5.2). We set the stage with Gauss's introduction to Section V (translation from [Gau 1]).

> Art. 153. In this section we shall treat particularly of functions in two unknowns x, y of the form
>
> $$ax^2 + 2bxy + cy^2$$
>
> where a, b, c are given integers. We will call these functions *forms of the second degree* or simply *forms*. On this investigation depends the solution of the famous problem of finding all the solutions of any indeterminate equation of the second degree in two unknowns where these unknown values are to be integers or rational numbers. This problem has already been solved in all generality by Lagrange, and many things pertaining to the nature of the *forms* were discovered by this great geometer and by Euler. They also furnished proofs for earlier discoveries of Fermat. However, a careful inquiry into the nature of the forms revealed so many new results that we decided it would be worthwhile to review the whole subject from the beginning - the more so because what these men have discovered is scattered in so many different places that few scholars are aware of them, further because the method we will use is almost entirely our own, and finally because the new things we add could not be understood without a new exposition of their discoveries. We have no doubt that many remarkable results still lie hidden and are

a challenge to the talents of others. In the proper places we will always report the history of important truths.

Note that Gauss writes his forms as $ax^2 + 2bxy + cy^2$, with even middle coefficient, whereas Lagrange and Legendre used $ax^2+bxy+cy^2$. Some theorems appear more natural in one notation and some in the other. This led to some debate in the 19th century as to which notation was the better. After Dedekind explained the relation between quadratic forms and quadratic fields, in which the form $ax^2+bxy+cy^2$ proves to be more natural, this latter form was generally, but by no means universally, accepted as the better notation.

Gauss made major contributions to the theory of quadratic forms, enriching the basic theory and introducing the deep and important concepts of composition and genus. This work strongly influenced the later development of number theory and abstract algebra.

First we review the basic setup that was presented in section 5.2. Matrix notation is used to simplify statements and proofs, but the proofs are essentially those given by Gauss or by Gustav Peter Lejeune Dirichlet (1805 - 1853). Dirichlet's lectures on the theory of numbers [Dir] made the *Disquisitiones* more accessible to the mathematical community and in many places we follow his exposition.

In addition to the references throughout the chapter, we also recommend [Dav] and for recent computational approaches [Bue].

2. Equivalence of Forms

Given a binary quadratic form

$$f(x, y) = ax^2 + bxy + cy^2 , \quad a, b, c \in \mathbf{Z} ,$$

also denoted by **f = (a, b, c)**, there are two basic problems:

1) find the integers n **representable** by the form, i.e., those n for which there exist integers x, y such that

$$n = f(x, y) ,$$

2) if n is representable by f, determine the number of ways this can be done.

Implicit in these questions is the desire for efficient algorithms. We concentrate on the first question and discuss aspects of the second later.

Two forms $f = (a, b, c)$ and $g = (a', b', c')$ are **equivalent, f ~ g**, if

$$g(x', y') = f(x, y) ,$$

where

$$\begin{pmatrix} x \\ y \end{pmatrix} = S \begin{pmatrix} x' \\ y' \end{pmatrix} , \quad S = \begin{pmatrix} s & t \\ u & v \end{pmatrix} ,$$

$\det S = \pm 1$ and $s, t, u, v \in \mathbf{Z}$; more explicitly

$$g(x', y') = f(sx' + ty', ux' + vy') .$$

Any integer matrix S with determinant ± 1 is called a **unimodular matrix**. We also write **g = Sf** to denote equivalence under the substitution S. Equivalence of forms is an equivalence relation.

The set of unimodular matrices under multiplication is the **general linear group $\mathbf{GL_2(Z)}$**, the multiplicative group of 2×2 matrices with integer entries, invertible over **Z**. (The operation Sf defines a group action of $GL_2(\mathbf{Z})$ on forms, but we will not use the general theory of group actions.)

In section 5.2, we saw that the unimodularity of S implies that *equivalent forms f and g represent the same integers* and that finding a "simple" form in each equivalence class might help solve the representation problem.

3. Matrix Notation and the Discriminant

Let M^t denote the transpose of the matrix M. Then, if

$$T = \begin{pmatrix} a & \frac{b}{2} \\ \frac{b}{2} & c \end{pmatrix} , \quad Z = \begin{pmatrix} x \\ y \end{pmatrix} ,$$

we have

$$f(x, y) = ax^2 + bxy + cy^2 = Z^t T Z . \tag{1}$$

The transformation of forms is then easy. Let

$$Z = SZ' , \quad S = \begin{pmatrix} s & t \\ u & v \end{pmatrix} , \quad Z' = \begin{pmatrix} x' \\ y' \end{pmatrix}$$

and substitute in (1) to get

$$\begin{aligned} f(x, y) =& Z^t T Z = (SZ')^t T(SZ') = Z'^t(S^t T S)Z' \\ =& Z'^t U Z' = g(x', y') = a'x'^2 + b'x'y' + cy'^2 , \end{aligned}$$

where

$$U = S^t T S = \begin{pmatrix} a' & \frac{b'}{2} \\ \frac{b'}{2} & c' \end{pmatrix} .$$

Multiplying out, we have the relation between the coefficients:

$$\begin{aligned} a' &= as^2 + bsu + cu^2 = f(s, u) , \\ b' &= 2ast + b(sv + tu) + 2cuv , \\ c' &= at^2 + btv + cv^2 = f(t, v) . \end{aligned} \tag{2}$$

Moreover

$$\det(U) = \det(S^t)\det(T)\det(S) = \det(T)(\det(S))^2$$

and, since we assume $\det(S) = \pm 1$,

$$\det(U) = \det(T) ,$$

i.e., $a'c' - \frac{b'^2}{4} = ac - \frac{b^2}{4}$ or

$$b'^2 - 4a'c' = b^2 - 4ac .$$

The **discriminant** of $f = (a, b, c)$, denoted by **disc(f)** or **D**, is $b^2 - 4ac$.

Warning: Some authors use $4ac - b^2$ for the discriminant and some call the discriminant the *determinant* of the form.

Why is the discriminant important ? First of all, we have just seen that

$$f \sim g \implies \text{disc}(f) = \text{disc}(g) , \tag{3}$$

i.e., the discriminant is an invariant of the equivalence classes and thus can sometimes be used to distinguish them. The converse of (3) is false; $x^2 + y^2$ and $-x^2 - y^2$ both have discriminant -4, but the former represents $2 = 1^2 + 1^2$ and $-x^2 - y^2 \leq 0$ does not. Second, we can associate to each form

$$f(x, y) = ax^2 + bxy + c = y^2 \left(a \left(\frac{x}{y} \right)^2 + b \left(\frac{x}{y} \right) + c \right) ,$$

the polynomial

$$P_f(z) = az^2 + bz + c \,,$$

with discriminant $D = b^2 - 4ac$ (this is why $b^2 - 4ac$ is called the discriminant of the form). The roots, $\frac{-b \pm \sqrt{D}}{2a}$, of $f(z) = 0$, will play an important role in our studies. We also see the first hint of a relation between quadratic forms and quadratic numbers.

Since the discriminant is an invariant under equivalence, we study the representation problem for a fixed discriminant D, i.e., at any given time we restrict our study to all forms of some fixed discriminant. From $D = b^2 - 4ac$ and the fact that a square is always congruent to 0 or 1 modulo 4, we have

$$D \equiv 0 \text{ or } 1 \pmod 4 \,.$$

For each such possible D there is at least one form of discriminant D, viz.,

$$D \equiv 0 \pmod 4: \quad f(x, y) = x^2 - \frac{D}{4} y^2 \,,$$

$$D \equiv 1 \pmod 4: \quad f(x, y) = x^2 + xy + \left(\frac{1 - D}{4}\right) y^2 \,.$$

If D is a square (so $D \geq 0$), then a form of discriminant D can be factored into a product of linear forms with rational coefficients. When $D = 0$, we have the square of a linear form and the representation problem reduces to the simple case of linear forms. When $D > 0$ is a square, the theory of representation and equivalence is not particularly deep and is worked out by Gauss in sections 206 - 212 of the *Disquisitiones* (there are interesting questions about bounds on the value of a product of two or more linear forms, for integer values of the variables, which will be treated using the geometry of number (chap. 22)). Hence, *we restrict our studies to quadratic forms with non square discriminants.* Note that if the discriminant of $ax^2 + bxy + cy^2$ is not a square, then $a, c \neq 0$.

4. Reduced Forms and the Number of Classes

The most important elementary result of the theory of quadratic forms is the

Reduction Theorem: *For every possible discriminant $D \neq$ square, there are only finitely many equivalence classes of forms of discriminant D.*

Proof: To prove the theorem, we show that every form is equivalent to a **reduced** form, i.e., a form (a, b, c) such that

$$|b| \le |a| \le |c| .$$

But for a given discriminant D, there are only finitely many reduced forms, i.e., finitely many triples satisfying

$$D = b^2 - 4ac , \quad |b| \le |a| \le |c| .$$

To see this, note that

$$\begin{aligned} |D| &= |b^2 - 4ac| \\ &\ge \big| |b|^2 - 4|a|\,|c| \big| , && \text{since } |x - y| \ge \big| |x| - |y| \big| , \\ &\ge 4|a|\,|c| - |b|^2 , && \text{since } |x| \ge x, -x , \\ &\ge 4|a|^2 - |b|^2 , && \text{since } |c| \ge |a| , \\ &\ge 4|a|^2 - |a|^2 , && \text{since } |b| \le |a| , \\ &= 3|a|^2 . \end{aligned}$$

Hence

$$|a| < \sqrt{\frac{|D|}{3}} , \quad |b| \le |a| \le \sqrt{\frac{|D|}{3}} , \quad c = \frac{b^2 - D}{4a} . \tag{1}$$

Thus, for a given D, there are only a finite number of choices for (a, b, c), i.e., there are a finite number of reduced forms of discriminant D. It follows that if every form is equivalent to a reduced form, there are only a finite number of equivalence classes.

To show that every form is equivalent to a reduced form, we need two special maps.

i) $\left\{ \begin{array}{l} x = -y' \\ y = x' \end{array} \right\}$ given by the matrix $S = \begin{pmatrix} 0 & -1 \\ 1 & 0 \end{pmatrix}$, which applied to $f = (a, b, c)$ yields, by (3.2),

$$(a, b, c) \sim (c, -b, a) .$$

ii) For $u \in \mathbf{Z}$, $\left\{ \begin{array}{l} x = x' + uy' \\ y = y' \end{array} \right\}$ given by the matrix $T_u = \begin{pmatrix} 1 & u \\ 0 & 1 \end{pmatrix}$, which yields

$$(a, b, c) \sim (a, b + 2ua, c + ub + u^2 a) .$$

The following algorithm starts with an arbitrary form f and produces a sequence of forms, equivalent to f, eventually arriving at a reduced form.

Algorithm: Starting with $f = (a, b, c)$

1) if $|a| < |b|$, choose $u \in \mathbf{Z}$ so that $-|a| < b + 2ua \leq |a|$ and apply T_u to (a, b, c) to get $(a, b' = b + 2ua, c')$. Thus $|\mathbf{b}'| \leq |\mathbf{a}|$,
2) if $|a| > |c'|$, apply S to (a, b', c') to get $(a' = c', -b', a)$, where $|c'| < |a|$ and go back to step 1),
3) if $|a| \leq |c'|$, stop and we have a reduced form.

This is illustrated by the flow diagram in figure 1.

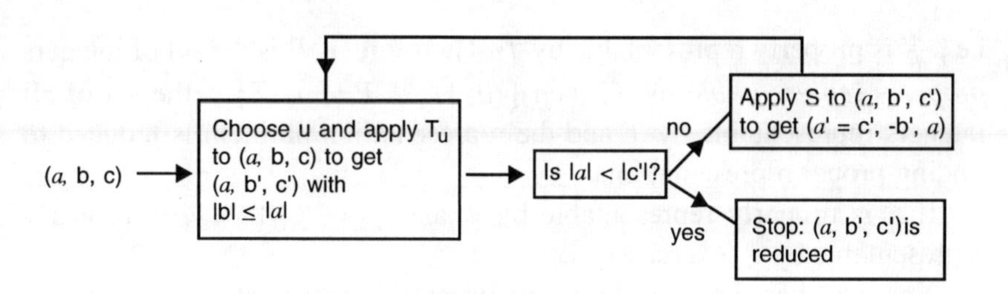

Figure 1

Note that step 1 leaves the coefficient of x^2 unchanged and step 2 makes it smaller. If we do not arrive at a reduced form after a finite number of steps, then we have generated a sequence of *positive* integers satisfying

$$|a| > |a'| > |a''| > \cdots ,$$

which is impossible. Hence the sequence of forms generated by the algorithm must stop at a reduced form and we are done.

Example: We find the number of classes of forms of discriminant -4. The coefficients of a reduced forms must satisfy (1), viz.,

$$|a| \leq \sqrt{\frac{4}{3}}, \quad |b| \leq \sqrt{\frac{4}{3}}, \quad b^2 - 4ac = -4 .$$

Hence $a = 1, -1$ ($a \neq 0$, since D is not a square) and $b = 0, 1, -1$.

If $b \neq 0$, then $c = \pm\frac{5}{4}$ is not an integer and thus is ruled out.

If $b = 0$, then $a = \pm 1$, $c = \pm 1$ and we have the forms $x^2 + y^2$, $-(x^2 + y^2)$. Since $x^2 + y^2$ represents $2 (= 1^2 + 1^2)$ and $-(x^2 + y^2) < 0$ does not, these forms are inequivalent.

Hence there are two reduced forms and correspondingly two classes.

5. Representation and Equivalence

The representation problem can be reduced to the study of equivalence.

The integer n is **properly representable** by the form $f = (a, b, c)$, if there exist $x, y \in \mathbf{Z}$, $(x, y) = 1$, such that $n = f(x, y)$. If $n = f(x, y)$ and $d = (x, y)$, then $\left(\frac{x}{d}, \frac{y}{d}\right) = 1$, $d^2 | n$ and

$$\frac{n}{d^2} = a\left(\frac{x}{d}\right)^2 + b\left(\frac{x}{d}\right)\left(\frac{y}{d}\right) + c\left(\frac{y}{d}\right)^2,$$

i.e., $\frac{n}{d^2}$ is properly representable by f. Therefore, if P is the set of integers *properly representable* by f, then $\{kd^2 \,|\, k \in P, d \in \mathbf{Z}\}$ is the set of all integers representable by f and the representation problem is reduced to finding proper representations.

If n is properly representable by f and $f \sim g$, then n is properly representable by g (exercise).

Now we characterize the integers properly representable by a form.

1) Recall if $f = (a, b, c) \sim g = (n, h, l)$ via the matrix $S = \begin{pmatrix} r & s \\ t & u \end{pmatrix}$, then $n = f(r, t)$. Since $\det S = ru - st = \pm 1$, we have $(r, t) = 1$, i.e., the representation is proper.

2) Suppose n is properly representable by f, say $n = f(r, t)$, $(r, t) = 1$. Then there exist $s, u \in \mathbf{Z}$ such that $ru - st = 1$. If we apply $S = \begin{pmatrix} r & s \\ t & u \end{pmatrix}$ to f, we get an equivalent form, where the coefficient of x^2 is n. Therefore we have proved

Theorem: *The set of integers properly representable by a form f equals the set of numbers appearing as coefficients of x^2 for all forms equivalent to f.*

6. Representations and Quadratic Residues

There is another criterion for representability, which connects with the theory of quadratic residues.

Theorem: *1) If n is properly representable by* $f = (a, b, c)$ *of discriminant D, then*

$$h^2 \equiv D(\text{mod}\, 4|n|) \tag{1}$$

is solvable, i.e., D is a quadratic residue modulo $4|n|$. *(Note that we choose* $|n|$ *so as to avoid negative moduli.)*

2) (partial converse) If $h^2 \equiv D(\text{mod}\, 4|n|)$ *is solvable, then n is properly representable by* **some** *form of discriminant D.*

Proof: 1) By the theorem of the last section, if n is properly representable by a form (a, b, c) of discriminant D, then there exists a form $(n, h, l) \sim (a, b, c)$. But equivalent forms have equal discriminants; thus $D = h^2 - 4nl$ or $h^2 \equiv D(\text{mod}\, 4|n|)$.

2) If there exists an h such that $h^2 \equiv D(\text{mod}\, 4|n|)$, then $D = h^2 - 4|n|k$, for some k. Then the form $g = |n|x^2 + hxy + ky^2$ has discriminant D and $g(1, 0) = n$ is a proper representation of n.

Corollary: *If all forms of discriminant D are equivalent, then the solvability of the congruence (1) is necessary and sufficient for n to be properly representable by the given form.*

This theorem was probably a strong motivation for Lagrange, Legendre and Gauss's interest in quadratic residues. Part 1 of the theorem generalizes our theorem about $x^2 - ay^2$ (sec. 6.4). Gauss presented part 1, with an elementary proof, as the first theorem about quadratic forms in Section V of the *Disquisitiones*. The second part of the theorem is the basis for Lagrange's study of representations in his 1773 paper "Recherches d'arithmetique" [Lag, vol. III, 695 - 795].

Using this theorem and the techniques of section 4, we now prove two of Fermat's claims made in his letter to Pascal (sec. 2. 6). Most of the other claims in the letter can be treated in a similar manner.

Example: *Sums of Two Squares*

Two Square Theorem: *Every prime* $p \equiv 1(\text{mod}\, 4)$ *can be properly represented as the sum of two squares, i.e., there exist x,y such that*

$$p = x^2 + y^2\,, \quad (x, y) = 1\,.$$

Proof: If there is a representation it must be proper since if $d = (x, y)$, then $d^2 | p$.

$x^2 + y^2$ has discriminant -4. By the previous theorem, p is properly representable by a form of discriminant -4 if the congruence $h^2 \equiv -4 (\mathrm{mod}\, 4p)$ is solvable. This congruence is solvable if $k^2 \equiv -1 (\mathrm{mod}\, p)$ is solvable, since if we let $h = 2k$, then $h^2 \equiv -4 (\mathrm{mod}\, 4p)$. But if $p \equiv 1 (\mathrm{mod}\, 4)$, then $\left(\frac{-1}{p}\right) = 1$ (sec. 11. 3) and the latter congruence is solvable.

In section 4, we saw that there are two classes of forms of discriminant -4, each containing one of the reduced forms $\pm(x^2 + y^2)$. Since $-(x^2 + y^2) < 0$, p must be properly representable by $x^2 + y^2$.

A characterization of all integers which are the sum of two squares is given by

Theorem: *Let n be a positive integer, $n = b^2 c$, c square free. Then*

$$n \text{ is a sum of two squares} \iff \text{no prime } p \equiv 3 (\mathrm{mod}\, 4) \text{ divides } c.$$

Proof: $\Longrightarrow$) If $p \equiv 3 (\mathrm{mod}\, 4)$ divides n and $n = x^2 + y^2$, then

$$x^2 + y^2 \equiv 0 (\mathrm{mod}\, p) \quad \text{or} \quad x^2 \equiv -y^2 (\mathrm{mod}\, p) .$$

If $y \not\equiv 0 (\mathrm{mod}\, p)$ then $(xy^{-1})^2 \equiv -1 (\mathrm{mod}\, p)$. But $p \equiv 3 (\mathrm{mod}\, 4)$ implies that $\left(\frac{-1}{p}\right) = -1$, which is not true (sec. 11. 3). Therefore, we have $y \equiv 0 (\mathrm{mod}\, p)$, $x \equiv 0 (\mathrm{mod}\, p)$, $p | x$, $p | y$, $p^2 | n$ and

$$\frac{n}{p^2} = \left(\frac{x}{p}\right)^2 + \left(\frac{y}{p}\right)^2 ,$$

which is a representation of $\frac{n}{p^2}$ as a sum of two squares. If $p | \frac{n}{p^2}$, then we can apply the same reasoning to $\frac{n}{p^2}$ and so on. Therefore, if $p \equiv 3 (\mathrm{mod}\, 4)$ divides n, it divides it an even number of times and cannot divide the square free factor c.

$\Longleftarrow$) The key to this proof is the identity

$$(a^2 + b^2)(c^2 + d^2) = (ac - bd)^2 + (ad + bc)^2 , \tag{2}$$

which is just another way of writing

$$|(a+bi)(c+di)|^2 = |a+bi|^2 |c+di|^2 ,$$

where $|z|$ denotes the absolute value of the complex number z. Thus we see that if two integers are each a sum of two squares, then so is their product.

Now factor n into primes,

$$n = 2^r \cdot q_1^{2s_1} \cdots q_k^{2s_k} \cdot p_1 \cdot p_m ,$$

where the p_i's are distinct primes, $p_i \equiv 1 (\bmod 4)$ and the q_i's are primes.

$2 = 1^2 + 1^2$, so, by repeated application of equation (2), 2^r = square + square. $q^{2s} = (q^s)^2 + 0^2$ and, by the two square theorem, each p_i = square + square. Thus we can use equation (2) to paste these representations together and obtain a representation of n as a sum of two squares.

Example: $x^2 + 2y^2$

Fermat claimed that if a prime $p \equiv 1, 3 (\bmod 8)$, then $p = x^2 + 2y^2$ is solvable. All such representations are proper since if $d = (x, y)$, then $d^2 | p$. We prove his claim.

The discriminant of $x^2 + 2y^2$ is -8. But p is properly representable by *some* form of discriminant -8 if $h^2 \equiv -8 (\bmod 4p)$ is solvable. This will be solvable if and only if $k^2 \equiv -2 (\bmod p)$ is solvable, i.e., if and only if $\left(\frac{-2}{p}\right) = 1$. Hence, if $\left(\frac{-2}{p}\right) = 1$, then p is properly representable by a form of discriminant -8.

A quick calculation with the inequalities (4.1) for reduced forms shows that every form of discriminant -8 is equivalent to $\pm(x^2 + 2y^2)$. But $-(x^2 + 2y^2) \leq 0$ cannot represent p. Therefore, if $\left(\frac{-2}{p}\right) = 1$, p is representable by $x^2 + 2y^2$. So we must decide when $\left(\frac{-2}{p}\right) = 1$.

Recall that $\left(\frac{-2}{p}\right) = \left(\frac{2}{p}\right)\left(\frac{-1}{p}\right)$ and

$$\left(\frac{-1}{p}\right) = 1 \iff p \equiv 1 (\bmod 4) ,$$

$$\left(\frac{2}{p}\right) = 1 \iff p \equiv 1, 7 (\bmod 8) .$$

We work mod 8:

$$p \equiv 1 (\bmod 8) \implies p \equiv 1 (\bmod 4) \implies \left(\frac{2}{p}\right)\left(\frac{-1}{p}\right) = 1 \cdot 1 = 1\,,$$

$$p \equiv 3 (\bmod 8) \implies p \equiv 3 (\bmod 4) \implies \left(\frac{2}{p}\right)\left(\frac{-1}{p}\right) = -1 \cdot -1 = 1\,,$$

$$p \equiv 5 (\bmod 8) \implies p \equiv 1 (\bmod 4) \implies \left(\frac{2}{p}\right)\left(\frac{-1}{p}\right) = -1 \cdot 1 = -1\,,$$

$$p \equiv 7 (\bmod 8) \implies p \equiv 3 (\bmod 4) \implies \left(\frac{2}{p}\right)\left(\frac{-1}{p}\right) = 1 \cdot -1 = -1\,.$$

Thus we have proved:

$$p \text{ is representable by } x^2 + 2y^2 \iff p \equiv 1, 3 (\bmod 8).$$

7. Proper Equivalence

Up to now, we have dealt primarily with Lagrange's contributions to the theory of quadratic forms. Gauss's main contribution to the more elementary part of the theory was a refinement of the notion of equivalence.

We have studied equivalence with respect to $GL_2(\mathbf{Z})$. However, in the proof that every form is equivalent to a reduced form, we used only the transformations

$$\begin{pmatrix} 0 & -1 \\ 1 & 0 \end{pmatrix}, \quad \begin{pmatrix} 1 & u \\ 0 & 1 \end{pmatrix}, \qquad u \in \mathbf{Z}\,.$$

These matrices have determinant one.

Hence, with Gauss, we say that two forms f and g are **properly equivalent**, denoted for the moment by $f \sim_p g$, if there exists an $A \in GL_2(\mathbf{Z})$ such that $g = Af$ and $\det A = 1$. Clearly $\sim_p$ is an equivalence relation. In fact, proper equivalence is equivalence under the action of the **special linear group** $\mathbf{SL_2(Z)} = \{A \in GL_2(\mathbf{Z}) | det A = 1\}$ on forms. The equivalence classes are called **proper equivalence classes.**

Exercise: Prove that a general equivalence class (i.e., under $\sim$) is either a proper equivalence class or the union of two proper equivalence classes.

As just noted, the proof of our reduction theorem actually proves

Theorem: *Every form is properly equivalent to a reduced form and the number of proper equivalence classes of forms of a fixed discriminant is finite.*

We call f and g **improperly equivalent**, denoted $f \sim_i g$, if there exists an $A \in GL_2(\mathbf{Z})$ such that $\det A = -1$ and $g = Af$.

Improper equivalence is *not* an equivalence relation. In fact,

$$f \sim_i g(g = Af)\ ,\ g \sim_i h(h = Bg) \implies f \sim_p h,$$

since $h = BAf$ and $\det BA = (-1)(-1) = 1$. Of course there may be another matrix C, $\det C = -1$, such that $h = Cf$, but in general this is not the case. Some, but not all, forms are improperly equivalent to themselves, as, e.g., $x^2 + y^2$ under the map $x = -x'$, $y = y'$.

Gauss reduced the problem of deciding equivalence of forms ($\sim$) to that of deciding proper equivalence ($\sim_p$) as follows:

1) any form $h = (a, b, c) \sim_i h' = (a, -b, c)$ under $x = -x'$, $y = y'$,
2) given two forms f and g, decide if $f \sim_p g$,
   i) if so, then $f \sim g$,
   ii) if not, decide if $f \sim_i g$ (which implies $f \sim g$) by noting that

$$f \sim_i g \iff f \sim_p g' = g(-x', y')\ .$$

Hence, in our further studies, we only consider proper equivalence and, *unless otherwise stated,* **f $\sim$ g will now denote proper equivalence**.

8. Definite and Indefinite Forms

Another distinction we shall make, due to Lagrange, concerns the sign of the discriminant.

Recall that with each form $f(x, y) = ax^2 + bxy + cy^2 = y^2\left(a\left(\frac{x}{y}\right)^2 + b\frac{x}{y} + c\right)$ of discriminant $D = b^2 - 4ac$, we can associate the quadratic polynomial $az^2 + bz + c$. Now we consider the parabola $w = az^2 + bz + c$.

1) If $D > 0$, then the equation $az^2 + bz + c = 0$ has two distinct roots and the parabola crosses the z-axis. In this case, the form $f(x, y)$

takes both positive and negative values for appropriate integer values of x and y (corresponding to *rational* values of w and z). A form with $D > 0$ is called **indefinite**.

2) If $D < 0$, then the roots of the equation are not real and the parabola lies completely above or completely below the z-axis. A form with $D < 0$ is called **definite**.
   i) If $a > 0$, then the parabola is above the axis and the corresponding form takes only positive values for $(x, y) \neq (0, 0)$. These are the **positive definite forms.**
   ii) If $a < 0$, the form takes only negative values and these are the **negative definite forms.**

We summarize with a diagram (fig. 2).

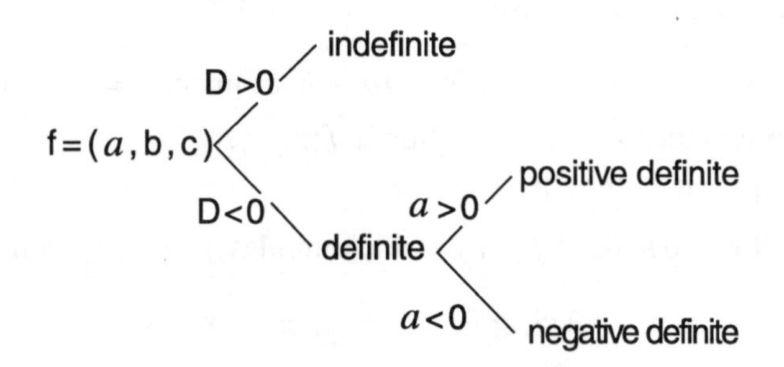

Figure 2

(Note that if $D = 0$, the parabola is tangent to the z-axis; then $f(x, y) \geq 0$ if $a > 0$ and $f(x, y) \leq 0$ if $a < 0$.)

Since the discriminant of a form is invariant under general equivalence, and therefore under proper equivalence, the property of a form being indefinite or definite is preserved under proper equivalence. By equations (3.2), the positive and negative definiteness is also preserved.

The representation and equivalence problems for definite and indefinite forms must be treated differently. A basic reason for this difference is that with definite forms we can easily choose a unique representative in each equivalence class, whereas with indefinite forms there is no such "natural" representative and we must choose a set of representatives for each class. Note also that for $f(x, y)$ a positive definite form and t a positive real number, $f(x, y) = t$ is the equation of an ellipse, whereas for indefinite forms it is the equation of a hyperbola. This is the basis for applying Minkowski's theory of the geometry of number to the study of these forms (chapter 22).

Now we will discuss definite forms, leaving a discussion of indefinite forms for the next chapter.

9. Positive Definite Forms

A positive integer n is representable by the positive definite form (a,b,c) if and only if $-n$ is representable by the negative definite form $(-a,-b,-c)$. Thus the representation problem for all definite forms is reduced to the problem for positive definite forms and we study the latter.

We have already shown that every positive definite form is *properly* equivalent to a reduced form (a, b, c) satisfying the conditions $|b| \leq |a| \leq |c|$. However there can be more than one reduced form in each equivalence class. The typical examples are

$$x^2 + xy + y^2 \quad \text{and} \quad x^2 - xy + y^2$$

which are reduced and equivalent under $x = y'$, $y = -x'$, and

$$x^2 + xy + 2y^2 \quad \text{and} \quad x^2 - xy + 2y^2$$

which are reduced and equivalent under $x = x' - y'$, $y = y'$. If we rule out this type of behavior by adding more conditions to the definition of a reduced form, we will have a unique form in each class.

If (a, b, c) is positive definite, we have $a = f(1, 0) > 0$ and $c = f(0, 1) > 0$. Hence a reduced form satisfies $|b| \leq a \leq c$, i.e.,

$$-a \leq b \leq a \leq c\,.$$

1) If $a = c$, then $(a, b, c) \sim (a, -b, c)$ by $x = y'$, $y = -x'$. To rule out one of these forms, we add the condition:

   if $a = c$, then $0 \leq b$.

2) If $a < c$, then $(a, -a, c) \sim (a, a, c)$ by $x = x' + y'$, $y = y'$. To rule out one of these forms, we add the condition:

   if $a < c$, then $-a < b$.

Adding these conditions to our definition of reduced form, we now say a **positive definite form is reduced** if either

$$0 \leq b \leq a = c \quad \text{or} \quad -a < b \leq a < c\ ,$$

and, from now on, *we mean this new definition when referring to reduced positive definite forms.*

As a corollary to our earlier reduction theorem, we have

Theorem: *Every positive definite form is (properly) equivalent to a reduced form. For each discriminant $D (< 0)$, there are only a finite number of equivalence classes of positive definite forms under proper equivalence.*

Example: In section 4, we proved that the only reduced forms (old definition) of discriminant -4 are $\pm(x^2 + y^2)$. Since $-(x^2 + y^2)$ is not positive definite and $x^2 + y^2$ satisfies condition 1), $x^2 + y^2$ is the only reduced positive definite form of discriminant -4 (new definition) .

Exercise: Show that the only reduced positive definite form of discriminant -3 is $x^2 + xy + y^2$.

Most importantly, it is now also true that

Theorem: *No two distinct reduced forms are equivalent.*

An arithmetic proof takes some work (see [Dic 2] or [Mat]). The basic idea of the proof is to find intrinsic interpretations for the coefficients of a reduced form (a, b, c). For example, it can be proved that a is the least positive integer represented by the form and therefore is unique. We shall prove the theorem geometrically in the next chapter.

Assuming the last theorem, we have solved the equivalence problem for positive definite forms. To decide if two such forms f and g are equivalent

i) transform f and g to equivalent reduced forms f' and g' (by our algorithm),

ii) $f \sim g \iff f' = g'$.

However, we have not solved the representation problem. The integer n is representable by f if and only if n appears as the coefficient of x^2 in a form equivalent to f, and we have not determined when this happens. We return to this question in the next chapter when we treat numbers of representations and automorphs of a form.

10. Primitive Forms and the Class Number

A form $f = (a, b, c)$ is **primitive** if $\gcd(a, b, c) = 1$.

Clearly, an integer n is properly representable by a form $f = (a, b, c)$, with $d = \gcd(a, b, c)$, if and only if $\frac{n}{d}$ is properly representable by the primitive form $\frac{f}{d} = \left(\frac{a}{d}, \frac{b}{d}, \frac{c}{d}\right)$. It also follows from the relation between coefficients of equivalent forms that

$$f \text{ primitive, } f \sim g \implies g \text{ is primitive .}$$

Hence the representation and equivalence problems for forms is reduced to the corresponding problems for primitive forms.

Recall that the discriminants of forms are those integers D congruent to 0 or 1 modulo 4. In each case there exists a primitive form, viz., $x^2 - \frac{D}{4}y^2$ and $x^2 + xy + \left(\frac{1-D}{4}\right)y^2$ respectively. A discriminant D is a **fundamental discriminant** if all forms of discriminant D are primitive. Such discriminants exist as we see by the following:

Theorem: *The fundamental discriminants are exactly those integers D such that*

1) $D \equiv 1 \pmod 4$, D square free, or

2) $D \equiv 0 \pmod 4$, $D = 4D'$, $D' \equiv 2$ or $3 \pmod 4$, D' square free.

Proof: a) Suppose that some D satisfying conditions 1 or 2 is the discriminant of an imprimitive form $f = (a, b, c)$, $\gcd(a, b, c) = k > 1$. Then k^2 divides D and$D$ must be of type 2), which implies $k = 2$. Then $a = 2a'$, $b = 2b'$, $c = 2c$, and $D' = \frac{D}{4} = b'^2 - 4a'c'$. But this means that $D' \equiv 0$ or $1 \pmod 4$, contradicting $D' \equiv 2$ or $3 \pmod 4$. Hence all forms of discriminant D are primitive and D is a fundamental discriminant.

b) Suppose D is a discriminant not satisfying 1) or 2). If $D \equiv 1 \pmod 4$ and not square free, then $D = D'k^2$, for some $k > 1$, and the form $\left(k, k, -k\frac{D'-1}{4}\right)$ has discriminant D and is not primitive. If $D \equiv 0 \pmod 4$, $D = 4D'$, then either D' is not square free or $D' \equiv 0{,}1 \pmod 4$. If some k^2 divides D' (which includes the case $D' \equiv 0 \pmod 4$, $\left(k, 0, -\frac{D'}{k}\right)$ has discriminant D and is not primitive. If $D' \equiv 1 \pmod 4$, the same is true for $\left(2, 2, -2\frac{D'-1}{4}\right)$. Thus D is not a fundamental discriminant.

One of the deepest and most basic quantities studied in the theory of quadratic forms is the class number. If D is a *fundamental discriminant*, then the **class number h(D)** is given by

$h(D)$ =number of *proper* equivalences classes of primitive forms (positive definite, if $D < 0$) of discriminant D.

Since primitive forms are a subset of all forms, then, by the theorem of section 7, $h(D)$ is finite.

Warning: Many writers call $h(D)$ the **strict** (or **narrow**) **class number** to indicate proper equivalence and use the term class number for general equivalence.

Dirichlet introduced analytic methods into the study of quadratic forms and in 1839 derived a formula for the class number [Ell, W - Ell, F], [Hec]. We state one case.

If $D < 0$ is a fundamental discriminant, then

$$h(D) = \frac{-w}{2}|D|^{-1/2} \sum_{n=1}^{|D|-1} n\left(\frac{D}{n}\right),$$

where

$$w = \begin{cases} 6, & \text{if } D = -3, \\ 4, & \text{if } D = -4, \\ 2, & \text{if } D < -4, \end{cases}$$

and $\left(\frac{D}{n}\right)$ is an extension of the Legendre symbol, which we will not define here. We shall see later that w is the number of proper transformations of a form of discriminant D which leave the form invariant.

A beautiful corollary of this is a formula first conjectured by Jacobi in 1832. If $D = -p$, where $p \neq 3$ is a prime $\equiv 3 (\bmod 4)$, then

$$h(-p) = \frac{B - A}{p},$$

where

A = sum of the quadratic residues mod p (between 0 and p) and
B = sum of the quadratic non residues mod p (between 0 and p).

Thus, e.g., if $D = -7$, then $A = 1 + 2 + 4 = 7$, $B = 3 + 5 + 6 = 14$, and $h(-7) = 1$, which can easily be checked from the inequalities for reduced forms.

There is no known elementary proof of this result nor even of the facts that $B - A$ is positive or that p divides $B - A$.

In article 303 of the *Disquisitiones*, Gauss conjectured that, for negative discriminants,

$$\lim_{D \to -\infty} h(D) = \infty .$$

This was proved by Heilbronn in 1933, based on earlier results of Hecke and Deuring.

Gauss also gave tables of discriminants having a given class number and, in particular, conjectured that there are only nine negative discriminants D for which $h(D) = 1$, viz.,

$$-3, -4, -7, -8, -11, -19, -43, -67, -163 .$$

Strictly speaking, since Gauss only considered the forms $ax^2 + 2bxy + cy^2$, with even middle coefficient, and defined the discriminant as $b^2 - ac$, his tables must be reinterpreted to give the conjecture as stated. This problem was first solved by Harold Stark in 1967, and shortly thereafter Alan Baker gave a different proof arising from his studies of constructive bounds for solutions of Diophantine equations (see Baker's talk upon receipt of the Fields medal [Bak]). Four years later, Baker and Stark independently proved that there are 18 negative D's for which $h(D) = 2$. The problem for class numbers $h(D) \geq 3$ remains open.

The history of the solutions to Gauss's two conjectures is quite interesting, but, since it involves a reformulation in terms of quadratic fields as well as methods of algebraic number theory, arithmetic on curves and analysis, we do not pursue it any further here. See Goldfeld [Gol, D] as well as Serre's lecture [Fla, appendix A] for more details.

On the other hand, for $D > 0$, Gauss conjectured that $\{D | h(D) = 1\}$ is infinite. Perhaps the use of the word 'conjecture' here, and often in referring to other well known conjectures, is a bit exaggerated. What Gauss says, in article 304 of the *Disquisitiones* (translation from [Gau 1]), is

> Upto the present, we cannot decide theoretically nor conjecture with any certainty of observation whether there is only a finite number of them (this hardly seems probable) or that they occur *infinitely rarely*, or that their frequency tends more and more to a fixed limit.

Whatever, conjecture or speculation, the problem remains open.

Chapter 13

Binary Quadratic Forms 2: Geometric Theory

1. Introduction

In the last chapter we studied quadratic forms using arithmetic methods. Now we turn to the geometric aspects of the theory.

H. J. S. Smith and Felix Klein gave the first geometric theory of the reduction of quadratic forms. Klein was a master of geometric ideas and a major player in the development of the theory of automorphic functions, where linear fractional transformations play a central role. He always attempted to give geometric interpretations to mathematical ideas; for example, he introduced lattices to give a geometric interpretation to the "ideal numbers" of algebraic number theory and to quadratic forms (see [Kle 3, Lecture VIII] and chapters 15 - 17). Klein's two volumes *Elementary Mathematics from an Advanced Viewpoint* [Kle 1] and his *Development of Mathematics in the 19th Century* [Kle 2] are full of insights into almost every area of 19th century mathematics.

In section 2 we discuss the key to the geometric theory: the properties of the roots of a form. This is followed by the geometric theory of the upper half planc, which gives both a new view of the reduction theory of positive definite forms and a method for counting the number of representations of an integer by a form. Finally, we round out our study of quadratic forms with a brief discussion of indefinite forms and the more advanced topics of composition and genus.

2. The Roots of a Form

Dirichlet [Dir - Ded] presented Gauss's number theoretical ideas to the mathematical community in a simplified and more motivated form. He approached the study of representation and equivalence of forms via the roots of a form. Some of these ideas were certainly known earlier (see [Wei 1] and [Wei 5]), but Dirichlet's is the first published account.

As in section 12.3, we associate to a form f of discriminant D,

$$f = ax^2 + bxy + cy^2 = y^2\left(a\left(\frac{x}{y}\right)^2 + b\frac{x}{y} + c\right),$$

the polynomial

$$P_f(z) = az^2 + bz + c\ .$$

We assume that D is not a square; thus $a \neq 0$. The roots of $P_f(z)$ are

$$r_1 = \frac{-b + \sqrt{D}}{2a}\ , \quad r_2 = \frac{-b - \sqrt{D}}{2a}\ ,$$

where $\sqrt{D}$ denotes the positive square root, if $D > 0$, and $i\sqrt{|D|}$, if $D < 0$. Thus we have

$$f(x, y) = a(x - r_1 y)(x - r_2 y)\ ,$$

where $\mathbf{r_1}$ is the **principal** or **first root** of the form and $\mathbf{r_2}$ is the **second root**. Since D is not a square, $\sqrt{D}$ is irrational and r_2 is determined by r_1.

First we examine to what extent the roots determine the form and then how the roots transform under equivalence of forms.

Theorem: *Let $f = (a, b, c)$ and $g = (a', b', c')$ with roots r_1, r_2 and r_1', r_2' respectively. If*

$$\text{disc}(f) = \text{disc}(g)(= D)\ ,$$

and

$$r_1 = r_1'\ ,$$

then

$$f = g$$

(in fact, the roots determine a form with leading coefficient one and the discriminant gives a).

Proof: Since $r_1 = r_1'$, we have $r_2 = r_2'$, i.e.,

$$\frac{-b+\sqrt{D}}{2a} = \frac{-b'+\sqrt{D}}{2a'}$$
$$\frac{-b-\sqrt{D}}{2a} = \frac{-b'-\sqrt{D}}{2a'} .$$

Subtracting the second equality from the first, we have $\frac{\sqrt{D}}{a} = \frac{\sqrt{D}}{a'}$ and $a = a'$. Hence $b = b'$ and

$$c = \frac{b^2 - D}{4a} = \frac{b'^2 - D}{4a'} = c' .$$

Now we consider equivalence of forms (sec. 12. 2). Let $f = (a, b, c) \sim g = (a', b', c')$ where

$$g = Af\ , \quad A = \begin{pmatrix} s & t \\ u & v \end{pmatrix} , \quad s, t, u, v \in Z\ , \quad \det A = \pm 1$$

($+1$ corresponds to proper equivalence ($\sim_p$) and -1 to improper equivalence ($\sim_i$)).

Then, regarding the corresponding polynomials $P_f(z)$ and $P_g(z')$ as polynomials in $z = \frac{x}{y}$ and $z' = \frac{x'}{y'}$, we have

$$z = \frac{x}{y} = \frac{sx' + ty'}{ux' + vy'} = \frac{s\frac{x'}{y'} + t}{u\frac{x'}{y'} + v}$$

or

$$z = \frac{sz' + t}{uz' + v} . \tag{1}$$

We use A to denote both the matrix and the corresponding linear fractional transformation (1). If (1) holds, we write $z = Az'$ and say that z is equivalent to z' (this is an equivalence relation).

Dirichlet's approach to equivalence of forms and representation by forms was to reduce equivalence of forms to equivalence of roots.

Theorem: *Using the above notation, let r_1, r_2 and r_1', r_2' be the roots of f and g respectively. Then*

i) $f \sim_p g, f = Ag \implies r_1 = Ar_1', r_2 = Ar_2'$,
ii) $f \sim_i g, f = Ag \implies r_1 = Ar_2', r_2 = Ar_1'$.

Proof: First we show that the roots of g map to the roots of f under the transformation A. Then we determine precisely which roots of g map to which roots of f. *Note that* $P_f(z) = f(z, 1)$.

By hypothesis, $f(sx' + ty', ux' + vy') = g(x', y')$. Let r' be one of the roots of g. Setting $x' = r'$ and $y' = 1$, we have

$$f(sr' + t, ur' + v) = g(r', 1) = P_{f'}(r') = 0\,,$$

i.e.,

$$a(sr' + t)^2 + b(sr' + t)(ur' + v) + c(ur' + v)^2 = 0\,. \tag{2}$$

We claim that $ur' + v \neq 0$. If not, $ur' + v = 0$ and, by (2), $sr' + t = 0$. So

$$\begin{aligned} sr' + t &= 0 \\ ur' + v &= 0 \end{aligned} \quad \text{or} \quad \begin{pmatrix} s & t \\ u & v \end{pmatrix} \begin{pmatrix} r' \\ 1 \end{pmatrix} = \begin{pmatrix} 0 \\ 0 \end{pmatrix},$$

which contradicts the fact that A is invertible, since $\det A \neq 0$.

Dividing both sides of (2) by $ur' + v$ yields

$$a\left(\frac{sr' + t}{ur' + v}\right)^2 + b\left(\frac{sr' + t}{ur' + v}\right) + c = 0\,;$$

so $\frac{sr'+t}{ur'+v}$ is a root of f. Hence the roots of g map to the roots of f.

Suppose that r' is a root of f corresponding to r_1, i.e., $r_1 = Ar'$ or

$$r_1 = \frac{sr' + t}{ur' + v}\,.$$

Solving for r' in terms of r_1, substituting $r_1 = \frac{-b+\sqrt{D}}{2a}$ and simplifying, we have

$$\begin{aligned} r' &= \frac{vr_1 - t}{-ur_1 + s} = \frac{v(-b + \sqrt{D}) - 2ta}{-u(-b + \sqrt{D}) + 2sa} \\ &= \frac{-vb - 2ta + v\sqrt{D}}{ub + 2sa - u\sqrt{D}}\,. \end{aligned}$$

Multiplying the numerator and denominator by $ub + 2sa + u\sqrt{D}$, multiplying out all the factors and simplifying, we have, after a somewhat lengthy computation

$$r' = \frac{-(2ast + b(sv + tu) + 2cuv) + (st - uv)\sqrt{D}}{2(as^2 + bsu + cu^2)}\,,$$

which, by the equations (12.3.2),

$$= \frac{-b' + \varepsilon\sqrt{D}}{2a'},$$

where $\varepsilon = \det A$.

If f is properly equivalent to $g(\varepsilon = +1)$, then $r' = r_1'$ and $r_1 = Ar_1'$, and if f is improperly equivalent to $g(\varepsilon = -1)$, then $r' = r_2'$ and $r_1 = Ar_2'$.

The converse of this theorem, which we only state for proper equivalence is

Theorem: *If* $\operatorname{disc}(f) = \operatorname{disc}(g)$, $\det A = \pm 1, r_1 = Ar_1'$ *and* $r_2 = Ar_2'$, *then* $f = Ag$.

Proof: (exercise: see [Hur-Kri])

Of course, by the previous theorem, it then follows that $\det A = +1$.

3. Positive Definite Forms and the Upper Half Plane

Since the roots r_1 and r_2 of a positive definite form $f = (a, b, c)$ are conjugate complex numbers ($r_2 = \overline{r_1}$), the form is determined by r_1 and its discriminant D, viz., $f = a(x - r_1 y)(x - \overline{r_1} y)$, where D determines a. We use **r** to denote the principal root. Since f is positive definite, $a = f(1, 0) > 0$. Therefore $\operatorname{Im} r = \frac{\sqrt{|D|}}{2a} > 0$ (**Re z** and **Im z** denote the real and imaginary parts of z respectively), i.e., r lies in the **upper half plane H** $= \{z \in C | \operatorname{Im} z > 0\}$ (the upper half of the complex plane - fig. 1).

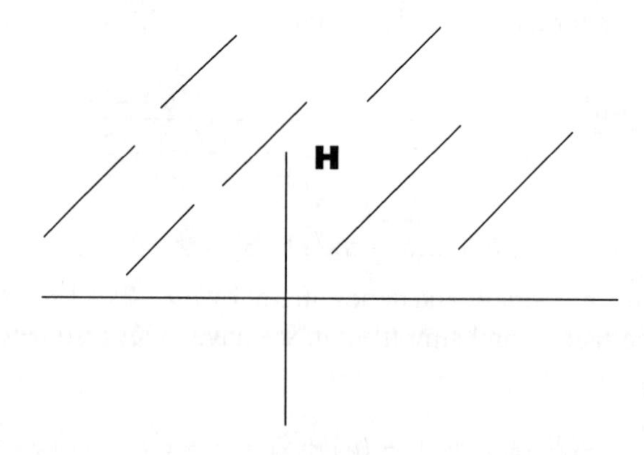

Figure 1

What additional restrictions guarantee that r is the principal root of a reduced form? Recall (sec. 12.9) that a reduced positive definite form is a form that satisfies one of the conditions

(1) $0 \leq b \leq a = c$,

(2) $-a < b \leq a < c$.

If r satisfies (1), then $0 \leq \frac{b}{2a} \leq \frac{1}{2}$ or $\frac{-1}{2} \leq \operatorname{Re} r = \frac{-b}{2a} \leq 0$.

If r satisfies (2), then $\frac{-1}{2} < \frac{b}{2a} \leq \frac{1}{2}$ or $\frac{-1}{2} \leq \operatorname{Re} r < \frac{1}{2}$.

In either case r must lie in the strip $\frac{-1}{2} \leq \operatorname{Re} z < \frac{1}{2}$ (fig. 2).

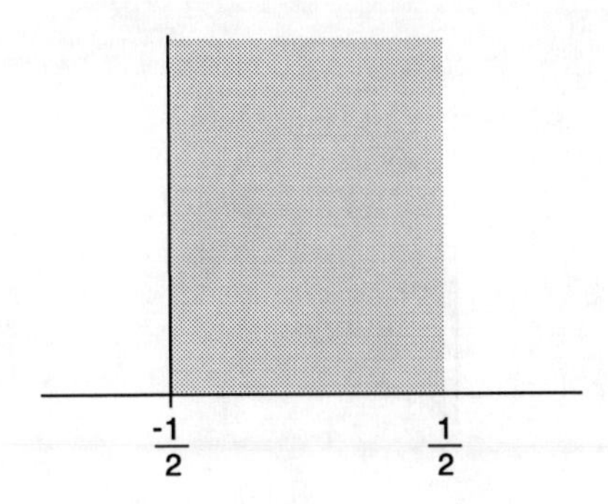

Figure 2

In both cases, taking the square of the absolute value of r, we have $|r|^2 = r\overline{r} = \frac{b^2-D}{4a^2} = \frac{4ac}{4a^2} = \frac{c}{a}$. In case (1), $\frac{c}{a} = 1$ and in case (2), $\frac{c}{a} > 1$. Thus $r \in \mathbf{H}$ must be outside the open disk $|z| < 1$ (fig. 3).

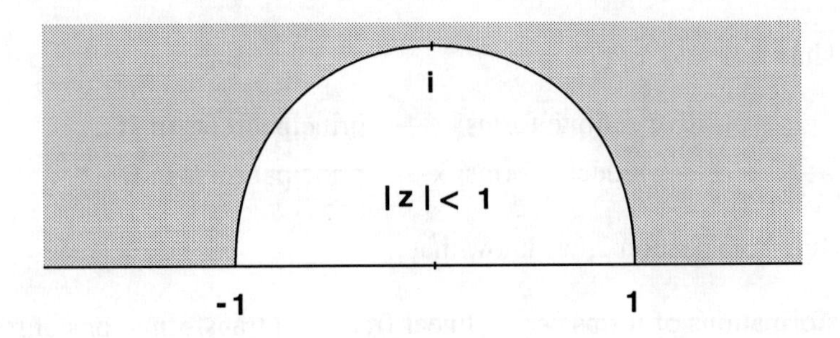

Figure 3

To summarize: the principal root r of a reduced form must satisfy one of the conditions

(1) $\frac{-1}{2} \le \operatorname{Re} r \le 0$, $|r| = 1$,

(2) $\frac{-1}{2} \le \operatorname{Re} r < \frac{1}{2}$, $|r| > 1$,

i.e., r must lie in the **modular domain D** (fig. 4) defined by

$$\mathbf{D} = \{z \in \mathbf{C} | \text{ either } -\frac{1}{2} \le \operatorname{Re} z \le 0, |z| = 1 \text{ or } \frac{-1}{2} \le \operatorname{Re} z < \frac{1}{2}, |z| > 1\}.$$

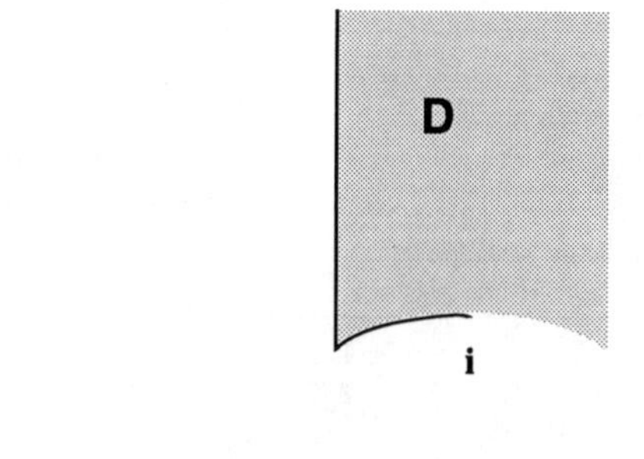

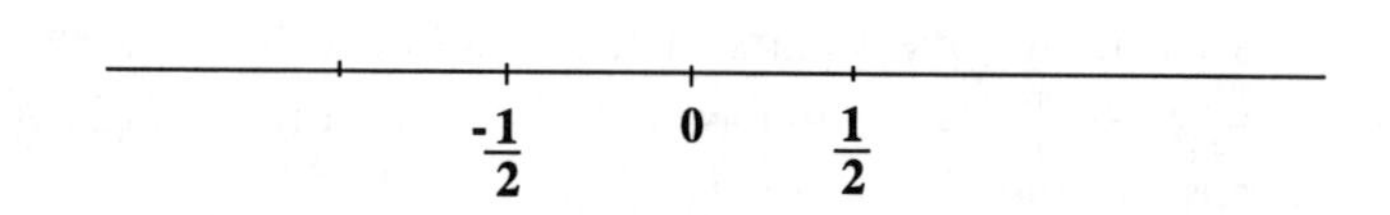

Figure 4

Thus

positive definite forms $\longleftrightarrow$ principal roots in **H** ,
reduced forms $\longleftrightarrow$ principal roots in **D** .

But, from section 2, we know that

transformations of forms $\longleftrightarrow$ linear fractional transformations of roots.

What do these transformations do to **H**?

Lemma: *If* $gz = \frac{az+b}{cz+d}$, $a, b, c, d \in \mathbf{R}$, $ad - bc = 1$, *then*

$$\mathrm{Im}(gz) = \frac{\mathrm{Im}(z)}{|cz+d|^2}\,.$$

Proof: (exercise)

In the following sections we study the geometry of the upper half plane. We show that every point in **H** can be mapped onto a unique point in **D** by a linear fractional transformation, with $a, b, c, d \in \mathbf{Z}$, $ad - bc = 1$, and that no two points in **D** can be mapped to each other by such a transformation. Translated to forms, this will imply that every positive definite form is equivalent to a unique reduced form. We will also deduce some results about forms with real coefficients.

4. Linear Fractional Transformations

Let $\overline{\mathbf{C}} = \mathbf{C} \cup \{\infty\}$, i.e., we adjoin a new symbol ∞ to the complex numbers, subject to the rules $\frac{1}{\infty} = 0$, $\frac{1}{0} = \infty$, and for $z \in C$, $z \pm \infty = \infty \pm z = \infty$. Geometrically $\overline{\mathbf{C}}$ can be modeled as a sphere (the so called "Riemann sphere") tangent to the complex plane at its south pole S, with $\overline{\mathbf{C}}$-{north pole N} bijective with **C** under projection from N (stereographic projection), and N representing ∞ (fig. 5).

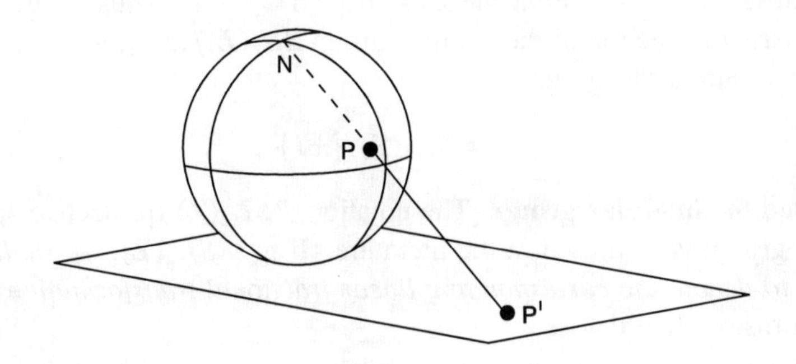

Figure 5

Recall that $SL_2(\mathbf{Z})$ denote the special linear group over **Z**, i.e.,

$$\mathbf{SL_2(Z)} = \left\{ g = \begin{pmatrix} a & b \\ c & d \end{pmatrix} \middle|\; a, b, c, d \in \mathbf{Z}\,,\, \det g = 1 \right\}.$$

(Note that in this section and the next, we use g to denote a matrix, not a form.) Each matrix g in $SL_2(\mathbf{Z})$ define a linear fractional transformation of $\overline{\mathbf{C}}$ to $\overline{\mathbf{C}}$ by

$$gz = \frac{az+b}{cz+d}, \tag{1}$$

where, since $gz = \frac{a+\frac{b}{z}}{c+\frac{d}{z}}$, we define $g\infty = \frac{a}{c}$ and $g\left(\frac{-d}{c}\right) = \infty$.

Exercise: Prove that

(1) $Iz = z$, where I is the identity matrix,

(2) $g_1(g_2z) = (g_1g_2)z$, for all g_1, g_2 in $SL_2(\mathbf{R})$ and all z in $\overline{\mathbf{C}}$,

(3) $z \mapsto gz$ is bijective on $\overline{\mathbf{C}}$ and on $\mathbf{H}$. (hint: for the latter part, use the lemma of section 3)

The first two properties state, in algebraic language, that we have a group action of $SL_2(\mathbf{Z})$ on $\overline{\mathbf{C}}$. For those familiar with projective geometry, $\overline{\mathbf{C}}$ is a model of the complex projective line, where the action of a matrix corresponds to a linear transformation of homogeneous coordinates (see chap. 18 for an introduction to projective geometry).

Since

$$gz = \frac{az+b}{cz+d} = \frac{-az-b}{-cz-d} = -gz\,,$$

the matrices g and $-g$ define the same map on $\overline{\mathbf{C}}$ and $\mathbf{H}$. Thus we are really considering the action of the quotient group $SL_2(\mathbf{Z})/\{\pm I\}$, where I is the identity matrix. The group

$$\Gamma = SL_2(\mathbf{Z})/\{\pm I\}$$

is called the **modular group**. The notation $PSL_2(\mathbf{Z})$ (projective special linear group) is also used in the literature. If $g \in SL_2(\mathbf{Z})$, *we shall also use* g *to denote the corresponding linear fractional transformation in* Γ; no confusion should arise.

5. The Fundamental Domain

We have Γ acting on $\mathbf{H}$. If $z, z' \in \mathbf{H}$, then we say that z is equivalent to z' modulo Γ, if $z = gz'$, for some g in Γ. Since we have a group action, this is an equivalence relation, and the equivalence classes are the orbits of **H** under Γ. A **fundamental domain** for Γ acting on **H** is a connected open set, with part of its boundary added to it, which is a set of distinct representatives for the equivalence classes.

Our first goal is to show that the modular domain **D** is such a fundamental domain. To do this, we introduce two special maps:

$$S = \begin{pmatrix} 0 & -1 \\ 1 & 0 \end{pmatrix} \longleftrightarrow \quad Sz = -\frac{1}{z},$$
$$T = \begin{pmatrix} 1 & 1 \\ 0 & 1 \end{pmatrix} \longleftrightarrow \quad Tz = z + 1.$$

Recall that these are the two matrices used to reduce quadratic forms (sec. 12.4). T is just a translation of **H** one unit to the right. S is inversion with respect to the unit circle ($z = re^{i\phi} \to w = \frac{e^{i\phi}}{r}$) followed by reflection through the y-axis (fig. 6). We also have the relations $S^2 = -I = I$ (in Γ) and $(ST)^3 = I$ (exercise—try to see what these relations mean geometrically).

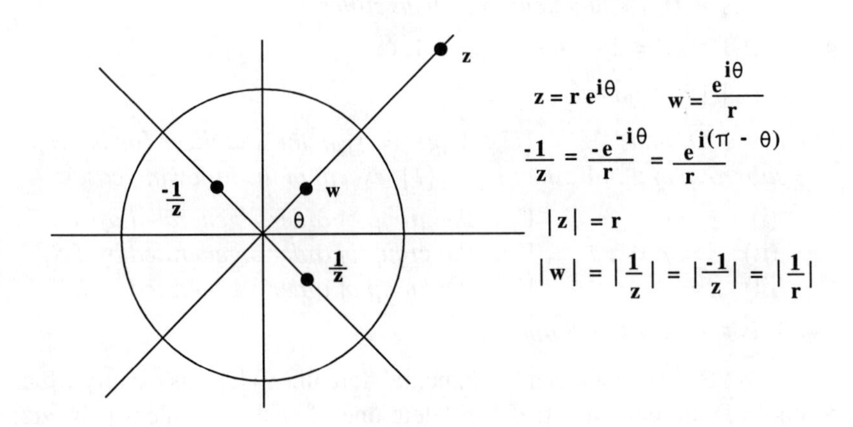

Figure 6

By adding the boundary of **D** to **D** we get $\overline{\mathbf{D}} = \{z \mid |z| \geq 1, |Re(z)| \leq \frac{1}{2}\}$ (fig. 7).

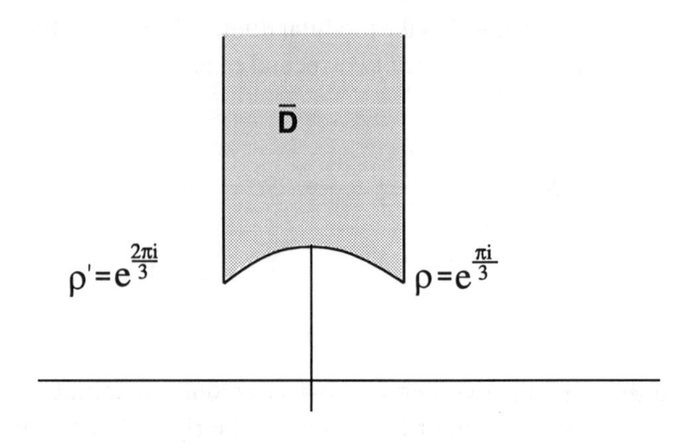

Figure 7

Now we determine a fundamental domain for Γ, following the exposition of Serre [Ser].

Theorem:

1) *For all z in **H**, there exists a g in Γ such that $gz \in \overline{\mathbf{D}}$,*
2) *if $z, z' \in \overline{\mathbf{D}}$, z equivalent to z', then either*
   (i) $\mathrm{Re}(z) = \pm\frac{1}{2}$ *and* $z = z' + 1$, or
   (ii) $|z| = 1$ *and* $z' = -\frac{1}{z}$,
3) *if $z \in \overline{\mathbf{D}}$, and $\Gamma z = \{g \in \Gamma | gz = z\}$ is the stabilizer (or isotropy) subgroup of z in Γ, then $\Gamma z = \{I\}$, except in the following cases:*
   (i) $z = i$, Γ_z *is the group of order 2 generated by S,*
   (ii) $z = \rho = e^{\frac{\pi i}{3}}$, Γ_z *is the group of order 3 generated by TS,*
   (iii) $z = \rho' = e^{\frac{2\pi i}{3}}$, Γ_z *is the group of order 3 generated by ST,*
4) *Γ is generated by S and T.*

1) says that $\overline{\mathbf{D}}$ contains a fundamental domain. 2) tells us exactly which points in $\overline{\mathbf{D}}$ are equivalent; if we delete one of each equivalent pair, viz., those points of $\overline{\mathbf{D}}$ on $\mathrm{Re}(z) = \frac{1}{2}$ or on the right hand side of the circular arc, then the resulting set **D** is a fundamental domain. 3) and 4) tell us about the structure of Γ. We will use 3) in section 7.

Proof: 1) The modular domain **D** was defined by the conditions satisfied by the principal root of a reduced form. We have seen that the action of $SL_2(\mathbf{Z})$ on forms corresponds to a transformation of the roots, which is just the action of $SL_2(\mathbf{Z})$ on **H**. Therefore by imitating the procedure used in section 12.4, for the reduction of forms, in the action of Γ on **H**, we have a procedure for moving a point into $\overline{\mathbf{D}}$.

Starting with any $z^{(0)}$ in **H**, we proceed as in figure 8.

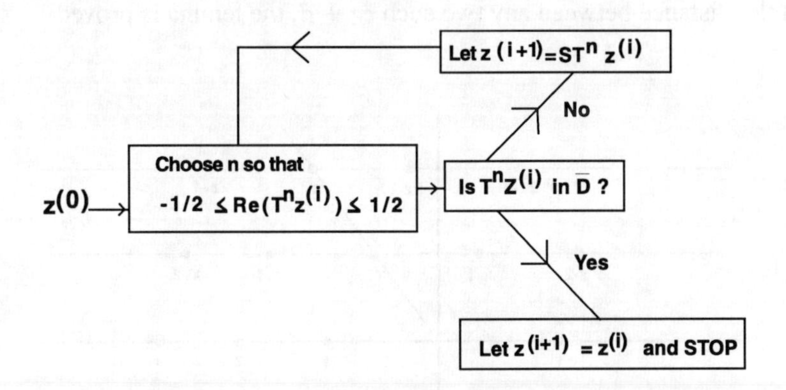

Figure 8

To prove 1) we show that the sequence $z^{(0)}, z^{(1)}, \ldots$ is finite, i.e., eventually stops at a point in $\overline{\mathbf{D}}$. Since we are dealing with all points in **H**, not just those corresponding to roots of forms with integer coefficients, the proof that our algorithm stops at a point in $\overline{\mathbf{D}}$ is different from the one for forms.

Note that each $z^{(i)} = g_i z^{(0)}$, where g is a product of powers of S's and T's. Therefore g_i corresponds to a matrix with integer entries. First, we show that

$$\max_i \operatorname{Im}(g_i z^{(0)}) \tag{1}$$

exists and is realized for some i; we then use this fact to show that the sequence is finite. Let $g_i = \frac{a_i z+b_i}{c_i z+d_i}$ and recall that $\operatorname{Im}(g_i z^{(0)}) = \operatorname{Im} \frac{z^{(0)}}{|c_i z+d_i|^2}$. Thus we can prove equivalently that

$$\min_i |c_i z^{(0)} + d_i|^2 \tag{2}$$

exists and is realized for some i. (Note that we cannot have $c_i = d_i = 0$, since $\det = 1$). We use the following lemma:

Lemma: *For any fixed z in* **H**, *the set $L_K = \{cz+d | c, d \in \mathbf{Z}, |cz+d| < K\}$ is finite for any $K > 0$, i.e., there are only a finite number of such $cz + d$ in any circle centered at the origin.*

Proof: L_K is a subset of all integer linear combinations of 1 and z (see the 'lattice' in fig. 9). Any two points of L_K on a horizontal line have distance > 1. Any two points on the same or different vertical lines have distance $> \mathrm{Im}(z)$, since $\mathrm{Im}(z)$ is the distance between successive horizontal lines, and since z is in **H**, $\mathrm{Im}(z) > 0$. Since there is a positive lower bound on the distance between any two such $cz + d$, the lemma is proved.

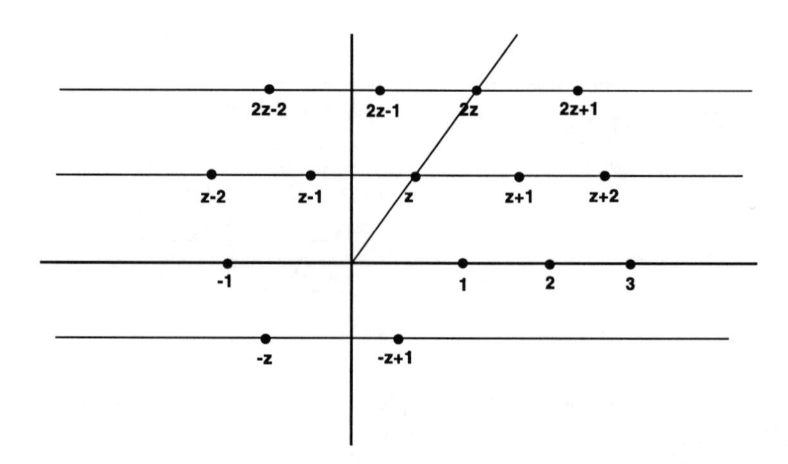

Figure 9

To prove that (2) achieves its minimum for some i and therefore (1) achieves its maximum, let $z = z^{(0)}$ and $K = 2|c_1 z^{(0)} + d_1|$ in the lemma. Then L_K contains $c_1 z^{(0)} + d_1$. By the lemma, L_K has a finite intersection with $T = \{c_i z^{(0)} + d_i\}$; therefore $\min_i |c_i z^{(0)} + d_i|$ is achieved on T.

Now assume that for some $z^{(0)}$, the sequence $z^{(0)}, z^{(1)}, \cdots$ is infinite, i.e., no point in the sequence is in $\overline{\mathbf{D}}$, and let $z^{(k)}$ be the point such that $\mathrm{Im}(z^{(k)}) = \max_i \mathrm{Im}(z^{(i)})$. Then $z^{(k+1)} = ST^n z^{(k)}$ for some n, where $-\frac{1}{2} < \mathrm{Re}(T^n z^{(k)}) < \frac{1}{2}$. But, by assumption, $T^n z^{(k)}$ is not in $\overline{\mathbf{D}}$ and thus $|T^n z(k)| < 1$. Since $\mathrm{Im}(Tz) = \mathrm{Im}(z)$, applying S reciprocates distances from the origin and we have

$$\mathrm{Im}(z^{(k+1)}) = \mathrm{Im}(ST^n z^{(k)}) > \mathrm{Im}(T^n z^{(k)}) = \mathrm{Im}(z^{(k)}) ,$$

contradicting the maximality of $\mathrm{Im}(z^{(k)})$. Hence the sequence $z^{(0)}, z^{(1)}, \cdots$ must stop at some point in $\overline{\mathbf{D}}$.

2) & 3) Let $z, gz \in \overline{\mathbf{D}}$, where $g = \begin{pmatrix} a & b \\ c & d \end{pmatrix}$. We show that this assumption strongly limits the possible values of g.

We can assume that $\mathrm{Im}(gz) > \mathrm{Im}(z)$; if not, replace (z, g) by (gz, g^{-1}) as the initial conditions. Since $\mathrm{Im}(gz) = \dfrac{\mathrm{Im}(z)}{|cz+d|^2} > \mathrm{Im}(z)$,

$$|cz+d| \leq 1 . \tag{3}$$

Suppose $|c| > 1$. Then, since $\mathrm{Im}(z) > \frac{1}{2}$, for $z \in \overline{\mathbf{D}}$,

$$|\,\mathrm{Im}(cz+d)| = |\,\mathrm{Im}(cz)| = |c|\,|\,\mathrm{Im}(z)| \geq 2\,\mathrm{Im}(z) \geq 1 .$$

But, by (3), $|\,\mathrm{Im}(cz+d)| \leq |cz+d| \leq 1$, a contradiction. *Hence* $|c| < 1$.

(I) $c = 0$: Then $|d| < 1$. But $d = 0$ contradicts $\det g = ad - bc = 1$; so $d = \pm 1$ and $ad - bc = a(\pm 1) - b(0) = 1$ implies that $a = d = \pm 1$. Therefore

$$g = \frac{az+b}{d} = z + e , \quad \text{a translation} .$$

If $|e| > 1$, then $z \in \overline{\mathbf{D}} \implies z + e \in \overline{\mathbf{D}}$; thus $|e| = 0, 1$.

$|e| = 0 \implies g = I$,

$|e| = 1 \implies g = z \pm 1$ and the only pairs (z, gz) for which $z, gz \in \overline{\mathbf{D}}$

are those for which $\mathrm{Re}(z) = \pm\frac{1}{2}$.

(II) $c = 1$: Then $|z + d| \leq 1$. But $|d| > 1$ implies $|z + d| > 1$; so $d = 0, 1, -1$.

i) $d = 0$: $ad - bc = a(0) - b(1) = 1 \implies b = -1$; hence $gz = a - \frac{1}{z}$. $|z+d| \leq 1 \implies |z| \leq 1$, and $z \in \overline{\mathbf{D}} \implies |z| = 1$; hence $|z| = 1$, i.e., z is on the circular arc. Then, if $|a| > 1$, $gz = a - \frac{1}{z}$ is not in $\overline{\mathbf{D}}$; hence $a = 0, 1, -1$.

$a = 0$: $gz = -\frac{1}{z}$; z and $-\frac{1}{z}$ are both on the circular arc (symmetric with respect to the y-axis).

$a = 1$: $gz = 1 - \frac{1}{z}$; $\rho \to \rho$, all other points in $\overline{\mathbf{D}}$ map outside $\overline{\mathbf{D}}$.

$a = -1$: $gz = -1 - \frac{1}{z}$; $\rho \to \rho$, all other points in $\overline{\mathbf{D}}$ map outside $\overline{\mathbf{D}}$.

ii) $d = 1 : |z+1| \leq 1$. From the diagram of $\overline{\mathbf{D}}$, it is easy to see that ρ' is the only z in $\overline{\mathbf{D}}$ such that $z+1 \in \overline{\mathbf{D}}$ and $|z+1| \leq 1$. We have $ad - bc = 1$ and $c = d = 1$ which implies $a - b = 1$. Thus $gz = \frac{az+(a-1)}{z+1} = \frac{a(z+1)-1}{z+1} = a - \frac{1}{z+1}$ and

$$g\rho' = a - \frac{1}{\rho' + 1} \in \overline{\mathbf{D}} \iff a = 0, 1 \, .$$

$a = 0$: $gz = \frac{-1}{z+1}$ fixes ρ'

$a = 1$: $gz = \frac{z}{z+1}$ pairs ρ and ρ' and fixes nothing in $\overline{\mathbf{D}}$.

iii) $d = -1$: The same procedure as for $d = 1$ yields

$gz = \frac{-1}{z-1}$ fixes ρ,

$gz = \frac{-z}{z-1}$ pairs ρ and ρ' and fixes nothing in $\overline{\mathbf{D}}$.

(III) $c = -1$: Since $\frac{az+b}{cz+d} = \frac{-az-b}{-cz-d}$, this reduces to the case $c = 1$.

We summarize these results in the following table.

g	gz	fixed points in $\overline{\mathbf{D}}$	pairs z, gz in $\overline{\mathbf{D}}$, $z \neq gz$
I	z	all	none
$T^{\pm 1}$	$z \pm 1$	none	$z = \frac{-1}{2} + si$ with $z' = \frac{1}{2} + si$
S	$\frac{-1}{z}$	i	$z, \frac{-1}{z}$ with $\lvert z\rvert = 1$
TS	$1 - \frac{1}{z}$	ρ	none
$(ST)^2$	$-1 - \frac{1}{z}$	ρ'	none
ST	$\frac{-1}{z+1}$	ρ'	none
	$\frac{z}{z+1}$	none	ρ with ρ'
$(TS)^2$	$\frac{-1}{z-1}$	ρ	none
	$\frac{-z}{z-1}$	none	ρ with ρ'

From this table, we can read off the results of parts 2) and 3).

4) Γ is generated by S and T: Let w be a point in the interior of $\overline{\mathbf{D}}$. Let $g \in \Gamma$ and $z' = g(w)$. Then, by the proof of part 1), there is a g' in Γ such that $g'z' \in \overline{\mathbf{D}}$, where g' is a product powers of S's and T's. Then $g'g(w) = w$. By part 3), $g'g = I$ and $g = g'^{-1}$ is a product of powers of S's and T's (with negative exponents allowed) and our proof is done.

By applying any g in Γ to $\mathbf{D}$ we get a new fundamental domain $g(\mathbf{D})$ for Γ. This set $\{g(\mathbf{D})|g \in T\}$ forms a **tessellation** of $\mathbf{H}$ (a decomposition of $\mathbf{H}$ into disjoint sets equivalent with respect to Γ - see fig. 10), where the domains $g(\mathbf{D})$ are denoted by g expressed as a product of powers of S's and T's.

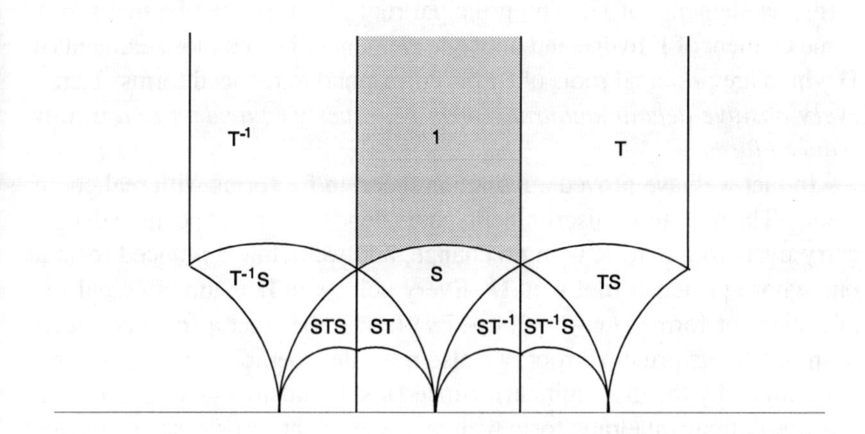

Figure 10

To obtain the figure it is necessary to prove that linear fractional transformations map generalized circles (circles or straight lines) to generalized circles. For elements of Γ, this only needs to be verified for T (obvious) and S (takes a little calculation), since they generate Γ. In particular, S maps lines not passing through 0 to circles passing through 0 (and vice versa), and it maps all other circles to circles.

Using this result, we derive the figure by regarding each $g(\mathbf{D})$ as a generalized triangle in $\mathbf{H} \cup \{\infty\}$ and find the image of each of the vertices of the triangle. Thus, e.g., to find $S(\mathbf{D})$ (labeled S in the figure), we note that S maps the three vertices of $\mathbf{D}$ as follows:

$$\infty \to -\frac{1}{\infty} = 0\,, \quad \rho = e^{\frac{\pi i}{3}} \to -e^{\frac{-\pi i}{3}} = e^{\frac{2\pi i}{3}} = \rho'\,, \quad \rho' \to \rho\,.$$

General linear fractional transformations are treated in most introductory books on complex variables. Sansone and Gerritsen [San-Ger, Vol. 2] present a very beautiful development and Siegel [Sie 1] is also recommended. Rademacher [Rad 1] gives a nice treatment using only elementary geometric arguments.

6. Forms and the Upper Half Plane Revisited

We now continue our discussion from section 3 of roots of a positive definite forms as elements of $\mathbf{H}$. The principal root of a form can be mapped by some element of Γ to one and only one element of $\mathbf{D}$, and those elements of $\mathbf{D}$ which are principal roots of forms correspond to reduced forms. Hence, *every positive definite quadratic form is properly equivalent to a* **unique** *reduced form.*

In fact we have proved a reduction theorem for forms with real coefficients. The notions of discriminant, equivalence, roots and positive definite carry over from $\mathbf{Z}$ to $\mathbf{R}$ with no change, and we define a reduced form as one whose principal root is in $\mathbf{D}$. Every point r in $\mathbf{H}$ is the principal root of a class of forms $t(x - ry)(x - \overline{r}y)$ (remember that a form is not determined by its principal root; we also need the coefficient of x^2, which is determined by the discriminant). Since $\mathbf{D}$ is a fundamental domain, every positive definite quadratic form with real coefficients is properly equivalent to a unique reduced form.

The modular group Γ plays a central role in the study of modular and automorphic functions and their applications in many areas of number theory. These ideas were first introduced by Jacobi with some early hints, as usual, by Euler. For the historical development see [Hou] and for more modern treatments [Shi], [Ser] and [Apo 1]. We present one example from this range of ideas.

Let $Q(x_1, x_2, \ldots, x_n) = \sum_{i,j} c_{i,j} x_i x_j$ be a positive definite ($Q > 0$, for all integer values of the variables) quadratic form with integer coefficients and let

$$F(z, Q) = \sum e^{2\pi i z Q(x_1,\ldots,x_n)}\,,$$

where the latter summation is over all integer values of $x_1, \ldots, x_n$. This type of series is called a theta series (or function). If $A(n)$ is the number of integer solutions of $Q(x_1, x_2, \ldots, x_n) = n$, then

$$F(z, Q) = \sum_{n=0}^{\infty} A(n) e^{2\pi i n z} .$$

It can be shown that

$$F\left(\frac{az+b}{cz+d}, Q\right) = \epsilon(d)(cz+d)^k F(z, Q) ,$$

where $a, b, c, d \in \mathbf{Z}$, $ad - bc = 1$, $c \equiv 0 (\operatorname{mod} N)$, N a natural number determined by Q and $\epsilon(d)$ is a "character mod N", i.e., F is a "modular form of weight $2k$ with respect to the congruence subgroup of Γ of level N".

Using analytic methods, Hecke [Hec] proved that

$$A(n) = A_0(n) + 0\left(n^{\frac{k}{2}}\right) ,$$

Where $A_0(n)$ is a number theoretic function determined by the "genus" of Q (genus is a rougher classification than equivalence classes — we will discuss it for binary forms in section 10).

$\mathbf{H}$ and Γ also provide a model for hyperbolic non Euclidean geometry [San-Ger, Vol. 2]. Generalizations of these ideas to three dimensions (upper half spaces) play a central role in Thurston's study of the classification of three manifolds (see [Thu], [Lan 2] and [Wee]).

As mentioned at the beginning of the chapter, there is also a geometric interpretation of quadratic forms due to Felix Klein, which uses lattices in the plane. The connection between 'reduced bases' of a lattice (which correspond to reduced forms) and our upper half plane interpretation is treated in Borevich and Shafarevich [Bor-Sha].

7. Automorphs and the Number of Representations

To count the number of representations of an integer by a form $f(x, y)$, we must find the automorphs of f, namely those $A \in SL_2(\mathbf{Z})$ such that $Af = f$. Again, we concentrate on positive definite forms.

Note that the maps in Γ corresponding to an automorph of f are those leaving the principal root r of f fixed, i.e., *the automorphs of f are the* *matrices corresponding to the stabilizer subgroup,* Γ_z *of* Γ. But we found the stabilizer subgroups of all points in $\overline{\mathbf{D}}$ in section 5, and we can translate these results to determine the automorphs of reduced positive definite forms.

1) The automorphs of the reduced forms $a(x^2+y^2)$, $a > 0$, with principal root i, are

$$\pm I, \pm S = \begin{pmatrix} 0 & \pm 1 \\ \mp 1 & 0 \end{pmatrix}.$$

2) The automorphs of the reduced forms $a(x^2 + xy + y^2)$, $a > 0$, with principal root $\rho' = e^{\frac{2\pi i}{3}}$, are

$$\pm I, \pm ST = \begin{pmatrix} 0 & \mp 1 \\ \pm 1 & \pm 1 \end{pmatrix}, \quad \pm (ST)^2 = \begin{pmatrix} \mp 1 & \mp 1 \\ \pm 1 & 0 \end{pmatrix}.$$

3) The automorphs of any other reduced positive definite forms are $\pm I$.

From these results we can find the automorphs of all forms with the help of the following lemma which relates automorphs of equivalent forms.

Lemma: *If* $f \sim g$*, then there are bijections between the sets* $S_f =$ {*automorphs of* f}, $S_g =$ {*automorphs of* g}, $S_{f,g} = \{A \in SL_2(\mathbf{Z}) | Af = g\}$*, and* $S_{g,f} = \{A \in SL_2(\mathbf{Z}) | Ag = f\}$. *In particular, if the sets are finite, then*

$$|S_f| = |S_g| = |S_{f,g}| = |S_{g,f}|.$$

Proof: The map $A \to A^{-1}$ is easily seen to be a bijection between $S_{f,g}$ and $S_{g,f}$. Hence, if we construct a bijection between S_f and $S_{f,g}$, then by symmetry there is a bijection between S_g and $S_{g,f}$.

Fix $B \in S_{f,g}$. Then

$$A \in S_f \implies BAf = g \implies BA \in S_{f,g},$$

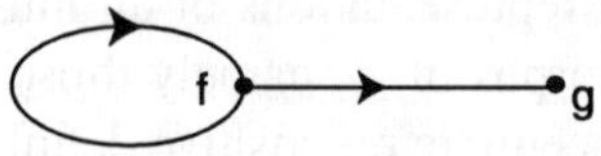

Figure 11

(fig. 11) and we have a map $F : S_f \to S_{f,g}$ given by $A \mapsto BA$. Since all maps in the S's are invertible, $BA = BA' \implies A = A'$, and F is one to one. If $C \in S_{f,g}$, then $C = B(B^{-1}C)$, where $B^{-1}C \in S_f$, and F is onto and thus a bijection.

Furthermore, since for any $E \in S_g$, $E = B(B^{-1}EB)B^{-1}$, $B^{-1}EB \in S_f$, we have $S_g = \{BAB^{-1} | A \in S_f\}$.

This is really a theorem about groups acting on sets and our proof made no use of the special action of matrices on forms.

Now we count automorphs. Recall that a form is primitive if its coefficients are relatively prime. Let

w(f) =the number of automorphs of the primitive positive definite quadratic form f.

Recall that equivalent forms have the same discriminant D, and the only reduced primitive forms of discriminants -3 and -4 are $x^2 + xy + y^2$ and $x^2 + y^2$ repectively (sec. 12.9). Combining all the previous results of this section, we have

Theorem:

$$w(f) = \begin{cases} 6 & \textit{if } D = -3\ , \\ 4 & \textit{if } D = -4\ , \\ 2 & \textit{if } D < -4\ . \end{cases}$$

Remarks: (1) We restrict to primitive forms to get a clean statement. Thus, e.g., $2(x^2+y^2)$ is a imprimitive form of discriminant -8 with 4 automorphs.

(2) $w(f)$, which is just a function of the discriminant of f, is the same w which appeared in the Dirichlet class number formula (sec. 12.10).

Now we can count representations. Recall that we proved (sec. 12.6):

A) if $n > 0$ is properly representable by a form f of discriminant D, then there exists a form $(n, h, k) \sim f$ such that

$$h^2 \equiv D(\bmod 4n)\ ,$$

B) if $h^2 \equiv D(\bmod 4n)$ is solvable, then n is representable by some form of discriminant D.

Furthermore, since the map $x = X + uY$, $y = Y$, yields

$$(n, h, k) \sim (n, h + 2un, k') ,$$

we can always choose the middle coefficient to be between 0 and $2n - 1$. A solution of

$$x^2 \equiv D(\text{mod}\, 4n) , \quad 0 < x < 2n ,$$

is called a *minimum root of the congruence.*

A careful analysis and refinement of the proofs of A and B (see [Lev, Vol. 2], [Mat] and [Dic 1]) leads to

Theorem: *Let f be a primitive positive definite quadratic form and $n > 0$. For each minimum root h of $h^2 \equiv D(\text{mod}\, 4n)$, let $k = \frac{h^2 - D}{4n}$. Then (n, h, k) is a form of discriminant D and*

the number of proper representations of n by f

$$= w(f) \cdot \text{(the number of such forms } (n, h, k) \text{ equivalent to } f) .$$

Corollary: *If the class number $h(D) = 1$, then every form (n, h, k) is equivalent to f and the number of proper representations of n by f*

$$= w(f) \cdot \text{(the number of minimum roots of } x^2 \equiv D(\text{mod}\, 4n)) .$$

Example: Sums of Two Squares

Let $f = x^2 + y^2$; so $D = -4$, $w(f) = 4$. If $h^2 \equiv -4(\text{mod}\, 4n)$, then h is even; set $h = 2h'$. Then h is a minimum root if and only if $h'^2 \equiv -1(\text{mod}\, n)$, $0 < h < n$, i.e., the number of minimum roots of $x^2 \equiv -4(\text{mod}\, 4n)$ equals the number of roots of $x^2 \equiv -1(\text{mod}\, n)$, $0 < x < n$.

Recall that if n is a prime $\equiv 1(\text{mod}\, 4)$, then n can be represented as the sum of two squares and every representation is proper (sec. 12.6). In this case, the number of minimum roots is 2 and therefore there are 8 representations. If (r, s) is a representation, then

$$(\pm r, \pm s) , \quad (\pm r, \mp s) , \quad (\pm s, \pm r) , \quad (\pm s, \mp r)$$

are all the representations. This justifies Fermat's statement that the representation is "essentially" unique.

8. Indefinite Forms, $D > 0$

The theory of indefinite forms is deeper and less complete than the theory of positive definite forms. Later, when we discuss the relation between forms and fields, we shall see that this translates a larger degree of ignorance about the structure of real quadratic fields as compared with the structure of imaginary quadratic fields. Now we present, without great precision or proofs, some of the basic results about indefinite forms. Some good references are [Mat], [Dic 2], [Zag], [Lev, Vol. 2] and [Hur-Kri].

Reduction

In the last chapter, we proved that there are only a finite number of classes of forms with discriminant D, under proper (or general) equivalence. This followed from the fact that every form is equivalent to some reduced form (a form (a, b, c) satisfying $|b| < |a| < |c|$) and that there are only finitely many reduced forms of discriminant D.

For positive definite forms, we refined the notion of a reduced form and showed that there is a unique reduced form in each class. However, for indefinite forms, although there are several notions of reduced forms which lead to fruitful developments, each of them yields many reduced forms in each equivalence class and there is no "natural" way to pick a unique reduced form (this can be made precise).

Using one of these definitions of a reduced indefinite form, given by Gauss, the reduction theory can be developed via the continued fraction expansion of the principal root of the form (which is a real number). In fact, in many ways the reduction theory of indefinite forms is equivalent to the theory of continued fractions of real quadratic numbers (in [Hur - Kri], they are treated as essentially one theory).

If we expand a form f with principal root r into a finite continued fraction then

$$r = \frac{p_n s_{n+1} + p_{n-1}}{q_n s_{n+1} + q_{n-1}}, \tag{1}$$

where $p_n q_{n-1} - q_n p_{n-1} = (-1)^{n-1}$, which equals 1 if n is odd (sec. 4.4). Therefore, for n odd, f is equivalent to a form g with root s_{n+1}. For n sufficiently large, g can be shown to be reduced in Gauss's sense. The s_i's corresponding to those b_i's in a period in the continued fraction will determine a set of reduced forms for f. The hard part is to prove that all reduced forms equivalent to f are determined in this way [Hur - Kri].

It is technically cleaner to use the negative continued fractions (sec. 4.10), since then $p_n q_{n-1} - q_n p_{n-1} = 1$, for all n [Zag].

Automorphs and Representations

The automorphs of a primitive indefinite form f, $\text{disc}(f) = D$, are determined by the solutions of the Pell type equations

$$x^2 - Dy^2 = 1 ,$$
$$x^2 - Dy^2 = 4 .$$

There are infinitely many automorphs of f and they form a cyclic group (recall that the solutions (x_n, y_n) of Pell's equation form a cyclic group, given by $x_n + y_n\sqrt{D} = (x_0 + \sqrt{y_0})^n$, where (x_0, y_0) is the minimal solution - sec. 4.9).

Correspondingly, a number n represented by f has infinitely many equivalent representations obtained by applying each automorph to the given representation. Lagrange introduced order into this theory by specifying how to choose one representation from each equivalence class, a primary representation, and proving:

(1) the number of primary representations of n by f is finite, and

(2) applying the automorphs to the primary representations yields all representations.

Geometric Methods

By adjoining the real axis to the upper half plane **H**, and representing an indefinite form with roots r_1, r_2 by a semicircle with endpoints r_1, r_2 on the real line (fig. 12), one can use the action of Γ and the modular domain **D** to study the reduction of indefinite forms (see [Mat] and [Lev, Vol. 2]).

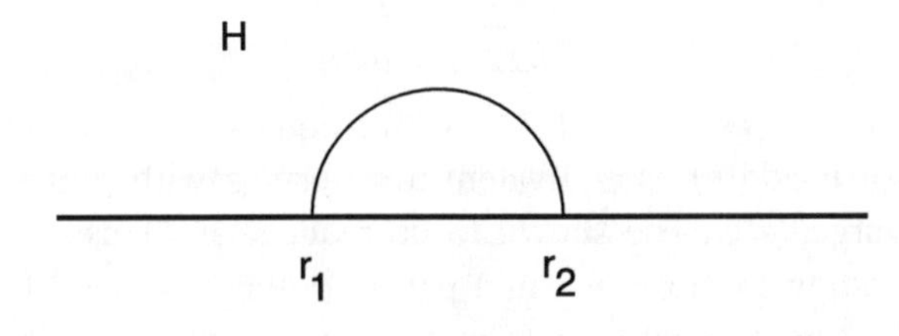

Figure 12

H. J. S. Smith [Smi, H 2], Hurwitz [Hur 1] and Humbert [Hum] replaced the tessellation of **H** by **D** and its images under Γ with the tessellation by $T = \{z = x + iy \in \mathbf{H} | 0 < x < 1, (x-1)^2 + y^2 > 1\}$ (fig. 13)

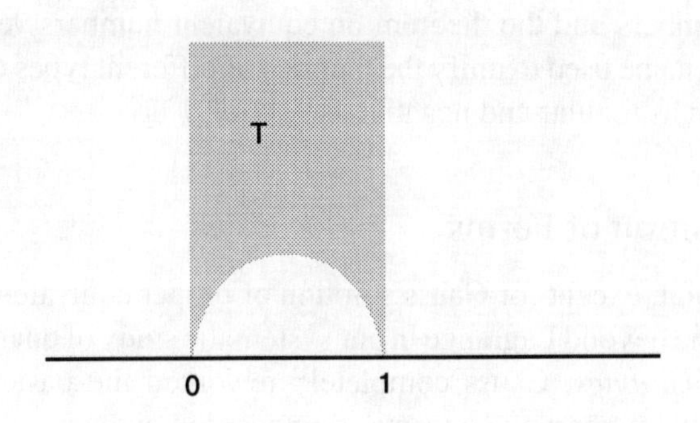

Figure 13

and its images under Γ (fig. 14 — actually, Hurwitz used an equivalent model in the projective plane).

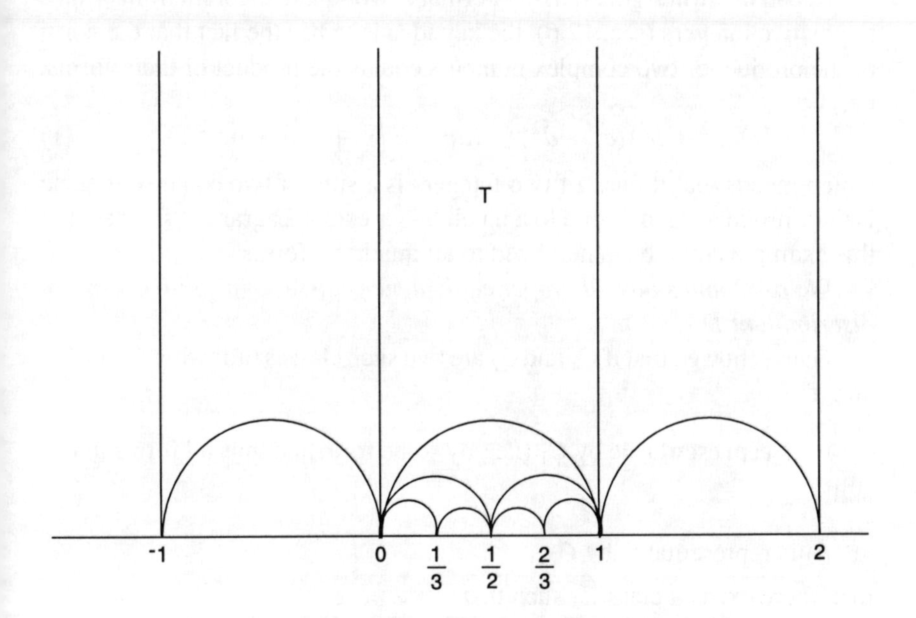

Figure 14

With this tessellation one can treat several reduction theories of indefinite forms and see the relations between them. Humbert gives a very thorough development. He also develops a geometric theory of continued fractions, which makes theorems such as the periodicity of the expansion of quadratic numbers and the theorem on equivalent numbers very intuitive. These ideas can be used to unify the theories of different types of continued fraction, such as regular and negative (see [Gol, J 1]).

9. Composition of Forms

Up to this point, except for Gauss's notion of proper equivalence, we have not really gone beyond Lagrange in our systematic study of quadratic forms. In the *Disquisitiones*, Gauss completely reworked the basic theory and introduced two major new concepts: composition and genus. He devoted a substantial part of the *Disquisitiones* to the study of these ideas.

We start with composition. As with many new ideas, this concept did not arise in a sudden flash of inspiration, but has an interesting history (see [Wei 1] and [Wei 5]).

Recall that when generalizing Fermat's two square theorem from primes to positive integers (sec. 12.6), the key idea is to use the fact that the norm of the product of two complex numbers equals the product of their norms, i.e.,

$$(a^2+b^2)(c^2+d^2) = (ac-bd)^2 + (ad+bc)^2\ , \tag{1}$$

which means that if each of two integers is a sum of two squares, then so is their product. Gauss (and to a much lesser extent Lagrange) showed that this example could be generalized to all quadratic forms.

We fix D and study all proper equivalence classes of primitive forms of discriminant D.

Gauss showed that if C_1 and C_2 are two such classes (allowing $C_1 = C_2$) and if

n_1 is **representable** by C_1 (i.e., by some form and thus all forms in C_1)

and

n_2 is representable by C_2,

then there exists a class C_3 such that

$n_1 n_2$ is representable by C_3.

Moreover, if $f_1(x_1, y_1) \in C_1, f_2(x_2, y_2) \in C_2$, then

$$f_1(x_1, y_1) f_2(x_1, y_2) = f_3(x_3, y_3) , \tag{2}$$

where f_3 is a quadratic form in C_3 and x_3, y_3 are bilinear functions of x_1, y_1, x_2, y_2.

This leads to a composition law for classes ($C_1 C_2 = C_3$), which Gauss called composition of forms. Although Gauss did not have the notion of an abstract group, he proved, in a more concrete language, that the classes are an Abelian group under composition and that this group is a direct product of cyclic groups. This group, the *class group*, which in the correspondence between forms and quadratic fields becomes the ideal class group (chapter 17), together with its generalizations to all algebraic number fields, is one of the central notions studied in algebraic number theory.

Gauss's proof of (2), as well as his whole theory of composition as presented in the *Disquisitiones*, is quite difficult to follow. Dirichlet's presentation, using the roots of a form, yields more insight.

Anticipating our study of the relation between forms and quadratic fields, $Q(\sqrt{d}) = \{a + b\sqrt{d} | a, b \in Q, d \in Z, d$ square free$\}$, we give some idea of how to develop a theory of composition. The *conjugate* of $\alpha = a + b\sqrt{d}$ is $\overline{\alpha} = a - b\sqrt{d}$ and the *norm* is

$$N(\alpha) = \alpha\overline{\alpha} = a - b^2 d .$$

Then

$$N(\alpha\beta) = N(\alpha)N(\beta) ,$$

for all $\alpha, \beta \in Q(\sqrt{d})$ (exercise).

If r is the principal root of a form $f = ax^2+bxy+cy^2$, then $r \in Q(\sqrt{d})$, where d is the square free part of the discriminant of f. The second root of f is $\overline{r}$ and

$$f = a(x - ry)(x - \overline{r}y) = aN(x - ry) .$$

If $f_1 = a_1x^2 + b_1xy + c_1y^2$ and $f_2 = a_2x^2 + b_2xy + c_2y^2$ have equal discriminant, then $f_1 f_2$ corresponds to a product of norms and we get a composition identity. This is best illustrated by an example.

Example: Since -24 is a fundamental discriminant, all forms of discriminant -24 are primitive (sec. 12.10). There are exactly two inequivalent reduced forms, viz., x^2+6y^2 and $2x^2+3y^2$ (exercise); hence the classes $C_1=\{x^2+6y^2\}$ and $C_2=\{2x^2+3y^2\}$ form a group of order two.

The identity

$$\left(x_1^2+6y_1^2\right)\left(x_2^2+6y_2^2\right)=(x_1x_2-6y_1y_2)^2+6\,(x_1y_2+x_2y_1)^2$$

tells us that $C_2^2=C_1$ and thus C_1 is the identity of our group. By trivial group theoretic arguments, we then know that $C_2^2=C_1$ and $C_1C_2=C_2C_1=C_2$; these equations also follow from the identities

$$\left(2x_1^2+3y_1^2\right)\left(2x_2^2+3y_2^2\right)=(2x_1x_2-3y_1y_2)^2+6\,(x_1y_2+y_1x_2)^2\ ,$$

and

$$\left(x_1^2+6y_1^2\right)\left(2x_2^2+3y_2^2\right)=2\,(x_1x_2-3y_1y_2)^2+3\,(2y_1x_2+x_1y_2)^2\ .$$

We prove the latter identity to illustrate the use of norms of complex numbers. Let

$$f_1=x_1^2+6y_2^2=\left(x_1-\sqrt{-6}y_1\right)\left(x_1+\sqrt{-6}y_1\right)=N\left(x_1-\sqrt{-6}y_1\right)$$

and

$$\begin{aligned}f_2=2x_2^2+3y_2^2&=\left(\sqrt{2}x_2-\sqrt{-3}y_2\right)\left(\sqrt{2}x_2+\sqrt{-3}y_2\right)\\&=N\left(\sqrt{2}x_2-\sqrt{-3}y_2\right)\ ;\end{aligned}$$

then

$$\begin{aligned}f_1f_2&=N\left(x_1-\sqrt{-6}y_1\right)\cdot N\left(\sqrt{2}x_2-\sqrt{-3}y_2\right)\\&=N\left(\left(x_1-\sqrt{-6}y_1\right)\left(\sqrt{2}x_2-\sqrt{-3}y_2\right)\right)\\&=N\left(\sqrt{2}\,(x_1x_2+3y_1y_2)-\sqrt{-3}\,(x_1y_2+2x_2y_1)\right)\\&=2\,(x_1x_2+3y_1y_2)^2+3\,(x_1y_2+2y_1x_2)^2\ .\end{aligned}$$

Gauss does not refer to this approach, but there are convincing arguments that he must have followed this line of reasoning and then reproved his results in the more opaque form presented in the *Disquisitiones*, in order to avoid the use of complex numbers (see [Wei 5] and [Edw 1, sec. 8.6]). Adams and Goldstein [Ada - Gol, sec. 11.5] have a nice discussion of composition, in the context of quadratic algebraic number theory (also see sec. 17.10).

10. Genus

We know from Lagrange that if $x^2 \equiv D(\text{mod}\, 4n)$ is solvable, then n is representable by *some* form of discriminant D. *But* we have no general criteria for deciding which form represents n. In some special cases, however, we can specify the form.

Example: Continuing our example of the last section, with discriminant $D = -24$, we know that $x^2 + 6y^2$ and $2x^2 + 3y^2$ are the only reduced forms. It can be easily proved, using our earlier results, that a prime p is representable by

$$x^2 + 6y^2 \iff p \equiv 1, 7(\text{mod}\, 24)\,,$$
$$2x^2 + 3y^2 \iff p \equiv 5, 11(\text{mod}\, 24) \quad \text{or} \quad p = 2, 3\,.$$

For general discriminant D, it can be proved that such results cannot exist.

Lagrange's result was refined by Gauss. He partitioned all forms of discriminant D into disjoint subsets (he called each subset a **genus**). Two forms f, g are in the same genus if there exists an integer $n \neq 0$ such that f and g both represent n. Since equivalent forms represent the same integers, each genus is a union of classes, and the partition into genera is coarser than the partition into classes. Gauss showed that the representability of n by a form in the genus G depends only on the residues of n modulo the primes that divide D.

For any form f, one can calculate a sequence of values of Legendre symbols called the *generic characters* of f. Two forms are in the same genus if and only if they have the same generic characters; thus the characters are a complete set of invariants for the genera. Gauss also proved that n is representable by some form in the genus G if the residues of n modulo the primes dividing D satisfy certain conditions depending only on the generic characters of G.

It is natural to ask if each genus can be further subdivided into disjoint subsets so that the representability of n by one of these subsets can be determined by further congruence conditions. In 1951, Helmut Hasse used class field theory to prove that if there are stronger criteria for representability, they cannot be given by congruence conditions. At the end of the nineteenth century, Herman Minkowski discovered that genera are equivalence classes under the action on forms of the group of linear fractional transformations with rational coefficients [Bor - Sha].

We see in this work the first appearance of the notion of "group characters" (as well as the use of the term "character"). This concept, which plays a key role in Dirichlet's work on primes in an arithmetic progression and the class number formula, and its generalization to 'group representations', is central to many areas of modern mathematics. See Mackey [Mac] for a beautiful treatment of the history of the theory of group representations and their unifying role in mathematics, including a very thorough coverage of number theory.

11. Sections V and VI of the Disquisitiones

We have discussed many of the ideas in sections 5 and 6 of the *Disquisitiones*, but far from all. Among the items we have omitted are

1) an algorithmic study of reduction and representation by indefinite forms,
2) a detailed theory of composition and genus,
3) the solution of the general Diophantine equation

$$Ax^2 + Bxy + Cy^2 + Dx + Ey + F = 0 ,$$

4) the three square theorem,
5) ternary quadratic forms,
6) a new proof of Legendre's theorem for deciding if there is a rational point on a conic,
7) primality testing.

We now move on to the last section of the Disquisitiones, the theory of cyclotomy.

Chapter 14

Cyclotomy

1. Introduction to Section VII of the *Disquisitiones*

Section VII of the **Disquisitiones**, "Equations Defining Sections of the Circle," presents Gauss's theory of constructibility of regular polygons by straightedge and compass (or equivalently, the problem of dividing a circle into n equal arcs with straightedge and compass). As we shall see, both here and in later chapters, the ideas introduced in this section have number theoretic implications that go far beyond this very special problem. **Cyclotomy** (= circle division) is the current term for these number theoretic studies.

Recall that Felix Klein divided Gauss's activity into three periods (sec. 7.1). The first, the experimental period, ended March 30, 1796, when Gauss started his mathematical diary with the entry stating that the circle could be divided into 17 equal parts using straightedge and compass (the endpoints of the arcs forming the vertices of a regular 17 sided polygon). Years later (Jan. 16, 1819), Gauss wrote to his former student C. L. Gerling about that day:

> By concentrated analysis of the connection of all roots (of the equation $1+x+\cdots+x^{p-1}=0$) according to arithmetical reasons, I succeeded, during a vacation in Braunschweig, in the morning of the day, before I got up, to see the connection clearly such that I was able to make the specific application to the 17-gon and to confirm it numerically right away.
>
> (from [Sch - Opo])

Gauss's first published work was the announcement of the constructibility of the 17-gon and he stated that his method worked for settling the question for all n-**gons** (regular n-sided polygons). His excitement over the solution of a problem open for almost 2000 years was decisive in his choosing a career in mathematics (at this time he was also considering philology).

The seventh section of the *Disquisitiones* presents these results as part of the broader theory of the nth roots of unity, i.e., the roots of the binomial equation $z^n = 1$ or $z^n - 1 = (z-1)(1+z+z^2+\cdots+z^{n-1}) = 0$. To see Gauss's point of view, we reproduce the first article of section VII:

EQUATIONS DEFINING SECTIONS OF A CIRCLE

335. Among the splendid developments contributed by modern mathematicians, the theory of circular functions without doubt holds a most important place. We often have occasion in a variety of contexts to refer to this remarkable type of quantity, and there is no part of general mathematics that does not depend on it in some fashion. Since the most brilliant modern mathematicians by their industry and shrewdness have built it into an extensive discipline, one would hardly expect any part of the theory, let alone an elementary part, could be significantly expanded. I refer to the theory of trigonometric functions corresponding to arcs that are commensurable with the circumference, i.e., the theory of regular polygons. Only a small part of this theory has been developed so far, as the present section will make clear. The reader might be surprised to find a discussion of this subject in the present work which deals with a discipline apparently so unrelated; but the treatment itself will make abundantly clear that there is an intimate connection between this subject and higher Arithmetic.

The principles of the theory which we are going to explain actually extend much farther than we will indicate. For they can be applied not only to circular functions but just as well to other transcendental functions. e.g. to those which depend on the integral $\int \frac{dx}{\sqrt{1-x^4}}$ and also to various types of congruences. Since, however, we are preparing a large work on those transcendental functions and since we will treat congruences at length in the continuation of these *Disquisitiones*, we have decided to consider only circular functions here. And although we could discuss them in all their generality, we reduce them to the simplest case in the following article, both for the sake of brevity and in order that the new principles of this theory may be more easily understood.

Remarks: 1) Although Gauss talks about the theory of circular or trigonometric functions, he is referring to the study of roots of unity written in the polar form $\cos x + i \sin x$, which he introduces in article 337. Gauss's doctoral thesis consisted of a proof of the fundamental theorem of algebra, and he developed significant parts of complex analysis before Cauchy, although he never published the latter work. Thus there is no question that Gauss's understood the power of complex notation, which he used formally in section VII, with no attempt at justification. Later, in his second memoir on biquadratic reciprocity, he gave one of the first presentations of complex numbers as points in the plane (chapter 15).

2) In the second paragraph, Gauss refers to generalizing his theory to other transcendental functions such as $\int \frac{dx}{\sqrt{1-x^4}}$. This integral arises in computing the arc length of a lemniscate. This remark tantalized many mathematicians and was part of Abel's motivation for developing his theory of elliptic functions. Only after some of his basic work was done, did Abel say that he finally understood Gauss's remark. As we now know, from his private papers, Gauss had developed, but never published, a significant part of the theory of elliptic functions of Abel and Jacobi. For an introduction to these ideas see Siegel [Sie 1]. M. Rosen [Ros] presents a beautiful discussion of Abel's theorem on the division of the arc of a lemniscate and some general background on elliptic functions and Houzel [Hou] presents a more general history of elliptic functions and Abelian integrals. We return to these ideas in chapter 19.

3) See W. K. Buhler [Kau, Interchapter 7] for an article by article summary of section VII.

In the *Disquisitiones*, Gauss states the following result:

Theorem: *A regular polygon with p sides (p a prime) is constructible by straightedge and compass if and only if $p = 2^{2^k} + 1$, for some k.*

Gauss proved the if part, but never published a proof of the only if part. This was first given by Wantzel [Wan]. It follows from the irreducibility of the "cyclotomic polynomials" and standard arguments on degrees of field extensions (see [Had], [Gol, L], [Wae]).

Gauss also studied the solvability of equations by radicals, i.e., the reduction of general polynomial equations (mixed equations, in Gauss's terminology) to a set of equations of the form $z^n = r$ (pure equations). In studying the solution of $z^p - 1 = 0$, Gauss introduced auxiliary equations (for the "periods" — see sec. 5). We quote from article 359 of the *Disquisitiones* :

359. The preceding discussion had to do with the *discovery* of auxiliary equations. Now we will explain a very remarkable property concerning their *solution*. Everyone knows that the most eminent geometers have been unsuccessful in the search for a general solution of equations higher than the fourth degree, or (to define the search more accurately) for the REDUCTION OF MIXED EQUATIONS TO PURE EQUATIONS. And there is little doubt that this problem is not merely beyond the powers of contemporary analysis but proposes the impossible (cf. what we said on this subject in Demonstratio nova. art.9[1]). Nevertheless it is certain that there are innumerable mixed equations of every degree which admit a reduction to pure equations, and we trust that geometers will find it gratifying if we show that our equations are always of this kind. But because of the length of this discussion we will present here only the most important principles necessary to show the reduction is possible; we reserve for another time a more complete consideration, which the topic deserves.

---

[1] This is Gauss' doctoral dissertation. Its full title is *Demonstratio nova theoremautis Omnem Functinoem Algebraicam Rationalem Integram unis variabilis in Factores Reales prima cel secundi gradus resolvi posse*, Helmstedt, 1799.

Although Gauss goes on to prove that $x^p - 1 = 0$ is solvable by radicals, he firmly states his belief that, in general, equations of degree greater than four are not solvable by radicals. Abel and Galois proved this for degree five, and Galois' more general studies of solvability by radicals led to modern Galois theory. Edward [Edw 2] provides an excellent history of these matters, as well as an English translation of Galois's fundamental memoir

2. Constructibility and the Theory of Equations

We first present a quick review of the procedure for transforming geometric constructions with straight edge and compass into the algebraic problem of solving polynomial equations using only rational operations and the extraction of square roots (see [Gol, L], [Had], [Wae], [Kle 4]). These ideas arose from the study of the classical Greek problems of trisecting the angle, squaring the circle, duplicating the cube, and constructing regular polygons.

We start with the Euclidean plane and a given segment (which we choose as the unit length). *A point P is constructible* if, beginning with the given segment, there exists a sequence of "allowable" operations which construct new points, such that P is constructed at the last step. The allowable operations are those that use a straightedge and compass to construct the intersection of two lines or the intersection of a line and a circle or the intersection of two circles. At any stage we produce lines using a straightedge to connect two points already constructed, and we draw the circles with a compass using an already constructed point as the center and another as a point on the circumference.

Now we identify the plane and the given segment with the complex numbers $\mathbf{C}$ such that the segment connecting (0,0) to (0,1) is the given segment. A complex number is **constructible** if the corresponding point in the plane is constructible; hence, by definition, 1 is constructible.

The set of constructible numbers is a subfield of $\mathbf{C}$ (exercise) and thus, since 1 is constructible, all rational numbers are constructible. *The constructible numbers are closed under the operation of square roots*, i.e., if z is constructible, then $\sqrt{z}$ is constructible (exercise; of course, both square roots are constructible).

Each new complex number (point) produced at each stage of the construction process is either in the subfield F of $\mathbf{C}$ generated by adjoining all previously constructed numbers (points) to $\mathbf{Q}$ or in a quadratic extension $F(\alpha)$, $\alpha^2 \in F$. The former case corresponds to obtaining new numbers (points) from old ones by rational operations (solutions of linear equations, representing the intersections of straight lines). The latter case corresponds to obtaining new numbers from old ones by rational operations and square roots (solutions of systems of linear and irreducible quadratic equations, representing intersections of lines and circles or circles and circles), i.e., to constructing rational functions of square roots of previously constructed numbers. Therefore, we have

z is constructible $\iff$ there exists a sequence $\mathbf{Q} \subseteq F_1 \subseteq \cdots \subseteq F_k$, of subfields of $\mathbf{C}$, such that each field is a quadratic extension of the previous one and $z \in F_k$

$\iff$ z is the root of a polynomial equation solvable by radicals, where all radicals are square roots.

For the problem of constructing regular n-sided polygons (**n-gons**), we choose our coordinate system so that the vertices of the n-gon lie on the unit circle in the complex plane with one vertex at the point 1. We say **the n-gon is constructible** if each of its vertices is a constructible point. These vertices are given by

$$\zeta_k = e^{\frac{2\pi i k}{n}}, \quad k = 1, 2, \ldots, n ,$$

which are the nth roots of unity (fig. 1), i.e., the solutions of

$$z^n - 1 = (z-1)\left(1 + z + z^2 + \cdots + z^{n-1}\right) = 0 .$$

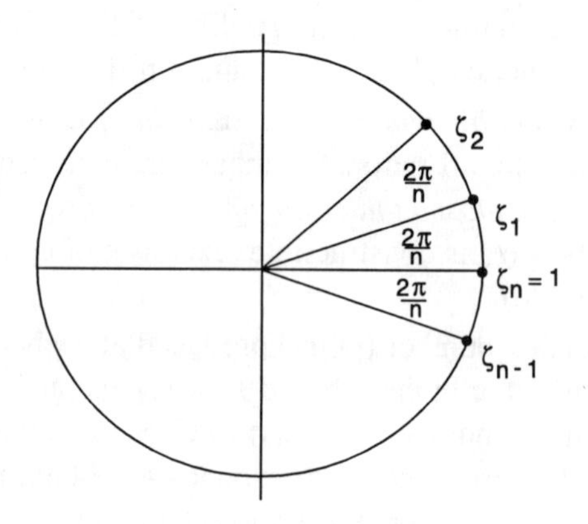

Figure 1

Thus constructibility of a regular n-gon is equivalent to constructibility of the roots of $1 + z + \cdots + z^{n-1} = 0$, since $z = 1$, which corresponds to the given segment, is constructible. Since

$$\zeta_1^k = \zeta_1^j \iff k \equiv j \,(\operatorname{mod} n) , \tag{1}$$

and

$$\zeta_k = \zeta_1^k , \tag{2}$$

the set of nth roots of unity form a cyclic group under multiplication generated by ζ_1. This group is isomorphic to $(\mathbf{Z}/n\mathbf{Z})^+$ by the map $\zeta_1^k \to k(\operatorname{mod} n)$. Any generator of the group is called a **primitive nth root of unity**.

Since constructible numbers are closed under multiplication, equation (2) implies that *the constructibility of the n-gon is equivalent to the constructibility of* ζ_1. Moreover, the construction of n-gons can be easily reduced to the construction of p-gons, for p a prime (see [Had]). So we must study the roots of the **cyclotomic polynomials**

$$\Phi_{\mathbf{p}}(\mathbf{z}) = 1 + z + \cdots + z^{p-1}\,.$$

$\Phi_p(z)$ is irreducible for all primes p (set $z = x + 1$ and apply Eisenstein's criterion), and every root is primitive since $(\mathbf{Z}/p\mathbf{Z})^+$ is cyclic of prime order. $\Phi_p(z) = 0$ is called a **cyclotomic equation.**

3. The 5-gon and Gaussian Periods

We use the construction of the 5-gon (the regular pentagon) to illustrate Gauss's procedures. Let $\zeta = \cos\frac{2\pi}{5} + i\sin\frac{2\pi}{5}$, a primitive 5th root of unity (fig. 2). Since $\zeta \neq 1$, it is a root of the cyclotomic equation $1 + \zeta + \zeta^2 + \zeta^3 + \zeta^4 = 0$.

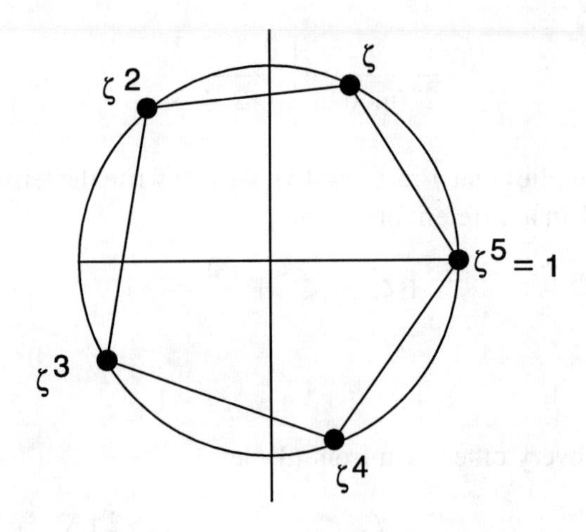

Figure 2

From $\zeta^5 = 1$, we have $\zeta^4 = \zeta^{-1}$ and $\zeta^3 = \zeta^{-2}$; thus $1 + \zeta + \zeta^2 + \zeta^{-2} + \zeta^{-1} = 0$ or

$$\zeta^2 + \zeta^{-2} = -1 - \left(\zeta + \zeta^{-1}\right). \tag{1}$$

But $\left(\zeta + \zeta^{-1}\right)^2 = \zeta^2 + 2 + \zeta^{-2}$ or

$$\zeta^2 + \zeta^{-2} = \left(\zeta + \zeta^{-1}\right)^2 - 2\,. \tag{2}$$

Equating (1) and (2), we have

$$\left(\zeta + \zeta^{-1}\right)^2 + \left(\zeta + \zeta^{-1}\right) - 1 = 0\,,$$

i.e., $\zeta + \zeta^{-1} = 2\cos\frac{2\pi}{5}$ satisfies the quadratic equation $x^2 + x - 1 = 0$. Since $2\cos\frac{2\pi}{5} > 0$,

$$2\cos\frac{2\pi}{5} = \frac{-1+\sqrt{5}}{2}\,.$$

Hence $\cos\frac{2\pi}{5}$ is constructible and so is $\sin\frac{2\pi}{5} = \sqrt{1-\cos^2\frac{2\pi}{5}}$; thus $\zeta = \cos\frac{2\pi}{5} + i\sin\frac{2\pi}{5}$ is constructible. Therefore the 5-gon is constructible.

This is a rather ad hoc procedure. Gauss systematized this line of reasoning as follows, so as to be able to apply it to the general case. Since 2 is a primitive root modulo 5 ($\bar{2}$ generates $\mathbf{F}_5^\times = (\mathbf{Z}/5\mathbf{Z})^\times$),

n	0	1	2	3
$2^n\,(\text{mod}\,5)$	1	2	4	3

and $\zeta^j = \zeta^k$ implies that $j \equiv k(\text{mod}\,5)$, we can write the terms of $\zeta + \zeta^2 + \zeta^3 + \zeta^4 = -1$ in a different order, viz.,

$$\zeta^{2^0} + \zeta^{2^1} + \zeta^{2^2} + \zeta^{2^3} = -1$$

or

$$\zeta + \zeta^2 + \zeta^4 + \zeta^3 = -1\,. \tag{3}$$

Choosing every other term from (3), let

$$\eta_1 = \zeta + \zeta^4 \left(= \zeta + \zeta^{-1} = 2\cos\frac{2\pi}{5}\right),$$
$$\eta_2 = \zeta^2 + \zeta^3\,.$$

Gauss called the η_i's **periods** (to some extent, they had been introduced earlier by Lagrange in his search for general solutions of polynomial equation

[Edw 2]). Then

$$\eta_1 + \eta_2 = -1\,,$$
$$\eta_1 \cdots \eta_2 = (\zeta + \zeta^{-1})(\zeta^2 + \zeta^3) = \zeta^3 + \zeta^4 + \zeta^1 + \zeta^2 = -1\,.$$

Hence η_1, η_2 are the roots of $x^2 + x - 1 = 0$, $\eta_1 = \frac{-1+\sqrt{5}}{2}$, and, as earlier, ζ is constructible.

Remarks: 1) Solving $\zeta + \zeta^{-1} = \frac{-1+\sqrt{5}}{2}$ for $\sqrt{5}$, we have

$$\begin{aligned}\sqrt{5} &= 1 + 2(\zeta + \zeta^{-1}) = 1 + \zeta + \zeta + \zeta^{-1} + \zeta^{-1}\\ &= 1 + \zeta + \zeta^{16} + \zeta^4 + \zeta^9 = 1 + \zeta + \zeta^4 + \zeta^9 + \zeta^{16}\,,\end{aligned}$$

i.e., $\sqrt{5}$ is the sum of consecutive square powers of ζ. Gauss generalized this to determine $1 + \zeta + \zeta^{2^2} + \cdots + \zeta^{(p-1)^2}$, where p is prime and $\zeta = e^{\frac{2\pi i}{p}}$, which he used for another proof of quadratic reciprocity (see the next section).

2) Since $\sqrt{5} = 1 + 2(\zeta + \zeta^4)$, $\mathbf{Q}(\sqrt{5}) = \{u + v\sqrt{5} | u, v \in \mathbf{Q}\}$ is a subfield of the "cyclotomic field" $\mathbf{Q}(\zeta)$, generated by ζ. In general, the **n^th^ cyclotomic field** is the subfield of $\mathbf{C}$ obtained by adjoining all n^{th} roots of unity to $\mathbf{Q}$. It is of the form $\mathbf{Q}(\zeta)$, where $\zeta = e^{\frac{2\pi i}{n}}$, a primitive n^{th} root of unity. In modern language, what Gauss did in section VII of the *Disquisitiones* was to work out the Galois theory of $\mathbf{Q}(\zeta)$, for n equal to a prime p, i.e., he determines all fields K such that $\mathbf{Q} \subseteq K \subseteq \mathbf{Q}(\zeta)$.

If $p - 1 = ma$, then there is exactly one subfield of degree m over $\mathbf{Q}$ generated by an arbitrary **Gaussian period of order m**

$$\eta_i = \sum_{j=0}^{v-1} \zeta^{ri+mj} \qquad (0 \le i \le m-1)$$

where r is a primitive root mod p. These periods are permuted by the elements of the Galois group (the automorphisms of $\mathbf{Q}(z)$ which leave $\mathbf{Q}$ fixed). The constructibility of the n-gon, when possible, can be treated analogously to our construction of the 5-gon, by finding quadratic equations satisfied by sets of periods. For a fuller treatment of these ideas see [Had], [Wae, Vol. 1], and [Rad 2].

4. Back to Quadratic Reciprocity

Recall that in his first diary entry of March 30, 1796, Gauss wrote that today he'd discovered that the 17-gon is constructible. By August 13th of that year, Gauss indicated, in the 23rd entry of the diary, that he saw the connection between the nth roots of unity and the "golden theorem" (quadratic reciprocity). He also wrote that he was ready to extend his endeavors beyond quadratic equations. Presumably he was beginning his studies of higher reciprocity laws.

Gauss's fourth and sixth proofs of quadratic reciprocity arose from his intimate knowledge of properties of sums of roots of unity gained in his work on cyclotomy. In studying the intermediate fields between $\mathbf{Q}$ and $\mathbf{Q}(\zeta)$, $\zeta = e^{\frac{2\pi i}{p}}$, and how they are generated by periods, Gauss was led to study special sums of roots of unity.

We introduce (in modern terminology) the **quadratic Gauss sum**

$$g_r = \sum_t \left(\frac{t}{p}\right) \zeta^{rt} ,$$

where $\zeta = e^{\frac{2\pi i}{p}}$, $r \in \mathbf{Z}$, p is a prime, $\left(\frac{t}{p}\right)$ is the Legendre symbol (with $\left(\frac{0}{p}\right) = 0$) and $\sum_t$ *denotes here, and in the rest of the section, summation from* $t = 0$ *to* $t = p - 1$. Our main interest, for now, is in $\mathbf{g} = g_1$, but it is convenient to look at all g_r to evaluate g.

Before discussing reciprocity, we digress to relate g to our formula $\sqrt{5} = \zeta + \zeta^4 + \zeta^9 + \zeta^{16}$. We have

$$g = \sum_u \left(\frac{u}{p}\right) \zeta^u + \sum_v \left(\frac{v}{p}\right) \zeta^v = \sum_u \zeta^u - \sum_v \zeta^v ,$$

where u runs through all quadratic residues and v through all quadratic non residues modulo p. But

$$1 + \sum_u \zeta^u + \sum_v \zeta^v = 0 ,$$

since this is just a rearrangement of the equation $1+\zeta+\zeta^2+\cdots+\zeta^{p-1} = 0$. Therefore

$$g = \sum_u \zeta^u - \sum_v \zeta^v = \sum_u \zeta^u - \left(-1 - \sum_u \zeta^u\right) = 1 + 2\sum_u \zeta^u .$$

Now consider the sum

$$\sum_t \zeta^{t^2}.$$

As t runs from 0 to $p-1$, t^2 takes on the values $0^2, 1^2, \ldots, (p-1)^2$. If $i \equiv j \pmod p$ then $\zeta^i = \zeta^j$, and if $x^2 \equiv d \pmod p$ then $(p-x)^2 \equiv d \pmod p$. Hence we see that $1^2, \ldots, (p-1)^2$ runs through each quadratic residue twice, and therefore

$$\sum_t \zeta^{t^2} = 1 + 2\sum_n \zeta^u = g.$$

In order to present Gauss' fourth proof of quadratic reciprocity, we need some elementary properties of Gauss sums.

Lemma:

$$\sum_t \zeta^{rt} = p, \quad \textit{if } r \equiv 0 (\textit{mod } p),$$

$$= 0, \quad \textit{otherwise}.$$

Proof: $r \equiv 0 \pmod p \Rightarrow \zeta^r = 1 \implies \zeta^{rt} = 1 \implies \sum_t \zeta^{rt} = p.$

$$r \not\equiv 0 \pmod p \implies \zeta^r \neq 1 \implies \sum_t \zeta^{rt} = \frac{\zeta^{rp}-1}{\zeta^r - 1} = \frac{0}{\zeta - 1} = 0.$$

Corollary:

$$\tfrac{1}{p}\textstyle\sum_t \zeta^{t(x-y)} = \delta(x,y) \quad = 1, \quad \textit{if } x \equiv y (\textit{mod } p)$$
$$= 0, \quad \textit{if } x \not\equiv y (\textit{mod } p).$$

Since half of the integers $1, 2, \ldots, p-1$ are quadratic residues and half are non residues, we have

Lemma: $\displaystyle\sum_t \left(\frac{t}{p}\right) = 0.$

Proposition: $g_r = \left(\dfrac{r}{p}\right) g.$

Proof: 1) If $r \equiv 0 \pmod p$, then $\zeta^{rt} = 1$ and therefore, by the last lemma, $g^r = \sum_t \left(\frac{t}{p}\right) = 0$. But $\left(\frac{r}{p}\right) = 0$ and therefore $\left(\frac{r}{p}\right) g_r = 0$.

2) If $r \not\equiv 0 (\text{mod } p)$, we introduce a trick that we will use more generally later. *Since rt runs through a complete system of residues as t does*, we have

$$\left(\frac{r}{p}\right) g_r = \sum_t \left(\frac{r}{p}\right)\left(\frac{t}{p}\right) \zeta^{rt} = \sum_t \left(\frac{rt}{p}\right) \zeta^{rt} = \sum_t \left(\frac{t}{p}\right) \zeta^t = g \,.$$

Hence

$$\left(\frac{r}{p}\right) g_r = g \implies \left(\frac{r}{p}\right)^2 g_r = \left(\frac{r}{p}\right) g \implies g_r = \left(\frac{r}{p}\right) g \,,$$

since $\left(\frac{r}{p}\right)^2 = 1$.

Key Theorem: $g^2 = \left(\frac{-1}{p}\right) p = (-1)^{\frac{p-1}{2}} p$.

Proof: We evaluate $\sum_{r=o}^{p-1} g_r g_{-r}$ in two ways.

i) If $r \not\equiv 0 (\text{mod } p)$, then $g_r g_{-r} = \left(\frac{r}{p}\right)\left(\frac{-r}{p}\right) g^2 = \left(\frac{-1}{p}\right) g^2$. If $r \equiv 0 (\text{mod } p)$, then $g_r = 0$. Hence $\sum_r g_r g_{-r} = \left(\frac{-1}{p}\right)(p-1) g^2$.

ii) $g_r g_{-r} = \sum_x \sum_y \left(\frac{x}{p}\right)\left(\frac{y}{p}\right) \zeta^{r(x-y)}$. Summing over r and applying the corollary to the first lemma, we have

$$\sum_r g_r g_{-r} = \sum_x \sum_y \left(\frac{x}{p}\right)\left(\frac{y}{p}\right) \sum_r \zeta^{r(x-y)}$$
$$= \sum_x \sum_y \left(\frac{x}{p}\right)\left(\frac{y}{p}\right) \delta(x, y) = (p-1)p \,.$$

Equating our two evaluations and multiplying by $\left(\frac{-1}{p}\right)$ yields

$$\left(\frac{-1}{p}\right)(p-1) g^2 = (p-1)p \implies \left(\frac{-1}{p}\right) g^2 = p \implies g^2 = \left(\frac{-1}{p}\right) g \,.$$

Corollary: *If $p \equiv 1 (\text{mod } 4)$, then $g^2 = p$ and $g = \pm\sqrt{p}$,*
If $p \equiv 3 (\text{mod } 4)$, then $g^2 = -p$ and $g = \pm\sqrt{-p}$.

Since $\sqrt{\pm p}$ is thus a polynomial in $\zeta = e^{\frac{2\pi i}{p}}$, we have

Corollary: *The quadratic fields* $\mathbf{Q}\sqrt{\pm p}$ *are subfields of the cyclotomic field* $\mathbf{Q}\left(e^{\frac{2\pi i}{p}}\right)$.

Later, we will discuss a significant generalization of this theorem, the Kronecker–Weber theorem (sec. 20.10).

For his fourth proof of quadratic reciprocity, Gauss had to determine the sign of this Gauss sum g. This is a very difficult problem and Gauss comments on it in a letter to Heinrich Olbers of Sept. 3, 1805:

> The determination of the sign of the root has vexed me for many years. This deficiency overshadowed everything that I found: over the last four years, there was rarely a week that I did not make one or another attempt, unsuccessfully, to untie the knot. Finally, a few days ago, I succeeded–but not as a result of my search but rather, I should say, through the mercy of God. As lightning strikes, the riddle has solved itself.
>
> (trans. from [Sch-Opo])

Gauss proved that

$$g = \sqrt{p}\,, \quad \text{if } p \equiv 1 (\bmod 4)\,,$$
$$g = i\sqrt{p}\,, \quad \text{if } p \equiv 3 (\bmod 4)\,,$$

or, equivalently,

$$g = i^{\left(\frac{p-1}{2}\right)^2} \sqrt{p}\,.$$

More generally, for any odd integer k, $g^2 = (-1)^{\frac{k-1}{2}} k$, where $\zeta = e^{\frac{2\pi i}{k}}$, $g = \sum_{t=0}^{k-1} \left(\frac{t}{k}\right) \zeta^t$ and $\left(\frac{t}{k}\right)$ is the Jacobi symbol; moreover

$$g = i^{\left(\frac{k-1}{2}\right)^2} \sqrt{k} \tag{1}$$

(see Rademacher [Rad 2, chap. 11] for a proof).

Following Rademacher, we now outline the essential idea of Gauss's fourth proof of quadratic reciprocity, assuming the evaluation of the Gauss sum.

For $\zeta = e^{\frac{2\pi i}{k}}$, k odd, let

$$g_{r,k} = \sum_{t=0}^{k-1} \left(\frac{t}{k}\right) \zeta^{rt}\,.$$

Our earlier g_r is just $g_{r,p}$ in this notation and g is $g_{1,p}$. Then one can show

1) $g_{r,k} = \left(\frac{r}{k}\right) g_{1,k}$,
2) If p, q are odd primes, $p \neq q$, then $g_{p,q} g_{q,p} = g_{1,pq}$. Note that different roots of unity are used in the each of the g's.

Combining 1) and 2), we have

$$\left(\frac{q}{p}\right) g_{1,p} \left(\frac{p}{p}\right) g_{1,q} = g_{1,pq} ,$$

and equation (1) yields

$$\left(\frac{p}{p}\right)\left(\frac{q}{p}\right) = g_{1,pq}/g_{1,p}g_{1,q}$$
$$= i^{\left(\frac{pq-1}{2}\right)^2 - \left(\frac{p-1}{2}\right)^2 - \left(\frac{q-1}{2}\right)^2} .$$

Working with residue classes of p and q modulo 4, it follows immediately that

$$\left(\frac{p}{p}\right)\left(\frac{q}{p}\right) = (-1)^{\frac{p-1}{2}\frac{q-1}{2}} .$$

In this proof we only used the fact that p and q are odd and thus we have proved the reciprocity law for the Jacobi symbol. For a slightly different proof, which avoids the Jacobi symbol but uses the same idea, see [Sch - Opo].

Gauss's sixth and last proof of reciprocity was published in 1817. Why so many proofs ? In the paper, Gauss said that for years he had searched for a proof of reciprocity which could be generalized to cubic and biquadratic residues (chap. 15) and that with the sixth proof he had finally succeeded. He presented these generalizations in 1828 and 1832 in his two memoirs on biquadratic residues. For discussion of the later development of the theory of Gauss sums, see [Ire - Ros] and [Wei 6].

What Gauss managed to do in the sixth proof was to use the ideas just discussed, while avoiding the evaluation of the Gauss sum. He avoided a direct use of complex numbers by working with congruences of polynomials modulo a prime (sec. 9.4). We present the proof in the later terminology of congruence in the ring $\mathbf{Z}[\zeta] = \{r_0 + r_1\zeta + \cdots + r_k\zeta^k \mid r_i \in \mathbf{Z}, \zeta = e^{\frac{2\pi i}{k}}\}$. For $k = p$, a prime, the field $\mathbf{Z}(\zeta)$ is isomorphic to $\mathbf{Q}[x]/(1 + \cdots + x^{p-1})$ (and to $\mathbf{Q}[x]$ modulo an appropriate polynomial, for k non prime), i.e., $\mathbf{Z}(\zeta)$ is constructed by congruences in polynomial rings. Thus, the two proofs are essentially the same. In fact, Gauss generalized congruence in $\mathbf{Z}$ to $\mathbf{Z}[i]$ in his second memoir on biquadratic residues.

Let $w_1, w_2, u \in \mathbf{Z}[\zeta]$. We say that **$\mathbf{w_1}$ is congruent to $\mathbf{w_2}$ modulo u**, written

$$\mathbf{w_1 \equiv w_2 (mod\, u)},$$

if there exists a $v \in \mathbf{Z}[\zeta]$ such that $w_1 - w_2 = vu$. Congruence is an equivalence relation in $\mathbf{Z}[\zeta]$. Of course, we are just considering equality in the quotient ring $\mathbf{Z}[\zeta]/u\mathbf{Z}[\zeta]$, analogous to describing congruence modulo p in $\mathbf{Z}$ by equality in $\mathbf{Z}/\mathbf{pZ}$.

We also need the following two properties, whose proofs are easy exercises:

1) If $a, b, c \in \mathbf{Z}$, then

$$a \equiv b (\operatorname{mod} c) \text{ in } \mathbf{Z} \iff a \equiv b (\operatorname{mod} c) \text{ in } \mathbf{Z}[\zeta] .$$

Thus, there is no ambiguity in using the same symbol $\equiv$ for both rings.

2) Binomial Theorem:

$$w_1, w_2 \in \mathbf{Z}[\zeta],\ p \text{ a prime} \implies (w_1 + w_2)^p \equiv w_1^p + w_2^p (\operatorname{mod} p) .$$

We also recall Euler's criterion: $\left(\frac{a}{p}\right) \equiv a^{\frac{p-1}{2}} (\operatorname{mod} p)$.

First, we compute $\left(\frac{2}{p}\right)$ as a prototype for the proof of reciprocity.

Computing $\left(\frac{2}{p}\right)$

Let $\zeta = e^{\frac{2\pi i}{8}}$, a primitive 8th root of unity. Then $\zeta^4 + 1 = 0$ and, multiplying by ζ^{-2}, $\zeta^2 + \zeta^{-2} = 0$. Thus $(\zeta + \zeta^{-1})^2 = \zeta^2 + \zeta^{-2} + 2 = 2$. Letting $\tau = \zeta + \zeta^{-1} (= \zeta - \zeta^3)$, we have $\tau \in \mathbf{Z}[\zeta]$ and $\tau^2 = 2$.

Let p be an odd prime. Then (*basic trick for all that follows*)

$$\tau^{p-1} = (\tau^2)^{\frac{p-1}{2}} = 2^{\frac{p-1}{2}} \equiv \left(\frac{2}{p}\right) (\operatorname{mod} p) ,$$

i.e., $\tau^{p-1} \equiv \left(\frac{2}{p}\right) (\operatorname{mod} p)$ or

$$\tau^p \equiv \left(\frac{2}{p}\right) \tau (\operatorname{mod} p) . \tag{2}$$

But, by the binomial theorem,

$$\tau^p = (\zeta + \zeta^{-1})^p \equiv (\zeta^p + \zeta^{-p})(\text{mod } p) . \tag{3}$$

Case 1: $p \equiv \pm 1(\text{mod } 8)$. Since $\zeta^8 = 1$, $\zeta^p + \zeta^{-p} = \zeta + \zeta^{-1} = \tau$.

Case 2: $p \equiv \pm 3(\text{mod } 8)$. Since $\zeta^3 = \frac{-1}{\zeta}$ and $-\zeta = \frac{1}{\zeta^3} = \zeta^{-3}$, we have $\zeta^p + \zeta^{-p} = \zeta^3 + \zeta^{-3} = -(\zeta + \zeta^{-1}) = -\tau$.

Hence

$$\begin{aligned} \zeta^p + \zeta^{-p} &= \tau , && \text{if } p \equiv \pm 1(\text{mod } 8) , \\ &= -\tau , && \text{if } p \equiv \pm 3(\text{mod } 8) . \end{aligned}$$

But

$$\begin{aligned} p = 8k \pm 1 &\implies \tfrac{p^2-1}{8} = 8k^2 \pm 2k && \text{is even,} \\ p = 8k \pm 3 &\implies \tfrac{p^2-1}{8} = 8k^2 \pm 6k + 1 && \text{is odd.} \end{aligned}$$

Substituting in (3) yields

$$\tau^p \equiv (-1)^{\frac{p^2-1}{8}} \tau (\text{mod } p) . \tag{4}$$

Equating (2) and (4), multiplying both sides by τ, and using $\tau^2 = 2$, we have

$$\left(\frac{2}{p}\right) 2 \equiv (-1)^{\frac{p^2-1}{8}} 2 (\text{mod } p)$$

$$\implies \left(\frac{2}{p}\right) \equiv (-1)^{\frac{p^2-1}{8}} (\text{mod } p)$$

which, since both sides equal ± 1,

$$\implies \left(\frac{2}{p}\right) = (-1)^{\frac{p^2-1}{8}} .$$

Part of this result, $p \equiv 1(\text{mod } 8) \implies \left(\frac{2}{p}\right) = 1$, is due to Euler, who assumed the existence of primitive roots. We digress to present this, using the current language of finite fields.

Let $p \equiv 1(\text{mod } 8)$, f be a generator of $\mathbf{F}_p^\times$, and $h = f^{\frac{p-1}{8}}$ (the analog of the primitive 8th root of unity $e^{\frac{2\pi i}{8}}$). Then h has order 8 and

$$h^8 = 1 \implies (h^4 - 1)(h^4 + 1) = 0 .$$

Since $h^4 = 1$ contradicts the order of h being 8, we have

$$\begin{aligned} h^4 = -1 &\implies h^2 = -h^{-2} \implies h^2 + h^{-2} = 0 \\ &\implies (h + h^{-1})^2 = h^2 + h^{-2} + 2 = 2 \\ &\implies 2 \text{ is a square in } \mathbf{F}_p^\times \implies \left(\frac{2}{p}\right) = 1\,. \end{aligned}$$

Now to Gauss's sixth proof of quadratic reciprocity. Analyzing our computation of $\left(\frac{2}{p}\right)$, we see that we found a sum τ of roots of unity, whose square was 2, and used Euler's criterion to derive equation (2), namely, $\tau^p \equiv \left(\frac{2}{p}\right)\tau \,(\mathrm{mod}\, p)$. Then, also computing τ^p by the binomial theorem, we equated the two expressions to arrive at the expression for $\left(\frac{2}{p}\right)$. To generalize this process to prove reciprocity, we must first express $\pm p$ as the square of a sum of roots of unity and try to imitate the above procedure. *But we have such a sum, the Gauss sum g.*

Proof of Quadratic Reciprocity: We work with $g = \sum_t \left(\frac{t}{p}\right)\zeta^t$, $\zeta = e^{\frac{2\pi i}{p}}$, and congruence modulo q in $\mathbf{Z}[e^{\frac{2\pi i}{p}}]$, where p and q are odd primes. Set $p^* = (-1)^{\frac{p-1}{2}} p$; thus $g^2 = p^*$, analogous to $\tau^2 = 2$. Then

$$g^{q-1} = (g^2)^{\frac{q-1}{2}} = p^{*\frac{q-1}{2}} \equiv \left(\frac{p^*}{q}\right) (\mathrm{mod}\, q)$$

or

$$g^q \equiv \left(\frac{p^*}{q}\right) g \,(\mathrm{mod}\, q)\,. \tag{5}$$

But, by the binomial theorem,

$$\begin{aligned} g^q &= \left(\sum_t \left(\frac{t}{p}\right)\zeta^t\right)^q \equiv \sum_t \left(\frac{t}{p}\right)^q \zeta^{qt} \,(\mathrm{mod}\, q) \\ &\equiv \sum_t \left(\frac{t}{p}\right)\zeta^{qt} \,(\mathrm{mod}\, q) \quad \left(q \text{ odd} \implies \left(\frac{t}{p}\right)^q = \left(\frac{t}{p}\right)\right) \\ &\equiv g_q \,(\mathrm{mod}\, q)\,. \end{aligned}$$

Thus $g^q \equiv g_q (\bmod q)$ and, since $g_g = \left(\frac{q}{p}\right) g$, we have

$$g^q \equiv \left(\frac{q}{p}\right) g (\bmod q) . \qquad (6)$$

Combining (5) and (6) and multiplying by g yields

$$\begin{aligned} \left(\frac{p^*}{q}\right) g^2 &\equiv \left(\frac{q}{p}\right) g^2 (\bmod q) \\ \Longrightarrow \left(\frac{p^*}{q}\right) p^* &\equiv \left(\frac{q}{p}\right) p^* (\bmod q) \\ \Longrightarrow \left(\frac{p^*}{q}\right) &\equiv \left(\frac{q}{p}\right) (\bmod q) \\ \Longrightarrow \left(\frac{p^*}{q}\right) &= \left(\frac{q}{p}\right) \quad \text{(since each side equals } \pm 1\text{).} \qquad (7) \end{aligned}$$

Now $\left(\frac{p^*}{q}\right) = \left(\frac{-1}{q}\right)^{\frac{p-1}{2}} \left(\frac{p}{q}\right) = (-1)^{\frac{q-1}{2}\frac{p-1}{2}} \left(\frac{p}{q}\right)$, and, combined with (7), we have

$$\left(\frac{p}{q}\right) = (-1)^{\frac{p-1}{2}\frac{q-1}{2}} \left(\frac{p}{q}\right) ,$$

the law of quadratic reciprocity.

5. Numbers of Solutions of Congruences; Equations over Finite Fields

There is another circle of ideas arising from section VII of the *Disquisitiones*. The Gaussian periods satisfy a multiplicative law

$$\eta_i \eta_j = \sum_k N_{ijk} \eta_k ,$$

where the N_{ijk} are natural numbers related to the number of solutions of the congruences

$$A\, x^m + B\, y^m \equiv C (\bmod p) .$$

This is the first deep idea connected with counting solutions of congruences, a topic developed primarily in the 20th century. In fact, Gauss's ideas seemed to have been completely overlooked until 1947 when A. Weil made an important advance. In [Wei 3], he recounts that in 1947 he read Gauss' two memoirs on biquadratic reciprocity. The first one deals with the number of solutions of $ax^4 - by^4 = 1 (\bmod p)$ and the relation between this problem and Gauss sums. Weil recognized that the same method was used by Gauss in the last section of the *Disquisitiones* to study $ax^3 - by^3 = 1 (\bmod p)$. He realized that he could apply the same ideas to $\sum a_i x_i^n = 0 (\bmod p)$ and from this he proved the "generalized Riemann hypothesis" for certain varieties over finite fields. This led Weil to a set of conjectures about varieties over a finite field.

These conjectures, the so called "Weil conjectures", guided the development of algebraic geometry for about 30 years. The last one was finally proved by DeLigne in 1973, for which he was awarded the Fields medal.

Ireland and Rosen [Ire - Ros] present an introduction to the study of equations over finite fields and their connection with Gauss sums, at the level of this book. Koblitz [Kob 1] is devoted to Dwork's proof of one of the Weil conjectures via p-adic analysis (chap. 23) and Katz [Kat] gives a more advanced overview of DeLigne's proof. Because of the thorough treatment by Ireland and Rosen, we shall not go into any systematic presentation of these matters, but we will discuss them later in the context of other topics.

6. Final Remarks on the *Disquisitiones*

We have presented, in a somewhat systematic manner, many of the important ideas in the *Disquisitiones*. Number theory branches in many directions after the *Disquisitiones*.

The theory of binary quadratic forms (and a bit on ternary forms) led to both the development of major parts of algebraic number theory and to the theory of forms in n variables, currently very active research areas. We shall not discuss forms in many variables, except for some results derived using the geometry of numbers (chapter 22). Scharlau and Opolka [Sch - Opo] give a broader historical treatment of this theory in the same spirit as this book.

Gauss's results on cyclotomy also led to later studies by Gauss, Jacobi and others on general exponential sums and higher reciprocity laws (see [Ire - Ros]). This has certainly justified Gauss's comment at the beginning of section VII, that although the reader might be curious about the inclusion of a study of division of the circle, and trigonometric functions (roots of unity), these studies have important arithmetical significance.

Although the *Disquisitiones* has been has been studied with great energy by many great mathematicians, Weil's work, discussed in the last section makes one hesitate to say that everything in the *Disquisitiones* is now well understood. Weil's work is a beautiful illustration of the value of reading original classic works.

Chapter 15

Algebraic Number Theory 1: The Gaussian Integers and Biquadratic Reciprocity

1. Gauss and Biquadratic Reciprocity

We have already encountered several instances where complex numbers were used to derive results about integers:

1) Euler and Lagrange factored Diophantine equations over **C** to obtain solutions (sec. 3.5).
2) Factoring binary quadratic forms over **C** led to results on the composition of forms (sec. 13.9),
3) Gauss sums were used to give two proofs of quadratic reciprocity (sec. 14.4).

In all of these applications, the complex numbers we used are **algebraic numbers**, namely, roots of polynomial equations with integer coefficients. Furthermore these applications are the first instances of what has become the vast edifice of algebraic number theory. The search for solutions of Diophantine equations and the search for higher reciprocity laws were the primary motivations for the development of algebraic number theory. The former, in the guise of attempts to prove Fermat's last theorem, tends to receive more attention than the latter, a distortion of the actual facts.

In this chapter, we will concentrate on the early generalizations of quadratic reciprocity. Euler made several conjectures about the cubic properties of 2, 3, 5, 7 and the fourth power properties of 2 modulo primes, but the first results of the search for higher reciprocity laws appear in Gauss's two memoirs on biquadratic residues [Gau 2, Gau 3]. We discuss these results briefly, including the introduction of the Gaussian integer complex numbers, a statement of the law of biquadratic reciprocity and an introduction to zeta functions. In the next two chapters, we treat some of the theory of quadratic fields as a prototype for general algebraic number theory and discuss the modern generalization of reciprocity laws. Finally, in chapter 23, we will indicate how the theory of valuations provides an alternate approach to the study of algebraic number fields.

Gauss's first memoir opens as follows (translation by George Brauer):

> "The theory of quadratic residues can be reduced to the most beautiful jewel among the fundamental theorems of higher arithmetic, which, as is known, were first discovered easily by inductive methods and then were proved in so many ways that nothing remains to be desired.
>
> However the theory of cubic and biquadratic residues is more difficult by far. In 1805, as we began to investigate these, except for the first results which gave several special theorems that stand out both because of their simplicity and because of the difficulty of their proofs, we soon recognized that the principles of arithmetic which were usable until then were in no way sufficient to build a general theory. Rather such a theory necessarily required an infinite enlargement to some extent of the field of higher arithmetic; in the course of this investigation it will become very clear how this is to be understood. As soon as we have described this new field, then at the same time access, by inductive methods, to the simplest theorems which exhaust the entire theory will be available to us; on the other hand, the proofs of these theorems are hidden so deeply that they could finally be brought to light only after many unsuccessful attempts.
>
> Proceeding now with the publication of these studies,we begin with the theory of quadratic residues and, in fact, in this paper we will present those investigations which can still be completely treated without an extension of the field of arithmetic, but which, to some extent, clear the way for such an extension and which yields several new extensions for the theory of cyclotomy."

Gauss begins with some elementary results. If r is a **biquadratic residue modulo p**, i.e., if there exists an x such that $x^4 \equiv r \pmod p$, then, since $(x^2)^2 \equiv r \pmod p$, we see that r is a quadratic residue mod p.

Gauss then proves that if r is a quadratic residue modulo a prime $p \equiv 3 \pmod 4$, then r is a biquadratic residue mod p. Then he says:

> "Since these obvious theorems exhaust the entire theory of biquadratic residues modulo primes of the form $4n+3$, we will completely exclude such moduli from our investigations and we will restrict our investigations to prime moduli of the form $4n+1$."

The rest of the memoir is taken up with special results, most importantly, the complete determination of the biquadratic character of 2, which had been correctly conjectured by Euler, viz., 2 is a fourth power residue modulo a prime $p \equiv 1 \pmod 4$ if and only if p can be written in the form $p = A^2 + 64B^2$, where $A, B \in \mathbf{Z}$. The second memoir continues with special results that can be treated over **Z**. But then Gauss writes:

> "In this way, the inductive method yields a rich harvest of special theorems which are related to the theorem for the number 2; but we lack a common tie; we lack rigorous proofs, as the methods by which we treat the number 2 in the first paper do not permit further application. True, there is no lack of various methods by which the proofs can be obtained for special cases......; meanwhile we will not be delayed by these as we must aim for a general theory which encompasses all cases. After we had begun to think about these matters, already in the year 1805, we soon arrived at the conviction that the natural course of a general theory is to be sought in an *extension of the field of arithmetic* as we already indicated in section one.
>
> For while in the problems treated until now, the higher arithmetic has been concerned only with real integers, the theorems appear in their full simplicity and natural beauty only if the domain of arithmetic is extended to the *imaginary* numbers, so that, without restriction, the numbers of the form $a+bi$, where, as usual, i denotes the imaginary quantity $\sqrt{-1}$ and a, b denote throughout real integers between $-\infty$ and $+\infty$, form the objects of the sum. We will call numbers of this sort *complex integers* and, in fact, we note that real numbers are not part of the complex numbers but are considered as a special form of the latter. The present paper will contain the elementary theory of complex quantities as well as the first beginnings of the theory of biquadratic residues, the complete development of which will be the task in following works.*

> * Only we wish to at least remark here that such an extension of the theory of biquadratic residues is the theory of cubic residues, which must be based, in a similar fashion, on the consideration of numbers of the form $a+bh$, where h is an imaginary root of the equation $h^3-1=0$, say $h = -\frac{1}{2} + i\sqrt{\frac{3}{4}}$, and likewise the theory of residues of higher powers requires the introduction of other imaginary quantities.

The footnote shows Gauss' realization that the study of higher reciprocity laws requires a theory of algebraic numbers.

Gauss goes on to give a formal presentation of the complex numbers and of what we now call the **Gaussian integers**,

$$\mathbf{Z[i]} = \{u + iv | u,\, v \in \mathbf{Z}\} \;.$$

He proves that Gaussian integers factor uniquely into "prime Gaussian integers". In section 39 of the paper, he introduces the geometric interpretation of complex numbers:

> "We now go over to the congruence of numbers with respect to complex moduli. However, at the beginning of this investigation, it is useful to indicate in which way one can give a clear picture of the complex number.
>
> In the same way that each real quantity can be expressed by a part of a straight line infinite in both direction with an arbitrary chosen origin, measured with respect to an arbitrary segment taken as unity, and, in this way, represented by the endpoint of the segment in such a way that the points on one side of the origin represent positive quantities, the points on the other side represent negative quantities, each complex quantity can also be represented by a point of an infinite plane, in which one fixed straight line serves to represent real quantities, namely, the complex quantity $x + iy$ can be represented by a point whose abscissa is equal to x and whose ordinate (taken positive on one side of the x-axis, negative on the other side) is equal to y."

After a detailed discussion of addition and multiplication of complex numbers. Gauss ends the section with:

> "However we will save an extensive treatment of this matter for another occasion. The difficulties with which the theory of imaginary quantities is believed to be encumbered, to a large extent, have their origin in the unsuitable naming (by several people they are named with the misnomer of impossible quantities). If one had called the positive quantities direct quantities, the negative quantities inverse quantities, and the imaginary quantities lateral quantities, . . ., then the result would be simplicity itself instead of complication, clarity instead of fog."

Gauss defines congruence modulo elements of $\mathbf{Z}[i]$, as a generalization of congruence in $\mathbf{Z}$, and proceeds with an extensive study of its properties, as well as their geometric interpretation as points in the plane with integer coordinates. Then the theory of quadratic residues modulo Gaussian primes is developed and ends with:

"If we extend this inductive method to other primes, then we find this highly elegant reciprocity theorem confirmed everywhere and in this way we arrive at the following *fundamental theorem* with respect to quadratic residues in the arithmetic of complex numbers.

If $a + bi$, $A + Bi$ are primes of such a character that a, A are odd, b, B are even, then either each number is a quadratic residue of the other or each number is a quadratic non residue of the other."

(Recall that when the moduli are in $\mathbf{Z}$, we only consider $p \equiv 1 (\text{mod}\, 4)$).

Gauss then discusses the theory of biquadratic residues over the Gaussian integers and states the law of biquadratic reciprocity. He never published a proof of this law. We will state the law and discuss its later history in section 3. Now we proceed with a quick introduction to the Gaussian integers as a model for the more general theory of quadratic fields.

2. The Gaussian Integers

We study the Gaussian integers $\mathbf{Z}[i]$ as a subring of the field $\mathbf{Q}(i) = \{a + bi | a, b \in \mathbf{Q}\}$, where they play the role of generalized integers. Although this would seem to be the obvious choice for the "integers" of $\mathbf{Q}(i)$, we will justify it in a deeper sense when we discuss algebraic integers. To avoid confusion we use the terms **rational integer** and **rational prime** for elements of $\mathbf{Z}$.

Divisibility and units in $\mathbf{Z}[i]$ are those for any ring, i.e., we say that α **divides** β (denoted $\alpha|\beta$) for $\alpha, \beta \in \mathbf{Z}[i]$, if there exists a $\gamma \in \mathbf{Z}[i]$ such that $\beta = \alpha\gamma$, and we say that α in $\mathbf{Z}[i]$ is a **unit** if $\alpha|1$, i.e., if $\frac{1}{\alpha} \in \mathbf{Z}[i]$. The units form a multiplicative subgroup. If $\alpha = x + iy$ is any complex number then $\alpha' = \mathbf{x} - \mathbf{iy}$ is the **conjugate** of α. *(From now on, we use a prime and not an over bar to denote conjugates.)* The **norm** of α, $\mathbf{N}(\alpha)$, is $\alpha\alpha' = x^2 + y^2$. Norms are a measure of the "size" of a complex number.

As noted by Gauss, we can represent the complex number $a + bi$ by the point (a, b) in the plane, and the set of all Gaussian integers correspond to the points of the "integer lattice" (all points in $\mathbf{R}^2$ with integer coordinates - figure 1). The conjugate α' then corresponds to the point $(a, -b)$, the reflection of (a, b) in the x-axis, and $N(\alpha)$ is the square of the distance from (a, b) to the origin.

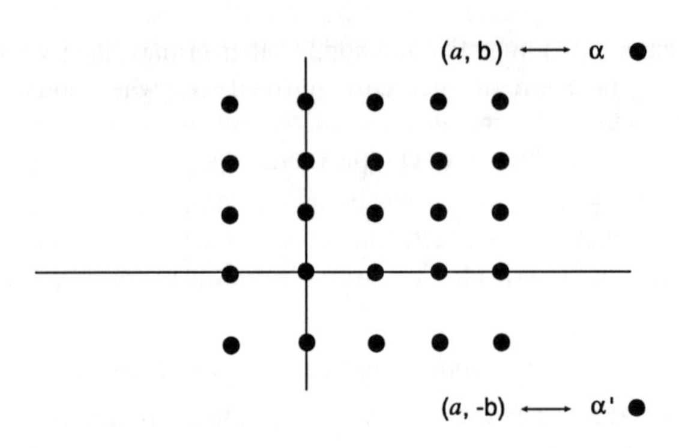

Figure 1

We will return to these geometric ideas in the next chapter, where we exploit them to develop a more general theory of quadratic fields.

Proposition: *If* $\alpha, \beta \in \mathbf{C}$, *then*

(i) $N(\alpha)$ *is a non negative real number,*

(ii) $N(\alpha) = 0 \iff \alpha = 0$,

(iii) $N(\alpha\beta) = N(\alpha)N(\beta)$,

(iv) *if* $\alpha \in \mathbf{Z}[i]$, *then* $N(\alpha)$ *is a non negative rational integer.*

Proof: (exercise).

Since $N(\alpha) = 1$ if and only if $\alpha\alpha' = 1$, and α in $\mathbf{Z}[i]$ divides 1 if and only if α' divides 1, we have

Proposition: α *is a unit in* $\mathbf{Z}[i] \iff N(\alpha) = 1$.

Thus the integer points in the plane which correspond to the units $x + iy$ are those points on the circle $N(x + iy) = x^2 + y^2 = 1$, and we have

Corollary: *The units of* $\mathbf{Z}[i]$ *are* $\pm 1, \pm i$.

Recall that Euler and Lagrange assumed unique factorization into "prime integers" for generalized integers such as the Gaussian integers when they used them for solving Diophantine equations. Gauss proved the unique factorization for the Gaussian integers. He first proved that $\mathbf{Z}[i]$ is a Euclidean ring, i.e., there exists a division algorithm with respect to the norm.

Theorem: *If $\alpha, \beta \in \mathbf{Z}[i]$, $\beta \neq 0$, then there exists $\gamma, \delta \in \mathbf{Z}[i]$ such that*

$$\alpha = \beta\gamma + \delta\ ,$$

where $0 \leq N(\delta) < N(\beta)$.

Proof: $\frac{\alpha}{\beta} = e + fi$, where $e, f \in \mathbf{Q}$. Choose $g, h \in \mathbf{Z}$ so that

$$|g - e| \leq \frac{1}{2}\ , \quad |h - f| \leq \frac{1}{2}$$

and let $\gamma = g + hi \in \mathbf{Z}[i]$. Then $\frac{\alpha}{\beta} = \gamma + (e - g) + (f - h)i$ or

$$\alpha = \beta\gamma + \{(e - g) + (f - h)i\}\beta\ .$$

Then $\delta = \{(e - g) + (f - h)i\}\beta$ is in $\mathbf{Z}[i]$ and $\alpha = \beta\gamma + \delta$. Moreover

$$\begin{aligned} N(\delta) &= N((e - g) + (f - g)i)N(\beta) \\ &= N(\beta)\{(e - g)^2 + (f - g)^2\} \leq N(\beta)\left(\frac{1}{4} + \frac{1}{4}\right) \\ &= \frac{1}{2}N(\beta) < N(\beta)\ . \end{aligned}$$

A standard result in algebra is that the existence of a division algorithm implies unique factorization. Of course, the algebraic proof is just a more or less straightforward generalization of Gauss's proof for Gaussian integers (which follows by analogy from the proof for $\mathbf{Z}$ - see chap. 1). We recall, without proof, the sequence of ideas (see almost any book on algebra, e.g., [Gol, L] or [Her]). All numbers are assumed to be in $\mathbf{Z}[i]$.

1) Division Algorithm $\Longrightarrow$ Euclidean algorithm $\Longrightarrow$ existence of greatest common divisors $\Longrightarrow$ if $(\alpha, \beta) = 1$, then there exist γ, δ such that

$$\alpha\gamma + \beta\delta = 1\ .$$

2) Definitions: α is **irreducible** if it is not the product of two non-units. A non-unit π (*not the real number* π) is a **(Gaussian) prime** if

$$\pi | \beta\gamma \implies \pi | \beta \text{ or } \pi | \gamma\ .$$

α and β are **associates** if $\alpha = u\beta$, where u is a unit.

Thus $\pm\alpha$, $\pm i\alpha$ are the associates of α. Associates behave the same under division, i.e., they divide and are divisible by the same integers.

3) π is prime $\Longleftrightarrow$ π is irreducible.

4) If $\alpha = \beta\gamma$, β and γ not units, then $N(\beta) < N(\alpha)$.

5) From 4) we derive the existence of a factorization of an integer into irreducibles (= primes).

6) From 3) we have the uniqueness of this factorization, up to order and associates of primes.

Thus we have

Unique Factorization Theorem: *Every Gaussian integer α is a product of Gaussian primes,*

$$\alpha = \pi_1^{n_1} \cdots \pi_k^{n_k} ,$$

unique up to order and associates.

BACK TO THE TWO SQUARE PROBLEM

Since $N(x+iy) = x^2 + y^2$, a rational integer is the sum of two squares if and only if it is the norm of a Gaussian integer. Thus the two square theorem (sec. 12.6) tells us exactly which rational integers m are norms. Conversely, we can also use the properties of the Gaussian integers to prove the two square theorem.

To determine other representations of m as a sum of two squares, we use the units. Let $\alpha = x + iy$ and $\beta = y + ix$; then $N(\alpha) = N(\beta) = x^2 + y^2 = m$. If u is a unit, then $N(u\alpha) = N(u\beta) = m$. Hence, if $x, y \neq 0$, $N(u\alpha)$ and $N(u\beta)$, $u = \pm 1, \pm i$, yield the eight representations, $(\pm x, \pm y)$, $(\pm x, +y)$, $(\pm y, \pm x)$, $(\pm y, +x)$, of m as a sum of two squares (sec. 13.7). Thus the units in $\mathbf{Z}[i]$ somehow correspond to the automorphs of $x^2 + y^2$. We shall see that this generalizes to arbitrary quadratic forms and fields, another example of the relation between the forms and fields (sec. 17.10).

Our immediate goals are to

(i) determine all Gaussian primes,

(ii) determine the factorization of all Gaussian integers into Gaussian primes.

Any Gaussian integer α divides a rational integer ($\alpha | N(\alpha) = \alpha\alpha'$) and thus any prime factor of α divides some rational prime dividing $N(\alpha)$, i.e.,

Lemma: *Every Gaussian prime divides some rational prime.*

Therefore, if we know how to factor all rational primes into Gaussian primes, we can factor α by first factoring $N(\alpha)$ into rational prime factors and then into Gaussian prime factors. These Gaussian primes are the possible factors of α; so we can test them to see which, if any, divides α, and to what power. Thus (ii) reduces to

(ii)′ determine the factorization of all rational primes into Gaussian primes.

We use two key lemmas for studying factorization.

Lemma: *If $N(\pi) = p$ is a rational prime, then π is a Gaussian prime.*

Proof: If $\gamma|\pi$, $\pi = \gamma\delta$, then

$$\begin{aligned} p = N(\pi) = N(\gamma)N(\delta) &\implies \text{one of } N(\gamma), N(\delta) = 1, \text{ say } N(\delta) = 1 \\ &\implies \delta \text{ is a unit} \\ &\implies \gamma \text{ is an associate of } \pi \implies \pi \text{ is a prime.} \end{aligned}$$

Lemma: *Every Gaussian prime π divides exactly one rational prime p.*

Proof: The existence was shown in the first lemma. Suppose $\pi\,|\,p$ and $\pi\,|q$, where p, q are distinct rational primes. Then $(p, q) = 1$ and there are rational integers x, y such that $1 = px + qy$. Therefore $\pi\,|1$ and is a unit, contradicting our assumption that π is prime.

Example: $2 = -i(1+i)^2$. But $-i$ is a unit and, since $N(1+i) = 2$, $1+i$ is prime. Therefore, by unique factorization, $1+i$ and its associates, $-1-i$, $i-1$ and $1-i$ are the only primes dividing 2.

Theorem: *Let p be a rational prime. Then p factors over $\mathbf{Z}[i]$ as follows:*

1) *$p = 2$; $p = -i(1+i)^2$, where π is a Gaussian prime and $N(\pi) = 2$,*
2) *$p \equiv 3 \pmod 4$; $p = \pi$ is a Gaussian prime and $N(\pi) = p^2$,*
3) *$p \equiv 1 \pmod 4$; $p = \pi\pi'$, where π and π' are non associated primes and $N(\pi) = N(\pi') = p$. π and π' are unique up to order and associates.*

Proof: (We assume the two square theorem.)

1) This was proved in the previous example.

2) Suppose $p = \alpha\beta, \alpha = a + bi, \beta = c + di$. Then $N(p) = N(\alpha)N(\beta)$ or $p^2 = (a^2 + b^2)(c^2 + d^2)$. If α and β are not units, so that $N(\alpha)$, $N(\beta) \neq 1$, then $a^2 + b^2 = p$. Since $p \equiv 3(\text{mod } 4)$, this contradicts the two square theorem; hence p is a Gaussian prime.

3) $p \equiv 1(\text{mod } 4)$ implies that p is the sum of two square, say $p = x^2 + y^2$. Setting $\pi = x + yi$, we have $p = \pi\pi'$ and $N(p) = p^2 = N(\pi)N(\pi')$. Since $N(\pi) = N(\pi')$, they both equal p. Hence, by our first lemma, π and π' are prime. Suppose $\pi = u\pi'$, u a unit. A straightforward case by case calculation shows that this leads to a contradiction (exercise); hence π and π' are not associates. Uniqueness follows from the unique factorization theorem.

This theorem answers the questions (i) and (ii)$'$ raised earlier. We can reformulate the theorem in terms of the Legendre symbol as follows:

Theorem: *Let p be a rational prime. Then p is a Gaussian prime if $\left(\frac{-1}{p}\right) = -1$, and p is the product of two non associated Gaussian primes if $\left(\frac{-1}{p}\right) = 1$.*

There is a corresponding theorem for arbitrary quadratic fields (sec. 17.6) and this is our first hint of a relation between factorization and the theory of quadratic residues.

3. Congruence and the Law of Biquadratic Reciprocity

Now we outline the ideas leading to biquadratic reciprocity and give a precise statement of the law. An excellent treatment with proofs is given by Ireland and Rosen [Ire - Ros, ch. 9], who also treat the law of cubic reciprocity.

Our main interest is deciding when $x^4 \equiv a(\text{mod } p)$ is solvable, i.e., deciding when a is a biquadratic residue modulo p. As we have seen, Gauss realized that for a deeper understanding of the problem, we must work over $\mathbf{Z}[i]$ (sec. 1).

To obtain a fruitful generalization of the theory of quadratic residues, we first reformulate our approach to the Legendre symbol. Recall that by

Fermat's little theorem we know that if $p \nmid a$,

$$a^{p-1} - 1 \equiv 0 (\bmod p) \implies \left(a^{\frac{p-1}{2}} - 1\right)\left(a^{\frac{p-1}{2}} + 1\right) \equiv 0 (\bmod p)$$
$$\implies a^{\frac{p-1}{2}} \equiv \pm 1 (\bmod p)\,,$$

and Euler's criterion (sec. 11.2) says that if $p \neq 2$, then

$$a^{\frac{p-1}{2}} \equiv \left(\frac{a}{p}\right) (\bmod p)\,. \tag{1}$$

Thus we could have defined $\left(\frac{a}{p}\right)$ by the following:

1) if $p \nmid a$, then $\left(\frac{a}{p}\right) = +1$ or -1, according to which value satisfies equation (1),
2) if $p | a$, then $\left(\frac{a}{p}\right) = 0$.

To generalize these ideas to $\mathbf{Z}[i]$, we first need a notion of congruence. If α, β and π are in $\mathbf{Z}[i]$, π a Gaussian prime, then α **is congruent to** β **modulo** π, $\alpha \equiv \beta (\textbf{mod}\, \pi)$, if $\pi | \beta - \alpha$. This is an equivalence relation compatible with addition and multiplication in $\mathbf{Z}[i]$, and in fact we are just studying the residue class ring (quotient ring) $\mathbf{Z}[i]/\pi\mathbf{Z}[i]$, where $\pi\mathbf{Z}[i]$ is an ideal in $\mathbf{Z}[i]$. $[\alpha]$, the *congruence* (equivalence) *class of* α, is just the coset $\alpha + \pi\mathbf{Z}[i]$ and we have

$$\alpha \equiv \beta (\bmod \pi) \iff [\alpha] = [\beta]\,.$$

This is a direct analogy with $\mathbf{Z}$, where the congruence classes are the elements of $\mathbf{Z}/p\mathbf{Z}$ and $p\mathbf{Z}$ is an ideal in $\mathbf{Z}$.

α is a **biquadratic (fourth power) residue modulo** π if there is an x in $\mathbf{Z}[i]$ satisfying

$$x^4 \equiv \alpha (\bmod \pi)\,.$$

Continuing the analogy with $\mathbf{Z}$, it can be proved that $\mathbf{Z}/\pi\mathbf{Z}[i]$ is a finite field with $N(\pi)$ elements; thus there are $N(\pi)$ congruence classes modulo π. Just as we proved Fermat's theorem using group theory (sec. 8.3), it is an easy exercise to prove that if $\pi \nmid \alpha$, then

$$\alpha^{N(\pi)-1} \equiv 1 (\bmod \pi)\,,$$

or equivalently, in $\mathbf{Z}[i]/\pi\mathbf{Z}[i]$,

$$[\alpha]^{N(\pi)-1} = [1] \, .$$

We can then prove an analog of Euler's result on quadratic residues, viz., if the prime $\pi \not| \alpha$, then there exists a rational integer j such that

$$\alpha^{\frac{N(\pi)-1}{4}} \equiv i^j \, (\mathrm{mod}\, \pi) \, . \qquad (2)$$

If $N(\pi) \neq 2$ and $\pi \not| \alpha$, we define the **biquadratic residue symbol,** $\left(\frac{\alpha}{\pi}\right)_4$, by

$$\left(\frac{\alpha}{\pi}\right)_4 = i^j ,$$

where j is determined by equation (2). Note that i^j is always a unit of $\mathbf{Z}[i]$. If $\pi | \alpha$, we set $\left(\frac{\alpha}{\pi}\right)_4 = 0$.

Then, if $\pi \not| \alpha$,

$$\left(\frac{\alpha}{\pi}\right)_4 = 1 \iff \alpha \text{ is a biquadratic residue modulo } \pi \, .$$

The biquadratic residue symbol is constant on congruence classes ($\alpha \equiv \beta (\mathrm{mod}\, \pi) \implies \left(\frac{\alpha}{\pi}\right)_4 = \left(\frac{\beta}{\pi}\right)_4$) and multiplicative $\left(\left(\frac{\alpha\beta}{\pi}\right)_4 = \left(\frac{\alpha}{\pi}\right)_4 \left(\frac{\beta}{\pi}\right)_4\right)$. Thus $[\alpha] \to \left(\frac{\alpha}{\pi}\right)_4$ defines a homomorphism of $\mathbf{Z}[i]/\pi\mathbf{Z}[i]$ to the multiplicative group of units of $\mathbf{Z}[i]$. In order to state the law of biquadratic reciprocity in general form, we generalize $\left(\frac{\alpha}{\pi}\right)_4$. Suppose $\alpha, \beta \in \mathbf{Z}[i]$, α not a unit, and $1 + i \not| \alpha$ (equivalently, $2 \not| N(\alpha)$). Let $\alpha = \prod_i \lambda_i$, where the λ_i are Gaussian primes. Then we define

$$\left(\frac{\beta}{\alpha}\right)_4 = \prod_i \left(\frac{\beta}{\lambda_i}\right)_4 \, .$$

$\left(\frac{\beta}{\alpha}\right)_4$ is independent of which prime decomposition of α we choose (i.e., up to associates) and, for fixed α, is constant on equivalence classes $[\beta]$.

Before stating the law of biquadratic reciprocity, it is convenient to eliminate the clumsiness arising from the fact that every element of $\mathbf{Z}[i]$ has four associates. This is done by defining a non-unit $\alpha = a + bi$ to be **primary** if $\alpha \equiv 1 (mod(1 + i)^2)$, which is equivalent to requiring that $a \equiv 1 (\mathrm{mod}\, 4)$, $b \equiv 0 (\mathrm{mod}\, 4)$ or $a \equiv 3 (\mathrm{mod}\, 4)$, $b \equiv 2 (\mathrm{mod}\, 4)$. Since $N(1 + i) = 2$, $1 + i$ plays the role of an 'even' prime. No primary element is divisible by $1+i$. Every set of associates $\{\pm\alpha, \pm i\alpha\}$ contains one primary element. Furthermore, every primary element is a product of primary Gaussian primes, a power of $1+i$ and a unit, unique up to order.

Law of Biquadratic Reciprocity: *Let α and β be relatively prime primary elements of $\mathbf{Z}[i]$. Then*

$$\left(\frac{\alpha}{\beta}\right)_4 = \left(\frac{\beta}{\alpha}\right)_4 (-1)^{\frac{N(\alpha)-1}{4}\cdot\frac{N(\beta)-1}{4}}.$$

As quoted in section 1, Gauss also found a law of cubic reciprocity, which involves the structure of the ring $\mathbf{Z}\left[e^{\frac{2\pi i}{3}}\right]$. Gauss's work may have been motivated to some extent by his studies of elliptic functions and the division of the arc of the lemniscate (sec. 14.1). In fact, Eisenstein published the first proofs of cubic and biquadratic reciprocity, first using elliptic functions and later with Gauss sums (for the latter see [Ire-Rose, ch. 9]). Jacobi claimed to have been the first to prove cubic reciprocity in his 1837 lectures and this led to a bitter priority dispute with Eisenstein.

Eisenstein also proved a reciprocity law for higher power residues [Ire-Ros, ch. 14]. We refer to Houzel [Hou] for a historical discussion of elliptic functions (including applications to reciprocity laws in section 15), to Weil [Wei 6] for the connections with cyclotomy and to H. J. S. Smith [Smi, H 1, vol. 1],who provides an extensive discussion of the study of reciprocity laws in the 19th century. Weil [Wei 7] presents one of Eisenstein's approaches to elliptic functions.

Before his premature death at the age of 29, Eisenstein published twenty five papers on reciprocity laws, which can be found in his recently published collected works [Eis 2]. Weil's review of the collected works [Wei 8] includes a delightful biography of this extraordinary mathematician, who Gauss regarded as a singular genius.

Note that we have not really answered our original question as to which rational integers are biquadratic residues modulo a given rational prime. Our study of Gauss's "natural" generalization of quadratic reciprocity concerns biquadratic residues modulo Gaussian primes. We saw in section 1 that for primes $p \equiv 3 \pmod 4$, the theory of biquadratic residues is equivalent to the theory of quadratic residues, but that this is not true for $p \equiv 1 \pmod 4$. In this latter case, where p and q are rational primes congruent to 1 modulo 4 and π is a Gaussian prime dividing p, one can prove

$$\left(\frac{q}{\pi}\right)_4 = 1 \iff x^4 \equiv q(\text{mod } p) \text{ has a solution } x \in \mathbf{Z}\,.$$

In 1969, K. Burde proved a beautiful biquadratic reciprocity law involving rational primes, and there has recently been renewed interest in such rational reciprocity laws (see [Ire-Ros, sec. 9.10]).

We have seen that the "natural" generalization of an important result (quadratic reciprocity) does not necessarily lead to a complete answer to our original question (which rational integers are biquadratic residues modulo a given rational prime). Often natural generalizations of important results do not answer the corresponding general questions.

Another example of this is Weil's generalization of Mordell's theorem about rational points on "curves of genus one" to rational sets of points on curves of higher genus, which did not provide the the desired characterization of rational points on curves of higher genus. The latter problem was only recently solved with Faltings' proof of Mordell's conjecture that curves of genus greater then one have only finitely many rational points. (These ideas will be discussed in chapters 18 - 20).

However, in no way should such natural generalizations be regarded as fruitless exercises. Although they may seem like the relatively easy direction of research, they often result in deep and beautiful new theories and sometimes lead to the development of techniques for solving the original questions (as was the case with Faltings work).

4. The Zeta Function and L Function of $\mathbf{Z}[i]$

We continue our discussion of $\mathbf{Z}[i]$, basically following the presentation of Scharlau and Opolka [Sch-Opo, pg. 73 ff.] by associating an "analytic object" to $\mathbf{Z}[i]$. Dirichlet and Jacobi published essentially the same results, expressed in the equivalent language of quadratic forms, in the 1830's and 40's and some of them were found in Gauss's unpublished notes.

The **zeta function of Z[i]**, $\zeta_{\mathbf{Z}[\mathbf{i}]}(\mathrm{s})$, is defined by

$$\zeta_{\mathbf{Z}[i]}(s) = \sum_{\alpha \in \mathbf{Z}[i], \alpha \neq 0} \frac{1}{N(\alpha)^s}\,,$$

where s is a variable. Note that, for now, we do not worry about convergence. We are dealing with formal Dirichlet series, i.e., expressions of the form $\sum_{n>0} \frac{a_n}{n^s}$, with real or complex coefficients, for which there is a clear cut algebra. Two such series $\sum \frac{a_n}{n^s}$ and $\sum \frac{b_n}{n^s}$ are defined to be equal if $a_n = b_n$ for all n, and we add and multiply by

$$\sum \frac{a_n}{n^s} + \sum \frac{b_n}{n^s} = \sum \frac{a_n + b_n}{n^s},$$
$$\sum \frac{a_n}{n^s} \times \sum \frac{b_n}{n^s} = \sum \frac{c_n}{n^s},$$

where $c_n = \sum_{j|n} a_{\frac{n}{j}} b_j$. The standard manipulations (including infinite products) can all be justified in the same way as for formal power series (see [Zag] for details).

By analogy with Euler's infinite product expansion for the Riemann zeta function (sec. 3.3),

$$\zeta(s) \left(= \sum \frac{1}{n^s} \right) = \prod_p \left(\frac{1}{1 - p^{-s}} \right), \tag{1}$$

we want to find an "Euler product" for $\zeta_{\mathbf{Z}[i]}(s)$.

We choose one of the four associates of each Gaussian prime as a representative, in such a way that if π is chosen and π' is not associated to π, then π' is also chosen (also let $1+i$ be chosen). If P denotes this set of prime representatives, then

$$\begin{aligned} \zeta_{\mathbf{Z}[i]}(s) \left(= \sum_{\alpha} \frac{1}{N(\alpha)^s} \right) &= 4 \prod_{q \in P} \left(1 + \frac{1}{N(q)^s} + \frac{1}{N(q)^{2s}} + \frac{1}{N(q)^{3s}} + \cdots \right) \\ &= 4 \prod_{q \in P} \frac{1}{1 - N(q)^{-s}}, \quad \text{(formal sum of a geometric series)} \end{aligned} \tag{2}$$

where each term of the product gives a unique factorization of an α in $\mathbf{Z}[i]$ and the 4 corresponds to the four units in $\mathbf{Z}[i]$, which, when multiplied by our terms, gives all associates of α.

The Euler product, $\zeta_{\mathbf{Z}[i]} = 4 \prod_{\pi \in P} \frac{1}{1-N(\pi)^{-s}}$, encodes information about the number of units and unique factorization. Analytic information about $\zeta_{\mathbf{Z}[i]}(s)$, regarded as an analytic functions of the complex variable s, e.g., the location and order of the poles, can be used to derive a great deal of arithmetic information about $\mathbf{Z}[i]$ (see [Zag]).

Exercise: Show that it follows from the proof of unique factorization in section 2 that the primes in P are of the form

$$\begin{aligned}
&1+i, \quad N(1+i)=2\,,\\
&p\equiv 3(\mathrm{mod}\,4), \quad N(p)=p^2\,,\\
&\pi \text{ and } \pi', \text{ where } \pi\pi' = p \equiv 1(\mathrm{mod}\,4),\ N(\pi)=p \text{ and}\\
&\qquad \pi \text{ and } \pi' \text{ are not associates.}
\end{aligned}$$

Applying the results of this last exercise in equation (2), we have

$$\begin{aligned}
\zeta_{\mathbf{Z}[i]}(s) &= 4\left(\frac{1}{1-N(1+i)^{-s}}\right)\prod_{p\equiv 3(4)}\left(\frac{1}{1-N(p)^{-s}}\right)\prod_{\pi}\left(\frac{1}{1-N(\pi)^{-s}}\right)\\
&= 4\left(\frac{1}{1-2^{-s}}\right)\prod_{p\equiv 3(4)}\left(\frac{1}{1-p^{-2s}}\right)\prod_{p\equiv 1(4)}\left(\frac{1}{1-p^{-s}}\right)^2,
\end{aligned}$$

where the square of the last product arises from the two non-associated factors π, π' of p with the same norm. Hence

$$\begin{aligned}
\zeta_{\mathbf{Z}[i]}(s) &= 4\left(\frac{1}{1-2^{-s}}\right)\prod_{p\equiv 3(4)}\left(\frac{1}{1-p^{-s}}\right)\left(\frac{1}{1+p^{-s}}\right)\prod_{p\equiv 1(4)}\left(\frac{1}{1-p^{-s}}\right)^2\\
&= 4\zeta(s)\prod_{p\equiv 3(4)}\left(\frac{1}{1+p^{-s}}\right)\prod_{p\equiv 1(4)}\left(\frac{1}{1-p^{-s}}\right),
\end{aligned}$$

where $\zeta(s)$ is the Riemann zeta function, in the form given in (1).

We call

$$L_{\mathbf{Z}[i]}(s) = \prod_{p\equiv 1(4)}\left(\frac{1}{1-p^{-s}}\right)\prod_{p\equiv 3(4)}\left(\frac{1}{1+p^{-s}}\right)$$

the **Dirichlet L-function** of $\mathbf{Z}[i]$ (so called because Dirichlet used L to denote this function); then

$$\zeta_{\mathbf{Z}[i]} = 4\zeta(s)L_{\mathbf{Z}[i]}(s)\,. \tag{3}$$

These notions were introduced by Dirichlet to prove his theorems on class numbers and on primes in an arithmetic progression. We present a simpler application in the vein of Jacobi's work.

Proposition: $L_{\mathbf{Z}[i]}(s) = \sum_{n=1}^{\infty} \frac{\chi(n)}{n^s} = 1 - \frac{1}{3^s} + \frac{1}{5^s} - \frac{1}{7^s} + \frac{1}{9^s} - \cdots$, *where*

$$\begin{aligned} \chi(n) = 0, &\quad \textit{if } n \textit{ is even}, \\ 1, &\quad \textit{if } n \equiv 1 \pmod 4, \\ -1, &\quad \textit{if } n \equiv 3 \equiv -1 \pmod 4. \end{aligned}$$

Proof: From the definition of $L_{\mathbf{Z}[i]}$, we have

$$L_{\mathbf{Z}[i]}(s) = \prod_p \left(\frac{1}{1 - \chi(p) p^{-s}} \right).$$

If we expand this product into a series $\sum a_n n^{-s}$, then $a_n \neq 0$ if and only if the factors of n are $\equiv 1$ or $3 \pmod 4$. Since $\chi(p) = 1$ for $p \equiv 1 \pmod 4$, the value of χ for these n is determined by the number of factors $\equiv 3 \pmod 4$. If $n \equiv 1 \pmod 4$, then the number of factors $\equiv 3 \pmod 4$ is even and $\chi(n) = 1$. If $n \equiv 3 \pmod 4$, then the number of such factors is odd, $\chi(n) = -1$, and the theorem is proved.

Applying the proposition to equation (3) yields

$$\begin{aligned} \zeta_{\mathbf{Z}[i]}(s) = 4\zeta(s)L(s) &= 4 \left(\sum_{k=1}^{\infty} \frac{1}{k^s} \right) \left(\sum_{m=1}^{\infty} \frac{\chi(m)}{m^s} \right) \\ &= 4 \sum_{n=1}^{\infty} \left(\sum_{m|n} \chi(m) \right) n^{-s}. \end{aligned} \tag{4}$$

Since $N(x + yi) = x^2 + y^2$, we also have

$$\begin{aligned} \zeta_{\mathbf{Z}[i]}(s) = \sum_{\alpha} \frac{1}{N(\alpha)^s} &= \sum_{(x,y) \neq 0} \frac{1}{x^2 + y^2} \\ &= \sum \frac{S_2(n)}{n^s}, \end{aligned} \tag{5}$$

where $S_2(n)$ denotes the number of representations of n as a sum of two squares. Equating the coefficients of n^{-s} in equations (4) and (5), we have

Theorem: $S_2(n) = 4 \sum_{m|n} \chi(m)$.

Remarks: 1) The proof rests on unique factorization in $\mathbf{Z}[i]$.

2) χ can be regarded as a character on $(\mathbf{Z}/4\mathbf{Z})^{\times}$, the reduced classes under multiplication (a character is a homomorphism of $(\mathbf{Z}/4\mathbf{Z})^{\times}$ to $\mathbf{C}$). Dirichlet introduced a general notion of character and systematically exploited it.

3) For a simple application of our theory to the derivation of a series expansion for $\pi/4$, see [Sch-Opo, pp.74-75 and pg.83].

The ideas of this section generalize to all quadratic number fields [Zag]. In fact, the generalization of the zeta function to all algebraic number fields is a central object of study in algebraic number theory (see [Hec] and [Shi]). Suitably generalized, the idea of encoding arithmetic information (in our case, the factorization of Gaussian integers into Gaussian primes) into an analytic function (Dirichlet series) and then applying the techniques of formal power series and complex analysis is one of the major techniques of twentieth century number theory.

Chapter 16

Algebraic Number Theory 2: Algebraic Numbers and Quadratic Fields

1. The Development of Algebraic Number Theory

Before studying quadratic fields in detail, we stand back and take a brief look at the development of algebraic number theory from Gauss to Hilbert. As described in the last chapter, the developers of this theory had two major goals in back of their minds: generalizing quadratic reciprocity and solving Diophantine equations, especially Fermat's last theorem. In order to continue our historical discussion we must introduce some general concepts.

An **algebraic number** is a complex number satisfying a polynomial equation $a_n x^n + a_{n-1} x^{n-1} + \cdots + a_0 = 0$, with rational coefficients not all equal to zero.

Proposition: *An algebraic number α is the root of a unique monic irreducible polynomial over* $\mathbf{Q}$.

Proof: The set I of polynomials in $\mathbf{Q}[x]$ which have α as a root is an ideal in $\mathbf{Q}[x]$. Since $\mathbf{Q}[x]$ is a principal ideal domain (all ideals are principal, i.e., generated by one element), I is a principal ideal, and hence I consists of all multiples of some polynomial $p(x)$ (a polynomial of *minimum degree* in I). Clearly $p(x)$ is irreducible, and dividing the coefficients of $p(x)$ by its leading coefficient yields the desired irreducible polynomial.

The degree of the irreducible polynomial over **Q** satisfied by α is called the **degree of** α. Thus, e.g., the rational number are the algebraic numbers of degree one.

An **algebraic number field** (or just **number field**) K is a subfield of the complex numbers **C** which is a finite extension of the rationals, i.e., K is finite dimensional as a vector space over the field **Q**. The **degree** of the extension is defined as the dimension of the vector space. For any β in K, finite dimensionality implies that the sequence $1, \beta, \beta^2, \dots, \beta^k$ must be linearly dependent for some k. Thus there exist $b_0, , \dots, b_k$ in **Q**, not all zero, such that $\sum_0^k b_i\beta^i = 0$, and β is algebraic. Therefore the elements of an algebraic number field of degree n are algebraic numbers of degree bounded by n.

Any number field K is of the form $K = \mathbf{Q}(\alpha_1, \alpha_2, , \dots, \alpha_n)$, with the α_i algebraic . In fact, by the primitive element theorem [Wae], it is of the form $K = \mathbf{Q}(\alpha)$, for some algebraic number α, i.e., every element of K is a rational function of α with rational coefficients (actually, a polynomial in α). It is an easy exercise to prove, using degrees of field extensions, that the set **A** of all algebraic numbers is a subfield of **C**. The fact that **A** is a proper subfield follows from the existence of *transcendental* (non algebraic) *numbers* (see chapter 21). **A** is not a number field, i.e. it is not a finite dimensional vector space over **Q** (just look at the degrees of the solutions of $x^n - 2 = 0$, for all n).

The first problem in developing an arithmetic theory of an algebraic number field K is choosing a subset of K, the 'algebraic integers' of K, which will play as fruitful a role in K as the rational integers play in **Q**.

Gauss studied the ring of Gaussian integers $\mathbf{Z}[i]$ in $\mathbf{Q}(i)$ as the 'algebraic integers' for his work on biquadratic reciprocity and mentioned the need for other rings for other reciprocity laws. In the 1830's and 40's, Jacobi and Eisenstein used the ring $\mathbf{Z}[\zeta] = \{u + v\zeta | u, v \in \mathbf{Z}\}$, where $\zeta = e^{\frac{2\pi i}{3}}$ is a primitive cube root of unity, $1 + \zeta + \zeta^2 = 0$, to study cubic reciprocity. Between 1840 and 1850, Dirichlet studied the rings $\mathbf{Z}[\alpha] = \{u_0 + \cdots + u_{n-1}\alpha^{n-1} | u_i \in \mathbf{Z}\}$, where α is the root of a monic irreducible equation $x^n + b_{n-1}x^{n-1} + \cdots + b_0 = 0$, $b_i \in \mathbf{Z}$. He characterized the units in these rings (references in [Edw 1] and [Ire - Ros]). Dirichlet's work on class number formulas, applying analytic techniques to study algebraic number fields, was also carried out during this period (see the appendix to chapter 17).

The most important contribution to the development of algebraic number theory was Kummer's work, in particular, his introduction of ideal numbers which, although restricted to cyclotomic fields, really signifies the beginning of a systematic theory of algebraic numbers. We quote at some length from H. Edwards [Edw 1, chap. 4], who has significantly clarified our understanding of these historical developments.

"4.1 The events of 1847

The proceedings of the Paris Academy and the Prussian Academy in Berlin for the year 1847 tell a dramatic story in the history of Fermat's Last Theorem. The story begins in the report of the March 1st meeting of the Paris Academy ([A1], p. 310), at which Lamé announced, evidently rather excitedly, that he had found a proof of the impossibility of the equation $x^n + y^n = z^n$ for $n > 2$ and had therefore completely solved this long outstanding problem. The brief sketch of a proof which he gave was, as he no doubt realized later, woefully inadequate, and there is no need to consider it in detail here. However, his basic idea was a simple and compelling one which is central to the later development of the theory. The proofs of the cases $n = 3, 4, 5, 7$ which had been found up to that time all depended on some algebraic factorization such as $x^3 + y^3 = (x + y)(x^2 - xy + y^2)$ in the case $n = 3$. Lamé perceived the increasing difficulty for large n as resulting from the fact that one of the factors in this decomposition has very large degree, and he noted that this can be overcome by decomposing $x^n + y^n$ completely into n linear factors. This can be done by introducing a complex number such that $r^n = 1$ and using the algebraic identity

$$x^n + y^n = (x + y)(x + ry)(x + r^2 y) \cdots (x + r^{n-1} y) \quad (n \text{ odd}) \qquad (1)$$

(For example, if $r = \cos\left(\frac{2\pi}{n}\right) + i \sin\left(\frac{2\pi}{n}\right) = e^{\frac{2\pi i}{n}}$ then the polynomial $x^n - 1$ has the n distinct roots $1, r, r^2, \ldots, r^{n-1}$ and by elementary algebra $X^n - 1 = (X - 1)(X - r)(X - r^2) \cdots (X - r^{n-1})$; setting $X = -\frac{x}{y}$ and multiplying by $-y^n$ then give the desired identity (1).) Put very briefly, Lamé's idea was to use the techniques which had been used in the past for the cruder factorization of $x^n + y^n$ (in special cases) to this complete factorization. That is, he planned to show that if x and y are such that the factors $x + y, x + ry, \ldots, x^{n-1} y$ are relatively prime then $x^n + y^n = z^n$ implies that each of the factors $x + y, x + ry, \ldots$ must itself be an nth power and to derive from this an impossible infinite descent. If $x + y, x + ry \ldots$ are not relatively prime he planned to show that there is a factor m common to all of them so that $\frac{x+y}{m}, \frac{x+ry}{m}, \ldots, \frac{x+r^{n-1}y}{m}$ are relatively prime and to apply a similar argument in this case as well.

Not doubting for a moment that the idea of introducing complex numbers in this way was the key that would unlock the door of Fermat's

Last Theorem, Lamé enthusiastically told the Academy that he could not claim the entire credit for himself because the idea had been suggested to him in a casual conversation by his colleague Liouville some months before. Liouville for his part, however, did not share Lamé's enthusiasm, and he took the floor after Lamé finished his presentation only to cast some doubts on the proposed proof. He declined any credit for himself in the idea of introducing complex numbers–pointing out that many others, among them Euler, Lagrange*, Gauss, Cauchy, and "above all Jacobi,"

---

* Liouville did not say so, and he may not have known it, but Lagrange actually *explicitly* mentioned the factorization $(x+y)(x+ry)\cdots(x+r^{n-1}y) = x^n+y^n$ in connection with Fermat's last theorem ([L3]).

---

had used complex numbers in similar ways in the past–and practically said that Lamé's brainchild was among the first ideas that would suggest themselves to a competent mathematician approaching the problem for the first time. What was more, he observed that Lamé's proposed proof had what appeared to him to be a very large gap in it. Would Lamé be justified, he asked, in concluding that each factor was an nth power if all he had shown was that the factors were relatively prime and that their product was an nth power? Of course this conclusion would be valid in the case of ordinary integers, but its proof depends** on the factorization

---

** The fact that Liouville spotted this gap instantly and saw instantly that it was related to the problem of proving unique factorization into primes for the complex numbers in question seems to indicate that he, and perhaps other mathematicians of the time as well, were well aware of the flaw in Euler's Algebra on this point (see Section 2.3). Nonetheless, I do not know of any writer of this period or earlier who criticizes Euler's argument.

---

of integers into prime factors and it is by no means obvious that the needed techniques can be applied to the complex numbers that Lamé needed them for Liouville felt that no enthusiasm was justified unless or until these difficult matters had been resolved.

Cauchy, who took the floor after Liouville, seemed to believe there was some likelihood that Lamé would succeed, because he hastened to point out that he himself had presented to the Academy in October of

1846 an idea which he believed might yield a proof of Fermat's Last Theorem but that he had not found the time to develop it further.

The proceedings of the meetings of the weeks following this one show a great deal of activity on the part of Cauchy and Lamé in pursuing these ideas. Lamé admitted the logical validity of Liouville's criticism, but he did not in the least share Liouville's doubts about the truth of the final conclusion. He claimed that his "lemmas" gave him a method of factoring the complex numbers in question and that all of his examples confirmed the existence of unique factorization into primes. He was certain that "there can be no insurmountable obstacle between such a complete verification and an actual proof."

In the meeting of March 15th, Wantzel claimed to have proved the validity of unique factorization into primes, but his arguments covered only the cases $n \leq 4$ which are easily proved ($n = 2$ is the case of ordinary integers, $n = 3$ is essentially the case proved in Section 2.5, and $n = 4$ was quickly proved by Gauss in his classic paper on biquadratic residues), beyond which he simply said that "one easily sees" that the same arguments can be applied to the cases $n > 4$. One doesn't, and Cauchy said so on March 22nd. Thereafter Cauchy launched into a long series of papers in which he himself attempted to prove a division algorithm for the complex numbers in question–"radical polynomials" as he called them–from which he could conclude that unique factorization was valid.

In the March 22nd proceedings it is recorded that *both* Cauchy and Lamé deposited "secret packets" with the Academy. The depositing of secret packets was an institution of the Academy which allowed members to go on record as having been in possession of certain ideas at a certain time–without revealing them–in case a priority dispute later developed. In view of the circumstances of March 1847, there is little doubt what the subject of these two packets was. As it turned out, however, there was no priority dispute whatever on the subject of unique factorization and Fermat's Last Theorem.

In the following weeks, Lamé and Cauchy each published notices in the proceedings of the Academy, notices that are annoyingly vague and incomplete and inconclusive. Then, on May 24, Liouville read into the proceedings a letter from Kummer in Breslau which ended, or should have ended, the entire discussion. Kummer wrote to Liouville to tell him that his questioning of Lamé's implicit use of unique factorization had been quite correct. Kummer not only asserted that unique factorization *fails*, he also included with his letter a copy of a memoir [K6] he had published* *three years earlier* in which he had demonstrated the failure of unique factorization in cases where Lamé had been asserting it was

---

* It must be admitted, however, that Kummer chose a very obscure place in which to publish it. Liouville republished it in his *Journal de Mathematiques Pures et Appliquees* in 1847 and this must have been the first time it reached a wide audience.

---

valid. However, he went on to say, the theory of factorization can be "saved" by introducing a new kind of complex numbers which he called "ideal complex numbers"; these results he had published one year earlier* in the proceedings of the Berlin Academy in resume form [K7] and a complete exposition of them was soon to appear in Crelle's Journal [K8]. He had for a long time been occupied with the application

---

* This notice was also reprinted in Crelle in 1847. (A flawed translation into English is contained in Smith's Source Book [S2].) The Crelle reprint gives the erroneous date 1845 for the original publication; the correct date is 1846.

---

of his new theory to Fermat's Last Theorem and said he had succeeded in reducing its proof for a given n to the testing of two conditions on n. For the details of this application and of the two conditions, he refers to the notice he had published that same month in the proceedings of the Berlin Academy (15 April 1847). There he in fact stated the two conditions in full and said that he "had reason to believe" that $n = 37$ did not satisfy them.

The reaction of the learned gentlemen of Paris to this devastating news is not recorded. Lamé simply fell silent. Cauchy, possibly because he had a harder head than Lamé or possibly because he had invested less in the success of unique factorization, continued to publish his vague and inconclusive articles for several more weeks. In his only direct reference to Kummer he said, "What little [Liouville] has said [about Kummer's work] persuades me that the conclusions which Mr. Kummer has reached are, at least in part, those to which I find myself led by the above considerations. If Mr. Kummer has taken the question a few steps further, if in fact he has succeeded in removing all the obstacles, I would be the first to applaud his efforts; for what we should desire the most is that the works of all the friends of science should come together to make known and to propagate the truth." He then proceeded to ignore–rather than to propagate–Kummer's work and to pursue his own ideas with only an occasional promise that he would eventually relate his statements to

Kummer's work, a promise he never fulfilled. By the end of the summer, he too fell silent on the subject of Fermat's Last Theorem. (Cauchy was not the silent type, however, and he merely began producing a torrent of papers on mathematical astronomy.) This left the field to Kummer, to whom, after all, it had already belonged for three years.

It is widely believed that Kummer was led to his "ideal complex numbers" by his interest in Fermat's Last Theorem, but this belief is surely mistaken. Kummer's use of the letter λ to represent a prime number, his use of the letter α to denote a "λ^{th} root of unity"–that is, a solution of $\alpha^{\lambda} = 1$ —and his study** of the factorization of prime numbers $p \equiv 1 \bmod \lambda$ into "complex numbers composed of λth roots

---

** Kummer's 1844 memoir dealt only with the factorization of such p. The general problem of factorization was covered by the succeeding papers in 1846 and 1847.

---

of unity" all derive directly from a paper of Jacobi [J2] which is concerned with *higher reciprocity laws*. Kummer's 1844 memoir was addressed by the University of Breslau to the University of Koenigsberg in honor of its jubilee celebration, and the memoir was definitely meant as a tribute to Jacobi, who for many years was a professor at Koenigsberg. It is true that Kummer had studied Fermat's Last Theorem in the 1830s and in all probability he was aware all along that his factorization theory would have implications for Fermat's Last Theorem, but the subject of Jacobi's interest, namely, higher reciprocity laws, was surely more important to him, both at the time he was doing the work and after. At the same time that he was demolishing Lamé's attempted proof and replacing it with his own partial proof, he referred to Fermat's Last Theorem as "a curiosity of number theory rather than a major item," and later, when he published his version of the higher reciprocity law in the form of an unproved conjecture, he referred to the higher reciprocity laws as "the principal subject and the pinnacle of contemporary number theory."

There is even an often told story that Kummer, like Lamé, believed he had proved Fermat's Last Theorem until he was told–by Dirichlet in this story–that his argument depended on the unproved assumption of unique factorization into primes. Although this story does not necessarily conflict with the fact that Kummer's primary interest was in the higher reciprocity laws, there are other reasons to doubt its authenticity. It first appeared in a memorial lecture on Kummer given by Hensel in 1910 and, although Hensel describes his sources as unimpeachable and gives their names, the story is being told at third hand over 65 years later. Moreover, the person who told it to Hensel was not apparently a

> mathematician and it is very easy to imagine how the story could have grown out of a misunderstanding of known events. Hensel's story would be confirmed if the "draft ready for publication" which Kummer is supposed to have completed and sent to Dirichlet could be found, but unless this happens the story should be regarded with great skepticism. Kummer seems unlikely to have assumed the validity of unique factorization and even more unlikely to have assumed it unwittingly in a paper he intended to publish."

Following Kummer's 1844 memoir, the main work until 1871 was the further study of the cyclotomic fields $\mathbf{Q}(\zeta)$, $\zeta^n = 1$, and their subrings $\mathbf{Z}[\zeta]$.

Dedekind transformed algebraic number theory from a theory of quadratic and cyclotomic fields to a study of general algebraic number fields. Dedekind published his notes of Dirichlet's lectures on number theory. In an appendix to the 2nd edition (1871), he presented a foundation for the theory of algebraic number fields which included the definition of an algebraic number field as well as most of the basic concepts and theorems which today form the first part of a course in algebraic number theory. This marks the beginning of algebraic number theory as a discipline. Dedekind's final definitive appendix appeared in 1893, in the 4th edition of Dirichlet-Dedekind [Dir].

Dedekind was the first to use axiomatics as a research tool and not just as a foundation for a subject. He had great skill in abstracting the essence of a result and he introduced many new concepts, e.g., subfields of the complex numbers (which he called *domains of rationality*), algebraic number fields, ideals, modules and lattices (in the sense of partially ordered sets). Dedekind looked for set theoretic constructions rather than introducing *ideal objects*, e.g., he used Dedekind cuts (pairs of sets of rational numbers) to construct the real numbers. Later we shall see how he introduced ideals in rings of algebraic integers to substitute for Kummer's ideal numbers.

At the same time, Leopold Kronecker, a student of Kummer's, was also making fundamental contributions to the theory. Kronecker's work did not use ideal theory, but was a more direct development of Kummer's approach. For a more detailed historical discussion of Kummer's and Kronecker's approaches, see [Ell, W - Ell, F]. Herman Weyl [Wey 1] contrasts Kronecker's and Dedekind's approach. Kronecker had a much more ambitious goal than Dedekind, namely a general theory that would encompass both algebraic number theory and algebraic geometry. However, Kronecker's papers were very difficult to understand and thus his work did not have as strong an impact as Dedekind's. Only in the mid twentieth century, due to the influence of A. Weil, has Kronecker's broad goal had an important impact on current research (see[Wei 9] and [Wei 10]). We will discuss the modern version of some of Kummer's and Kronecker's ideas, the so called 'valuation theoretic approach' to algebraic number theory, in chapter 23.

In 1893, the German Mathematical Society asked David Hilbert and Hermann Minkowski to prepare a survey report on the state of the theory of numbers. Hilbert and Minkowski divided the work, deciding that the former would report on algebraic number theory and the latter on rational number theory. Distracted by his intensive work on the geometry of numbers (chapter 22), Minkowski never finished his report. Hilbert's report, the "Zahlbericht" appeared in 1897 [Hil]. This volume, based upon the revolutionary work of Kummer, Kronecker and Dedekind, presented the ideal theoretic foundations of algebraic number theory and included many deep and important new contributions. (See Constance Reid's biography of Hilbert [Rei, C] for a more complete version of the personal aspects of this story.)

Subsequent to the appearance of the Zahlbericht, Hilbert published a series of papers which opened up a new approach to reciprocity laws in algebraic number fields. This work was a decisive turning point in the development of algebraic number theory. First, Hilbert changed the emphasis of algebraic number theory from a tool for for studying the rational integers to a deep and beautiful theory worth studying in its own right. Second, his papers set the direction of research in algebraic number theory for the next several decades, culminating in Emil Artin's general reciprocity law (sec. 17.13).

Herman Weyl, a student of Hilbert's and one of the great mathematicians of the twentieth century, wrote an obituary of Hilbert [Wey 2], including both a personal account and a detailed discussion of his mathematical work. We quote from Weyl as to Hilbert's influence and style:

> "If examples are wanted let me tell my own story. I came to Gottingen as a country lad of eighteen, having chosen that university mainly because the director of my high school happened to be a cousin of Hilbert's and had given me a letter of recommendation to him. In the fullness of my innocence and ignorance I made bold to take the course Hilbert had announced for that term, on the notion of number and the quadrature of

the circle. Most of it went straight over my head. But the doors of a new world swung open for me, and I had not sat long at Hilbert's feet before the resolution formed itself in my young heart that I must by all means read and study whatever this man had written. And after the first year I went home with Hilbert's Zahlbericht under my arm, and during the summer vacation I worked my way through it–without any previous knowledge of elementary number theory or Galois theory. These were the happiest months of my life, whose shine, across years burdened with our common share of doubt and failure, still comforts my soul.

.

.

.

Before giving a more detailed account of Hilbert's work, it remains to characterize in a few words the peculiarly Hilbertian brand of mathematical thinking. It is reflected in his literary style walk through a sunny open landscape; you look freely around, demarcation lines and connecting roads are pointed out to you, before you must brace yourself to climb the hill; then the path goes straight up, no ambling around, no detours. His style has not the terseness of many of our modern authors in mathematics, which is based on the assumption that printer's labor and paper are costly but the reader's effort and time are not. In carrying out a complete induction Hilbert finds time to develop the first two steps before formulating the general conclusion from n to $n + 1$. How many examples illustrate the fundamental theorems of his algebraic papers–examples not constructed ad hoc, but genuine ones worth being studied for their own sake!"

The best presentation of algebraic number theory in the spirit of Dedekind and Hilbert was given by Hecke [Hec]. For more modern presentations, see [Mar], [Rib], [Bor - Sha] and [Lan 3].

The rest of this chapter is concerned with the general notion of an algebraic integer and a detailed studied of quadratic fields. In the next chapter, we will present Dedekind's ideal theory for quadratic fields.

2. Algebraic Integers

The search for a good generalization of the rational integers to algebraic number fields followed a long and circuitous route. The most natural first approach, as followed by Gauss, Kummer and others, was to take the ring

$\mathbf{Z}[\alpha]$ in the field $\mathbf{Q}(\alpha)$. Although, as we now know, this suffices for the Gaussian integers and cyclotomic fields, we shall see that even for quadratic fields a larger ring is desirable in many cases. Gauss and Kummer were both lucky in treating cases where $\mathbf{Z}[\alpha]$ suffices. Weil [Wei 8] points out that Eisenstein used the currently accepted definition of algebraic integer in 1851, but he wrote as though it was not his own original idea. It is not clear whether Dedekind, who presented the definition in 1871 [Dir], along with a deep study of its implications, knew of Eisenstein's work.

We do not know how Eisenstein and Dedekind were led to their definition of an algebraic integer but, following Hecke [Hec], we present one possibility, whose general plausibility is reinforced by Dedekind's axiomatic attitudes. We list a set of properties that we would like algebraic integers to satisfy. Assuming these properties, we force out a condition for an algebraic number to be an algebraic integer which we then take as our definition. Then we will study those algebraic integers in a fixed number field K in order to generalize arithmetic from $\mathbf{Q}$ to K.

The properties that we would like the set Ω of all algebraic integers (as a subset of the algebraic numbers) to satisfy are

a) $\Omega \cap \mathbf{Q} = \mathbf{Z}$,
b) Ω is a ring,
c) if $\alpha \in \Omega$, then the **conjugates** of α (the other roots of the monic irreducible polynomial for α over $\mathbf{Q}$) belong to Ω.

Now suppose that the algebraic number α, which is a root of the irreducible equation $x^n + b_{n-1}x^{n-1} + \cdots + b_0 = 0$, $b_i \in \mathbf{Q}$, is an algebraic integer. Then, since each b_i is a symmetric function of α and its conjugates, we see by properties b) and c) that the b_i must be algebraic integers and therefore, by property a), they are rational integers. Hence we are led to the following:

Definition: An algebraic number α is an **algebraic integer** if the monic irreducible equation for α over $\mathbf{Q}$, $p(x) = 0$, has rational integer coefficients. The **degree** of α is the degree of $p(x)$.

It follows immediately from the definition that the set Ω of all algebraic integers satisfies property a), which says that the algebraic integers of degree one are the rational integers, and property c). It is also true that Ω is a ring. We will not prove this, but refer the reader to [Hec] for the classical proof using symmetric functions and [Ire-Ros] for the modern proof via modules. We shall see that the ring property is trivial for quadratic number fields

Note that the algebraic numbers of degree one are the rational integers.

Proposition: *If α is the root of a (not necessarily irreducible) monic polynomial with coefficients in* $\mathbf{Z}$, *then α is an algebraic integer.*

Proof: Exercise - (Hint: look at the kernel of the homomorphism $\mathbf{Z}[x] \to \mathbf{Z}$, where $f(x) \mapsto f(\alpha)$, and recall that $\mathbf{Z}[x]$ is a principal ideal domain.) See [Hec] for a more classical proof.

If α in Ω is a root of $f(x) = x^n + b_{n-1}x^{n-1} + \cdots + b_0 = 0$, $b_i \in \mathbf{Z}$, then any k^{th} root of α is a root of $f(x^k) = 0$. Thus k^{th} roots of elements of Ω are in Ω, and there are no natural irreducible (prime) elements of Ω, i.e., the arithmetic of $\mathbf{Z}$ does not generalize to Ω. Nevertheless, we shall see that there is a rich factorization theory for the integers of a fixed number field.

We define **the ring of integers $\mathbf{I_K}$ of an algebraic number field K** by $\mathbf{I_K} = \Omega \cap \mathbf{K}$ (since Ω and K are rings, I_K is also a ring). The elements of I_K are called the **integers of K**. Note that this notion is independent of field extensions, i.e., if α is an integer of K and L is an extension of K ($K \subseteq L$), then α is an integer of L.

From now on, we concentrate on quadratic fields as a prototype for the more general theory.

3. Quadratic Fields

The **quadratic fields** are the fields

$$\mathbf{Q}(\sqrt{D}) = \{x + y\sqrt{D} \mid x, y \in \mathbf{Q}\}\,, \quad \text{where } D \in \mathbf{Q}\,,\ D \neq \text{ square}\,.$$

Exercises: 1) Prove that the quadratic fields are all the extensions of $\mathbf{Q}$ of degree 2. Hence the elements of $\mathbf{Q}(\sqrt{D})$ are rational numbers (algebraic numbers of degree one) or **quadratic numbers** (algebraic numbers of degree two).

2) Prove that $\mathbf{Q}\left(\sqrt{\frac{x}{y}}\right) = \mathbf{Q}\left(\sqrt{xy}\right)$, when $x, y \in \mathbf{Z}$.

3) Prove that $\mathbf{Q}\left(\sqrt{xy}\right) = \mathbf{Q}(\sqrt{d})$, $x, y \in \mathbf{Z}$, where d is the square free integer derived from xy by removing all square factors.

From 2), we see that the quadratic fields $\mathbf{Q}(\sqrt{D})$ are the fields $\mathbf{Q}(\sqrt{d})$, where $d \in \mathbf{Z}$, and, by 3), we can assume d is square free. Thus the quadratic fields are the fields

$$\mathbf{Q}(\sqrt{d}) = \{x + y\sqrt{d} \mid x, y \in \mathbf{Q}\ ,\ d \in \mathbf{Z}\ ,\ d \text{ square free}\}\ .$$

From now on, when we write $\mathbf{Q}(\sqrt{d})$, we assume that d is a square free integer.

$\mathbf{Q}(\sqrt{d})$ is a **real quadratic field** if $d > 0$, and an **imaginary quadratic field** if $d < 0$. If $\alpha = x + y\sqrt{d} \in \mathbf{Q}(\sqrt{d})$, then

$$\begin{aligned}
&\boldsymbol{\alpha}' = x - y\sqrt{d} \text{ is the } \textbf{conjugate of } \alpha \textbf{ in } \mathbf{Q}(\sqrt{d})\ ,\\
&\mathbf{N}(\boldsymbol{\alpha}) = \alpha\alpha' = x^2 - dy^2 \text{ is the } \textbf{norm of } \alpha\ , \text{ and}\\
&\mathbf{Tr}(\boldsymbol{\alpha}) = \alpha + \alpha' = 2x \text{ is the } \textbf{trace of } \alpha\ .
\end{aligned}$$

The proofs of the following propositions are simple exercises.

Proposition: *If $\alpha, \beta \in \mathbf{Q}(\sqrt{d})$, then*

1) $(\alpha \pm \beta)' = \alpha' \pm \beta'$,
2) $(\alpha\beta)' = \alpha'\beta'$,
3) $\left(\dfrac{\alpha}{\beta}\right)' = \dfrac{\alpha'}{\beta'}$
4) $\alpha = \alpha' \iff \alpha \in \mathbf{Q}$.

Proposition: *If $\alpha, \beta \in \mathbf{Q}(\sqrt{d})$, then*

1) $N(\alpha), \mathrm{Tr}(\alpha) \in \mathbf{Q}$,
2) $N(\alpha\beta) = N(\alpha)N(\beta)$, $\mathrm{Tr}(\alpha + \beta) = \mathrm{Tr}(\alpha) + Tr(\beta)$,
3) $N(\alpha) = 0 \iff \alpha = 0$,
4) *α is a root of $x^2 - \mathrm{Tr}(\alpha)x + N(\alpha) = 0$ and if $\alpha \notin \mathbf{Q}$, this polynomial is irreducible.*

Some excellent references on quadratic fields are [Hil], [Ada -Gol], [Rei, L], [Coh], [Hec], [Bor - Sha], [Zag], [Sch - Opo], and [Inc].

4. Quadratic Integers

We now determine the set $\mathbf{I_d}$ $(= I_{\mathbf{Q}(\sqrt{d})})$ of algebraic integers of $\mathbf{Q}(\sqrt{d})$.

Let $\alpha = u + v\sqrt{d} \in I_d$, where $u, v \in \mathbf{Q}$.

If $\alpha \in \mathbf{Q}$ $(v = 0)$, then it satisfies the irreducible equation $x - \alpha = 0$ and, by the definition of an algebraic integer, $\alpha \in \mathbf{Z}$; hence $I_d \cap \mathbf{Q} \subseteq \mathbf{Z}$. The converse is obvious and we see that $I_d \cap \mathbf{Q} = Z$.

If $\alpha \notin \mathbf{Q}$ $(v \neq 0)$, then it is a **quadratic integer** (algebraic integer of degree two), i.e., a root of the irreducible equation

$$x^2 - \mathrm{Tr}(\alpha)x + N(\alpha) = 0 ,$$

where, by the definition of an algebraic integer, $\mathrm{Tr}(\alpha) = \alpha + \alpha' = 2u$ and $N(\alpha) = u^2 - dv^2$ are in $\mathbf{Z}$.

Hence, if we set

$$u_1 = 2u , \quad v_1 = 2v ,$$

then

$$\mathrm{Tr}(\alpha) = u_1, \;\; N(\alpha) = \frac{u_1^2 - dv_1^2}{4} \in \mathbf{Z} ,$$

i.e.,

$$u_1 \in \mathbf{Z}, \;\; u_1 \equiv dv_1^2 (\mathrm{mod}\, 4) .$$

Claim: $v_1 \in \mathbf{Z}$.

From the last equation, we have

$$dv_1^2 = 4k + u_1^2 , \quad \text{for some } k \in \mathbf{Z} .$$

Since the right hand side is an integer, we have $dv_1^2 \in \mathbf{Z}$. Let $v_1 = \frac{a}{b}$, where $a, b \in \mathbf{Z}$, $(a, b) = 1$. If $|b| > 1$, then, since d is square free, $d\left(\frac{a}{b}\right)^2$ is not an integer, which is a contradiction. Thus the denominator of v_1 is $+1$, i.e., $v_1 \in \mathbf{Z}$.

To summarize

$$\alpha \in I_d \implies u_1, v_1 \in \mathbf{Z} \;\text{ and }\; u_1^2 \equiv dv_1^2 (\mathrm{mod}\, 4) .$$

We examine different cases modulo 4, noting that since d is square free, $d \not\equiv 0 (\mathrm{mod}\, 4)$.

i) $d \equiv 2, 3 \pmod 4$:

Checking all cases modulo 4 (exercise), we find that the only solutions of $u_1^2 \equiv dv_1^2 \pmod 4$ are those for which u_1, v_1 are even. Thus

$$\alpha \in I_d \Longrightarrow \alpha = \frac{u_1 + v_1\sqrt{d}}{2},$$

$$u_1, v_1 \in 2\mathbf{Z} \Longrightarrow \alpha = u + v\sqrt{d}, u_1, v_1 \in \mathbf{Z}.$$

So $I_d \subseteq \mathbf{Z}[\sqrt{d}]$. But every number of the form $u + v\sqrt{d}$, where $u, v \in \mathbf{Z}$, has trace and norm in $\mathbf{Z}$, i.e., $\mathbf{Z}[\sqrt{d}] \subseteq I_d$. Therefore, for $d \equiv 2, 3 \pmod 4$,

$$I_d = \{u + v\sqrt{d} | u, v \in \mathbf{Z}\} = \mathbf{Z}[\sqrt{d}],$$

just what everyone expects.

ii) $d \equiv 1 \pmod 4$:

Checking all cases modulo 4, we find that the only solutions of $u_1^2 \equiv dv_1^2 \pmod 4$ are those for which $u_1 \equiv v_1 \pmod 2$, i.e., u_1 and v_1 are both odd or both even.

If u_1 and v_1 are even, then we have the same conditions as in case i) and $\mathbf{Z}[\sqrt{d}] \subseteq I_d$.

If u_1 and v_1 are both odd, then

$$\alpha = \frac{u_1}{2} + \frac{v_1}{2}\sqrt{d} = u + v\sqrt{d},$$

where u, v are half integers, i.e., $u, v \in \mathbf{Z} + \frac{1}{2} \left(= \left\{n + \frac{1}{2} \mid n \in \mathbf{Z}\right\}\right)$. Conversely any such number has trace and norm in $\mathbf{Z}$. Therefore, for $d \equiv 1 \pmod 4$

$$I_d = \left\{u + v\sqrt{d} \mid u, v \in \mathbf{Z} \quad \text{or} \quad u, v \in \mathbf{Z} + \frac{1}{2}\right\}.$$

Hence for the first time we see that the set of algebraic integers of a number field $\mathbf{Q}(\theta)$ can be larger than $\mathbf{Z}[\theta]$.

Putting these results together, we have

$$\mathbf{I_d} = \Big\{\mathbf{u} + \mathbf{v}\sqrt{\mathbf{d}} \mid \mathbf{u}, \mathbf{v} \in \mathbf{Z} \text{ if } \mathbf{d} \equiv \mathbf{2}, \mathbf{3} \pmod{\mathbf{4}}, \text{ and}$$

$$\mathbf{u}, \mathbf{v} \in \mathbf{Z} \text{ or } \mathbf{u}, \mathbf{v} \in \mathbf{Z} + \frac{\mathbf{1}}{\mathbf{2}} \text{ if } \mathbf{d} \equiv \mathbf{1} \pmod{\mathbf{4}}\Big\}.$$

From this result, it follows immediately, by direct computation, that

Corollary: *I_d is a ring.*

We reformulate our characterization of I_d slightly. Note that if $d \equiv 2, 3 \pmod 4$, then $I_d = \mathbf{Z}[\sqrt{d}]$ is the set of all integer linear combinations of $\{1, \sqrt{d}\}$, and that 1 and $\sqrt{d}$ are linearly independent over $\mathbf{Z}$. The case $d \equiv 1 \pmod 4$ can be cast in a similar form. Consider the numbers

$$u + v\left(\frac{1}{2} + \frac{1}{2}\sqrt{d}\right) = \left(u + \frac{v}{2}\right) + \frac{v}{2}\sqrt{d}\,, \quad u, v \in \mathbf{Z}\,. \tag{1}$$

If v is even, $\frac{v}{2} \in \mathbf{Z}$, then, as u takes all values in $\mathbf{Z}$, we can make $u + \frac{v}{2}$ equal to any given integer. Hence, by (1),

$$\left\{u + v\left(\frac{1}{2} + \frac{1}{2}\sqrt{d}\right) \middle| u \in \mathbf{Z}, v \in 2\mathbf{Z}\right\} = \left\{u' + v'\sqrt{d} | u, v \in \mathbf{Z}\right\}\,. \tag{2}$$

If v is odd, $\frac{v}{2} \in \mathbf{Z} + \frac{1}{2}$, then as u takes all values in $\mathbf{Z}$, $u + \frac{v}{2}$ takes all values in $\mathbf{Z} + \frac{1}{2}$. Thus, by (1),

$$\left\{u + v\left(\frac{1}{2} + \left(\frac{1}{2}\right)\sqrt{d}\right) \middle| u \in \mathbf{Z}, v \in 2\mathbf{Z} + 1\right\}$$
$$= \left\{u' + v'\sqrt{d} \,\middle|\, u, v \in \mathbf{Z} + \frac{1}{2}\right\}\,. \tag{3}$$

Combining (2) and (3), we see that if $d \equiv 1 \pmod 4$, then

$$I_d = \left\{u + v\left(\frac{1}{2} + \frac{1}{2}\sqrt{d}\right) \middle| u, v \in \mathbf{Z}\right\} = \mathbf{Z}\left[\frac{1}{2} + \frac{1}{2}\sqrt{d}\right]\,,$$

i.e., I_d consists of all integer linear combinations of 1 and $\frac{1}{2} + \frac{1}{2}\sqrt{d}$. Moreover 1 and $\frac{1}{2} + \frac{1}{2}\sqrt{d}$ are linearly independent over $\mathbf{Z}$,

If we set

$$\eta = \begin{cases} q\sqrt{d}\,, & \text{if } d \equiv 2, 3 \pmod 4\,, \\ \frac{1}{2} + \frac{1}{2}\sqrt{d}\,, & \text{if } d \equiv 1 \pmod 4\,, \end{cases}$$

then

$$\mathbf{I_d} = \mathbf{Z}[\eta] = \{\mathbf{u} + \mathbf{v}\eta | \mathbf{u}, \mathbf{v} \in \mathbf{Z}]\,.$$

We say that $\{1, \eta\}$ is an **integral basis** for I_d, i.e., a finite set, linearly independent over $\mathbf{Z}$, such that every element of I_d is a rational integer linear combination of elements of the set.

Note that the irreducible polynomials for η are

$$x^2 - d\,, \qquad \text{if } d \equiv 2, 3 \pmod 4\,,$$

$$x^2 - x + \frac{1-d}{4}\,, \qquad \text{if } d \equiv 1 \pmod 4\,.$$

The corresponding forms $x^2 - dy^2$ and $x^2 - xy + \frac{1-d}{4}y^2$ were essentially the examples used in section 12.3 to show that for every discriminant d, there exists a quadratic form with this discriminant ($x^2 - xy + \frac{1-d}{4}y^2$ plays the same role as $x^2 + xy + \frac{1-d}{4}y^2$ for this purpose). This is another hint of a connection between forms and fields.

5. Geometric Representation; Divisibility and Units

We generalize our geometric representation of Gaussian integers as integer points in the plane (sec.15.2) to all quadratic fields.

The number $\alpha = a + b\sqrt{d} \in \mathbf{Q}(\sqrt{d})$ is represented by the point $(a, b\sqrt{|d|})$ in the plane. When no confusion can arise, we *use the terms number and point interchangeably.* For imaginary quadratic fields, it is usually convenient to identify the plane $\mathbf{R}^2$ with the complex plane and refer to the complex number $a + b\sqrt{d}$ as a point in the plane.

In the last section, we saw that the algebraic integers I_d are $\mathbf{Z}$ linear combinations of 1 and η. Thus the corresponding points in the plane are the integer linear combinations of

$$(1,0), (0, \sqrt{|d|})\,, \qquad \text{if } d \equiv 2, 3 \pmod 4\,,$$

$$(1,0), \left(\tfrac{1}{2}, \left(\tfrac{1}{2}\right)\sqrt{|d|}\right)\,, \qquad \text{if } d \equiv 1 \pmod 4\,.$$

Figures 1 and 2 illustrate two special cases.

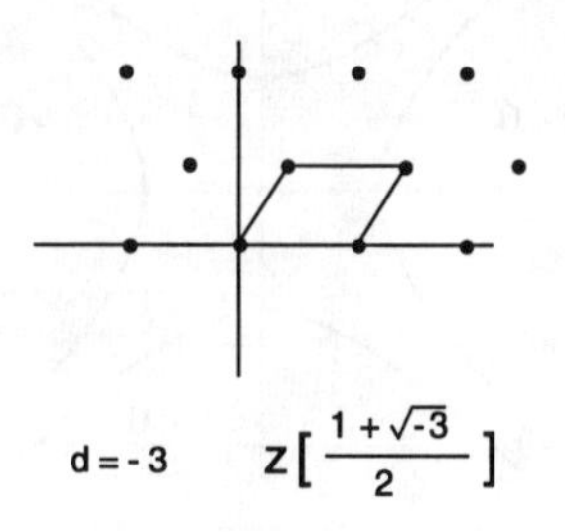

Figure 1

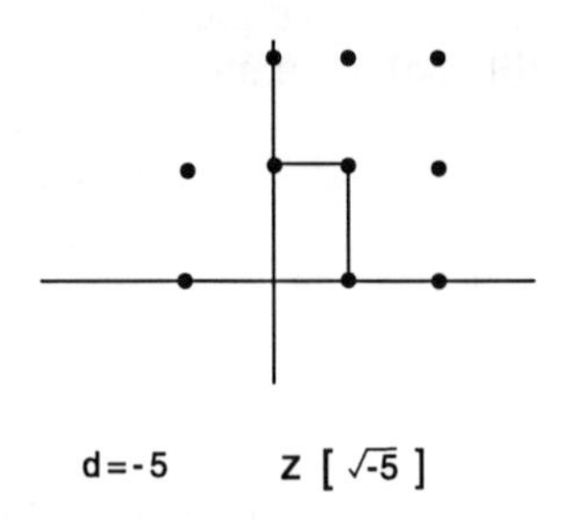

Figure 2

These regular patterns of points in the plane are called *lattices* (not to be confused with lattices in the theory of partially ordered sets). When we study ideals, we will make these notions more precise. For now we stress geometric intuition. Note that if $d \equiv 2, 3 \pmod 4$, the pattern is formed by rectangles and if $d \equiv 1 \pmod 4$, by non rectangular parallelograms.

The geometry is quite different for real and imaginary quadratic fields. Let $\alpha = a + b\sqrt{d}$.

1) If $d < 0$, then

$$N(\alpha) = a^2 + |d|b^2 = a^2 + (b\sqrt{|d|})^2$$

is just the square of the distance from the corresponding point $(a, b\sqrt{|d|})$ to the origin. The curve $N(\alpha) = c > 0$ is a circle centered at the origin, and every element in $\mathbf{Q}(\sqrt{d})$ lies on exactly one such circle.

2) If $d > 0$, $N(\alpha) = a^2 - db^2$ and $N(\alpha) = c$ is a rectangular hyperbola centered at the origin (fig. 3) and every element of $\mathbf{Q}(\sqrt{d})$ lies on exactly one such hyperbola.

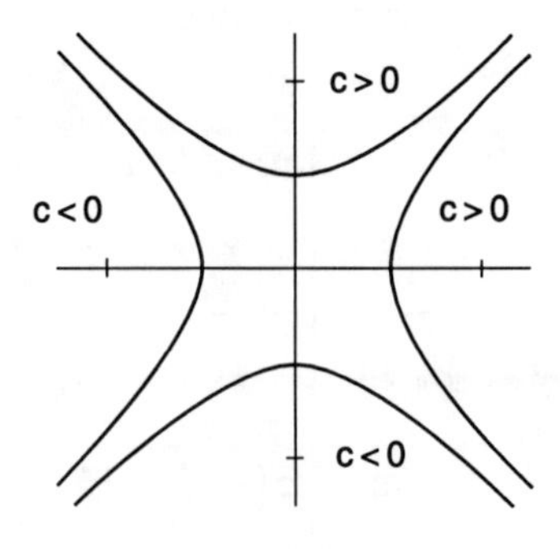

Figure 3

DIVISIBILITY AND UNITS

Since I_d is a commutative ring with identity 1, we recall some notions from ring theory (sec. 9.2). If $\alpha, \beta \in I_d$ then α **divides** β, $\alpha|\beta$, if there exists a $\gamma \in I_d$ such that $\beta = \alpha\gamma$. A **unit** in I_d is an element dividing 1. The units of I_d are a subgroup of the multiplicative group of the ring (exercise). Let $\mathbf{U_d}$ denote this group of units.

Lemma: *If $\alpha \in I_d$, then*

$$\alpha \text{ is a unit} \iff N(\alpha) = \pm 1 .$$

Proof: Exercise - essentially the same proof as for the Gaussian integers (sec.15.2), but norms can be negative if $d > 0$.

By the lemma, if $d < 0$, then the unit $x + y\sqrt{d}$ satisfies $x^2 + |d|y^2 = 1$. It is a trivial exercise to find all solutions of $x^2 + |d|y^2 = 1$, where $x, y \in \mathbf{Z}$ if $d \equiv 2, 3 (\bmod 4)$ and $x, y \in \mathbf{Z}$ or $x, y \in \mathbf{Z} + \frac{1}{2}$ if $d \equiv 1 (\bmod 4)$. Solving the equation then yields

$$U_d = \begin{cases} \{\pm 1, \pm i\}, & \text{if } d = -1, \\ \{\pm 1, \pm\omega, \pm\omega'\}, & \text{if } d = -3, \text{ where } \omega = \frac{1}{2} + (\frac{1}{2})\sqrt{-3} \\ & \text{and } \omega' \text{ is the complex conjugate of } \omega, \\ \{\pm 1\}, & \text{if } d = -2 \text{ or } d < -3. \end{cases}$$

Note that for $d = -1$, the units are the fourth roots of unity, for $d = -3$, they are the sixth roots of unity, and for $d = -2$ or $d < -3$, they are the square roots of unity. Thus U_d, with $d < 0$, consists of roots of unity. Obviously, if α in some I_d is a root of unity, say $\alpha^k = 1$, then $\alpha\alpha^{k-1} = 1$, and α is a unit. Therefore we have found all roots of unity which appear in any I_d.

Observe that the number of units in I_d equals the number of automorphisms of positive definite forms of discriminant d (sec. 13.7), indicating another connection between forms and fields.

If $d > 0$ then, by our lemma, $\alpha = x + y\sqrt{d}$ is in U_d if and only if

$$N(\alpha) = x^2 - dy^2 = \pm 1. \qquad \text{(Pell's equation)}$$

Solving these equations for $x, y \in \mathbf{Z}$ and $d \equiv 2, 3 (\bmod 4)$ yields all units of I_d. If $d \equiv 1 (\bmod 4)$, then $x, y \in \mathbf{Z}$ or $x, y \in \mathbf{Z} + \frac{1}{2}$, and we let $x = \frac{x_1}{2}$ and $y = \frac{y_1}{2}$, with $x_1, y_1 \in \mathbf{Z}$ and solve $x_1^2 - dy_1^2 = \pm 4$.

We have seen that Pell's equation can be solved using the continued fraction for $\sqrt{d}$, and the solutions form a group. We also noted that the equation $x_1^2 - dy_1^2 = \pm 4$ can be treated in a similar way (sec. 4.9). These results yield the following:

Theorem: *If $d > 0$, then there is a unit κ in I_d, the* **fundamental unit**, *which is the smallest unit greater than 1. This unit corresponds to the minimum positive solution of Pell's equation and U_d is a cyclic group generated by κ, i.e.,*

$$U_d = \{\pm\kappa^n \mid n \in \mathbf{Z}\} .$$

This continued fraction approach includes techniques for computing fundamental units. For a more direct, but less constructive approach, see [Ire - Ros, sec. 13.1].

6. Factorization in Quadratic Fields

In line with our historical discussion in section 1, we now consider factorization in I_d. We first take a brief look at fields with unique factorization, and then we study examples of non unique factorization.

The norm plays a key role in these studies as it measures the size of integers. Recall that $N(\alpha\beta) = N(\alpha)N(\beta)$, and if $\alpha \in I_d$ then $N(\alpha) \in \mathbf{Z}$. Hence, if α, β are in I_d and $\alpha|\beta$, then $N(\alpha)|N(\beta)$ and we have some control on the size of the norms of all divisors of β.

Again recall the distinction (in any commutative ring) between *irreducible* (α is not the product of two non-units) and *prime* ($\alpha|\beta\gamma \implies \alpha|\beta$ or $\alpha|\gamma$). The key to proving the uniqueness of prime factorization, when such a factorization exists, is showing that all irreducibles are primes.

Recall that a **unique factorization domain** (**UFD**) is a ring in which every non unit is a product of primes, unique up to order and associates. The general problem of determining for which d, I_d is a UFD, is still unsolved. However, it is true that *for all d, every non-unit of I_d can be factored into irreducibles in I_d.*

For suppose that a non unit α is in I_d and

$$\alpha = \alpha_1\alpha_2\cdots\alpha_n , \quad \alpha_i \text{ non-units (so } |N(\alpha)| \geq 2) .$$

Then

$$|N(\alpha)| = |N(\alpha_1)|\cdots|N(\alpha_n)| \geq 2^n$$

and

$$n \leq \log_2 |N(\alpha)| \ .$$

Thus any factorization of α into non-units has at most $\log_2 |N(\alpha)|$ factors. Now if α is not irreducible, then $\alpha = \beta_1\beta_2$, for some non-units β_1 and β_2. If either of the β's are not irreducible then they can also be factored into non-units and so on. But we just saw that α can be factored into at most $\log_2 |N(\alpha)|$ non-units. Hence this factorization process must stop and produces a factorization of α into irreducible.

When the irreducibles are prime, we have unique factorization. We consider special classes of rings since, as we shall see, not all I_d's are UFD's. Note that since all I_d are subrings of **C**, they have no zero-divisors and thus are integral domains.

7. Euclidean Domains and Unique Factorization

We recall that I_d is a **Euclidean domain with respect to its norm** or **norm Euclidean** if it has a "division algorithm", i.e., if for every pair $\alpha, \beta \in I_d$, $\beta \neq 0$, there exist $\gamma, \delta \in I_d$ such that

$$\alpha = \beta\gamma + \delta \ ,$$

with $|N(\delta)| < |N(\beta)|$ (we use absolute values since the norm may be negative in a real quadratic field).

While studying the Gaussian integers (sec. 15.2), we outlined a proof that

$$\text{norm Euclidean} \implies \text{UFD} \ .$$

To determine which I_d are norm Euclidean, we reformulate our definition.

Theorem:

I_d is norm Euclidean $\iff$ For every $\mu \in \mathbf{Q}(\sqrt{d})$ there exists a $\gamma \in I_d$ such that $|N(\mu - \gamma)| < 1$.

Proof: $\Leftarrow$) Let $\alpha, \beta \in I_d$. Then $\frac{\alpha}{\beta} \in \mathbf{Q}(\sqrt{d})$ and, by assumption, there is a $\gamma \in I_d$ such that $|N(\frac{\alpha}{\beta} - \gamma)| < 1$. Let $\mu = \frac{\alpha}{\beta} - \gamma$ and $\delta = \beta\mu$. Then

$\alpha = \beta\gamma + \beta\mu = \beta\gamma + \delta$. Moreover $\delta = \alpha - \beta\gamma$ is in I_d and

$$|N(\delta)| = |N(\beta\mu)| = \left|N\left(\beta\left(\frac{\alpha}{\beta} - \mu\right)\right)\right| =$$

$$|N(\beta)|\left|N\left(\frac{\alpha}{\beta} - \mu\right)\right| < |N(\beta)|;$$

hence I_d is norm Euclidean.

$\Longrightarrow$) If $\mu \in \mathbf{Q}(\sqrt{d})$ then $\mu = \frac{\alpha}{\beta}$, for some $\alpha, \beta \in I_d$ (in particular, if $\mu = \frac{r}{s} + \frac{r'}{s'}\sqrt{d} = \frac{rs'+r's\sqrt{d}}{ss'}$, let $\alpha = rs' + r's\sqrt{d}$, $\beta = ss'$). The rest of the proof is straightforward and is left as an exercise.

Now we use the theorem to determine all norm Euclidean *imaginary* quadratic fields.

i) $d < 0$, $d \equiv 2$ or $3 \pmod 4$: In this case, the elements of I_d form a rectangular lattice in the *complex* plane (fig. 4).

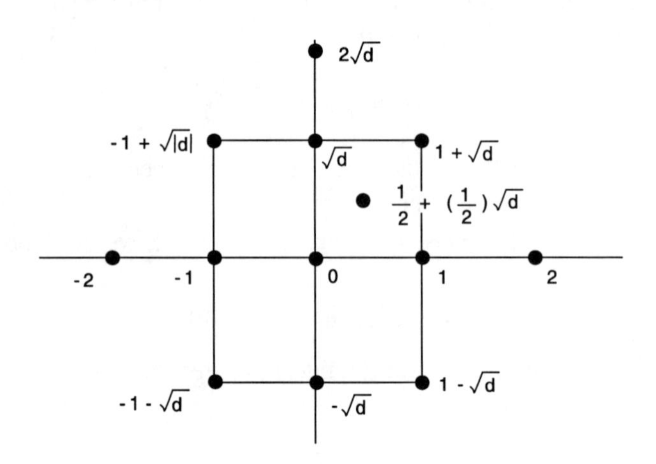

Figure 4

The distance between numbers in $\mathbf{Q}(\sqrt{d})$ refers to the distance between the corresponding points in the complex plane.

It is clear from the symmetry of the lattice that the maximum distance between a number in $\mathbf{Q}(\sqrt{d})$ and an integer is realized by the distance from

$0 \in I_d$ to $\frac{1}{2} + \frac{1}{2}\sqrt{d} \in \mathbf{Q}(\sqrt{d})$. Thus, by the previous theorem

$$I_d \text{ is norm Euclidean } \iff N\left(\frac{1}{2} + \frac{1}{2}\sqrt{d}\right) = \frac{1}{4}(1 + |d|) < 1 .$$

But since $d < 0$, square free and $\equiv 2, 3 (\text{mod } 4)$, the only solutions to this inequality are $d = -1$ and $d = -2$.

ii) $d < 0$, $d \equiv 1 (\text{mod } 4)$: In this case the lattice is shown in figure 5.

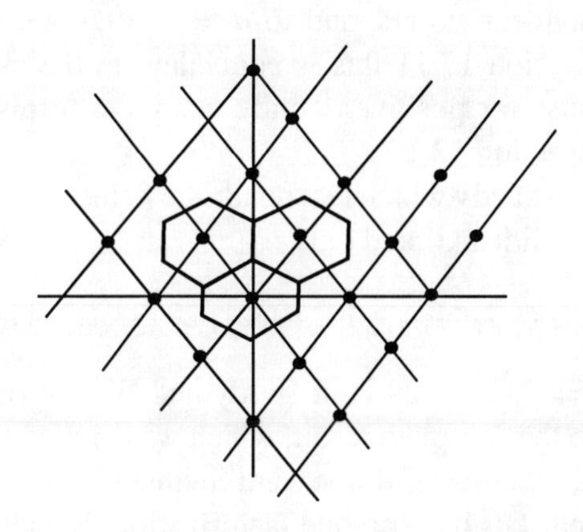

Figure 5

Drawing the hexagons so that the sides bisect lines connecting the integers, it follows from the symmetry of the lattice that the maximum distance from a point in $\mathbf{Q}(\sqrt{d})$ to an integer is realized by the distance between the top vertex of the centermost hexagon and the origin.

This top point is the intersection of the vertical axis with the line passing through $\frac{1}{2}\left(\frac{1}{2} + \frac{1}{2}\sqrt{d}\right)$ with slope $\left(\frac{-1}{\frac{1}{2}\sqrt{|d|}}\right)/\frac{1}{2}$, viz., the point $\frac{1}{4}\left(\sqrt{d} + \frac{1}{\sqrt{d}}\right)$. Thus

$$I_d \text{ is norm Euclidean } \iff \left|N\left(\frac{1}{4}\left(\sqrt{d} + \frac{1}{\sqrt{d}}\right)\right)\right|$$
$$= \frac{1}{16}\left(|d| + 2 + \frac{1}{|d|}\right) < 1 .$$

The only square free $d < 0, d \equiv 1 (\bmod 4)$, satisfying this inequality are $d = -3, -7, -11$.

To summarize:

Theorem: *If $d < 0$, then*

$$I_d \text{ is norm Euclidean} \iff d = -1, -2, -3, -7, -11 .$$

The other imaginary quadratic fields that have unique factorization but are not norm Euclidean correspond to $d = -19, -43, -67, -163$. As we shall see in section 17.11 this is equivalent to the class number one conjecture of Gauss for positive definite quadratic forms, whose history was discussed in section 12.10.

It is not known exactly which real quadratic fields, are UFD's. If $d > 0$, then I_d is norm Euclidean if and only if

$$d = 2, 3, 5, 6, 7, 11, 13, 17, 19, 21, 29, 33, 37, 47, 57, 73 .$$

The proof for $d = 2, , \dots, 29$ is in Hardy and Wright [Har - Wri] along with references to the complete result, which was proved in the 1940's and 50's by Chatland, Davenport, Barnes and Swinnerton-Dyer.

Many other real I_d's have unique factorization. Recall that Gauss conjectured that there are infinitely many $D > 0$ such that the indefinite forms of discriminant D have class number one. As we shall later see, this is equivalent to the conjecture that I_d is a UFD for infinitely many $d > 0$, which is still unsolved.

8. Non-Unique Factorization and Ideals

Not all I_d's have unique factorization.

Theorem: *Let $d < 0, d \neq -1, -2$, d square free and $d \equiv 2, 3 (\bmod 4)$. Then I_d is not a unique factorization domain.*

Proof: i) $d \equiv 3 (\bmod 4)$: Assume that I_d is a UFD. We show that there are two factorizations of $1 - d$ into non associated irreducible elements, viz., $1 - d = 2\left(\frac{1-d}{2}\right) = (1 + \sqrt{d})(1 - \sqrt{d})$, contradicting the assumption. If we show that 2 is irreducible then, since $\frac{1 \pm \sqrt{d}}{2}$ is not in I_d, there is no refinement of the first factorization which will make 2 an associate of an irreducible factor of the second factorization.

Since $N(2) = 4$, any irreducible factor of 2 must have norm 2. But

$$N(x + y\sqrt{d}) = x^2 + |d|y^2 = 2$$

has no integer solution for $|d| \geq 5$.

(ii) $d \equiv 2 (\mathrm{mod}\, 4)$: The same type of reasoning as in (i), applied to $-d = (-2)\left(\frac{d}{2}\right) = (-\sqrt{d})(\sqrt{d})$, will yield the result.

Stark [Sta] has a nice elementary treatment of unique factorization, together with applications to the solution of Diophantine equations along the lines of Euler's approach (sec. 3.5).

Now we follow Hecke [Hec, sec. 23] in analyzing an example of non unique factorization in more detail to try to understand how Dedekind may have proceeded in restoring unique factorization. Keep in mind that Kummer had already introduced "ideal elements" for cyclotomic fields and Dedekind's goal was to make this notion more precise and to generalize to other number fields.

Example: In $I_{-5} = \mathbf{Z}[\sqrt{-5}]$, we have the factorization

$$21 = 3 \cdot 7 = (1 + 2\sqrt{-5})(1 - 2\sqrt{-5}) .$$

Let $\alpha = (1 + 2\sqrt{-5})$; so $\alpha' = (1 - 2\sqrt{-5})$. The norm can now be used, as we did for Gaussian primes, to show that α, α', 3 and 7 are irreducible and hence factorization in I_{-5} is not unique. We carry out the proof for 3. Similar arguments work for α, α' and 7.

If $3 = \delta\rho$, δ and ρ not units, then $N(3) = 9 = N(\delta)N(\rho)$. Thus $N(\delta) = N(\rho) = 3$. Since $-5 \equiv 3 (\mathrm{mod}\, 4)$, we have $\delta = x + y\sqrt{-5}$, for some $x, y \in \mathbf{Z}$, and $N(\delta) = x^2 + 5y^2 = 3$. Clearly there are no integer solutions to the last equation and therefore 3 is irreducible.

Thus 3 and α are irreducible and clearly not associates; they have no common factor in I_{-5}. To understand why factorization is not unique and how to still find a fruitful arithmetic theory, we must work in the ring Ω of all algebraic integers.

In section 2, we saw that if λ is an algebraic integer, then so is $\sqrt{\lambda}$. If we let $\lambda = 2 + \sqrt{-5}$ and $\chi = 2 + 3\sqrt{-5}$, then it is easy to check that

$$\alpha^2 = \lambda(-\chi') , \quad \alpha'^2 = \lambda'(-\chi) ,$$
$$3^2 = \lambda\lambda' , \quad 7^2 = \chi\chi' .$$

But the square roots of $\lambda, \lambda', \chi, \chi', -\chi, -\chi'$ (which are in Ω) are not in I_{-5}. By an appropriate choice of square roots, it can be shown that

$$\begin{aligned} \alpha &= \sqrt{\lambda}\sqrt{-\chi'}\,, & \alpha' &= \sqrt{\lambda'}\sqrt{-\chi}\,, \\ 3 &= \sqrt{\lambda}\sqrt{\lambda'}\,, & 7 &= \sqrt{\chi'} = (\sqrt{-\chi})(-\sqrt{-\chi})\,, \end{aligned}$$

and the two factorizations of 21 have the common refinement

$$21 = \sqrt{\lambda}\sqrt{\lambda'}\sqrt{-\chi}\sqrt{-\chi'}\,.$$

It can also be shown [Hec, sec.23] that $\sqrt{\lambda}$ acts like a greatest common divisor for α and β, viz.,

$$\sqrt{\lambda}|\alpha,\ \sqrt{\lambda}|\beta \ \text{ and if } \ \rho \in \Omega,\ \rho|\alpha, \beta, \ \text{ then } \ \rho|\lambda\,.$$

We could try to adjoin $\sqrt{\lambda}$, $\sqrt{\lambda'}$ etc. to $\mathbf{Q}(\sqrt{-5})$ but the resulting number field would have new integers that might not have unique factorization. Instead Dedekind took the set J of all numbers in I_{-5} divisible by $\sqrt{\lambda}$ as a new entity "which represents" $\sqrt{\lambda}$ in I_{-5}. So he stayed in the ring by regarding certain subsets of I_{-5} as "new numbers". If $\sqrt{\lambda}|\alpha,\beta \in I_{-5}$, then

$$\sqrt{\lambda}|(\mu\alpha + \rho\beta)\,,$$

for all $\mu, \rho \in I_{-5}$. Thus J is an ideal in the ring I_{-5}.

Dedekind introduced the term 'ideal' in analogy with the ideal objects Kummer adjoined to cyclotomic fields. These ideals play the role of greatest common divisors. A number ρ in I_{-5} will be represented by all the number in I_{-5} which it divides, viz., the ideal ρI_{-5}.

Thus for Dedekind new objects should be constructed as sets of known objects rather than by adding new symbols satisfying special conditions (as, for example, Kummer did with his ideal objects). Dedekind used the same philosophy in constructing the real numbers from the rationals by defining real numbers as Dedekind cuts, i.e., certain pairs of sets of rational numbers. For more details on Kummer's ideal numbers, see [Ell, W - Ell, F].

We now turn to the study of ideals in the rings I_d with the goal of restoring unique factorization by proving that every ideal is a unique product of "prime" ideals. When I_d is a unique factorization domain this will correspond to ordinary factorization.

Chapter 17

Algebraic Number Theory 3: Ideals in Quadratic Fields

1. Arithmetic of Ideals in I_d

We start with two examples of ideals in ring I_d of integers of the quadratic field $\mathbf{Q}(\sqrt{d})$ (sec 16.4).

1) For any finite set $\{\alpha_1, \ldots, \alpha_n\}$ in I_d, the set $A = (\alpha_1, \ldots, \alpha_n) = \{\sum_{i=1}^{n} \alpha_i x_i | x_i \in I_d\}$, which represents the range of a linear form, is an ideal. We say that A is generated by $\alpha_1, \ldots, \alpha_n$. A **principal ideal** is an ideal generated by one element, i.e., an ideal of the form $(\alpha) = \alpha I_d = \{\mu\alpha | \mu \in I_d\}$.

Exercise: The point of this exercise is to show that no confusion should arise from using (α, β) to denote both ideals and a greatest common divisors. If $m, n \in \mathbf{Z}$ and $d = (m, n)$ is their greatest common divisor, then there exist $x, y \in \mathbf{Z}$ such that $d = mx + ny$. Show that this implies $(d) = (m, n)$ as ideals in $\mathbf{Z}$. Prove that in any principal ideal domain (all ideals principal) a generator of the ideal (α, β) is a greatest common divisor of α and β.

2) Given a ring I_d and an algebraic integer γ, *not necessarily in* I_d, then $\{\alpha \in I_d | \gamma | \alpha\}$ is an ideal in I_d, where divisibility is in the ring of all algebraic integers.

In fact, every ideal of I_d is generated by two elements and is also of the form 2) for some γ. We must develop more machinery before proving this.

To avoid confusion, we shall use capital Latin letters to denote ideals, lower case Greek letters for elements of I_d and lower case Latin letters for rational integers. Unless otherwise stated, *our ideals are not the zero ideal (0).*

Following Dedekind, we define the **product AB** of the ideals A and B in I_d as the ideal *generated* by all products of elements of A and elements of B, viz., the smallest ideal containing these products. Therefore

$$\begin{aligned} AB &= (\{\alpha\beta | \alpha \in A, \beta \in B\}) \ . \\ &= \{\text{finite sums } \sum_i \alpha_i \beta_i | \alpha_i \in A, \beta_i \in B\} \ . \end{aligned}$$

Proposition: (exercise)

1) $(\alpha I_d)(\beta I_d) = (\alpha\beta) I_d, [(\alpha)(\beta) = (\alpha\beta)]$

2) $(\alpha I_d) B = \alpha B,$

3) if $A = (\alpha_1, \ldots, \alpha_m)$ *and* $B = (\beta_1, \ldots, \beta_n)$ *then*

$$AB = (\alpha_i \beta_j; \ i = 1, \ldots, m \ , \quad j = 1, \ldots, n) \ ,$$

4) $AB = BA,$

$A(BC) = (AB)C,$

$AI_d = I_d A = A; \ I_d = (1)$ *is called the* **unit ideal**.

Example: We return to our factorization example from section 16.8, viz., $21 = 3 \cdot 7 = \left(1 + 2\sqrt{-5}\right) \cdot \left(1 - 2\sqrt{-5}\right)$, to find a common refinement of these factorizations in terms of ideals in I_{-5}; more precisely to find a common refinement of the factorizations

$$21 I_{-5} = 3 I_{-5} \cdot 7 I_{-5} = \left(1 + 2\sqrt{-5}\right) I_{-5} \cdot \left(1 - 2\sqrt{-5}\right) I_{-5} \ .$$

Let $A = \left(3, 1 + 2\sqrt{-5}\right)$, $\overline{A} = \left(3, 1 - 2\sqrt{-5}\right)$, $B = \left(7, 1 + 2\sqrt{-5}\right)$ and $\overline{B} = \left(7, 1 - 2\sqrt{-5}\right)$ be the ideals generated by taking one element from each factorization. We shall see later that these ideals act as greatest common divisors of their generators.

By part 3 of the proposition

$$A\overline{A} = \left(3 \cdot 3, 3\left(1 + 2\sqrt{-5}\right), 3\left(1 - 2\sqrt{-5}\right), 3 \cdot 7\right) \ .$$

Since 3 divides each generator, $A\overline{A} \subseteq (3)$. But $3 = 3 \cdot 7 - 2(3 \cdot 3)$ implies that $(3) \subseteq A\overline{A}$ and therefore $A\overline{A} = (3)$.

Similarly $AB = \left(3 \cdot 7, 3\left(1+2\sqrt{-5}\right), 7\left(1+2\sqrt{-5}\right), \left(1+2\sqrt{-5}\right)^2\right)$, and $1+2\sqrt{-5}$ divides each generator; so $\left(1+2\sqrt{-5}\right) \supseteq AB$. But, since $1+2\sqrt{-5} = 7\left(1+2\sqrt{-5}\right) - 2\left(3\left(1+2\sqrt{-5}\right)\right)$, $\left(1+2\sqrt{-5}\right) \subseteq AB$ and therefore $AB = \left(1+2\sqrt{-5}\right)$.

In the same way , we can prove that

$$\overline{AB} = \left(1 - 2\sqrt{-5}\right), \quad B\overline{B} = (7).$$

Thus

$$(21) = (3)(7) = A\overline{A} \cdot B\overline{B}$$

and

$$(21) = \left(1+2\sqrt{-5}\right)\left(1-2\sqrt{-5}\right) = AB \cdot \overline{A}\,\overline{B},$$

and we have a common refinement. There is no finer refinement, but we have not yet developed the techniques to prove this.

The following exercises reduce the divisibility of integers to relations between principal ideals and give a hint about how to generalize the arithmetic of integers to ideals.

Exercises: Let R be a commutative ring (e.g., $\mathbf{Z}$ or I_d). Prove that

1) $\alpha R \supseteq \beta R \iff \alpha|\beta$,
2) $\alpha R = \beta R \iff \alpha$ is an associate of β,
3) $\alpha R = R \iff \alpha$ is a unit of R.

We say that the principal ideal αI_d **divides** the principal ideal βI_d, denoted by $\alpha I_d | \beta I_d$, if $\alpha|\beta$. From exercise 1), it follows that

Proposition: $\alpha R | \beta R \iff \alpha R \supseteq \beta R$.

Remark: The proposition may seem confusing at first glance, since in $\mathbf{Z}$ if m divides n, then m is smaller than n. The way to think about this is that since 3 divides 6, the multiples of 3 contain the multiples of 6. Appropriately generalized, this will be the key to studying divisibility of arbitrary ideals in R.

2. Lattices and Ideals

We present a fairly geometric approach via lattices to ideal theory in quadratic fields. This approach is strongly influenced by M. Artin's treatment [Art, M], but he discusses some topics in more detail. In section 16.5, we set up a correspondence between $\mathbf{Q}(\sqrt{d})$ and points in the plane, where I_d corresponds to a lattice. Before studying ideals using these geometric ideas, we must be more precise about the notion of a lattice. The following three definitions are equivalent.

(I) A **lattice** L (actually a 2-dimensional lattice) is a discrete subgroup of $\mathbf{R}^2$ containing two elements which are linearly independent over $\mathbf{R}$. By **discrete** we mean that the intersection of L with any bounded subset of the plane is finite.

(II) A **lattice** L is a set of the form

$$\mathbf{Z}\alpha + \mathbf{Z}\beta = \{m\alpha + n\beta | m, n \in \mathbf{Z}\} ,$$

where $\alpha, \beta \in \mathbf{R}^2$ are linearly independent over $\mathbf{R}$.

Clearly this set is a subgroup of $\mathbf{R}^2$. We shall see in a moment that it is discrete and thus (II) $\Rightarrow$ (I). The proof that (I) $\Longrightarrow$ (II) is given in the appendix to this section.

(III) A **lattice** L is the image of the integer lattice $\{x = (x_1, x_2) \in \mathbf{Z}^2\}$ under a non singular linear transformation, $y = Mx$, $M = (m_{ij})$, $\det(M) \neq 0$, of the plane.

The equivalence of (II) and (III) is immediate by taking α and β to be the column vectors of M.

The set $\{\alpha, \beta\}$ of (II) is called an **integral basis** (or **lattice basis**) of L. L has infinitely many such bases and we will discuss their relations later. Of course algebraically, we are just considering the theory of torsion-free finitely generated abelian groups (**Z**-modules) with two generators.

A **fundamental parallelogram** of L (fig. 1) for the integral basis $\{\alpha, \beta\}$ is the set

$$T = \{r\alpha + s\beta | r, s \in \mathbf{R}, 0 \leq r, s < 1\} .$$

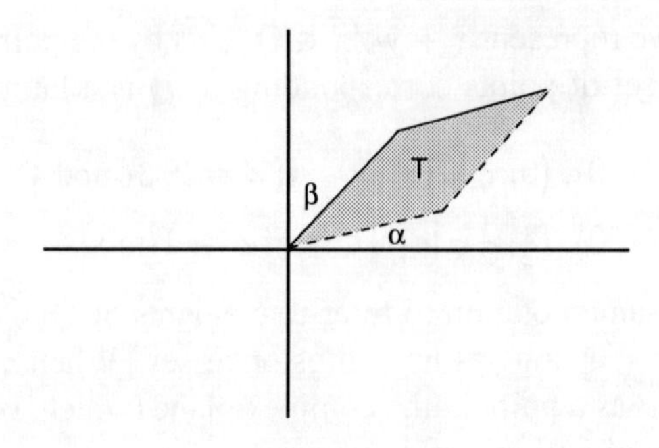

Figure 1

It is clear, from the linear independence of α and β, that the only point of L in T is the vertex $(0, 0)$.

We say $\gamma, \delta \in \mathbf{R}^2$ are equivalent modulo L if $\gamma = \delta + \mu$, for some $\mu \in L$. This is an equivalence relation, and every element of $\mathbf{R}^2$ is equivalent to a unique element of T - i.e., T is a system of distinct representatives for the equivalence relation (exercise). Geometrically, this means that the parallelograms $\{T + \mu | \mu \in L\}$ are disjoint and **tile** the plane (fig. 2); $\mathbf{R}^2 = \cup_\mu \{T + \mu\}$ (*disjoint union*).

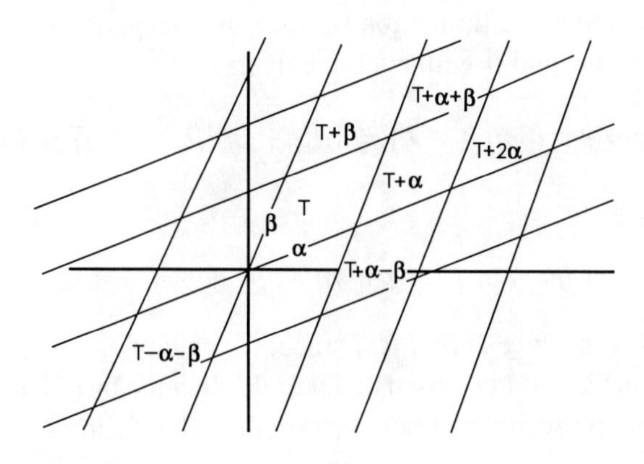

Figure 2

L is the set of vertices of these parallelograms. We see immediately from figure 2 that any L given by definition (II) is discrete; hence (II) $\Longrightarrow$ (I) (exercise-make this precise).

As earlier, we represent $x + y\sqrt{d} \in \mathbf{Q}(\sqrt{d})$ by the point $(x, y\sqrt{|d|})$ in the plane. The set of points corresponding to I_d is a lattice with integral basis

$$\left\{(1,0), \left(0, \sqrt{|d|}\right)\right\}, \qquad \text{if } d \equiv 2, 3 (\text{mod } 4)\ ,$$
$$\left\{(1,0), \left(\tfrac{1}{2}, \tfrac{1}{2}\sqrt{|d|}\right)\right\}, \qquad \text{if } d \equiv 1 (\text{mod } 4)\ .$$

We use the same notation to refer to elements of $\mathbf{Q}(\sqrt{d})$ and the corresponding points, as long as no confusion arises. When $d < 0$, we often refer to $x + y\sqrt{d}$ as a point in the complex plane (which, of course, is just $\mathbf{R}^2$ with a notion of multiplication of points).

If $L' \subseteq L$, where L' and L are lattices, then we say that L' is a **sublattice** of L. By (I), a subgroup of a lattice which contains two elements that are linearly independent over $\mathbf{R}$ is a lattice, since it inherits the discreteness of L.

An ideal A of I_d is a subgroup of I_d. If $x + y\sqrt{d} \neq 0$ is in A, then so is $\sqrt{d}(x + y\sqrt{d})$ and the corresponding points are linearly independent over $\mathbf{R}$. Thus by (I) A is a sublattice of I_d, and by (II) A *has an integral basis*,

$$A = \{m\alpha + n\beta | m, n \in \mathbf{Z}\}$$

for some $\mathbf{Z}$ linearly independent α, β in I_d.

Since A is generated by all integer linear combinations of α and β and is closed under multiplication by I_d, it is also generated by all linear combinations of α and β with coefficients in I_d, i.e.,

$$A = \{m\alpha + n\beta | m, n \in \mathbf{Z}\} = \{\mu\alpha + \rho\beta | \mu, \rho \in I_d\} = (\alpha, \beta)\ ,$$

and we have

Theorem: *Every ideal in I_d is generated by two elements.*

Not every sublattice is an ideal since, for example, the sublattice L of $\mathbf{Z}[i]$ with basis $\{2, i\}$ is not closed under multiplication by i ($i \cdot i = -1 \notin L$).

The following result will be a key tool in what follows. Recall that if G is a subgroup of the group H, then the **index, [H : G]**, is the number of cosets of G in H.

Proposition: *Let L' be a sublattice of L with lattice basis $L' = \{\alpha, \beta\}$. Then the index $[L : L']$ is finite and equal to the number of points of L in the fundamental parallelogram of L' (with respect to α, β), i.e., the number of L-points which are in the parallelogram whose vertices are $0, \alpha, \beta, \alpha+\beta$, and not on the edges $\overline{\alpha, \alpha+\beta}$, $\overline{\beta, \alpha+\beta}$. In particular, the quotient group L/L' is finite.*

Proof: Exercise - Show that the L points in the fundamental parallelogram for L' are a set of distinct representatives for the cosets.

Example: If $L = \mathbf{Z}[i]$, $L' = \{2, 2i\}$, then $[L : L'] = 4$ (see fig. 3).

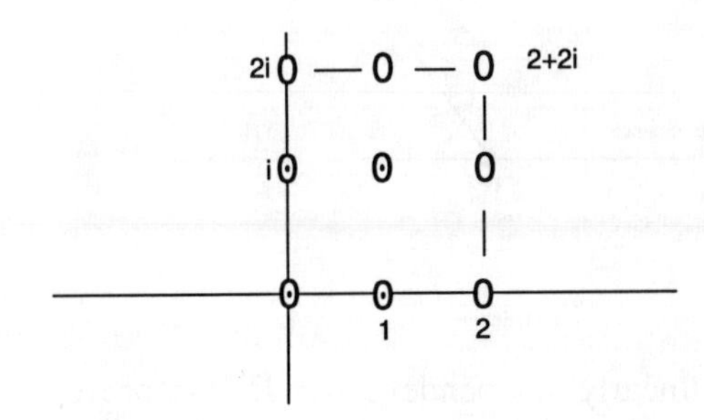

Figure 3

We will assume the following

Proposition: *The area of the fundamental parallelogram determined by a basis of a lattice L is independent of the choice of the basis. We denote this area by $\Delta(L)$.*

For a simple geometric proof, see Hilbert and Cohn-Vossen [Hil - Coh, pp. 32-35]. More generally, one can prove

Proposition: *Let $L' \subseteq L$ be lattices. Then*

$$[L : L'] = \frac{\Delta(L')}{\Delta(L)}\,.$$

Appendix: Lattices

We now prove the result stated at the beginning of section 2.

Proposition: *If L is a discrete 2-dimensional subgroup of the plane, then there exist* α_1, α_2 *in L, linearly independent over* $\mathbf{R}$*, such that* $L = \{m_1\alpha_1 + m_2\alpha_2 | m, n \in \mathbf{Z}\}$.

Proof: By discreteness, we can choose an $\alpha_1 \neq 0$ in L so that the interior of the segment $\overline{\mathbf{0}\alpha_1}$ contains no points of L. Choose any $\beta \in L$ not on the line M through by $\mathbf{0}$ and α_1 and let P be the parallelogram spanned by α_1 and β. Choose $\alpha_2 \in P \cap L$ with minimum positive distance to M (fig. 4).

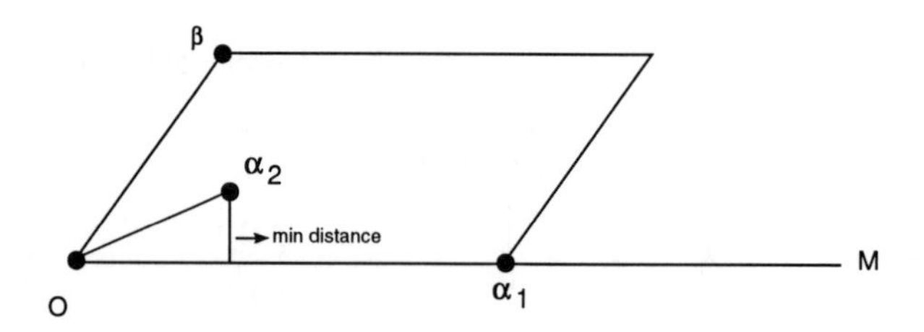

Figure 4

By choice, α_1, α_2 are linearly independent over $\mathbf{R}$. We prove they are a basis for L. Let $\Gamma \in L$, $\Gamma = r_1\alpha_1 + r_2\alpha_2$, $r_1, r_2 \in \mathbf{R}$. Our goal is to prove that the r_i's are integers. Let $s_i = r_i - [r_i]$, $0 \leq s_i < 1$. Then $\gamma = s_1\alpha_1 + s_2\alpha_2$ is in L and also in P', the parallelogram spanned by α_1 and α_2 (fig. 5). We prove that $s_1 = s_2 = 0$, which means that the r_i's are integers.

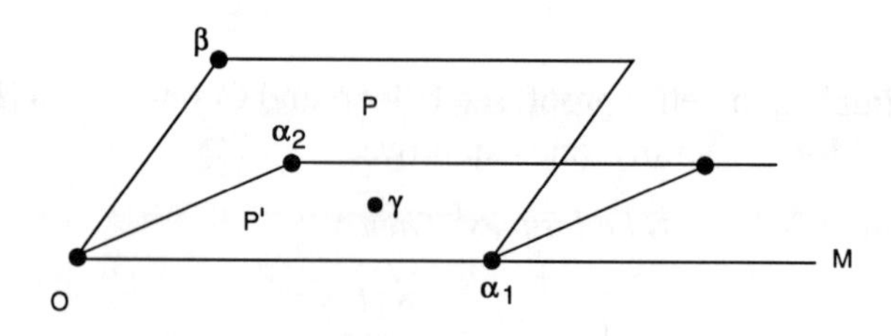

Figure 5

i) Suppose γ is also in P (fig. 5). Since α_2 was chosen with minimum positive distance to M, every point inside P' is closer than α_2 to M. Now $\gamma \in P'$ is not on the sides $\overline{\alpha_1, \alpha_1 + \alpha_2}$ or $\overline{\alpha_2, \alpha_1 + \alpha_2}$ (since $s_1, s_2 < 1$); therefore $s_2 = 0$, for otherwise $\gamma \in L \cap P$ has smaller positive distance than α_2 to M. Thus $\gamma \in \overline{\mathbf{0}\alpha_1}$ and $\gamma \neq \alpha_1$ (since $s_1 < 1$). But $\overline{\mathbf{0}\alpha_1}$ has no points of L in its interior; therefore $\gamma = (0, 0)$ and $s_1 = 0$.

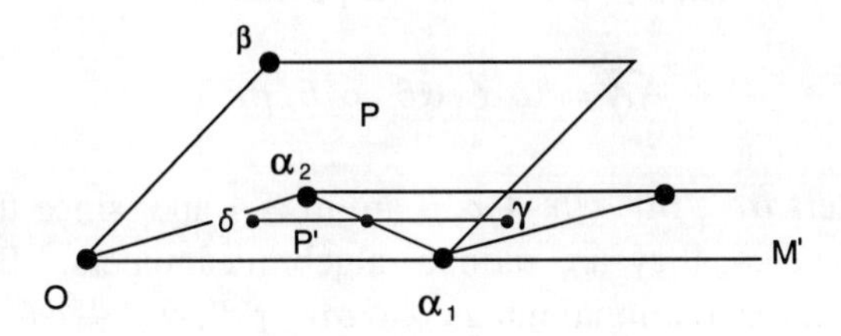

Figure 6

ii) Suppose that γ is not in P (fig. 6). Then δ, the reflection of γ through the midpoint of $\overline{\alpha_1, \alpha_2}$, is in P. Moreover, $\frac{\gamma+\delta}{2} = \frac{\alpha_1+\alpha_2}{2}$ or $\delta = \alpha_1+\alpha_2-\gamma$; hence $\delta \in L$. Applying the reasoning in case i) to δ, we see that $\delta = (0, 0)$, $\gamma = \alpha_1 + \alpha_2$ and $s_i = 1$. But this contradicts the assumption that $s_1, s_2 < 1$; hence γ must be in P.

3. More Arithmetic of Ideals

If A and B are ideals in I_d, we say that A **divides** B, denoted by $\mathbf{A|B}$, if there is an ideal C in I_d such that $B = AC$. If $A|B$ and $A \neq B$, then A is a **proper divisor** of B.

The following lemma is a primary tool for studying the divisibility of ideals. If A is an ideal in I_d, then the **conjugate** of A is $\overline{\mathbf{A}} = \{\alpha' \in I_d | \alpha \in A\}$, where α' is the conjugate of α. $\overline{A}$ is an ideal in I_d, and if $A = (\alpha_1, \alpha_2)$, then $\overline{A} = (\alpha_1', \alpha_2')$.

Key Lemma: *If A is an ideal in I_d, then, for some $n \in \mathbf{Z}$,*

$$A\overline{A} = nI_d \, .$$

Remark: More generally, in the ring of integers of any number field, for every ideal A there is an ideal $B \neq (0)$ such that AB is principal. The different proofs of this theorem correspond to the different approaches to ideal theory in algebraic number fields.

Proof: The proof is similar to our analysis of the factorizations of the ideal $(21)I_{-5}$ in section 1. By our last theorem, A is generated by two elements; say $A = (\alpha, \beta)$. Then $\overline{A} = (\alpha', \beta')$ and

$$A\overline{A} = (\alpha\alpha', \alpha\beta', \alpha'\beta, \beta\beta') .$$

The integers $\alpha\alpha', \beta\beta', \alpha\beta' + \alpha'\beta$ are in $A\overline{A}$ and, since they are equal to their conjugates, they are rational algebraic integers. But a rational algebraic integer is a rational integer; so $\alpha\alpha', \beta\beta', \alpha'\beta + \alpha\beta' \in \mathbf{Z}$. If

$$n = \gcd(\alpha\alpha', \beta\beta', \alpha'\beta + \alpha\beta') ,$$

then n is a $\mathbf{Z}$-linear combination of these integers; hence $n \in A\overline{A}$, and since $A\overline{A}$ is an ideal,

$$nI_d \subseteq A\overline{A} .$$

We prove that $A\overline{A} \subseteq nI_d$ and thus $nI_d = A\overline{A}$, by showing that n divides each of the generators, i.e., $\frac{\alpha\alpha'}{n}, \frac{\alpha\beta'}{n}, \frac{\alpha'\beta}{n}, \frac{\beta\beta'}{n} \in I_d$. Recall that a quadratic algebraic number is a quadratic integer if its norm and trace are in $\mathbf{Z}$.

By the definition of n, we have $n|\alpha\alpha'$ and $n|\beta\beta'$. Now $\frac{\alpha\beta'}{n}, \frac{\alpha'\beta}{n} \in \mathbf{Q}(\sqrt{d})$ and are conjugate to each other. Therefore

$$\mathrm{Tr}\left(\frac{\alpha\beta'}{n}\right) = \mathrm{Tr}\left(\frac{\alpha'\beta}{n}\right) = \frac{\alpha'\beta + \alpha\beta'}{n} ,$$

$$N\left(\frac{\alpha\beta'}{n}\right) = N\left(\frac{\alpha'\beta}{n}\right) = \frac{\alpha\alpha'}{n} \cdot \frac{\beta\beta'}{n} .$$

Again, by the definition of n, the trace and norm are in $\mathbf{Z}$ and thus $\frac{\alpha\beta'}{n}, \frac{\alpha'\beta}{n} \in I_d$.

As an immediate application we have

Proposition: *If A, B, C are ideals in I_d and $AB = AC$, then $B = C$.*

Proof: If $A\overline{A} = nI_d$, then

$$\begin{aligned} AB = AC &\implies \overline{A}AB = \overline{A}AC \\ &\implies (nI_d)B = (nI_d)C \\ &\implies nB = nC \\ &\implies B = C\,. \end{aligned}$$

Just as for principal ideals (sec. 1) we have

Theorem: $A|B \iff A \supseteq B$.

Proof: $\Longrightarrow$) If $A|B$ then $B = AC$, for some C. Let $\beta \in B$. Then $\beta = \sum_i \alpha_i\gamma_i$, for some $\alpha_i \in A$, $\gamma_i \in C$. But $\gamma_i \in C$ implies $\gamma_i \in I_d$; so, by the definition of an ideal, $\beta \in A$.

$\Longleftarrow$) First assume that A is principal, say $A = \alpha I_d$. Then

$$A = \alpha I_d \supseteq B \implies \alpha|\beta, \text{ for all } \beta \in B \implies \beta = \alpha\beta_0, \text{ for some } \beta_0 \in I_d\,.$$

Let $C = \{\text{all such } \beta_0\}$. Then C is an ideal and $B = \alpha C = (\alpha I_d)C$. Now if A is any ideal and $A\overline{A} = nI_d$, then

$$A \supseteq B \implies \overline{A}A \supseteq \overline{A}B \implies nI_d \supseteq \overline{A}B\,.$$

By our proof for principal ideals, this means that $nI_d|\overline{A}B$, i.e., $(nI_d)C = \overline{A}B$, for some C; thus

$$\overline{A}AC = (nI_d)C = \overline{A}B \implies AC = B \implies A|B\,.$$

Corollary: *A is a proper divisor of B* $\iff A \supset B$.

(Recall that $\supset$ means proper containment.)

Corollary: *For every ideal B in I_d,*

1) $B|I_d \iff B = I_d$,
2) $AB = I_d \iff A = I_d$ *and* $B = I_d$.

From now on it will be convenient to identify α with (α). Since

$$A|(\alpha) \iff A \supseteq (\alpha) \iff \alpha \in A\,,$$

writing $\mathbf{A}|\alpha$ for $A|(\alpha)$ should not cause confusion.

4. Unique Factorization of Ideals

Now that we have introduced ideals into I_d as new objects and represented integers of I_d as principal ideals, our goal is to prove unique factorization of ideals into products of "prime ideals".

Proceeding by analogy with the rational integers, or any principal ideal domain, we define irreducible and prime ideals and show that they are the same.

$A \neq I_d$ is **irreducible** if $A = BC \implies B = I_d$ or $C = I_d$.

$P \neq I_d$ is **prime** if $P|AB \implies P|A$ or $P|B$.

Exercise: P is prime $\iff \{\alpha\beta \in P \implies \alpha \in P$ or $\beta \in P\}$.

The condition in the exercise is the definition of prime ideal usually given in algebra texts. However our definition is the natural generalization of the notion of a prime integer.

The proof that primes are irreducible is a direct analog of the proof for integers. For if P is prime, $P = AB$, then, since $P|AB$, we have $P|A$ or $P|B$; say $P|A$. Thus, for some C,

$$A = PC$$

and

$$PI_d = P = AB = (PC)B \ .$$

But

$$PI_d = PCB \implies I_d = CB \implies B = I_d \ .$$

To prove that irreducibles are primes requires a bit more work. First we give a different characterization of irreducible ideals. Recall that an ideal A in I_d is *maximal* if there are no ideals *strictly* between A and I_d, i.e., there is no S such that $A \subset S \subset I_d$.

Proposition: *A is irreducible $\iff$ A is maximal.*

Proof: $\implies$) Suppose A is not maximal; then there is an S such that $A \subset S \subset I_d$. Now

$$A \subset S \implies S|A \implies A = ST \ ,$$

for some T. But A is irreducible and $S \neq I_d$; thus $T = I_d$ and $A = S$, a contradiction.

$\Longleftarrow$) Suppose A is maximal and $A = ST$, $S \neq I_d$. Then $S|A$, which implies that $S \supseteq A$, and thus $I_d \supseteq S \supseteq A$. But A is maximal; so $S = A$ and $A = AT$. Now $A = AI_d = AT$ implies $T = I_d$ and A is irreducible.

All maximal ideals are prime. This follows by standard algebraic results, viz.,

$$A \text{ maximal} \implies I_d/A \text{ is a field} \implies I_d/A \text{ is an integral domain} \implies A \text{ is prime}.$$

Combining these results, we have

Proposition: *P irreducible $\implies$ P is prime.*

We present another proof that uses less machinery and is closer to to the proof usually given for principal ideal domains.

Theorem: *If A, B are ideals of I_d, not both (0), then there exists a unique* **greatest common divisor** *of A and B, namely, an ideal D such that*

1) $D|A$, $D|B$

2) if $C|A$, $C|B$, then $C|D$.

In fact, **D = (A, B)**, the ideal generated by $A \cup B$, and thus, just as for principal ideals, using the same notation for 'generates' and 'greatest common divisor' is consistent.

Since $A = (\alpha, \beta)$ if and only if $A = ((\alpha), (\beta))$, we see that *A is the greatest common divisor of its generators.*

Warning: One must be a little careful in using the notation gcd $(\alpha, \beta) = (\alpha, \beta)$. If α, β are integers in I_d then gcd (α, β) is only unique up to units, whereas if they represent the principal ideals $(\alpha), (\beta)$, then gcd $((\alpha), (\beta))$ is unique. Thus, e.g., in $\mathbf{Z}$, ± 1 are greatest common divisors of 3 and 5, whereas the ideal $(1) = (-1) = \mathbf{Z}$ is the unique greatest common divisor of the ideals (3) and (5). Generally the meaning is clear from context.

proof of theorem. Letting $D = (A, B)$, we have

1) $D \supseteq A, B \implies D|A, D|B$.

2) $C|A, B \implies C \supseteq A, B \implies C|(A, B)$, since C is an ideal $\implies C|D$.

If D' is also a greatest common divisor then, by 2), $D|D'(D \supseteq D')$ and $D'|D(D' \supseteq D)$; thus $D = D'$ and greatest common divisors are unique.

Corollary: $C \cdot (A, B) = (CA, CB)$

Proof: Exercise (Use the theorem and the definition of multiplication of ideals.)

Corollary: *P irreducible $\Longrightarrow$ P prime.*

Proof: If $P|AB$ and $P \nmid B$ then $(P, B) = (1)(= I_d)$, since, as an irreducible, P has no factors other than P and (1). By the last corollary, $A = A \cdot (1) = A \cdot (P, B) = (AP, AB)$. Hence

$$\begin{aligned} P|AP, P|AB &\Longrightarrow P \supseteq AP, AB \\ &\Longrightarrow P \supseteq (AP, AB) = A \Longrightarrow P|A \,. \end{aligned}$$

We can now prove the

Fundamental Theorem of Ideal Theory (DEDEKIND): *If A is an ideal in I_d, $A \neq (0), (1)$, then A is a product of prime ideals. The product is unique up to the order of the primes.*

Proof: First we deal with the uniqueness. Assume that

$$P_1 P_2 \cdots P_r = Q_1 Q_2 \cdots Q_s$$

are two factorizations of A into prime (= irreducible) ideals. Then

$$\begin{aligned} P_1| \text{ left hand side } &\Longrightarrow P_1| \text{ right hand side} \\ &\Longrightarrow P_1|Q_i, \text{ for some } i, \text{ (since } P_1 \text{ is prime)} \\ &\Longrightarrow P_1 = Q_i, \text{ (since both are irreducible)} \,. \end{aligned}$$

We cancel P_1 from both sides, reducing the number of factors, and repeat the argument with P_2 and so on.

The existence of a factorization into prime ideals requires more work. If A is prime, we are done; so assume A is not prime. The key idea is to prove that for any ideal A, there are only a finite number of ideals dividing it.

Recall that if L' is a sublattice of L, then L/L' is a finite group (sec. 2). Thus, since A is a sublattice of I_d, I_d/A is finite. If $B|A$ then $A \subseteq B \subseteq I_d$. By one of the of the elementary homomorphism theorems of group theory, there is a one to one correspondence between subgroups of I_d containing A and subgroups of I_d/A (see [Art, M]).

Since I_d/A is finite, it has only a finite number of subgroups. Since ideals are subgroups, the number of ideals containing A is finite. Thus we can find a maximal (= irreducible = prime) ideal P among them. But $P|A$; so $A = PA'$, for some A'. Every divisor of A' is a divisor of A; but $A|A$ and $A \nmid A'$ and thus A' has fewer divisors than A. Now we can apply the same procedure to find a prime divisor of A', and so on. The number of available divisors is reduced at each step and the process must eventually stop.

5. Applications of Unique Factorization

Divisibility and Diophantine Equations

Factoring ideals into prime ideals immediately yields a criterion for the divisibility of integers. If $\alpha, \beta \in I_d$ and

$$(\alpha) = P_1^{m_1} \cdots P_k^{m_k}, \qquad m_i \geq 0,$$
$$(\beta) = P_1^{n_1} \cdots P_k^{n_k}, \qquad n_i \geq 0,$$

then $\alpha|\beta$ if and only if $m_i \leq n_i$, for all i. To apply this in specific cases, we must know how to factor ideals. We will discuss this shortly.

Recall that Euler solved the Diophantine equation $y^2 = x^3 - 2$ by assuming that I_{-2} is a unique factorization domain (sec. 3.5), which is true (sec. 16.7). A similar approach can be used to study $y^2 = x^3 + d$ (when d is square free). We sketch the main ideas. From the factorization of $x^3 = y^2 - d$,

$$x^3 = (y + \sqrt{d})(y - \sqrt{d}),$$

we have the factorization of principal ideals, in I_d,

$$(xI_d)^3 = (y + \sqrt{d})I_d \cdot (y - \sqrt{d})I_d .$$

For certain values of d, the ideals $(y + \sqrt{d})I_d$ and $(y - \sqrt{d})I_d$ are relatively prime. Hence, by unique factorization, they are the cubes of ideals. A detailed analysis, for special d, actually leads to the solutions. This was first studied by Mordell ([Mor 1], [Mor 2, chapter 2, with references to the original papers]) and there is a nice treatment in [Ada - Gol, chapter 10]. The analysis requires the notion of class number of I_d which we will soon discuss.

UNIQUE FACTORIZATION DOMAINS

We know that any principal ideal domain is a unique factorization domain. For I_d, the converse is also true.

Proposition: I_d *a* $UFD \implies I_d$ *a* PID.

Proof: If A is an ideal in I_d, then we know that A is generated by two elements, $A = (\alpha, \beta)$, for some $\alpha, \beta \in I_d$. But then A is the greatest common divisor of αI_d and βI_d (sec. 4). Since we assumed I_d is a UFD, there is an integer δ in I_d which is a greatest common divisor of α and β, unique up to units. Thus δI_d is a greatest common divisor of αI_d and βI_d as ideals and, by the uniqueness of greatest common divisors of ideals, $A = \delta I_d$ and is principal.

6. The Factorization of Rational Primes

The procedure for factoring an ideal $A \subseteq I_d$ into prime ideals is a direct analog of our procedure for factoring Gaussian integers (sec. 15.2).

Since $A\overline{A} = nI_d$ for some $n \in \mathbf{Z}$, a prime ideal factor of A must divide some rational prime factor of n. Thus the problem of factoring ideals in I_d reduces to the problem of factoring pI_d, p a rational prime, into prime ideals of I_d. We study this problem in some detail, both here and in later sections, and show its fundamental relation to quadratic residues. For a more algorithmic approach to factoring (p), without any theoretical characterization, see Stewart and Tall [Ste - Tal, chapter 10].

First we prove

Proposition: *Every prime ideal P of I_d divides exactly one rational prime.*

Proof: The existence of such a rational prime was shown above.

Assume that $P|p$ and $P|q$, where $p \neq q$ are rational primes. Then, since $(p, q) = 1$, there exist $s, t \in \mathbf{Z}$ such that $sp + tq = 1$. This means that $1 \in (p, q)$, *the ideal in* I_d generated by p and q; thus the ideal $(p, q) = (1) = I_d$. Therefore

$$P|p, q \implies P|I_d \implies P = I_d ,$$

a contradiction since P is prime.

Since there are infinitely many rational primes and the fundamental theorem of ideal theory guarantees that every rational prime has a prime ideal factor, we have the

Corollary: *Every I_d has infinitely many prime ideals.*

The proof of the proposition can be used to further analyze the factorization of pI_d in I_d, where p a rational prime.

We need to know that if P is prime, then $\overline{P}$ is prime, which is an immediate consequence of the following easy exercises:

a) $\overline{AB} = \overline{A} \cdot \overline{B}$,

b) $\overline{\overline{A}} = A$,

c) $A|B \implies \overline{A}|\overline{B}$, in particular, if $A|n$, $n \in \mathbf{Z}$, then $\overline{A}|n$.

Let P be a prime ideal. Then $P\overline{P} = nR$ for some $n \in \mathbf{Z}$. If $n = p_1 \cdots p_k$ is the prime factorization of n over $\mathbf{Z}$, then

$$P\overline{P} = nI_d = (p_1 I_d) \cdots (p_k I_d) .$$

Since $\overline{P}$ is prime, $P\overline{P}$ is the factorization of nI_d into prime ideals in I_d. After refining the factorization on the right hand side into prime ideals, we must have two prime ideal factors (by uniqueness); thus $k \leq 2$.

If $k = 1$ then $P\overline{P} = nI_d = p_1 I_d$.

If $k = 2$ then $p_1 I_d$ and $p_2 I_d$ must be prime ideals or there would be too many factors. Hence, by uniqueness, $P\overline{P} = (p_1 I_d)(p_2 I_d)$ and $P = p_1 I_d$ or $P = p_2 I_d$; thus P is a rational prime ideal $p_i I_d$.

Conversely, let q be a rational prime. Then qI_d has a prime ideal factor P. By the above discussion, either

i) $P = pI_d$, for some rational prime p, in which case $pI_d | qI_d$ and, since P can only divide one rational prime, $p = q$, i.e., qI_d is a prime ideal, or

ii) $P\overline{P} = pI_d$, p a rational prime. Then $P|p$ and $P|q$. But P can only divide one rational prime and, as in case i), $p = q$ and $qI_d = P\overline{P}$.

To summarize

Theorem: *Let p be a rational prime. Then either*

1) pI_d is a prime ideal (we say p is **inert** *in I_d) or*

2) $pI_d = P\overline{P}$, *for some prime ideal P in I_d. We distinguish two cases:*

a) $P \neq \overline{P}$ *(we say p **splits** in I_d),*

b) $P = \overline{P}$ *(we say p **ramifies** in I_d).*

Furthermore, if P is a prime ideal, then $P\overline{P} = pI_d$, for some rational prime p.

We will provide a more detailed analysis after developing some new concepts.

7. Class Structure and the Class Number

We introduce an equivalence relation for ideals and a multiplication of equivalence classes which will lead us to the notion of the 'class group'.

Kummer introduced these ideas for cyclotomic fields and it was generally believed that they were motivated by an unsuccessful attempt to prove Fermat's last theorem. Then Dedekind, in his 11th supplement to Dirichlet's lectures [Dir], generalized the concepts to all algebraic number fields and formalized the relation between quadratic forms and quadratic fields, especially the relation between the class structure of forms and fields. However, as Harold Edwards discovered in his research on Kummer, the story of these developments is more complex.

The following quote from section 5.1 of Edwards book [Edw 1] discusses these ideas as well as the genesis of the relation between forms and fields . When Edwards talks about divisors, ideal divisors or ideal complex numbers, he is referring to Kummer's "ideal objects" which are related to the valuation theoretic discussion of division in algebraic number fields (chapter 23). However if one interprets these ideal objects as ideals in the ring of integers of an algebraic number field, everything makes sense and little is lost.

> "It frequently happens that great innovations are made not by revolutionaries bent on change but by men who have a great respect for what has gone before and who are motivated by the wish to conserve and fulfill the traditions of their predecessors. This was certainly the case with Kummer. As K.-R. Biermann points out in his biography of Kummer in the *Dictionary of Scientific Biography*, Kummer was by nature very conservative, not in any narrow political sense, but in the sense that he was dedicated to building on the basis of existing traditions. In understanding the motivation of Kummer's work, it is important to realize that

he had no wish to introduce new abstract structures for their own sake but rather, as he said at the beginning of his announcement [K7] of the new theory, his goal was "to complete and simplify" existing structures.

This chapter is devoted to Kummer's proof of Fermat's Last Theorem for a large class of prime number exponents λ which are now known as the *regular* primes. This proof requires another important innovation of Kummer's, namely, the notion of *equivalence* between two divisors (two ideal complex numbers). In keeping with Kummer's personality, it is to be expected that this new notion of equivalence was motivated by some very compelling consideration; Kummer would not have introduced it simply because it was an "interesting" possibility. Although it is tempting to suppose that it was the very attempt to prove Fermat's Last Theorem that motivated the definition of equivalence of divisors, the fact is that this definition was included as a very prominent part of Kummer's initial announcement of the theory of ideal divisors in 1846,well before Lamé's premature announcement prodded him into working out in detail the consequences of his theory for Fermat's Last Theorem. Thus it seems unlikely that this work had a major role in motivating the original definition of equivalence.

There appear to be at least two considerations which did motivate the definition. The first is that in applying divisor theory to actual problems – for example to the problems in cyclotomy which Kummer considers in his 1846 announcement–one is very soon confronted with the problem "when is a given divisor the divisor of an actual cyclotomic integer?" The solution of this problem leads, as will be shown in the next section, very naturally to the notion of equivalence. But the second motive appears from Kummer's statements to be almost as important to him as the first. It is the fact that this notion of equivalence is very closely related to Gauss's notion of the equivalence of binary quadratic forms.

Here again a "conservative" interpretation can be placed on Kummer's innovation. The study of Pell's equation and of other equations of degree two in two variables leads very naturally to the concept of *equivalence* of binary quadratic forms (see Section 8.2). Gauss in the *Disquisitiones Arithmeticae* was led to introduce the stronger notion of *proper equivalence* (see Section 8.3). Kummer observes that this notion always seems forced and artificial in the theory of binary quadratic forms – it requires, for example, that one regard the two forms $ax^2+2bxy+cy^2$ and $cx^2+2bxy+ay^2$ as not being properly equivalent – despite the fact that "it must be recognized that the Gaussian classification corresponds more closely to the essential nature of the matter." Thus, Kummer implicitly concludes, the Gaussian notion of proper equivalence is something which needs to be *saved* from this appearance of artificiality. He says that the theory of ideal factorization accomplishes this because "the entire theory of binary quadratic forms can be interpreted as the theory of complex numbers of the form $x + y\sqrt{D}$ (D is Gauss's notation for the *determinant* $b^2 - ac$ of the quadratic form $ax^2 + 2bxy + cy^2$) and as a result of this interpretation leads necessarily to ideal complex numbers (divisors) of the same type. He then goes on to say, in effect, that the notion of equivalence which he has defined for ideal complex numbers of the form $a_0 + a_1\alpha + a_2\alpha^2 + \cdots + a_{\lambda-1}\alpha^{\lambda-1}$ also applies to the ideal complex numbers of the form $x+y\sqrt{D}$ and that, when these latter are interpreted as binary quadratic forms, the notion of equivalence coincides with Gauss's notion of proper equivalence. This, he then concludes, is the "true basis" of the Gaussian notion of equivalence.

Quite mysteriously, Kummer *never published* any details on this connection between binary quadratic forms and ideal complex numbers (divisors) of the form $x + y\sqrt{D}$. The few informal remarks in his 1846 announcement and a few more sketchy indications in later treatises ([K8, p. 366] and [K11, p. 114]) constitute all that he said on the subject or, at any rate, all that has survived. Thus, although it seems quite certain that the analogy with Gauss's theory played some role in the genesis of the notion of equivalence of divisors, its exact role cannot be known. "

Kummer's definition of equivalence translates to ideals in I_d as follows:

a) **A is equivalent to B, A $\sim$ B**, if there is an ideal C such that AC and BC are both principal.

We will use the following equivalent definition (exercise - prove the equivalence using the key lemma of section 3):

b) $\mathbf{A} \sim \mathbf{B}$ if there is a $\lambda \in \mathbf{Q}(\sqrt{d})$ such that $\mathbf{A} = \lambda\mathbf{B}$.

For example, in I_{-5}, let $A = (3, 1 + 2\sqrt{-5})$ and $B = (3\sqrt{-5}, -10+\sqrt{-5})$. Then $A = \left(\frac{1}{\sqrt{-5}}\right) B$ and $C = \left(\frac{1}{\sqrt{-5}}\right)(-3\sqrt{-5}, -10 - \sqrt{-5})$ makes AC and BC principal.

The relation $\sim$ is an equivalence relation and the equivalence classes are called **ideal classes**. We will denote the class of A by [**A**]. In an imaginary quadratic field, $A = \lambda B$ implies that the lattices corresponding to A and B are geometrically similar, since multiplication by the complex number λ is a rotation followed by a dilation. For real quadratic fields the geometry is not as simple (see [Bor - Sha]).

Any principal ideal αI_d in I_d is equivalent to I_d since $\alpha I_d = \alpha \cdot I_d$; thus any two principal ideals are equivalent. Conversely, if $A \sim I_d$, then, for some some $\lambda \in I_d$, $A = \lambda I_d = \lambda(1) = (\lambda)$, and A is a principal ideal of I_d. Therefore we have

Proposition: $[I_d]$ *is the set of principal ideals.*

Proposition: *Multiplication of ideals is compatible with equivalence.*

Proof:

$$A \sim A', B \sim B' \implies A = \lambda A', B = \lambda' B' \implies AB = (\lambda\lambda')A'B'$$
$$\implies AB \sim A'B'.$$

Hence we can define the **product of ideal classes** by

$$[\mathbf{A}][\mathbf{B}] = [\mathbf{AB}].$$

Theorem: *With this definition of multiplication, the ideal classes of* I_d *form an abelian group.*

Proof: Associativity and commutativity follow from the same properties for products of ideals. $[A][I_d] = [AI_d] = [A]$ implies that $[I_d] = [(1)]$ is the identity. Since $A\overline{A} = nI_d \sim I_d$, for some n in $\mathbf{Z}$, we have $[A][\overline{A}] = [(1)]$; so $[\overline{A}]$ is the inverse of A.

This group, the **ideal class group** (or just **class group**) of $\mathbf{Q}(\sqrt{d})$, will be denoted by $\mathbf{C_d}$.

In section 5, we proved that I_d is a UFD if and only if I_d is a PID; hence

Corollary: I_d *is a UFD* $\iff |C_d| = 1$.

One of the main theorems of algebraic number theory is the following.

Theorem: C_d *is a finite group.*

$|C_d|$, the order of C_d, is the **class number** of $\mathbf{Q}(\sqrt{d})$ or I_d, to be denoted by $\mathbf{h_d}$ (or $\mathbf{h}$).

For now we will assume that this theorem is true. Later we will see that it follows easily from the relation between forms and fields, which we will discuss in section 10 and from the finiteness of the class number for forms.

From the finiteness of the class number, we have

Theorem: *If* h *is the class number of* I_d*, and* A *is an ideal in* I_d*, then* A^h *is principal.*

Proof: Since the order of a subgroup divides the order of the group, we have

$$\begin{aligned}[A]^h = [I_d] &\implies [A^h] = [I_d] \\ &\implies A^h \sim I_d;\ \text{hence } A^h \text{ is principal}\,.\end{aligned}$$

Now we prove that ideals in I_d can be represented by algebraic integers, which may belong to other fields. In some sense, this justifies Dedekind's use of ideals as "ideal factors" of integers in I_d.

Theorem: *If A is an ideal in I_d, then there exists an algebraic integer γ, not necessarily in I_d, or even quadratic, such that*

$$A = \{\alpha \in I_d \mid \gamma | \alpha\},$$

where divisibility is over the ring of all algebraic integer.

Proof: By the previous theorem, A^h is principal, say $A^h = \beta I_d$. Let $\gamma = \sqrt[h]{\beta}$ (choose any h^{th} root).

The ring Ω of all algebraic integers is closed under the operation of taking all roots (sec. 16.2). Therefore

$$\begin{aligned}\alpha \in A \implies \alpha^h \in A^h = \beta I_d &\implies \frac{\alpha^h}{\beta} \in I_d \\ &\implies \frac{\alpha}{\sqrt[h]{\beta}} = \frac{\alpha}{\gamma} \in \Omega \implies \gamma|\alpha \text{ (division in } \Omega)\,.\end{aligned}$$

Conversely suppose $\alpha \in I_d$ and $\gamma|\alpha$ (division in Ω). Then $\frac{\alpha}{\gamma} \in \Omega$, which implies that $\frac{\alpha^h}{\gamma^h} = \frac{\alpha^h}{\beta} \in \Omega$. Since α^h and β are in I_d, we have

$$\begin{aligned}
\frac{\alpha^h}{\beta} &\in \mathbf{Q}(\sqrt{d}) \\
&\Longrightarrow \frac{\alpha^h}{\beta} \in I_d \text{ (since } \tfrac{\alpha^h}{\beta} \text{ is an algebraic integer)} \\
&\Longrightarrow \alpha^h \in \beta I_d \subseteq A^h \\
&\Longrightarrow (\alpha^h) = (\alpha)^h \subseteq A^h \\
&\Longrightarrow A^h | (\alpha)^h \text{ (division of ideals - to contain is to divide)} \\
&\Longrightarrow A | \alpha \text{ (by unique factorization of ideals; count the} \\
&\qquad \text{multiplicity of prime powers)} \\
&\Longrightarrow (\alpha) \subseteq A \Longrightarrow \alpha \in A\,.
\end{aligned}$$

It can also be shown that the integers needed to represent ideals in $\mathbf{Q}(\sqrt{d})$ can be chosen so that they all belong to a field L of degree h over $\mathbf{Q}(\sqrt{d})$. Therefore there exists a field L containing $\mathbf{Q}(\sqrt{d})$ such that every ideal of I_d is a principal ideal in L (see [Hec, section 33]).

8. Finiteness of the Class Number; Norm of an Ideal

Our definition of equivalence of ideals and the class group carries over to the ring of integers of any algebraic number field, and the corresponding class number is always finite. Kummer introduced these concepts and proved the finiteness for cyclotomic fields. As we saw in the quote from Edwards in the last section, Kummer probably had these ideas for quadratic fields but they were first published by Dedekind. Minkowski proved finiteness for all algebraic number fields using his theory of the Geometry of Numbers (chapter 22).

One of the central problems of algebraic number theory is the study of the structure of the class groups. This is still a wide open area.

A typical application of the class number arises in Kummer's attack on Fermat's last theorem. In the cases he can prove, he assumes that the prime exponent p in $x^p + y^p = z^p$ does not divide the class number of the corresponding cyclotomic field $\mathbf{Q}(e^{\frac{2\pi i}{p}})$. Such primes are called *regular*. It is not known if there are infinitely many regular primes, but it is known that there are infinitely many irregular (not regular) primes. See [Bor - Sha], [Edw 1] and [Rib] for details.

In sec. 12.10, we stated Gauss's conjecture that there are infinitely many discriminants $D > 0$ such that the corresponding class number of the quadratic forms of discriminant D is one. It will follow from the results in section 10 that this is equivalent to the conjecture that there are infinitely many real quadratic fields with class number one (= UFD'S). In fact, it is not even known if there are infinitely many algebraic number fields of class number one.

Although we will prove the finiteness of the class number as a corollary of the relation between forms and fields (sec. 10), we now outline Minkowski's geometric proof in the case of quadratic fields since it generalizes to all number fields.

First we must generalize the notion of norms to ideals. This will allow us to measure the "size" of an ideal.

For any ideal A in I_d, we have $A\overline{A} = nI_d$ for some $n \in \mathbf{Z}$, and since $nI_d = -nI_d$, *we can choose* $n > 0$. We call n the **norm of A**, denoted **N(A)**, i.e.,

$$N(A) = n, \quad \text{if} \quad A\overline{A} = (n)\,, \quad n > 0\,.$$

Since $\overline{\overline{A}} = A$, we have $N(A) = N(\overline{A})$.

If A is principal, $A = \alpha I_d$, then $\overline{A} = \alpha' I_d$ and $A\overline{A} = \alpha\alpha' I_d = N(\alpha)I_d$. Since, by definition, we require the norm of A to be positive, we have $N(\alpha I_d) = |N(\alpha)|$.

If $N(A) = n$, $N(B) = m$, then $(AB)(\overline{AB}) = (A\overline{A})(B\overline{B}) = (mn)$, and we have

Theorem: $N(AB) = N(A)N(B)$.

The norm of an ideal is as useful as the norm of an integer and for the same reasons, namely,

Proposition: *1)* $A|B \implies N(A)|N(B)$.

2) $A = I_d \iff N(A) = 1$.

3) $N(A) = p$*, a rational prime* $\implies$ *A is a prime ideal.*

4) If P is a prime ideal, then for some rational prime p, either

a) $P = pI_d$ *and* $N(P) = p^2$*, or*

b) $P\overline{P} = pI_d$ *and* $N(P) = N(\overline{P}) = p$.

Proof. (exercise)

Example: Recall that in I_{-5}, $21 = \left(1+2\sqrt{-5}\right)\left(1-2\sqrt{-5}\right)$ as integers, and in section one we introduced the ideals $A = \left(3, 1+2\sqrt{-5}\right)$ and $B = \left(7, 1+2\sqrt{-5}\right)$ to refine the ideal factorization of both sides to $A\overline{A}B\overline{B}$. Since $N(A) = N(\overline{A}) = 3$ and N(B) = $N(\overline{B}) = 7$, this is a factorization into prime ideals.

There is an equivalent definition of the norm, used to prove the finiteness of the class number, which generalizes to all number fields.

Recall that if A is an ideal of I_d, then, as viewing I_d and A as groups, the index $[I_d : A] = |I_d/A|$ is finite. With the propositions of section 2, one can show (with non-trivial work) that

Theorem: $N(A) = [I_d : A]$.

Since I_d is a commutative ring and A an ideal, I_d/A is not just a finite group, but also a finite ring (the quotient or *residue class ring*).

Another more number theoretic way of looking at this is to define congruence of integers in I_d modulo A by

$$\alpha \equiv \beta (\mathrm{mod}\, A) \quad \text{if} \quad A \mid \alpha - \beta\ .$$

By the construction of quotient rings, congruence is an equivalence relation compatible with addition and multiplication in I_d, the elements of I_d/A are the congruence classes and $N(A)$ is the number of such classes.

In **Z** and **Z**$[i]$, which are both principal ideal domains, we've introduced congruence modulo an integer (sec. 15.3). In these cases, if $A = (\gamma)$, then

$$\alpha \equiv \beta (\mathrm{mod}\, \gamma) \iff \alpha \equiv \beta (\mathrm{mod} (\gamma))\ .$$

If P is a prime (= maximal) ideal, then I_d/P is a finite field. Since $P\overline{P} = pI_d$, for some rational prime p (section 6), $N(P) = p$ or p^2, and I_d/P has p or p^2 elements. This makes the theory of finite fields a useful tool for studying number fields as we shall see in our further analysis of the factorization of rational primes.

To prove the finiteness of the class number requires more geometry of lattices (a la Minkowski). We outline the basic ideas, without proofs.

Proposition: (Minkowski) *Let L be a lattice. Then there is a point $\alpha \neq (0,0)$ in L of minimal length (denoted for the moment by $|\alpha|$) and any such point satisfies*

$$|\alpha|^2 \leq 2\frac{\Delta(L)}{\sqrt{3}},$$

where we recall that $\Delta(L)$ is the area of a fundamental parallelogram of L.

Using the bases we derived for I_d (sec. 16.4), it follows immediately that

$$\Delta(I_d) = \sqrt{d}\,, \text{ if } d \equiv 2, 3 (\text{mod}\, 4)\,,$$
$$\sqrt{\frac{d}{2}}\,, \text{ if } d \equiv 1 (\text{mod}\, 4)\,.$$

From the proposition, one can prove

Proposition: (Minkowski) *Every ideal class of I_d contains an ideal A such that*

$$N(A) \leq k = \sqrt{\frac{|d|}{3}}\,, \text{ if } d \equiv 1 (\text{mod}\, 4)\,,$$
$$2\sqrt{\frac{|d|}{3}}\,, \text{ if } d \equiv 2, 3 (\text{mod}\, 4)\,.$$

To prove finiteness of the class number, we must show that there are only finitely many ideals with norm $\leq k$. Then every ideal class will contain an ideal in the finite set. More generally we show that there are only finitely many sublattices of L of I_d such that $[I_d : L] = n$.

The order of the equivalence class $\overline{\alpha} \in I_d/L$ divides $n = |I_d/L|$; so $n\overline{\alpha} = \overline{0} = L$ (using additive notation for lattice groups). Then $\overline{n\alpha} = n\overline{\alpha} = L$, i.e., $n\alpha \in L$, for every $\alpha \in I_d$. Hence $nI_d \subseteq L \subseteq I_d$, for every L such that $[I_d : L] = n$. But the number of L satisfying this is in one-one correspondence with the number of subgroups of I_d/nI_d. Since I_d/nI_d is finite, we are done.

9. Bases and Discriminants

Before discussing the correspondence between forms and fields, we look at the relation between different bases and introduce the notion of a discriminant.

Proposition: *Let $\{\alpha_1, \beta_1\}$ be an integral basis for the ideal $A \subseteq I_d$ as a lattice (sec. 2) and let $\{\alpha_2, \beta_2\} \subseteq A$. Then*

$$\begin{pmatrix} \alpha_2 \\ \beta_2 \end{pmatrix} = M \begin{pmatrix} \alpha_1 \\ \beta_1 \end{pmatrix}, \tag{1}$$

for some

$$M = \begin{pmatrix} r & s \\ t & u \end{pmatrix}, \quad r, s, t, u \in \mathbf{Z}.$$

Furthermore

$$\{\alpha_2, \beta_2\} \text{ is a basis} \iff \det M = \pm 1. \tag{2}$$

The proof is essentially the same as the proof that the forms $F(x, y)$ and $F(rx + sy, tx + uy)$ represent the same integers if and only if the matrix M transforming one to the other has determinant ± 1 (sec. 5.2).

We know that if $\{\alpha_1, \beta_2\}$ and $\{\alpha_2, \beta_2\}$ are bases of A, then their fundamental parallelograms tile the plane and, by (2), they have the same area. Conversely, if α and β span the parallelogram P and $P \cap A$ equals the vertices of P, then P tiles the plane and $\{\alpha, \beta\}$ is a basis.

Given $\alpha, \beta \in \mathbf{Q}(\sqrt{d})$, form the determinant

$$\delta(\alpha, \beta) = \begin{vmatrix} \alpha & \alpha' \\ \beta & \beta' \end{vmatrix}.$$

If $\{\alpha_1, \beta_1\}$ and $\{\alpha_2, \beta_2\}$ (not necessarily bases) are related as in (1), it follows that

$$\delta(\alpha_2, \beta_2) = \det M \cdot \delta(\alpha_1, \beta_1).$$

If they are both bases, then by (2)

$$\delta(\alpha_2, \beta_2) = \pm\, \delta(\alpha_1, \beta_1).$$

If $\{\alpha, \beta\}$ is a basis for A, let

$$\mathbf{D}(\alpha, \beta) = \delta(\alpha, \beta)^2.$$

Then $D(\alpha, \beta)$, the **discriminant of A**, is independent of the basis chosen for A; we denote it by $\mathbf{D_A}$. D_A is a rational integer since it is an integer of I_d and is equal to its conjugate.

Warning: Some authors use the notation δ for the discriminant of A.

The **discriminant of the field $\mathbf{Q}(\sqrt{\mathbf{d}})$**, denoted by $\mathbf{D_d}$, is the discriminant of the ideal I_d. It is a very important invariant of $\mathbf{Q}(\sqrt{d})$.

Using the bases derived for I_d, it follows immediately that

$$D_d = \begin{cases} d\ , & \text{if } d \equiv 1 (\text{mod } 4)\ , \\ 4d\ , & \text{if } d \equiv 2, 3 (\text{mod } 4)\ . \end{cases}$$

Exercise: Prove that $I_d = \{\frac{u+v\sqrt{D_d}}{2} \mid u, v \in \mathbf{Z}, u \equiv vD_d (\text{mod } 2)\}$.

There are geometric interpretations for the discriminant. E.g., if $d < 0$, then

$$|D_d| = 4 \cdot (\Delta(I_d))^2\ ,$$

where $\Delta(I_d)$ is the area of a fundamental parallelogram for the lattice I_d (exercise - see [Ada -Gol] or [Bor - Sha]).

It can also be shown ([Hec, Theorem 76; *beware of the different notation*]) that if $\{\alpha, \beta\}$ is a basis for A, then

$$N(A) = \left| \frac{\delta(\alpha, \beta)}{\sqrt{D_d}} \right| . \qquad (3)$$

We shall just use D in place of D_d to denote the discriminant of $\mathbf{Q}(\sqrt{d})$, when the subscript is clear from context.

10. The Correspondence between Forms and Fields

Now we formalize the relation between quadratic fields and binary quadratic forms that has appeared in special cases over the last several chapters. First we make a slight change of notation for quadratic fields.

In the last section, we saw that the discriminant D of $\mathbf{Q}(\sqrt{d})$, d square free, is given by

$$D = \begin{cases} d\ , & \text{if } d \equiv 1 (\text{mod } 4)\ , \\ 4d\ , & \text{if } d \equiv 2, 3 (\text{mod } 4)\ . \end{cases}$$

In both cases $\mathbf{Q}(\sqrt{D}) = \mathbf{Q}(\sqrt{d})$; so we will denote our fields by $\mathbf{Q}(\sqrt{D})$, with discriminant D, and let $\mathbf{I_D} = I_d$ denote the integers.

Note that these discriminants are fundamental discriminants, i.e., discriminants of primitive forms (sec. 12.10).

Among the relations we have seen between forms and fields is the fact that the roots of a form of discriminant D are elements of $\mathbf{Q}(\sqrt{D})$ and the form factors over the field (sec. 13.2). Thus, e.g.,

$$x^2 + 3y^2 = (x + \sqrt{-3}y)(x - \sqrt{-3}y) = N(x + \sqrt{-3}y) ,$$

for all $x, y \in \mathbf{Z}$. To make this relation more precise, we are going to set up a correspondence between proper equivalence classes of forms and equivalence classes of ideals. First we must restrict the notion of equivalence of ideals.

If A, B are ideals in I_D, then **A is strictly equivalent to B** (or **equivalent in the narrow sense**), denoted by $\mathbf{A} \approx \mathbf{B}$, if there exists a $\lambda \in \mathbf{Q}(\sqrt{D})$, $N(\lambda) > 0$, such that $A = \lambda B$. This is an equivalence relation and *the strict ideal classes are subsets of the ideal classes.* This notion of equivalence corresponds to proper equivalence of forms that Kummer described in the excerpt in section 7. In the case of imaginary quadratic fields, equivalence and strict equivalence are the same, since all norms are > 0.

Multiplication of strict equivalence classes is defined as for ideal classes and we have a strict class group and strict class number. We denote the strict class number of $\mathbf{Q}(\sqrt{D})$ by $\mathbf{h'_D}$ (later we prove it is finite).

To set up our correspondence, we start with ideals and bases and associate forms to them.

If $\{\alpha, \beta\}$ is an integral basis for the ideal A in I_D, we regard it as an ordered basis. Recall that the determinant $\delta(a, b)$ is defined by

$$\delta = \begin{vmatrix} \alpha & \alpha' \\ \beta & \beta' \end{vmatrix} = \alpha\beta' - \alpha'\beta .$$

Since $\delta' = -\delta$, $\delta = s\sqrt{D}$ for some $s \in \mathbf{Z}$. If $s < 0$, then the ordered basis $\{\beta, \alpha\}$ of A will have $s > 0$. We only allow those bases for which $s > 0$ and call them **oriented bases** (this is just a choice of orientation of a basis as a pair of vectors in $\mathbf{R}^2$). When we want to designate an ideal together with an oriented basis, we use the notation $(\mathbf{A}, \alpha, \beta)$.

We define a function φ from triples (A, α, β), for all ideals A in a *fixed* I_D, to forms by

$$\varphi(A, \alpha, \beta) = f_{\alpha,\beta}(x, y) = ax^2 + bxy + cy^2 ,$$

where

$$f_{\alpha,\beta} = \left(\frac{1}{N(A)}\right)(\alpha x + \beta y)(\alpha' x + \beta' y)$$
$$= \left(\frac{1}{|N(A)|}\right) N(\alpha x + \beta y) ,$$

for all $x, y \in \mathbf{Z}$.

We *outline* the properties of φ. Good references include [Ada-Gol], [Hec], and [Nar, Chap. VIII, Sec. 2].

I) *$f_{\alpha,\beta}$ has rational integer coefficients and is primitive.*

Proof: $N(\alpha x + \beta y) = \alpha\alpha' x^2 + (\alpha\beta' + \alpha'\beta)xy + \beta\beta' y^2$. The coefficients equal their conjugates and therefore are in $\mathbf{Z}$. By our proof in section 3 that $A\overline{A} = nI_D$, for some n in $\mathbf{Z}$, $n = N(A)$ is the greatest common divisor of the coefficients; hence $f_{\alpha,\beta}$ is primitive.

II) *discriminant* $(f_{\alpha,\beta}) = D(= D_d)$.

Proof:

$$\text{discriminant } (f_{\alpha,\beta}) = \left(\frac{1}{N(A)^2}\right)(\alpha\beta' + \alpha'\beta)^2 - 4\alpha\alpha'\beta\beta'$$
$$= \left(\frac{1}{N(A)^2}\right)(\alpha\beta' - \alpha'\beta)^2 = \frac{\delta(a, b)^2}{N(A)^2} ,$$

which by equation (9.3) = D_d.

If $D < 0$, $f_{\alpha,\beta}$ is a definite form and, since the coefficient of x^2 is $\frac{\alpha\alpha'}{N(A)} > 0$, $f_{\alpha,\beta}$ is positive definite. Therefore, from I and II, we have

$$\text{Image } (\varphi) \subseteq \{\text{primitive forms (positive definite if } D < 0)\}.$$

III) Image $(\varphi) = \{$primitive forms (positive definite if $D < 0)\}$, i.e., φ is surjective..

Proof: Let $f = ax^2 + bxy + cy^2$ be primitive with discriminant D (positive definite if $D < 0$).

An obvious approach would be to generalize our example at the beginning of the section by representing forms as norms of elements. Thus we have

$$ax^2 + bxy + cy^2 = a(x - r_1 y)(x - r_2 y) = aN(x - r_1 y) ,$$

where $r_1, r_2 = \frac{-b\pm\sqrt{D}}{2a}$ are the roots of the form (sec. 13.2). Therefore it would seem natural to try the ideals with integral basis $\{1, \frac{-b\pm\sqrt{D}}{2a}\}$. However $\frac{-b\pm\sqrt{D}}{2a}$ is not an integer unless $a = \pm 1$, a problem that can be avoided by essentially clearing denominators.

There are two cases according to the sign of a.

i) Let $a < 0$, $A = (2a, b + \sqrt{D})$ and assume for the moment that $\alpha = 2a$ and $\beta = b + \sqrt{D}$ is an integral basis for A. Then

$$\begin{aligned}\varphi(A, \alpha, \beta) &= \left(\frac{1}{N(A)}\right)(\alpha\alpha' x^2 + (\alpha\beta' + \alpha'\beta)xy + \beta\beta' y^2)\\ &= \left(\frac{1}{N(A)}\right) N(2ax + (b + \sqrt{D})y)\\ &= \left(\frac{1}{N(A)}\right)(4a^2x^2 + 4abxy + (b^2 - D)y^2)\\ &= \left(\frac{1}{N(A)}\right)(4a^2x^2 + 4abxy + 4acy^2)\,.\end{aligned}$$

From the proof of $A\overline{A} = nI_D$ (sec. 3), we have

$$n = \gcd(\alpha\alpha', \alpha\beta' + \alpha'\beta, \beta\beta') = -(4a)\gcd(a, b, c) = -4a\ ,$$

since $a < 0$ and f is primitive. As we assumed that $a < 0$, $N(A) = n = -4a$ and we have $\varphi(A, \alpha, \beta) = f$. A quick calculation shows that the basis $\{\alpha, \beta\}$ is oriented.

To finish the proof, we must show that $\{\alpha, \beta\}$ is an integral basis for A. It is sufficient to prove that $B = \{\alpha x + \beta y | \alpha, \beta \in \mathbf{Z}\}$ is an ideal since it must then be the ideal $A = (\alpha, \beta)$.

By the exercise in section 9, an integer in I_D has the form $\lambda = \frac{u+v\sqrt{D}}{2}$, where $u, v \in \mathbf{Z}$ and $u \equiv vD (\mathrm{mod}\, 2)$. Thus, for any $\mu = 2ax + (b + \sqrt{D})y \in B$, we must show that $\lambda\mu \in B$. Now

$$\begin{aligned}\lambda\mu &= \left(\frac{u + v\sqrt{D}}{2}\right)(2ax + (b + \sqrt{D})y)\\ &= xua + \frac{ybu}{2} + \frac{yvD}{2} + \left(xva + \frac{yvb}{2} + \frac{uy}{2}\right)\sqrt{D}\\ &= \left(x\frac{u - vb}{2} - yvc\right)2a + \left(xva + y\frac{u + vb}{2}\right)(b + \sqrt{D})\,.\end{aligned}$$

We must show that the coefficients of $2a$ and $b+\sqrt{D}$ are integers. But

$$D = b^2 - 4ac \implies D \equiv b^2 (\bmod 4a) \implies D \equiv b^2 (\bmod 2)$$
$$\implies D \equiv b (\bmod 2) ,$$

and thus

$$u \equiv vD (\bmod 2) \implies u \equiv vb (\bmod 2) \implies u \equiv -vb (\bmod 2) .$$

Hence $\frac{u-vb}{2}$ and $\frac{u+vb}{2}$ are integers and we are done.

ii) If $a > 0$, then the basis $\{2a, b+\sqrt{D}\}$ is *not* oriented. In this case, let $A = (2a, b-\sqrt{D})$ and repeat the proof in case i) (exercise).

IV) $$A = \{\alpha_1, \beta_1\} = \{\alpha_2, \beta_2\} \iff f_{\alpha_1,\beta_2} \sim f_{\alpha_2,\beta_2} , \quad (1)$$

where $\sim$ denotes proper equivalence of forms. Moreover, the matrix M which connects the bases of A also defines the equivalence of the corresponding forms.

From IV, we see that φ induces a surjective map

$$\varphi' : \text{ideals} \to \{\text{proper equivalence classes, } \overline{f}\text{, of primitive forms (positive definite, if } D < 0)\} ,$$

defined by

$$\varphi'(A) = \overline{f}_{\alpha,\beta}, \text{ where } \alpha, \beta \text{ is any basis of } A .$$

V) If $A \approx B$, then $\varphi'(A) = \varphi'(B)$.

Thus φ' is constant on strict ideal classes and induces a surjective map

$$\overline{\varphi} : \text{ strict ideal classes } \to \{\text{proper classes of primitive forms (positive definite, if } D < 0)\} ,$$

defined by

$$\overline{\varphi}([\mathbf{A}]) = \overline{f}_{\alpha,\beta} ,$$

for any basis $\{\alpha, \beta\}$ of any ideal in $[A]$.

By IV), $\overline{\varphi}$ is injective and thus bijective. Restricting ourselves to $\mathbf{Q}(\sqrt{D})$, we summarize:

The Correspondence Theorem: *$\overline{\varphi}$ is a bijection between the strict equivalence classes of ideals in I_D and proper equivalence classes of primitive forms of discriminant D (positive definite, if $D < 0$).*

To generalize to primitive forms whose discriminant is not the discriminant of a quadratic field, i.e., not a fundamental discriminant, as, for example, $x^2 - 3^2 \cdot 11^4 \cdot 19y^2$ with discriminant $4 \cdot 3^2 \cdot 11^4 \cdot 19$, we must look at subrings of I_D, the "orders" of Dirichlet, and **Z**-modules in these rings. In the example above, one considers

$$\{r + \sqrt{4 \cdot 3^2 \cdot 11^4 \cdot 19}\} = \{r + s(2 \cdot 3 \cdot 11^2)\sqrt{19}\} \subseteq I_{19} .$$

The whole theory of ideals, including unique factorization and our correspondence, carries over (see [Ada-Gol] or [Bor-Sha] for details).

This correspondence between equivalence classes of primitive forms and strict ideal classes essentially provides a dictionary for translating concepts and results between forms and fields. The next section is devoted to several examples of this.

11. Applications of the Correspondence

Class Numbers

Let D be a fundamental discriminant. We know that $h(D)$, the number of proper equivalence classes of primitive forms of discriminant D (positive definite, if $D < 0$), is finite (sec. 12.9). By the correspondence theorem, this equals h'_D, the strict class number of $Q(\sqrt{D})$. Hence

Corollary: *h'_D is finite.*

The theory of reduced forms, which provided an algorithm for computing class numbers for forms (chapter 12) can now be applied to compute strict class numbers of fields. The corollary also allows us to prove

Theorem: *h_D, the class number of $\mathbf{Q}(\sqrt{D})$ is finite.*

Proof: Every ideal class splits into at most two strict ideal classes. To see this, assume that the ideals A, B, C are equivalent, but no two are strictly equivalent. Then $A = \lambda B$, $B = \mu C$, $N(\lambda), N(\mu) < 0$, and thus $A = \lambda\mu C$, $N(\lambda\mu) = N(\lambda)N(\mu) > 0$, contradicting A not strictly equivalent to B. Therefore $h'_D \leq 2h_D$ and we are done.

The relation between h_D and h'_D can be made more precise.

Proposition: *1) If $D < 0$, then $h'_D = h_D$,*

2) *If $D > 0$ and κ is the fundamental unit of I_D (sec. 16.5), then*

(i) $N(\kappa) = -1 \implies h'_D = h_D$,

(ii) $N(\kappa) = 1 \implies h'_D = 2h_D$.

We do not prove this but note that an algorithm for finding fundamental units would then let us compute h_D from h'_D (see [Ada - Gol] and our discussion of units and their relation to Pell's equation in sec. 16.5).

Example: $h(-24)$, the class number of primitive forms of fundamental discriminant -24, is 2 (sec. 13.9). Hence, since $-24 < 0$, $h'_{-24} = h_{-24} = 2$.

From the relations between various class numbers, we see that Gauss' conjecture that there are infinitely many fundamental discriminants $D > 0$ with $h(D) = 1$ is equivalent to the conjecture that there are infinitely many real quadratic fields with $h_D = 1$ (i.e., with unique factorization).

Units and Automorphs of Forms

We show that, under our correspondence, units correspond to automorphs of forms. To see this, we look more closely at the map φ of the last section.

Suppose $A \approx B = \lambda A$. Then

1) $\{\alpha, \beta\}$ is a basis for A if and only if $\{\lambda\alpha, \lambda\beta\}$ is a basis for B.

2) $\varphi(A, \alpha, \beta) = \varphi(B, \lambda\alpha, \lambda\beta)$.

Now let $\lambda = u$, a unit. Then $A = uA$, since each side divides the other. By 1) and 2), $f_{\alpha,\beta} = f_{\lambda\alpha,\lambda\beta}$ and, by 1) and property IV of section 10, there is a matrix M transforming $f_{\alpha,\beta}$ to $f_{\lambda\alpha,\lambda\beta}(= f_{\alpha,\beta})$, i.e., M is an automorph. Thus units correspond to automorphs and this explains why we obtained the same answer in counting each of them separately (sec. 13.7 and sec. 16.5).

The Class Group, Composition of Forms, and the Representation of Numbers by Forms

In section 13.9, we discussed Gauss's theory of composition of forms and how Gauss and his precursors may have been led to these ideas by using norms of quadratic integers. We want a more precise statement of the relation between these two ideas.

For a given fundamental discriminant D, the proper equivalence classes of primitive forms of a discriminant D form a group under composition of classes, the strict ideal classes of $\mathbf{Q}(\sqrt{D})$ form a group under multiplication of classes, and $\overline{\varphi}$ is a bijection between these groups. In fact, $\overline{\varphi}$ *is an isomorphism between these groups*, and thus questions about composition translate to questions about the structure of the strict class group. In general, it is much easier to work with the strict class group than the group of forms.

Recall that if an integer n is represented by a form $f(x, y)$, i.e., $m = f(u, v)$ for some $u, v \in \mathbf{Z}$, then m is represented by all forms equivalent to f and we say that m is represented by the proper equivalence class $\overline{f}$. Under our correspondence, it can be shown that the representation of m by a class of primitive forms $\overline{f}$ (where f has fundamental discriminant D) is equivalent to the existence of an ideal of norm m in I_D.

If D is a fundamental discriminant, f_1 and f_2 have discriminant D, m is represented by $\overline{f}_1$, and n is represented by $\overline{f}_2$, then mn is represented by the composition $\overline{f}_1 \cdot \overline{f}_2$ (sec. 13.9). A beautiful partial converse is due to Gauss.

Theorem: *If mn is represented by $\overline{f}$, disc $(f) = D$, and* $\gcd(m, n) = 1$, *then there are classes $\overline{f}_1$ and $\overline{f}_2$ such that $\overline{f} = \overline{f}_1 \cdot \overline{f}_2$, m is represented by $\overline{f}_1$, and n is represented by $\overline{f}_2$.*

The proof is given in the language of fields and is not very hard.

A more detailed discussion of composition and representation, including primitive forms of any discriminant, is given with proofs in [Ada - Gol]. Borevich and Shafarevich [Bor - Sha] treat both the representation theory and the genus theory of forms in the language of fields.

12. Factorization of Rational Primes Revisited

In section 6, we proved that $(p) = pI_d$, p a rational prime, factors into at most two prime ideals in I_d. More specifically, either (p) is inert ((p) is prime), (p) splits ($(p) = P\overline{P}$, $P \neq \overline{P}$, P and $\overline{P}$ prime) or (p) ramifies ($(p) = P^2$, P prime, so $P = \overline{P}$).

Now we will provide more information about this factorization and its connection to quadratic residues. Recall that the discriminant D of $\mathbf{Q}(\sqrt{d})$ is given by $D = d$, for $D \equiv 1 (\bmod 4)$ and $D = 4d$, for $d \equiv 2, 3 (\bmod 4)$. If p is an odd prime, then we will frequently use that fact that $p \nmid D$ if and only if $p \nmid d$.

Theorem: *Let p be an odd prime and D the discriminant of* $\mathbf{Q}(\sqrt{d})$. *Then*

(i) if $p \nmid D$ *and* $x^2 \equiv d (\bmod p)$ *has no solution in* $\mathbf{Z}$, *then* (p) *is inert,*
(ii) if $p \nmid D$ *and* $x^2 \equiv d (\bmod p)$ *is solvable in* $\mathbf{Z}$, *then* (p) *splits,*
(iii) if $p | D$, *then* (p) *ramifies.*

We give two proofs, the first is constructive and the second gives a deeper insight into the connection with quadratic residues.

First Proof: (i) Note that in this case the condition $p \nmid D$ is superfluous, since if $p | D$, then $p | d$ and $0^2 \equiv d (\bmod p)$.

Now suppose that (p) factors into two primes, $(p) = P\overline{P}$. Then I_d/P is a finite field with $N(P) = |I_d/P| = p$ or p^2 elements (sec. 8). Since $p^2 = N((p)) = N(P)N(\overline{P}) = N(P)^2$, we must have $N(P) = p$. We show that this leads to a contradiction and thus P is inert.

Consider the map $\varphi : \mathbf{Z}/p\mathbf{Z} \to I_d/P$, defined by $\overline{n} \to n + P$, for $n \in \mathbf{Z}$. To see that φ is well defined, we note that if $\overline{m} = \overline{n}$, then $m \equiv n (\bmod p)$ and, since $P | (p)$, $m \equiv n (\bmod P)$; hence $m + P = n + P$. φ is injective since

$$\begin{aligned} \varphi(\overline{m}) = \varphi(\overline{n}) &\implies m + P = n + P \implies m - n \in P \implies P | m - n \\ &\implies \overline{P} | m - n \implies P\overline{P} = (p) | m - n \\ &\implies m \equiv n (\bmod p) \implies \overline{m} = \overline{n} . \end{aligned}$$

Thus every coset of I_d/P contains integers and, in particular, there is an $a \in \mathbf{Z}$ such that $a \in \sqrt{d} + P$.

Therefore $a \equiv \sqrt{d} (\bmod P)$, $a^2 \equiv d (\bmod P)$ and, since $P | p$, $a^2 \equiv d (\bmod p)$, contradicting the assumption that $x^2 \equiv d (\bmod p)$ in not solvable.

(ii) Let $a \in \mathbf{Z}$ satisfy $a^2 \equiv d(\bmod p)$. Since $p \nmid d$, we have $p \nmid a$ and $(a, p) = 1$.

Let $A = (p, a+\sqrt{d})$, so $\overline{A} = (p, a-\sqrt{d})$. We claim that $A\overline{A} = (p)$, $A \neq \overline{A}$, and thus, by the discussion in the first paragraph of this section, A is prime and (p) splits.

First we have

$$A\overline{A} = \left(p^2, p(a+\sqrt{d}), p(a-\sqrt{d}), a^2-d\right)$$
$$= (p)(p, (a+\sqrt{d}), (a-\sqrt{d}), \frac{a^2-d}{p})$$

where $p|(a^2-p)$. But this latter ideal contains p and $2a = (a+\sqrt{d}) + (a-\sqrt{d})$, which are relatively prime (p odd, $(a, p) = 1$) and thus contains 1 and equals I_d. Therefore $A\overline{A} = (p)$.

If $A = \overline{A}$, then, as above, A contains p and $2a$ and $A = I_d$, a contradiction.

(iii) Let $A = (p, \sqrt{d})$. Then $A = \overline{A}$ and $A\overline{A} = (p)(p, \sqrt{d}, \frac{d}{p})$, where, by assumption, $p|d$. But, since d is square free, $\gcd(p, \frac{d}{p}) = 1$; therefore $(p, \sqrt{d}, \frac{d}{p}) = I_d$, $(p) = A^2$, and (p) ramifies.

Second Proof: We assume a reasonable familiarity with quotient rings.

1) Suppose $d \equiv 2, 3(\bmod 4)$. Under the canonical homomorphism of $I_d \to I_d/P$, there is a bijection between

$$\{\text{prime ideals in } I_d \text{ dividing (p)}\} = \{\text{prime ideals in } I_d \supseteq (p)\}$$

and (1)

$$\{\text{prime ideals of } I_d/P\} .$$

Now we introduce a sequence of isomorphisms. First we have

$$I_d = \{u + v\sqrt{d} | u, v \in \mathbf{Z}\} \cong \mathbf{Z}[x]/(x^2-d) .$$

Hence

$$I_d/P \cong (\mathbf{Z}[x]/(x^2-d))/(p) \cong \mathbf{Z}[x]/(x^2-d, p) .$$

Since

$$\mathbf{Z}[x]/p\mathbf{Z}[x] \cong (Z/p\mathbf{Z})[x] = F_p[x] ,$$

and the order in which the relations $x^2 - d = 0$ and $p = 0$ are introduced into the ring $\mathbf{Z}[x]$ is irrelevant to the final result, we have

$$I_d/P \cong \mathbf{Z}[x]/(x^2 - d, p) \cong (\mathbf{Z}[x]/p\mathbf{Z}[x])/(x^2 - \overline{d}) = F_p[x]/(x^2 - \overline{d}) ,$$

where $\overline{d}$ is the residue class of d modulo p.

Under the isomorphism $I_d/P \cong F_p[x]/(x^2 - \overline{d})$, prime ideals in one ring correspond to prime ideals in the other. Combining this with (1), we see that

$$\{\text{prime ideals in } I_d \text{ dividing (p)}\}$$

corresponds to

$$\{\text{prime ideals in } F_p[x]/(x^2 - \overline{d})\} ,$$

which corresponds to

$$\{\text{prime ideals in } F_p[x] \text{ containing } (x^2 - \overline{d})\}$$

and, since $F_p[x]$ is a principal ideal domain, this corresponds to

$$\{\text{polynomials in } F_p[x] \text{ dividing } x^2 - \overline{d}\} .$$

But $F_p[x]$ is a unique factorization domain and thus $x^2 - \overline{d}$ is a prime polynomial or it splits uniquely into two linear factors.

$x^2 - \overline{d}$ is a prime polynomial (and correspondingly (p) is inert in I_d) if and only if it has no roots in F_p, which is true if and only if $x^2 \equiv d (\text{mod } p)$ is unsolvable.

$x^2 - \overline{d}$ factors if and only if

$$\begin{aligned} x^2 - \overline{d} = (x - \overline{a})(x - \overline{b}) &\iff \overline{a} + \overline{b} = 0 \iff \overline{b} = -\overline{a} \\ &\iff \overline{d} = \overline{a}^2 \iff a^2 \equiv d (\text{mod } p) \\ &\iff x^2 \equiv d (\text{mod } p) \text{ is solvable} . \end{aligned}$$

If $p \nmid D$ (and thus $p \nmid d$), then $p \nmid a$, $\overline{a} \neq -\overline{a}$ (since p is odd) and thus there are two prime ideals, namely $(x - \overline{a})$ and $(x - \overline{b})$, dividing $(x^2 - \overline{d})$. Correspondingly, there are two ideals dividing (p) in I_d and (p) splits.

If $p | D$ (and thus $p | d$), then $p | a, \overline{a} = -\overline{a}$, and $(x - \overline{a}) = (x + \overline{a})$; so the corresponding prime ideals in I_d are equal and (p) ramifies.

2) Suppose $d \equiv 1 (\text{mod } 4)$. Then $I_d = \mathbf{Z}[\frac{1}{2} + \frac{\sqrt{d}}{2}] \cong \mathbf{Z}[x]/(x^2 - x + \frac{1-d}{4})$, where $x^2 - x + \frac{1-d}{4}$ is the irreducible polynomial over $\mathbf{Q}$ for $\frac{1}{2} + \frac{\sqrt{d}}{2}$ (sec. 16.4). Following the same procedure as in the previous case leads to

$$\{\text{ideals in } I_d \text{ dividing } (p)\}$$

corresponds to

$$\{\text{polynomials in } F_p[x] \text{ dividing } x^2 - x + (1 - \overline{d})\overline{4}^{-1}\}\,.$$

Applying the techniques of section 11.1 reduces the factorization of $x^2 - x + (1 - \overline{d})\overline{4}^{-1}$ to the factorization of $y^2 = \overline{d}\,\overline{4}^{-1}$ or $z^2 = \overline{d}$, where $z = \overline{2}y$. Thus we are reduced to the same polynomial as in case 1 and the same results apply.

Recall that the Legendre symbol is defined for $p \nmid a$ as $\left(\frac{a}{p}\right) = 1$, if $x^2 \equiv a (\text{mod } p)$ is not solvable in $\mathbf{Z}$, $\left(\frac{a}{p}\right) = 1$, if it is solvable, and $\left(\frac{a}{p}\right) = 0$ if $p|a$. Since we always have $\left(\frac{D}{p}\right) = \left(\frac{d}{p}\right)$, for the discriminant D of I_d and any prime p, we can reformulate our theorem as follows.

Theorem: *Let p be an odd prime and D the discriminant of* $\mathbf{Q}(\sqrt{d})$ *. Then*

i) if $\left(\frac{D}{p}\right) = -1$, (p) *is inert,*

ii) if $\left(\frac{D}{p}\right) = +1$, (p) *splits,*

iii) if $\left(\frac{D}{p}\right) = 0$, (p) *ramifies.*

A similar set of conditions can be given for $p = 2$ (see [Ire - Ros]). Using the law of quadratic reciprocity yields a deeper and more elegant formulation of our criteria.

Theorem: *The form of the factorization of* (p) *in* I_d, *p odd, is determined by the residue class of p modulo D. More specifically, if* $p \equiv q (\text{mod } D)$, *q an odd prime, then* (p) *is inert (resp. splits, ramifies) if and only if the same is true for* (q).

Proof: Only a finite number of odd primes ramify (the prime divisors of D). If (p) ramifies and $p \equiv q(\bmod D)$, then $p|D$, $p = q$ and (q) ramifies. Hence we must only consider the cases where (p) splits or is inert, i.e., $p\nmid D$.

i) Suppose $d \equiv 2, 3(\bmod 4)$, $D = 4d$, and $p\nmid D$.

If $d > 0$ and $p \equiv q(\bmod 4d)$, then, by form $1'$ of the law of quadratic reciprocity (sec. 11.5),

$$p \equiv q(\bmod 4d) \Longrightarrow \left(\frac{d}{p}\right) = \left(\frac{d}{q}\right) \Longrightarrow \left(\frac{D}{p}\right) = \left(\frac{d}{q}\right)$$

and we are done.

If $d < 0$, then

$$\left(\frac{D}{p}\right) = \left(\frac{-|D|}{p}\right) = \left(\frac{-1}{p}\right)\left(\frac{|D|}{p}\right) = (-1)^{\frac{p-1}{2}}\left(\frac{|D|}{p}\right)$$

and similarly $\left(\frac{D}{q}\right) = (-1)^{\frac{q-1}{2}}\left(\frac{|D|}{q}\right)$. If $p \equiv q(\bmod 4d)$, then $p \equiv q(\bmod 4|d|)$ and, by the reciprocity law, $\left(\frac{|d|}{p}\right) = \left(\frac{|d|}{q}\right)$ and thus $\left(\frac{|D|}{p}\right) = \left(\frac{|D|}{q}\right)$. But $p \equiv q(\bmod 4d)$ also implies $p \equiv q(\bmod 4)$ and thus $(-1)^{\frac{p-1}{2}} = (-1)^{\frac{q-1}{2}}$ and we are done.

ii) Suppose $d \equiv 1(\bmod 4)$, $D = d$, and $p\nmid D$. Then D is odd and, by the law of reciprocity for Jacobi symbols (sec. 11.8), which is also true for $D < 0$ (exercise),

$$\left(\frac{D}{p}\right) = (-1)^{\frac{p-1}{2}\cdot\frac{D-1}{2}}\left(\frac{p}{D}\right) = \left(\frac{p}{D}\right),$$

since $\frac{D-1}{2}$ is even; hence $\left(\frac{D}{p}\right)$ depends only on the residue class of p modulo D.

Therefore the question of how rational primes factor in a given I_d depends on only a finite number of congruence conditions. The latter form of our decomposition law is essentially equivalent to the law of quadratic reciprocity (see [Edw 1, sec. 7.8]) and thus we see an intimate connection between reciprocity laws and the splitting of rational primes in number fields. In the next section, we discuss general reciprocity laws which are related to a generalization of these ideas.

We must also note that under the correspondence between forms and fields (sec. 10), the ideal class of (p) in I_d corresponds to the proper equivalence class of $x^2 - dy^2$, if $d \equiv 2, 3 (\mathrm{mod}\, 4)$, and to the class of $x^2 - xy + \left(\frac{1-d}{4}\right) y^2$, if $d \equiv 1 (\mathrm{mod}\, 4)$. Our law for the splitting of rational primes in quadratic fields then leads to significant results on the representation of numbers by forms, via composition of forms, and this circle of ideas leads naturally to Gauss' genus theory of forms (see [Ada - Gol, sec. 11.5] and [Bor - Sha chapter 3, sec. 8]).

13. General Reciprocity Laws

Now we present a 20th century view of reciprocity laws following the very beautiful paper by B. Wyman [Wym].

We are interested in the solvability of $f(x) = 0 (\mathrm{mod}\, p)$, where $f(x)$ is a polynomial with integer coefficients and we want $x \in \mathbf{Z}$. If $f(x)$ is quadratic, then in section 11.1 we saw how to reduce the problem to polynomials of the form $x^2 - d \equiv 0 (\mathrm{mod}\, p)$, which led to the law of quadratic reciprocity.

More generally, suppose that the monic polynomial $f(x) = a_0 + a_1 x + \cdots + a_{n-1} x^{n-1} + x^n$ is irreducible over $\mathbf{Z}$ and let $f_p(x) = \Sigma \overline{a}_i x_i$, where $\overline{a}_i$ is the residue class of a_i modulo the prime p. If $f_p(x)$ factors into distinct linear factors over the finite field F_p, we say that $f(x)$ splits completely mod p. Let

$$\mathbf{Spl(f)} = \{p | f \text{ splits completely mod } p\} .$$

Higher reciprocity laws can be described, in a vague sense, as characterizing $\mathrm{Spl}(f)$ in terms of f in some "nice" way.

Example: (Quadratic reciprocity) Let p be an odd prime, $f = x^2 - d$, $f_p = x^2 - \overline{d}$, $d \neq 0$. Then

$$\begin{aligned} \left(\frac{d}{p}\right) = 1 &\iff b^2 \equiv d (\mathrm{mod}\, p), \text{ for some } b \\ &\iff f_p \text{ has the distinct roots } \overline{b} \text{ and } -\overline{b} \\ &\iff p \in \mathrm{Spl}(f) . \end{aligned}$$

Then the first form of quadratic reciprocity can be reformulated as

$$p \in \mathrm{Spl}(f) \iff p \nmid d \text{ and } p \text{ is congruent to one of a set of numbers } \bmod 4d\ ,$$

i.e., $\mathrm{Spl}(f)$ can be described by congruence conditions $\bmod 4d$.

Before stating a very general reciprocity law, we quickly recall some notions from Galois theory.

The splitting field of f is the field $\mathbf{Q}(r_1, \ldots, r_n)$, where the r_i are the complex roots of f (not to be confused with $\mathrm{Spl}(f)$). $\mathrm{Gal}(f)$, the Galois group of f, is the group of automorphisms of $\mathbf{Q}(r_1, \ldots, r_n)$ which leave each element of $\mathbf{Q}$ fixed. Galois theory is concerned with the relation between the subgroups of $\mathrm{Gal}(f)$ and the subfields of $\mathbf{Q}(r_1, \ldots, r_n)$. We say that f is an **abelian polynomial** (and $\mathbf{Q}(r_1, \ldots, r_n)$ is an **abelian extension** of $\mathbf{Q}$) if $\mathrm{Gal}(f)$ is an Abelian group.

Abelian Polynomial Theorem: *If f is a monic polynomial with rational integer coefficients, irreducible over $\mathbf{Z}$, then*

$\mathrm{Spl}(f)$ *is described by congruence conditions relative to a modulus depending only on* f $\iff$ f *is an abelian polynomial.*

$\Leftarrow$) is equivalent to the Artin reciprocity law, the most general reciprocity law known.

The proof of the theorem requires a substantial amount of algebraic number theory, including class field theory (the study of abelian extensions of the rationals).

All of the earlier reciprocity laws, quadratic, cubic, biquadratic and Eisenstein reciprocity, are special cases of the Artin reciprocity law. For a broader and deeper discussion of general reciprocity laws, see [Tat].

This finishes our introduction to algebraic number theory, with a strong emphasis on quadratic fields. The reader who wishes to explore the deeper aspects of the theory must study a general introduction including the role of Galois theory. Beyond this are the analytic theory (including zeta functions), and class field theory.

Probably the best place to begin is with Harold Stark's chapter "Galois Theory, Algebraic Number Theory and Zeta Functions" in the survey volume *From Number Theory to Physics* [W-M-L-I] (this book contains many other gems, e.g., Cartier's chapter on zeta functions). Hecke [Hec] is the best book for the classical approach, and we have mentioned a variety of more modern books including Borevich and Shafarevitch [Bor - Sha], Marcus [Mar], and Lang [Lan 3]. We will discuss another approach to algebraic number theory, the so called "local theory" in the last section of the book.

Appendix: Dirichlet and 19th Century Number Theory

We have had frequent occasion to refer to the work of Gustav Peter Lejuene Dirichlet (1805-1859), but we have not described the man nor put him in historical context.

Dirichlet was one of the most influential number theorists of the 19th century and he is also famous for his work in analysis and applied mathematics. He was the first to master all of Gauss' *Disquisitiones Arithmeticae*. According to Eduard Kummer, Dirichlet did not just read the *Disquisitiones* several times, but throughout his life it was always on the table where he was working and was a source for continual study. Dirichlet reworked, simplified, and added to Gauss' number theory and lectured on this work during his years in Berlin and Göttingen. Among the students in his classes were Eisenstein, Kronecker, Dedekind, and Riemann.

After his death, Dirichlet's lectures were edited by Dedekind and appeared in 1863 as *Vorlesungen über Zahlentheorie*, referred to ever since in the number theory community as Dirichlet - Dedekind. Dedekind added appendices in the subsequent three editions and the final fourth edition [Dir - Ded] contains the definitive version of Dedekind's fundamental contributions to algebraic number theory (see our discussion in sec. 16.1). Even today, Dirichlet - Dedekind remain a clear and fascinating introduction to number theory.

Dirichlet contributed to several parts of number theory: Diophantine equations, the law of biquadratic reciprocity, algebraic number theory (the Dirichlet unit theorem [Hec, sec. 3.5]), Diophantine approximation and the Dirichlet box principle (sec. 21.2), the class number formula for quadratic forms (sec. 12.10), and the theorem on primes in an arithmetic progression (sec. 8.4). In his proofs of the latter two theorems, he introduced techniques of analysis, including what are now called Dirichlet series (sec. 16.4), into number theory. Many regard this work, especially his two part paper *Recherches sur diverses applications de l'analyse infinitésimale à la théorie des nombre* [Dir], as the true beginning of analytic number theory. See [Ell, W - Ell, F] for an introduction to Dirichlet's ideas and [Hec] and [Ire - Ros] for more formal proofs.

The only articles in English about Dirichlet seem to be a short biographical essay by Oystein Ore in the Dictionary of Scientific Biography and a very interesting article by David Rowe [Roe] stressing Dirichlet's work on biquadratic reciprocity, and including some correspondence with Gauss and an informative letter from Dirichlet to his mother. There is a good discussion of Dirichlet and an English translation of a small part of Kummer's lengthy memorial lecture [Dir] in [Sch - Opo].

Chapter 18

Arithmetic on Curves 1: Rational Points and Plane Algebraic Curves

1. Introduction

Now we turn from the theme of algebraic number theory, whose intensive development began in the 19th century, to arithmetic on curves, one of the main topics of 20th century number theory. Arithmetic on curves or arithmetic algebraic geometry refers to the applications of methods of algebraic geometry to number theory.

Our two main goals are the Mordell theorem on the finite generation of 'rational points on elliptic curves' and the Nagel-Lutz theorem characterizing such 'points of finite order'. We will prove some special cases of these theorems and discuss these topics more generally. In order to do so we will need to assume results from algebraic geometry and complex function theory. We will outline these results, without proof, and try to make them as intuitive as possible. Our description of results in algebraic geometry follows classical ideas; the difficulty of formulating them precisely in great generality and proving them led to the modern abstract algebraic treatment. Our exposition has been strongly influenced by the lectures of John Tate [Tat 2], who showed that some deep parts of the theory can be treated by elementary methods. Tate's lecture notes have since been expanded and updated into a book [Sil - Tat].

For a beautiful treatment of algebraic curves, with a mixture of the classical and modern ideas and tracing some of the historical development, see Brieskorn and Knorrer [Bri - Kno]. Walker [Wal] is an excellent introduction from the classical viewpoint, using algebraic methods, and Fulton [Ful] presents an introduction to curves in a modern algebraic setting. Other good references for algebraic geometry include Shafarevich [Sha], which includes a historical appendix, Jenner [Jen], Kendig [Ken], Reid [Rei, M], Mumford [Mum], Kirwan [Kir] and Hartshorne [Har].

Arithmetic algebraic geometry is mostly a development of the 20th century, beginning with Poincaré's fundamental paper in 1901 [Poi]. However, there were earlier results going back as far as Diophantus, who studied rational solutions of polynomial equations. We will delay the history for a while since it can be discussed more sensibly after we've developed some of the language and basic results. Instead we begin with the ideas in Poincaré's paper.

Modern treatments of the subject include [Sil], [Hus], [Lan 4], [Lan 5] and [Kob 2]. More elementary discussions are included in [Nag], [Mor 2], [Lev 2], [Ros], [Cha], [Kna] and [Sil - Tat]. [Wei1] and [Bas] contain substantial aspects of the early history.

General Problem: Given a polynomial $f(x, y) \in \mathbf{Q}[x, y]$ (i.e., with rational coefficients), find all rational solutions of $f(x, y) = 0$.

Geometrically, we are asking for all **rational points** (points with rational coordinates) on the curve $C : f(x, y) = 0$. These are **plane algebraic curves**. For arithmetic purposes, we assume that $f(x, y)$ has rational coefficients. Since one can always clear the denominators of the coefficients in $f = 0$, there is no loss of generality in restricting $f(x, y)$ to be in $\mathbf{Z}[x, y]$. The **degree** of the curve $f = 0$ (called the order, in some books) is the degree of the polynomial f.

Finding rational solutions of $f = 0$ is equivalent to finding the integer solutions of an associated *homogeneous* polynomial equation (all terms of the same degree). For example, if $\left(\frac{r}{u}, \frac{s}{u}\right)$ is a rational point on $y^2 = x^3 + 1$ (written with common denominator u), then

$$\left(\frac{s}{u}\right)^2 = \left(\frac{r}{u}\right)^3 + 1 \implies us^2 = r^3 + u^3 , \tag{1}$$

and we have an integer solution of

$$ZX^2 = Y^3 + Z^3 . \tag{2}$$

Conversely, any integer solution of (2) yields, after division by Z^3, a rational solution of (1). The fact that rational solutions of (2) also yield rational solutions of (1), after division by Z, will play an important role when we discuss projective curves.

Generally it is better to study rational points instead of integer points because it is more flexible to work in the field $\mathbf{Q}$ as opposed to the ring $\mathbf{Z}$.

Notation: If R is a ring then the pairs (x, y), $x, y \in R$, are called R points. Thus we have $\mathbf{Z}$ points (integer points), $\mathbf{Q}$ points (rational points), $\mathbf{R}$ points (real points) and $\mathbf{C}$ points (complex points).

We will be primarily concerned with the study of cubic curves (degree three), but first we look at lines and conics. For the moment, *we regard our curves as defined in the real plane* $\mathbf{R}^2$, so that we can easily draw pictures.

2. Lines

We call $L : ax + by = c$, where $a, b, c \in \mathbf{Q}$, $(a, b) \neq (0, 0)$, a **rational line**. Clearly x is rational if and only y is rational, and thus we have all the rational points on L. For later generalizations, it is convenient to look at these rational points in a more geometric way. It is an easy exercise to prove that two rational points determine a rational line, and that two rational lines are either parallel or intersect in a rational point.

If we consider the graph of $y = -\frac{a}{b}x - \frac{c}{b}$ (assume $b \neq 0$), then we can think of the x-axis as a parameter space for the parametrization $x \to \left(x, -\frac{a}{b}x - \frac{c}{b}\right)$. The rational points on the line correspond to rational values of the parameter (fig. 1).

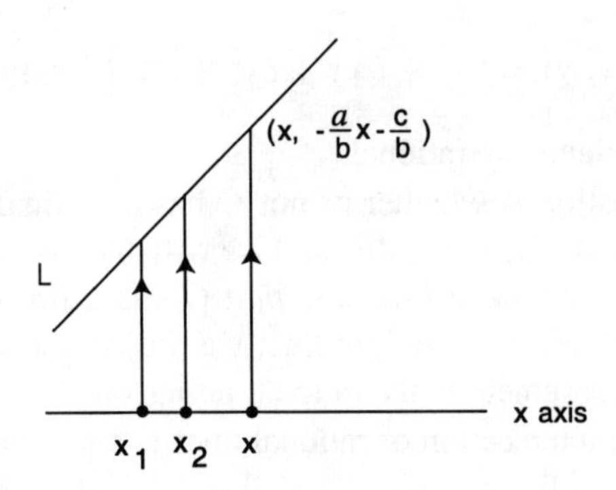

Figure 1

More generally, if L' is any other rational line and P is a rational point not on L or L', we have a projection of L' to L from P (fig. 2). If Q is rational, then the line PQ is rational and intersects the rational line L' in a rational point Q' or PQ is parallel to L'. Conversely if Q' is rational, then PQ' is rational and intersects L in the rational point Q or PQ' is parallel to L. Thus the rational points on any rational line L' can be used as the parameter space for the rational points on L by projection from a rational point not on L or L'. The projection is, in general, not one to one because some lines may be parallel. In fact, it is only one to one when L and L' are parallel. This will be remedied by adding "rational points at infinity" to L and L' when we introduce projective geometry (sec. 5).

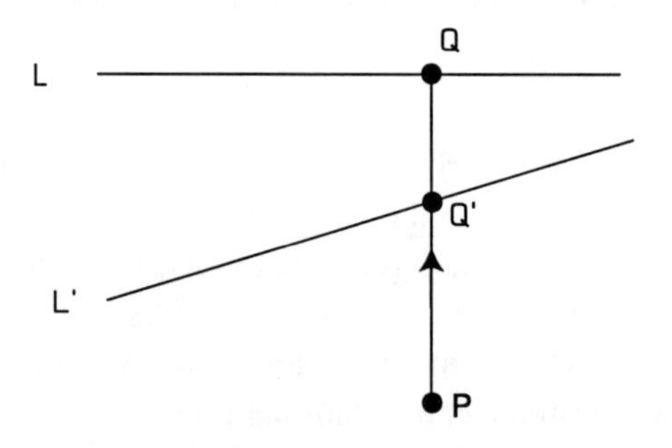

Figure 2

3. Conics

Let C be a **rational conic**:

$$C : f(x, y) = ax^2 + bxy + cy^2 + dx + ey + f = 0 ,$$

where the coefficients are rational.

Our first question is whether or not C has a rational point. This was completely answered by Legendre, and we will discuss his result at the end of the section. For now, *we assume that C has a rational point R*. By analogy with the case of a straight line, we project the conic from R to a rational line and parametrize the rational points on C.

Although the intersection of rational lines is a rational point, the intersection of a rational line $ax + by + c = 0$ and the rational conic C does not necessarily contain rational points. For the x coordinates of the intersection points satisfy the quadratic equation

$$f\left(x, -\frac{a}{b}x - \frac{c}{b}\right) = 0\,, \tag{1}$$

with rational coefficients, and we can choose the line and conic so that the roots are irrational. For example, the x coordinates of the points of intersection of $x^2 + y^2 = 1$ and $y = x$ are $\pm\sqrt{\frac{1}{2}}$.

However, *if* one of the roots of (1) is rational, then the other roots is rational (the sum of the roots = − coefficient of x, which is rational) and the corresponding y values, $y = -\frac{a}{b}x - \frac{c}{b}$, are also rational. To summarize

Proposition: *If C is a rational conic containing a rational point R, then every rational line through R intersects C in another rational point. When the line is tangent to C, we say that the line intersects C in a point with 'multiplicity two' and that R is also the second point of intersection.*

Now we can parametrize the rational points on C. Let L be a *rational* line and project C to L from R (fig. 3). By our proposition (and since rational lines intersect in a rational point), we have P is rational if and only if Q is rational. Thus the rational points on L serve as a parameter space for the rational points on **Q**.

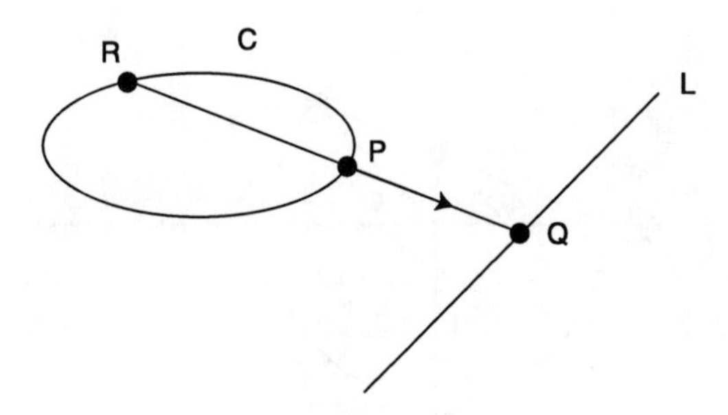

Figure 3

The parametrization is not bijective since the line RP or the tangent to C at R might be parallel to L (fig. 4). As in the case of lines, we will add a "rational point at infinity" to L to restore bijectivity (sec. 5).

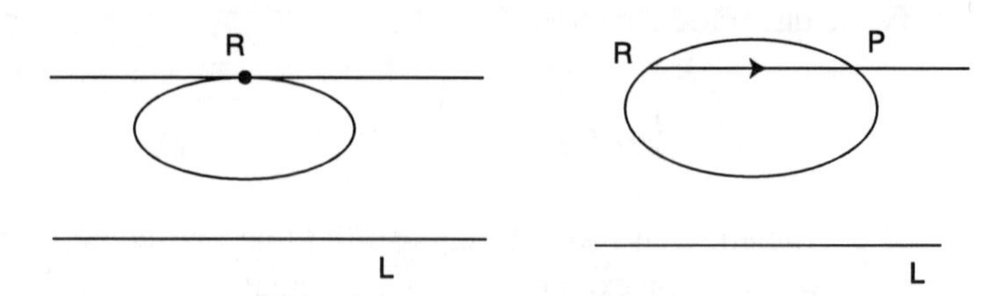

Figure 4

Example: We project the circle $C : x^2 + y^2 = 1$ to the y-axis from the points $R = (-1, 0)$ (see fig. 5). The line connecting a point (x, y) on C to $(-1, 0)$ has equation $y = t(x + 1)$. Every point on C, except $(-1, 0)$ corresponds to a unique value of t. To find x, y given t, we have

$$(1 + x)(1 - x) = 1 - x^2 = y^2 = t^2(1 + x)^2 \ .$$

Since $x = -1$ is a root of the equation, we cancel $1 + x$, and thus $1 - x = t^2(1 + x)$ or

$$x = \frac{1 - t^2}{1 + t^2}\,, \quad y = \frac{2t}{1 + t^2}\,, \tag{2}$$

which is a parametrization of $C - \{(0, -1)\}$ by rational functions. Since $t = \frac{y}{1+x}$, we see that (x, y) is a rational point if and only if t is rational.

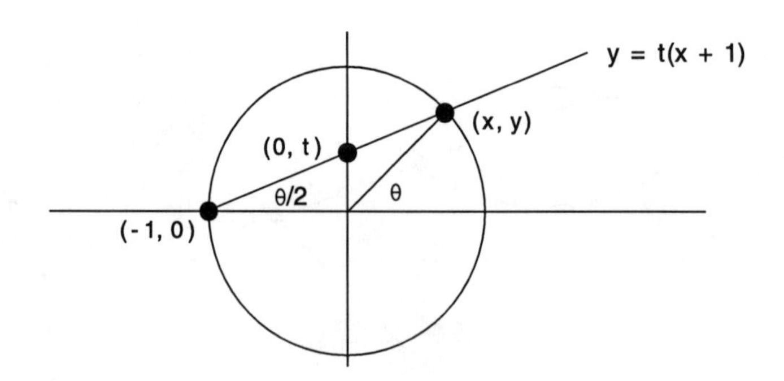

Figure 5

Exercise: Setting $x = \frac{X}{Z}$, $y = \frac{Y}{Z}$, the rational points on $x^2 + y^2 = 1$ correspond to integer points on $X^2 + Y^2 = Z^2$. Use the parametrization (2) to derive the formula of section 1.2 for Pythagorean triples.

Our parametrization of the circle has other interesting applications; we note that it is true for the real and complex points on the unit circle, as well as for the rational points.

Example: *Trigonometric identities.*

Considering the *real* points on the circle, we see from the figure 3 that

$$x = \cos\theta = \frac{1-t^2}{1+t^2}, \quad y = \sin\theta = 2\frac{t}{1+t^2}, \tag{3}$$

and we have a parametrization of the sine and cosine functions by rational functions. To verify any trigonometric identity in which one angle is the argument for all the functions, substitute the rational functions for sine and cosine, clear denominators to get a polynomial in t, and check whether it is the zero polynomial.

Example: *Integration of trigonometric functions.*

We consider integrals of the form $I = \int R(\sin\theta, \cos\theta)d\theta$, where R is a rational function of $\sin\theta$ and $\cos\theta$. From figure 3 we have $\frac{\theta}{2} = \arctan t$; thus $d\theta = \frac{2dt}{1+t^2}$. Substituting this and the parametrization (3) into the integral, we have $I = \int$ (rational function of t) dt, which is integrable in terms of elementary functions of t, using partial fractions. Substituting $t = \frac{y}{1+x} = \frac{\sin\theta}{1+\cos\theta}$ proves that the integral of any rational functions of trigonometric functions of θ can be expressed in terms of elementary functions of θ.

Exercise: Find the rational points on $x^2 + y^2 = 2$ by projecting from $(1, 1)$ and picking the line carefully to simplify the formulas.

If a rational conic has a rational point, then we have shown that we can find all of them, *at least in principle*. However, not all conics have rational points.

Example: Let $C : x^2 + y^2 = 3$. Then, setting $x = \frac{X}{Z}$, $y = \frac{Y}{Z}$, we have the homogeneous equation $X^2 + Y^2 = 3Z^2$, and integer solutions of the latter equation correspond to rational points on the former.

Assume that (X, Y, Z) is an integer solution (with no loss of generality we may assume $\gcd(X, Y, Z) = 1$). If $3|X$, then $9|X^2$, $9|Y^2$ and $3|Y$. From $9|X^2$ and $9|Y^2$, we have $3|Z^2$ which implies $3|Z$. Hence $3|X, Y, Z$, contradicting $\gcd(X, Y, Z) = 1$. Hence $X \not\equiv 0 \pmod 3$ and similarly $Y \not\equiv 0 \pmod 3$. Therefore $X, Y \equiv \pm 1 \pmod 3$, $X^2, Y^2 \equiv 1 \pmod 3$ and

$X^2 + Y^2 \equiv 2(\text{mod } 3)$. But $X^2 + Y^2 = 3Z^2 \equiv 0(\text{mod } 3)$, a contradiction, and there are no integer solutions of $X^2 + Y^2 = 3Z^2$ or rational points on $x^2 + y^2 = 3$.

Legendre [Leg] found an algorithm which allows us to decide, in a finite number of steps, if a rational conic has a rational point. He starts with a general rational conic and shows that by elementary algebraic manipulation the problem can be reduced to finding a rational point on a curve of the form

$$ax^2 + by^2 + c = 0\ ,$$

where a, b, c in $\mathbf{Z}$ are square free, not all of the same sign and $abc \neq 0$. Setting $x = \frac{X}{Z}$, $y = \frac{Y}{Z}$, this is equivalent to finding a non zero integer solution to $aX^2 + bY^2 + cZ^2 = 0$. This latter problem is solved by

Legendre's Theorem: *Let a, b, c in $\mathbf{Z}$ be square free, not all of the same sign, with $abc \neq 0$. Then*

$$F(X, Y, Z) = aX^2 + bY^2 + cZ^2 = 0$$

has a non trivial integer solution ($(X, Y, Z) \neq (0, 0, 0)$), if and only if $-bc, -ca, -ab$ are quadratic residues modulo a, b, and c respectively.

Of course, these latter conditions can be checked in a finite number of steps and thus we have an algorithm. For direct proofs of the theorem see [Leg], [Dav 1], [Gro]. Legendre's theorem is also equivalent to a special case of the Hasse-Minkowski theorem, viz.,

Under the same conditions as in Legendre's theorem, $F(X, Y, Z) = 0$ has a non trivial solution in $\mathbf{Q}$ if and only if

i) *it has a non trivial solution in $\mathbf{R}$ and*

ii) $F \equiv 0(\text{mod } p^n)$ *has a non trivial solution* ($(X, Y, Z) \not\equiv (0, 0, 0)$ mod p^n)*, for all odd primes p and all $n > 0$.*

This latter form of the theorem is intimately tied to the notion of p-adic numbers and we will discuss it further in that context (chapter 23).

4. Cubics and the Geometric Form of Mordell's Theorem

From now on, we will be primarily concerned with the set of rational points on a **cubic curve**:

$$C : f(x, y) = \sum_{i+j\leq 3} a_{i,j}x^i y^j = 0 \,,$$

where $a_{i,j} \in \mathbf{Q}$. (Note that we do not call this a rational cubic - in algebraic geometry, that term is usually reserved for curves which can be parametrized by rational functions.)

If P is a point on the curve C, then **x(P)** and **y(P)** will denote the x and y coordinates of P; so $P = (x(P), y(P))$.

Example: Let C be the curve $x^3+y^3 = 1$. Then, setting $x = \frac{X}{Z}$, $y = \frac{Y}{Z}$ and multiplying by Z^3, the rational points on C correspond to integer solutions of $X^3 + Y^3 = Z^3$. By Fermat's last theorem for $n = 3$, there are no non trivial integer solutions and hence no rational points on C.

We try to proceed as for conics. Consider a *rational* secant line L : $ax + by = c$ ($a, b, c \in \mathbf{Q}$) connecting a rational point P of C to a point Q of C. Then, as opposed to conics, Q need not be rational and, 'in general', L intersects C in a third point R (fig. 6).

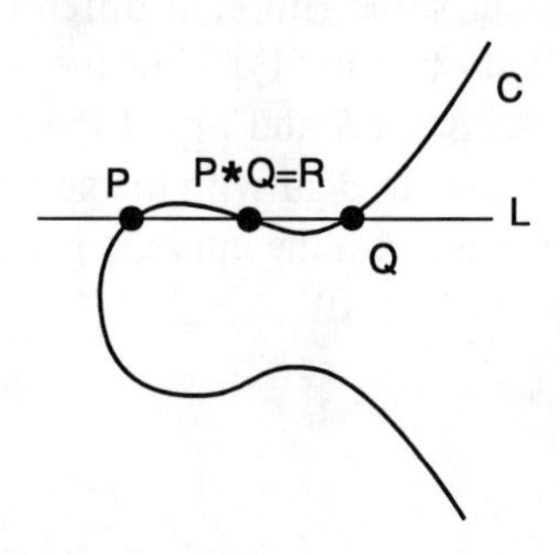

Figure 6

However, if Q rational, then so is R. For $x(P)$, $x(Q)$, and $x(R)$ are roots of

$$f\left(x, -\frac{a}{b}x - \frac{c}{b}\right) = 0 \,, \tag{1}$$

which is, 'in general', a cubic equation with rational coefficients. Since $x(P)$ and $x(Q)$ are rational, f has two rational roots. Hence the third root

$x(R)$ is rational (sum of the roots $=$ – coefficient of x^2) and, since R is on the rational line L, $y(R)$ is rational. Therefore, if we set $P * Q = R$, we have a law of composition for the set of rational points on C (fig. 1).

When we say 'in general', we mean that in some cases problems can arise. For example, if $y^2 = x^3 + 1$, the line $x = 0$ passes through the rational point $(0, -1)$ (fig. 7). Since (1) becomes the *quadratic* equation $y^2 = 0^3 + 1 = 1$, the only other point of intersection with the curve is the rational point $(0, 1)$. In the next section we will fix this problem by adding a 'rational point at infinity' to the curve.

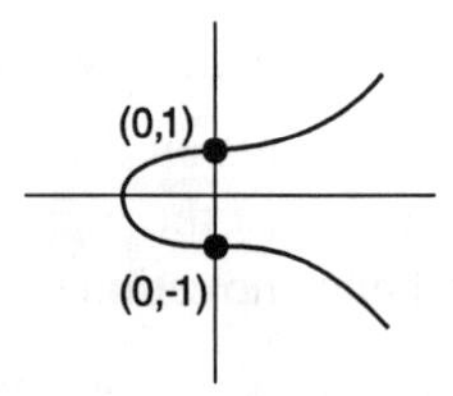

Figure 7

If P is a rational point on C, then the tangent to C at P is a rational line since it has a rational slope (implicit differentiation) and $x(P)$ is a rational root of multiplicity two in (1). Thus the third root is rational and corresponds to the rational point R and we set $P * P = R$ (fig. 8). As in the case of secants, we will have to deal with the special cases where there is no third point of intersection with the curve, e.g., the tangent line $x = -1$ to $y^2 = x^3 + 1$ at the point $(-1, 0)$.

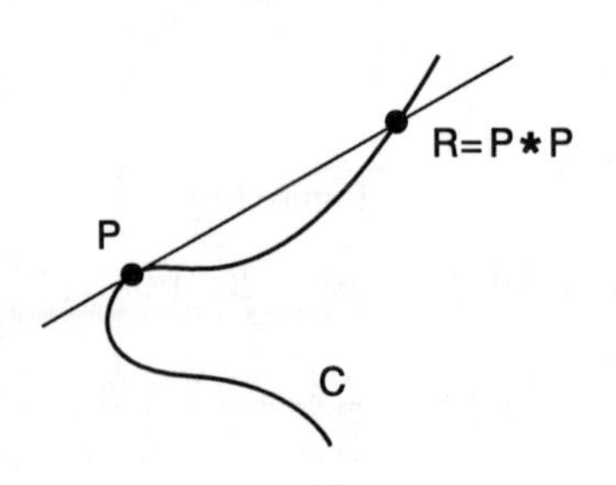

Figure 8

Starting with any set of rational points on C, one can iterate these constructions to get new rational points. This is called the **tangent-secant process**. One of the first deep results about rational points was

Mordell's Theorem: *(Geometric version) Let C be a 'nonsingular' cubic curve defined by an equation with rational coefficients. Then there exists a finite set of rational points on C such that every rational point on the curve can be derived from them by a finite number of applications of the tangent-secant process.*

Speaking informally, a point P on a curve is '**nonsingular**' if the curve has a unique tangent line at P and **singular** otherwise. A curve is **nonsingular** if all of its points are nonsingular. A more precise analytic treatment will be given in section 19.5. We must also specify whether we are talking about real or complex points; for the moment we are only concerned with real points.

Example: Over $\mathbf{R}$, $\mathbf{0} = (0, 0)$ is a singular point of $C : y^2 = x^2(x + a)$, since there are two tangents at this point (fig. 9).

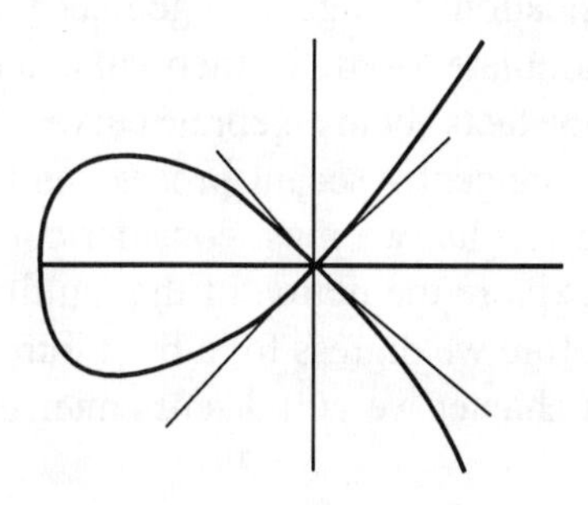

Figure 9

This example also gives some idea of the significance of singular points. A line through $\mathbf{0}$ intersects the curve "twice" at this point and then once more at some point P. Hence using rational lines M through $\mathbf{0}$, we can project the rational points of C to the rational points Q on a fixed rational line L (fig. 10).

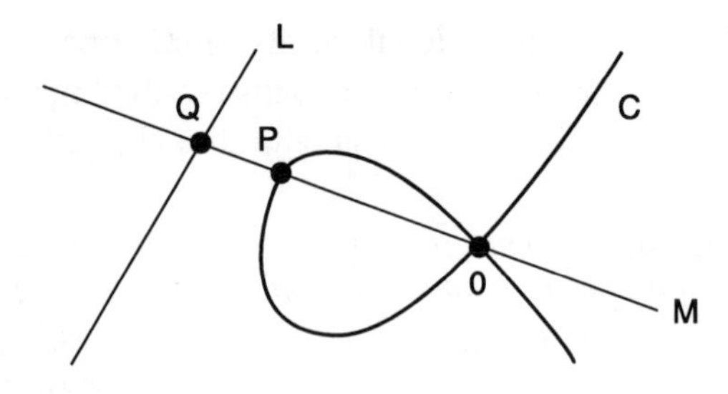

Figure 10

Thus the existence of the singular point allows us to find rational points on the cubic using the same method as for conics, and thereby reduces the difficulty of the problem.

Exercise: Derive formulas describing the projection of $y^2 = x^2(x+a)$ from $(0,0)$ to the line $x = 1$. For $a = 0$, $y^2 = x^3$, show that you get the parametrization $t \to (t^2, t^3)$.

Mordell's theorem has much broader implications than may be obvious at first sight. We shall discuss some of them from a classical viewpoint and give a computational proof of a special case of Mordell's theorem. To penetrate more deeply into the subject really requires a familiarity with the modern algebraic foundations of algebraic geometry.

Later we will reformulate Mordell's theorem in a more algebraic form, but first we present some facts about algebraic curves. To avoid the problem cases indicated in the tangent - secant process and to have natural and uniform statements of results, we must consider curves embedded in the projective plane and explore the notion of the multiplicity of intersection of these curves. Therefore we digress for a brief introduction to projective geometry. In the next chapter we will discuss intersection theory in more depth.

5. The Need for Projective Geometry

In considering the projection of one line to another line (sec. 2), the projection of a circle to a line (sec. 3) and the tangent - secant process for generating rational points on cubics (sec. 4), we ran into problems which prevented the maps from being bijective in the first two cases and which precluded a uniform process applying to all pairs of rational points in the third. To get around this problem, we are going to enlarge the plane $\mathbf{R}^2$ by adding new points.

There is a basic asymmetry in the axioms of Euclidean geometry which originally motivated the enlargement of the plane. Two points always determine a line, but two lines will either determine a point (their intersection) *or* be parallel (no intersection). This asymmetry often results in excluding special cases from general statements of theorems, as the following example shows.

Desargues' Theorem: *If the lines A_1A_2, B_1B_2, C_1C_2 joining the vertices of the triangles $A_1B_1C_1$ and $A_2B_2C_2$ intersect in a common point, then, 'in general', the points of intersection, $R = A_1B_1 \cap A_2B_2$, $Q = A_1C_1 \cap A_2C_2$, $P = B_1C_1 \cap B_2C_2$, of the lines determined by corresponding sides of the triangles exist and lie on a line (fig. 11).*

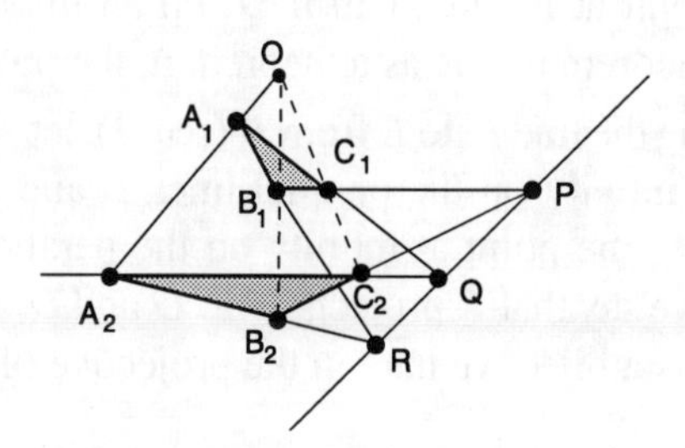

Figure 11

Here, 'in general' means that we avoid special cases where lines are parallel and the conclusion of the theorem makes no sense. For example, if A_1B_1 and A_2B_2 are parallel, there is no point R of intersection (fig. 12). (Note however, that if there is no other parallelism, then P and Q exist and, in this case, A_1B_1 and A_2B_2 are parallel to PQ.)

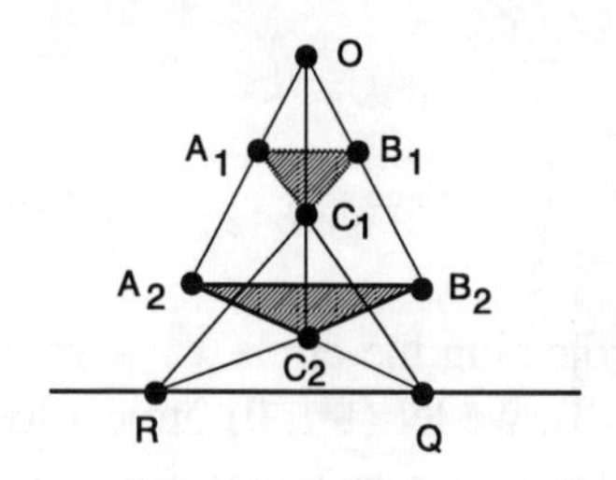

Figure 12

The problem of avoiding special cases due to parallelism of lines will be solved by adding a new point to each line, the so called *point at infinity on the line*, with the condition that parallel lines have the same point at infinity (intersect at that point). The set of new points will assumed to lie on a line, the so called *line at infinity*. The set of points in the plane, together with the new points will constitute the *projective plane* (or *real projective plane*). We will make these ideas more precise after showing their power to resolve our earlier problems.

First of all, symmetry is restored in the axioms. Not only do two points determine a unique line in the projective plane, but now two lines determine a unique point (which is their common point at infinity, if they are parallel in the ordinary plane).

For Desargues' theorem, where A_1B_1 and A_2B_2 are parallel to PQ, we have a common point at infinity, call it O, on all three lines and, in this case, Desargue's theorem is true as a theorem in the projective plane.

When projecting the line L' to L from P (sec. 2), let R' on L' correspond to O, the point at infinity on the parallel lines L and PR', and let S on L correspond to O', the point at infinity on the parallel lines L' and PS (fig. 13). Then, if we say that P projects R' to O and O' to S, the projection of L' to L from P is a bijective map in the projective plane.

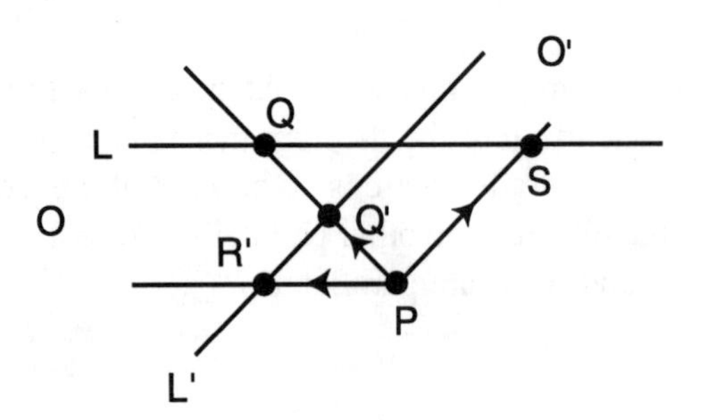

Figure 13

Similarly, when projecting the circle $x^2 + y^2 = 1$ to the y axis from the point $(-1, 0)$ (sec. 3), we let $(-1, 0)$ project to O, the common point at infinity on the parallel lines $x = 0$ and $x = -1$, and the map becomes bijective in the projective plane (fig. 14).

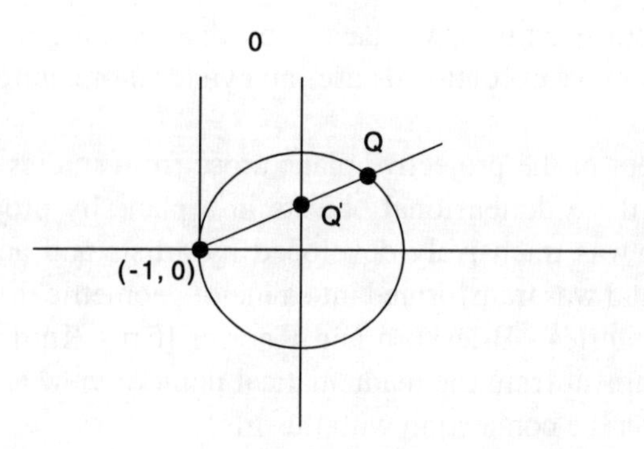

Figure 14

In the case of the secant process on $C : y^2 = x^3 + 1$ (fig. 15), we note that the slope of the tangent line to C approaches $\pm\infty$, as $y \to \pm\infty$, (so the tangents get closer and closer to to being parallel to the y axis). We say that O, the point at infinity on the y axis, is a point at infinity on the curve C. Hence the y axis intersects C at O as well as at the points $(0, -1)$, $(0, 1)$. If we call O a rational point on the curve, the secant process applied to the latter two rational points now yields another rational point. Similarly if the line L connects any two rational points of the form $(a, -\sqrt{a^3+1})$ and $(a, \sqrt{a^3+1})$, $a > -1$ (fig. 5), then L is parallel to the y-axis, contains O, and O is the third rational point of intersection of L with C. If $a = -1$, the tangent to C at $(-1, 0)$ intersects C with multiplicity two at $(-1, 0)$ (i.e., $y^2 = (-1)^3 + 1 = 0$ has a double root at $y = 0$) and O is the third point of intersection of T with the curve. Hence, by working in the projective plane, the secant process, applied to two rational points on C (and the tangent process applied to one rational point), always yields another rational point.

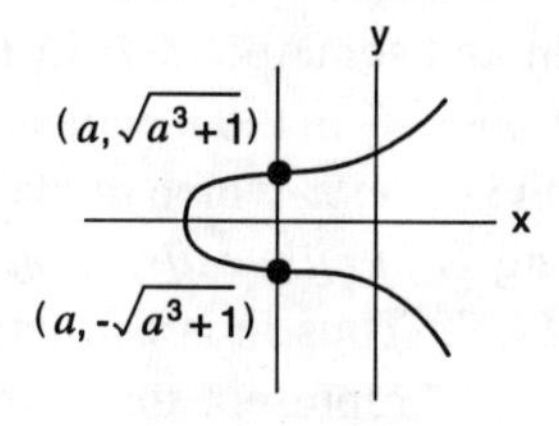

Figure 15

In all of our examples, we have seen that working in the projective plane lets us remove exceptional cases and yields more uniform statements of results.

The concept of the projective plane arose from studies of perspective (representing three dimensional objects in a plane by projection from a point), which was intensively developed by artists and architects in the Renaissance and was transformed into a major geometric theory during the 16th - 18th centuries. Brieskorn and Knorrer [Bri - Kno] present a nice historical treatment from the mathematical point of view and Pedoe [Ped] concentrates on the connection with the arts.

It is not difficult to develop projective geometry synthetically (axiomatically) for studying lines or conics (see [Cox, H], [Cox, H 2] and [Veb - You]). For more general algebraic curves, an approach via analytic geometry is preferable, and we present a more precise introduction to the fundamentals from this point of view. In any case, we shall always stress geometric intuition.

6. The Real Projective Plane; Homogeneous Coordinates

The Real Projective Plane and its Models

We call $\mathbf{R}^2$ the **affine plane**, where the term affine signifies that we are only studying the properties of intersections of points and lines as, e.g., in Desargue's theorem ($\mathbf{R}^2$ with the usual inner product, which yields angle measurement and distance, is generally called the Euclidean plane). We shall represent the affine plane $\mathbf{R}^2$ as the plane $z = 1$ in $\mathbf{R}^3$, identifying the point (x, y) with $(x, y, 1)$. Thus $(x, y, 1)$ are the *affine coordinates* of the points in $z = 1$. The point $P = (x, y, 1)$ can also be represented by the line through the origin $\mathbf{0}$ and P (fig. 16).

A line in the affine plane, which is identified with a line L in $z = 1$, under our correspondence, can be represented by the set of all lines L_P, passing through the origin and some point P in L. This set $\{L_P | P \in L\}$ of lines, together with the line L_∞ in the x-y plane which contains $\mathbf{0}$ and is parallel to L, constitutes the plane π which contains L and $\mathbf{0}$ (fig. 17).

If we let P move along L *in either direction*, then "as P approaches infinity, L_P approaches L_∞". Thus in constructing the projective plane, it seems reasonable to let L_∞ represent the point at infinity on L. It is

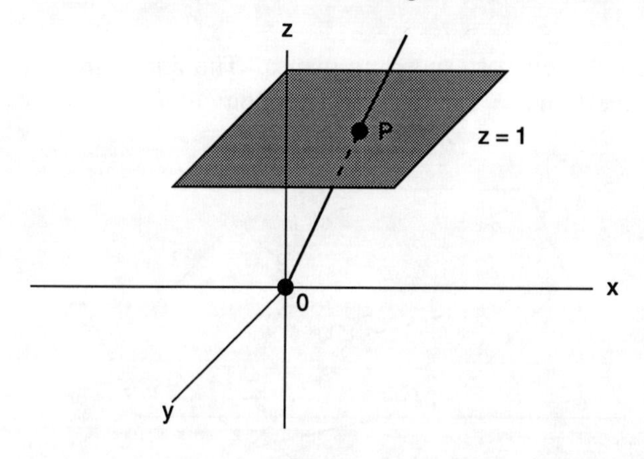

Figure 16

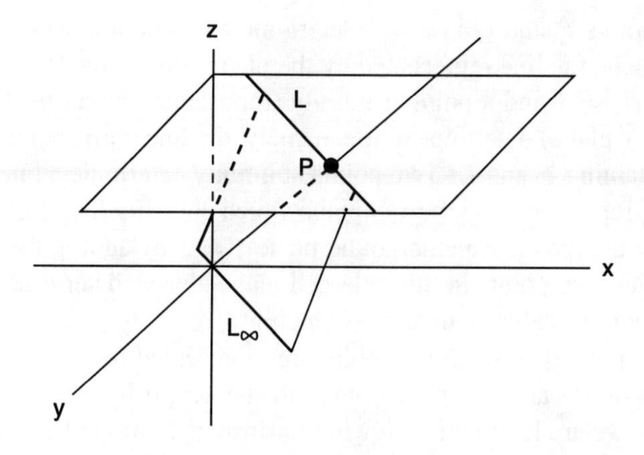

Figure 17

intuitively clear that if L and L' are parallel lines in $z = 1$, then the planes π (resp. π') containing $\mathbf{0}$ and L (resp. $\mathbf{0}$ and L') intersect in the x-y plane, i.e., $L_\infty = L'_\infty$ (fig. 18). Thus two lines in $z = 1$ have the same point at infinity if and only if they are parallel. The set of all points at infinity on all lines in $z = 1$ is represented by the set of all lines in the x-y plane, which contain $\mathbf{0}$, and thus we let the x-y plane represent the *line at infinity*.

If two lines in $z = 1$ intersect in a point P, then their corresponding planes intersect in the line through $\mathbf{0}$ and P. If two lines in $z = 1$ are parallel, then they intersect in their common points at infinity, which is the intersection of their corresponding planes. The line at infinity intersects any other line L in $z = 1$ at the point at infinity on L.

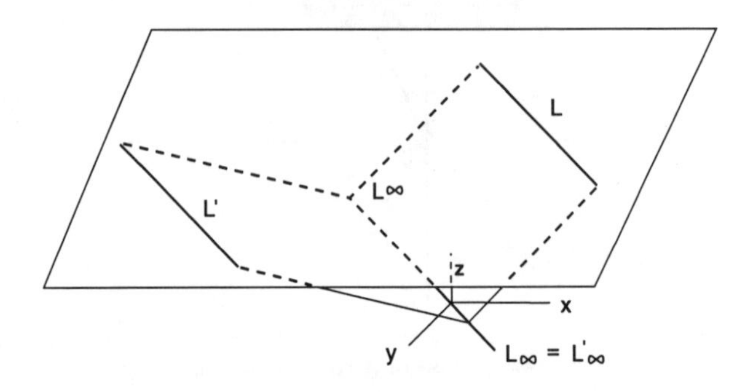

Figure 18

Two points P and Q in $z = 1$ determine a unique line which contains them, namely, the line represented by the plane containing P, Q and $\mathbf{0}$. A point P in $z = 1$ and a point at infinity (represented by a line L through $\mathbf{0}$ in the x-y plane) determine a line, namely, the line corresponding to the plane containing P and L. Two points at infinity determine a line, the line at infinity, represented by the x-y plane which contains L and L'.

To summarize: if we enlarge the plane $z = 1$ by adding the points at infinity, then two points in this enlarged plane always determine a unique line and two lines always intersect in a point.

Since the lines through the origin are represented by one dimensional subspaces of $\mathbf{R}^3$ and the planes through the origin by two dimensional subspaces, we are led to the following intrinsic definition of the projective plane which is independent of the choice of the special plane $z = 1$.

Definition: *The* **real projective plane** $\prod_2(\mathbf{R})$ *(or just* $\prod_2$*) is the set of one dimensional subspaces of* $\mathbf{R}^3$*. Each such subspace is a* **point** *in the projective plane (or a* **projective point***). A* **projective line** *(or just a* **line** *in* $\prod_2$*) is a two dimensional subspace of* $\mathbf{R}^3$.

Two projective points determine a unique projective line, viz., the two dimensional subspace they span. Two projective lines determine a projective point, viz., their intersection.

Our use of the plane $z = 1$ in $\mathbf{R}^2$ (with points at infinity adjoined) may now be regarded as a **model** of the projective plane. The correspondence

$(x, y) \leftrightarrow (x, y, 1)$ is an **embedding** of the affine plane in the projective plane, and $z = 1$ is then a model for the affine plane.

Our definition of $\prod_2$ does not single out any special points or lines at infinity. *The line at infinity is a property of our model and will change for different models.*

The plane $x = 1$ can also be used as a model. Here the points $(y, z) \in \mathbf{R}^2$ are represented by $(1, y, z)$ and, in this model, the point $P = (1, y, z)$ corresponds to the projective point, which is the line in $\mathbf{R}^3$ passing through $\mathbf{0}$ and P (fig. 19). The points at infinity are represented by the lines through $\mathbf{0}$ which are parallel to $x = 1$ (the lines through $\mathbf{0}$ in the y-z plane).

Similarly, we could use any other plane *not containing the origin* as a model. E.g., in the plane $x + y + z = 1$ (fig. 20), the affine point (x, y) is represented by $(x, y, 1 - x - y)$ and the points at infinity by the lines through $\mathbf{0}$, parallel to this plane (i.e., in the plane $x + y + z = 0$).

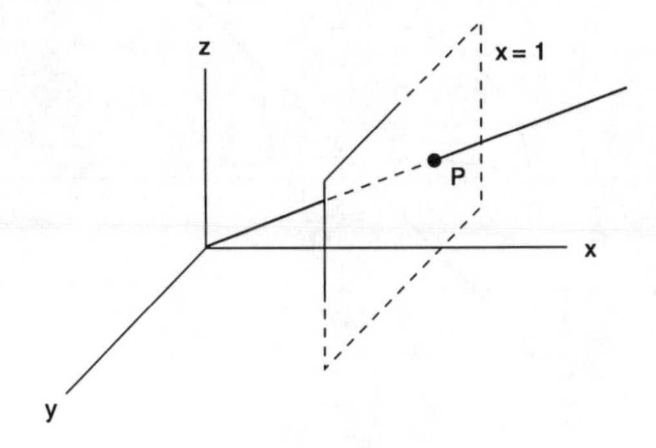

Figure 19

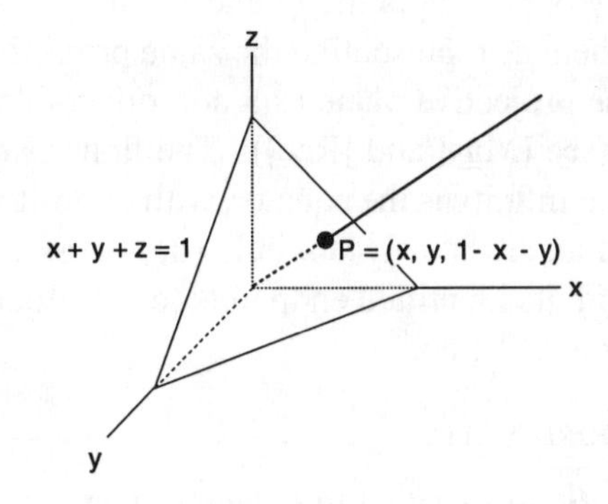

Figure 20

In any model, the points not at infinity are often called the **finite** or **affine points** of the model; they are the points in the **finite** or **affine part** of the projective plane.

Note that in our examples of models, the correspondence between points in $\mathbf{R}^2$ and the finite part of the model preserve lines. This is not a requirement for a model, but it is desirable for our geometric intuition. Surfaces other then the plane can also be used for models. A favorite of the topologists is $S^+ = \{(x, y, z)|x^2 + y^2 + z^2 = 1, z > 0\}$, the northern hemisphere of the unit sphere ($z > 0$) and its its boundary (the equator, $z = 0$) (fig. 21). The northern hemisphere represents the affine plane by the correspondence $(x, y) \leftrightarrow (x, y, \sqrt{1 - x^2 - y^2})$.

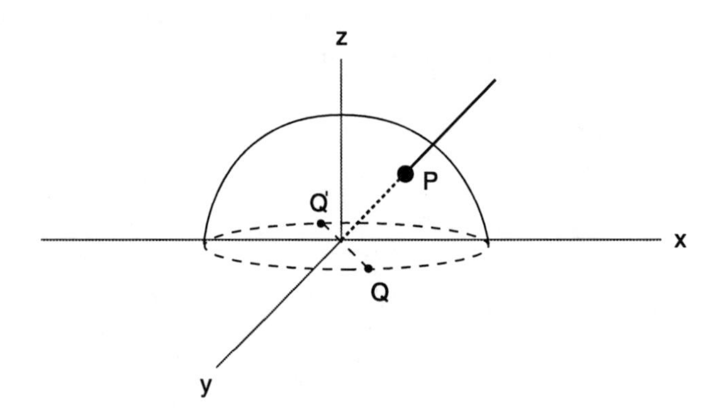

Figure 21

A line through the origin intersects S^+ at one point P, if $z > 0$, and at two diametrically opposite points, Q and Q', if $z = 0$. If we 'identify' Q and Q' (regard them as representing the same projective point), we then have a model of the projective plane as a non-orientable compact surface without boundary (see [Mas] and [Ree]). The finite part of our model is $z > 0$, and the line at infinity is the equator, with opposite points identified. A projective line is an arc of a great circle (the intersection of S^+ with a plane through $\mathbf{0}$) with its identified endpoints (on the equator) representing the point at infinity on the line.

Homogeneous Coordinates

From now on, coordinates written in capital letters denote coordinates in $\mathbf{R}^3$, *and coordinates written in lower case letters refer to the affine plane or the affine part of a model.. When it clarifies the discussion, we place a bar over a letter which denotes a projective object.*

Since a projective point $\overline{P}$ is a 1-subspace of $\mathbf{R}^3$ (a line through $\mathbf{0}$) and such a line is determined by any *non-zero* point on it, we call the coordinates of any such non-zero point **homogeneous** or **projective coordinates** of $\overline{P}$. If (X, Y, Z) are homogeneous coordinates of $\overline{P}$, then so are $\lambda(X, Y, Z) = (\lambda X, \lambda Y, \lambda Z)$ for any $\lambda \neq 0$. Two such triples are said to be equivalent, denoted by $(\mathbf{X}, \mathbf{Y}, \mathbf{Z}) \sim (\lambda\mathbf{X}, \lambda\mathbf{Y}, \lambda\mathbf{Z})$. This is an equivalence relation with the equivalence classes representing projective points, and we abuse notation by writing $\overline{P} = (X, Y, Z)$ or $\overline{P} = (\lambda X, \lambda Y, \lambda Z)$, when no confusion can arise. $\mathbf{0} = (0, 0, 0)$ *is not* the homogeneous coordinates of any point.

In the plane $Z = 1$, the affine coordinates $(x, y, 1)$ of a point P (corresponding to (x, y) in $\mathbf{R}^2$) are also homogeneous coordinates for the corresponding projective point. Conversely, if $(X, Y, Z), Z \neq 0$, are any homogeneous coordinates for $\overline{P}$, then setting $\lambda = \frac{1}{Z}$, we have $\lambda(X, Y, Z) = \left(\frac{X}{Z}, \frac{Y}{Z}, 1\right) = (x, y, 1)$ represents the corresponding affine point, where $x = \frac{X}{Z}$ and $y = \frac{Y}{Z}$. The points at infinity in this model (the 1-subspaces in the X-Y plane) have homogeneous coordinates of the form $(X, Y, 0)$.

Choosing a different model just requires us to specialize the homogeneous coordinates in a different way. For example, in the model $X = 1$, if $X \neq 0$ then $(X, Y, Z) \sim (1, \frac{Y}{X}, \frac{Z}{X}) = (1, y, z)$, where $y = \frac{Y}{X}, z = \frac{Z}{X}$. The points at infinity in this model (the 1-subspaces in the Y-Z plane) have homogeneous coordinates of the form $(0, Y, Z)$.

In the model $X + Y + Z = 1$, since

$$(X, Y, Z) \sim \left(\frac{X}{X+Y+Z}, \frac{Y}{X+Y+Z}, \frac{Z}{X+Y+Z}\right),$$

the latter coordinates also represent points in the affine part of the model (their sum is one). The points at infinity in this model (the 1-subspaces of $\mathbf{R}^3$ contained in $X + Y + Z = 0$) have homogeneous coordinates of the form $(X, Y, -X - Y)$.

7. Algebraic Curves in the Projective Plane

A projective line $\overline{L}$ is a 2-subspace of $\mathbf{R}^3$ and thus has an equation of the form

$$aX + bY + cZ = 0 \,. \tag{1}$$

Since (X, Y, Z) satisfies the equation if and only if $(\lambda X, \lambda Y, \lambda Z)$ does, for $\lambda \neq 0$, the equation is satisfied by any homogeneous coordinates of any point on $\overline{L}$. We say that (1) is an equation for the projective line $\overline{L}$ in homogeneous coordinates. Dividing (1) by Z, we have

$$a\left(\frac{X}{Z}\right) + b\left(\frac{Y}{Z}\right) + c = 0$$

or

$$ax + by + c = 0 \,, \tag{2}$$

where $x = \frac{X}{Z}$, $y = \frac{Y}{Z}$.

In the model $Z = 1$, $(x, y, 1)$ are coordinates of the points on the corresponding affine line $\overline{L} \cap (Z = 1)$ in the affine part of the model and (2) is the usual equation for the corresponding line in $\mathbf{R}^2$ (where $(x, y, 1) \leftrightarrow (x, y)$). Note that, *equivalently*, we can also go from the homogeneous to the affine equation by setting $Z = 1, x = X$, and $y = Y$ in (1). We shall use whichever procedure is more convenient. The point at infinity on $\overline{L}$ in this model is $(-b, a, 0)$.

If we divide (1) by X and let $y = \frac{Y}{X}$, $z = \frac{Z}{X}$ (or, equivalently, set $X = 1, y = Y, z = Z$), we get $a + by + cz = 0$, the equation in $\mathbf{R}^2$ of the affine part of the line $\overline{L}$ in the model $X = 1$ (where $(1, y, z) \leftrightarrow (y, z)$).

Conversely, we could start with the line $L : ax + by + c = 0$ in $\mathbf{R}^2$ (or the corresponding line in $Z = 1$ via $(x, y) \leftrightarrow (x, y, 1)$), set $x = \frac{X}{Z}$, $y = \frac{Y}{Z}$, and multiply by Z to get $aX + bY + cZ = 0$. Now also allowing $Z = 0$ (for the point at infinity in the model $Z = 1$), we have the equation for a projective line, which consists of the embedded affine line and its point at infinity. We say that (1) is the *homogeneous* or *projective* equation corresponding to (2).

We proceed similarly with the affine circle $x^2 + y^2 = 1$ (again in the model $Z = 1$). The set of all lines through $\mathbf{0}$ and points on the circle (i.e., the corresponding projective points) form a right circular cone (fig. 22). The set of lines (projective points) in the cone form the *corresponding circle* in $\prod_2$. Setting $x = \frac{X}{Z}$, $y = \frac{Y}{Z}$ in $x^2 + y^2 = 1$ and multiplying by Z^2, we have

$$X^2 + Y^2 = Z^2 \,. \tag{3}$$

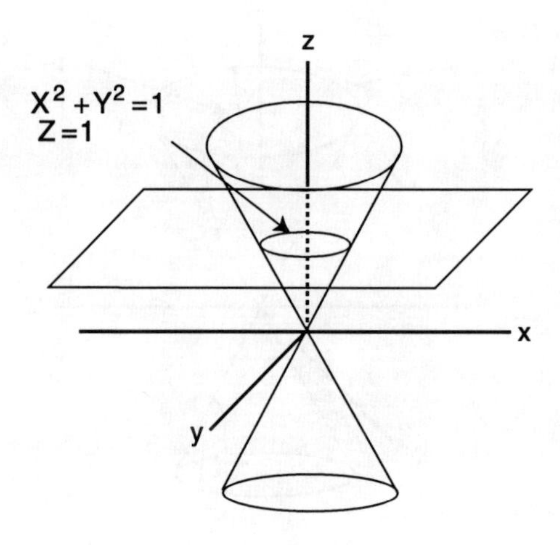

Figure 22

Since all triples $(x, y, 1)$, with $x^2 + y^2 = 1$, satisfy equation (3) and (X, Y, Z) satisfies it if and only if $(\lambda X, \lambda Y, \lambda Z)$ does (for $\lambda \neq 0$), (3) is an equation for the projective circle. If we set $Z = 1$, $x = X$ and $Y = y$ in equation (3) we are back to the affine circle. The circle has no points at infinity, so the affine and projective circle are represented by the same points in our model. $X^2 + Y^2 = Z^2$ is the *homogenized* or *projective equation* corresponding to $x^2 + y^2 = 1$.

Now consider the projective circle (the cone $X^2 + Y^2 = Z^2$) and the model $X = 1$. Geometrically, we are cutting the cone with the plane $X = 1$ and, in the affine part of the model, the intersection is a hyperbola H (fig. 23). Moreover, if we project each point P on the hyperbola H to a point P' on the affine circle C, from the origin ($P' = C \cap$ (line $\mathbf{0}P$)), we have a bijection between the whole affine hyperbola and the circle minus the points $(0, 1, 1)$ and $(0, -1, 1)$. These latter two points are the points at infinity on the hyperbola, in the model $X = 1$ (since they are on the plane $X = 0$, which is the line at infinity for this model). What we accomplish by changing models is to move the points at infinity on a curve in one model ($X = 1$) to the finite part of the corresponding curve in another model ($Z = 1$), a technique which is very useful in visualizing pieces of a projective curve.

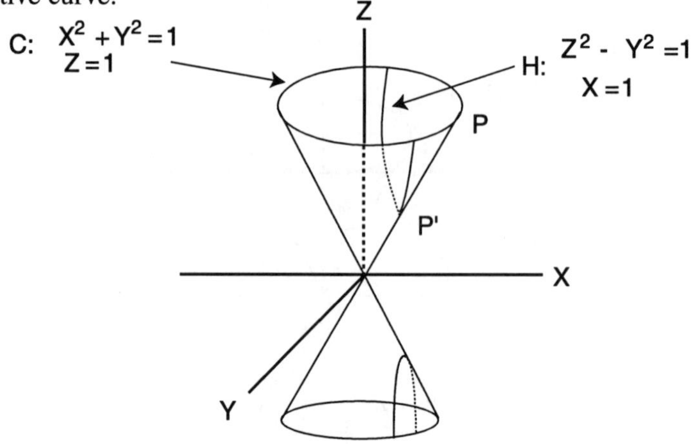

Figure 23

Algebraically, we have gone from the affine circle, $f(x, y) = x^2 + y^2 - 1 = 0$, to the projective circle, $F(X, Y, Z) = X^2 + Y^2 - Z^2 = 0$, by homogenizing f, and then to the affine hyperbola, $h(y, z) = z^2 - y^2 - 1 = 0$, by *dehomogenizing F with respect to X* (dividing $F = 0$ by X^2 and setting $y = \frac{Y}{X}$, $z = \frac{Z}{X}$). (This dehomogenizing is equivalent to setting $X = 1$, $y = Y$ and $z = Z$ in $X^2 + Y^2 - Z^2 = 0$.) The points at infinity on the hyperbola can be found by setting $X = 0$ (the line at infinity for the model $X = 1$) in $X^2 + Y^2 - Z^2 = 0$, yielding $Y = \pm Z$. Therefore the points are $(0, Z, Z) \sim (0, 1, 1)$ and $(0, -Z, Z) \sim (0, -1, 1)$, the same result we derived geometrically. Note that these points are the asymptotes to $z^2 - y^2 = 1$ in the model $X = 1$. Hence the hyperbola, which by definition approaches an asymptote as we move out along any branch, has the same points at infinity as its asymptotes. This makes the idea that the point at infinity lies on the curve quite intuitive.

It must be emphasized that the projective equation $X^2 + Y^2 = Z^2$ is the central one and that by adding the restriction that (X, Y, Z) lie in some plane (a model), we get various representations of the projective curve. Since we can get any conic by cutting the cone with an appropriate plane, we do not distinguish between different conics in projective geometry (this is related to the notion of change of homogeneous coordinates, which are given by linear transformations of $\mathbf{R}^3$ - see [Har 2] and [Sch - Spe] for more details).

Now to more general algebraic curves. If $C : f(x, y) = 0$, $f(x, y) \in \mathbf{Z}[x, y]$, is an algebraic curve in the affine plane, we want to embed it in the projective plane, i.e., add appropriate points at infinity to get an extended curve $\overline{C}$ in the projective plane.

Geometrically, if the slope of the curve approaches the slope of a line L as we move out to infinity along some branch, we want the point at infinity on L to be a point at infinity on the curve. We use the slope as a measure because we are interested in the directions of lines.

In the following examples, the model of the projective plane is always $Z = 1$.

Example: $C : y = x^2$. As we move out along either branch of this parabola, $(x \to \pm\infty)$, the slope of the tangent line $2x \to \pm\infty$, the slope of the y axis. Thus $O = (0, 1, 0)$, the point at infinity on the y-axis, should be on the extended projective curve $\overline{C}$.

Example: $C : x^2 - y^2 = 1$. As we move out along different branches, the slope of the hyperbola approaches the slope of one of its asymptotes, $x = \pm y$. Thus $O' = (1, 1, 0)$, the point at infinity on $x = y$, and $O'' = (-1, 1, 0)$, the point at infinity on $x = -y$, should both be on the corresponding projective curve (fig. 24).

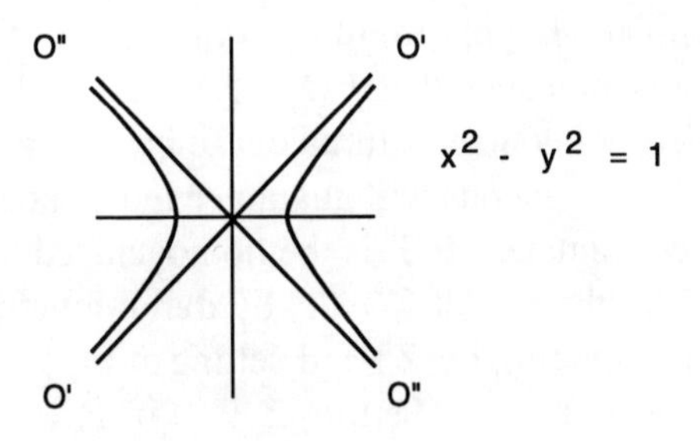

Figure 24

Example: $C : y^2 = x^3 + 1$. As $x \to \infty$, the slope on the branches $y > 0$ and $y < 0$ approaches ∞ and $-\infty$ respectively and thus $O = (0, 1, 0)$, the point at infinity on the y axis, should be on the corresponding projective curve (fig. 25).

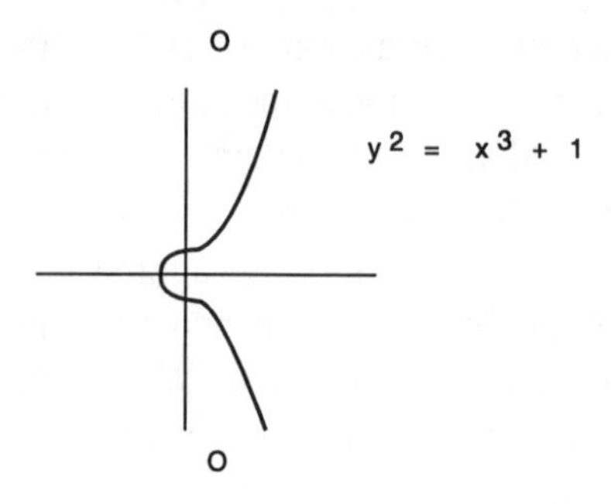

Figure 25

Algebraically, we want an equation $F(X, Y, Z) = 0$, for the extended curve $\overline{C}$ in the projective plane, as a function of homogeneous coordinates (X, Y, Z). F should be independent of the choice of representative coordinates, i.e., $F(X, Y, Z) = 0$ if and only if $F(\lambda X, \lambda Y, \lambda Z) = 0$, for $\lambda \neq 0$, and we want to be able to recover our original equation $f(x, y) = 0$ of the affine curve when restricting our equation to the affine part of an appropriate model. Our treatment of the line and circle illustrate our approach.

In the cases of the line and the circle, we started with the affine equation $f(x, y) = 0$ (thought of as referring to points $(x, y, 1)$ in $Z = 1$), homogenized the equation by setting $x = \frac{X}{Z}, y = \frac{Y}{Z}$, multiplied by Z^n, where n is the degree of f, to clear denominators and arrived at an equation $F(X, Y, Z) = Z^n f\left(\frac{X}{Z}, \frac{Y}{Z}\right) = 0$. This process of **homogenizing** f results in a **homogeneous polynomial** F (all terms of the same degree - equivalently, there is an n such that $F(\lambda X, \lambda Y, \lambda Z) = 0 = \lambda^n F(X, Y, Z)$, for all $\lambda \neq 0$, where, of course, n turns out to be the degree of F). Thus, requiring F to be homogeneous will guarantee independence of the choice of homogeneous coordinates. If F is the homogenized version of f, then we can recover f (in the model $Z = 1$) by **dehomogenizing** F (with respect to Z), i.e., by dividing by Z^n and setting $x = \frac{X}{Z}, y = \frac{Y}{Z}$ or just by setting $Z = 1, X = x, Y = y$. Setting $Z = 0$ in $F = 0$ gives the points points at infinity on the curve in the model $Z = 1$. By dehomogenizing with respect to the other variables or by putting linear conditions on them, we can derive the equation for the affine version in other models.

These considerations lead to the

Definition: *A* **plane algebraic curve in** $\prod_2$ *(or a* **projective plane algebraic curve***) is the set of all points in* $\prod_2$ *whose homogeneous coordinates* *satisfy*

$$F(X, Y, Z) = 0,$$

*where $F(X, Y, Z)$ is a homogeneous polynomial in $\mathbf{Z}[X, Y, Z]$ (equivalently in $Q[X, Y, Z]$). We could, of course, allow the coefficients to be real, but **Z** and **Q** are sufficient for our number-theoretic goals.*

This definition is consistent with the requirements of our last three examples (again, the model is $Z = 1$).

i) If $y = x^2$, then the homogenized equation is $YZ = X^2$ and $Z = 0$ implies that $X = 0$ and Y can take any non zero value. Therefore $(0, Y, 0) \sim (0, 1, 0)$ is the point at infinity. Note that if we dehomogenized with respect to X (divide by X^2 and set $y = \frac{Y}{X}$, $z = \frac{Z}{X}$), then we get the hyperbola $yz = 1$, in the model $X = 1$. If we let $X = 0$ (the line at infinity for $X = 1$) in $YZ = X^2$, then $Y = 0$ or $Z = 0$. Hence $(0, 0, Z) \sim (0, 0, 1)$ and $(0, Y, 0) \sim (0, 1, 0)$ are the points at infinity on the hyperbola as well as on the z and y axes respectively (which are also the points at infinity on the asymptotes for $yz = 1$).

ii) If $x^2 - y^2 = 1$, then the projective equations is $X^2 - Y^2 = Z^2$. Setting $Z = 0$, we have $X = \pm Y$ and thus $(Y, Y, 0) \sim (1, 1, 0)$ and $(-Y, Y, 0) \sim (-1, 1, 0)$ are the points at infinity on the hyperbola, as well as on its asymptotes $x = \pm y$.

iii) If $y = x^3 + 1$, the homogeneous equation is $Y^2 Z = X^3 + Z^3$ and setting $Z = 0$ implies $X = 0$; thus $(0, Y, 0) \sim (0, 1, 0)$, the point at infinity on the y axis (in the model $Z = 1$) is also the point at infinity on the curve.

8. Geometry Over a Field; Higher Dimensions and Duality

General Fields

Our discussion of the real projective plane can be generalized to geometry over any other field F. The vector space $F^n = \{(f_1, ..., f_n) | f_i \in F\}$, over F, is **affine n-space over F**, where F^2 is the affine plane. The **projective plane over F, $\prod_2$(F)**, is the set of one dimensional subspaces of affine 3-space F^3. Each such subspace is a **projective point**. A **projective line** is a two dimensional subspace of F^3.

When F is a finite field, F^2 and $\prod_2(F)$ contain only finitely many points and we have a natural setting for studying the number of solutions of congruences in two variables. Thus we can interpret the problem of counting solutions of $f(x, y) \equiv 0 (\text{mod } p)$ as counting points on the curve $f(x, y) = 0$ in the affine plane F_p^2; going to the corresponding projective curve in $\prod_2(F_p)$ leads to more uniform results (see sec. 20.9 and [Ire - Ros, chapters 10, 11]). Finite projective planes are also a very active research area in combinatorics, related to coding theory and block designs [VanL - Wil].

In our further study of rational points on curves, we shall work with the complex projective plane $\prod_2(\mathbf{C})$. Since $\mathbf{R}^2 \subset \mathbf{C}^2$, $\prod_2(\mathbf{R}) \subset \prod_2(\mathbf{C})$, and our earlier discussion over the reals carries over to the enlarged space. The main advantage of this more general view is that $\mathbf{C}$ is algebraically closed (every polynomial equation, $p(x) = 0$, has a root in $\mathbf{C}$). This is particularly useful when dealing with the intersection of curves.

Example: A line $y = c$ intersects the circle $x^2 + y^2 = 1$ in 0, 1 or 2 points in $\mathbf{R}^2$ (fig. 26).

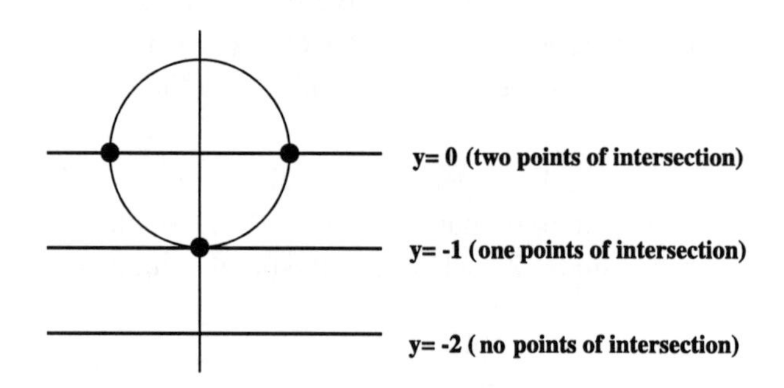

Figure 26

However, the situation changes over $\mathbf{C}$. If $y = -2$, then $(\pm i\sqrt{3}, -2)$ are the two points of intersection in $\mathbf{C}^2$. If $y = -1$, then $x^2 = 0$, $x = 0$ is a root of multiplicity two, and we say that the line intersects the circle twice or with multiplicity two at $(0, -1)$. Thus, with a proper interpretation of 'intersection multiplicity', a line $y = c$ always intersects the circle twice in $\mathbf{C}^2$.

Although the circle has no points at infinity in $\prod_2(\mathbf{R})$, when we look at the projective equation $X^2 + Y^2 = Z^2$ in $\prod_2(\mathbf{C})$, the circle intersects the line at infinity, $Z = 0$, at the points $(\pm i, 1, 0)$. In fact, as we shall see in the next chapter, any line in $\prod_2(\mathbf{C})$ intersects the projective circle in two points.

Higher Dimensions

Projective n-space, $\prod_n(F)$, over a field F is defined as the set of one dimensional subspaces of F^{n+1}. A projective points is a 1-subspace and a projective k-hyperplane ($k = 1, \ldots, n$) is a $k+1$ dimensional subspace. Notions such as homogeneous coordinates and the embedding of affine space can be easily extended to $\prod_n(F)$. Projective spaces are the natural setting for studying algebraic varieties (solution sets of a *finite* number of polynomial equations), as can be seen in books about algebraic geometry.

The sphere, as a model of the extended complex plane $\mathbf{C}$, given by stereographic projection (sec. 13.4), is also a model of projective 1-space $\prod_1(\mathbf{C})$, where ∞ (the north pole) represents the point at infinity (in the model $Y = 1$ in $\mathbf{C}^2$).

Duality

We finish our brief introduction to projective geometry with a discussion of duality, a subject we will not use here, but which is too beautiful to skip.

Recall that one of the problems that led to the need for the projective plane is the asymmetry of Euclid's axioms, viz., two points determine a line but two lines either determine a point or are parallel. In the projective plane two lines always intersect in a point. The projective plane can be axiomatized so that there is a 'duality' between points and lines in the following sense:

> *if, in every axiom, the word "point" is replaced by the word "line" and "line" by "point" and if every statement "the point p is on (intersects) the line L" is replaced by the dual statement "the line L′ contains (intersects) the point p′" and "L intersects p" is replaced by "p′ intersects L′", then the resulting statement is true.*

For example, the dual of the axiom "given any two points, there is a unique line which intersects them (two points determine a line)" is "given any two lines, there is a unique point which intersects them (two lines determine a point)". This duality propagates to all theorems via the **duality principle**:

if, in each step of the derivation of a theorem, we replace all statements by their duals, as described above, then we have a derivation of the dual theorem.

If you look at some older books on projective geometry, e.g., Cremona [Cre], you will find the pages set up in two columns, one containing the proof of a theorem and the other the proof of the dual theorem. Some theorems and their duals, which are both important and non-trivial, were proved independently (in the affine plane, with special cases when lines are parallel) before the full development of projective geometry and the duality principle. Duality is well illustrated by

Pappus's Theorem: *If $p_1, \ldots, p_6$ are points in $\prod_2(\mathbf{R})$ such that p_1, p_3, p_5 lie on a line M_1 and p_2, p_4, p_6 lie on a line M_2, then the points of intersection*

$$r_1 = p_1p_4 \cap p_3p_2, \; r_2 = p_1p_6 \cap p_5p_2, \; r_3 = p_3p_6 \cap p_5p_4$$

lie on a line (fig. 27).

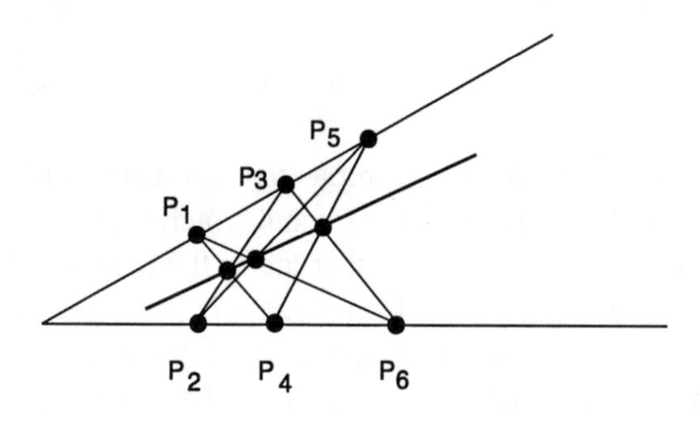

Figure 27

If we think of the points $p_1, p_4, p_5, p_2, p_3, p_6$ as the consecutive vertices of a (generalized) hexagon, then Pappus's theorem says that, under our condition, the points of intersection of opposite sides lie on a line. The dual theorem states:

If $L_1, \ldots, L_6$ are lines in $\prod_2(\mathbf{R})$ such that L_1, L_3, L_5 meet in a point M_1 and L_2, L_4, L_6 meet in a point M_2, then the lines

$$N_1 = (L_1 \cap L_2)(L_4 \cap L_5), \; N_2 = (L_2 \cap L_3)(L_5 \cap L_6)$$
$$N_3 = (L_3 \cap L_4)(L_6 \cap L_1)$$

meet in a point (draw a picture using colored pencils).

In some sense, the lines form a 'dual hexagon'. There are two other dual theorems which place Pappus's theorem and its dual in a more natural light.

Pascal's Theorem: *If a hexagon is inscribed in a conic then the points of intersection of opposite sides lie on a line (fig. 28).*

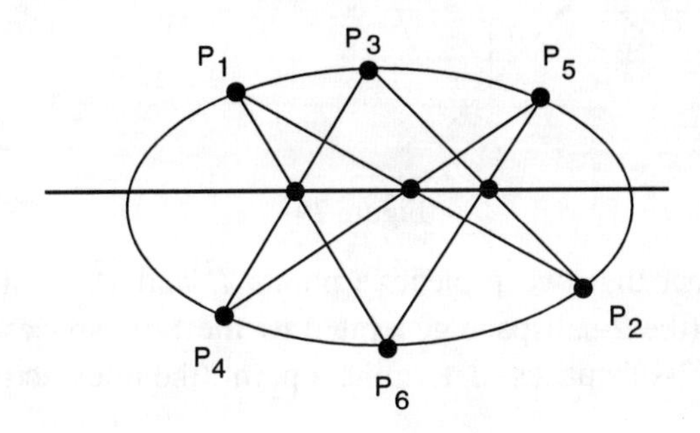

Figure 28

A **reducible** (degenerate) conic is a pair of straight lines (the product of two linear equations set equal to zero). Hence, in the degenerate case, Pascal's theorem reduces to Pappus's theorem.

To dualize the theorem requires a notion of dual of a curve (a conic), which we will describe in a moment (see [Brie - Kno] for more details). For now, we just state the dual.

Branchion's Theorem: *If a hexagon is circumscribed about an irreducible conic, then the lines connecting opposite vertices intersect in a point.*

Duality can be interpreted very simply in our construction of $\prod_2(\mathbf{R})$. Just think of $aX + bY + cZ$ as the ordinary dot product of (a, b, c) and (X, Y, Z). Then we say that the projective point $\overline{P}$ (1-subspace) with homogeneous coordinates (a, b, c) and the projective line, $\overline{L} : aX + bY + cZ = 0$ (the set of all points in $\mathbf{R}^3$ perpendicular to (a, b, c)), are dual to each other. We can visualize this in terms of the model $Z = 1$ (fig. 29).

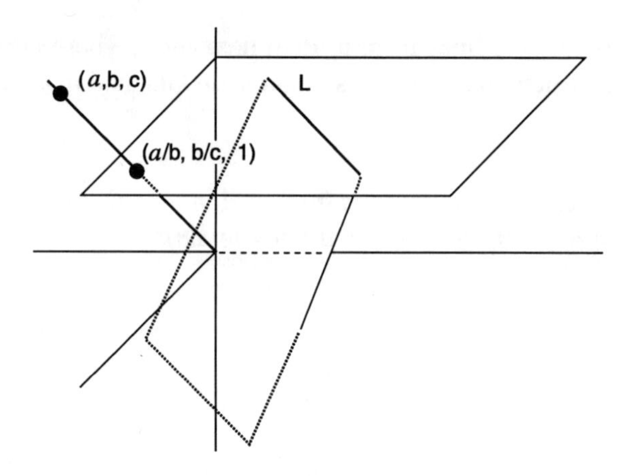

Figure 29

Thus the fact that two projective points $\overline{P}$ and $\overline{P}'$ (1-subspaces) determine a line (the 2-subspace generated by the 1-subspaces) implies that the dual lines (2-subspaces) determine a point (the intersection of the two 2-subspaces).

If we consider the dual lines of all points on a curve C, then these lines will form the 'envelope' of another curve, the dual of C. Thus, e,g., it is not hard to visualize, in the model $Z = 1$, that the dual of a circle is a circle. In general, the dual lines of all points on any conic C form the 'envelope' of another conic, the dual of C.

Chapter 19

Arithmetic on Curves 2: Rational Points and Elliptic Curves

1. Introduction

We now treat a variety of topics needed for a precise group theoretic formulation of Mordell's theorem and to understand its scope. In particular, we discuss the central role of elliptic curves for arithmetic questions as well as the relation between complex analysis and elliptic curves. In the last section we presented the early history of the study of rational points which ties together some of the diverse topics of this chapter and the previous one. Many results are stated without proof. In addition to the references given at the beginning of the last chapter, we recommend the papers by Stroeker [Str, R], Horowitz [Hor] and [Coh].

2. Intersection of Curves; Bezout's Theorem

We work with algebraic curves $F(X, Y, Z) = 0$ in the complex projective plane $\prod_2(\mathbf{C})$ where $F \in \mathbf{Q}[X, Y, Z]$ is a homogeneous polynomial. The **degree** of the curve $F = 0$ is the degree of F. As in chapter 18, let

$$f(x, y) = F(x, y, 1) = 0$$

be the corresponding affine curve (*in the model* $Z = 1$). When discussing points at infinity on the affine curve, we are referring to the points at infinity on the corresponding projective curve. We shall also draw pictures of the real affine points on the curve (the intersection of the curve with $\mathbf{R}^2$).

The main result in the intersection theory of plane curves is

Bezout's Theorem: *Let $F = 0$ and $G = 0$ be be algebraic curves of degrees m and n respectively, such that F and G have no non-constant factors over* **C** *in common. Then F and G intersect in mn points 'counted appropriately according to their multiplicities'.*

The assumption that F and G have no factors in common prevents curves from having a whole piece in common as, for example,

$$y(y - x^2) = 0 \text{ and } y = 0 ,$$

both of which contain the y axis.

A complete treatment of multiplicity of intersection and a proof of the theorem would require the development of an algebraic machine (see [Bri - Kno], [Wal] or [Ful]). Instead, we shall only present several examples which give a general idea of how to interpret Bezout's theorem.

Example: The x axis is tangent to $y = x^2$ at the origin (fig. 1). It intersects the parabola at the origin with multiplicity two. We interpreted this earlier by noting that if we set $y = 0$ in $y = x^2$, then 0 is a root of multiplicity two of $x^2 = 0$. Geometrically, if we move the line $y = 0$ 'up a little' to $y = c$, $c > 0$, then it cuts $y = x^2$ in two points.

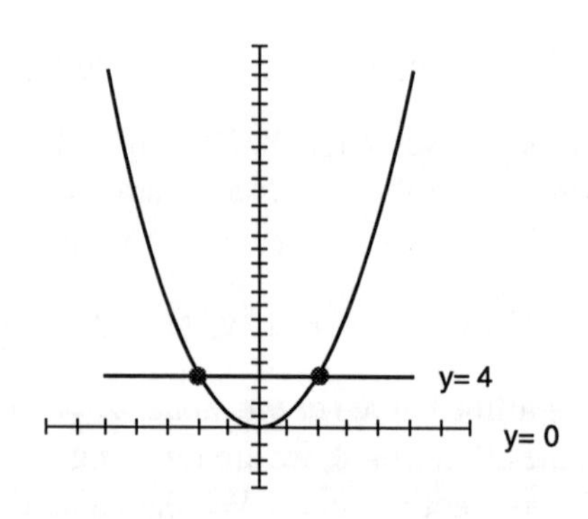

Figure 1

Example: The two ellipses (curves of degree two) in figure 2 intersect in

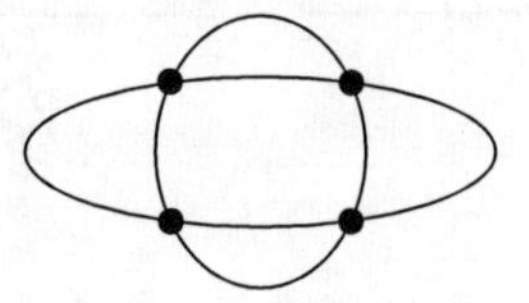

Figure 2

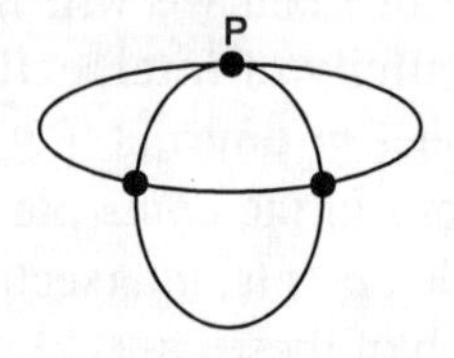

Figure 3

four points. But the ellipses in figure 3 meet in three points since the curves are tangent at P. However, if we move the horizontal ellipse 'down a little' (fig. 4), we now get four points. P is a point of intersection of the ellipses of multiplicity two.

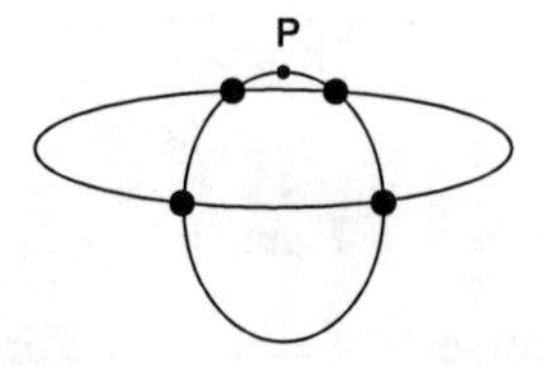

Figure 4

Example: $y = 0$ and $y = x^3$ intersect at the origin (fig. 5a). If we rotate $y = x^3$ 'a little' (fig. 5b), we get the three intersections required by Bezout's theorem (note that 0 is a root of multiplicity three of $x^3 = 0$).

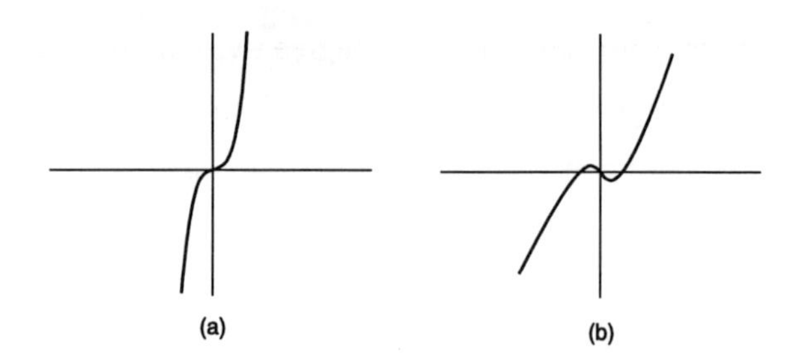

Figure 5

In general, if two curves intersect at a point P, then by looking at all 'small movements' of one of them we will find a maximum number m of intersections, the **multiplicity** of **intersection** of the curves at P. We also say that the curves meet m times at P. Our examples all involved curves with *real* intersections in the *affine* plane. More generally, we can have intersections at complex points, intersections at points at infinity and multiple intersection in each of these cases.

Example: The curves $y = 3$ and $x^2 + y^2 = 1$ have no intersections in the real affine plane (fig. 6). Solving the pair of equations, we have $(x, y) = (\pm i\sqrt{8}, 3)$, the two points in $\mathbf{C}^2$ required by Bezout's theorem.

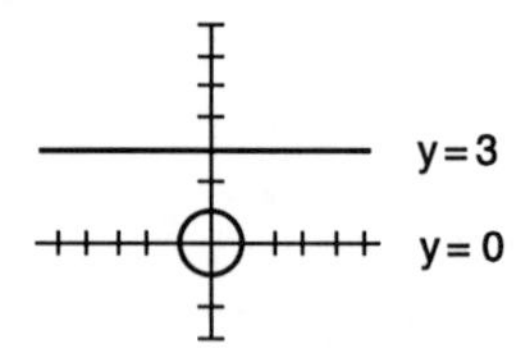

Figure 6

Example: The y-axis intersects $y = x^2$ at the origin. However, as we saw in section 18.7, $y = x^2$ also contains the point at infinity on the y axis; hence we have two points of intersection. More formally, homogenizing our equations yields

$$YZ = X^2 \text{ and } X = 0\ ,$$

the equations for the corresponding projective curves. Solving simultaneously, we have $YZ = 0$, which implies $Y = 0$ or $Z = 0$, and the solutions are

$$(0,0,Z) \sim (0,0,1) \text{ and } (0,Y,0) \sim (0,1,0)\,,$$

where $(0,0,1)$ corresponds to $(0/1, 0/1) = (0,0)$ in the affine plane and $(0,1,0)$ is the point at infinity on the y axis.

Example: The concentric circles $x^2 + y^2 = 1$ and $x^2 + y^2 = 2$ have no points in common in the real affine plane. Homogenizing the equations and solving we have

$$X^2 + Y^2 = Z^2, X^2 + Y^2 = 2Z^2$$
$$\implies Z^2 = 2Z^2 \implies Z = 0 \implies X = \pm iY\,.$$

If $X = iY$, the point is $(iY, Y, 0) \sim (i, 1, 0)$. If $X = -iY$, the point is $(-iY, Y, 0) \sim (-i, 1, 0)$. So in $\prod_2(\mathbf{C})$, we have two points at infinity lying on both curves. But Bezout requires four points. In fact, the circles are tangent at these points and the multiplicity of each intersection is two.

To see this, we dehomogenize both equations by taking the model $X = i$. Thus, setting $X = i, Y = y, Z = z$, we have the affine equations

$$z^2 - y^2 = -1\,, \quad 2z^2 - y^2 = -1\,,$$

which are hyperbolas in the y-z plane (fig. 7). The points $(z, y) = (0, \pm 1)$ are on both curves and sketching the curves makes it easy to see the tangency at these points

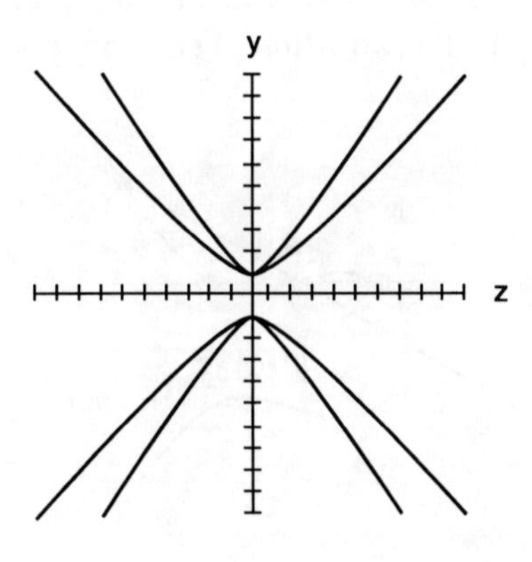

Figure 7

3. The Group Law and the Algebraic Form of Mordell's Theorem

In section 18.4, we saw how, using the tangent-secant process, two rational points on a non-singular affine cubic determine a third rational point, with certain exceptions. Moreover, we eliminated these exceptions for the curve $y^2 = x^3 + 1$ by working in the projective plane (sec. 18.7). Now we extend this procedure to general cubics.

Recall that a point on a curve is a *rational point* if one of its sets of homogeneous coordinates (X, Y, Z) is rational. For example, $(0, 1, 0)$, the point at infinity on the y axis and on $y^2 = x^3 + 1$, is rational. If $Z \neq 0$, then this is equivalent to requiring that the corresponding affine point $\left(\frac{X}{Z}, \frac{Y}{Z}\right)$ (in the model $Z = 1$) is rational.

Bezout's Theorem guarantees that a non-singular cubic and a line will intersect in three points in $\prod_2(\mathbf{C})$. By the same reasoning as in section 18.4, if two of these points are rational, then so is the third. Thus, in the projective plane, the tangent-secant process has no exceptions. Two distinct rational points P and Q on a non-singular cubic C determine a third rational point $P * Q$, the third point of intersection of C with the line through P and Q. If $P = Q$, then $P * P$ is the third point of intersection of the tangent at P with C (see the diagrams in sec. 18.4). This gives us a commutative law of composition for the rational points on a cubic (if any exist). Unfortunately, this law does not define a group since there is no identity element, but we can modify it a little to obtain a group.

Let C be a non-singular cubic with rational points and fix one such rational point O on C. Define **addition of rational points** on C by $\mathbf{P}+\mathbf{Q} = (P * Q) * O$ (fig. 8).

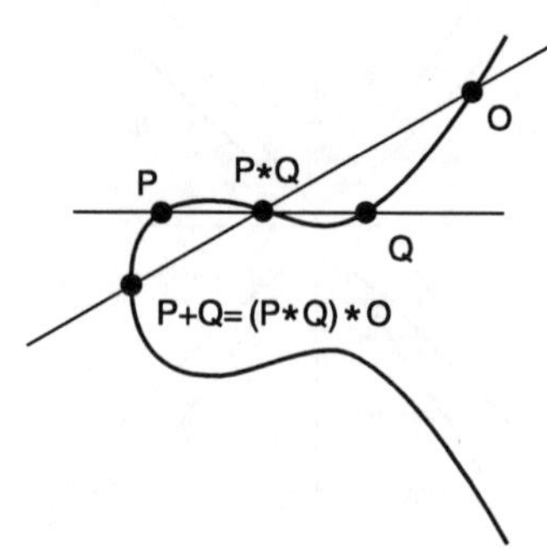

Figure 8

This is a law of composition and, with this law, we have a commutative group. For now, we only illustrate this geometrically.

i) *identity:* O is the identity element; $P + O = O$ (fig. 9). Joining P to O gives the third point of intersection $P * O$ and then joining $P * O$ to O gives the third point of intersection $P = (P * O) * O$.

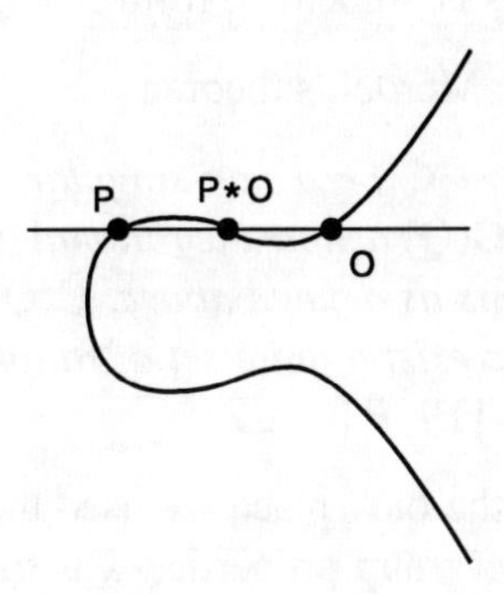

Figure 9

ii) *inverses:* Let $S = O * O$, the third point of intersection of the tangent at O with the curve (figure 10). We claim that for any Q, $R = Q * S$ is the inverse $-Q$ of Q, i.e., $Q + R = (Q * R) * O = O$. To see this note that $Q * R = S$ (by the construction of S) and $S * O = O$, since the line joining S to O meets the curve once at S and with multiplicity two at O (it is the tangent at O).

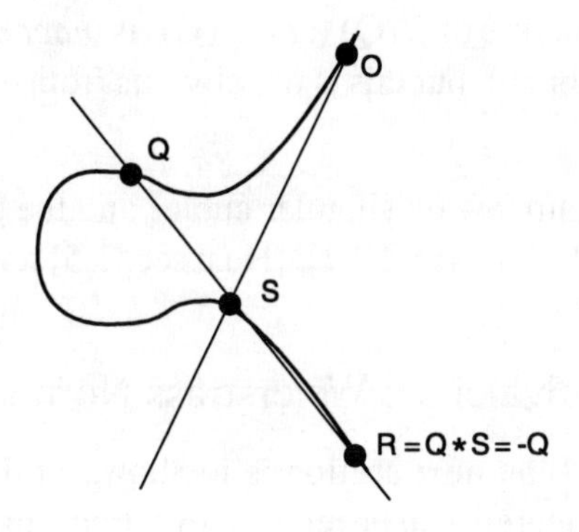

Figure 10

iii) *associativity*: There is a somewhat intricate geometric proof by cases, using Bezout's theorem (see [Sil - Tat, sec. I.2] and [Ful, pp. 124-125]). For now, we assume associativity. Later, after reducing our study to a special class of cubics and deriving formulas for the addition of points on these curves in terms of coordinates, the proof for these special cubics becomes a routine exercise (sec. 6).

iv) *commutativity:* This is immediate from $P * Q = Q * P$.

Now we can reformulate Mordell's theorem.

Mordell's Theorem: *Let C be a non-singular cubic in $\prod_2(\mathbf{C})$ with a rational point O and let $C(\mathbf{Q})$ denote the rational points on the curve. With addition of rational points as defined above, $C(\mathbf{Q})$ is a finitely generated Abelian group, i.e., there exist a finite set of rational points $P_1, \cdots, P_k \in C(\mathbf{Q})$ such that $C(\mathbf{Q}) = \{\Sigma r_i P_i | r_i \in \mathbf{Z}\}$.*

Remarks: 1) Note that the only place we used the rationality in defining addition and proving the group properties was to prove that the rational points are closed under addition. Thus, if we fix any point O on the curve, our definition of addition works for the set $C(\mathbf{C})$ of all complex points of C (in $\prod_2(\mathbf{C})$) and $C(\mathbf{C})$ is an Abelian group with identity O.

If K is a subfield of $\mathbf{C}$, let $C(K)$ be the points on C which have some set of homogeneous coordinates in K. Then, if the O we choose for $C(\mathbf{C})$ is in $C(K)$, $C(K)$ is a subgroup of $C(\mathbf{C})$ (closure under addition is proved just as for rational points). In particular, if we choose a rational point O (if it exists) as the identity for all our groups, then all $C(K)$ are subgroups of $C(\mathbf{C})$.

2) The group structure of $C(\mathbf{Q})$ (and $C(\mathbf{C})$) is *independent* of the choice of O (up to isomorphism), but this is not obvious from our formulation (see [Sil]).

3) There is a group law on singular cubics and the groups can be fully determined (see [Sil - Tat, sec. III.7], [Hus, sec. 1.5] and [Sil, sec.3.2]).

4. Birational Equivalence; Weierstrass Normal Form

Our goal in this and the next section is to show, on the one hand, that a proof of Mordell's theorem can be reduced to a proof for a very special class of cubics, and, on the other hand, that Mordell's theorem for non-singular cubics determines the structure of the rational points on a wider class of curves, the curves of 'genus one'.

Recall that we simplified our study of the representation of integers by quadratic forms by defining the equivalence of forms and showing that equivalent forms represent the same integers (chap. 12). Then we used special forms in each class to help solve our problem pertaining to the whole class. We use a similar idea to study rational points on curves.

Example: In section 18.3, we set up a correspondence between the real points on $x^2 + y^2 = 1$ and $y = 0$ by a parametrization of the circle. If we let (x, y) denote coordinates on the circle and $(u, v) = (0, t)$, where t is a coordinate for the line, then we showed that

$$u = 0\,, \quad v = t = \frac{y}{1+x}$$

and

$$x = \frac{1-t^2}{1+t^2} = \frac{1-v^2}{1+v^2}\,,$$
$$y = \frac{2t}{1+t^2} = \frac{2v}{1+v^2}\,.$$

Except for the point $(-1, 0)$, these relations define a bijection between points on the two curves (in affine space). This bijection is given by *rational functions* of the coordinates with rational coefficients and rational points on the circle correspond to rational points on the line.

If we pass to projective coordinates (X, Y, Z) and (U, V, W), where $x = \frac{X}{Z}$, $y = \frac{Y}{Z}$, $u = \frac{U}{W}$ and $v = \frac{V}{W}$, then the correspondence is given by polynomial, namely,

$$\begin{aligned} U &= 0 \\ V &= Y \\ W &= Z + X \end{aligned} \qquad \text{and} \qquad \begin{aligned} X &= W^2 - V^2 \\ Y &= 2VW \\ Z &= W^2 + V^2 . \end{aligned}$$

To check directly when the maps are inverse to each other in projective form, we have

$$\begin{aligned} (X, Y, Z) \longrightarrow (U, V, W) &= (0, Y, Z + X) \\ &= (0, 2VW, W^2 + V^2 + W^2 - V^2) \\ &= 2W(0, V, W) \sim (0, V, W)\,, \end{aligned}$$

if $W \neq 0$. $W = 0$ corresponds to the point at infinity on the y-axis. The map from the line to the circle sends the point at infinity $(U, V, W) \longrightarrow (-1, 0, 1)$, i.e., the affine point $(-1, 0)$ on the circle, but the map from the circle to the line sends $(X, Y, Z) = (-1, 0, 1) \longrightarrow (0, 0, 0)$, which is not on the line. Hence the maps are inverse to each other except for these points.

Note that the curves have different degrees.

Of course, extending the maps to make the point $(-1, 0)$ on the circle correspond to the point at infinity on the line yields a complete bijection, but this bijection is not completely described by our rational functions.

Definition: *Let* $C_1 : f(x, y) = 0$ *and* $C_2 : g(u, v) = 0$, *where* f *and* g *have rational coefficients, be algebraic curves in the affine plane* $\mathbf{C}^2$. *Then* $\mathbf{C_1}$ **is birationally equivalent to** $\mathbf{C_2}$ **(over** $\mathbf{Q}$**)**, *if there exist rational functions of two variables* P, Q, R, T, *with rational coefficients*

$$\begin{aligned} x &= P(u, v) \\ y &= Q(u, v) \end{aligned} \quad \text{and} \quad \begin{aligned} u &= R(x, y) \\ v &= T(x, y) \end{aligned}$$

which define mappings of points of C_1 *to* C_2 *and of* C_2 *to* C_1, *and which are inverse to each other with the possible exception of a finite number of points on each curve. The exceptions refer to both the domains of the functions and of the inverse relations.*

Thus, *except for a finite number of points*, we have bijective mappings. We run into trouble if the denominators of the functions become zero, but then we just go to the projective plane. If we take the curves in the projective plane, then it is equivalent to assume that *the functions relating the homogeneous coordinates are homogeneous polynomials, but, as we saw in the last example, we must still allow a finite number of exceptions.*

It follows immediately that birational equivalence is an equivalence relation. Thus birational equivalence sets up a correspondence between complex points on the curves. Under this correspondence *rational points correspond to rational points*, since everything is given by rational functions with rational coefficients. Therefore, the Diophantine problem of finding rational points on the curves is basically the same for both curves, *even though they may have different degrees*, as in the example of a circle and a line.

The need for allowing finitely many exceptions to the bijection is best seen in the following example.

Example: Consider the curves

$$C_1 : x^2 + y^2 = 1$$
$$C_2 : u^2 + v^2 = u^2 v^2$$

and, in projective form,

$$X^2 + Y^2 = Z^2 ,$$
$$V^2 W^2 + U^2 W^2 = U^2 V^2 .$$

They are birationally equivalent under the mappings

$$x = \frac{1}{u}, \quad u = \frac{1}{x}$$
$$y = \frac{1}{v}, \quad v = \frac{1}{y}$$

and, in projective form,

$$X = VW, \quad U = YZ$$
$$Y = UW, \quad V = XZ$$
$$Z = UV, \quad W = XY .$$

The map from C_1 to C_2 in projective form is easily guessed from the form of the equations. Dehomogenizing yields $x = \frac{1}{u}$, $y = \frac{1}{v}$. Ignoring, for the moment, zero values in the denominator, we have $u = \frac{1}{x}$, $v = \frac{1}{y}$. Homogenizing these equations yields $\frac{U}{W} = \frac{1}{\frac{X}{Z}}$, $\frac{V}{W} = \frac{1}{\frac{Y}{Z}}$ or $U = \frac{Z}{X} \cdot W$, $V = \frac{Z}{Y} \cdot W$. Since W is arbitrary in homogeneous coordinates, and since we want polynomials to express our maps, we set $W = XY$ to clear denominators, and we have $U = YZ$, $V = XZ$.

To check exactly when these maps are inverse to each other, we have

$$\begin{aligned}(U, V, W) \longrightarrow (X, Y, Z) &= (VW, UW, UV) \\ &= (XZXY, YZXY, YZXZ) \\ &= XYZ(X, Y, Z) \sim (X, Y, Z) ,\end{aligned}$$

if $XYZ \neq 0$. The only points for which $XYZ = 0$ and which lie on $X^2 + Y^2 = Z^2$ are $(0, 1, 1)$ and $(1, 0, 1)$.

Mapping from C_1 to C_2, using homogeneous coordinates, we have

$$(1, 0) \longleftrightarrow (1, 0, 1) \longrightarrow (0, 1, 0) ,$$
$$(-1, 0) \longleftrightarrow (-1, 0, 1) \longrightarrow (0, -1, 0) .$$

But $(0, 1, 0)$ and $(0, -1, 0)$ are homogeneous coordinates for the same point and the map is not globally one to one. Again, we see that birational equivalence need not preserve the degree of a curve.

Even though birational equivalence does not necessarily preserve lines (and therefore addition of points on cubics), the groups of rational points on birationally equivalent non-singular cubics are isomorphic. The proof requires an intrinsic definition of the addition of points (see [Sil] or [Str]). Thus if we prove Mordell's theorem for non-singular cubics, it will be true for all non-singular curves birationally equivalent to them. The key to a proof of Mordell's theorem is the following.

Theorem: *A non-singular cubic C, with a rational point, is birationally equivalent to a non-singular cubic of the form*

$$y^2 = x^3 + ax^2 + bx + c \tag{1}$$

and, also, to a non-singular cubic of the form

$$y^2 = 4x^3 - g_2 x - g_3 \,. \tag{2}$$

The latter form is usually called the **Weierstrass normal form** of C, but we will use this term to refer to either type. We skip the proof of this theorem, which requires techniques we have not developed (see [Sil], [Str], or [Hus]).

Non-singularity of either of the normal forms is equivalent to requiring that the discriminant of the cubic polynomial in x on the right hand side of the equation be non-zero. Recall that the discriminant of a polynomial $f(x)$ is $D = \prod_{i \neq j}(e_i - e_j)^2$, where the e's are the roots of $f(x) = 0$. It 'discriminates' between the roots, i.e., $D \neq 0$ if and only if the roots are distinct. For the curve defined by equation (2), $D = g_2^3 - 27g_3^2$. The proof is by computation using the fact that the coefficients of a monic polynomial f(x) are symmetric functions of the roots.

Our theorem, together with the fact that birationally equivalent curves have isomorphic groups of rational points, implies that *to prove Mordell's theorem, it is only necessary to prove it for non-singular cubics in normal form.*

Example: $u^3 + v^3 = 1$ is birationally equivalent to $y^2 = x^3 - 432$ by

$$x = \frac{12}{u+v}, \quad y = 36\frac{u-v}{u+v}$$

and

$$u = \frac{36+y}{6x}, \quad v = \frac{36-y}{6x}.$$

It is a good exercise to carry out the details in homogeneous coordinates to show that there are only finitely many exceptions.

Thus Fermat's last theorem for $n = 3$ implies that the only affine rational points on $y^2 = x^3 - 432$, are those corresponding to the trivial solutions $(\pm 1, 0)$ and $(0, \pm 1)$ of $u^3 + v^3 = 1$. Homogenizing the equations we see that $U^2 + V^2 = 1$ contains $(1, -1, 0)$ and $Y^2 Z = X^3 - 432 \cdot Z^3$ contains $(0, 1, 0)$, their respective rational points at infinity

Wiles proof of Fermat's last theorem starts with the theorem conjectured by G. Frei and proved by K. Ribet [Rib 2] that Fermat's last theorem for any exponent is equivalent to questions about non-singular cubics.

5. Singular Points and the Genus

Since birationally equivalent curves can have different degrees, the degree of a curve is not a good measure of the structure of its rational points. However, there is an integer determined by a curve, its 'genus', which has proved to be a very good measure of the Diophantine properties of the equation defining the curve.

If $C : f(x, y) = 0$ is an algebraic curve, then to every **C** point $(x_1 + ix_2, y_1 + iy_2)$ on the curve we associate the point (x_1, x_2, y_1, y_2) in $\mathbf{R}^4$ and we let $S(C)$ be the set of all such points in $\mathbf{R}^4$. Equating the real and imaginary parts of $f(x_1 + ix_2, y_1 + iy_2)$ to zero yields two equations whose graphs in $\mathbf{R}^4$ intersect in $S(C)$. Each equation puts one condition on the points of $\mathbf{R}^4$ and thus the dimension of $S(C)$ is two (it is a surface).

Example: $C : f(x, y) = y = 0$; then $f(x_1 + ix_2, y_2 + iy_2) = y_1 + iy_2$ and $S(C)$ is given by the equations $y_1 = 0$ and $y_2 = 0$, each of which defines a three dimensional subspace of $\mathbf{R}^4$. Their intersection is the two dimensional subspace $\{(x_1, x_2, 0, 0)\}$. Since the fourth coordinate is zero, this can also be visualized as the plane $\{(x_1, x_2, 0)\}$ in the 3-subspace $y_2 = 0$.

Example: $C : f(x, y) = y - x = 0$; then $S(C)$ is given by the equations $y_1 = x_1, y_2 = x_2$. If we fix $y_2 = c$, we are taking a cross section of the surface. We represent the point (x_1, x_2, y_1, c) by the point (x_1, x_2, y_1) in $\mathbf{R}^3$ (i.e., we are looking at the three dimensional hyperplane $y_2 = c$ in $\mathbf{R}^4$). Then the points in $S(C)$ are represented by points in $\mathbf{R}^3$ satisfying the equations $y_1 = x_1, x_2 = c(= y_2)$. This is a line L_c in the plane $x_1 = y_1$, parallel to the x_1-y_1 plane and c units above it (fig. 11). As c varies, L_c sweeps out the plane $x_1 = y_1$. Hence $S(C)$ is represent by the plane $x_1 = y_1$ in $\mathbf{R}^3$.

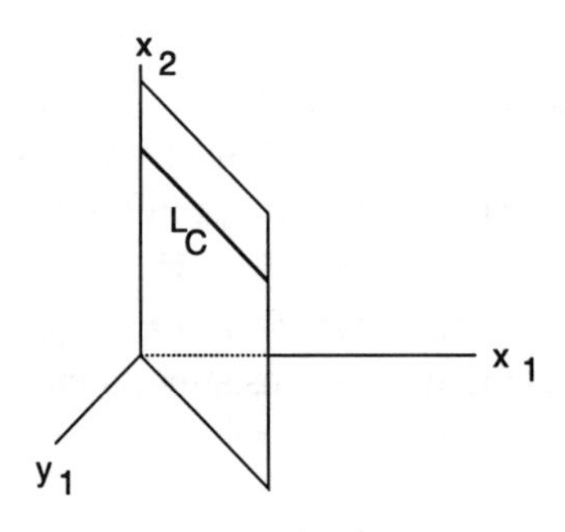

Figure 11

Kendig [Ken, chap. 1] presents many such examples of how to visualize higher degree curves in $\mathbf{R}^3$ by taking cross sections.

If C is a *non-singular* affine curve, then the set of points at infinity on C (in the projective plane) is finite, since it is the intersection of C with the line at infinity. Then a finite number of points in $\mathbf{R}^4$ (corresponding to the points at infinity on C) can be adjoined to $S(C)$, in a way that yields a set $S'(C)$ which is a 'compact connected orientable two dimensional manifold without boundary' (a 'compact surface' - see [Mas]).

If C has singular points, then C is birationally equivalent to a non-singular curve C'. This curve C' will, in general, not be a plane curve, but will sit in some higher dimensional space, say $\mathbf{C}^k$, and be described by a set of polynomial in k variables. I will not elaborate on the meaning of birational equivalence in this context. See [Bri -Kno], [Ful], or [Wal] for more details. The set of points $(x_1, y_1, \cdots, x_k, y_k)$ in $\mathbf{R}^{2k}$ corresponding to the points $(x_1 + iy_1, \cdots, x_k + iy_k)$ on C' (again adding points at infinity) is a compact surface, also denoted by $S'(C)$.

It is a well known result in topology that every compact surface is homeomorphic to a doughnut with g holes (equivalently a sphere with g handles attached) for some g [Mas]. For example, when $g = 3$ we have the surface of figure 12. The number g is the genus of the surface and

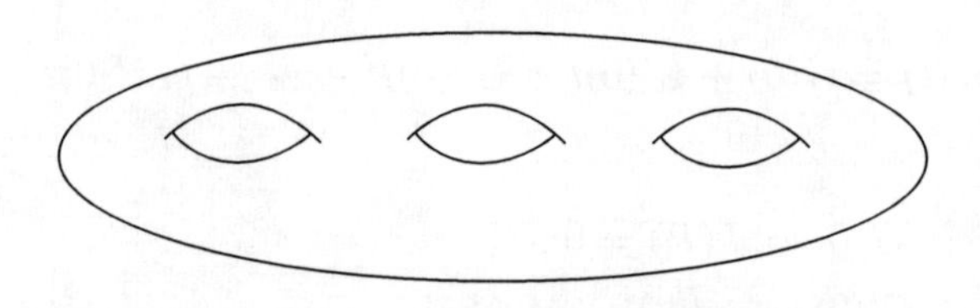

Figure 12

characterizes it up to homeomorphism. The **genus, g = g(C)** of the curve C is the genus of the corresponding surface $S'(C)$.

Surprisingly, the structure of the set of rational (and integral) points on a curve depends on this topological notion of the genus of the curve, rather than its degree. To get some idea of the reason for this, we need a more arithmetic description of the genus. The genus g is an invariant of C as a topological space. However C has additional structure, since it is given by a polynomial equation. In fact, g can be described in terms of the geometry of the curve. The idea of expressing a topological invariant of a space in terms of additional structure, such as algebraic or differential geometric, is a very powerful notion, e.g., it is one of the key ideas behind the Atiyah - Singer Index Theorem in partial differential equations. For an interesting discussion of these ideas, see Atiyah [Ati].

There was a hint of a relation between rational points and geometry in our example of $y^2 = x^2(x+1)$ in section 18.4, where the existence of a singular point simplified the problem of finding all rational points. The genus of a curve can be described in terms of its singular points and the degree of its defining equation. To understand this,we must look more closely at the structure of singular points.

Let $C : f = 0$ be a curve in the affine plane and let L be a line through the point $P = (x_0, y_0)$ on the curve. Let

$$x = x_0 + at \ , \quad y = y_0 + bt \ ,$$

$(a, b) \neq (0, 0)$, be a parametrization for L. The intersections of L and C are determined by the roots of the equation

$$G(t) = f(x_0 + at, y_0 + bt) = 0 .$$

Since G is a polynomial, it has a finite Taylor series in t,

$$G(t) = G(0) + G'(0)t + G''(0)t^2 + \cdots + G^{(n)}(0)t^n ,$$

where

$$\begin{aligned} G(0) &= f(P) = 0 , \\ G'(0) &= f_x(P)a + f_y(P)b , \\ G''(0) &= f_{xx}(P)a^2 + 2f_{xy}(P)ab + f_{yy}(P)b^2 , \end{aligned}$$

and so on (f and its partial derivatives are all evaluated at P).

A) If $f_x(P)$, $f_y(P)$ are not both zero, then P is a **non-singular** (or **regular**) point of C. Then

   i) if $f_x(P)a + f_y(P)b \neq 0$, L intersects C at P once.

   ii) if $f_x(P)a + f_y(P)b = 0$, then, since b/a is the slope of L, the equation of L is

   $$f_x(P)(x - x_0) + f_y(P)(y - y_0) = 0 ,$$

   and L is, by definition, the **tangent line** to C at P. This is unique since a and b are determined up to a constant factor.

B) If $f_x(P) = 0 = f_y(P)$ and not all second partial derivatives are zero, we call P a **singular point of order two** (or a **double point**). Examples are shown in figure 13, where a simple crossing of two arcs, as in figure figure 13(a), is an **ordinary double point**.

Figure 13

C) More generally, if all the partial derivatives up to order $r-1$ vanish at P but some r^{th} order partial derivative is not zero at P, we call P a **singular point of order r**.

Thus the **singular points** of a curve are the points P for which $f_x, f_y = 0$. Let $\mathbf{r_P}$ denote the order of the singular point P.

In *projective space*, where $F(X, Y, Z) = 0$ is the corresponding homogeneous equation for C, we proceed in a similar manner. A *singular point of order r* is one for which all partial derivatives of F up to order $r-1$ vanish and some partial of order r is not equal to zero. There can be singular points on the line at infinity, which is why the following theorem must be stated for curves in projective space.

Theorem: *If $C : F(X, Y, Z) = 0$ is an* **irreducible** *curve in $\prod_2(\mathbf{C})$ (F is the power of an irreducible polynomial), and the degree of F is n, then the genus is given by*

$$g(C) = \frac{(n-1)(n-2)}{2} - \sum \frac{r_P(r_P - 1)}{2},$$

where the sum is over all singular points P of C.

The number $\frac{(n-1)(n-2)}{2}$ is the maximum possible number of singular points on a curve of degree n. (Classically the genus was called the deficiency, meaning a measure of the deviation from this maximum.)

Example: The Fermat curves are defined by $C_n : F(X, Y, Z) = X^n + Y^n - Z^n = 0$. Since the only solution of $F_X = F_Y = F_Z = 0$ is $(0, 0, 0)$, which does not represent a projective point, the curves are non-singular and $g(C_n) = \frac{(n-1)(n-2)}{2}$.

Theorem: *If C_1 is birationally equivalent to C_2, then $g(C_1) = g(C_2)$.*

The converse is not true, e.g., $x^2 + y^2 = 1$ and $x^2 + y^2 = 3$ are non-singular curves of genus zero, but the former has rational points and the latter does not (sec. 18.3).

As noted earlier, to find a non-singular curve birationally equivalent to a given curve, we must consider curves in higher dimensional space. The best we can do, always staying in the plane, is

Theorem: *Every plane curve in $\prod_2(\mathbf{C})$ is birationally equivalent to a curve whose only singularities are ordinary double points.*

To see how this is done, consider a curve with a third order singularity at P. One can 'lift' the curve to three space, separating the arcs intersecting at P, and then project back to the plane from a different direction. This yields three ordinary double points in place of the original triple point. One then repeats the process for other higher order singularities. This can be done in a birational way. For more details about this process of 'resolving singularities', see Walker [Wal].

Theorem: *A curve of genus zero with a rational point is birationally equivalent to a line (the complex projective line $\prod_1(\mathbf{C})$). These curves can be parametrized by rational functions of one variable with rational coefficients.*

Conversely any such parametrizable curve has genus zero. Restricting the correspondence to the rational values of the parameter yields a rational parametrization of the rational points on the curve.

Curves of genus zero are also called **rational curves**.

Example: The curve $y = x^3$ has the rational parametrization $x = t$, $y = t^3$ and thus it has genus zero. We can also compute the genus from the singularity formula. The curve has no singular points in the affine plane. In the projective plane, the homogenized equation is $YZ^2 - X^3 = 0$, and the point at infinity $(0, 1, 0)$ is a singular point of order 2 on the curve. Thus by the genus formula, the genus is $\frac{(3-1)(3-2)}{2} - \frac{2(2-1)}{2} = 0$. Thus topologically, the complex points of the curve are a sphere (a doughnut with zero holes). The parametrization comes from projecting the curve from $(0, 1, 0)$, the point at infinity on the y axis, to the x axis (i.e., by using lines parallel to the y axis) .

Example: Since, as we saw in an earlier example, for $n \geq 3$, the Fermat curves have genus greater than zero, they are not parametrizable. Shafarevich [Sha, Chap. 1] presents an elegant proof of this using techniques of algebraic function theory, rather than our theorems about the genus.

6. Elliptic Curves and the Group Law

Curves of genus one will be our main concern.

Theorem: (i) *A non-singular cubic has genus one (since $\frac{(3-1)(3-2)}{2} = 1$).*

(ii) *Every irreducible curve $C : f = 0$ of genus one, with a rational point, is birationally equivalent to a non-singular cubic and thus to one in Weierstrass normal form.*

Note: A non-singular curve $F = 0$ of genus one is irreducible since if $F = GH$, $F \neq G$, then, by Bezout's theorem, the curves $G = 0$ and $H = 0$ will have a point in common, which would be a singular point.

An **elliptic curve** is a non-singular curve of genus one containing a rational point O. We will use E to denote an elliptic curve and $E(K)$ for the points on E with coefficients in the field $K \subseteq C$. The reason for the term 'elliptic' is that these curves are parametrized by 'elliptic functions' (sec. 7). *An ellipse has genus zero and is not an elliptic curve.*

In section 4 we stated that to prove Mordell's theorem for non-singular cubics, it suffices to prove it for those in Weierstrass normal form. Now we will see that if we prove it for curves in normal form, then *it is also true for all elliptic curves.*

Since we have reduced the problem of determining the rational points on elliptic curves to the study of non-singular cubics in normal form

$$E : y^2 = f(x) = x^3 + ax^2 + bx + c ,$$

until otherwise stated, we shall always mean curves of the latter form when referring to **elliptic curves**. For now *E means $E(\mathbf{C})$*, the complex points. The following proposition is useful for graphing elliptic curves.

Proposition: *E is non-singular $\iff$ $f(x)$ has distinct roots.*

(Recall that f has distinct roots if and only if the discriminant of f is non-zero (sec. 4).)

Proof: As we shall see in a moment, the point at infinity on the projective form of the curve is non-singular, so we only need to consider the curve in affine form.

If (x_0, y_0) is a singular point of $E : F(x, y) = y^2 - f(x) = 0$, then

$$F_y(x_0, y_0) = 2y_0 = 0 \implies f(x_0) = 0$$

and

$$F_x(x_0, y_0) = -f'(x_0) = 0 \implies f'(x_0) = 0 ,$$

which are the criteria for x_0 to be a multiple root of $f(x)$. The converse is a simple exercise.

Since the degree of $f(x)$ is three, $f(x)$ has one or three distinct real roots (the complex conjugate of a root is a root). If E is non-singular, these roots are distinct. Using standard curve sketching methods, we find that the real points on the curve have one of the shapes shown in figures 14 or 15. In both cases, the curves are symmetric with respect to the x axis.

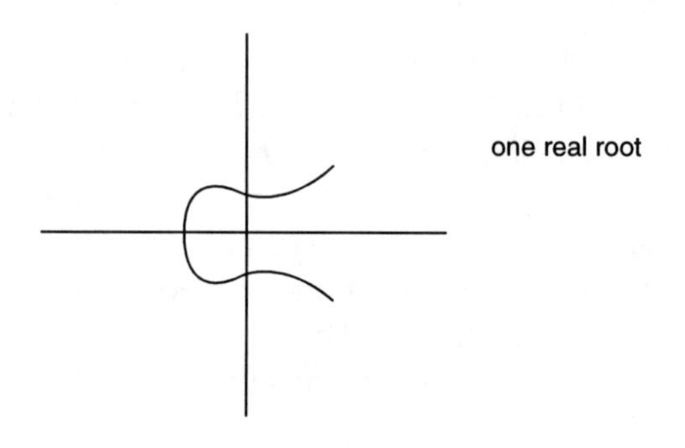

Figure 14

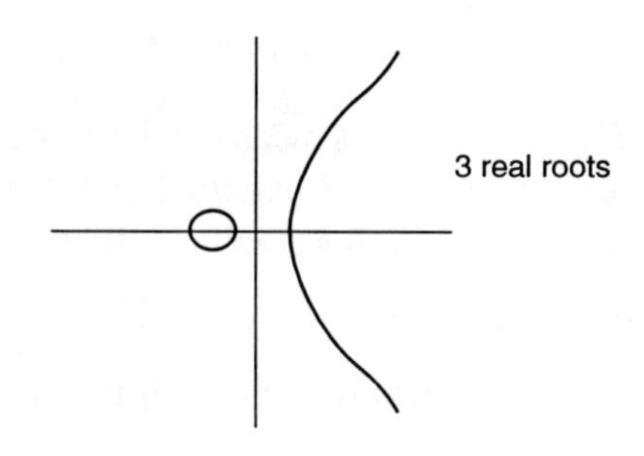

Figure 15

Points at infinity

Homogenizing the equation for E, we have the projective equation

$$Y^2Z = X^3 + aX^2Z + bXZ^2 + cZ^3 .$$

If we set $Z = 0$, we have $X^3 = 0$, so $X = 0$, and Y can take any value. Thus the line at infinity ($Z = 0$) meets E at the rational point

$$(0, Y, 0) \sim (0, 1, 0) .$$

Since $X^3 = 0$ has $X = 0$ as a triple root, $Z = 0$ meets the curve at $(0, 1, 0)$ with *multiplicity three* and is the *tangent* line to E at this non-singular point. A non-singular point P on a curve C is an **inflection point of C** if the tangent to C at P is not part of the curve and meets P at C with multiplicity three; so $(0, 1, 0)$ is an inflection point of E. (Note that $(0, 1, 0)$ also is on the y-axis and the tangent to the curve approaches the vertical direction as y becomes large.)

The Group Law

We take the rational point $O = (0, 1, 0)$ at infinity as our fixed point for defining addition (we are discussing the group $E(\mathbf{C})$, not just $E(\mathbf{Q})$ — recall the comments at the end of section 3). First we consider the subgroup $E(\mathbf{R})$ of real points, in the real affine plane, where lines connecting O to a real point on E can be visualized as lines parallel to the y axis and addition is illustrated in figure 16.

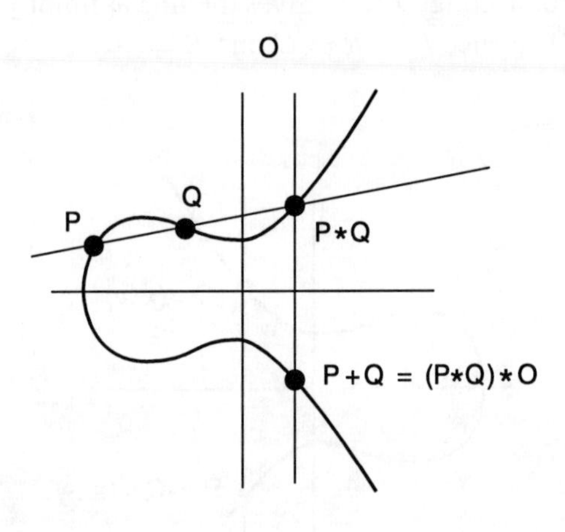

Figure 16

Identity: O is the identity for the group. $P + O = P$ is illustrated in figure 17. Note that $O + O = O$ follows from the fact that the tangent line at O is the line at infinity and the third intersection point of this line with E is again O (inflection point-triple intersection). Connecting O to O again gives the lines at infinity which meets the curve at O.

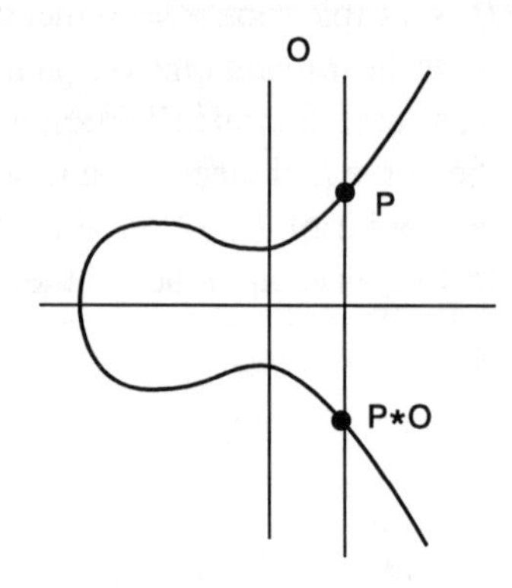

Figure 17

Inverses: If $Q = (x, y)$, then $R = (x, -y)$ is the inverse of Q. To see this, note that the line through Q and R is parallel to the y axis and $Q * R$, its third point of intersection with E, is O (fig. 18). As in the proof that O is the identity, connecting O to O gives the line at infinity which meets E a third time at O. Hence $Q + R = O$ and $R = -Q$.

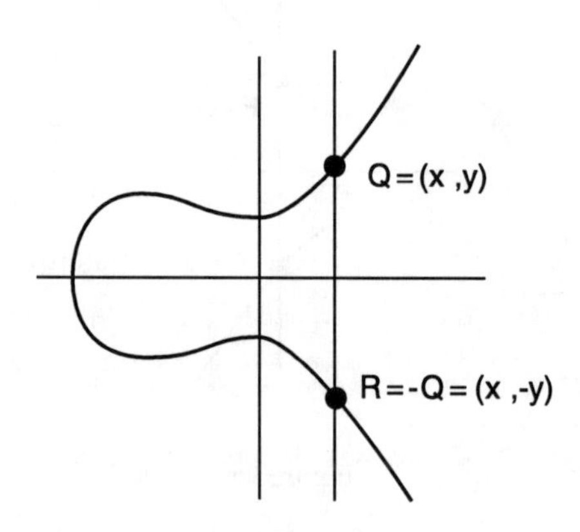

Figure 18

Example: We determine the real points P of order two, i.e., $P + P = O$, on the elliptic curve $E : y^2 = f(x)$, where $f(x)$ has three real roots, corresponding to the points Q_1, Q_2 and Q_3 (fig. 19). By definition, O has order 1; so assume $P \neq O$.

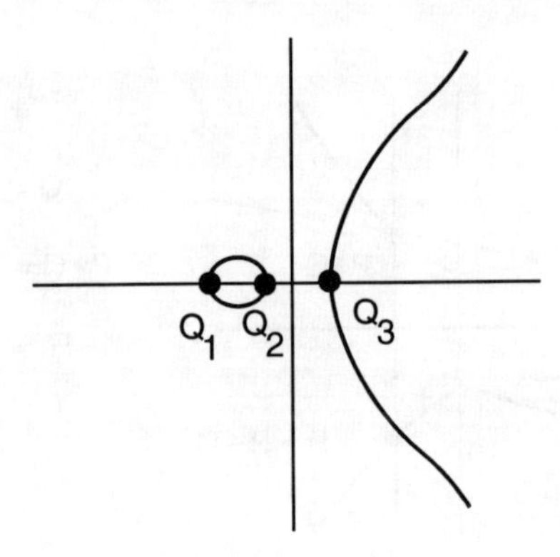

Figure 19

For any Q_i, the tangent to Q_i is parallel to the y-axis and intersects E at O. Connecting O to O gives a third point O and we have $2Q_i = O$. If $P \neq Q_1, Q_2, Q_3$, then the tangent line at P is not parallel to the y-axis and clearly $2P \neq O$. Thus Q_1, Q_2 and Q_3 are the real points of order two (using the formula for adding points, to be derived in a moment, one can show there are no complex points of order two). Note that $Q_i + Q_j = Q_k$, when i, j and k are distinct.

The points of order *dividing two* in an Abelian group form a subgroup. In our example we have $\{O, Q_1, Q_2, Q_3\}$, which is the well known Klein four group, a product of two cyclic groups of order two. Even if two of the roots of $f(x)$ are complex, the points corresponding to these roots, together with zero, are all the points of order dividing two and again they form the four group. We will discuss the complex points of order n as an application of elliptic functions in section 9.

Now we derive algebraic equation for the addition of points in $E(\mathbf{C})$. Although we draw pictures in the real affine plane, the derivation holds for $E(\mathbf{C})$.

Given $R = P + Q$, with coordinates as shown in figure 20, we want equations for x_3 and y_3. The line through P and Q is

$$y = \lambda x + v \,, \tag{1}$$

where

$$\lambda = \frac{y_1 - y_2}{x_1 - x_2}, \quad v = y_1 - \lambda x_1 = y_2 - \lambda x_2 .$$

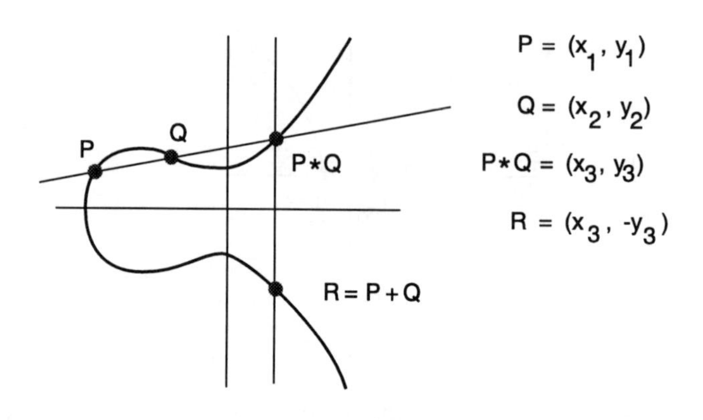

Figure 20

To find $P * Q$, we substitute (1) in the equation for the curve to obtain

$$(\lambda x + v)^2 = x^3 + ax^2 + bx + c$$

or

$$x^3 + (a - \lambda^2)x^2 + (b - 2\lambda v)x + c - v^2 = 0 .$$

This is a cubic with roots x_1, x_2, x_3. Since $x_1 + x_2 + x_3 = \lambda_2 - a$ (the sum of the root is - (coefficient of x^2)), we have

$$\begin{aligned} \mathbf{x_3} &= \lambda^2 - \mathbf{a} - \mathbf{x_2} - \mathbf{x_1}, \\ \mathbf{y_3} &= \lambda \mathbf{x_3} + \mathbf{v} . \end{aligned} \tag{2}$$

To derive a formula for doubling of a point, $P + P = 2P$, we proceed just as above, except that instead of using the equation for the secant line through P and Q, we need the equation $y = \lambda x + v$ for the tangent line at $P = (x_1, y_1)$ (fig. 21). We find λ by implicit differentiation,

$$y^2 = f(x) \Longrightarrow y' = \frac{f'(x)}{2y} \Longrightarrow \lambda = \frac{f'(x_1)}{2y_1},$$

and $v = y_1 - \lambda x_1$. Then (2) holds with these values of λ and v.

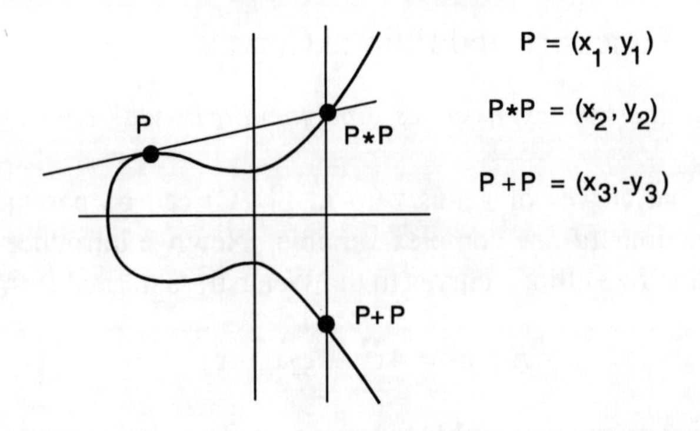

Figure 21

With these formulas for adding points, associativity of addition can now be verified directly (exercise).

Exercise: For $P, Q, R \in E(\mathbf{C})$,

$$P + Q + R = 0 \iff P, Q, R \text{ are collinear.}$$

(This is also true on any non-singular cubic if O is chosen to be an inflection point [Bri - Kno, sec. 7.4])

Everything we derived for curves in the normal form $y^2 = x^3 + ax^2 + bx + c$ is essentially the same for curves in the Weierstrass normal form $E : y^2 = 4x^3 - g_2x - g_3$.

E is non-singular if and only if $4x^3 - g_2x - g_3$ has no multiple roots. (0, 1, 0) is the only point at infinity on the curve (an inflection point), and if we choose this rational point as the O for the group law, we have, for $P_1 \neq P_2$,

$$P_1 + P_2 = P_3 \,,$$
$$(x_1, y_1) + (x_2, y_2) = (x_3, -y_3) \,,$$

where (3)

$$x_3 = \frac{\lambda^2}{4} - x_1 - x_2 \,,$$
$$y_3 = \lambda x_3 + v \,,$$

and

$$\lambda = \frac{y_1 - y_2}{x_1 - x_2} \,, \quad v = y_1 - \lambda x_1 = \frac{x_1 y_2 - x_2 y_1}{x_1 - x_2} \,.$$

There is a similar formula for $P + P$.

7. Elliptic Functions and Elliptic Curves

This section and the next assumes some familiarity with complex function theory.

Recall that curves of genus zero in $\prod_2(\mathbf{C})$ can be parametrized by rational functions of one complex variable. Now we introduce functions which parametrize elliptic curves in the Weierstrass normal form

$$E : y^2 = 4x^3 - g_2 x - g_3 \, .$$

This is done with **meromorphic** functions on $\mathbf{C}$ (analytic functions, whose only singularities are poles). A complex number λ is a **period** for a meromorphic function f if $f(z+\lambda) = f(z)$ for all z, where z is a pole if and only if $z + \lambda$ is a pole. Clearly, the set of periods of $f(z)$ form an additive subgroup of $\mathbf{C}$.

Let $\lambda_1, \lambda_2 \in \mathbf{C} - \{0\}$ and let $L = [\lambda_1, \lambda_2] = \{m\lambda_1 + n\lambda_2 | m, n \in \mathbf{Z}\}$ be a two dimensional lattice with integral basis λ_1, λ_2 (fig. 22 - see sec. 17.2 for basic lattice terminology). For reasons that will not concern us here, many authors assume that $\text{Im}\frac{\lambda_1}{\lambda_2} > 0$, so that $\frac{\lambda_1}{\lambda_2} > 0$ is in the upper half plane.

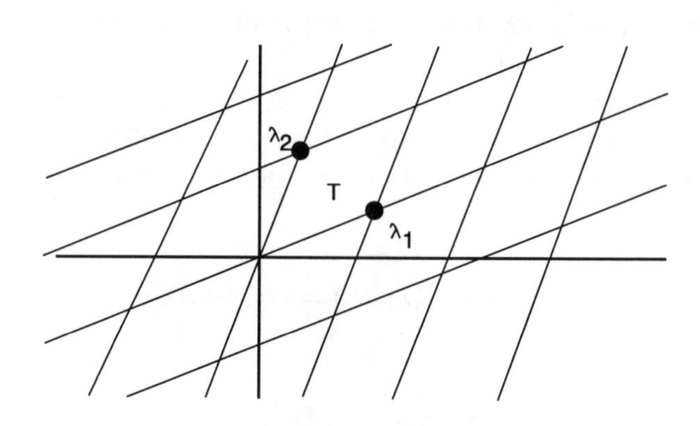

Figure 22

An **elliptic function** f is a meromorphic function whose set of periods form a two dimensional lattice L. We say that f is **doubly periodic** and has **period lattice L. M(L)**, the set of elliptic functions with period lattice L, is a field under addition and multiplication of functions, the so called **elliptic**

function field with period lattice L. The reason for the term 'elliptic' will become clear when we discuss the historical development in section 9.

Since $f(z)$ in $M(L)$ is periodic with respect to all λ in L, its values for any z in **C** is determined by its values in the fundamental parallelogram $T = \{t_1\lambda_1 + t_2\lambda_2 \,|\, 0 \leq t_1, t_2 < 1\}$.

Another way to look at this is to note that since L is an additive subgroup of **C**, the elements of the quotient group $\mathbf{C}/L$ are equivalence classes with respect to the equivalence relation $z_1 \sim z_2$ if and only if $z_1 - z_2 \in L$. Then T is a system of distinct representative for this equivalence relation. One often refers to the group operation in $\mathbf{C}/L$ as adding points modulo L, i.e., one can add two numbers in T as complex numbers and then move the sum back into T by subtracting an appropriate lattice point.

Topologically, if we identify equivalent (opposite) points on the boundary of T, then T becomes a torus (fig. 23).

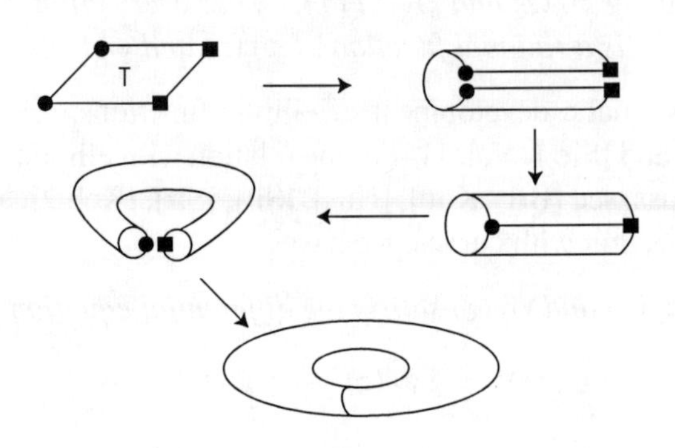

Figure 23

A *Riemann surface* is a one dimensional complex analytic manifold (roughly speaking, around each point there is a neighborhood homeomorphic to an open disc in **C** via holomorphic functions, such that there exists a holomorphic relation between overlapping neighborhoods). **C** is a Riemann surface and induces a complex analytic structure on T. We will identify each p in T with the equivalence class $p + L$ in $\mathbf{C}/L$ and write $T = \mathbf{C}/L$. The Riemann surface T is often called a **complex torus** (it is a complex Lie group). See [Spr], [Ber], [Jon] and [Wey 3] for an introduction to Riemann surfaces from a more classical viewpoint and [For] for a modern treatment. An elliptic function can now be thought of as a meromorphic function on the torus. Both points of view are useful at various times.

A central role in our study is played by a special elliptic function, the **Weierstrass $\wp$ function** (pronounced 'pay') **with respect to the lattice L**, defined by the series

$$\wp_{\mathbf{L}}(\mathbf{z}) = \frac{1}{z^2} + \sum_{\lambda \in L-\{0\}} \left\{ \frac{1}{(z-l)^2} - \frac{1}{\lambda} \right\} .$$

Theorem: *1) The series for $\wp_L(z)$ converges absolutely and uniformly on every compact subset of $\mathbf{C} - L$. It defines a meromorphic function on $\mathbf{C}$ with a double point at each lattice point and no other poles. $\wp_L(z)$ is an elliptic function with period lattice L, i.e., $\wp_L(z) \in M(L)$.*

2) $\wp_L'(z) = \dfrac{d\wp_L(z)}{dz} = -2\sum_{\lambda \in L} \dfrac{1}{(z-l)^3} \in M(L).$

3) The elliptic function field $M(L) = \mathbf{C}(\wp_L(z), \wp_L'(z))$, the field generated by adjoining $\wp_L(z)$ and $\wp_L'(z)$ to $\mathbf{C}$. Thus every elliptic function with period lattice L is a rational function of $\wp_L(z)$ and $\wp_L'(z)$.

For a systematic development of elliptic functions, see [Ahl], [San - Ger, Vol. 2] and [Sie 1, Vol. 1]. For the relation with elliptic curves, which we now discuss, see [Bri - Kno], [Hus], [Sil], [Str], [Kob 2], and [Kna]. To see this connection with curves, we have

Theorem: *$\wp_L(z)$ and $\wp_L'(z)$ satisfy the differential equation*

$$\wp_L'(z)^2 = 4\wp_L(z)^3 - g_2\wp_L(z) - g_3 ,$$

where g_2 and g_3 are functions of the lattice (in fact, if we let $G_{2k}(L) = \sum_{\lambda \in L-\{0\}} \frac{1}{\lambda^{2k}}$, the 'Eisenstein series of weight $2k$', then $g_2 = 60G_4$ and $g_3 = 140G_6$).

Thus we see that $(\wp_L(z), \wp_L'(z))$ is a point on the curve

$$E : y^2 = 4x^3 - g_2x - g_3$$

(E is, in fact, non-singular and thus an elliptic curve). The map

$$\varphi : \mathbf{C} - L \to E(\mathbf{C}) ,$$
$$z \mapsto (\wp_L(z), \wp_L'(z)) ,$$

is surjective and we have a parametrization of E as an affine curve. If we let $\varphi(\lambda)$, $\lambda \in L$, be the point at infinity $(0, 1, 0)$ on E, then φ is surjective to the curve in the projective plane. *Hence elliptic functions parametrize elliptic curves.*

Conversely, every elliptic curve in Weierstrass normal form determines a lattice L such that $(\wp_L(z), \wp_L'(z))$ parametrizes the curve. The fundamental parallelogram of L is often called the **period parallelogram of the curve E**.

Since $\wp_L(z)$ and $\wp_L'(z)$ both have L as their period lattice, we see that φ is constant on the equivalence classes of C/L, i.e., $\varphi(z+l) = \varphi(z)$ for $\lambda \in L$. Hence the induced map

$$\begin{aligned} \overline{\varphi} : \mathbf{C}/L &\to E(\mathbf{C}) , \\ [z] &\mapsto (\wp_L(z), \wp_L'(z)) , \end{aligned}$$

where $[z] = z + L$, is a bijection between the complex torus $\mathbf{C}/L$ and the curve $E(\mathbf{C})$. Therefore $\overline{\varphi}$ provides a bijective parametrization of $E(\mathbf{C})$ with T, the fundamental parallelogram of L (which is identified with $\mathbf{C}/L$), as the parameter space ($\overline{\varphi}$ is also a complex analytic isomorphism of $\mathbf{C}/L$ and $E(\mathbf{C})$ regarded as Riemann surfaces).

How does this map relate to the group structure on the curve?

Addition Theorem: *If $u, v \in C$, $u \neq v$, then*

$$\begin{aligned} \wp_L(u+v) &= -\wp_L(u) - \wp_L(v) + \frac{1}{4}\lambda(u,v)^2 , \\ -\wp_L'(u+v) &= \lambda(u,v)\wp_L'(u+v) + \mu(u,v) \end{aligned}$$

where

$$\begin{aligned} \lambda(u,v) &= \frac{(\wp_L'(u) - \wp_L'(v))}{(\wp_L(u) - \wp_L(v))} , \\ \mu(u,v) &= \frac{\wp_L(u)\wp_L'(v) - \wp_L(v)\wp_L'(u)}{\wp_L(u) - \wp_L(v)} . \end{aligned}$$

Letting $v \to u$, we get similar equations for $\wp_L(2u)$ and $\wp_L'(2u)$.

These equations have exactly the same form as the equations (3) in section 6, for adding points on the curve. Hence if

$$\begin{aligned} P_1 &= (x_1, y_1) = (\wp_L(u), \wp_L'(u)) , \\ P_2 &= (x_2, y_2) = (\wp_L(v), \wp_L'(v)) , \\ P_3 &= (x_3, y_3) = (\wp_L(u+v), \wp_L'(u+v)) , \end{aligned}$$

we have $\varphi(u+v) = \varphi(u)+\varphi(v)$, where $u+v$ denotes addition of complex numbers and $\varphi(u) + \varphi(v)$ denotes addition of points on the curve. This implies that $\overline{\varphi}([u+v]) = \overline{\varphi}([u])+\overline{\varphi}([v])$ and $\overline{\varphi}$ is an isomorphism between the groups $\mathbf{C}/L$ and $E(\mathbf{C})$ (in fact, it is a complex isomorphism between complex Lie groups).

We have now looked at our cubics in normal form from a geometric perspective as the set of points satisfying $y^2 = 4x^3 - g_2x - g_3$ and from a complex analytic perspective as a Riemann surface (a complex torus). In general, there are three ways to study an algebraic curve:

i) geometric — the set of points satisfying an equation $f(x, y) = 0$, $F \in \mathbf{Q}[x, y]$,
ii) complex analytic — a Riemann surface,
iii) algebraic — an algebraic function field in one variable.

We have not discussed the third point of view. See Stroeker [Str] and Shafarevich [Sha, chap. 1] for an introduction and Koblitz [Kob 2, sec. 1.6] for a special example. For a proof of the equivalence of the three points of view, see Bers [Ber]..

8. Complex Points of Finite Order

We now use our new point of view to find the subgroup $\mathbf{E_n}(\mathbf{C})$ of $E(\mathbf{C})$ of points of order dividing n on an elliptic curve in the form $y^2 = 4x^3 - g_2x - g_3$. By definition, the identity has order one. Let $L = [\lambda_1, \lambda_2]$ be the lattice whose function $\wp_L$ parametrizes the curve E, T the fundamental parallelogram of L, and recall that adding points on a curve E corresponds to adding the corresponding points in T modulo L.

$n = 2$: Clearly the four points marked in figure 24 are the only z in T such that $[z] + [z] = [0]$ (i.e., $z + z \in L$). This is the Klein four group, a direct product of of cyclic groups C_1, C_2 of order two (we derived this result by a different method in section 6).

$n = 3$: The points marked in figure 25 are exactly those points z in T satisfying $3[z] = [0]$ (i.e., $3z \in L$) and are a product of two cyclic groups of order dividing three.

For arbitrary n, the n^2 points $k\frac{\lambda_1}{n} + m\frac{\lambda_2}{n}$, $k = 0, \cdots, n-1, m = 0, \cdots, n-1$, in T are the points of order dividing n and this is a product of the cyclic groups $C_i = \{\frac{k\lambda_i}{n} | k = 0, \cdots, n-1\}$, $i = 1, 2$, of order n.

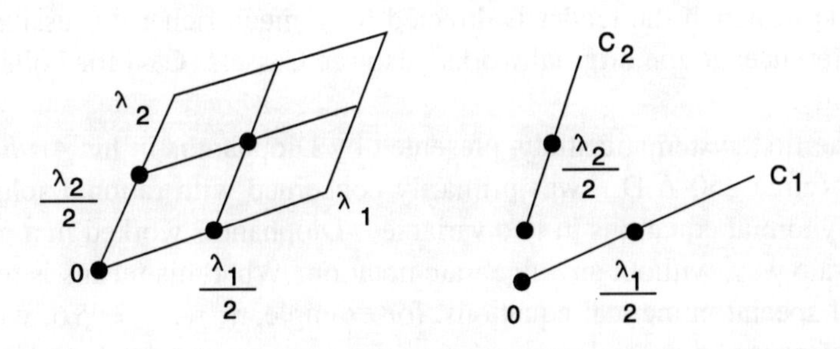

Figure 24

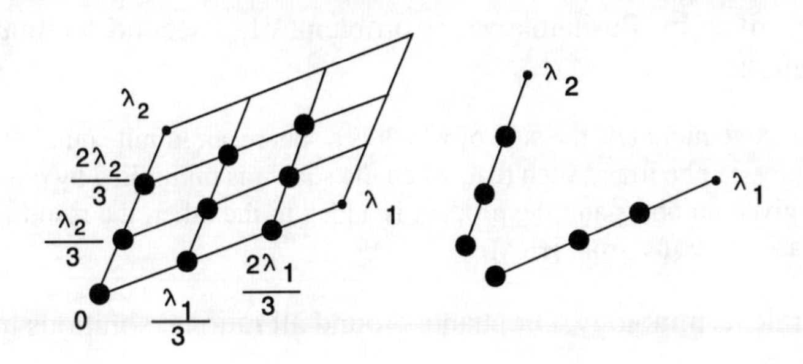

Figure 25

Remarks: 1) The coordinates of points of finite order on E are always algebraic numbers.

2) There is no known way of determining which $u \in T$ correspond to *rational points* on E. Nevertheless, this points of view is still essential in studying rational points on elliptic curves, e.g., Weil used it to give a proof of Mordell's theorem [Wei 11].

3) In the next chapter, we will provide an algorithm for finding rational points of finite order on E.

9. The Early History

Since Poincaré's fundamental memoir [Poi] represents the beginning of the modern theory of arithmetic on curves, we shall regard the work leading up to and including Poincaré's work as the early history. Our discussion is based mostly on the historical work of Bashmakova [Bas] and Weil

[Wei 1] to which the reader is directed for a much richer discussion and for references to the original works (also see Cassels [Cas] for Poincaré's memoir).

The first systematic study, presented by Diophantus in his *Arithmetic* [Hea] (circa 350 A.D.), was primarily concerned with rational solutions of polynomial equations in two variables. Diophantus worked in a purely algebraic way, without any algebraic notation. What this means is that he solved special numerical equations, for example, $x^2 + y^2 = 16$, without any reference to special properties of the numbers involved, and solved sufficiently many examples to make it clear that he understood the general methods we would today formulate with algebraic notation. However, there are two places where he formulated his problems in a general way. One example, given by Bashamkova, is problem VI_{12}, second lemma (in the *Arithmetic*):

> Given two numbers, the sum of which is a square, an infinite number of squares can be found such that, when the square is multiplied by one of the given numbers and the product is added to the other, the result is a square. (trans. from [Hea])

In algebraic terminology, Diophantus found all rational solutions to

$$ax^2 + b = y^2 \ , \ \text{where } a + b = \text{square.}$$

When given a conic with a rational point O, Diophantus generally understood how to represent all the rational points by rational functions of a parameter. His method is equivalent to our geometric method, using the straight lines passing through O (sec. 18.3).

After treating second degree equations, Diophantus considered special equations of the third, fourth and sixth degrees. Some of these examples are curves of genus zero, with a rational point (and thus birationally equivalent to a line). These led to solutions represented by rational functions of a parameter, as in our example of $y^2 = x^2(x + a)$ (sec. 18.4). Diophantus also attacked some equations of genus one. When he already knows one rational point for such an equation, he used an algebraic formulations of the tangent-secant process (sec. 18.4) to find a second rational point.

An example of the tangent process appears in problem VI_{18}, applied to the equation

$$y^2 = x^3 - 3x^2 + 3x + 1 \ ,$$

with the obvious rational solution (0, 1). Substituting $y = \frac{3}{2}x - 1$ (which is the tangent to the curve at (0, 1)) into the equation leads to a cancellation of the constant and x term and to a second rational solution $\left(\frac{21}{4}, \frac{71}{8}\right)$.

In some examples Diophantus used a non-tangent line through a rational point P to find a second rational point of intersection. In these cases, he was using the secant line connecting P to the rational point at infinity on the curve. He only uses the secant process when one of the rational points was a point at infinity (of course, he didn't think in these terms). Note that, given a rational point on a curve, Diophantus only used the tangent-secant process to find a second rational point. There is no indication that he ever thought of iterating this method to find more rational points. See Weil [Wei 1, chapter 1] for other methods of Diophantus.

For the next major advances we must jump to the 16th century when Bombelli, in his *Algebra*, and Viete, in his *Zetetica*, used the new methods of algebra to make many of Diophantus' methods more clear.. However, it was Fermat who really mastered Diophantus' methods by studying Bachet's translation of the *Arithmetic* [Hea] and added major new advances.

Fermat realized that iterating Diophantus' tangent and secant methods could yield infinitely many rational points on a curve, and he developed systematic techniques for applying this in special cases. Fermat also realized that there could be points of finite order, such as (0, 1) on the curve $y^3 = x^3 + 1$. However, as with Diophantus, Fermat only applied the secant process when one of the rational points was a point at infinity on the curve. It has only recently been discovered that Newton realized that the secant process could be generalized to two rational points on a curve in the affine plane to yield a third rational point [New]. He also gave the geometric interpretation of the tangent-secant process (sec. 18.4). Newton's successors, including Euler, seemed to have been unaware of this work; thus it played no role in later developments.

The next significant advances were by Euler. He systematized and extended the study of

$$y^2 = f_3(x), \; y^2 = f_4(x), \text{ and } y^3 = f_3(x)\,,$$

where $f_n(x) \in \mathbf{Z}[x]$, $\deg f_n = n$. Among his results is a proof that a sufficient condition for $y^2 = f_3(x)$ to be parametrizable by rational functions is that $f_3(x)$ have a double root. Euler also applied the general secant process to special curves.

In what would seem to be a totally different direction, but one which is intimately related to arithmetic on curves, Euler studied elliptic integrals. With the advent of the integral calculus, a major direction of interest was the question of which functions could be integrated in closed form, in terms of a specified set of functions. Thus, e.g., polynomials can be integrated in terms of polynomials. Euler proved that rational functions can be integrated in terms of rational functions, logarithms and inverse trigonometric functions. Using this result, one can then integrate rational functions $f(x, y)$, where y is of the form $\sqrt{ax^2+bx+c}$. Weil [Wei 1, chap. 3, sec. XV] notes that the substitutions needed for the latter result are the same as those needed to find rational solutions of $y^2 = ax^2 + bx + c$, when a is a square. He also notes that the problem of integrating $\sqrt{1-x^4}$ (which Euler unsuccessfully attempted in his younger days) is almost the same as finding non trivial rational solutions of $y^2 = 1 - x^4$ (or, equivalently, non trivial integral solutions of $Y^2 = Z^4 - X^4$). Since Euler had reconstructed Fermat's proof that the latter equation has no such solutions, Weil argues that this must have helped convince Euler that $\sqrt{1-x^4}$ could not be integrated in terms of known functions. Of course, this also implies that Euler might have guessed at some deep connection between integration problems and Diophantine problems (also see [Wei 3]).

Elliptic integrals, which arise from the attempt to calculate the length of an arc of an ellipse, were of great interest in Euler's time and were seen as the next order of difficulty after the integration of rational functions. Thus, to find the length of an arc of the ellipse $\frac{x^2}{a^2} + \frac{y^2}{b^2} = 1$, one must integrate the element of arc length

$$\begin{aligned} ds &= \sqrt{(1+(dy/dx)^2)}dx \\ &= \frac{(b^2-a^2)x^2+a^4}{a\sqrt{(b^2-a^2)x^2+a^4)(a^2-x^2)}}\,dx\ . \end{aligned}$$

An **elliptic integral** is an integral of the form $\int R(x, y)dx$, where R is a rational function of the variables x, y and $y = \sqrt{P(x)}$, P a polynomial of degree 3 or 4 with no multiple roots. (We omit any discussion of the complex plane as the natural setting for these questions and the related multivalued square root function — see [Sie 1, Vol. 1] for more details.)

For the arc length of the ellipse we have

$$\int \left((b^2 - a^2)x^2 + a^4\right) \frac{dx}{y} ,$$

where $y = \sqrt{(b^2 - a^2)x^2 + a^4)(a^2 - x^2)}$, and for $\sqrt{1 - x^4}$, we have $\int \frac{dx}{y}$, where $y = \sqrt{1 - x^4}$.

On December 23, 1752, the Berlin Academy of Sciences gave Euler the recently received two volume *Produzioni Mathematiche* by Fagnano for examination. Euler was obviously stimulated by Fagnano's work on the lemniscate for on January 27, 1753 he presented to the Academy the first of a series of papers which culminated in the general addition and multiplication theorems for elliptic integrals [Wei 1].

Fagnano was concerned with the length of an arc of the lemniscate. Recall that a lemniscate is the locus of a point $P = (x, y)$ in a plane such that the product of its distances from two fixed points, of distance $2c$ apart, has constant value c^2 (fig. 26). If we let the two fixed points be $\left(\frac{\pm 1}{\sqrt{2}}, 0\right)$ and set

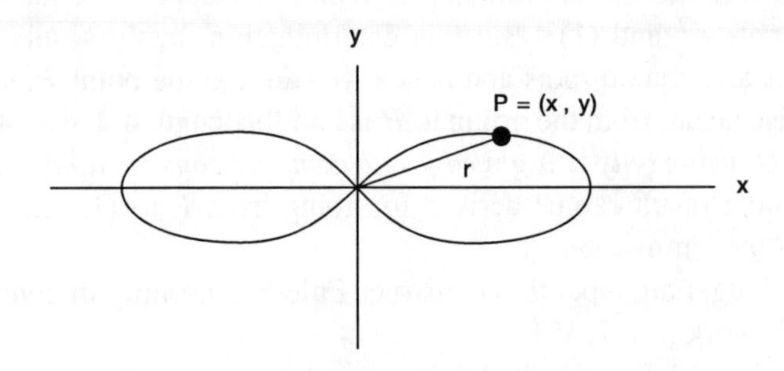

Figure 26

$c^2 = \frac{1}{2}$, then this lemniscate can be parametrized by

$$x = \pm\frac{1}{\sqrt{2}}\sqrt{r^2 + r^4} , \tag{1}$$

$$y = \pm\frac{1}{\sqrt{2}}\sqrt{r^2 - r^4} ,$$

where r is the distance from P to the origin ($r^2 = x^2 + y^2$) and the signs of the square roots are chosen according to which quadrant we are studying.

If we restrict ourselves to an arc in the first quadrant and let $s(r)$ be the length of the arc from the origin to the point P with parameter r, then

$$s(r) = \int_0^r \frac{dt}{\sqrt{1-t^4}}\,, \quad 0 \le r \le 1\,,$$

where we take the positive square root (one of our examples of an elliptic integral).

Fagnano, in the course of trying to evaluate this integral, found that if the parameter value u corresponding to the point $Q = (x', y')$ satisfies

$$r^2 = 4u^2 \frac{1-u^4}{(1+u^4)^2}\,, \tag{2}$$

then

$$\int_0^r \frac{dt}{\sqrt{1-t^4}} = 2\int_0^u \frac{dt}{\sqrt{1-t^4}}\,. \tag{3}$$

Thus the arc from the origin to Q is twice the length of the arc from the origin to P. Therefore, if we are given the coordinates x', y' of Q, we can determine the coordinates x, y of P by applying equations (2), $u^2 = x'^2 + y'^2$ and (1). But this determination involves only rational operations and square roots and hence we can find the point P, such that the lemniscate arc from the origin to P is half the length of the arc from the origin to Q, *using only straight edge and compass constructions* (sec. 14.1-2). A similar result can be derived for going from P to Q, i.e., doubling the arc of the lemniscate.

C. L. Siegel attempts to reconstruct Euler's reasoning in generalizing Fagnano's work [Sie 1, Vol. 1].

There is a relation for doubling a circular arc, analogous to equations (2) and (3). Let

$$r = 2u\sqrt{1-u^2}\,;$$

then, by direct algebraic manipulation,

$$\int_0^r \frac{dr}{\sqrt{1-r^2}} = 2\int_0^u \frac{du}{\sqrt{1-u^2}}\,.$$

By the subsitution $u = \sin x$, these equations yield

$$\sin(2x) = 2\sin x \cos x\,.$$

However, this is a special case of

$$\sin(x+y) = \sin x \cos y + \cos x \sin y \,,$$

which leads to the more general equation

$$\int_0^u \frac{du}{\sqrt{1-u^2}} + \int_0^v \frac{dv}{\sqrt{1-v^2}} = \int_0^r \frac{dr}{\sqrt{1-r^2}}, \; r = u\sqrt{1-v^2} + v\sqrt{1-u^2}\,,$$

where $u = \sin x$, $v = \sin y$, $u\sqrt{1-v^2} + v\sqrt{1-u^2} = \sin(x+y)$.

Euler found a corresponding addition theorem for the lemniscate integral, viz.,

$$s(u) + s(v) = s(r)\,,$$

where

$$r = \frac{u\sqrt{1-v^4} + v\sqrt{1-u^4}}{1+u^2v^2}\,.$$

He then generalized it to

$$\int_0^r \frac{dt}{\sqrt{P(t)}} = \int_0^u \frac{dt}{\sqrt{P(t)}} + \int_0^v \frac{dt}{\sqrt{P(t)}}\,,$$

where

$$r = \frac{u\sqrt{P(v)} + v\sqrt{P(u)}}{1+u^2v^2}, \quad P(t) = 1 + ct^2 - t^4\,,$$

(c an arbitrary constant). Euler's work made elliptic integrals one of the hot topics of the day. We will return to the connection between the problems of rational points and of integration shortly.

Legendre made the next major contributions. He found an algorithm for deciding if a conic has a rational point a (sec. 18.3) and continued Euler's work on elliptic integrals.

The most significant advance in the study of elliptic integrals was found by Gauss, Abel and Jacobi in the early part of the 19th century. They followed the analogy between the lemniscate integral and the integral $\int_0^r \frac{dt}{\sqrt{1-t^2}}$ for circular arcs. The latter integral can be used to define the arc sine function and then *inverting* it yields the sine function. The whole theory of trigonometric functions can be based on this inversion. What Gauss, Abel and Jacobi did was to invert elliptic integrals and develop a theory of these inverse functions. These functions, naturally called elliptic functions, are best studied in the complex plane, and were the first examples of elliptic functions as defined in section 7. Making the elliptic functions, rather than elliptic integrals, the basic objects of studies simplified the theory and led to major advances. Of course, we know that the fact that the elliptic functions $\wp$ and $\wp'$ parametrize elliptic curves is an example of a connection between arithmetic and analysis, but this was not discovered until later in the 19th century.

The history of elliptic functions is rather complicated. Gauss never published his work, but found many results about twenty years before Abel and Jacobi. It is unclear to what extent Jacobi's work depended on a prior knowledge of Abel's. We refer to [Hou] for the systematic historical development as well as to illuminating discussions in [Kle 2], [Bel 1], [Bel 2], [Sha, historical appendix], and [Wei 7]. However, it is interesting to note that Abel stated that he finally understood Gauss' cryptic remark on the lemniscate function, at the beginning of section VII of the *Disquisitiones* (sec. 14.1), after developing his theory of elliptic functions. See Rosen [Ros] for a very clear modern exposition of some of Abel's work. Abel also treated more general integrals, the so called Abelian integrals and their inverses, and proved a general addition theorem. See R. Cooke [Coo] for the developments connected with Abel's theorem. The theory of elliptic functions motivated some of the most beautiful work in analysis, geometry, algebra and number theory of the 19th and 20th centuries and its influence continues to the present day.

In 1835, Jacobi published a paper "On the Application of the Theory of Elliptic Integrals and Abelian Integrals to Diophantine Analysis" [Jac], in which he explicated the connection between Euler's work on rational points and his addition theorem for elliptic integrals. He showed that given rational points on the curve $y^2 = P(x)$, $P(x)$ a fourth degree polynomial, one can use the addition theorem for the elliptic integral $\int_0^r \frac{dx}{\sqrt{P(x)}}$ to determine new rational points (see [Bas] for more details). This was presented from a more modern viewpoint in section 6. If we restrict the isomorphism between a complex torus $\mathbf{C}/L$ and the elliptic curve parametrized by $(\wp_L(z), \wp_L'(z))$ to the subgroup of rational points on the latter and the subgroup of parameter values corresponding to rational points on the former, then the addition theorems for the elliptic functions $\wp_L(z)$ and $\wp_L'(z)$ correspond to the algebraic formulas for adding rational points on the curve. Hence an analytic addition theorem for functions and an algebraic addition theorem for rational points are different ways of looking at the same phenomenon. Jacobi provided a deep study of the structure of rational points on elliptic curves (without any group theoretic notion of adding points), but he did not guess the finite generation of the rational points or have the notion of birational equivalence. He also studied curves of higher genus and outlined a plan for constructing a theory of them. Unfortunately, Jacobi's successors, particularly Poincaré, were ignorant of this work and it had no influence in the development of the subject. However, Jacobi's related work in many directions was of critical influence.

For the next forty-five years there was no significant progress on the study of rational points, although the requisite fields (algebraic geometry, complex analysis, elliptic and modular functions, ...) were actively pursued. In 1890, Hilbert and Hurwitz [Hil - Hur] characterized rational points on curves of genus zero, using birational equivalence.

Then, in 1901, Poincaré's fundamental memoir "Sur les propertiétés arithmétiques des courbes algébriques" [Poi] appeared. Unaware of the arithmetic work of Jacobi and Hilbert and Hurwitz, but a master of the requisite analysis, geometry and algebra, Poincaré presented a program for the systematic study of rational points on algebraic curves. (In fact, he did not mention any earlier work.) First he introduced birational equivalence for projective algebraic curves. Then he rediscovered the Hilbert-Hurwitz result for curves of genus zero.

The main body of the paper is concerned with non-singular plane cubics, and Poincaré assumed that they could be parametrized by elliptic functions; he also assumed knowledge of the properties of these functions. (Although Poincaré assumed that all curves of genus one can be so parametrized, he never stated that when such curves have a rational point, they are birationally equivalent to curves in Weierstrass normal form.) Assuming the curve has a rational point O, Poincaré introduced the addition of rational points and showed that this is an Abelian group with O as the identity element. He presented the equivalence of the tangent-secant process, the algebraic equations for adding points, and the addition theorems for elliptic functions (sec. 5, 6). Thus the geometric, algebraic and analytic viewpoints, and their connections, were always present. Poincaré defined the rank of a curve as the minimal number of rational points needed to generate the group (today we reserve the term for the number of generators of the torsion free part of the group) and asked about the properties of this integer. Thus he implicitly assumed that the rank is finite, i.e., the group is finitely generated. This came to be known as Poincaré's hypothesis and later as Mordell's theorem.

He also discussed conditions for curves of genus one, with a rational point, to be birationally equivalent.

The memoir ends by discussing two possible generalizations of these results. The first is to extend to points whose coordinates lie in some algebraic number field and the second is to consider curves of higher genus. In the latter case, Poincaré realized that there is no natural way of adding points but that one can generalize to the addition of 'rational groups of points'. Both of these directions led to great advances in the hands of Weil. Later, we will discuss these twentieth century developments. It suffices to say for now that Poincaré's memoir, together with the subsequent work of Mordell and Weil, opened up an enormous area of arithmetic research, with beautiful results being discovered up to the present day.

Chapter 20

Arithmetic on Curves 3: The Twentieth Century

1. From Poincaré to Weil

In the last chapter, we recounted the study of rational points on curves from Diophantus to Poincaré. This culminated in Poincaré's conjecture, which almost twenty years later became Mordell's theorem, the first truly general statement about Diophantine equations (where we allow rational as well as integer solutions.). Now we continue with the history from Poincaré's 1901 paper to about 1930, closely following the exposition by Cassels [Cas], who has been a significant contributor to the subject and was a long time friend of Mordell's. Davenport's obituary of Mordell [Dav 2] and Mordell's autobiographical essays [Mor 3], [Mor 4] offer interesting insights into Mordell's personality and to his mathematical work.

The first significant work after Poincaré was by Thue, who revealed the fundamental connection between Diophantine equations and Diophantine approximation, which resulted in his theorem that $f(x, y) = d$ has only finitely many integer solutions for any homogeneous irreducible $f(x, y)$ in $\mathbf{Z}[x, y]$ of degree greater than two (sec. 21.5).

Then in 1917, Thue [Thue 2] (and independently Landau and Ostrowski [Lan - Ost]) used his theorem to prove that

$$ax^2 + bx + c = dy^n \tag{1}$$

has only finitely many solutions when $a, b, c, d \in \mathbf{Z}$, $n \geq 3$, $a \neq 0$, and $b^2 - 4ac \neq 0$. We sketch the idea of the proof, which is the key to later results, and which uses the arithmetic of quadratic fields (see chapters 16 and 17).

We first multiply equation (1) by a and let $x' = ax$, $b' = ab$, $c' = ac$, $d' = ad$. Then $x'^2 + b'x + c' = d'y^n$, and factoring $x'^2 + b'x + c'$ yields

$$N(x' - \theta) = d'y^n ,$$

where θ is a root of $x'^2 + b'x + c' = 0$ and N is the norm (sec. 16.3). By unique factorization of ideals in the integers of $\mathbf{Q}(\theta)$ (sec. 17.4), one can show that

$$x' - \theta = \mu\eta^n ,$$

where $\eta = u + v\theta \in \mathbf{Z}[\theta]$ and where $\mu \in \mathbf{Q}(\theta)$ is restricted to a finite set. Equating coefficients of θ on both sides of the last equations yields $g_\mu(u, v) = 1$, where g_μ is a homogeneous polynomial with rational coefficients depending only on μ. Multiplying by an appropriate constant e to clear denominators, we have $f_\mu(u, v) = e$, where f_μ satisfies the conditions of Thue's theorem and thus, for each of the finitely many μ, there are only finitely many integer solutions.

Meanwhile, Mordell, an American born mathematician who had gone to Cambridge University to study, began his serious research in number theory about 1910. In his own words [Mor 4]

> "There was then no Ph. D. degree; it came in only after the First World War, so the present day mathematicians are much more learned than we were... There were two Smith Prizes in those days for which B. A.'s could compete. Neville got the first one, and I the second for an essay on *The Diophantine equation* $y^2 = x^3 + k$, a topic which has played a prominent part in my research career even in the very latest years. I might mention that my Cambridge inaugural lecture, "A Chapter in the Theory of Numbers" [Mor 1] dealt with this topic.
>
> I continued my studies, staying on in Cambridge, and wrote another paper entitled, *Indeterminate equations of the 3rd and 4th degrees* [Mor 5]. I was very unfortunate with this paper. It was rejected by the London Mathematical Society; I really don't know why. Perhaps they did not approve of my style, but it was a really important paper and has played a prominent part in the progress of number theory even in the present day.
>
> I hope you will bear with me if I mention one of my results. I had proved that the integer solutions of $y^2 = ax^3 + bx^2 + cx + d$ could be found from the representations of unity by binary quartics. Neither I nor the referee was aware that in 1909 Thue had proved there could be only a finite number of representations. This meant that the cubic had only a finite number of solutions, a really important result.

By 1918, Mordell knew of Thue's work and in 1922 he published a proof that

$$Ey^2 = Ax^3 + Bx^2 + Cx + D$$

has only finitely many integer solutions. Mordell thought he had extended the result to quartics in x, but found an error and tried in vain to correct it for six months. Suddenly he realized that he had in fact proved Poincaré's conjecture (Mordell's theorem) [Mor 3]. Cassels attempts to reconstruct Mordell's chain of reasoning [Cas].

At the end of Mordell's fundamental paper there are five conjectures, the last stating that a curve of genus greater than one has only finitely many rational points [Mor 5]. This became known and famous as Mordell's conjecture and was proved by Gerd Faltings [Fal] in 1983.

Mordell continues his earlier narration [Mor 4] about integer solutions of $Ey^2 = Ax^3 + Bx^2 + Cx + D$ with the next major discovery:

> " Some years later, Professor C. L. Siegel, one of the worlds foremost mathematicians, generalized this result and communicated his results to me. I asked him whether he would not object to this being published by the London Mathematical Society. He did not reply and so I took it for granted that he did not object. Proof sheets were sent to him, and he was then very annoyed because the mathematicians of Frankfurt had agreed not to publish anything for a few years. However, he agreed to let it appear anonymously as due to X. When I saw him a few years ago, he said it need no longer be anonymous."

This incident occurred around 1925. The result states that the equation $ay^2 = f(x)$, where $f(x) \in \mathbf{Z}[x]$, $\deg(f) \geq 3$, has only finitely many integer solutions. Shortly thereafter, Siegel generalized this and using Mordell's theorem and his own generalization of Thue's theorem in Diophantine approximation went on to prove part of his famous integer point theorem, viz., a curve $f(x, y) = 0$, $f \in \mathbf{Z}[x, y]$, of genus one has only finitely many integer points.

Meanwhile in the early twenties, André Weil, then a student at the École Normale, studied the works of Riemann and Fermat with the strong conviction that reading papers of the great masters of the past was as rich a source of inspiration as the papers of modern day authors [Wei 12, vol. 1, pp. 524 - 526]. Wanting to make a contribution to Diophantine analysis, he began with the study of Fermat's method of infinite descent (sec. 2.2) and the point of view of birational equivalence. Weil's interest in history was stimulated by his visits to Dehn's seminar on the history of mathematics at the University of Frankfurt. Here he met and received great encouragement from Siegel. See Siegel [Sie 2] for a beautiful and touching recollection of the Frankfurt mathematical group.

In his thesis [Wei 13], Weil generalized Mordell's theorem by proving that if K is an algebraic number field, then $E(K)$, the group of K-rational points (coordinates in K) on an elliptic curve E, is a finitely generated abelian group. To quote Cassels [Cas]:

> "From now on, Weil's generalization of Mordell's theorem (and subsequent generalizations) was usually referred to as the Mordell - Weil Theorem. Mordell himself strongly disapproved of this usage and frequently insisted (in public and in private) that what he had proved should be called Mordell's Theorem and that everything else could, for his part, be called simply Weil's Theorem. I would tell him that then, even when I was working with rational points on curves of genus 1, what I was using was a special case of Weil's Theorem. He would not, or could not, see the point.
>
> Weil also published a much simpler and more perspicuous proof of Mordell's original result [Wei 11]. He considered a curve of genus 1 with rational point in the shape
>
> $$y^2 = x^3 - Ax - B \quad (A, B \in \mathbf{Z})$$
>
> the rational point being that 'at infinity'. Now
>
> $$x - \theta = \mu\eta^2$$
>
> where θ is a root of $\theta^2 - A\theta - B$ (supposed irreducible for this discussion), $\eta \in \mathbf{Q}(\theta)$ and μ from a finite set. Weil starts by showing that μ depends only on the coset of x, y in the group $E(\mathbf{Q})/2E(\mathbf{Q})$, where $E(\mathbf{Q})$ is the group of rational points. The argument concludes with a descent."

This proof we shall give is essentially a variation of this latter proof of Weil.

Shortly after Weil's work, Siegel [Sie 3] used the Mordell - Weil theorem to complete his proof of the

Integer Point Theorem: *A curve* $f(x, y) = 0$, $f \in \mathbf{Z}[x, y]$, *of genus* $g > 0$ *has only finitely many integer points.*

He also characterized those equations of genus zero with infinitely many integer points. The most accessible exposition of this proof is probably that given by Robinson and Roquette [Rob - Roq] using the language of nonstandard analysis.

In his thesis, Weil also studied curves of genus greater than one. Poincaré realized that there seemed to be no reasonable way of adding rational points on such curves and introduced the addition of 'rational sets'. A set of g k-rational points (coordinates in a number field k) on a curve C of genus g is a *rational set* if every symmetric function of their coordinates is rational. In modern language, these are 'rational divisors on the Jacobian of C' (for a discussion of Jacobi's original ideas see [Bas]). Weil introduced a notion of 'adding' rational sets of g points and proved that the rational sets on a curve of genus g are a finitely generated abelian group. (When $g = 1$ and the field is $\mathbf{Q}$, this reduces to Mordell's theorem.)

Weil's goal was to prove Mordell's conjecture. He told this to Hadamard who responded "Work on; you should not publish a half-result"; fortunately, he did not follow this advice since we had to wait about 50 years for Faltings' proof.

We summarize these results over $\mathbf{Q}$ and $\mathbf{Z}$ in table 1:

	genus 0	genus 1	genus > 1
rational points	0 or ∞ many	0 or finitely generated abelian group	finite
integer points	complete characterization	finite	finite
rational sets of points – genus g – 0 or finitely generated abelian group			

Table 1.

The proofs of these theorems, as given by Mordell, Siegel, and Weil, are ineffective, i.e., they do not provide a computable method to find the solutions for any given equation of positive genus. Based on the fundamental work of Baker which makes Thue's theorem effective (sec. 21.6), Baker and Coates found a method for computing a bound for the size of the coordinates of the integer points on a curve of genus one [Bak - Coa]. Lutz and Nagell found an algorithm to find points of finite order on a curve of genus one (elliptic curve) in an appropriate normal form (sec. 2). This is the extent of our knowledge of effective methods. For example, although Legendre's theorem (sec. 18.3) provides an algorithm for deciding if a conic (genus zero) has a rational point and there is an algorithm for finding integer solutions on such a curve [Gau 1, art. 216, 222], there are conjectures but no proven method for deciding if an elliptic curve has a rational point.

Arithmetic on curves has been such an active area of research during the last 50 years with important new results coming at such a fast and furious pace that a coherent presentation of the history is certainly premature. The main emphasis in recent years has been on the study of elliptic curves. In the remainder of this chapter, we prove that the set of points of finite order on an elliptic curve is finite, give an algorithm for finding these points, prove a special case of Mordell's theorem and give short discussions of some other topics.

General references for the material in this chapter were given at the end of section 18.1.

2. Points of Finite Order; The Lutz–Nagell Theorem

As discussed in the last two chapters, Mordell proved that if E is an elliptic curve, then $E(\mathbf{Q})$ is a finitely generated abelian group. By the fundamental theorem for finitely generated abelian groups, $E(\mathbf{Q})$ is a direct sum of a free group F and $E(\mathbf{Q})_{\mathbf{tors}}$, the torsion subgroup of $E(\mathbf{Q})$, i.e., the subgroup of points of finite order ($mP = O$ for some $m > 0$).

The study of the **rank** of $E(\mathbf{Q})$ (the number of independent generators of F) is a very active area of research and we will discuss some results and open problems later. On the other hand, the structure of $E(\mathbf{Q})_{\text{tors}}$ is completely known.

It is an amazing and very deep theorem, proved by Barry Mazur, that the possibilities for $E(\mathbf{Q})_{\text{tors}}$ are very limited ([Maz 1], [Maz 2]) .

Mazur's Theorem: *If $E(\mathbf{Q})$ is an elliptic curve, then $E(\mathbf{Q})_{\text{tors}}$ is one of the following 15 groups :*

$$\mathbf{Z}/n\mathbf{Z}\,, \quad 1 \geq n \geq 10 \quad or \quad n = 12\,,$$
$$(\mathbf{Z}/2\mathbf{Z}) \times (\mathbf{Z}/2n\mathbf{Z})\,, \quad 1 \leq n \leq 4\,.$$

Each of these groups is the torsion subgroup of some elliptic curve.

The proof of this result is far beyond the scope of this book, but we shall prove a more limited theorem discovered independently by Elisabeth Lutz [Lut] and Trygve Nagel [Nag 2].

First we need yet another normal form for elliptic curves. We apply the birational map $x = X - \frac{a}{3}$, $y = Y$ to $y^2 = x^3 + ax^2 + bx + c$ to eliminate the x^2 term, and we have $Y^2 = X^3 + AX + B$.

Lutz–Nagell Theorem: *Let*

$$E : y^2 = x^3 + Ax + B$$

be an elliptic curve, with $A, B \in \mathbf{Z}$, *and let* $P = (x, y)$ *be a rational point of finite order on* E ($P \in E(\mathbf{Q})_{\text{tors}}$). *Then*

i) P *is an integer point* ($x, y \in \mathbf{Z}$) *and*

ii) $y = 0$ *(points of order two) or*
$y^2 | D$, *where* $D \in \mathbf{Z}$ *is the 'discriminant' of the curve.*

Therefore there are only a finite number of possible choices for y and for each such y at most three values of x; hence $E(\mathbf{Q})_{\text{tors}}$ *is finite.*

In the projective plane, the point at infinity $(0, 1, 0)$ *on* E, *which serves as the identity for the group law, also has finite order.*

Remarks:

1) The **discriminant D** of E: $y^2 = f(x)$ is defined to be the discriminant D_f of $f(x)$. Recall that if $f(x) = x^3 + Ax + B = (x - r_1)(x - r_2)(x - r_3)$, then

$$D_f = (r_1 - r_2)^2 (r_1 - r_3)^2 (r_2 - r_3)^2 .$$

In sec. 19.4, we saw that

$$D \neq 0 \iff \text{the } r_i \text{ are distinct} \iff E \text{ is non-singular.}$$

For our curve, $D = -4A^3 - 27B^2$. The proof is a straightforward computation using the fact that the coefficients of f are symmetric functions of the roots.

2) The restriction to $A, B \in \mathbf{Z}$ is not a real restriction. If

$$E : y^2 = x^3 + Ax + B ,$$

with $A, B \in \mathbf{Q}$, then multiply the equation by u^6 and group terms as follows:

$$(u^3 y)^2 = (u^2 x)^3 + (u^4 A)(u^2 x) + u^6 B .$$

Let $X = u^2 x$, $Y = u^3 y$ and fix u to be an integer such that $A' = u^4 A$ and $B' = u^6 B$ are integers. Then $x = \frac{X}{u^2}$, $y = \frac{Y}{u^3}$ is a birational map transforming E to

$$E' : Y^2 = X^3 + A'X + B' ,$$

with A', $B' \in \mathbf{Z}$.

Since our birational map is linear, straight lines map to straight lines and three points are collinear on E if and only if the corresponding points on E' are collinear. Hence, in this case, the isomorphism of $E(\mathbf{Q})_{\text{tors}}$ and $E'(\mathbf{Q})_{\text{tors}}$ (the points of finite order) is immediate from the geometric definition of adding points..

3) Although the theorem gives necessary but not sufficient conditions for (x, y) to be of finite order, it immediately yields an **algorithm** for finding all rational points of finite order on E. Such a point must be an integer point with either the y coordinate equal to zero or to a divisor of D. Substituting one of these y values in $x^3+Ax+(B-y^2) = 0$ yields a polynomial equation with integer coefficients and leading coefficient 1. Since any integer root must divide $B - y^2$ we can find all integers x corresponding to this y in a finite number of steps. If O denotes the identity of the group $E(\mathbf{Q})$, then, by the theorem, the set

$$S = \{(x, y) | (x, y) \in E(\mathbf{Z}) \ , \ y = 0 \text{ or } y|D\} \cup \{O\}$$

contains all non-zero points of finite order. Since the order of a point of finite order is *bounded* by the size of S, we can use our formulas for adding points (sec. 19.6) to test which points of S have finite order.

4) Our proof closely follows that given by Tate ([Tat 2] and [Sil - Tat]), but with a simpler form of the equation of the curve.

Example: $E : y^2 = x^3 + 1$, $D = -27$. $y = 0$ yields the point $(-1, 0)$. If $y^2| - 27$, then either $y = \pm 1$, which yields the points $(0, \pm 1)$, or $y = \pm 3$, which yields the points $(2, \pm 3)$. Adding the point at infinity and using the formulas for adding points, it is easy to check that $E(\mathbf{Q})_{\text{tors}}$ is a cyclic group of order 6 with (2, 3) as a generator.

3. The Easy Part of the Theorem

Since the proof of (ii) is easy, but (i) will take work, we first prove (ii) assuming (i).

Using general theory one can prove that D is always in the ideal of $\mathbf{Z}[t]$ generated by $f(t) = t^2 + At + B$ and $f'(t)$, but now we prove a stronger result using straightforward computation.

Lemma: *For some* $p(t), q(t) \in \mathbf{Z}[t]$,

$$D = p(t)f(t) + q(t)f'(t)^2$$

is a polynomial identity.

Proof: Recall that $D = -4A^3 - 27B^2$. Let $q(t) = 3t^2 + 4A$ and use the division algorithm to show $f(t) = t^3 + At + B$ divides $f'(t)^2 q(t) - D = (3t^2 + A)^2(3t^2 + 4A) + (4A^3 + 27B^2)$.

Proof of (ii): Let $P = (x, y) \in E(\mathbf{Q})_{\text{tors}}$; by (i), x and y are integers. Recall that $y = 0$ if and only if P has order two (sec. 19.6); so assume $2P \neq O$.

$y^2 = f(x)$ implies $y^2 | f(x)$. If we can show that $y^2 | f'(x)^2$, then, by the lemma $y^2 | D$. But 2P = (X, Y) has finite order and thus, by (i), Y is an integer. By the duplication formula for points (sec. 19.6),

$$2x + X = \lambda^2 , \quad \lambda = \frac{f'(x)}{2y} .$$

Since $x, X \in \mathbf{Z}$, we have $\lambda^2 \in \mathbf{Z}$. But $\lambda \in \mathbf{Q}$; therefore $\lambda \in \mathbf{Z}$, $y | f'(x)$ and $y^2 | f'(x)^2$.

4. The Hard Part of the Theorem

Now to the proof of (i), namely, if (x, y) is a rational point of finite order, then x and y are integers.

The rational points with $y = 0$ (points of order two) are easy to treat. In this case $x = \frac{a}{b}$ satisfies $x^3 + Ax + B = 0$ and b must divide the coefficient of x^3; so $b = \pm 1$ and x is an integer.

In general, to show that a rational point (x, y) of finite order on E has integer coordinates, we will prove that when x and y are written in lowest terms no prime divides the denominator of x or y. To do this, we study those rational points whose denominators are divisible by some prime p and show that this set has no points of finite order. Our techniques are just those of elementary p-adic number theory (chap. 23) and the Lutz - Nagell theorem can be generalized in that context. However our presentation will be self-contained and can be regarded as one motivation for introducing p-adic numbers..

Let $\frac{s}{t} \in \mathbf{Q}$, where $0 \neq \frac{s}{t} = \left(\frac{m}{n}\right) p^{v}$, $(m, p) = (n, p) = 1$ and $m > 0$. Then v is the **order of $\frac{s}{t}$ at p**, denoted by $\mathbf{ord_p}\left(\frac{s}{t}\right) = \mathbf{v}$; v is also called the p-value or **p-adic valuation** of $\frac{s}{t}$ at p. The following properties are immediate from the definition: if $(s, t) = 1$, then

1) $p|$ denominator of $\frac{s}{t} \iff \operatorname{ord}_p\left(\frac{s}{t}\right) < 0$,
2) $p|$ numerator of $\frac{s}{t} \iff \operatorname{ord}_p\left(\frac{s}{t}\right) > 0$,
3) $p \nmid$ numerator or denominator $\iff \operatorname{ord}_p\left(\frac{s}{t}\right) = 0$.

Now we characterize rational points on E in terms of their divisibility by p.

Lemma: *Let $(x, y) \in E(\mathbf{Q})$ and assume that p divides the denominator of x or y. Then p divides both denominators, i.e.,*

$$x = \frac{m}{np^r}, \quad y = \frac{d}{ep^s}, \tag{1}$$

where $r, s > 0$, $(m, p) = (n, p) = (d, p) = (e, p) = 1$, and, more specifically, there exists a $q > 0$ such that

$$\operatorname{ord}_p(x) = -2q, \quad \operatorname{ord}_p(y) = -3q,$$

i.e., $r = 2q$, $s = 3q$.

Proof: We write x and y in the form given in equation (1) with $(m, p) = (n, p) = (d, p) = (e, p) = 1$ and assume that p divides the denominator of x, i.e., $r > 0$. Substituting (1) in the equation for E, we have

$$\frac{d^2}{e^2 p^{2s}} = \frac{m^3 + Amn^2 p^{2r} + Bn^3 p^{3r}}{n^3 p^{3r}}. \tag{2}$$

Since p does not divide d or e, we have $\operatorname{ord}_p\left(\frac{d^2}{e^2 p^{2s}}\right) = -2s$. Since $r > 0$ and $p \nmid m$, we see that $p \nmid (m^3 + Amn^2 p^{2r} + Bn^3 p^{3r})$; so

$$\operatorname{ord}_p \frac{m^3 + Amn^2 p^{2r} + Bn^3 p^{3r}}{n^3 p^{3r}} = -3r.$$

By (2), the two orders are equal, i.e., $2s = 3r$. Therefore $s > 0$, p divides the denominator of y, and, since $3|s$ and $2|r$, there exists a q such that $r = 2q$ and $s = 3q$.

If we assume that p divides the denominator of y $(s > 0)$ then, by the same reasoning, we arrive at the same result. This proves the lemma.

We continue the study of divisibility of rational points by introducing

$$E(p^r) = \{(x, y) \in E(\mathbf{Q}) | ord_p(x) \leq -2r, ord_p(y) \leq -3r) \cup \{(0, 1, 0)\},$$

recalling that $(0, 1, 0)$ is the identity element of $E(\mathbf{Q})$. Namely, we are considering those points where p^{2r} divides the denominator of x and p^{3r} divides the denominator of y.

Clearly

$$E(\mathbf{Q}) \supseteq E(p) \supseteq E(p^2) \supseteq \cdots$$

Proposition: *For any prime p, there are no points of finite order in $E(p)$.*

It follows, from this proposition and the previous lemma, that the rational points of finite order have no primes in the denominators of their coordinates; thus they are integer points. Therefore, as soon as we prove this proposition, the proof of (i) and of the whole theorem will be done.

Proof of the proposition: This is a long and involved proof and will occupy the rest of this section.

We will introduce a birational map φ of E to a curve E', which maps the point at infinity to the origin. By analyzing the arithmetic of rational points on E', we will show that no point of finite order on $E'(\mathbf{Q})$ is in $\varphi(E(p))$. This will imply that no point of finite order in $E(\mathbf{Q})$ is in $E(p)$, for any prime p, and we will be done.

In affine coordinates, the map φ of the (x, y) plane to the (t, s) plane is defined by

$$t = \frac{x}{y}, \quad s = \frac{1}{y}.$$

The inverse map is given by

$$x = \frac{t}{s}, \quad y = \frac{1}{s}.$$

φ maps $E : y^2 = x^3 + Ax + B$ to

$$\mathbf{E}' : s = t^3 + Ats^2 + Bs^3 ,$$

Since the images of the point at infinity and those points where $y = 0$ (points of order two) are not easy to see in affine coordinates, we lift the map to the projective plane, where coordinates will be denoted by (X, Y, Z) and (T, S, U) respectively. Then we have

$$\frac{T}{U} = t = \frac{x}{y} = \frac{X/Z}{Y/Z} = \frac{X}{Y},$$
$$\frac{S}{U} = s = \frac{1}{y} = \frac{1}{Y/Z} = \frac{Z}{Y};$$

so

$$S = Z, T = X, U = Y$$

is the corresponding map of projective planes. This map is bijective and the point at infinity $(X, Y, Z) = (0, 1, 0)$ maps to $(S, T, U) = (0, 0, 1)$, which corresponds to the origin in the affine t-s plane.

The only points in the x-y plane which map to the line at infinity in the t-s plane are those for which $y = 0$ (points of order two). Since we have already shown that all such rational points are integer points, nothing is lost by now restricting our study to the curve E' in the affine t-s plane.

φ maps lines to lines since if $y = ax + b$, then, dividing by b times y, we have $\frac{1}{b} = \frac{a}{b}\frac{x}{y} + \frac{1}{y} = \frac{a}{b}t + s$ or $s = -\frac{a}{b}t + \frac{1}{b}$. The line at infinity meets the point at infinity $(0, 1, 0)$ on E with multiplicity three (inflection point). The image of the line is the t-axis which is tangent to $\pm E$ at the inflection point $(0, 0)$, and this latter point is the image of $(0, 1, 0)$ in the t-s plane (fig. 1).

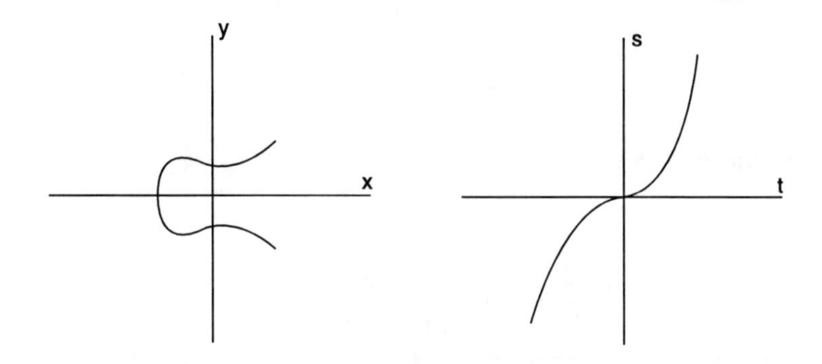

Figure 1

Since lines map to lines, a tangent-secant configuration for adding points on E maps to the configuration for adding the images on E'; thus φ (viewed as a map of projective planes) is an isomorphism between $E(\mathbf{Q})$ and $E'(\mathbf{Q})$. $\mathbf{O} = (0, 0)$ is the identity element of $E(\mathbf{Q})$. To derive explicit formulas for adding points on E' in the t-s plane,we use the tangent-secant definition of addition (sec. 19.3) and proceed as in section 19.6.

Let $P_1 = (t_1, s_1)$ and $P_2 = (t_2, s_2)$ be points on $E' : s = t^3 + Ats^2 + Bs^3$ and let $P_1P_2 = (t_3, s_3)$ be the third point of intersection of P_1P_2 with E' (fig. 2). Then $P_3 = P_1 + P_2 = (P_1 * P_2) * O$, the third point of intersection of the line through $P_1 * P_2$ and O with E', and $P_1 + P_2 = (-t_3, -s_3)$, since E' is symmetric with respect to the origin. (Note that $P_1 * P_2 = (t_3, s_3) = -P_3$.)

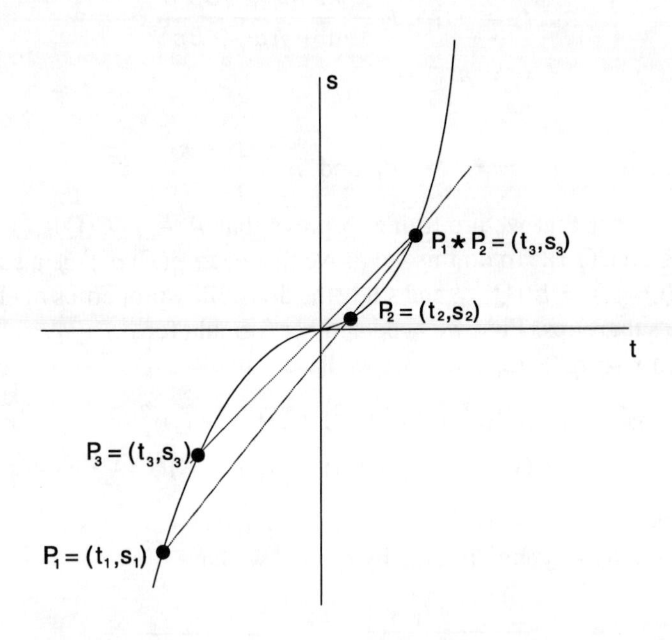

Figure 2

Let $s = \rho t + \sigma$ be the equation of the line through P_1 and P_2. Substituting this into the equation for E', we have

$$\rho t + \sigma = t^3 + At(\rho t + \sigma)^2 + B(\rho t + \sigma)^3 ;$$

so

$$t^3(1 + A\rho^2 + B\rho^3) + t^2(2A\rho\sigma + 3B\rho^2\sigma) + \cdots = 0$$

or

$$t^3 + t^2 \frac{2A\rho\sigma + 3B\rho^2\sigma}{1 + A\rho^2 + B\rho^3} + \cdots = 0 .$$

The sum of the roots

$$\begin{aligned} t_1 + t_2 + t_3 &= -(\text{coefficient of } x^2) \\ &= -\frac{2A\rho\sigma + 3B\rho^2\sigma}{1 + A\rho^2 + B\rho^3} ; \end{aligned}$$

so we have

$$\begin{aligned} t_3 &= -t_1 - t_2 - \frac{2A\rho\sigma + 3B\rho^2\sigma}{1 + A\rho^2 + B\rho^3} , \\ s_3 &= \rho t_3 + \sigma , \end{aligned} \tag{3}$$

where

$$\sigma = s_1 - \rho t_1 \text{ and } \rho = \frac{s_1 - s_2}{t_1 - t_2} .$$

Remember that we are trying to prove that $P \in \varphi(E(\mathbf{Q})_{\text{tors}})$ implies that $P \in \varphi(E(p))$. To do this we prove that each $\varphi(E(p^r))$ is a subgroup of $\varphi(E(\mathbf{Q})_{\text{tors}}) = E'(\mathbf{Q})_{\text{tors}}$ and study the divisibility properties of elements of these subgroups. First we need another formula for ρ.

From $s_i = t_i^3 + At_i s_i^2 + Bs_i^3$, we have

$$\begin{aligned} s_1 - s_2 = (t_1 - t_2)(t_1^2 + t_1 t_2 + t_2^2) + A\big((t_1 - t_2)s_2^2 \\ + t_1(s_1 - s_2)(s_1 + s_2)\big) + B(s_1 - s_2)(s_1^2 + s_1 s_2 + s_2) , \end{aligned}$$

where we used $t_1 s_1^2 - t_2 s_2^2 = (t_1 - t_2)s_2^2 + t_1(s_1^2 - s_2^2)$ to simplify. Now solving for $s_1 - s_2$ and dividing by $t_1 - t_2$, we have

$$\rho = \frac{t_1^2 + t_1 t_2 + t_2^2 + As_2^2}{1 - B(s_1^2 + s_1 s_2 + s_2) - At_1(s_1 + s_2)} . \tag{4}$$

The one in the denominator for ρ makes is easy to compute $\text{ord}_p(\rho)$. To find ρ for $P_1 = P_2$, just let $t_1 = t_2$ and $s_1 = s_2$ in (4). For the remainder of the proof, we shall only deal with the case where $P_1 \neq P_2$; *the case* $P_1 = P_2$ *is left as an exercise.*

To prove that $\varphi(E(p^r))$ is a subgroup of $E'(\mathbf{Q})_{\text{tors}}$, we first need some divisibility properties of rational points. It is convenient to introduce the sets

$$\mathbf{R_p} = \{0\} \cup \{x | \text{ord}_p(x) \geq 0\} .$$

Lemma:

1) R_p *is a ring.*
2) $\frac{m}{n} \in U(R_p)$, *the units of* $R_p \iff \mathrm{ord}_p\left(\frac{m}{n}\right) = 0$,
3) p *is the only prime in* R_p,
4) R_p *is a unique factorization domain.*

Proof: 1) - straightforward exercise.

2) If $\frac{m}{n} \neq 0$, $(m, n) = 1$, then $\frac{m}{n}$ is a unit if and only if $\frac{1}{\left(\frac{m}{n}\right)} = \frac{n}{m} \in R_p$, which is true if and only if $\mathrm{ord}_p\left(\frac{m}{n}\right) = 0$.

3) and 4) follow from the fact that if $\frac{m}{n}$ is not a unit, then $\frac{m}{n} = \left(\frac{u}{v}\right) p^r$, with $(u, p) = (v, p) = 1$ and $r > 0$.

Since we will be dealing with a fixed prime p, we let $R = R_p$ for the rest of the proof. Then $p^r R = \{p^r x | x \in R\} = \{x \in R | \mathrm{ord}_p(x) > r\}$ is the set of all rationals in lowest terms whose numerator is divisible by p^r.

Now we examine the divisibility properties of rational points (t, s) on E'. Let $\mathbf{E'_r} = \varphi(E(p^r))$. If $(t, s) \in E'_r$, where $(t, s) = \varphi((x, y))$, $(x, y) \in E(p^r)$, then $x = \frac{m}{np^{2r}}$, $y = \frac{d}{ep^{3r}}$, and

$$t = \frac{x}{y} = \left(\frac{me}{nd}\right) p^r , \quad s = \frac{1}{y} = \left(\frac{e}{d}\right) p^{3r} .$$

Therefore

$$(t, s) \in E'_r \iff t \in p^r R , \quad s \in p^{3r} R , \qquad (5)$$
$$(\mathrm{ord}_p(t) \geq r, \ \mathrm{ord}_p(s) \geq 3r) .$$

Lemma: E'_r *is a subgroup of* $E'(\mathbf{Q})$.

Proof: We show that if $P_1, P_2 \in E'_r$, $P_i = (t_i, s_i)$ then $P_1 \pm P_2 = (\pm t_3, \pm s_3) \in E'_r$. By equation (5),

$$t_1, t_2 \in p^r R , \quad s_1, s_2 \in p^{3r} R .$$

Hence the numerator in equation (4) for ρ belongs to $p^{2r} R$. The denominator is a unit since it equals one plus terms divisible by p; thus $\rho \in p^{2r} R$. From these results we see that $\sigma = s_1 - \rho t_1 \in p^{3r} R$. Recall that

$$t_1 + t_2 + t_3 = -\frac{2A\rho\sigma + 3B\rho^2\sigma}{1 + A\rho^2 + B\rho^3} .$$

The denominator is a unit and the numerator is in $p^{3r}R$. Hence

$$t_1 + t_2 + t_3 \in p^{3r}R \tag{6}$$

and, since $t_1, t_2 \in p^r R$, we have $\pm t_3 \in p^r R$. From $s_3 = \rho t + \sigma$, we see that $\pm s_3 \in p^{3r}R$; therefore $P_1 \pm P_2 = (\pm t_3, \pm s_3) \in E'_r$ and we are done.

We are now on the last leg of our proof of the proposition and thus of (i). In order to finish the proof, we exploit equation (6). If $P = (t, s)$, let $t(P) = t$; so (6) can be restated as

$$t(P_1) + t(P_2) \equiv t(P_3) (\text{mod } p^{3r}R) , \tag{6'}$$

where $a \equiv b (\text{mod } p^u R)$ means $a - b \in p^u R$ (equivalently, $\text{ord}_p(a - b) \geq u$).

Now we can prove that $P \in E'(\mathbf{Q})_{\text{tors}}$ implies $P \notin E'_1$. Suppose that $P \in E'_1$ and P has finite order m. Then, since arbitrarily large powers of p cannot divide a fixed integer, there exists an $r > 0$ such that $P \in E'_r$, $P \notin E'_{r+1}$ (so $t(P) \notin p^{r+1}R$). There are two cases:

1) $p \nmid m$ — By induction on (6′), it is easy to see that

$$t(mP) \equiv mt(P) (\text{mod } p^{3r}R) .$$

But $mP = O$, $(m, p) = 1$ and $t(O) = 0$; hence $t(P) \equiv 0 (\text{mod } p^{3r}R)$, contradicting our assumption that $t(P) \not\equiv 0 (\text{mod } p^{r+1}R)$.

2) $p | m$ — The proof is similar to the proof of 1) after removing the common factors. Let $m = pn$ and $P' = nP$; then P' is in E'_1, since E_1 is a group and there exists an $r > 0$ such that $P' \in E'_r$, $P' \notin E'_{r+1}$ (so $t(P') \notin p^{r+1}R$). But $pP' = pnP = mP = O$, so we have

$$0 = t(pP') \equiv pt(P') (\text{mod } p^{3r})$$

and

$$t(P') \equiv 0 (\text{mod } p^{3r-1}R) .$$

But $3r - 1 \geq r + 1$, which contradicts $P' \notin E'_{r+1}$. Therefore, there are no points of finite order in E'_1.

Suppose that (x, y) is a point of finite order in $E(p)$. Then x and y are of the form

$$x = \frac{m}{np^{2i}} , \quad y = \frac{u}{vp^{3i}} ,$$

for some $i > 0$. Hence

$$t = \frac{x}{y} = \frac{mvp^i}{nu}, \quad s = \frac{1}{y} = \frac{vp^{3i}}{u},$$

i.e., $\varphi((x, y)) = (t, s)$ is a point of finite order in E'_1. This contradicts our result; hence there are no points of finite order in E(p), for any prime p, and therefore all points of finite order have integer coordinates. Our proof of the Lutz–Nagell theorem is done.

5. Mordell's Theorem; An Outline of the Proof

Without taking a break from the long proof of the Lutz–Nagell theorem, we move on to the proof of a special case of Mordell's theorem. This proof is also rather long, but its overall structure is easier to grasp. We work with elliptic curves in the same form as for the Lutz–Nagell theorem.

Mordell's Theorem: *Let*

$$E : y^2 = x^3 + Ax + B \tag{1}$$

be an elliptic curve, with $A, B \in \mathbf{Z}$, such that the roots of $f(x) = x^3 + Ax + B = 0$ are rational. Then $E(\mathbf{Q})$, the group of rational points on E, is a finitely generated abelian group.

Our proof, based on Weil's second proof of the theorem [Wei 11], follows closely the exposition by Chahal [Cha], with elements taken from the expositions by Silverman and Tate [Sil - Tat], Chowla [Cho], Mordell [Mor 2] and Ireland and Rosen [Ire - Ros]. Mordell, Rose [Ros, H] and Ireland and Rosen present the full theorem with no conditions on the roots of f(x). See Cassels [Cas] for an insightful reconstruction of Mordell's discovery of his proof of the theorem.

Remarks: (1) Since E is, by definition, non-singular, the roots of $f(x)$ are distinct.

(2) In fact, the roots of $f(x)$ must be integers. For suppose that $\frac{s}{t}$, $t > 0$, is a root in lowest terms, $(s, t) = 1$. Then $\left(\frac{s}{t}\right)^3 + A\left(\frac{s}{t}\right) + B = 0$ or $s^3 + Ast^2 + Bt^3 = 0$: hence $t | s^3$. If $t > 1$, then $(s, t) > 1$, a contradiction.

(3) Requiring the roots of $f(x)$ to be rational allows us to work over $\mathbf{Q}$. If we drop the assumption that the roots of $f(x)$ are rational, then it is not hard to generalize our proof to a proof of the full Mordell theorem by working over an algebraic number field containing the roots.

The proof of Mordell's theorem is lengthy so we first outline the main steps and show how they imply the theorem.

We will define the height $H(P)$ of a rational point (rational coordinates) on E, a positive integer which in some sense measures the "complexity" of P and which satisfies the following properties:

Lemma A: *For any real number $K > 0$, the set $\{P \in E(\mathbf{Q}) | H(P) < K\}$ is finite.*

Lemma B: *Fix a point $Q = (x_0, y_0) \in E(\mathbf{Q})$. Then there is a constant c depending only on Q and E (i. e., on x_0, y_0, A and B) such that*

$$H(P + Q) \leq c(H(P))^2 ,$$

for all $P \in E(\mathbf{Q})$.

Lemma C: *There is a constant d depending only on E (i.e., on A and B) such that*

$$H(P) \leq d(H(2P))^{\frac{1}{4}} ,$$

for all $P \in E(\mathbf{Q})$.

Lastly we need the

Weak Mordell–Weil Theorem: *The quotient group $E(\mathbf{Q})/2E(\mathbf{Q})$ is finite.*

As we now show, these four results imply Mordell's theorem. Let $Q_1, Q_2, \ldots, Q_n$ be a set of representatives of the cosets of $2E(\mathbf{Q})$ in $E(\mathbf{Q})$. Then, since any P in $E(\mathbf{Q})$ must be in some coset, there is an $i(1)$, $1 \leq i(1) \leq n$, depending on P, such that $P \in Q_{i(1)} + 2E(\mathbf{Q})$ (we use additive notation for the group $E(\mathbf{Q})$). Hence

$$P = Q_{i(1)} + 2P_1 ,$$

for some $P_1 \in E(\mathbf{Q})$. Similarly, there is an $i(2)$, $1 \leq i(2) \leq n$, such that $P_1 \in Q_{i(1)} + 2E(\mathbf{Q})$ or $P_1 = Q_{i(1)} + 2P_2$, for some $P_2 \in E(\mathbf{Q})$. Repeating this process m times we have the series of equations

$$\begin{aligned} P &= Q_{i(1)} + 2P_1 \\ P_1 &= Q_{i(2)} + 2P_2 \\ &\vdots \\ P_{m-1} &= Q_{i(m)} + 2P_m , \end{aligned}$$

where the Q's are all coset representatives, not necessarily all distinct.

Substituting the second equation in the first we get $P = Q_{i(1)} + 2Q_{i(2)} + 2^2 P_2$. Now substituting the third equation for P_2 in this expression, the fourth equation for P_3 in the resulting equation and so on, we have

$$P = Q_{i(1)} + 2Q_{i(2)} + 2^2 Q_{i(3)} + \cdots + 2^{m-1} Q_{i(m)} + 2^m P_m \,. \tag{2}$$

To prove that $E(\mathbf{Q})$ is finitely generated, we will define *a finite* set S such that for every $P \in E(\mathbf{Q})$, there is an m such that $P_m \in S$. Then, by equation (2), $\{Q_1, Q_2, \ldots, Q_n\} \cup S$ generates $E(\mathbf{Q})$.

Let $c_i = c(-Q_i)$ be the constant of lemma B for $Q = -Q_i$, d the constant of lemma C, $K = \max(dc_1, \ldots, dc_n)$, and $S = \{P \in E(\mathbf{Q}) | H(P) \leq K^2\}$. By lemma A, S is finite. Suppose our claim is not true, i.e., there exists a $P \in E(\mathbf{Q})$ such that $H(P_m) > K^2$, for all $m \geq 1$. Then, since

$$P_{j-1} + (-Q_{i(1)}) = 2P_j \,, \tag{3}$$

$1 \leq j \leq m$, where $P = P_0$, we have

$$\begin{aligned} H(P_j) &\leq dH(2P_j)^{\frac{1}{4}} && \text{by lemma C} \\ &= dH(P_{j-1} + (-Q_{i(j)}))^{\frac{1}{4}} && \text{by equation (3)} \\ &\leq d[c_{i(j)} H(P_{j-1})^2]^{\frac{1}{4}} && \text{by lemma B} \\ &\leq KH(P_{j-1})^{\frac{1}{2}} && \text{by the definition of } K. \end{aligned}$$

We assumed that $H(P_{j-1}) > K^2$, i.e., $K < H(P_{j-1})^{\frac{1}{2}}$; so

$$H(P_j) \leq KH(P_{j-1})^{\frac{1}{2}} < H(P_{j-1})^{\frac{1}{2}} H(P_{j-1})^{\frac{1}{2}} = H(P_{j-1}) \,,$$

for $1 \leq j \leq m$. Since the height is a positive integer, it follows that $H(P_j) \leq H(P_{j-1}) - 1$ and thus

$$H(P_m) \leq H(P_{m-1}) - 1 \leq H(P_{m-2}) - 2 \leq \cdots \leq H(P_1) - (m-1) \,.$$

If $m > H(P_1) + 1$, then $H(P_m) < 0$, contradicting the positiveness of the height function. Thus, for some m, $H(P_m) \leq K^2$, i.e., $P_m \in S$, and $\{Q_1, Q_2, \ldots, Q_n\} \cup S$ generates $E(\mathbf{Q})$. Weil saw this as an infinite descent argument since we are using the fact that there cannot be an infinite decreasing sequence of positive integers.

6. Some Preliminary Results

Before proving the lemmas and the weak Mordell–Weil theorem, we need some preliminary results.

First, as in our proof of the Lutz–Nagell theorem, we will show that any rational point on E can be written in a special way.

Lemma 1: *If* $P = (x, y) \in E(\mathbf{Q})$*, then there exist* $X, Y, Z \in \mathbf{Z}$*, with* $(X, Z) = 1$*,* $(Y, Z) = 1$ *and* $Z > 0$ *such that*

$$x = \frac{X}{Z^2}, \quad y = \frac{Y}{Z^3}. \tag{1}$$

Proof: Let $x = \frac{X}{M}$, $y = \frac{Y}{N}$ be in lowest terms, with $M, N > 0$. Substituting this into the equation $y^2 = x^3 + Ax + B$ for the curve and multiplying by M^3N^2 yields

$$M^3Y^2 = N^2X^3 + AN^2M^2X + BN^2M^3. \tag{2}$$

We will show that $M^3 = N^2$. For then, setting $Z = \frac{N}{M}$, we will have $Z^2 = \frac{N^2}{M^2} = \frac{M^3}{M^2} = M$ and $Z^3 = \frac{N^3}{M^3} = \frac{N^3}{N^2} = N$; hence $x = \frac{X}{Z^2}$ and $y = \frac{Y}{Z^3}$.

Since N^2 divides each term on the right hand side of (2), it divides M^3Y^2. But $(N, Y) = 1$; so $N^2|M^3$. Again from (2), we have M^2 divides M^3Y^2, AN^2M^2X, and BY^2M^3; so $M^2|N^2X^3$ and, since $(X, M) = 1$, $M^2|N^2$. This last result implies that $M^3|AN^2M^2X$ and since M^3 divides M^3Y^2 and BY^2M^3, we have $M^3|N^2X^3$. Again, since $(X, M) = 1$, $M^3|N^2$, and we are done.

Now we derive several formulas for the addition of points.

Given a point R, we let $\mathbf{x}(R)$ denote its x coordinate. Now for our curve E, given by equation (5.1), we need to find formulas for $x(P + Q)$, where P and Q are any points on the curve. Proceeding as in sec. 19.6, let $P = (x_1, y_1)$, $Q = (x_2, y_2)$ and $P + Q = (x_3, -y_3)$, where $P * Q = (x_3, y_3)$ is the third point of intersection of the line through P and Q with E (see figure 7 of sec. 19.6); so $x_3 = x(P + Q) = x(P * Q)$.

First assume that $P \neq \pm Q$ (if $P = -Q$, then $P + (-P)$ is the point at infinity on E which is the identity element of $E(\mathbf{Q})$). The equation of the line through P and Q is

$$y = y_1 + \left(\frac{y_2 - y_1}{x_2 - x_1}\right)(x - x_1). \tag{3}$$

Substituting this in equation (5.1) of E yields a cubic equation in x,

$$\left[y_1 + \frac{y_2 - y_1}{x_2 - x_1}(x - x_1)\right]^2 = x^3 + Ax + B\,, \tag{4}$$

with roots x_1, x_2 and x_3. Since the sum of the roots $= -$(coefficient of x_2), we obtain

$$x_3 = x(P+Q) = \left(\frac{y_2 - y_1}{x_2 - x_1}\right)^2 - (x_1 + x_2)$$

and, since $y_1^2 = x_1^3 + Ax_1 + B$, $y_1^2 = x_2^3 + Ax_2 + B$, this can be written as

$$x(P+Q) = \frac{(x_1 + x_2)(x_1 x_2 + A) + 2B - 2y_1 y_2}{(x_1 - x_2)^2}\,. \tag{5}$$

Later in the proof we will need yet another formula for adding points. Let r_1, r_2, r_3 be the roots of $f(x) = x^3 + Ax + B = 0$. For any one of the roots r, we will find a formula for $(x_1 - r)(x_2 - r)(x_3 - r)$, where x_1, x_2, x_3 again represent the roots of equation (4). To find $\prod(x_i - r)$ we need to "shift the roots by r ". Setting $X = x - r$, $X_1 = x_1 - r$, and substituting into equation (4) yields

$$\left[y_1 + \frac{y_2 - y_1}{x_2 - x_1}(X - X_1)\right]^2 = X^3 + CX^2 + DX\,, \tag{6}$$

where the right hand side has no constant term because $f(r) = 0$, and thus $X = 0$ is a root of $f(X + r)$. Since x_1, x_2, x_3 are the roots of (4), $x_1 - r$, $x_2 - r$, $x_3 - r$ are the roots of (6) and the product of the roots is the constant term on the left hand side of (6), viz.,

$$(x_1 - r)(x_2 - r)(x_3 - r) = \left[y_1 - \frac{y_2 - y_1}{x_2 - x_1}X_1\right]^2\,. \tag{7}$$

Substituting $x(P+Q) = x_3$, $X_1 = x_1 - r$ in the latter equation yields our final formula

$$X(P+Q) - r = \frac{1}{(x_1 - r)(x_2 - r)}\left[\frac{y_1(x_2 - r) - y_2(x_1 - r)}{x_2 - x_1}\right]^2\,. \tag{8}$$

There is an analogous formula for $x(2P) - r$. If we replace the slope $\frac{y_2 - y_1}{x_2 - x_1}$ of the line through P and Q by the slope $\frac{3x_1^2 + A}{2y_1}$ of the tangent line to the curve at P, then (7) becomes

$$(x(2P) - r)(x_1 - r)^2 = \left[y_1 - \frac{(3x_1^2 + A)(x_1 - r)}{2y_1}\right]^2 \tag{9}$$

$$= \left[\frac{2y_1^2 - (3x_1^2 + A)(x_1 - r)}{2y_1}\right]^2 .$$

Since $f(r) = r^3 + Ar + B = 0$, dividing $x^3 + Ax + B$ by $x - r$ yields

$$x^3 + Ax + B = (x - r)(x^2 + rx + A + r^2) .$$

Thus the numerator on the right hand side of (9) is the square of

$$2y_1^2 - \left(3x_1^2 + A\right)(x_1 - r) = (x_1 - r)\left(-x_1^2 + 2rx_1 + A + 2r^2\right)$$

and for each $r = r_1, r_2, r_3$, we have

$$x(2P) - r = \left[\frac{-x_1^2 + 2rx_1 + A + 2r^2}{2y_1}\right]^2 . \tag{10}$$

7. The Height Function

Now we move on to the notion of the height of a point and prove lemmas A, B and C of section 5.

First we define the height of a rational number. The **height of a rational number** $x = \frac{m}{n} \neq 0$, with $(m, n) = 1$, is given by

$$\mathbf{H(x)} = H\left(\frac{m}{n}\right) = \max\{|m|, |n|\} ,$$

with $H(0) = 1$. Thus the height is always a positive integer. Note that for any positive real number K, $\{x \in \mathbf{Q} | H(x) \leq K\}$ is finite. Now if $P = (x, y)$ is a rational point, we define the **height of** P by

$$\mathbf{H(P)} = H(x) ,$$

where, for $P = O$, the point at infinity, we let $H(P) = 1$.

Lemma A of section 5, which says that $\{P \in E(\mathbf{Q}) | H(P) \leq K\}$ is finite, now follows immediately, since for every x coordinate of a rational point on $y^2 = x^3 + Ax + B$ there are at most two y values.

To prove lemma B, we first need

Lemma 1: *If $P = (x, y) \in E(\mathbf{Q})$, then there is a constant K, depending only on A and B, such that*

$$|y| \leq K(H(P))^{\frac{3}{2}}.$$

Proof: By lemma 6.1, we can write $P = (x, y) = \left(\frac{X}{Z^2}, \frac{Y}{Z^3}\right)$, with $(X, Z) = (Y, Z) = 1$ and $Z \geq 1$. Since P is on the curve, we have

$$Y^2 = X^3 + AXZ^4 + BZ^6.$$

By the definition of height, $|X|$ and Z^2 are both $\leq H(x)$; so, by the triangle inequality

$$\begin{aligned} Y^2 = |Y^2| &\leq |X^3| + |A|\,|X|\,|Z^4| + |B|\,|Z^6| \\ &\leq H(x)^3 + |A|H(x)H(x)^2 + |B|H(x)^3 \\ &= (1 + |A| + |B|)H(x)^3. \end{aligned}$$

Therefore

$$|Y| \leq (1 + |A| + |B|)^{\frac{1}{2}} H(x)^{\frac{3}{2}}.$$

Since $H(P) = H(x)$, we let $K = (1 + |A| + |B|)^{\frac{1}{2}}$ and we are done.

Now we prove lemma B.

Lemma B: *Fix a point $Q = (x_0, y_0) \in E(\mathbf{Q})$. Then there is a constant c depending only on Q and E (i. e., on x_0, y_0, A and B) such that*

$$H(P + Q) \leq c(H(P))^2,$$

for all $P \in E(\mathbf{Q})$.

Proof: The lemma is clearly true for $Q = O$, the point at infinity, since it is the identity of the group $E(\mathbf{Q})$. It also suffices to prove the lemma for all but a finite set $\{P_1, \ldots, P_n\}$ of rational points for we can always increase the constant to be larger than $\frac{H(P_i+Q)}{H(P_i)^2}$, for $i = 1, \ldots, n$. So from now on, we assume that $P \neq Q, -Q, O$.

By lemma 6.1, we know that P can be written in the form $P = \left(\frac{X}{Z^2}, \frac{Y}{Z^3}\right)$ and we let $P + Q = (x, y)$. By equation (6.5) we have

$$x = \frac{\left(\frac{X}{Z^2} + x_0\right)\left(\frac{Xx_0}{Z^2} + A\right) + 2B - \frac{2Yy_0}{Z^3}}{\left(x_0 - \frac{X}{Z^2}\right)^2}.$$

Multiplying the numerator and denominator on the right hand side by Z^4, we obtain

$$x = \frac{aYZ + bX^2 + cXZ^2 + dZ^4}{eX^2 + fXZ^2 + gZ^4} \tag{1}$$

where $a, b, \ldots, g$ are functions of x_0, y_0, A and B. Hence we have

$$H(x) \leq \max\{|aYZ + bX^2 + cXZ^2 + dZ^4|, |eX^2 + fXZ^2 + gZ^4|\},$$

since cancellation on the right hand side of (1) can only reduce the height.

Now we proceed as in the proof of lemma 1. By the definition of the height and by lemma 1, we have

$$|X| \leq H(P), \quad Z \leq H(P)^{\frac{1}{2}}, \quad \text{and} \quad |Y| \leq K\left(H(P)^{\frac{3}{2}}\right),$$

where K is a function of A and B. Applying the triangle inequality, we have

$$\begin{aligned} |aYZ + bX^2 + cXZ^2 + dZ^4| &\leq |a|K\,H(P)^{\frac{3}{2}}H(P)^{\frac{1}{2}} + |b|H(P)^2 \\ &\quad + |c|H(P)H(P) + |d|H(P)^2 \\ &= (|a|K + |b| + |c| + |d|)H(P)^2, \end{aligned}$$

and

$$\begin{aligned} |eX^2 + fXZ^2 + gZ^4| &\leq |e|H(P)^2 + |f|H(P)H(P) + |g|H(P)^2 \\ &= (|e| + |f| + |g|)H(P)^2. \end{aligned}$$

Therefore

$$H(P+Q) = H(x) \leq cH(P)^2,$$

for all $P \neq \pm Q, O$, where c = $\max\{|a|K + |b| + |c| + |d|, |e| + |f| + |g|\}$ depends only on x_0, y_0, A and B.

We now prove lemma C.

Lemma C: *There is a constant d depending only on E (i.e., on A and B) such that*

$$H(P) \leq d(H(2P))^{\frac{1}{4}},$$

for all $P \in E(\mathbf{Q})$.

Proof: Recall that r_1, r_2, and r_3 are the roots of $x^3 + Ax + B = 0$, and assume that $P \neq O$ or $(r_j, 0)$. We can always adjust our constant to take care of these four cases.

By lemma 1, we let $P = \left(\frac{X_1}{Z_1^2}, \frac{Y_1}{Z_1^3}\right)$ and $2P = \left(\frac{X}{Z^2}, \frac{Y}{Z^3}\right)$. Substituting these values in equation (6.9), multiplying the equation by $Z^2 Z_1^4$, and taking the square root of both sides yields, for $j = 1, 2, 3$,

$$(X - r_j Z^2)^{\frac{1}{2}} = \left(\frac{Z}{2Y_1 Z_1}\right)\left(-X_1^2 + 2r_j X_1 Z_1^2 + AZ_1^4 + 2r_j Z_1^4\right) . \quad (2)$$

Collecting terms on the right hand side according to powers of r_j and setting

$$\lambda_1 = \left(-X_1^2 + AZ_1^4\right)\frac{Z}{2Y_1 Z_1}, \quad \lambda_2 = \frac{X_1 Z_1 Z}{Y_1}, \quad \text{and } \lambda_3 = \frac{Z_1^4 Z}{Y_1 Z_1} \quad (3)$$

transforms (2) into a set of three linear equations in the λ_i:

$$\alpha_j = \lambda_1 + r_j \lambda_2 + r_j^2 \lambda_3 , \quad j = 1, 2, 3 , \quad (4)$$

where $\alpha_j = (X - r_j Z^2)^{\frac{1}{2}}$. By (2), α_j is rational, and $\alpha_j^2 = X - r_j Z^2$ is an integer since r_j is an integer; hence α_j is an integer. Thus the set of equations (4) has all integer coefficients.

The determinant D of this system is the well known *van der Monde determinant*

$$D = \begin{vmatrix} 1 & r_1 & r_1^2 \\ 1 & r_2 & r_2^2 \\ 1 & r_3 & r_3^2 \end{vmatrix} = \prod_{i>j}(r_i - r_j) \in \mathbf{Z} .$$

Since the roots are distinct, $D \neq 0$. By Cramer's rule, each $D\lambda_j$ is a linear combination of the α_j's, with integer coefficients; hence *$D\lambda_j$ is an integer.*

To estimate the height of P, we first use our results to get estimates for X_1^2 and Z_1^4. From (3), an immediate calculation gives

$$D(A\lambda_3 - 2\lambda_1) = X_1^2 \frac{DZ}{Y_1 Z_1} \quad (5)$$

and

$$D\lambda_3 = Z_1^4 \frac{DZ}{Y_1 Z_1} . \quad (6)$$

Since $D\lambda_j$ is an integer, the left hand sides of (5) and (6) are integers; so $X_1^2 \frac{DZ}{Y_1 Z_1}$ and $Z_1^4 \frac{DZ}{Y_1 Z_1}$ are integers. Therefore

$$(X_1, Z_1) = 1 \implies Z_1 | DZ \text{ and } (Y_1, Z_1) = 1 \implies Y_1 | DZ .$$

Thus, from $(Y_1, Z_1) = 1$, we have $\frac{DZ}{Y_1 Z_1} \in \mathbf{Z}$. Since the left hand sides of (5) and (6) are integers, we also have

$$X_1^2 | D(A\lambda_3 - 2\lambda_1) \text{ and } Z_1^4 | D\lambda_3 ,$$

or

$$X_1^2 \leq |D(A\lambda_3 - 2\lambda_1)| \text{ and } Z_1^4 \leq |D\lambda_3| . \tag{7}$$

The $D\lambda_i$ are integer linear combinations of the $\alpha_j = (X - r_j Z^2)^{\frac{1}{2}}$, with coefficients independent of P. Recalling that $\frac{x+y}{2} \leq \max\{x, y\}$, we have

$$\begin{aligned}(X - r_j Z^2)^{\frac{1}{2}} &\leq |X| + |r_j| Z^2 \leq 2 \max\{|X|, |r_j| Z^2\} \\ &\leq 2 \max\{|r_j||X|, |r_j| Z^2\} = 2|r_j| \max\{|X|, Z^2\} \\ &= 2|r_j| H(2P) .\end{aligned}$$

So

$$\alpha_j = (X - r_j Z^2)^{\frac{1}{2}} \leq C^{\frac{1}{2}} H(2P)^{\frac{1}{2}} ,$$

for all j, with $C = \max\{2|r_1|, 2|r_2|, 2|r_3|\}$, which is independent of P. Therefore, by (7) and the fact that the $D\lambda_i$ are linear combinations of the α_j, we have

$$X_1^2 \leq C_1 H(2P)^{\frac{1}{2}} \text{ and } Z_1^4 \leq C_2 H(2P)^{\frac{1}{2}} ,$$

for appropriate constants C_1 and C_2 independent of P. Hence

$$H(P)^2 = \max\{X_1^2, Z_1^4\} \leq C_3 H(2P)^{\frac{1}{2}}$$

for $C_3 = \max\{C_1, C_2\}$, and

$$H(P) \leq d H(2P)^{\frac{1}{4}}$$

for $d = \sqrt{C_3}$, and we are done.

8. The Weak Mordell–Weil Theorem

We have proved lemmas A, B, and C which, together with the Weak Mordell–Weil theorem, imply Mordell's theorem. Hence we have only one more step to go.

Weak Mordell–Weil Theorem: *The quotient group $E(\mathbf{Q})/2E(\mathbf{Q})$ is finite.*

Proof: Let $\mathbf{Q}^*$ denote the *multiplicative* group of non-zero rational numbers and $\mathbf{Q}^{*2}$ the subgroup of squares of element of $\mathbf{Q}^*$. *We use multiplicative notation for the group $\mathbf{Q}^*$ and additive notation for $E(\mathbf{Q})$.*

The idea of the proof is to construct a *homomorphism* ϕ from $E(\mathbf{Q})$ to a product of copies of the quotient group $\mathbf{Q}^*/\mathbf{Q}^{*2}$, whose *image* $\phi(E(\mathbf{Q}))$ is finite and whose *kernal* is $2E(\mathbf{Q})$. Then, by the fundamental theorem on group homomorphisms, $E(\mathbf{Q})/2E(\mathbf{Q})$ is isomorphic to $\phi(E(\mathbf{Q}))$ and thus is finite.

We construct ϕ in several steps. First we let $\beta : \mathbf{Q}^* \longrightarrow \mathbf{Q}^*/\mathbf{Q}^{*2}$ be the natural homomorphism that maps each $x \in \mathbf{Q}^*$ to its coset, i.e., $\beta(x) = x\mathbf{Q}^{*2}$. Then we define $\phi_{\mathbf{j}} : E(\mathbf{Q}) \longrightarrow \mathbf{Q}^*/\mathbf{Q}^{*2}$, $j = 1, 2, 3$, by

$$\begin{aligned}\phi_j(P) &= 1\,, \quad \text{if } P = O\,,\\ &= \beta[(x(P) - r_i)(x(P) - r_k)]\,, \quad i \neq j \neq k\,, \quad \text{if } P = (r_j, 0)\,,\\ &= \beta(x(P) - r_j)\,, \quad \text{otherwise.}\end{aligned}$$

Note that the points $(r_j, 0)$, $j = 1, 2, 3$ are the points of order two in $E(\mathbf{Q})$.

Finally we define $\phi : E(\mathbf{Q}) \longrightarrow \mathbf{Q}^*/\mathbf{Q}^{*2} \times \mathbf{Q}^*/\mathbf{Q}^{*2} \times \mathbf{Q}^*/\mathbf{Q}^{*2}$ by

$$\phi(\mathbf{P}) = (\phi_1(P), \phi_2(P), \phi_3(P))\,.$$

First we show that ϕ is a homomorphism. Assume that $P = (x_1, y_1)$ and $Q = (x_2, y_2)$ are in $E(\mathbf{Q})$ and not equal to O or $(r_j, 0)$. We must show that $\phi(P+Q) = \phi(P)\cdot\phi(Q)$. For any point $R = (x, y) \in E(\mathbf{Q})$, ϕ is, by definition, independent of y. Hence $\phi(-R) = \phi(x, -y) = \phi(R)$ (where $-R$ is the inverse of R in $E(\mathbf{Q})$). Moreover, $\phi(R)\phi(-R) = \phi(R)^2 = I$, the identity element of $\mathbf{Q}^*/\mathbf{Q}^{*2}$, since every non identity element of $\mathbf{Q}^*/\mathbf{Q}^{*2}$ has order 2. So we have

$$\phi(P+Q) = \phi(P)\phi(Q) \text{ if and only if } \phi(P+Q)\phi(P)\phi(Q) = I\,.$$

Coordinate-wise, this is equivalent to

$$\beta(x(P+Q) - r_i)\beta(x(P) - r_i)\beta(x(Q) - r_i) = I\,,$$

or

$$(x_3 - r_i)(x_1 - r_i)(x_2 - r_i) \in \mathbf{Q}^{*2} ,$$

which follows immediately from equation (6.8), since $\prod_j (x_j - r_i)$ equals a rational square.

The case where at least one of $P, Q = O$ is trivial and the case where they are both of the form $(r_i, 0)$, i.e., elements of order two in $E(\mathbf{Q})$, is left as an exercise.

Proposition: $\phi(E(\mathbf{Q}))$, *the image of* ϕ, *is finite.*

Proof: Let $P = (x, y) = \left(\frac{X}{Z^2}, \frac{Y}{Z^3}\right) \in E(\mathbf{Q})$. To show that the image of ϕ is finite, we will show that each $\phi_i(P) = (x - r_i)\mathbf{Q}^{*2}$ can take only a finite number of values, which are independent of P. We work with ϕ_1; the reasoning is identical for ϕ_2 and ϕ_3.

Substituting $x = \frac{X}{Z^2}$, $y = \frac{Y}{Z^3}$ into the equation for E, where $(X, Z) = (Y, Z) = 1$, yields

$$\frac{Y^2}{Z^6} = \left(\frac{X}{Z^2} - r_1\right)\left(\frac{X}{Z^2} - r_2\right)\left(\frac{X}{Z^2} - r_3\right)$$

$$Y^2 = (X - r_1 Z^2)\left(X^2 - XZ^2(r_2 + r_3) + r_2 r_3 Z^4\right) .$$

Let

$$\mathbf{S} = X - r_1 Z^2 , \quad \mathbf{T} = X^2 - XZ^2(r_2 + r_3) + r_2 r_3 Z^4 , \quad \mathbf{M} = \gcd(S, T);$$

then $S = MS_1$, $T = MT_1$, where $(S_1, T_1) = 1$. Therefore $Y^2 = M^2 S_1 T_1$ and $S_1 T_1 = \left(\frac{Y}{M}\right)^2$, where $\frac{Y}{M}$ is an integer. Since S_1 and T_1 are relatively prime and their product is a square, each of them is a square. So we have $S_1 = S_2^2$, $T_1 = T_2^2$ and $X - r_1 Z^2 = S = MS_2^2$.

Now

$$x - r_1 = \frac{X}{Z^2} - r_1 = \frac{X - r_1 Z^2}{Z^2} = M\left(\frac{S_2}{Z}\right)^2 ;$$

therefore $\phi_1(P) = M(\mathbf{Q}^{*2})$. If we can prove that M can take only a finite number of values, independent of P, then we are done.

Lemma: M *divides* $3r_1^2 + A$.

The expression $3r_1^2 + A$ was not pulled out of the air. It is the *resultant* of S and T regarded as polynomials (see [Wal] for the theory of resultants). We avoid resultant theory and use straightforward algebraic calculations.

Proof of the lemma: First we change $T = X^2 - XZ^2(r_2+r_3)+r_2r_3Z^4$ to a slightly different form. The relations between the roots and coefficients of $x^3 + Ax + B = (x - r_1)(x - r_2)(x - r_3) = 0$ yield

$$r_1 + r_2 + r_3 = 0 \,, \tag{1}$$

and

$$r_1r_2 + r_1r_3 + r_2r_3 = A \,. \tag{2}$$

From (2) we have $r_2r_3 = A - r_1(r_2 + r_3)$. But by (1), $-r_1 = r_2 + r_3$; so $r_2r_3 = A + r_1^2$ and

$$T = X^2 + r_1XZ^2 + (A + r_1^2)Z^4 \,.$$

The identities

$$-S(X + 2r_1Z^2) + T = (3r_1^2 + A)Z^4$$

and

$$S\left[\left(r_1^2 + A\right)X + \left(r_1^2 + A\right)r_1Z^2\right] = \left(3r_1^2 + A\right)X^2$$

follow by straightforward calculation (or by the theory of resultants). Since $(X, Z) = 1$, any common divisor of S and T, including the greatest common divisor M, divides $3r_1^2 + A$ and we are done.

We are now on the last leg of a long proof.

Proposition: $\mathrm{Ker}(\phi) = 2E(\mathbf{Q})$.

Proof: If $2P \in 2E(\mathbf{Q})$, then, by equation (6.9), the $x(2P) - r_i$, $i = 1, 2, 3$, are squares of rational numbers and $\phi(2P) = I = \mathbf{Q}^{*2}$; so $2E(\mathbf{Q}) \in \ker(\phi)$.

Now we prove the converse, viz., if $P = (x, y) \in \ker(\phi)$, $\phi(P) = I$, then $P = 2Q$, for some $Q \in E(\mathbf{Q})$, i.e., $P \in 2E(\mathbf{Q})$. Some of our reasoning is very similar to that used in the proof of lemma C.

If there is such a point $Q = (x_1, y_1)$, then x_1 must satisfy equation (6.9),

$$x - r_j = \left[\frac{-x_1^2 + 2r_jx_1 + A + 2r_j^2}{2y_1}\right]^2 , \tag{3}$$

$j = 1, 2, 3$, where $x = x(P) = x(2Q)$. But our assumption that $\phi_j(P) = (x - r_j)\mathbf{Q}^{*2} = I = \mathbf{Q}^{*2}$ means that $x - r_j$ is the square of a rational number, say

$$x - r_j = \rho_j^2 \ , \tag{4}$$

i.e., $(x - r_j)^{\frac{1}{2}} = \rho_j$ is rational. Thus, collecting terms according to powers of r_j, the equations (3) become

$$\rho_j = \lambda_1 + r_j\lambda_2 + r_j^2\lambda_3 \ , \quad j = 1, 2, 3 \ , \tag{5}$$

with

$$\lambda_1 = \frac{-x_1^2 + A}{2y_1} \ , \quad \lambda_2 = \frac{x_1}{y_1} \text{ and } \lambda_3 = \frac{1}{y_1} \ . \tag{6}$$

As in the proof of lemma C, the determinant of this linear system is the van der Monde determinant of value $\prod_{i>j}(r_i - r_j) \neq 0$. Therefore, by Cramer's rule, we have a unique rational solution of these equations. From $\lambda_3 = \frac{1}{y_1}$, we see that y_1 is rational, and from $\lambda_2 = \frac{x_1}{y_1}$, we see that x_1 rational; so the point $\mathbf{Q}= (x_1, y_1)$ given by our equations is a rational point satisfying the doubling formula (3), but we still have to prove that it lies on the curve.

Substituting equation (5) into (4), multiplying out, using the relations $r_j^3 = -Ar_j - B$ and $r_j^4 = -Ar_j^2 - Br_j$ to get rid of r_j^3 and r_j^4, and collecting terms according to powers of r_j, we obtain the vector equation

$$\left(\lambda_1^2 - 2B\lambda_2\lambda_3 - x\right)\mathbf{v_0} + \left(1 + 2\lambda_1\lambda_2 - 2A\lambda_2\lambda_3 - B\lambda_3^2\right)\mathbf{v_1} + \left(\lambda_2^2 + 2\lambda_1\lambda_3 - A\lambda_3^2\right)\mathbf{v_2} = 0 \ , \tag{7}$$

where

$$\mathbf{v_0} = (1, 1, 1) \ , \quad \mathbf{v_1} = (r_1, r_2, r_3) \text{ and } \mathbf{v_2} = \left(r_1^2, r_2^2, r_3^2\right) \ .$$

The determinant of the matrix formed by the vectors is again a non-zero van der Monde determinant; so the $\mathbf{v_i}$ are linearly independent. Hence the coefficients of the $\mathbf{v_i}$ in (7) equal zero, i.e.,

$$\lambda_1^2 - 2B\lambda_2\lambda_3 = x \ ,$$

$$2\lambda_1\lambda_2 - 2A\lambda_2\lambda_3 - B\lambda_3^2 = -1 \ , \tag{8}$$

$$\lambda_2^2 + 2\lambda_1\lambda_3 - A\lambda_3^2 = 0 \ . \tag{9}$$

Solving for λ_1 in (9) and substituting this in (8) yields

$$\lambda_2^3 + A\lambda_2\lambda_3^2 + B\lambda_3^3 = \lambda_3 . \tag{10}$$

By (8) and (9), $\lambda_3 \neq 0$; so dividing (10) by λ_3^2 gives

$$\frac{1}{\lambda_3^2} = \left(\frac{\lambda_2}{\lambda_3}\right)^3 + A\left(\frac{\lambda_2}{\lambda_3}\right) + B .$$

By (6), $\frac{1}{\lambda_3} = y_1$ and $\frac{\lambda_2}{\lambda_3} = x_1$; hence

$$y_1^2 = x_1^3 + Ax_1 + B .$$

Therefore, our rational point $Q = (x_1, y_1)$ lies on E, and we are done with the proof of the proposition, the proof of the weak Mordell–Weil theorem, and thus with the proof of Mordell's theorem.

9. Equations over Finite Fields; the Zeta and L Functions of a Curve

To talk about the currently very active study of elliptic curves, we must introduce the ζ and L functions of these curves. See Ireland and Rosen [Ire - Ros] and Koblitz [Kob 2] for a more extensive discussion of the theory and examples.

Again we study elliptic curves in the normal form

$$E : y^2 = x^3 + Ax + B ,$$

with $A, B \in \mathbf{Z}$ and discriminant D= $-4A^3 - 27B^2 \neq 0$. For each prime p, we associate to E the congruence

$$y^2 \equiv x^3 + Ax + B (\text{mod } p),$$

which can also be regarded as an equation, over the finite field $\mathbf{F_p} = \mathbf{Z}/p\mathbf{Z}$, of the curve

$$\mathbf{E_p} : y^2 = x^3 + A_p x + B_p ,$$

where $A_p = A(\text{mod } p)$, $B_p = B(\text{mod } p)$, $0 \leq A_p, B_p \leq p-1$. If $p \nmid D$ and if the curve has at least one point with coordinates in $\mathbf{F_p}$, then we say that $\mathbf{E_p}$ is an **elliptic curve over $\mathbf{F_p}$**. E_p is also called the *reduction of E modulo p*. (If $p|D$, then the curve modulo p can degenerate in bad ways.)

Let $\mathbf{E_p(F_{p^m})}$ be the set of points on E_p with coordinates in $\mathbf{F}_{p^m}$, the unique field with p^m elements which contains $\mathbf{F}_p$, and let

$$\mathbf{N_{p^m}} = |E_p(\mathbf{F}_{p^m})| .$$

The **zeta function of $\mathbf{E_p}$** is defined as

$$\mathbf{Z(E_p, u)} = \exp\left(\sum_{m=1}^{\infty} N_{p^m} \frac{u^m}{m!}\right) .$$

It can be proved that

$$Z(E_p, u) = \frac{1 - a_p u + pu^2}{(1-u)(1-pu)} , \tag{1}$$

for some integer a_p. Hasse proved that

$$\left(1 - a_p u + pu^2\right) = (1 - \lambda u)(1 - \overline{\lambda} u) ,$$

where $\overline{\lambda}$ is the complex conjugate of λ (so $\lambda\overline{\lambda} = p$ and $a_p = \lambda + \overline{\lambda}$) and

$$|\lambda| = |\overline{\lambda}| = \sqrt{p} .$$

This latter result is known as the *Riemann hypothesis for elliptic curves over finite fields.*

Setting $u = p^{-s}$ in (1), we obtain the zeta function as a function of s, viz.,

$$\zeta(E_p, s) = \frac{1 - a_p p^{-s} + p^{1-2s}}{(1-p^{-s})(1-p^{1-s})} .$$

This zeta function is a way of coding information about the points on E over the finite fields $\mathbf{F}_{p^m}$ into one function. It was introduced in analogy with the classical Riemann zeta function (sec. 3.3) and the zeta functions of algebraic number fields (sec. 15.4). The use of zeta functions to encode information and then apply the methods of analysis starts with Euler and has a long history. See [Ell, W - Ell, F] and chapter 3 for details.

For convenience, when $p|D$ we define

$$\zeta(E_p, s) = \frac{1}{(1-p^{-s})(1-p^{1-s})} .$$

The $\zeta(E_p, s)$, for all p, are called the **local zeta functions of E at p.** Now we put together all the information about all primes. The **global zeta**

function of E (or just the **zeta function of E**) is given by the product of all the local zeta functions;

$$\zeta(\mathbf{E}, \mathbf{s}) = \prod_p \zeta(E_p, s)\ .$$

By simple manipulation, it can be shown that

$$\zeta(E, s) = \frac{\zeta(s)\zeta(s-1)}{L(E, s)}\ ,$$

where $\zeta(s)$ is the classical Riemann zeta function and

$$L(E, s) = \prod_{p|D} \frac{1}{\left(1 - a_p p^{-s} + p^{1-2s}\right)}$$

is the **L function of E.**

From Hasse's result, it can be shown that $L(E, s)$, as a function of the complex variable s, converges when $Re(s) > \frac{3}{2}$. Hasse conjectured that $\zeta(E, s)$ (and therefore $L(E, s)$) can be analytically continued to all of **C** as a meromorphic function. From the work of Weil, Deuring and Shimura we know that this is true for curves with 'complex multiplication' (defined in the next section) and for other special cases.

Combining extensive computer studies of curves of the form $y^2 = x^3 - ax$ with some heuristic arguments (see [Kob 2; pg. 91]), Birch and Swinnerton–Dyer made several conjectures which have since guided a large part of the research in this subject. We state their principal conjectures in weak and strong forms under the assumption that L can be analytically continued to **C**, so that we can talk about the analytic behavior of L at $s = 1$.

Weak Birch and Swinnerton–Dyer Conjecture: *If E is an elliptic curve defined over Q, then*

$$L(E, 1) = 0 \iff E \textit{ has infinitely many rational points.}$$

Recall that the rank of E is the number of free generators of $E(\mathbf{Q})$.

Strong Birch and Swinnerton–Dyer Conjecture: *$rank(E)$ = the order of the zero of $L(E, s)$ at $s = 1$.*

This is an amazing conjecture. Somehow knowing local information (the number of points over finite fields) determines global information (structure of the rational points). This is very similar to the Hasse principle we discussed in section 9.7, where solutions of equations modulo p^n determine solutions over $\mathbf{Q}$.

We have discussed equations over finite fields only in the context of studying rational points. But the studies of the number of points on equations over finite fields, especially the so called "Weil Conjectures", have been a major theme of twentieth century number theory. We do not present a systematic discussion of these ideas because it would be difficult to improve the exposition of Ireland and Rosen [Ire - Ros]. The more limited treatments by Silverman and Tate [Sil - Tat] and Chahal [Cha] are also recommended. A knowledge of these topics is essential to understanding some of the most important work in number theory and algebraic geometry of the past 50 years.

10. Complex Multiplication

Our principal knowledge about rational points on elliptic curves are for curves with 'complex multiplication'. Recall the parametrization of an elliptic curve by the Weierstrass $\wp$ functions defined on lattices (sec. 19.7). Every elliptic curve E determines a lattice L such that the group $E(\mathbf{C})$ is isomorphic to $\mathbf{C}/L$.

Now consider the endomorphisms End $(\mathbf{C}/L)$, i.e., the homomorphisms of $\mathbf{C}/L$ to $\mathbf{C}/L$. It is easy to show the group isomorphism

$$End\,(\mathbf{C}/L) \cong \{\alpha \in \mathbf{C} | \alpha L \in L\}\ ,$$

i.e., the endomorphisms are given by multiplication by a complex number (geometrically, by rotating and stretching the lattice). Trivially, we always have $\mathbf{Z} \subseteq \mathrm{End}\,(\mathbf{C}/L)$. The important question is whether there are any other endomorphisms.

Definition: If End $(\mathbf{C}/L) \neq \mathbf{Z}$, then we say that the elliptic curve $E = \mathbf{C}/L$ has **complex multiplication** or is a **CM curve**.

So a CM curve has at least one endomorphism which is given by multiplication by a complex number (which is not real). This is a very strong symmetry condition, namely L can be rotated and stretched in such a way that it is still contained in itself.

In 1977 Coates and Wiles proved half of the weak Birch and Swinnerton–Dyer conjecture for CM curves, viz., if E is an CM elliptic curve over $\mathbf{Q}$, then

$$E(\mathbf{Q}) \text{ infinite } \implies L(E,1)=0\,.$$

Complex multiplication traces back to the Euler–Fagnano work on doubling the arc of the lemniscate (see sec. 19.9 and [Sie 1]). It is also deeply connected with the development of algebraic number theory. In the latter half of the 19th century, Kronecker and Weber proved the

Kronecker–Weber Theorem: *Every finite abelian extension of* $\mathbf{Q}$ *(i.e., with abelian Galois group) is a subfield of a cyclotomic field* $\mathbf{Q}(\zeta)$*, where* $\zeta = e^{2\pi i/m}$ *is a primitive* m*th root of unity for some positive integer* m*.*

We proved this for the quadratic fields $Q(\sqrt{\pm p})$, p a prime, in section 14.4. See [Gol, L 2] for a proof of the full theorem.

Kronecker raised the question of finding, for any algebraic number field K, analytic functions (like $e^{2\pi i/z}$ for $\mathbf{Q}$) such that special values of these functions when adjoined to K generate all abelian extensions of K.

The solution of this problem was Kronecker's 'Jugentraum' (the dream of his youth) and became the twelfth in Hilbert's famous list of problems [Hil]. When K is imaginary quadratic, the problem was solved by the work of Kronecker, Weber, Takagi and Hasse (the last two in the twentieth century) using special elliptic functions. Complex multiplication is central to this work. See [Shi] for an advanced update on the general problem, Silverman and Tate [Sil - Tat; chap. 6] for a more elementary discussion, and Edwards [Edw 3] for a sketch of Kronecker's life.

Thus we see that complex multiplication of elliptic curves is tightly connected to the structure of imaginary quadratic fields. This was recently reinforced by the work of Goldfeld and Gross-Zagier on effective methods for class structure of the quadratic fields (see Goldfeld [Gol, D] for a survey of these results).

For an extensive and deep survey of many of the topics in this chapter at a more advanced level, see Cassels [Cas 2].

Chapter 21

Irrational and Transcendental Numbers, Diophantine Approximation

The three topics in the title of the chapter are so intertwined that to trace their development requires treating them together.

1. The Early History

Recall that a complex number α is **irrational** if it is not rational, **algebraic** if it is the root of a polynomial equation with integer coefficients (the **degree** of α is then the degree of an irreducible equation satisfied by α), and **transcendental** if it is not algebraic.

The oldest known result in the subject is due to the Pythagorean school ($\sim$ 500 B.C.), whose philosophy was built on the idea that everything in mathematics and nature can be explained by whole numbers and their ratios. Thus, according to Plato, their discovery that the ratio of the length of the diagonal of a square to the length of a side is irrational, i.e., $\sqrt{2}$ is irrational, destroyed the Pythagorean's most fundamental belief and stunned the Greek mathematical community.

We present two proofs that $\sqrt{2}$ is irrational. If $\sqrt{2}$ were rational, then we could write $\sqrt{2} = \frac{a}{b}$, $(a, b) = 1$. Thus the irrationality is equivalent to saying that $a^2 = 2b^2$ has no non-zero relatively prime integer solutions.

1) (assumed to be essentially the first proof but presented in algebraic notation rather than the geometric language of the Greeks)

$$a^2 = 2b^2 \implies a^2 \text{ is even } \implies a \text{ is even, say } a = 2c$$
$$\implies (2c)^2 = 2b^2 \implies b^2 = 2c^2 \implies b \text{ is even,}$$

contradicting $(a, b) = 1$.

2) If $a^2 = 2b^2$, then a^2 has an even number of prime factors and $2b^2$ an odd number, contradicting unique factorization in $\mathbf{Z}$.

Exercise: Prove that if m, n are positive integers and $n \neq m^{\text{th}}$ power of an integer, then $\sqrt[m]{n}$ is irrational.

In his work *Theaetus*, Plato states that his teacher Theodorus ($\sim$ 400 B.C.) proved that $\sqrt{3}, \sqrt{5}, \ldots, \sqrt{17}$ (except for $\sqrt{4}, \sqrt{9}, \sqrt{16}$) are irrational. See [Har-Wri] for a discussion of his possible methods of proof.

The constant π was introduced in ancient times and the question of its transcendentality arose indirectly (from our modern point of view, not from the Greek's) in the problem of squaring the circle. Actually, to prove that the circle cannot be squared we *only need to know that π is not constructible*, i.e., not in a field extension of $\mathbf{Q}$ of degree a power of 2 (see sec. 14.2 and [Gol, L]). However, the only known way to prove that π is not constructible is as a corollary to the fact that π is transcendental

Now we must jump over 2000 years for any further progress in the study of irrational and transcendental numbers.

2. From Euler to Dirichlet

The first real progress comes with Euler.

The number e was introduced by Napier ($\sim$ 1614) as the base of his natural logarithms. Euler showed that $e^{ix} = \cos x + i \sin x$, which yields $e^{\pi i} = -1$. Since that time, the study of e and π has been closely related. In fact, the proofs of irrationality and transcendency of π essentially generalize those for e.

In 1737, Euler proved that e is irrational. To prove this we work with the Taylor expansion $e = \sum_0^\infty \frac{1}{k!}$. Consider the tail of this series, namely,

$$e - \sum_1^n \frac{1}{k!}, \tag{1}$$

and assume that e is rational, $e = \frac{a}{b}$, $b > 0$. If we pick $n \geq b$ so that b divides $n!$, then multiplying (1) by n! yields an integer α. But

$$0 < \alpha = n!\left(e - \sum_{1}^{n} \frac{1}{k!}\right) = \frac{1}{n+1} + \frac{1}{(n+1)(n+2)} + \cdots$$
$$< \frac{1}{(n+1)} + \frac{1}{(n+1)^2} + \frac{1}{(n+1)^3} + \cdots$$
$$= \frac{1}{n} < 1\,,$$

i.e., α is an integer satisfying $0 < \alpha < 1$, a contradiction, and therefore e is irrational.

Euler conjectured that e and π are transcendental.

In his *Introductio in Analysis Infinitorium* of 1748 [Eul 3], Euler conjectured that for $a, b \in \mathbf{Q}$, b not a rational power of a, $\log_a b$ is transcendental. This was only proved in the 20th century.

Next we come to Lambert ($\sim$ 1770) who proved that e^x and $\tan x$ are irrational for $x \in \mathbf{Q}$, $x \neq 0$. Since $\tan \frac{\pi}{4} = 1$, we have , as a corollary, that $\frac{\pi}{4}$ and thus π are irrational. See [Sie 3] or [Har - Wri] for the proof for e^x, and [Sie 4] or the English translation of Lambert's paper [Str] for $\tan x$.

From Euler and Lambert to the 19th century, the main technique for studying e and the approximation of irrational numbers by rationals (**Diophantine approximation**) was continued fractions. Before discussing this, we prove two very general but elementary results.

Since the rationals are dense in the reals, any real irrational number can be approximated arbitrarily closely by rationals. To ask non-trivial questions about such approximations, we use a function of the denominator as a measure of approximation, i.e., for some $f(q)$ and real α, we ask for solutions of

$$\left|\alpha - \frac{p}{q}\right| < f(q) \text{ or } \left|\alpha - \frac{p}{q}\right| > f(q)$$

or with $<, >$ replaced by $\leq, \geq$. For convenience in stating results, *we shall always assume q is a positive integer, unless otherwise stated.*

Theorem: *If $\alpha \in \mathbf{R}$, there exist infinitely many distinct solutions of*

$$\left|\alpha - \frac{p}{q}\right| \leq \frac{1}{q}\,.$$

Proof: $\left|\alpha - \frac{p}{q}\right| < \frac{1}{q}$ is equivalent to $|q\alpha - p| < 1$. If $\alpha \notin \mathbf{Z}$, then there are infinitely many $q \in \mathbf{Z}$ for which $q\alpha$ is not an integer. For each such q, let p be the integer nearest to $q\alpha$ (if $q\alpha = n + \frac{1}{2}$, $n \in \mathbf{Z}$, choose n); thus $\alpha - \frac{p}{q} \neq 0$, $\left|\alpha - \frac{p}{q}\right| \leq \frac{1}{q}$, and these inequalities guarantee infinitely many distinct solutions. If $\alpha = n \in \mathbf{Z}$, let $\frac{p}{q} = n - \frac{1}{q} = \frac{nq-1}{q}$, for each q.

We want to be able to distinguish different types of numbers (rational, irrational, algebraic, transcendental). Our second result is a move in that direction.

Let $\alpha \in \mathbf{Q}$, $\alpha = \frac{a}{b}$, $(a, b) = 1$, $b > 0$. If $\frac{a}{b} \neq \frac{p}{q}$ $(aq - pb \neq 0)$, then

$$\left|\frac{a}{b} - \frac{p}{q}\right| = \left|\frac{aq - pb}{bq}\right| \geq \frac{1}{bq},$$

and we have the following.

Theorem: *If $\alpha \in \mathbf{Q}$, there exists a positive rational $c(\alpha)$, depending only on α, such that*

$$\left|\alpha - \frac{p}{q}\right| \geq \frac{c(\alpha)}{q},$$

for all rational $\frac{p}{q} \neq \alpha$.

Note that $c(\alpha)$ is just $\frac{1}{b}$. We use the artifice $c(\alpha)$, instead of just writing $\frac{1}{b}$, to guide us in generalizing the result.

Hence, for α rational, we can find infinitely many rational approximations $\frac{p}{q}$ within $\frac{1}{q}$ of α, but we can never do better than $\frac{c(\alpha)}{q}$. So there is a limit on how well rationals can be approximated by other rationals.

For α irrational, we can do better. Recall that in section 4.7 we used continued fractions to prove

Lagrange's Theorem: *If α is irrational, then there exist infinitely many solutions to*

$$\left|\alpha - \frac{p}{q}\right| < \frac{1}{q^2}.$$

We also stated, without proof, Hurwitz's theorem that

$$\left|\alpha - \frac{p}{q}\right| < \frac{1}{\sqrt{5}}q^2$$

has infinitely many solutions for each irrational α and that the constant $\frac{1}{\sqrt{5}}$ is the best constant for the theorem to be true for all irrationals.

We present another proof of Lagrange's theorem, due to Dirichlet, and based on the

Dirichlet Box Principle: *If $n+1$ objects are placed in n boxes, then some box contains at least 2 objects.*

This principle, also known as the **pigeonhole principle**, leads to many results in Diophantine approximation and in other areas of mathematics.

Proof of Lagrange's theorem: Let $(\mathbf{x}) = \mathbf{x} - [\mathbf{x}]$, be the **fractional part of x**; so $0 \leq (x) < 1$. Let Q be a positive integer and break the interval $[0, 1]$ into Q subintervals $\left[\frac{s}{Q}, \frac{s+1}{Q}\right]$, $s = 0, 1, \ldots, Q-1$.

$0 \quad \frac{1}{Q} \quad \frac{2}{Q} \quad .... \quad \frac{(Q-1)}{Q} \quad 1$

Since α is irrational, the $Q+1$ numbers $0, (\alpha), (2\alpha), \ldots (Q\alpha)$ are distinct and $0 < (n\alpha) < 1$. Therefore, by the box principle, two of them, say $(q_1\alpha)$, $(q_2\alpha)$, are in the same subinterval and thus differ in absolute value by $< \frac{1}{Q}$ ($(q_i\alpha) \neq$ endpoint for $q_i > 0$). Then

$$|(q_2\alpha) - (q_1\alpha)| < \frac{1}{Q}$$

or

$$|(q_2 - q_1)\alpha - ([q_2] - [q_1])| < \frac{1}{Q}\,.$$

Let $q = q_2 - q_1$ and $p = [q_2] - [q_1]$; then $|q\alpha - p| < \frac{1}{Q}$ or $|\alpha - \frac{p}{q}| < \frac{1}{q}Q$. Without loss of generality, we assume that $q_2 > q_1$. Since $0 < q_1, q_2 < Q$, we have $0 < q < Q$. Therefore

$$|\alpha - \frac{p}{q}| < \frac{1}{qQ} < \frac{1}{q^2}\,.$$

Thus, for each Q, $\left|\alpha - \frac{p}{q}\right| < \frac{1}{Q} < \frac{1}{q^2}$ has a solution.

Now suppose there are only a finite number of solutions, $\frac{p_1}{q_1}, \ldots, \frac{p_k}{q_k}$ to $\left|\alpha - \frac{p}{q}\right| < \frac{1}{q^2}$. Then the set $\left\{\left|\alpha - \frac{p_i}{q_i}\right|\right\}$ is bounded away from zero, i.e., there is a positive integer Q' such that

$$\left|\alpha - \frac{p_i}{q_i}\right| > \frac{1}{Q'} . \tag{2}$$

For this Q', there exist p, q such that $\left|\alpha - \frac{p}{q}\right| < \frac{1}{qQ'} < \frac{1}{q^2}$, i.e., $\frac{p}{q}$ is a solution. But $\frac{1}{qQ'} < \frac{1}{Q'}$ and $\left|\alpha - \frac{p}{q}\right| < \frac{1}{Q'}$, contradicting (2). Hence there are infinitely many solutions.

In section 22.6, we will present yet another proof of Lagrange's theorem, based on geometric ideas.

3. Liouville to Hilbert; The Beginning of Transcendental Number Theory

In 1844, Liouville showed that there is a limit to how well any algebraic number can be approximated and he established the first connection between Diophantine approximation and transcendental numbers.

Liouville's Theorem: *If α is algebraic of degree $n \geq 2$, then there exists a $c(\alpha) > 0$, depending only on α, such that*

$$\left|\alpha - \frac{p}{q}\right| > \frac{c(\alpha)}{q^n} ,$$

for all $\frac{p}{q} \in \mathbf{Q}$.

Proof: Let $f(x) = a_0 + a_1 x + \cdots a_n x^n$, $a_i \in \mathbf{Z}$, be an irreducible equation for α. To make f unique, assume that $a_n > 0$ and gcd $(a_0, \ldots, a_n) = 1$.

Since f is irreducible, we have $f\left(\frac{p}{q}\right) \neq 0$, for all $\frac{p}{q} \in \mathbf{Q}$. Furthermore,

$$\left|q^n f\left(\frac{p}{q}\right)\right| = \left|a_n p^n + a_{n-1} p^{n-1} q + \cdots + a_0 q^n\right| \geq 1 ,$$

since $a_n p^n + a_{n-1} p^{n-1} q + \cdots + a_0\, q^n$ is a non-zero integer. Hence

$$\left|f\left(\frac{p}{q}\right)\right| \geq \frac{1}{q^n} . \tag{1}$$

By the mean value theorem and our assumption that $f(\alpha) = 0$, we have

$$\left| f\left(\frac{p}{q}\right) \right| = \left| f\left(\frac{p}{q}\right) - f(\alpha) \right| = \left| \left(\alpha - \frac{p}{q}\right) f'(\xi) \right| ,$$

for some ξ between $\frac{p}{q}$ and α . We may assume that $\alpha - 1 < \frac{p}{q} < \alpha + 1$, for otherwise $\left|\alpha - \frac{p}{q}\right| > 1$ and our theorem is true.

$|f'(\xi)|$ is bounded on the closed interval $[\alpha - 1, \alpha + 1]$ by a function of α, *since α uniquely determines $f(x)$*. Say $|f'(\xi)| < d(\alpha)$; so

$$\left| f\left(\frac{p}{q}\right) \right| < d(\alpha) \left| \alpha - \frac{p}{q} \right| .$$

Then

$$\begin{aligned} \left| \alpha - \frac{p}{q} \right| &> \frac{1}{d(\alpha)} \left| f\left(\frac{p}{q}\right) \right| \\ &> \frac{1}{d(\alpha)} \frac{1}{q^n} , \text{ by (1)} \\ &= \frac{c(\alpha)}{q^n} , \end{aligned}$$

where $c(\alpha) = \frac{1}{d}(\alpha)$.

This theorem allowed Liouville to be the first to prove the existence of transcendental numbers by constructing numbers that can be so well approximated that they cannot be algebraic of any degree.

We call α a **Liouville number** if there exists a sequence of distinct rationals $\frac{p_1}{q_1}, \frac{p_2}{q_2}, \ldots$ such that

$$\left| \alpha - \frac{p_r}{q_r} \right| < \frac{K}{q_r^r} ,$$

for some constant $K > 0$.

Note that there must be infinitely many q's; for if some q occurs infinitely often in the sequence, then there are infinitely many corresponding p's and thus the corresponding $\frac{p_r}{q_r}$ can be made arbitrarily large and will not satisfy the inequality.

Suppose that the Liouville number α is algebraic of degree n. Then, by Liouville's theorem, there exists a $c(\alpha) > 0$ such that

$$\frac{c(\alpha)}{q_r^n} < \left|\alpha - \frac{p_r}{q_r}\right|,$$

and we have

$$\frac{c(\alpha)}{q_r^n} < \frac{K}{q_r^r},$$

for all q_r. For $r > n$, this means that

$$\frac{c(\alpha)}{K} < \frac{1}{q_r^r} - n.$$

But there are infinitely many q_r and thus arbitrarily large q_r, contradicting a lower bound $\frac{c(\alpha)}{K}$. Therefore we have

Theorem: *All Liouville numbers are transcendental.*

Example: Let

$$\alpha = \sum_{1}^{\infty} \frac{1}{10^{k!}} = .110001000\cdots$$

and let

$$\frac{p_r}{q_r} = \sum_{1}^{r} \frac{1}{10^{k!}} = \frac{p_r}{10^{r!}}.$$

Then

$$\begin{aligned}
\left|\alpha - \frac{p_r}{q_r}\right| &= \sum_{r+1}^{\infty} 10^{k!} \\
&= \frac{1}{10^{(r+1)!}}\left(1 + \frac{1}{10^{r+2}} + \cdots\right) \\
&< \frac{2}{10^{(r+1)!}} = \frac{2}{10^{(r+1)r!}} \\
&< \frac{2}{(10^{r!})^r} = \frac{2}{q_r^r},
\end{aligned}$$

i.e., α is a Liouville number. Hence α is transcendental and we have Liouville's important result that **transcendental numbers exist**.

It is easy to vary this construction to show that infinitely many transcendental numbers exist (for example, replace 10 by any positive integer). One only has to make sure that the terms of the series converge to zero very rapidly.

This result marked the beginning of transcendental number theory, which is one of the most difficult areas of mathematics, and which has strong connections to algebraic number theory and Diophantine equations.

Liouville's results led to several questions.

1) Are 'most' numbers transcendental?

2) Are all transcendental numbers Liouville numbers?

3) Are 'interesting' numbers such as π and e, as well as values of classical functions at rational values of x (e.g., e^x and $\log_e x$) transcendental?

The answers came over a long period of time.

1) 30 years after Liouville's paper, Cantor (1874) proved that 'most' number are transcendental. We quickly review Cantor's approach.

Recall that Cantor introduced the idea of infinite cardinals and how to compare them. Cantor proved that **C**, the set of complex numbers, is uncountable. He then showed that the set of polynomials in one variable with integer coefficients (i.e., the set of n-tuples of integers, for all n) is countable. Every such equation has a finite number of roots and the set of these roots is the set Ω of algebraic numbers; thus *the algebraic numbers are countable.* Therefore **C** - Ω, *the set of transcendental numbers, is uncountable.* See Birkhoff and MacLane [Bir - Mac] for more details.

2) No! In fact, in 1932 Mahler proposed a classification of transcendental numbers by how well they can be approximated. This is a very technical and difficult topic and we will not discuss it (see [Lev, Vol. 2]).

As to 3), nothing was known about 'interesting numbers' until 1873 when Hermite [Her, C] proved that e is transcendental. In 1882, Lindemann proved that π is transcendental [Lin] and thus proved, as a corollary, that the circle cannot be squared using only compass and straightedge (see sec. 14.2 and [Gol, L]). In fact he proved the following more general result:

Lindemann's Theorem: *If* $\alpha_1, \ldots, \alpha n$ *are algebraic numbers and a is an integer, then*

$$a + \sum e^{\alpha_i} \neq 0 .$$

Corollary: π *is transcendental.*

Proof: If π is algebraic then $i\pi$ is algebraic since the product of algebraic numbers is algebraic. But $e^{i\pi}+1=0$, contradicting Lindemann, where $n=1$, $a=1$, and $\alpha_1=i\pi$.

It follows from Lindeman's Theorem that, except for the point (0, 1), the curve $y=e^x$ has no points with both coordinates algebraic [Klei 3].

Three years later, Weierstrass generalized Lindemann's theorem.

Weierstrass's Theorem: *If* $\alpha_1,\ldots,\alpha_n$ *are distinct algebraic numbers and* $A_1,\ldots,A_n$ *are distinct non-zero algebraic numbers, then*

$$\sum A_i e^{\alpha_i} \neq 0\,,$$

i.e., the e^{α_i} *are linearly independents over the field of algebraic numbers.*

As we shall see, this type of linear independence result proves to be very fruitful.

For many years mathematicians worked over these results to simplify them and find alternate proofs. We present an outline of Hilbert's proof that *e is transcendental* [Hil] following an exposition by Felix Klein [Kle 3].

Theorem: *e is transcendental.*

Proof: Suppose that *e* is algebraic; thus there exist $a_1,\ldots,a_n\in\mathbf{Z}$, not all zero, such that

$$a_0+a_1e+\cdots+a_ne^n=0\,. \tag{2}$$

The idea of the proof is to show that we can find integers d and I_k and real numbers δ_k, $k=0,\ldots,n$, where δ_k is very small (and $\neq 0$ for $k\neq 0$) such that $de^k=I_k+\delta_k$. Then, multiplying equation (2) by d and substituting for de^k, we have

$$\underbrace{(a_0I_0+a_1I_1+\cdots+a_nI_n)}_{\textstyle \| \atop \textstyle P_1}+\underbrace{(a_1\delta_1+\cdots+a_n\delta_n)}_{\textstyle \| \atop \textstyle P_2}=0\,. \tag{3}$$

We then show that the I's , δ's and d can be chosen so that $0\neq P_1\in\mathbf{Z}$ and $0<P_2<1$, contradicting (3).

To achieve this, Hilbert, motivated by Hermite's proof, introduced the integral

$$J = \int_0^\infty z^p[(z-1)(z-2)\cdots(z-n)]^{p+1}e^{-z}dz\,,$$

where the integer p will be chosen later. Let **g(z)** denote the integrand. Our earlier d will be $\frac{J}{p!}$.

Multiplying (2) by J, splitting up the integrals and dividing by $p!$, we have

$$\left(a_0\int_0^\infty \frac{g(z)}{p!}dz + a_1e\int_1^\infty \frac{g(z)}{p!}dz + \cdots + a_ne^n\int_n^\infty \frac{g(z)}{p!}dz\right)$$
$$+\left(a_1e\int_0^1 \frac{g(z)}{p!}dz + \cdots + a_ne^n\int_0^n \frac{g(z)}{p!}dz\right) = 0\,.$$

Our earlier P_1 will be the expression in the first parenthesis and P_2 the expression in the second.

Now Hilbert chose p so that $0 \neq P_1 \in \mathbf{Z}$ and $0 < P_2 < 1$. To do this, recall the value of the Gamma function $\Gamma(t)$, for $t = p$ an integer, is

$$\Gamma(p) = \int_0^\infty z^pe^{-z}dz = p!\,. \tag{4}$$

For $k > 0$, introducing the substitution $z = z' + k$ in $e^k\int_k^\infty g(z)dz$ yields

$$e^k\int_0^\infty (z'+k)^p\big[(z'+k-1)(z'+k-2)\cdots(z'+k-n)\big]^{p+1}e^{-z'-k}dz' =$$

$$\int_0^\infty (z'+k)^p\big[(z'+k-1)\cdots(z'+k-k)\cdots(z'+k-n)\big]^{p+1}e^{-z'}dz'$$

$$\|$$

$$z'$$

$$= \int_0^\infty (\text{sum of powers of } z'\text{, each} > p+1)\, e^{-z'}dz'\,.$$

By (4) the integral of each term $= cr!$, where $c \in \mathbf{Z}$ and $r \geq p+1$; thus the complete integral is an integer divisible by $(p+1)!$.

Similarly, in evaluating J, the lowest power of z in the integrand contributes $((-1)^n n!)^{p+1} p!$ to the integral and all the other terms are integers divisible by $p+1$. Hence P_1 is an integer and $P_1 \equiv \pm a_0 (n!)^{p+1} (\text{mod } p+1)$. We can choose p to be, say, a prime greater than both a_0 and n, so that $P_1 \not\equiv 0 (\text{mod } p+1)$ and therefore $P_1 \neq 0$.

Now, applying the mean value theorem for integrals, the kth integral in P_2 can be replace by c_k^p, for some constant c_k, $0 \leq c_k \leq k$, yielding

$$a_k e^k \int_0^k \frac{g(z)}{p!} dz = \frac{a_k e^k c_k^p}{p!} \leq \frac{a_k e^k k^p}{p!},$$

which can be made as small as we want by choosing p sufficiently large (consistent with our earlier restriction on p). Hence we are done.

We have now seen Liouville's on approximating algebraic numbers and constructing transcendental numbers, Cantor's proof that the transcendental numbers are uncountable and Hilbert's proof that e is transcendental.

In 1900, Hilbert presented a now famous list of unsolved problems to the International Congress of Mathematicians in Paris [Hil 3]. These problems have guided the development of a good deal of 20th century mathematics. He presents his seventh problem as follows:

> "7. IRRATIONALITY AND TRANSCENDENCE OF CERTAIN NUMBERS.
>
> Hermite's arithmetical theorems on the exponential function and their extension by Lindemann are certain of the admiration of all generations of mathematicians. Thus the task at once presents itself to penetrate further along the path here entered, as A. Hurwitz has already done in two interesting papers,* "Ueber arithmetische Eigenschaften gewisser tranzendenter Funktionen." I should like, therefore, to sketch a class of problems which , in my opinion, should be attacked as here next in order. That certain special transcendental functions, important in analysis, take algebraic values for certain algebraic arguments, seems to us particularly remarkable and worthy of thorough investigation. Indeed, we expect transcendental functions to assume, in general, transcendental values for even algebraic arguments; and, although it is well known that there exist integral transcendental functions which even have rational values for all algebraic arguments, we shall still consider it highly probable that the exponential function $e^{i\pi z}$, for example, which evidently has algebraic values for all rational arguments z, will on the other hand always take transcendental values for irrational algebraic values of the argument z. We can also give this statement a geometrical form , as follows:

If, in an isosceles triangle, the ratio of the base angle to the angle at the vertex be algebraic but not rational, the ratio between base and side is always transcendental.

In spite of the simplicity of this statement and its similarity to the problems solved by Hermite and Lindemann, I consider the proof of this theorem very difficult; as also the proof that

The expression α^β, for an algebraic base α and an irrational algebraic exponent β, e.g., the number $2^{\sqrt{2}}$ or $e^\pi = i^{-2i}$, always represents a transcendental or at least an irrational number.

It is certain that the solution of these and similar problems must lead us to entirely new methods and to a new insight into the nature of special irrational and transcendental numbers."

[trans. in [Hil 3]]

---

* Math. Annalen, vols. 22, 32 (1883, 1888)

The second problem became known as Hilbert's α^β conjecture. As Hilbert notes, corollaries of this conjecture include the transcendence of $2^{\sqrt{2}}$ and of $e^\pi = (e^{\pi i})^{-i} = (-1)^{-i}$.

An amusing incident concerning this conjecture is related in C. Reid's biography of Hilbert [Rei, C]. Carl Ludwig Siegel came to Gottingen as a student in 1919. He always remembered a lecture by Hilbert who, wanting to give his audience examples of problems in the theory of numbers which seem simple at first glance but which are, in fact, incredibly difficult, mentioned the Riemann Hypothesis, Fermat's Last Theorem and the transcendence of $2^{\sqrt{2}}$. Hilbert said that given recent progress he hoped to see the proof of the Riemann Hypothesis in his lifetime. Fermat's problem required totally new methods and possibly the youngest members of the audience would live to see it solved. As for $2^{\sqrt{2}}$, Hilbert said that no one at the lecture would live to see its proof. Hilbert was wrong! Siegel proved the transcendence of $2^{\sqrt{2}}$ about 10 years later (unpublished) and the solution of the α^β conjecture came shortly afterwards. He was right about Fermat's theorem and the Riemann Hypothesis is still unproved.

Reid's biography provides a fascinating portrait of Hilbert and his contemporaries (especially Minkowski) as well as historical and anecdotal information on the transition from 19th to 20th century mathematics. Her biography of Courant [Rei, C 2], who was part of Hilbert's school, is also recommended.

4. Simultaneous Approximation; Kronecker's Theorem

During the 19th century work also proceeded on the simultaneous approximation of several real numbers by rationals (with the same denominator).

By a slight variation of his earlier arguments using the box principle, Dirichlet proved the following (see Har-Wri] for a proof):

Theorem: *If $\alpha_1, \dots, \alpha_k \in \mathbf{R}$, then the system of inequalities*

$$\left| \alpha_i - \frac{p_i}{q} \right| < \frac{1}{q^{1+\frac{1}{k}}}, \quad i = 1, \dots, k,$$

has at least one solution. If at least one α_i is irrational, then there are infinitely many solutions.

Since our inequalities are equivalent to $|q\alpha_i - p_i| < \frac{1}{q^{\frac{1}{k}}}$, we have

Corollary: *Given $\alpha_1, \dots \alpha_k$, where at least one α_i is irrational, and any $\epsilon > 0$, there exists a q in $\mathbf{Z}$ such that $q\alpha_i$ differs from an integer by $< \epsilon$ for all i.*

Kronecker [Kro] proved a similar but much deeper theorem. First we state the one dimensional version in two equivalent ways.

Theorem: *i) If θ is irrational, α arbitrary, $N > 0$, $\epsilon > 0$, then there exist integers $n(> N)$ and p such that*

$$|n\theta - p - \alpha| < \epsilon .$$

ii) Recall that $(\mathbf{x}) = x - [x]$ denotes the fractional part of x. If θ is irrational, then $(n\theta_i)$, $n = 1, 2, \dots$ is dense in $(0, 1)$.

One corollary of this theorem is the billiard ball problem or equivalently the problem of reflected light rays, solved by Konig and Szucs. If a ball bounces off the side of a billiard table, then the angle of incidence equals the angle of reflection.

Suppose we have a *square* billiard table and we give a ball an initial push (fig. 1). Then the path of the ball is either closed and periodic or it is dense in the square (i.e., passing arbitrarily close to every point in the square).

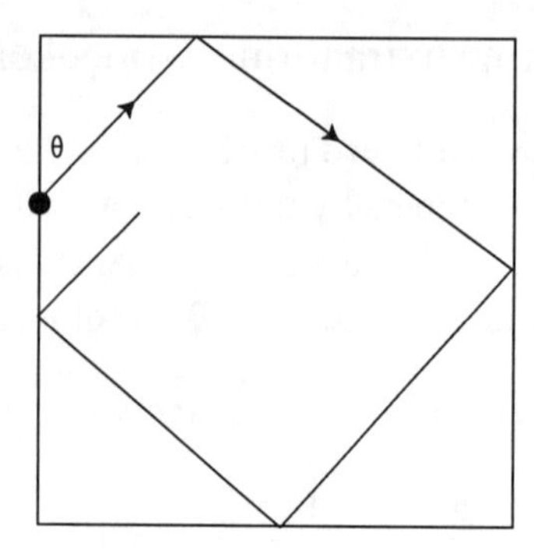

Figure 1

The path is periodic if and only if $\tan\theta$ is rational, where θ is the angle between a side of the table and the initial direction of the ball. See [Har - Wri] for a proof of this.

We now state the full theorem two different equivalent forms.

Kronecker's Theorem:

i) *If* $1, \theta_1, \ldots, \theta_k$ *are linearly independent over* $\mathbf{Z}$ *(equivalently over* $\mathbf{Q}$*),* $\alpha_1, \ldots, \alpha_k$ *arbitrary,* $N > 0, \epsilon > 0$*, then there exist integers* $n (> N)$ *and* $p_1, \ldots, p_k$ *such that*

$$|n\theta_i - p_i - \alpha_i| < \epsilon \ , \quad \textit{for all } i \ .$$

ii) *If* $1, \theta_1, \ldots, \theta_k$ *are linearly independent over* $\mathbf{Z}$*, then the points*

$$((n\theta_1), (n\theta_2), \ldots, (n\theta_k)) \ , \quad n = 1, 2, \ldots \ ,$$

are dense in the unit k-cube $(= \{(x_1, \ldots, x_k) | 0 \leq x_i < 1\})$.

See [Har - Wri] for three proofs of this theorem and [Sie 5] and [Cas 3] for a slightly more general lattice theoretic formulation.

5. Thue: Diophantine Approximation and Diophantine Equations

We have now surveyed the major work up to 1900. The first important result in the the 20th century was Axel Thue's discovery of the relation between Diophantine approximation and Diophantine equations.

Thue (1863–1922) was a Norwegian mathematician who made original contributions to number theory and logic. He read very little and thus rediscovered much. His papers are written in a strictly logical order, with no motivation, and are difficult to read. There is an excellent biography of Thue at the beginning of his collected works [Thue].

Thue studied binary forms of arbitrary degree, i.e. homogeneous polynomials in two variables

$$\sum_{i=o}^{n} a_i x^{n-i} y^i \ , \quad a_i \in \mathbf{Z}, a_0 \neq 0 \, .$$

and proved the first truly general theorem about Diophantine equations [Thue 2]. The method of proof was even more significant than the theorem.

Thue's Theorem: *Let* $f(z) = a_n z^n + a_{n-1} z^{n-1} + \cdots + a_0$ *with* $n \geq 3$, $a_i \in \mathbf{Z}$, *and assume that* f *is irreducible over* $\mathbf{Q}$. *Let*

$$F(x, y) = y^n f\left(\frac{x}{y}\right) = a_n x^n + a_{n-1} x^{n-1} y + \cdots + a_0 y^n$$

be the corresponding homogeneous irreducible polynomial. Then, for every $d \in \mathbf{Z}$,

$$F(x, y) = d$$

has at most a finite number of integer solutions.

Proof: 1) $d = 0$: If $y = 0$, then $x = 0$. If $y \neq 0$, then $f\left(\frac{x}{y}\right) = 0$ and $z - \frac{x}{y}$ is a factor of $f(z)$, contradicting irreducibility. Hence the only solution is $x = y = 0$.

2) $d \neq 0$: Assume the theorem is false. If $y = 0$, there are at most two integers x such that $a_n x^n = d$. Therefore, if there are infinitely many solutions, then there are infinitely many with $y \neq 0$. We can then assume that there are infinitely many with $y > 0$ (otherwise consider $F(x, -y)$, which is also irreducible).

Since, for each such y, there are at most n integers x with (x, y) a solution, we have a sequence of solutions

$$(x_1, y_1), (x_2, y_2), \ldots,$$

$y_i > 0$, $y_i \to \infty$. Thus we have solutions with arbitrarily large y.

Now we factor

$$f(z) = a_n \prod_{n=1}^{n} (z - \theta_i) \ ;$$

so

$$F(x, y) = a_n \prod (x - \theta_i y) .$$

If $F(x, y) = d$, we have

$$|a_n| \prod |x - \theta_i y| = |d| . \tag{1}$$

For at least one of the θ's, say θ_1, we have

$$|x - \theta_1 y| \leq \left(\frac{|d|}{|a_n|} \right)^{\frac{1}{n}} = C_1 ,$$

where C_1 is independent of y. Also, for $j \neq 1$,

$$\begin{aligned} |x - \theta_j y| &= |(\theta_1 - \theta_j) y + (x - \theta_1 y)| \\ &\geq \big| |\theta_1 - \theta_j| y - |x - \theta_1 y| \big| . \end{aligned} \tag{2}$$

Note that since f is irreducible, the roots of f are distinct; hence $\theta_j \neq \theta_1$ for$j \neq 1$. Let $C_2 = \min_{j \neq 1} |\theta_1 - \theta_j| > 0$.

So we have C_2 which bounded $|\theta_1 - \theta_j|$ from below and C_1 which bounds $|x - \theta_k y|$ from above. For $y > \frac{C_1}{C_2}$, we see by (2) that

$$\begin{aligned} |x - \theta_j y| &\geq \big| |\theta_1 - \theta_j| y - |x - \theta_1 y| \big| \\ &\geq C_2 y - C_1 . \end{aligned}$$

| |
smaller larger (keeping everything positive)

If we also require that $y > 2\frac{C_1}{C_2}$, then $C_2 y - C_1 > (\frac{1}{2}) C_2 y$; hence

$$|x - \theta_j y| > \left(\frac{1}{2} \right) C_2 y .$$

From (1), we have

$$|a_n| \, |x - \theta_1 y| \left| \left(\frac{1}{2} \right) C_2 y \right|^{n-1} < |d|$$

or

$$\left| \frac{x}{y} - \theta_1 \right| < \frac{1}{y^n} \left(\frac{|d|}{|a_n|} \left(\frac{2}{C_2} \right)^{n-1} \right) .$$

Combining all the constants into one, C_3, this says that

$$\left|\frac{x}{y} - \theta_1\right| < \frac{C_3}{y^n}.$$

To summarize: for each $y > 2\frac{C_1}{C_2}$ there is a θ such that $\left|\frac{x}{y} - \theta\right| < \frac{C_3}{y^n}$. But there are infinitely many such y; so, for some θ, there are infinitely many solutions to $\left|\frac{x}{y} - \theta\right| < \frac{C_3}{y^n}$.

Thue finished the proof by showing that this result contradicts the following theorem, which he proved.

Theorem: *Let θ be a root of an irreducible equation of degree ≥ 3 with integer coefficients and let $A > 0$. Then*

$$\left|\frac{x}{y} - \theta\right| < \frac{A}{y^n}$$

has only a finite numbers of solutions $x, y (y > 0)$.

This relation between Diophantine equations and approximation was a totally new idea which has proved to be of great significance for further studies.

The last approximation theorem was generalized in two stages. First by Siegel [Sie 3], who used it as an essential tool to prove his theorem about integer points on curves (sec. 20.1). Then K. F. Roth [Rot 2] gave the final form in 1955, for which he received a Field's medal (see [Dav 3] and [Rot 3]). The final form of the theorem carries the names of all three originators.

Thue-Siegel-Roth Theorem: *Let θ be an irrational algebraic number and $\epsilon > 0$. Then there are only finitely many solutions of*

$$\left|\frac{x}{y} - \theta\right| < \frac{1}{q^{2+\epsilon}}, \quad q > 0.$$

Remarks:: (1) The degree of θ plays no role in the theorem.

(2) If $\epsilon = 0$, then we have infinitely many solutions. So this is the best possible result using a power of q as the measure of approximation. See Lang [Lan 1] for a discussion of possible results with other measures.

In 1970 Schmidt proved a multidimensional version of the theorem [Sch], i.e., one about simultaneous approximation.

6. The 20th Century

In 1929, Gelfond proved Hilbert's α^β conjecture for β an imaginary quadratic number. Then Kuzman extended the result to include β a real quadratic irrationality. In the same year, Siegel introduced important new methods to study the transcendentality of values of Bessel functions. In 1932, he proved that for the Weierstrass function $\wp(z)$ (sec. 19.7), where

$$\wp'(z)^2 = 4\wp(z)^3 - g_2\wp(z) - g_3\,,$$

with g_2, g_3 algebraic, one of the periods of $\wp(z)$ is transcendental.

Finally, in 1934, Gelfond and Schneider independently proved the α^β conjecture using Siegel's techniques. In fact, they proved the following equivalent form of the theorem.

Theorem: *Let $\alpha_1, \alpha_2, \beta_1\,, \beta_2$ be non zero algebraic numbers. If $\log\alpha_1$ and $\log\alpha_2$ are linearly independent over* **Q**, *then*

$$\beta_1 \log\alpha_1 + \beta_2 \log\alpha_2 \neq 0\,,$$

i.e., they are linearly independent over the algebraic numbers Ω.

It does not matter which branch of the logarithm is chosen as long as the same choice is used throughout the proof.

Corollary: *If α and β are algebraic numbers, $\alpha \neq 0, 1$ and β is irrational, then α^β is transcendental.*

Proof: Let $\gamma = \alpha^\beta$; so $\beta\log\alpha - \log\gamma = 0$. Assume for the moment that $\log\alpha$ and $\log\gamma$ are linearly independent over **Q**. Then if γ is algebraic, we have, by the theorem, $\beta\log\alpha - \log\gamma \neq 0$, a contradiction. Hence γ is transcendental.

To see the linear independence, assume that $r\log\alpha + s\log\gamma = 0$, where $r, s \in \mathbf{Q}$, $r, s \neq 0$. Then $\beta = \frac{\log\gamma}{\log\alpha} = -\frac{r}{s} \in \mathbf{Q}$, contradicting our assumption that β is irrational.

As another example, let $p \neq q$ be primes. Then

$$\begin{aligned} &r\log p + s\log q = 0,\ \ r, s \in \mathbf{Q} \text{ and non zero} \\ &\implies p^r q^s = 1 \implies r = s = 0\,, \end{aligned}$$

i.e., $\log p$ and $\log q$ are linearly independent over **Q**, and thus by the theorem over Ω.

By analogy with Weierstrass's theorem (sec. 3), Gelfond conjectured that if $\alpha_1, \ldots, \alpha_n$ are non zero algebraic number and $\log \alpha_1, \ldots, \log \alpha_n$ are linearly independent over $\mathbf{Q}$, then they are linearly independent over Ω.

Independent of the conjecture's intrinsic interest, Gelfond and others showed that if one can find an effective (computable) lower bound for

$$|\beta_1 \log \alpha_1 + \cdots + \beta_n \log \alpha_n| , \tag{1}$$

where $\beta_1, \ldots, \beta_n$ are algebraic and not all zero, then one can solve many important problems. These include

i) finding all imaginary quadratic fields of class number one (i.e., with unique factorization — see sec. 17.7),
ii) finding a computable upper bound for the solutions of Thue's equation $F(x, y) = k$ (sec. 5).

In 1966, Alan Baker [Bak 2] found such an effective lower bound for (1), and thus solved these problems ((i) had been solved earlier by Stark using other methods (see [Gol, D])). Baker received the Fields medal for this work which involved the use of analytic functions of several complex variables (see [Tur] and [Bak]). In particular, he proved

Baker's Theorem: *Under the conditions of Thue's theorem (sec. 5), the solutions of $F(x, y) = k$ satisfy*

$$\max(|x|, |y|) < e^C \ , \ \text{where } C = (nH)^{10^5} \ ,$$

where n is the degree of F and H is the largest absolute value of the numbers of the set consisting of k and the coefficients of F.

Although the bound in Baker's theorem is huge, it has been possible to apply Baker's methods, for special classes of Diophantine equations of classical interest, to reduce the bound to a manageable size and actually find all solutions.

Baker and Coates [Bak - Coa] generalized Baker's work to find effective bounds on the size of integral points on curves of 'genus one' (see sec. 19.5 for the notion of genus), but the question remains open for higher genus. See [Tij] and [Bak] for a discussion of the Gelfond-Baker work.

Much other work has been done on transcendental numbers in the past fifty years, but none, I believe, as remarkable in its implications as Baker's work.

7. Other Results and Problems

Before discussing the various problems, we must review some basic definitions. The complex numbers $\alpha_1, \ldots, \alpha_n$ are **algebraically independent (over Q)** if

$$f(\alpha_1, \ldots, \alpha_n) = 0,\ f \in \mathbf{Q}[x_1, \ldots, x_n] \implies f \text{ is the zero polynomial.}$$

The **transcendence degree** of a field $E \subseteq \mathbf{C}$ is the size of the largest algebraically independent set in E. **The transcendence degree of** $\{\boldsymbol{\alpha_1, \ldots, \alpha_n}\}$ is the transcendence degree of $\mathbf{Q}(\alpha_1, \ldots, \alpha_n)$.

1) Are e and π algebraically independent? In particular, are $e + \pi$ and $e\pi$ transcendental or even irrational? At least one of them must be irrational for otherwise $(x - e)(x - \pi) = x^2 - (e + \pi)x + e\pi$ is an equation with rational coefficients and roots e and π, contradicting the transcendence of e and π.

2) Is **Euler's constant** γ,

$$\gamma = \lim_{n \to \infty} \left(\sum_{i=1}^{n} \frac{1}{i} - \log n \right),$$

irrational or transcendental? Nothing is known. (See [Gra - Knu - Pta] for a discussion of Euler's constant.)

3) A major direction of study is still the question of transcendentality of values of classical functions defined by differential equations ([Lan 6], [Sie 4]).

4) Schanuel's Conjecture: if $\alpha_1, \ldots, \alpha_n$ are linearly independent over $\mathbf{Q}$, then

$$\{\alpha_1, \ldots, \alpha_n, e^{\alpha_1}, \ldots, e^{\alpha_n}\}$$

has transcendence degree at least n.

This conjecture implies many of the major results already proved or conjectured such as the α^β theorem. As an example we prove the following.

Proposition: *Schanuel's conjecture* $\implies$ *e and π are algebraically independent.*

Proof: Let $\alpha_1 = \pi i$, $\alpha_2 = 1$. Then $\{\pi i, 1, e^{\pi i} = -1, e^1\}$ has transcendence degree at least 2. Since $\mathbf{Q}(\pi i, 1, -1, e) = \mathbf{Q}(\pi i, e)$, we see that e and πi are algebraically independent.

To see that this implies our result, suppose that e and π are algebraically dependent, i.e., there is a $P(x, y) \in \mathbf{Q}[x, y]$ such that $P(e, \pi) = 0$. Let

$$R(x, y) = P(x, -iy)\ ,$$
$$\overline{R}(x, y) = \overline{P}(x, -iy)\ ,$$

Where the overbar denotes complex conjugation. Then

$$R(e, \pi i) = P(e, \pi) = 0\ ,$$
$$\overline{R}(e, \pi i) = \overline{P}(\pm e, \pm\pi) = P(e, \pi) = 0\ ;$$

therefore

$$(R + \overline{R})(e, \pi i) = 0\ .$$

But $R + \overline{R}$ has rational coefficients, contradicting the algebraic independence of e and πi, and we are done.

See [Lan 6, pp.30 ff.] for more on Schanuel's conjecture.

5) In 1978, Apéry proved that $\zeta(3)\ \left(= \sum \frac{1}{n^3}\right)$ is irrational. What is particularly astonishing is that Apéry's methods do not connect to any other work in the subject and might well have been found much earlier. A. van der Poorten's article [Poo] presents the proof together with some interesting history.

Apéry's work is leading to new attempts to classify irrational numbers as well as to attacks on the irrationality of other values of the zeta function.

8. The Literature

Many of the basic results concerning irrational and transcendental numbers, as well as Diophantine approximation, are proved in Hardy and Wright [Har - Wri]. To understand the deeper aspects of the subject, the best references are the books by many of the most significant researchers. For irrational and transcendental numbers we recommend the books of Siegel [Sie 5], Gelfond [Gel] , Baker [Bak 3], Lang [Lan 6], and Mahler [Mah, K], and for Diophantine approximation the books of Cassels [Cas 3] and Lang [Lan 1], as well as the more elementary exposition by Niven [Niv 2].

Chapter 22

Geometry of Numbers

1. The Motivating Problem; Quadratic Forms

The geometry of numbers deals with the use of geometric notions, especially convexity and lattice, to solve problems in number theory, usually via the solutions of inequalities in integers. Its genesis lies in the problem of minimizing the values of a quadratic form for integer values of the variables.

After Gauss, the study of binary quadratic forms was generalized in two directions — first to algebraic number theory (chapters 15–17) and then to general quadratic forms in any number of variables. The theory of general quadratic forms begins by generalizing such notions as representations of integers, equivalence under groups of matrices and reduction theory from the binary case (chapters 12, 13) to the general case. Among the pioneers in these studies were Jacobi, Dirichlet, Eisenstein, Hermite, H. J. S. Smith, H. Minkowski and C. L. Siegel. The theory is active to the present day and is deeply tied to both algebraic theories, e.g., the theory of arithmetic subgroups of Lie groups, and to analytic studies such as the theory of modular functions. Scharlau and Opolka [Sch - Opo] provide a brief and moderately elementary survey from a modern point of view and Borevich and Shafarevich [Bor - Sha] is a good systematic introduction.

Here we will only be concerned with inequalities involving quadratic forms, since trying to solve them led Minkowski to create the geometry of numbers. We will concentrate on **positive definite quadratic forms in n variables**, i.e., the forms

$$Q(x) = \sum_{i,j=1}^{n} a_{i,j} x_i x_j \ , \tag{1}$$

where $x = (x_1, \ldots, x_n)$, $a_{i,j} \in \mathbf{R}$, $a_{ij} = a_{ji}$, and $Q(x) > 0$ for all $x \neq \mathbf{0} = (0, \ldots, 0)$. $\mathbf{D = det(a_{i,j})}$ is called **the determinant of the form**.

Hermite asked the question raised earlier: given such a form, how small can we make its value for integer values of the variables which are not all zero? A point $g = (g_1, \ldots, g_n) \in \mathbf{R}^n$, with all $g_i \in \mathbf{Z}$, will be called an **integer point**. The set $\mathbf{Z^n}$ of integer points is called the **integer lattice**. Thus we are studying the values of the forms on $\mathbf{Z}^n$ and later we will show, from the geometry, that a positive definite form defined on $\mathbf{Z}^n - \{\mathbf{0}\}$ actually achieves its minimum value at one of these points. Hermite generalized Lagrange's reduction theory for two variables to prove the following:

Theorem: ($\sim$ 1845) *Let Q be a positive definite quadratic form in n variables with determinant D. Then there exists a non-zero integer point g, such that*

$$Q(g_1, \ldots, g_n) \leq \left(\frac{4}{3}\right)^{\frac{n-1}{2}} D^{\frac{1}{n}}.$$

Hermite realized that many important parts of the theory depend on his theorem (see his letter of 1845 to Jacobi — translated in [Sch - Opo]). Thus, for example, he and Eisenstein used it to prove that the number of equivalence classes of positive definite forms in n variables with determinant D, under the action of $Gl_n(\mathbf{Z})$, is finite. These results were proved using arithmetic methods and tools of matrix theory (but without the powerful modern notation of matrices).

Later in the century Hermann Minkowski (1864–1909) and H. J. S. Smith carried the theory much further. There is an interesting story about this latter work as related by C. Reid in her biography of Hilbert [Rei, C].

> "Although Minkowski was still only 17 years old, he was involved in a deep work with which he hoped to win the Grand Priz des Sciences Mathematiques of the Paris Academy.
>
> The Academy had proposed the problem of the representation of a number as the sum of five squares. Minkowski's investigations, however, had led him far beyond the stated problem. By time the deadline of June 1, 1882 arrived, he still had not had his work translated into French as the rules of the competition required. Nevertheless, he decided to submit it. At the last minute, at the suggestion of his older brother Max, he wrote a short prefatory note in which he explained that his neglect had been due to the attractions of his subject and expressed the hope that the Academy would not think "I would have given more if I had given less." ...

> Then, in the spring of 1883, came the announcement that this boy, still only 18 years old, had been awarded jointly with the well-known English mathematician Henry Smith the Grand Priz des Sciences Mathematiques. ...
>
> For a while it seemed, though, that Minkowski might not actually receive his prize. The French newspapers pointed out that the rules of the competition had specifically stated that all entries must be in French. The English mathematicians let it be known that they considered it a reflection upon their distinguished countryman, who had since died, that he should be made to share a mathematical prize with a boy. ("It is curious to contemplate at a distance," an English mathematician remarked some forty years later, " the storm of indignation which convulsed the mathematical circles of England when Smith, bracketed after his death with the then unknown German mathematician, received a greater honor than any that had been paid to him in life.") In spite of the pressures upon them, the members of the prize committee never faltered. From Paris, Camille Jordan wrote to Minkowski: " Work, I pray you, to become a great mathematician." "

Minkowski's close friend Hilbert also wrote about this work

> "The seventeen year old student attacked this topic with all his energy and solved it brilliantly, developing, far beyond the original question, a general theory of quadratic forms, specifically their division in orders and genera, even for arbitrary rank. It is remarkable how well versed Minkowski was in the theory of elementary divisors, and in transcendental tools, such as Dirichlet series and Gauss sums."
>
> (trans. from [Sch -Opo])

Minkowski continued to work on these ideas and on November 6, 1889, he wrote to Hilbert

> " Perhaps you or Hurwitz are interested in the following theorem (which I can prove in half a page): in a positive definite form of determinant D with $n(\geq 2)$, one can always assign such values to the variables that the form is $< nD^{\frac{1}{n}}$ "
>
> (trans. from [Sch - Opo])

This theorem was based on geometric reasoning and it revolutionized the subject. We now explain these ideas, first systematically presented in Minkowski's fundamental book [Min 1].

Unless otherwise stated, all quadratic forms are assumed to be positive definite.

2. Minkowski's Fundamental Theorem

We will not hesitate to assume various geometric properties which are intuitively clear (the reader may first want to think about the cases $n = 2$ or 3). A set S in $\mathbf{R}^n$ is **symmetric with respect to the origin (or has center at 0** or **is centered at the origin**) if $x \in S$ implies that $-x \in S$. The volume of S will be denoted by **V(S)** (we are only dealing with nice sets where all the reasonable definitions of volume coincide).

Minkowski reformulated the problem of minimizing forms in a geometric language. For any positive definite quadratic form and positive real number λ, the set

$$Q_\lambda = \{(x_1, \ldots, x_n) \in \mathbf{R}^n | Q(x_1, \ldots, x_n) < \lambda\}$$

is a set in $\mathbf{R}^n$. Since for $n = 2$ the sets are ellipses and for $n = 3$ they are ellipsoids, we call the general Q_λ **n-dimensional ellipsoids** or just **ellipsoids**. Our diagrams will illustrate the case of ellipses in $\mathbf{R}^2$, which are symmetric with respect to the coordinate axes ($n = 2$ and $a_{12} = a_{21} = 0$ in equation 1.1). As we shall see, nothing is lost by visualizing this very special case.

As λ varies, the Q_λ are all similar and symmetric with respect to the origin (fig. 1).

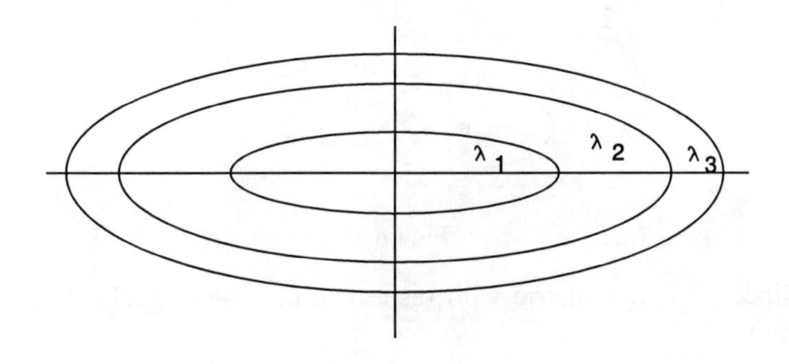

Figure 1

The minimization problem for Q is now equivalent to the question: how big must λ be to guarantee that the ellipsoid Q_λ contains a non-zero integer point?

Minkowski first proved the following theorem.

Theorem: *If the volume of the ellipsoid E is* $> 2^n$, *then it contains a non-zero integer point.*

This yielded a new proof of Hermite's theorem and, as we shall see, led to far reaching generalizations. We first follow Minkowski's approach to the geometry and delay the applications to quadratic forms.

Proof: We assume that E does not contain any non-zero integer points and show that $V(E) \leq 2^n$.

Let $E' = \frac{1}{2}E = \left\{\frac{1}{2}x | x \in E\right\}$. Then $V(E') = \frac{V(E)}{2^n}$. For each integer point g, consider the ellipsoid $E'_g = E' + g$, which is centered at g (fig. 2).

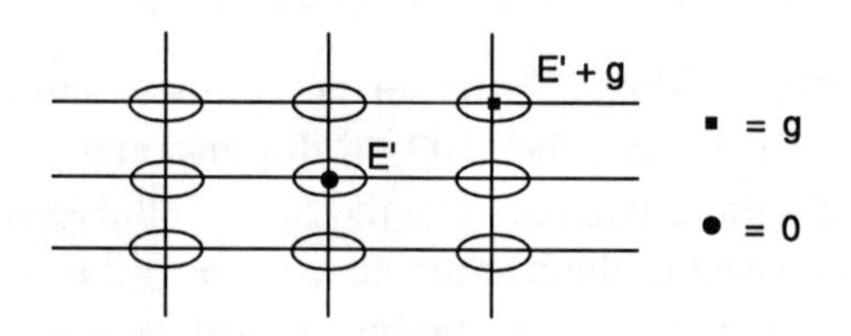

Figure 2

a) The ellipsoids do not overlap.

Suppose $p \in E'_g \cap E'_{g'}$, $g \neq g'$ (fig. 3). Then $p - g \in E'$, $p - g' \in E'$

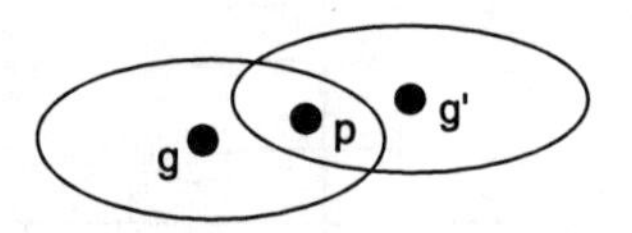

Figure 3

and, since E' is symmetric with respect to 0, $g - p \in E'$. Hence the midpoint

$$\frac{1}{2}[(p - g') + (g - p)] = \frac{1}{2}(g - g') \in E' .$$

Therefore

$$2\left[\frac{1}{2}(g - g')\right] = g - g' \in 2E' = E ,$$

contradicting the assumption that E has no non-zero integer points.

b) The 'density' of the ellipsoids is ≤ 1.

Intuitively, what we are saying is that for a large hypercube C, the ratio of the total volume of the E'_g's, with $g \in C$, to the volume of C is ≤ 1, i.e. if there are m such E'_g's with centers in C, and C is sufficiently large, then we must show that $\frac{mV(E')}{V(C)} \leq 1$. The problem is that pieces of some of the ellipsoids stick out of the sides of C and we must prove that the volume of these pieces is insignificant when C is large.

Now we make these ideas precise. Let N be a positive integer and $C = \{x \in \mathbf{R}^n | 0 \leq x_i \leq N\}$. There are $(N+1)^n$ ellipsoids E'_g, $g = (g_1, \ldots, g_n) \in \mathbf{Z}^n$, with $g \in C$, since there are $N+1$ choices for each integer coordinate with $0 \leq g_i \leq N$. By part a), these ellipsoids do not overlap and thus their total volume is $(N+1)^n V(E')$.

Since E is bounded, so is E' and therefore E' is contained in some hypercube, say $\{x | |x_i| < k\}$ (fig. 4). Then all the E'_g, $g \in C$, are contained

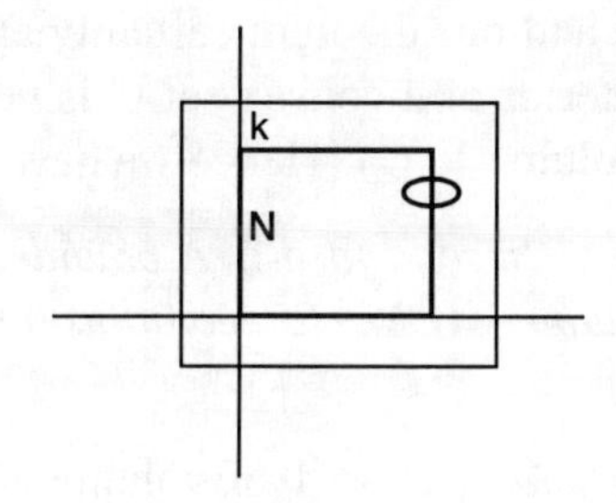

Figure 4

in the hypercube $C' = \{x | -k \leq x_i \leq N+k\}$, where $V(C') = (N+2k)^n$. Hence

$$(N+1)^n V(E') \leq (N+2k)^n$$

and

$$V(E') \leq \left(\frac{N+2k}{N+1}\right)^n < \left(\frac{N+2k}{N}\right)^n$$

$$= \frac{1+2k}{N} \to 1\,, \quad \text{as } N \to \infty\,.$$

Therefore $V(E') = \frac{V(E)}{2^n} \leq 1$ or $V(E) \leq 2^n$, and the theorem is proved.

Minkowski then observed that the only properties of E used in the proof were the following:

i) E is convex — A set $S \subseteq \mathbf{R}^n$ is **convex** if $x, y \in S$ implies that the line segment $\overline{xy} \in S$, where, analytically, $\overline{xy} = \{\lambda x + \mu y | \lambda, \mu \in \mathbf{R}, \lambda, \mu \geq 0, \lambda + \mu = 1\}$. In fact, we only used the fact that if $x, y \in S$, then $\frac{1}{2}(x + y) \in S$, but this implies convexity (exercise).

ii) E is symmetric with respect to the origin.

iii) E is bounded — E is contained in a box $\{(x_1, \ldots, x_n) | \, |x_i| < k\}$, for some $k > 0$.

(The techniques for proving that ellipsoids are convex and bounded will be discussed later.)

Thus Minkowski actually proved the following general theorem:

Minkowski's Theorem: *(First form): A bounded convex set C in* $\mathbf{R}^n$*, with center at* $\mathbf{0}$ *and volume* $V(C) > 2^n$*, contains a non-zero integer point.*

But we really have a stronger result. Our ellipsoids $\{Q(x_1, \ldots, x_n) < \lambda\}$ are open sets (the pre-image of an open set under the continuous map from $\mathbf{R}^n$ to $\mathbf{R}$ defined by Q) and our theorem certainly applies to open convex set. However, if the interior of a convex set C is not empty, then it is an open set and also has volume $V(C)$. Hence we have the following.

Minkowski's Theorem: *(Second form): A bounded convex set C in* $\mathbf{R}^n$*, with center at* $\mathbf{0}$ *and volume* $V(C) > 2^n$*, contains a non-zero integer point g in its interior.*

Remarks: i) The hypercube $|x_i| < 1$ of volume 2^n doesn't contain any non-zero integer points, so our constant 2^n is best possible.

ii) A convex set with volume $< 2^n$ *can* contain lattice points. If $n = 2$, consider the rectangle with sides parallel to the axes, centered at the origin, of width k and height $\frac{1}{2k}$. By choosing k sufficiently large, we can make the rectangle contain any preassigned number of lattice points on the x axis. This example obviously generalizes to any dimension.

Minkowski's Theorem: *(Third form): A bounded convex set C in* $\mathbf{R}^n$*, with center at* $\mathbf{0}$ *and volume* $V(C) \geq 2^n$*, contains a non-zero integer point in its interior or on its boundary.*

Proof: Let $C_k = \left(1 + \frac{1}{k}\right) C, k = 1, 2, \ldots$. Then $V(C_k) > 2^n$ and C_k contains an integer point $g^{(k)} \neq \mathbf{0}$. Hence we have a sequence of non-zero integer points, $g^{(1)}, g^{(2)}, \ldots$, which must all lie in $2C$. But $2C$ is bounded and contains only finitely many integer points. Therefore the sequence takes a constant value g from some point on, and g must be in C or on its boundary (C, together with its boundary, equals the intersection of all the C_k).

This type of continuity argument will sometimes be used in our applications without specific elaboration.

This seemingly innocuous but very pretty theorem would not necessarily impress one as something of great depth. It was Minkowski's genius to recognize that the implications of this theorem and the related geometric ideas went far beyond the application to bounds for quadratic forms; he made it a working tool for the number theorist. In his hands, the theorem blossomed into an important theory with many applications and led to the solutions of outstanding unsolved problems. This work also led to Minkowski's deep contributions to the geometry of convex sets [Min 2]. We shall concentrate on the applications of Minkowski's theorem and only mention other developments.

In 1914, H. F. Blichfeldt gave a new proof of Minkowski's Theorem [Bli] which has led to further research on the geometry of more general sets. The proof proceeds via a more general theorem.

Blichfeldt's Theorem: *Let M be an bounded set in $\mathbf{R}^n$ with $V(M) > 1$. Then M contains two distinct points x, y such that $x - y$ is a non-zero integer point (not necessarily in M).*

Corollary: *Blichfeldt $\Longrightarrow$ Minkowski (First form).*

Proof: Let $K \in \mathbf{R}^n$ be a bounded convex set, centered at the origin, with $V(K) > 2^n$. Let $K' = \frac{1}{2}K$; so $V(K') > 1$. By Blichfeldt's theorem, there exist $x, y \in K'$ such that $g = x - y$ is a non-zero integer point. Then $2x, 2y \in K$ and, by symmetry, $-2y \in K$. Hence, by convexity, the midpoint of $2x$ and $-2y$, namely, $\frac{1}{2}(2x + (-2y)) = x - y$, is a non-zero integer point in K.

Proof of Blichfeldt's theorem: Let A be the hypercube $\{0 < x_i < 1\}$ and $A_g = A + g$, for each integer point g. Since M is bounded, only a finite number of the sets $M_g = A_g \cap M$ are non empty and $M = \cup_g M_g$ (fig.5).

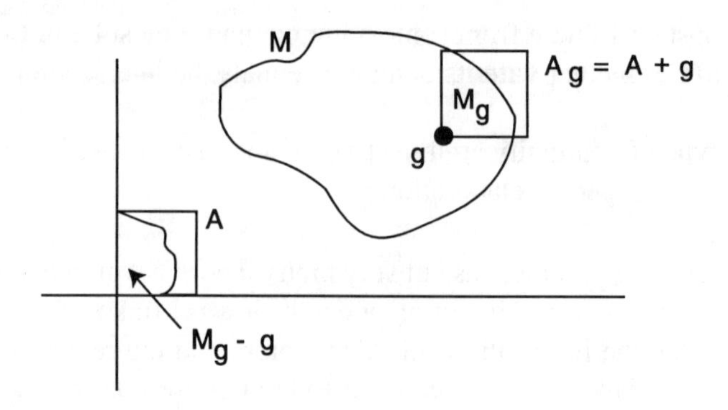

Figure 5

If we translate M_g back inside A (i.e., consider $M_g - g$), then $V(M_g - g) = V(M_g)$ and

$$\sum_g V(M_g - g) = \sum_g V(M_g) \geq V(M) > 1 .$$

Therefore, since $\cup_g (M_g - g) \subseteq A$ and $V(A) = 1$, two of the sets must overlap, and there exist integer points g, g' and a point p such that $p \in (M_g - g) \cap (M_{g'} - g')$. Hence

$$x = p + g \in M_g \quad \text{and} \quad y = p + g' \in M_{g'} ,$$

and

$$x - y = g - g'$$

is an integer point.

Moreover, if M is an open set, then all the other sets in the proof are open; in this case we can choose a p with rational coordinates and then x and y also have *rational* coordinates.

3. Minkowski's Theorem for Lattices

Lattices

Now we can use Minkowski's theorem to attack our original problem of minimizing quadratic forms (sec. 1).

Example: $n = 2$ — Consider the class of positive definite quadratic forms $Q(x, y) = ax^2 + by^2, a, b \in \mathbf{R}, a, b > 0$, with determinant $D = ab$. The sets $\{Q < k\}$ are ellipses with area $\frac{k\pi}{\sqrt{ab}} = \frac{k\pi}{D^{\frac{1}{2}}}$. By Minkowski's theorem, if

$$\frac{k\pi}{D^{\frac{1}{2}}} \geq 2^2 \quad \text{or} \quad k \geq \frac{4}{\pi} D^{\frac{1}{2}} ,$$

then there is an integer point $(g_1, g_2) \neq (0, 0)$ in the interior or on the boundary of the ellipse; hence $Q(g_1, g_2) \leq k$. Reformulating this, we see that there exist integers g_1, g_2, not both zero, such that

$$Q(g_1, g_2) \leq \frac{4}{\pi} D^{\frac{1}{2}} .$$

We thus have a result similar to Hermite's theorem (sec. 1) with $\left(\frac{4}{3}\right)^{\frac{1}{2}}$ replaced by the weaker bound $\frac{4}{\pi}$.

We could now study bounds for positive definite quadratic forms in n variables by computing volumes of ellipsoids, but instead we will reformulate Minkowski's theorem in terms of lattices in $\mathbf{R}^n$, a generalization of the two dimensional lattices of section 17.2, in order to reduce the problem to the study of n-dimensional spheres. This also leads to an easier way of applying Minkowski's theorem to a variety of number theoretic results.

It is a standard result in linear algebra that a positive definite quadratic form $Q(x)$ can be transformed into a sum of squares by a non-singular linear transformation, $y = Ax$, of its variables, i.e., there exists an $A = (a_{ij}), a_{ij} \in R, \det(A) \neq 0$, such that

$$Q'(y) = Q(A^{-1}y) = \sum y_i^2 .$$

The matrix A maps the set $\mathbf{Z}^n$ to the set $\{y = Ax | x \in \mathbf{Z}^n\}$. Recall that $\mathbf{Z}^n$ is called the **integer lattice**.

Definition: (I) An **n-dimensional lattice** Λ is the image of the integer lattice in $\mathbf{R}^n$ under a non-singular linear transformation $y = Ax$. The points of Λ are called Λ-**lattice points** or, when no confusion can arise, just **lattice points**. Thus, e.g., integer points are $\mathbf{Z}^n$-lattice points by the identity transformation.

Since the ellipsoid $\{Q < \lambda\}$ in x-space is mapped by our matrix A to the sphere $\{Q' = \sum y_i^2 < \lambda\}$ in y-space, our minimization problem for Q

is transformed to the problem of minimizing $\sum y_i^2$ on the lattice Λ. Thus the minimization problem for all positive definite Q of determinant D is changed to the problem of deciding how big λ must be so that the sphere $\{\sum y_i^2 < \lambda\}$ contains a point of every lattice corresponding to the Q's.

The problem of minimizing a class of forms over the integer lattice has now been transformed to the problem of minimizing a single form for a class of lattices, and we shall see that to solve the latter problem we only have to compute the volumes of spheres.

First we study lattices and then we generalize Minkowski's theorem from $\mathbf{Z}^n$ to arbitrary lattices. There are two other equivalent ways of defining a lattice:

(II) An n-dimensional lattice Λ is a set of the form

$$\Lambda = \mathbf{Z}\alpha_1 + \cdots + \mathbf{Z}\alpha_n = \{m_1\alpha_1 + \cdots + m_n\alpha_n | m_i \in \mathbf{Z},\ \alpha_i \in \mathbf{R}^n\}$$

where $\alpha_1, \ldots, \alpha_n$ are linearly independent over $\mathbf{R}$. The α_i can be taken as the column vectors of the matrix A in our original definition.

(III) An n-dimensional lattice Λ is a discrete subgroup of $\mathbf{R}^n$. By **discrete** we mean that the intersection of Λ with any bounded subset of the plane is finite.

Since the proofs of equivalence of definitions I – III are basically the same as those given for the case of two dimensional lattices in section 17.2, we omit them. For a deeper study of lattices, see Lekkerkerker [Lek] or Cassels [Cas 4].

A **sublattice** of a lattice Λ in $\mathbf{R}^n$ is a subset of Λ which is also a lattice in $\mathbf{R}^n$ (so, in particular, it contains n points linearly independent over $\mathbf{R}$).

The set $\{\alpha_1, \ldots, \alpha_n\}$ of (II) is called an **integral** or **lattice basis** of Λ. Λ has infinitely many such bases and we will study them in a moment.

$\Delta = |\det(A)|$ is the **determinant** of the lattice $\Lambda : y = Ax$; we shall prove it is independent of the basis chosen for Λ. Geometrically Δ is the volume of the parallelepiped, $T = \{m_1\alpha_1 + \cdots + m_n\alpha_n | m_i \in \mathbf{Z}$, $0 \le m_i \le 1\}$, whose edges are the column vectors $\alpha_1, \ldots, \alpha_n$ of A. This is clear since T is the image of the cube $\{0 \le x_i \le 1\}$, of volume one, under $y = Ax$ and thus $V(T) = |\det(A)| \times V(\text{cube})$. ($T$ is called a **fundamental parallelepiped** for Λ given by the basis $\alpha_1, \ldots, \alpha_n$.)

Δ can also be interpreted as the reciprocal of the "density" of the lattice Λ. We give a very brief sketch of this idea.

Let M be a bounded subset of $\mathbf{R}^n$ which has a volume and let $f(\Lambda) = |\Lambda \cap M|$ be the number of Λ-lattice points in M. Then we claim that

$$d = \lim_{\lambda \to \infty} \frac{f(\lambda M)}{V(\lambda M)}$$

exists and is independent of M; d is the **density** of Λ. First we note that by the map $y = Ax$, λM in y-space corresponds to the set $A^{-1}(\lambda M)$ in x-space, with volume $\frac{V(\lambda M)}{\Delta}$, and lattice points in λM correspond to integer points in $A^{-1}(\lambda M)$. Therefore $f(\lambda M)$ is the number of integer points in $A^{-1}(\lambda M)$ which, as λ gets large, is approximately $\frac{V(\lambda M)}{\Delta}$ (the proof of this is similar to our density argument in the first proof of Minkowski's theorem). Hence $\lim_{\lambda \to \infty} \frac{f(\lambda M)}{V(\lambda M)} = \frac{1}{\Delta}$; so Λ has a density, and $\Delta = \frac{1}{d}$ is independent of the basis.

CHANGE OF BASIS

Now we give a more rigorous proof that Δ is independent of the basis by studying how to change bases.

Let $\{\alpha_1, \ldots, \alpha_n\}$ and $\{\beta_1, \ldots, \beta_n\}$ be bases of Λ. Then Λ is given by $\{y = Ax\} = \{y = Bx\}$, where A (resp. B) is the matrix with column vectors α_i (resp. β_i). Since every β is an integral linear combination of the α's, we have $\beta_k = \sum_i \alpha_i p_{ik}$, for some $p_{ik} \in \mathbf{Z}$, and thus $B = AP$, where $P = (p_{ij})$ is an *integral* matrix ($p_{ij} \in \mathbf{Z}$). Similarly $A = BQ$, for some integral matrix Q. Therefore $A = APQ$ and, since $\det(A) \neq 0$, $PQ = I$. But $\det(P) = \frac{1}{\det(Q)}$ and $\det(P)$, $\det(Q)$ are integers. Hence $\det(P) = \det(Q) = \pm 1$ and we have

Theorem: *If the columns of A and B are bases for the lattice Λ, then $B = AP$, where P is a* **unimodular** *matrix (integral with determinant ± 1).*

Corollary: *Δ is independent of the basis chosen for Λ.*

Proof: $B = AP \implies |\det(B)| = |\det(A)|\,|\det(P)| = |\det(A)|$.

The converse of the last theorem is also true.

Theorem: *If the columns of A are a basis for Λ and P is unimodular, then the columns of $B = AP$ are a basis for Λ.*

Proof: (exercise — this follows from the fact that the inverse of a unimodular matrix is unimodular, which is proved by using the formula for the inverse of a matrix in terms of cofactors.)

MINKOWSKI'S THEOREM REFORMULATED

Let $\Lambda : y = Ax$ be a lattice of determinant Δ. Then A defines a linear transformation from $\mathbf{R}^n$ to $\mathbf{R}^n$ (as we did before, we think of A as a map from x-space with integer points to y-space with Λ-points). Let K be a bounded, convex set in y-space, centered at the origin, with $V(K) > 2^n\Delta$. Then $A^{-1}K$ is a bounded, convex set in x-space, centered at the origin, with $V(A^{-1}K) > 2^n$ (a linear transformation preserves convexity, symmetry and boundedness). By Minkowski's theorem, $A^{-1}K$ contains a non-zero integer point g, and thus the Λ-point Ag is in K. Hence we have

Minkowski's Theorem for Lattices:

1) *A bounded convex set K centered at the origin with $V(K) > 2^n\Delta$ contains a non-zero lattice point of every lattice of determinant Δ.*

2) *Similarly, if $V(K) \geq 2^n\Delta$, then there is a non-zero lattice point in K or on its boundary.*

This form of the theorem suggested new questions to Minkowski. A lattice Λ is said to be **admissible** for a bounded symmetric convex set K if there are no non-zero Λ- points in K. Minkowski's theorem tells us

$$\det \text{ (admissible lattice for K) } \geq \frac{V(K)}{2^n}. \tag{1}$$

$\Delta(\mathbf{K})$, **the critical determinant of K**, is the greatest lower bound of $\det(\Lambda)$, taken over all admissible lattices for K. An admissible lattice Λ for K with $\det(\Lambda) = \Delta(K)$ is a **critical lattice** for K. For a given convex K (and for more general non-convex sets), the exact evaluation of $\Delta(K)$ and finding a critical lattice are central problems of the geometry of numbers, with strong number theoretic consequences. We do not pursue this theme here but refer to [Cas 4] and [Lek].

4. Back to Quadratic Forms

As described at the beginning of the last section, the problem of minimizing all positive definite quadratic forms $Q(x_1, \ldots, x_n)$ of determinant D over the integer lattice is changed to the problem of deciding how big λ must be so that the sphere $\{\sum y_i^2 < \lambda\}$ contains a point of every lattice $\Lambda : y = Ax$, for which A transforms a Q to $\sum y_i^2$.

By Minkowski's theorem for lattices, we need two pieces of information: (i) the volume of $\{\sum_{i=1^n} y_i^2 < \lambda\}$ and (ii) the determinant of Λ.

(i) Let $S_\lambda^n = \{\sum y_i^2 < \lambda\}$. A point y satisfies $\sum y_i^2 < 1$ if and only if $\sqrt{\lambda}\, y$ satisfies $\sum(\sqrt{\lambda} y_i)^2 = \lambda \sum y_i^2 < \lambda$. Hence $S_\lambda^n = \sqrt{\lambda} S_1^n$ and $V(S_\lambda^n) = \lambda^{\frac{n}{2}} V(S_1^n)$. But

$$V(S_1^n) = \frac{2\pi^{\frac{n}{2}}}{n\Gamma\left(\frac{n}{2}\right)}$$

where Γ is the gamma function (see Siegel [Sie 5, pp. 25 - 26]).

(ii) If $y = Ax$, with $\det(A) = \Delta$, transforms Q to $\sum y_i^2$, i.e., $\sum y_i^2 = Q(A^{-1}y)$, then it is known from linear algebra [Bor - Sha, supplement A] that

$$\det\left(\sum y_i^2\right) = \det(Q) \cdot \left(\det(x = A^{-1}y)\right)^2 ;$$

thus

$$1 = D \cdot (\Delta^{-1})^2$$

and

$$\det(\Lambda) = \Delta = D^{\frac{1}{2}} .$$

Hence we have $V(S_\lambda^n) \geq 2^n \Delta = 2^n D^{\frac{1}{2}}$, if $\frac{\lambda^{\frac{n}{2}} \pi^{\frac{n}{2}}}{\Gamma(1+\frac{1}{n})} \geq 2^n D^{\frac{1}{2}}$ or

$$\lambda \geq \frac{4}{\pi}\left(\Gamma\left(1+\frac{1}{n}\right)\right)^{\frac{2}{n}} D^{\frac{1}{n}} .$$

By Minkowski's theorem for lattices, this means that every lattice of determinant $D^{\frac{1}{2}}$ contains a non-zero lattice point y such that

$$\sum y_i^2 \leq \frac{4}{\pi}\left(\Gamma\left(1+\frac{1}{n}\right)\right)^{\frac{2}{n}} D^{\frac{1}{n}} .$$

Transforming back to x-space, we have

Theorem (Minkowski): *A positive definite quadratic form Q in n variables with real coefficients and determinant D assumes a value*

$$Q(x) \leq \frac{4}{\pi}\left(\Gamma\left(1+\frac{1}{n}\right)\right)^{\frac{2}{n}} D^{\frac{1}{n}},$$

for some non-zero integer point x.

The constant in Hermite's theorem (sec. 1) was $\left(\frac{4}{3}\right)^{\frac{n-1}{2}}$. Minkowski's is better for $n \geq 4$.

5. Sums of Two and Four Squares

In section 12.6, we proved

Theorem: *A prime* $p \equiv 1 \pmod 4$ *is the sum of two integer squares, i.e.,* $p = \lambda_1^2 + \lambda_2^2$, *for some* $\lambda_1, \lambda_2 \in \mathbf{Z}$,

and then generalized this to decide which positive integers are the sum of two squares.

Now we give a geometric proof of this theorem as a prototype for the corresponding theorem about four squares.

Proof: (i) For the moment, we assume that there exists a sublattice Λ of the integer lattice in $\mathbf{R}^2$, with $\det(\Lambda) = p$, such that

$$y_1^2 + y_2^2 \equiv 0 \pmod p ,$$

for all (y_1, y_2) in Λ.

(ii) The circle $y_1^2 + y_2^2 < 2p$ has area

$$4\pi p^2 > 4p = 4 \cdot \det(\Lambda) ,$$

and thus, by Minkowski's theorem for lattices, there is a Λ point $(\lambda_1, \lambda_2) \neq (0, 0)$ in the circle. Therefore we have

$$0 < \lambda_1^2 + \lambda_2^2 < 2p$$

as well as

$$\lambda_1^2 + \lambda_2^2 \equiv 0 \pmod p .$$

Since $\lambda_1^2 + \lambda_2^2$ is a multiple of p which is strictly between 0 and $2p$, it must equal p.

Now we construct Λ and then we are done. For any integer u, let Λ_u be the lattice

$$y_1 = px_1 + ux_2$$

$$y_2 = \qquad\quad x_2 ,$$

where $(x_1, x_2) \in \mathbf{Z}^2$; then $\det(\Lambda_u) = p$. Since $y_1 \equiv uy_2 \pmod p$, we have

$$y_1^2 + y_2^2 \equiv (u^2 + 1)y_2^2 \pmod p ,$$

for all $(y_1, y_2) \in \Lambda_u$. Since $p \equiv 1(\mathrm{mod}\, 4)$, -1 is a quadratic residue modulo p and there is an integer w such that $w^2+1 \equiv 0(\mathrm{mod}\, p)$. Therefore $\Lambda = \Lambda_w$ is the desired lattice.

Now we use the same idea to prove

Theorem: *(Lagrange): Every positive integer m is the sum of four integer squares,*

$$m = \lambda_1^2 + \lambda_2^2 + \lambda_3^2 + \lambda_4^2\,, \quad \lambda_i \in \mathbf{Z}\,.$$

As in the case of two squares, we can reduce the proof of the theorem to the result for primes by noting that if two integers are each the sum of four squares, then so is their product. This follows from the algebraic identity

$$\begin{aligned}
&(x_1^2 + x_2^2 + x_3^2 + x_4^2)(y_1^2 + y_2^2 + y_3^2 + y_4^2) \\
&\quad = (x_1y_1 - x_2y_2 - x_3y_3 - x_4y_4)^2 + (x_1y_2 + x_2y_1 + x_3y_4 - x_4y_3)^2 \\
&\quad + (x_1y_3 - x_2y_4 + x_3y_1 + x_4y_2)^2 + (x_1y_4 + x_2y_3 - x_3y_2 + x_4y_1)^2\,.
\end{aligned}$$

Just as the analogous identity for sums of two squares expresses the multiplicative property of the norm (absolute value) for complex numbers, this identity expresses the multiplicative property of the norm for the quaternions [Har - Wri].

Now we prove

Theorem (Lagrange): *Every prime p is the sum of four integer squares,*

$$p = \lambda_1^2 + \lambda_2^2 + \lambda_3^2 + \lambda_4^2\,, \lambda_i \in \mathbf{Z}\,,$$

Proof: (i) For the moment, we assume there exists a sublattice Λ of the integer lattice in $\mathbf{R}^4$, with $\det(\Lambda) = p^2$, such that

$$y_1^2 + y_2^2 + y_3^2 + y_4^2 \equiv 0(\mathrm{mod}\, p)\,,$$

for all (y_1, y_2, y_3, y_4) in Λ.

(ii) From section 4, we know that the solid sphere $S_{2p}^{(4)} = \{y_1^2 + y_2^2 + y_3^2 + y_4^2 < 2p\}$ has volume

$$\frac{1}{2}\pi^2(2p)^2 = 2p^2\pi^2 > 2^4p^2 = 2^4 \cdot \det(\Lambda)\,,$$

and thus, by Minkowski's theorem for lattices, there is a non-zero Λ-point $(\lambda_1, \lambda_2, \lambda_3, \lambda_4)$ in this sphere. Therefore we have

$$0 < \lambda_1^2 + \lambda_2^2 + \lambda_3^2 + \lambda_4^2 < 2p$$

as well as

$$\lambda_1^2 + \lambda_2^2 + \lambda_3^2 + \lambda_4^2 \equiv 0 (\text{mod } p) .$$

Since $\lambda_1^2 + \lambda_2^2 + \lambda_3^2 + \lambda_4^2$ is a multiple of p which is strictly between 0 and $2p$, it must equal p, i.e., $p = \lambda_1^2 + \lambda_2^2 + \lambda_3^2 + \lambda_4^2$.

Now we construct Λ and then we are done. For any integers u, v, let $\Lambda_{u,v}$ be the lattice

$$\begin{aligned} y_1 &= \quad\quad\quad\quad + x_3 \\ y_2 &= \quad\quad\quad\quad\quad\quad + x_4 \\ y_3 &= p x_1 \quad\quad + u x_3 + v x_4 \\ y_4 &= \quad\quad p x_2 - v x_3 + u x_4 , \end{aligned}$$

where $(x_1, x_2, x_3, x_4) \in \mathbf{Z}^4$; then $\det(\Lambda_{u,v}) = p^2$. A direct computation, which can be simplified by noting that $y_3 \equiv u y_1 + v y_2 (\text{mod } p)$ and $y_4 \equiv -v y_1 + u y_2 (\text{mod } p)$, yields

$$y_1^2 + y_2^2 + y_3^2 + y_4^2 \equiv (u^2 + v^2 + 1)(y_1^2 + y_2^2)(\text{mod } p) ,$$

for all $(y_1, y_2, y_3, y_4) \in \Lambda_{u,v}$. If we can find integers t, w such that $t^2 + w^2 + 1 \equiv 0 (\text{mod } p)$, then $\Lambda = \Lambda_{t,w}$ is the required lattice.

By Chevally's theorem (sec. 9.6), $x^2 + y^2 + z^2 \equiv 0 (\text{mod } p)$ has a non-zero solution $(a, b, c)(\text{mod } p)$; say $c \not\equiv 0 (\text{mod } p)$. Then c^{-1} exists modulo p and $(ac^{-1})^2 + (bc^{-1})^2 + 1 \equiv 0 (\text{mod } p)$. Setting $t = ac^{-1}$ and $w = bc^{-1}$, we have $t^2 + w^2 + 1 \equiv 0 (\text{mod } p)$ and we are done.

There is also a quick direct proof of this last result. If $p = 2$, then $1^2 + 0^2 + 1 \equiv 0 (\text{mod } 2)$. If p is odd then $\{u^2 \mid 0 \le u \le \frac{p}{2}\}$ and $\{-1 - v^2 \mid 0 \le v \le \frac{p}{2}\}$ are sets of $\frac{p+1}{2}$ integers such that the elements of each set are not congruent modulo p. But there are only p congruence classes modulo p; so there is an integer s congruent to an integer of each set, i.e., there are t, w such that $t^2 \equiv s \equiv -1 - w^2 (\text{mod } p)$ or $t^2 + w^2 + 1 \equiv 0 (\text{mod } p)$.

There is a common technique behind the proofs of the two and four square theorem, namely, if we take the integer lattice $\mathbf{Z}^n$ and impose m homogeneous linear congruence conditions on the coordinates modulo

$k_1, \ldots, k_n$ respectively, then the set of integer points satisfying these conditions is a sublattice of determinant $\le k_1 k_2 \cdots k_n$. Thus for the two square theorem we set $y_1 \equiv u y_2 (\text{mod } p)$ to get the lattices Λ_u. For the four square theorem, we set $y_3 \equiv u y_1 + v y_2 (\text{mod } p)$ and $y_4 \equiv -v y_1 + u y_2 (\text{mod } p)$.

This technique can also be used to prove Legendre's theorem, which yields an algorithm for deciding if a conic contains a point with rational coordinates (sec. 18.3 and [Cas 4, sec. III.7]). However, although we can prove three important theorems using this technique, it has not yet proved capable of yielding more general theorems on the representation of integers by quadratic forms. Thus, for now, it remains a special trick and not a general method.

6. Linear Forms

Minkowski crafted the geometry of numbers into a valuable tool for number theorists. After applying the theory to quadratic forms, he then applied his fundamental theorem to systems of linear forms yielding basic results in algebraic number theory such as Dirichlet's characterization of units in a number field and the finiteness of the class number (see [Hec]). More importantly, he proved a conjecture of Kronecker on 'discriminants' of number fields, which we shall treat in the next section. First we discuss linear forms.

The **box,** $B = \{|y_1| < \lambda_1, \ldots, |y_n| < \lambda_n\}$, is a convex set in $\mathbf{R}^n$ symmetric about the origin, with $V(B) = 2^n \lambda_1 \cdots \lambda_n$. This is intuitive when viewed geometrically and not difficult to prove directly (exercise). We shall provide a general method of proving such results in section 8.

By Minkowski theorem for lattices (sec. 3), a lattice Λ of determinant Δ has a point inside the box if

$$2^n \lambda_1 \cdots \lambda_n > 2^n \Delta \,,$$

i.e.,

$$\lambda_1 \cdots \lambda_n > \Delta \,.$$

Thus, recalling that a lattice, $\Lambda : y = Ax$, is given by linear forms in x_i, we have

Minkowski's Linear Form Theorem: *If* $y = Ax$, $A = (a_{ij})$, $a_{ij} \in \mathbf{R}$, $\Delta = |\det(A)| \neq 0$, *and if* $\lambda_1, \ldots, \lambda_n > 0$ *satisfy* $\lambda_1 \cdots \lambda_n > \Delta$, *then there*

is a non-zero integer point $(x_1, \ldots, x_n)$ *such that*

$$|y_1| < \lambda_1, \ldots, |y_n| < \lambda_n ,$$

i.e.,

$$|a_{i1}x_1 + a_{i2}x_2 + \cdots + a_{in}x_n| < \lambda_i , \quad \textit{for all } i.$$

If $\lambda_1 \cdots \lambda_n \geq \Delta$, the result holds with $|y_i| < \lambda_i$ replaced by $|y_i| \leq \lambda_i$, for all i. However, we can do a little more.

Corollary: *If* $\lambda_1 \cdots \lambda_n = \Delta$, *then there is a non-zero integer point such that*

$$|y_1| \leq \lambda_1, |y_2| < \lambda_2, \ldots, |y_n| < \lambda_n ,$$

and similarly for any other λ_i.

Proof: Apply the second form of Minkowski's theorem (sec.2) to the box $\{|y_1| < (1+\epsilon)\lambda_1, |y_2| < \lambda_2, \ldots, |y_n| < \lambda_n\}$ and let $\epsilon \to 0$. This is the same continuity argument used to prove the third form of Minkowski's theorem (sec. 2).

Example: Diophantine Approximation — In sections 4.8 and 21.2, we proved Lagrange's theorem, viz., that for α irrational, there are infinitely many distinct rational solutions $\frac{p}{q}$ of $\left|\alpha - \frac{p}{q}\right| < \frac{1}{q^2}$. Now we give yet another proof using the linear forms theorem.

Proof: Let

$$y_1 = x_1 - \alpha x_2$$
$$y_2 = \qquad x_2 ;$$

thus $\Delta = 1$. Let $\lambda_1 = \frac{1}{k}$ and $\lambda_2 = k$; so $\lambda_1\lambda_2 = 1$. By the corollary, there is an integer point $(x_1, x_2) \neq (0, 0)$ such that

$$|x_1 - \alpha x_2| < \frac{1}{k}$$
$$|x^2| \leq k .$$

Now if we let $k = 1, 2, \ldots$, there are infinitely many integer solutions to the pair of inequalities; otherwise$|x_1 - \alpha x_2| > k'$ for some k' and $x_1, x_2 \in \mathbf{Z}$, contradicting $|x_1 - \alpha x_2| < \frac{1}{k}$ for k sufficiently large. Furthermore, if x_2 takes any fixed value, say m, there are only finitely many integer solutions, since only finitely many integers x_1 can satisfy $|x_1 - \alpha m| < 1$ (or $\leq \frac{1}{k}$). Hence there is a sequence $\left(x_1^{(1)}, x_2^{(1)}\right), \left(x_1^{(2)}, x_2^{(2)}\right), \ldots$, with $x_2^{(i)} \to \infty$, such that

$$\left| \frac{x_1^{(i)}}{x_2^{(i)}} - \alpha \right| < \frac{1}{\left|x_2^{(i)}\right|^k} \leq \frac{1}{\left(x_2^{(i)}\right)^2} .$$

Since $x_2^{(i)} \to \infty$, there are infinitely many distinct rationals among the $\frac{x_1^{(i)}}{x_2^{(i)}}$ and we are done.

Minkowski also derived the theory of continued fractions (and the associated theory of indefinite binary quadratic forms) geometrically with the linear form theorem (see [Min 1]). Starting with Klein's interpretation of continued fractions in terms of lattice points in the plane (sec. 4.7), he made a detailed study of the distribution of lattice points on the boundaries of rectangles whose sides are defined by the linear forms $|ax + by| = k_1$ and $|cx + dy| = k_2$. Minkowski expressed considerable pleasure about his proof of periodicity for quadratic irrationalities, which he regarded as the most natural proof of this theorem. This geometric approach was also the basis of Minkowski's generalization of continued fractions [Min 2].

7. Sums and Products of Linear Forms; The Octahedron

The set

$$K = \left\{ \sum_1^n |y_i| < \lambda \right\}$$

is a bounded convex set in $\mathbf{R}^n$, symmetric with respect to the origin (see the next section for a method of proof). For $n = 2$, we have a square (fig. 6), and for $n = 3$, an octahedron. For general n, we also call K the (generalized) **octahedron**. To find its volume, note that the octahedron consists of 2^n congruent pieces, one in each octant. The volume of the piece in the positive octant (all $y_i > 0$) is

$$\lambda^n \int_0^1 \int_0^{1-y_1} \cdots \int_0^{1-y_1-\cdots-y_n} dy_1 dy_2 \cdots dy_n = \frac{\lambda^n}{n!} ;$$

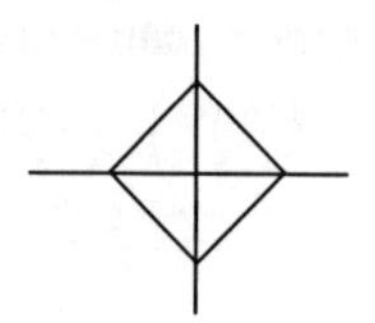

Figure 6

hence $V(K) = 2^n \frac{\lambda^n}{n!}$. If a lattice Λ, with determinant Δ, satisfies

$$2^n \frac{\lambda^n}{n!} \geq 2^n \Delta \quad \text{or} \quad \lambda \geq (n!\Delta)^{\frac{1}{n}} ,$$

then K contains a non-zero point of Λ. Thus, by Minkowski's theorem for lattices, we have

Theorem (Sums of Linear Forms): *If* $y = Ax$, $A = (a_{ij}), a_{ij} \in \mathbf{R}$, $\Delta = |\det(A)| \neq 0$, *then there exists a non-zero integer point* $(x_1, \ldots, x_n)$ *such that*

$$|y_1| + \cdots + |y_n| \leq (n!\Delta)^{\frac{1}{n}} .$$

Example: For $n = 2$, the theorem guarantees the existence of a non-zero integer point (x_1, x_2) satisfying

$$|ax_1 + bx_2| + |cx_1 + dx_2| \leq \sqrt{2\Delta} ,$$

where $\Delta = |ad - bc| \neq 0$. As Siegel points out [Sie 5], the case $a = \sqrt{7}$, $b = \sqrt{6}, c = \sqrt{15}, d = \sqrt{13}$ shows that this is a non trivial result.

Products of linear forms can be reduced to sums by means of the inequality for the arithmetic and geometric means, namely,

$$|y_1 \cdots y_n|^{\frac{1}{n}} \leq \frac{1}{n}(|y_1| + \cdots + |y_n|)$$

or

$$|y_1 \cdots y_n| \leq \frac{1}{n^n}(|y_1| + \cdots + |y_n|)^n .$$

Applying the last theorem, we now have

Theorem (Products of Linear Forms): *Under the same assumptions as in the last theorem, there is a non-zero integer points* $(x_1, \ldots, x_n)$ *such that*

$$|y_1 \cdots y_n| \leq \frac{n!\Delta}{n^n} .$$

Example: Discriminants of Algebraic Number Fields — As we mentioned earlier, Minkowski used the geometry of numbers to prove a conjecture of Kronecker, viz.,

the 'discriminant' of a number field ($\neq \mathbf{Q}$) is greater than one.

We defined the notions needed to understand this result in the context of quadratic fields (chapters 16 and 17), where the result is trivial (sec. 17.8). Now we quickly review the basic ideas needed to understand the general conjecture and prove a special case of Minkowski's result.

A complex number θ is an **algebraic number** (of **degree n**) if it is the root of an irreducible polynomial $\sum_0^n a_i x^i$, with integer coefficients. If $a_n = 1$, θ is an **algebraic integer**. An **algebraic number field** K (or just **number field**) is a subfield of **C** which is a finite extension of the rationals, i.e., $n = \deg(K/\mathbf{Q}) < \infty$.

Every number field K of degree n over **Q** is **generated** by an algebraic number θ of degree n, in the sense that

$$K = \mathbf{Q}(\theta) = \{u_0 + u_1\theta + \cdots + u_{n-1}\theta^{n-1} | u_i \in \mathbf{Q}\} ,$$

i.e., if $\omega \in K$, then $\omega = q(\theta)$, for some $q(x) \in \mathbf{Q}[x]$, $\deg q \leq n$. If $p(x)$ is the irreducible equation for θ, with roots $\theta^{(0)} = \theta, \theta^{(1)}, \ldots, \theta^{(n-1)}$ (the **conjugates** of θ), then $\omega^{(0)} = \omega = q(\theta)$, $\omega^{(1)} = q(\theta^{(1)}), \ldots, \omega^{(n-1)} = q(\theta^{(n-1)})$ are the **conjugates of ω in K** (they are independent of the choice of which element generates K).

$\mathbf{I_K}$, the set of algebraic integers of K, is a ring with an **integral basis**, i.e., there exist $\omega_0, \ldots, \omega_{n-1} \in I_K$ such that

$$I_K = \left\{ \sum_0^{n-1} v_i \omega_i | v_i \in \mathbf{Z} \right\} .$$

With this basis, let $\omega_i^{(j)}$, $j = 0, \ldots, n-1$, be the n conjugates of ω_i, with $\omega_i^{(0)} = \omega_i$, and consider the linear forms

$$\xi^{(0)} = x_0\omega_0^{(0)} + \cdots + x_{n-1}\omega_{n-1}^{(0)}$$
$$\xi^{(1)} = x_0\omega_0^{(1)} + \cdots + x_{n-1}\omega_{n-1}^{(1)}$$
$$\vdots$$
$$\xi^{(n-1)} = x_0\omega_0^{(n-1)} + \cdots + x_{n-1}\omega_{n-1}^{(n-1)} .$$

The **discriminant of K** is defined by $\mathbf{d_K} = [\det(\omega_i^{(j)})]^2$. The following properties will prove useful:

(i) d_K is a non-zero rational integer, independent of the choice of integral basis,

(ii) if $K = \mathbf{Q}(\theta)$ is **totally real** (θ and all its conjugates are real) then d_K is a positive integer,

(iii) the product $\xi^{(0)} \cdots \xi^{(n-1)}$ is a non-zero integer for any non-zero integer point $(x_0, \ldots, x_{n-1})$.

Now we prove Kronecker's conjecture for totally real fields by finding a lower bound for d_K. The proof for all number fields, given in [Min 1] and [Rib], is not much harder.

By the theorem for products of linear forms applied to the $\xi^{(i)}$, with $\Delta = \sqrt{d_K}$, there exists a non-zero integer point $(x_0, \ldots, x_{n-1})$ such that

$$|\xi^{(0)} \cdots \xi^{(n-1)}| \leq \frac{n!\sqrt{d_K}}{n^n} .$$

By (iii), the left hand side is a positive integer and therefore greater than or equal to one. Hence $1 \leq \frac{n!\sqrt{d_K}}{n^n}$ or $d_K \geq \left(\frac{n^n}{n!}\right)^2$, and $d_K > 1$ for $n > 1 (K \neq \mathbf{Q})$.

Corollary: *The discriminant of a totally real number field $K (\neq \mathbf{Q})$ is divisible by at least one rational prime.*

This has the same significance for general number fields as in the quadratic case (sec. 17.6, 17.12), namely, the existence of 'ramified primes'.

It should be noted that the only known general construction for finding n linear forms whose product is not very small is by taking a form, whose coefficients are in a number field, and constructing the other forms by taking conjugates, as in our example.

8. Gauge Functions; The Equation of a Convex Body

Minkowski introduced the gauge function of a convex set which provides an analytic description of the set. By characterizing which functions are gauge functions, he also produced a general procedure which can often be used to prove that a set is convex. The gauge function also provides a direct way of stating our arithmetic results on minimization of functions (or sets of functions) over integer points, instead of having to translate from geometry to arithmetic. As before, this new point of view leads to yet another formulation of Minkowski's theorem and to new generalizations.

We shall only consider **convex bodies**, i.e., bounded open convex sets containing the origin (*warning*: there is some variation in this terminology among other authors). Since the interior of a convex set is an open convex set, this is not a real restriction. ∂**B** will denote the **boundary** of B.

Let B be a convex body in $\mathbf{R}^n$. The **gauge function** of B is a function $f : \mathbf{R}^n \to [0, \infty)$ defined as follows:

(i) $f(\mathbf{0}) = 0$,

(ii) for $x \neq \mathbf{0}$, $f(x) = \frac{\|x\|}{\|x'\|}$, where x' is the intersection of the ray from $\mathbf{0}$ to x with the boundary of B (fig. 7), and $\|y\| = \sqrt{\sum y_i^2}$ denotes the length of $y = (y_1, \ldots, y_n)$.

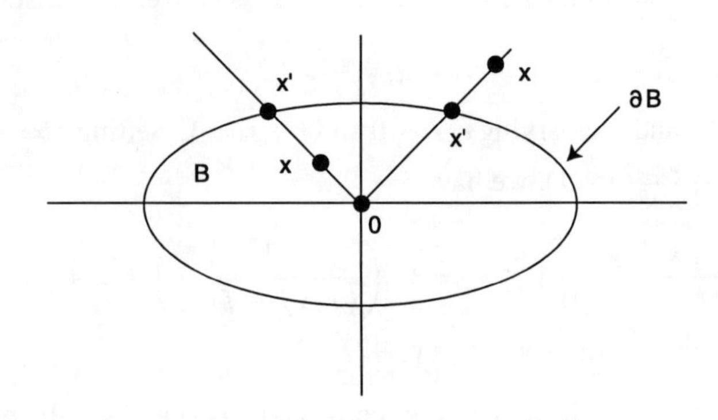

Figure 7

Equivalently, if $x = \mu x'$, $\mu > 0$, $x' \in \partial B$, then $f(x) = \mu$. Note that

(a) $f(x) < 1 \iff x \in B$,
(b) $f(x) = 1 \iff x \in \partial B$,
(c) $f(x) \leq 1 \iff x \in \overline{B}$,

where $\overline{\mathbf{B}} = B \cup \partial B$, the **closure** of B.

Thus (a) can be regarded as an equation (or, more precisely, an inequality) for B and (b) is an equation for the boundary.

The gauge function measures the distance from $\mathbf{0}$ to x with $\|x'\|$ as the unit. If we set $\rho(x, y) = \|x - y\|$, then ρ is a metric on $\mathbf{R}^n$. These metrics are a special class of the so called 'projective metrics', and the 'Minkowski geometry' induced by them is studied in Busemann and Kelly [Bus - Kel].

Theorem: *If B is a convex body with gauge function $f(x)$, then*

(i) $f(x) > 0$ *for* $x \neq \mathbf{0}$, $f(\mathbf{0}) = 0$,
(ii) $f(\lambda x) = \lambda f(x)$, *for* $\lambda \in \mathbf{R}$, $\lambda \geq 0$,
(iii) $f(x + y) \leq f(x) + f(y)$.

Moreover f is continuous.

Proof: (i) follows immediately from the definition of f. To prove (ii), suppose that $x = \mu x'$, $\mu > 0$, $x' \in \partial B$; so f$(x) = \mu$. Then $f(\lambda x) = f(\lambda \mu x') = \lambda \mu = \lambda f(x)$.

The triangle inequality (iii) takes more work. If x or y is $\mathbf{0}$, then it is obvious. Otherwise, we normalize by considering $x' = \frac{x}{f(x)}$, $y' = \frac{y}{f(y)}$; so $f(x') = f(y') = 1$ and $x', y' \in \partial B$. Since $\overline{B}$ is convex (we assume this), we have

$$f(rx' + sy') \leq 1\,, \tag{1}$$

for any real r and s satisfying $r, s > 0$ and $r + s = 1$. Setting $r = \frac{f(x)}{f(x)+f(y)}$ and $s = \frac{f(y)}{f(x)+f(y)}$ in (1), we have

$$f\left(\frac{x + y}{f(x) + f(y)}\right) \leq 1 \implies \left(\frac{1}{(f(x) + f(y))}\right) f(x + y) \leq 1$$
$$\implies f(x + y) \leq f(x) + f(y)\,.$$

The continuity follows from properties (i) – (iii) and will be proved as part of the next theorem.

Symmetry is incorporated into the gauge function by

Proposition: *B is symmetric $\iff$ f is an even function* $(f(x) = f(-x))$.

Proof: $\Longrightarrow$) $x = \lambda x', x' \in \partial B \implies -x = \lambda(-x'), -x \in B$ (by symmetry) $\implies f(x) = f(-x) = \lambda$.

$\Longleftarrow$) $x \in B \implies f(x) < 1 \implies f(-x) = f(x) < 1 \implies -x \in B$.

Exercises: (I) Let B be the square in $\mathbf{R}^2$, centered at the origin with sides of length 2 parallel to the axes. Show that the gauge function of B is $f(x) = \max\{|x_1|, |x_2|\}$, where $x = (x_1, x_2)$. Find the gauge function for the general box $\{|x_1| < \lambda_1, |x_2| < \lambda_2\}$.

(II) Let B be the interior of the ellipse $\frac{x_1^2}{a^2} + \frac{x_2^2}{b^2} = 1$. Show that the gauge function $f(x) = \sqrt{\frac{x_1^2}{a^2} + \frac{x_2^2}{b^2}}$.

The general idea for finding a gauge function of a convex body B is to find an equation for the boundary of B of the form $g(x) = 1$ and then, if necessary, fiddle with g to make it homogeneous (i.e., to satisfy part (ii) of the theorem). Thus in exercise (II), we have $g(x) = \frac{x_1^2}{a^2} + \frac{x_2^2}{b^2}$ and taking the square root makes it homogeneous. $\sqrt{g(x)}$ clearly satisfies part (i) of the theorem, but it is still necessary to verify (iii), the triangle inequality.

In many of our earlier applications we proceeded in the opposite direction, beginning with a function f and assuming that the set $\{f < \lambda\}$ is a convex body. The convexity can often be proved by the converse to the last theorem.

Theorem: *If a function $f : \mathbf{R}^n \to [0, \infty)$ satisfies*

(i) $f(x) > 0$ *for* $x \neq \mathbf{0}$, $f(\mathbf{0}) = 0$,

(ii) $f(\lambda x) = \lambda f(x)$, *for* $\lambda \in \mathbf{R}$, $\lambda \geq 0$,

(iii) $f(x + y) \leq f(x) + f(y)$,

then f is continuous, $B = \{x | f(x) < 1\}$ is a convex body, and f is the gauge function of B.

Proof: 1) f is continuous — First we prove continuity at $\mathbf{0}$. If $e_1 = (1, 0, \ldots, 0), \ldots, e_n = (0, \ldots, 0, 1)$ denote the unit vectors, then every x

in $\mathbf{R}^n$ can be written in the form

$$x = \sum_1^n \lambda_i e_i = \sum_1^n |\lambda_i|(\pm e_i) .$$

Hence, by (i) – (iii),

$$0 \leq f(x) \leq \sum f(|\lambda_i|(\pm e_i)) = \sum |\lambda_i| f(\pm e_i) .$$

Therefore, as $x \to \mathbf{0}$, $|\lambda_i| \to 0$, $f(x) \to 0 = f(\mathbf{0})$, and f is continuous at $\mathbf{0}$.

To prove continuity at $x \neq \mathbf{0}$, we write $x = (x + y) + (-y)$, for any $y \neq \mathbf{0}$. Then $f(x) \leq f(x + y) + f(-y)$ or $-f(-y) \leq f(x + y) - f(x)$. From $f(x + y) \leq f(x) + f(y)$, we have $f(x + y) - f(x) \leq f(y)$. Combining these inequalities, we have

$$-f(-y) \leq f(x + y) - f(x) \leq f(y) .$$

By continuity at $\mathbf{0}$, $\lim_{y \to \mathbf{0}} f(y) = \lim_{y \to \mathbf{0}} f(-y) = 0$ and thus f is continuous at x.

2) B is open since it is the pre-image of an open set under the continuous map f.

3) B is convex — Let $x, y \in B$; so $f(x) < 1$, $f(y) < 1$. For $\lambda, \mu > 0$, $\lambda + \mu = 1$, we have $f(\lambda x + \mu y) \leq f(\lambda x) + f(\mu y) = \lambda f(x) + \mu f(y) < \lambda + \mu = 1$. Hence $\lambda x + \mu y \in B$ and B is convex.

4) B is bounded — Assume B is unbounded. Then there exists a sequence of points in B, say $x_0, x_1, \ldots$, such that

$$f(x_n) < 1 \text{ and } \|x_n\| \to \infty, \text{ as } n \to \infty .$$

Let $\lambda_n = \frac{1}{\|x_n\|}$. Then as $n \to \infty$,

$$f(\lambda_n x_n) = \frac{f(x_n)}{\|x_n\|} < \frac{1}{\|x_n\|} \to 0 .$$

But $\|\lambda_n x_n\| = 1$ and thus $\lambda_n x_n$ lies on the unit sphere in $\mathbf{R}^n$, which is a compact set. Since f is a continuous function on a compact set, it attains a minimum $f(x')$ at some point x' on the set and, by (i), $f(x') > 0$. This contradicts $f(\lambda_n x_n) \to 0$.

5) f is the gauge function of B (exercise).

Exercise: Show that $f(x) = (\sum |x_i|^r)^{\frac{1}{r}}$, $x \in \mathbf{R}^n$, $r > 1$, is an even gauge function. The triangle inequality is known as the Minkowski inequality (see [Sie 5] or [Har - Lit - Pol]) and this setting is probably the reason Minkowski studied this inequality. If

$r = 1$, we have the generalized octahedron,

$r = 2$, we have the ellipsoid,

$r \to \infty$, we have the square box (all sides equal — why?).

Why does the convexity of the square box imply the convexity of any box?

Now we reformulate Minkowski's fundamental theorem for a symmetric convex body B in $\mathbf{R}^n$ with gauge function f. For any $x \in \partial(\lambda B)$, $f(x) = \lambda$. First we choose λ small enough so that λB doesn't contain any non-zero integer points. Then we increase λ until λB contains an integer point on its boundary and none ($\neq \mathbf{0}$) inside; suppose this occurs at $\lambda = \mu$. Then, since $f(\partial(\lambda B))$ is an increasing function of λ, we have

$$\mu = \operatorname{Min} f(g) ,$$

where the minimum is over all *non-zero* integer points g. Thus we have shown that the minimum is actually achieved at an integer point. We call μ the **first minimum of f** (or B). Since $V(\lambda B) = \lambda^n V(B)$, we have

Minkowski's Theorem:

a) *(geometric form) If B is a bounded symmetric convex body in $\mathbf{R}^n$ centered at the origin, and μ is its first minimum, then*

$$\mu^n V(B) \leq 2^n .$$

b) *(arithmetic form) If $f : \mathbf{R}^n \to [0, \infty)$ is a function satisfying the conditions of the last theorem, then both the volume $V(B)$ of $B = \{x | f(x) < 1\}$ and the minimum μ of f over all non-zero integer points exist, and $\mu^n V(B) \leq 2^n$.*

The same reformulation can be carried out for convex bodies with respect to any lattice L of determinant Δ. We define the first minimum μ_L of B (with respect to L) in a similar way and μ_L is the minimum of f over all non-zero points of L. Then Minkowski's theorem for lattices becomes $\mu_L^n V(B) \leq 2^n \Delta$ and there is a similar arithmetic form.

Minkowski's theorem allows us to state our arithmetic theorems directly, since bounds on the first minimum have always been our arithmetic objective. For example, in the theorem on sums of linear forms (sec. 7), L is given by $y = Ax$, $f(y) = \sum |y_i|$, the volume of $B = \{y | f(y) < 1)$ is $\frac{2^n}{n!}$, and thus $\mu_L \leq \left(\frac{2^n \Delta}{V(B)}\right)^{\frac{1}{n}} = (n!\Delta)^{\frac{1}{n}}$.

9. Successive Minima

The formulation of Minkowski's theorem as presented in the last section leads to a natural generalization, which was also given by Minkowski.

Let f be an even gauge function defined on $\mathbf{R}^n$ and $B = \{x | f(x) < 1\}$ the corresponding symmetric convex body. We have defined the first minimum of B; for the moment we denote it by α_1. Let $g^{(1)}$ be an integer point in $\partial(\alpha_1 B)$. Suppose there are a maximum of k_1 integer points on $\partial(\alpha_1 B)$, say $g^{(1)}, \ldots, g^{(k_1)}$, which are *linearly independent* over $\mathbf{R}$ (the choice is not necessarily unique); so their f values are all α_1. Now let λ increase from α_1 until $\partial(\lambda B)$ contains an integer point independent of $g^{(1)}, \ldots, g^{(k_1)}$, say for $\lambda = \alpha_2$. Choose a maximum number of integer points on $\partial(\alpha_2 B)$, say $g^{(k_1+1)}, \ldots, g^{(k_2)}$, so that $g^{(1)}, \ldots, g^{(k_2)}$ are linearly independent; the f values of the new points are all α_2. Continue this process by again letting λ increase from α_2 and so on. Since, for λ sufficiently large, λB will contains n linearly independent vectors, our process must stop with the selection of n independent integer points $g^{(1)}, \ldots, g^{(n)}$. Setting $\mu_i = f(g^{(i)})$, we have, by our construction,

$$\mu_1 \leq \mu_2 \leq \cdots \leq \mu_n .$$

The μ_i are called the **successive minima** of$B$ (μ_1 is the first minimum).

Minkowski's Theorem for Successive Minima: *If $\mu_1, \ldots, \mu_n$ are the successive minima of the symmetric convex body B, then*

$$\mu_1 \mu_2 \cdots \mu_n V(B) \leq 2^n .$$

This result is considerably stronger than the basic Minkowski theorem, $\mu_1^n V(B) \leq 2^n$, and has correspondingly stronger arithmetic implications (see [Min 1], [Cas 4] and [Lek]).

We shall not prove the theorem. Minkowski's original proof [Min 1] is quite difficult. Weyl [Wey 4] and Davenport [Dav 4] later gave simpler proofs, but the theorem remains quite deep. Siegel [Sie 5] gives a proof, together with a nice discussion as to why the obvious approach to such a proof doesn't work.

10. Other Directions

We have presented the genesis, proof and many applications of Minkowski's fundamental theorem. Under his brilliant hand these results led to a general theory carried on in the 20th century by such luminaries as Mordell, Weyl, Mahler and Davenport. The theory now includes deeper studies of quadratic forms (including indefinite forms), Mordell's generalizations to non convex sets, packings of $\mathbf{R}^n$ by convex sets and applications to Diophantine approximation, where, e.g., it plays an important role in Schmidt's generalization of the Thue–Siegel–Roth theorem (chap. 21).

Siegel's lectures [Sie 5] provides a good introduction, as well as some of the deeper results on reduction of forms. The books by Cassels' [Cas 4] and Lekkerkerker [Lek] provide comprehensive treatments with exhaustive bibliographies. Minkowski [Min 1] is quite difficult but his more elementary "Diophantische Approximationen" [Min 3] is somewhat more readable and a rich source of the ideas and conjectures which have guided the development of the field. Hancock [Han] gives English translations of substantial parts of Minkowski's books and papers (without explicitly saying so). Unfortunately, his confusing commentaries are often interweaved into Minkowski's text.

Chapter 23

p-adic Numbers and Valuations

1. History

In chapters 15 – 17 we looked at Dedekind's generalization of Kummer's work (chapters 15 – 17). Another line of development of algebraic number theory begins with Kronecker, who in many ways was more faithful to Kummer's ideas. Kronecker had a more majestic goal than Dedekind, viz., a general theory which would encompass both algebraic number theory and algebraic geometry. Kronecker's goal is still only partially realized and remains a challenge to mathematicians [Wei 9]. His papers are very difficult to read, which may account for the early domination of Dedekind's approach, but are still an unworked gold mine of information.

Kronecker's work was continued by his student Hensel. Its foremost advocate in this century was Hensel's student Hasse. In the twentieth century, the rich analogy between number fields and geometry over **R** and **C** has been extended to geometry over finite fields. In the course of extending Kronecker's work, Hensel was led to the creation of p-adic numbers which have proved to be a powerful tool of independent interest for modern number theory. From a current point of view, the notion unifying these ideas is that of a 'valuation on a field'.

In sections 2 – 4 we first introduce the p-adic numbers from Hensel's point of view, using both direct construction and valuations. In section 5, we discuss the relation between p-adic integers and congruences over the rationals, which provides a more elementary motivation for the construction of p-adic numbers. In section 6, we discuss the deeper question of the relation between p-adic solutions and rational and integer solutions of polynomial equations. Finally, in section 7, we briefly discuss a modern version of the the Kummer - Kronecker theory of factorization in number fields, based on valuations.

A short discussion of the historical development is in [Ell, W - Ell, F]. Some good general references for p-adic numbers are [Bor - Sha], [Kob 1], [Kob 3], [Bac], and [Cas 5].

2. The p-adic Numbers; An Informal Introduction

First we introduce the p-adic numbers in a way that gives some idea of how the basic notions were discovered. In the next section, we formalize these ideas.

While studying the factorization of the ideal (p) in a number field K, where p is a rational prime, Hensel found criteria which depend on the factorization modulo p^k of polynomials with integer coefficients, for some $k \geq 1$. Moreover, he found it useful to write the coefficients of these polynomials in the base p, i.e., in the form $\pm(a_0 + a_1 p + \cdots + a_n p^n)$, where $0 \leq a_i \leq p-1$. A key observation is to note that for a positive integer

$$x = a_0 + a_1 p + \cdots + a_n p^n ,$$

$0 \leq a_i \leq p-1$, we have

$$x \equiv a_0 + a_1 p + \cdots + a_i p^i \,(\mathrm{mod}\, p^{i+1}) , \quad i = 0, 1, \ldots , \tag{1}$$

and these latter conditions uniquely characterize x (of course, for $i \geq n$ the same expression always appears on the right hand side of the congruence).

By a stroke of genius, whose motivation is hard to see, Hensel then considered all formal infinite expressions

$$a_0 + a_1 p + a_2 p^2 + \cdots + a_i p^i + \ldots ,$$

with $0 \leq a_i \leq p-1$, as new objects to be studied. He called these formal power series **p-adic integers**. Perhaps he was led to this by realizing that negative integers have a representation in this form. For example, if $p = 3$ we have, by *formally* summing a non convergent geometric series,

$$\sum_0^\infty 2 \cdot 3^i = 2 \cdot \left(\frac{1}{1-3} \right) = -1 ,$$

where we give meaning to this equation by noting that by (1),

$$2 \cdot 1 + 2 \cdot 3 + 2 \cdot 3^2 + \cdots + 2 \cdot 3^i = 2\left(\frac{1-3^{i+1}}{1-3}\right)$$
$$= 3^{i+1} - 1 \equiv -1 (\bmod 3^{i+1}) .$$

We say that -1 is represented by the 3-adic integer $\sum_0^\infty 2 \cdot 3^i$. In general, we say that the rational integer **x is represented by the p-adic integer** $\sum \mathbf{a_i p^i}$, denoted by $\mathbf{x} = \sum \mathbf{a_i p^i}$, if

$$x \equiv a_0 + a_1 p + \cdots + a_i p^i (\bmod p^{i+1}) , \quad i = 0, 1, \ldots \tag{2}$$

(we will justify the use of the equal sign shortly). Of course, as we saw in (1), the base p expansion of a positive integer is also its representation as a p-adic integer. We shall see that not every p-adic integer represents a rational integer, so we have really extended the rational integers.

We define two p-adic integers $\alpha = \sum a_i p^i$ and $\beta = \sum b_i p^i$, $0 \le a_i$, $b_i \le p - 1$, to be **equal** if

$$\sum_0^i a_k p^k \equiv \sum_0^i b_k p^k (\bmod p^{i+1}) , \quad i = 0, 1, \ldots .$$

It follows immediately from (2) that if $\alpha = \beta$, then $a_i = b_i$, for all i. In particular, the representation of a rational integer by a p-adic integer is unique.

Hensel then introduced addition and multiplication of p-adic integers (for a fixed p) by first adding and multiplying the formal power series:

$$\sum a_i p^i + \sum b_i p^i = \sum (a_i + b_i) p^i ,$$
$$\sum a_i p^i \cdot \sum b_i p^i = \sum c_i p^i ,$$

where $c_i = \sum a_k b_{i-k}$. *However* the resulting sum or product may not satisfy the condition that its coefficients lie between 0 and $p - 1$, so we must *shift* the coefficients of the series to reduce it to the proper form.

This is best illustrated by an example. Let $p = 3$; then

$$\begin{array}{lr} 2 \cdot 1 + 2 \cdot 3 + 2 \cdot 3^2 + \cdots + 2 \cdot 3^i + \cdots & (= -1) \\ + \ 2 \cdot 1 + 2 \cdot 3 + 2 \cdot 3^2 + \cdots + 2 \cdot 3^i + \cdots & (= -1) \\ \hline 4 \cdot 1 + 4 \cdot 3 + 4 \cdot 3^2 + \cdots + 4 \cdot 3^i + \cdots & (= -2). \end{array}$$

We can make the coefficients less than 3 as follows:

$$\begin{aligned}
&4 \cdot 1 + 4 \cdot 3 + 4 \cdot 3^2 + \cdots + 4 \cdot 3^i + \cdots \\
&= (1+3) \cdot 1 + 4 \cdot 3 + 4 \cdot 3^2 + 4 \cdot 3^3 + \ldots \\
&= 1 \cdot 1 + (1 \cdot 3 + 4 \cdot 3) + 4 \cdot 3^2 + 4 \cdot 3^3 + \cdots \\
&= 1 \cdot 1 + (2 \cdot 3 + 3 \cdot 3) + 4 \cdot 3^2 + 4 \cdot 3^3 + \cdots \\
&= 1 \cdot 1 + 2 \cdot 3 + (1 \cdot 3^2 + 4 \cdot 3^2) + 4 \cdot 3^3 + \cdots \\
&= 1 \cdot 1 + 2 \cdot 3 + (2 \cdot 3^2 + 3 \cdot 3^2) + 4 \cdot 3^3 + 4 \cdot 3^4 + \cdots \\
&= 1 \cdot 1 + 2 \cdot 3 + 2 \cdot 3^2 + (2 \cdot 3^3 + 3 \cdot 3^3) + 4 \cdot 3^4 + \cdots \\
&\ \vdots \\
&= 1 \cdot 1 + 2 \cdot 3 + 2 \cdot 3^2 + \cdots + 2 \cdot 3^i + \cdots .
\end{aligned}$$

Therefore we have $-2 = 1 \cdot 1 + \sum_{i=1}^{\infty} 2 \cdot 3^i$, which is consistent with our earlier method of summing geometric series, viz., $1 + \sum_{i=1}^{\infty} 2 \cdot 3^i = 1 + 2 \cdot \frac{3}{1-3} = -2$. A little thought will show that addition is done just as with the decimal representation of rational integers except that we write the digits (coefficients of 3^i) and do our carrying from *left to right*, and reduce the sum of digits modulo 3 rather than modulo 10.

Thus in our example, -1 is represented by its sequence of digits $2\,2\,2 \cdots$ and we are adding

$$\begin{array}{r} 222\cdots \\ +\ \underline{222\cdots} \ . \end{array}$$

Adding the digits in the units (first) column, we have $2+2=4=1+3$, which exceeds 3 by 1, so we enter a one and carry a one:

$$\begin{array}{l} \ \ 2\ {}^{1}22\cdots \\ +\ \underline{2\ \ 22\cdots} \\ \ \ 1 \qquad\ \ . \end{array}$$

Adding the second column (the threes column), we have $2+2+1=2+3$, which exceeds 3 by 2, so we enter a 2 and carry a 1:

$$\begin{array}{l} \ \ 2\,{}^{1}2\,{}^{1}2\cdots \\ +\ \underline{2\ 2\ 2\cdots} \\ \ \ 1\ 2 \qquad . \end{array}$$

The subsequent columns behave just as the second and we have

$$\begin{array}{r} 2\,{}^{1}2\,{}^{1}2\,{}^{1}2\,{}^{1}2\cdots \\ +\ \underline{2\ \ 2\ \ 2\ \ 2\ \ 2\cdots} \\ 1\ \ 2\ \ 2\ \ 2\ \ 2\cdots . \end{array}$$

To illustrate multiplication, we let $p = 5$ and multiply two finite series, $7 = 2 \cdot 1 + 1 \cdot 5$ and $40 = 0 \cdot 1 + 3 \cdot 5 + 1 \cdot 5^2$, to keep calculations simple. Then

$$\begin{aligned} &(2 \cdot 1 + 1 \cdot 5) \cdot (0 \cdot 1 + 3 \cdot 5 + 1 \cdot 5^2) \\ &= 0 \cdot 1 + (2 \cdot 3 + 1 \cdot 0) \cdot 5 + (2 \cdot 1 + 1 \cdot 3) \cdot 5^2 + (1 \cdot 1) \cdot 5^3 \\ &= 0 \cdot 1 + 6 \cdot 5 + 5 \cdot 5^2 + 1 \cdot 5^3 \\ &= 0 \cdot 1 + 1 \cdot 5 + 6 \cdot 5^2 + 1 \cdot 5^3 \\ &= 0 \cdot 1 + 1 \cdot 5 + 1 \cdot 5^2 + 2 \cdot 5^3, \end{aligned}$$

i.e., $7 \cdot 40 = 280$. Using the digit notation, we multiply as in ordinary multiplication, but proceeding left to right and reducing digits modulo 5, viz.,

$$\begin{array}{rl} & 21 \\ \times & \underline{031} \\ & 000 \\ & 14 \\ + & \underline{21} \\ & 0112\,. \end{array}$$

If $\mathbf{Z_p}$ denotes the p-adic integers, then, under addition and multiplication, $\mathbf{Z_p}$ is a commutative ring with no divisors of zero (exercise), i.e., $\mathbf{Z_p}$ is an integral domain (= entire ring). The map $\varphi : \mathbf{Z} \to \mathbf{Z_p}$, given by

$$n \mapsto \text{ the } p\text{-adic representation of } n,$$

is an isomorphism of $\mathbf{Z}$ with a subring of $\mathbf{Z_p}$. This justifies our use of the equal sign, as in $-1 = \sum 2 \cdot 3^i$, if we identify -1 with its image $\varphi(-1) = \sum 2 \cdot 3^i$ under the isomorphism.

Divisibility is defined as for any commutative ring, viz., $\alpha | \beta$ if there exist a γ such that $\beta = \alpha\gamma$. The units are those p-adic integers $\sum a_i p^i$ with $a_0 \neq 0$ (exercise — the proof is analogous to the proof that the invertible formal power series, i.e., those $\sum a_i x^i$ for which there exists a $\sum b_i x^i$ with $\sum a_i x^i . \sum b_i x^i = 1$, are exactly the power series with $a_0 \neq 0$).

Clearly every p-adic integer $\alpha = \sum a_i p^i$ can be factored in the form

$$\alpha = p^m (a_m + a_{m+1} p + a_{m+2} p^2 + \cdots) , \tag{3}$$

where a_m is the smallest non zero coefficient. The only prime in $\mathbf{Z_p}$ is p (exercise) and thus (3) represents a factorization of α into a unit and a product of primes. Since the representation as a series in p is unique, (3) is the only prime factorization of α in $\mathbf{Z_p}$, and we see that $\mathbf{Z_p}$ is a unique factorization domain with p as the only prime.

We prove that $\mathbf{Z_p}$ is a proper extension of $\mathbf{Z}$ by showing that $\sum p^i$ does not represent a rational integer. If $\sum p^i$ does represent a rational integer x, then $x < 0$ since positive integers are represented by finite sums and the representation is unique. Hence $-x > 0$ must be represented by a finite sum; but

$$\begin{aligned}
(-1) \sum p^i &= \sum (-1) p^i \\
&= -1 \cdot 1 + (-1) \cdot p + (-1) \cdot p^2 + (-1) \cdot p^3 + \cdots \\
&= (-p + p - 1) \cdot 1 + (-1) \cdot p + (-1) \cdot p^2 + (-1) \cdot p^3 + \cdots \\
&= (p - 1) \cdot 1 + (-2) \cdot p + (-1) \cdot p^2 + (-1) \cdot p^3 + \cdots \\
&= (p - 1) \cdot 1 + (-p + p - 2) \cdot p + (-1) \cdot p^2 + (-1) \cdot p^3 + \cdots \\
&= (p - 1) \cdot 1 + (p - 2) \cdot p + (-2) \cdot p^2 + (-1) \cdot p^3 + \cdots \\
&= (p - 1) \cdot 1 + (p - 2) \cdot p + (p - 2) \cdot p^2 + (-2) \cdot p^3 + \cdots \\
&\ \ \vdots \\
&= (p - 1) \cdot 1 + (p - 2) \cdot p + (p - 2) \cdot p^2 + (p - 2) \cdot p^3 + \cdots ,
\end{aligned}$$

an infinite sum, and we have a contradiction. Therefore $\sum p^i$ does not represent a rational integer.

Of course, as Hensel must have realized, since $\sum p^i = \frac{1}{1-p}$ as a formal series in p, $\sum p^i$ represents the rational number $\frac{1}{1-p}$ in the sense that

$(1-p)\sum p^i$ represents 1, i.e.,

$$(1-p)\sum_{i=0}^{n} p^i = (1-p)\frac{1-p^{n+1}}{1-p}$$
$$= 1 - p^{n+1} \equiv 1(\text{mod } p^{n+1}) \,,$$

for $n = 0, 1, \ldots$. Furthermore, $-\sum p^i = (p-1) + \sum_1^\infty (p-2)p^i$ represents $\frac{1}{p-1}$ in the sense that

$$(p-1)\left((p-1) + \sum_1^n (p-2)p^i\right) \equiv 1(\text{mod } p^{n+1}) \,,$$

for $n = 0, 1, \cdots$.

More generally, we will say that a p-adic integer $\delta = \sum d_i p^i$ **represents a rational number** $\frac{a}{b}$ if $\delta b = a$ or equivalently if

$$b\left(\sum_{i=0}^{n} d_i p^i\right) \equiv a(\text{mod } p^{n+1}), \quad n = 0, 1, \cdots .$$

If $\gcd(a, b) = 1$ and $p \nmid b$, then $\frac{a}{b}$ is represented by a p-adic integer. For if $p \nmid b$, then b is a unit in $\mathbf{Z_p}$ and has an inverse b^{-1}. Hence $\frac{a}{b} = \frac{ab^{-1}}{bb^{-1}} = ab^{-1}$, i.e., the p-adic integer ab^{-1} represents $\frac{a}{b}$ (more formally $(ab^{-1})b = a(b^{-1}b) = a$).

A rational number $\frac{a}{b}$ with $\gcd(a, b) = 1$ and $p \nmid b$ is called a **p-integer**. The p-integers form a subring $\mathbf{Z(p)}$ of $\mathbf{Q}$ which is a unique factorization domain with one prime p. Under the map

$$\varphi : \mathbf{Z}(p) \to \mathbf{Z}_p \,,$$

given by

$$\frac{a}{b} \mapsto \text{the } p\text{-adic representation of } \frac{a}{b} \,,$$

$\mathbf{Z}(p)$ is isomorphic to a subring of $\mathbf{Z}_p$ and, as for integers, we will identify $\frac{a}{b}$ with its image $\varphi\left(\frac{a}{b}\right) = \sum_{i=0} d_i p^i$ and write $\frac{a}{b} = \sum_{i=0} d_i p^i$. It can be shown that $\varphi(\mathbf{Z}(p))$ consists of those p-adic integers whose sequence of coefficients eventually become periodic, and thus there are p-adic integers which do not represent p-integers [Mah, K 2].

Now we consider rational numbers $\frac{a}{b}$, $(a, b) = 1$, which are not p-integers. Thus $p|b$ and, since $(a, b) = 1$, $p \nmid a$. If we let $b = \sum b_i p^i$, then, for some $m > 0$, $b = p^m(b_m + b_{m+1}p + \cdots) = p^m b'$, where b' is a unit with inverse b'^{-1}. Therefore, if we set $ab'^{-1} = \sum d_i p^i$, we have

$$\begin{aligned}\frac{a}{b} &= \frac{a}{p^m b'} = \frac{1}{p^m} ab'^{-1} \\ &= \frac{1}{p^m}(d_0 + d_1 p + d_2 p^2 + \cdots) \\ &= \frac{d_0}{p^m} + \frac{d_1}{p^{m+1}} + \cdots + \frac{d_{m-1}}{p} + d_m + d_{m+1}p + \cdots ,\end{aligned}$$

a formal *Laurent series* in p with only a finite number of non zero negative powers of p and with $0 \le d_i \le p-1$. We call this Laurent series the *p-adic representation of* $\frac{a}{b}$.

Hensel then defined $\mathbf{Q_p}$, the **p-adic numbers**, to be those formal Laurent series in p with only a finite number of non-zero negative powers and with all coefficients between 0 and $p - 1$, i.e.

$$\mathbf{Q}_p = \left\{ \sum_{i=r}^{\infty} a_i p^i \,|\, 0 \le a_i \le p-1 \text{ with } r, a_i \in \mathbf{Z} \right\} .$$

As with $\mathbf{Z}_p$, we define addition and multiplication as the corresponding operation on formal Laurent series, with a suitable rearrangement of the sum (or product) to get the required condition on the coefficients. Under these operations, $\mathbf{Q}_p$ *is a field.* In fact, it is the fraction field of $\mathbf{Z}_p$. To see that $\mathbf{Q}_p$ is closed under division, let

$$\begin{aligned}\alpha &= \sum_{i=m}^{\infty} a_i p^i = p^m \sum_{i=0}^{\infty} a_{m+i} p^i = p^m \alpha' , \\ \beta &= \sum_{i=k}^{\infty} a_i p^i = p^k \sum_{i=0}^{\infty} a_{k+i} p^i = p^k \beta' ,\end{aligned}$$

where $k, m \in \mathbf{Z}$ and α' and β' are units. Therefore $\frac{\alpha'}{\beta'} \in \mathbf{Z}_p$, and $\frac{\alpha}{\beta} = p^{m-k}\frac{\alpha'}{\beta'} \in \mathbf{Q}_p$.

Just as for $\mathbf{Z}$ and $\mathbf{Z}_p$, we can embed $\mathbf{Q}$ in $\mathbf{Q}_p$ by an isomorphism mapping every rational to its p-adic representation. It is easy to show that a p-adic number represents a rational number if and only if its sequence of coefficients eventually becomes periodic. Hence $\mathbf{Q}$ is properly contained in $\mathbf{Q}_p$.

We summarize our construction with a diagram (fig. 1).

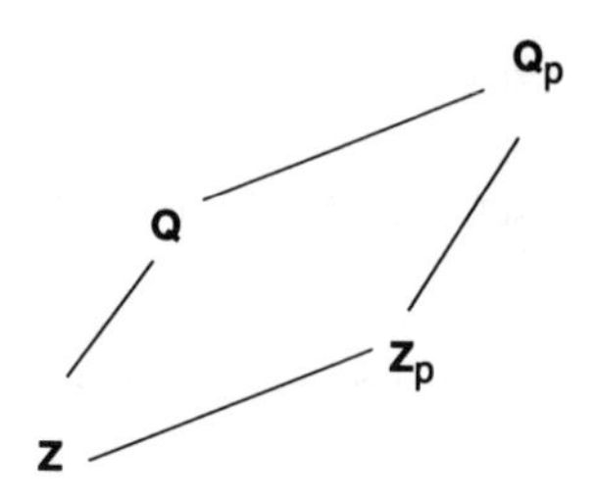

Figure 1

3. The Formal Development

In dealing with p-adic integers, the main obstacle to a clean treatment was the condition $0 \leq a_i < p$ on the coefficients of a formal power series $\sum_{i=0}^{\infty} a_i p^i$. When adding or multiplying such series, we had to manipulate the sum to recover this condition. To formalize our construction, we deal with the set of *all* formal series in p (with integer coefficients), define an equivalence relation on this set and concentrate on the *sequences* $\{s_n\} = \left\{\sum_{i=r}^{n} a_i p^i\right\}$ of partial sums of our series. Our treatment closely follows the one in Borevich and Shafarevich [Bor - Sha].

Note that $s_n - s_{n-1} = a_n p^n \equiv 0(\text{mod } p^n)$, i.e.,

$$s_n \equiv s_{n-1}(\text{mod } p^n) . \tag{1}$$

Conversely, suppose that $\{s_n\}$ is a sequence of integers satisfying (1). Let $a_0 = s_0$. Then, by equation (1) for $n = 1$, there is an integer a_1 such that $s_1 = s_0 + a_1 p = a_0 + a_1 p$. Setting $n = 2$ in equation (1), there is an integer a_2 such that $s_2 = s_1 + a_2 p^2 = a_0 + a_1 p + a_2 p^2$. Continuing with $n = 3, 4, \ldots$, we see that there is a sequence of integers such that $s_n = \sum_{i=r}^{n} a_i p^i$, i.e., $\{s_n\}$ is the sequence of partial sums of a power series $\sum_{i=r}^{\infty} a_i p^i$. Hence we shall consider all sequences of integer satisfying (1); we call such sequences **p-sequences**.

We call two p-sequences $\{s_n\}$ and $\{s'_n\}$ **equivalent**, written $\{s_n\} \sim \{s'_n\}$, if

$$s_n \equiv s'_n(\text{mod } p^{n+1}) , \quad n = 0, 1, \cdots . \tag{2}$$

It is easily checked that this is an equivalence relation. We define a **p-adic integer** as an equivalence class of p-sequences and, as before, we let $\mathbf{Z_p}$ denote the set of p-adic integers.

A **canonical sequence** is a p-sequence $\{\overline{s_n}\}$ which is the sequence of partial sums of a series $\sum_{i=0}^{\infty} \overline{a_i} p^i$ such that $0 \le \overline{a_i} < p$; the series is called a **canonical series.** Hence *a canonical series is a p-adic integer in the sense used in section two.* To connect with our new definition we prove the following.

Proposition: *A p-adic integer α (an equivalence class of p-sequences) contains a unique canonical sequence.*

Proof: First we show that $\{\overline{s_n}\}$ is a canonical sequence if and only if it is a p-sequence satisfying

$$0 < \overline{s_n} < p^{n+1} \,. \tag{3}$$

If $\{\overline{s_n}\}$ is a canonical sequence, then $\overline{s_n} = \sum_{i=0}^{n} \overline{a_i} p^i \le \sum_{i=0}^{n} (p-1)p^i = p^{n+1} - 1 < p^{n+1}$; so $\{\overline{s_n}\}$ satisfies (3). Conversely, if $\{\overline{s_n}\}$ satisfies (3), then, by our earlier construction of the series $\sum_{i=0}^{\infty} \overline{a_i} p^i$ corresponding to $\{\overline{s_n}\}$, we have

$$0 \le \overline{s_{n+1}} = \overline{s_n} + \overline{a_{n+1}} p^{n+1} < p^{n+2} \,.$$

If $\overline{a_{n+1}} > p$, then $\overline{a_{n+1}} p^{n+1} > p^{n+2}$, a contradiction, and if $\overline{a_{n+1}} < 0$, then $p^{n+1} > \overline{s_n} > (-\overline{a_{n+1}}) p^{n+1} > p^{n+1}$, another contradiction. Hence $0 \le \overline{a_i} < p$ and $\{\overline{s_n}\}$ is canonical.

To see that α contains a canonical sequence, let $\{s_n\}$ be any p-sequence determining α and let $\overline{s_n}$ be the smallest non-negative integer congruent to s_n modulo p^{n+1}; hence

$$\overline{s_n} \equiv s_n (\text{mod}\, p^{n+1}) \,. \tag{4}$$

Therefore, since $\{s_n\}$ is a p-sequence, we have

$$\overline{s_n} \equiv s_n \equiv s_{n-1} \equiv \overline{s_{n-1}} (\text{mod}\, p^n) \,.$$

Thus $\{\overline{s_n}\}$ is a p-sequence and, by its construction, satisfies (3); hence $\{\overline{s_n}\}$ is canonical. By equation (4), $\{\overline{s_n}\} \sim \{s_n\}$; so α contains the canonical sequence $\{\overline{s_n}\}$.

If α contains canonical sequences $\{\overline{s_n}\}$ and $\{\overline{t_n}\}$, then, since $\overline{s_n} \equiv \overline{t_n}$ $(\text{mod}\, p^{n+1})$ and $0 \le \overline{s_n}, \overline{t_n} < p^{n+1}$, we have $\overline{s_n} = \overline{t_n}$, for all n, and we are done.

Therefore we see that a p-adic integer, as an equivalence class of p-sequences, contains a unique canonical sequence whose corresponding canonical series is a p-adic integer in the sense of section 2. For the moment, we use 'canonical series' to refer to our earlier definition of p-adic integers and 'p-adic integer' to refer to our new definition using equivalence classes.

Addition and multiplication can now be defined quite naturally. It is trivial to prove that if $\{s_n\}$ and $\{t_n\}$ are p-sequences, then $\{s_n+t_n\}$ and $\{s_nt_n\}$ are p-sequences. Suppose that $\{s_n\}$ and $\{t_n\}$ determine the p-adic integers α and β respectively. Then the **sum** $\alpha + \beta$ (resp., the **product** $\alpha\beta$) is the p-adic integer determined by the p-sequence $\{s_n + t_n\}$ (resp., $\{s_nt_n\}$). It follows immediately, from the definition of equivalence, that the sum and product are well defined, i.e., independent of the choice of p-sequences. However to see that our sum and product are the same as the sum and product of canonical sequences, as discussed in section 2, takes some work (exercise).

We now summarize some of the basic properties of $\mathbf{Z}_p$, most of which were discussed for canonical series in section 2. We state them in an order which makes their proofs easy exercises.

Theorem:

(1) *Z_p is a commutative ring. Divisibility in $\mathbf{Z}_p$ is defined as for any commutative ring.*

(2) *If $\{s_n\}$ determines α, then α is a unit if and only if $s_0 \not\equiv 0 (\text{mod } p)$, i.e., if and only if $p \nmid \alpha$.*

(3) *The map which associates to each rational integer n the p-adic integer determined by the constant p-sequence $\{n, n, \cdots\}$ is an isomorphism between $\mathbf{Z}$ and a subring of $\mathbf{Z}_p$.*

(4) *Every $\alpha \in \mathbf{Z}_p$, $\alpha \neq 0$, has a unique representation of the form $\alpha = p^m\epsilon$, where ϵ is a unit in Z_p and m is a positive integer.*

(5) *$\mathbf{Z}_p$ is a unique factorization domain with one prime p.*

(6) *$\mathbf{Z}_p$ has no divisors of zero and thus it is an integral domain.*

(7) *Congruence in $\mathbf{Z}_p$ is defined as in $\mathbf{Z}$, i.e., if $\alpha, \beta, \gamma \in \mathbf{Z}_p$, then $\alpha\beta \equiv (\text{mod } \gamma)$ if $\gamma | (\alpha - \beta)$. As in $\mathbf{Z}$, congruence modulo γ in $\mathbf{Z}_p$ is an equivalence relation and corresponds to equality in the quotient ring $\mathbf{Z}_p/(\gamma)$, where (γ) is the principal ideal generated by*

γ. Every p-adic integer α is congruent to a rational integer modulo p^n (if α is determined by the partial sums of $\sum_{i=0}^{\infty} a_i p^i$, then $\alpha = \sum_{i=0}^{n} a_i p^i \pmod{p^n}$). Rational integers are congruent modulo p^n in $\mathbf{Z}$ if and only if they are congruent modulo p^n in $\mathbf{Z}_p$.

(8) *There are p^n residue classes modulo p^n in $\mathbf{Z}_p$, i.e., $\left|\mathbf{Z}_p/(p^n)\right| = p^n$.*

The **p-adic numbers $\mathbf{Q_p}$** are defined to be the elements of the fraction field of $\mathbf{Z}_p$. As discussed in section 2, they are easily shown to be in one to one correspondence with the set $\left\{\sum_{i=r}^{\infty} a_i p^i \,|\, 0 \le a_i \le p-1, r \in \mathbf{Z}\right\}$ of formal Laurent series in p satisfying the designated conditions. Every $\alpha \in \mathbf{Q}_p$ can be uniquely represented as $\alpha = p^m \varepsilon$, where ε is a unit in $\mathbf{Z}_p$ and m is a *rational* integer. If $r \in \mathbf{Q}$, $r = p^m\left(\frac{a}{b}\right)$, $m \in \mathbf{Z}$, $\gcd(a, b) = 1$, and $p \not| b$, then b is a unit in $\mathbf{Z}_p$, and the map which associates to r the p-adic number $p^m ab^{-1}$ is an isomorphism between $\mathbf{Q}$ and a subfield of $\mathbf{Q}_p$. If $\alpha \in \mathbf{Q}_p$, $\alpha = \sum_{i=r}^{\infty} a_i p^i$, $a_r \neq 0$, then

$$\alpha = p^r\left(\sum_{i=r}^{\infty} a_i p^{i-r}\right) = p^r\left(\sum_{j=0}^{\infty} a_{j+r} p^j\right);$$

thus every non-zero p-adic integer can be represented in the form $p^r u$, where $r \in \mathbf{Z}$ and u is a unit.

As mentioned at the beginning of section 2, Hensel was led to the construction of the p-adic numbers by the study of the factorization of the ideals (p), p a rational prime, in the ring I_K of integers of a number field K. He then generalized the p-adic numbers in order to study the factorization of a prime ideal of I_K into ideals in I_E, where E is a finite extension of K. To give some idea of this, we rephrase our construction of p-adic numbers.

A p-adic number can be viewed as a formal Laurent series in p,

$$\sum_{i=r}^{\infty} a_i p^i, \; r \in \mathbf{Z},$$

where the coefficients a_i are chosen from the set $\{0, 1, \ldots, p-1\}$, which can be regarded as a complete set of representatives for the equivalence classes of $\mathbf{Z}/p\mathbf{Z}$. To generalize to a number field K, Hensel picked a prime ideal P in I_K, an element π in P such that π^2 is not in P, and a complete set S of representatives for I_K/P. He then introduced the set of formal

Laurent series in π,

$$\mathbf{K_P} = \left\{ \sum_{i=r}^{\infty} a_i \pi^i , a_i \in S , \quad r \in \mathbf{Z} \right\} .$$

K_P is the set of **P-adic numbers** and with appropriate definitions of addition and multiplication of P-adic numbers and of representation of an element of K by a P-adic number, Hensel proved that K_P is a field, that every element of K can be represented by a P-adic number and that K is isomorphic to a proper subfield of K_P. In addition to using the P-adic numbers to study prime ideal factorization, Hensel also used them to solve the then open problem of finding the highest power of a rational prime which divides the discriminant of a number field.

4. Convergence

Most good mathematicians doing research in a specific area do not work in a vacuum. They are aware of developments in other parts of mathematics that may provide useful analogies or results for their work. About ten years after the introduction of p-adic numbers, the concurrent development of topological ideas motivated Hensel to introduce the notion of convergence for p-adic numbers. We state many results without proof. See Borevich and Shafarevich [Bor - Sha] for details.

Representing a p-adic integer α by one of its p-sequences, say $\alpha = \sum_{i=0}^{\infty} a_i p^i$, is analogous to representing a real number r, $0 < r < 1$, by its decimal expansion, $r = \sum_{i=1}^{\infty} r_i \cdot 10^{-i}$, where $r_i \in \mathbf{Z}$, $0 \leq r_i \leq 10$. The partial sums $t_n = \sum_{i=1}^{n} r_i 10^{-i}$ are a sequence of rational numbers, $t_1, t_2, \ldots$, converging to r since $|t_n - r| < \frac{1}{10^n}$. Similarly, we would like to formulate a notion of convergence for p-adic numbers so that the sequence $\{s_n\} = \left\{\sum_{i=0}^{n} a_i p^i\right\}$ of rational numbers converge to α. The key to this is the observation that for large n, $\alpha - s_n = \sum_{i=n+1}^{\infty} a_i p^i$ is divisible by a high power of p, namely by p^{n+1}. We call two p-adic numbers "p-close" if their difference is divisible by a high power of p. To make these ideas more precise, we introduce a measure of closeness.

Earlier, we saw that every non zero p-adic number is uniquely representable in the form $\alpha = p^n \varepsilon$, where $n \in \mathbf{Z}$ and ε is a unit in $\mathbf{Q}_p$. The exponent n is the **p-value of** α or the **order of** α **at p** and is denoted by $\mathbf{ord_p}(\alpha)$, where we set $\mathrm{ord}_p(0) = \infty$ (since 0 is divisible by any power of

p). Note that if $\alpha \in \mathbf{Z}$, then $\mathrm{ord}_p(\alpha)$ is just the power of p appearing in the prime factorization of α. Hence the sequence $\{\mathrm{ord}_{p_i}(\alpha)\}$, where $\{p_i\}$ is the sequence of rational primes, completely determines the prime factorization of α in $\mathbf{Z}$. This can be used to provide a new way of constructing algebraic number fields (sec. 7). The function $\mathrm{ord}_p : \mathbf{Q}_p \to \mathbf{Z}$ is also called the **p-adic valuation of $\mathbf{Q}_p$**.

It is an easy exercise, using the representation of p-adic numbers by formal Laurent series, to prove the following properties of the p-value:

(1) $\mathrm{ord}_p(\alpha\beta) = \mathrm{ord}_p(\alpha) + \mathrm{ord}_p(\beta)$,

(2) $\mathrm{ord}_p(\alpha + \beta) \geq \min(\mathrm{ord}_p(\alpha), \mathrm{ord}_p(\beta))$,

(3) $\mathrm{ord}_p(\alpha + \beta) = \min(\mathrm{ord}_p(\alpha), \mathrm{ord}_p(\beta))$, if $\mathrm{ord}_p(\alpha) = \mathrm{ord}_p(\beta)$.

The p-adic valuation is useful for arithmetic in $\mathbf{Z}_p$ since, if $\alpha, \beta \in \mathbf{Z}_p$, then $\alpha | \beta$ if and only if $\mathrm{ord}_p(\alpha) \leq \mathrm{ord}_p(\beta)$.

Now to convergence. A sequence $\alpha_1, \alpha_2, \cdots$ of p-adic numbers **converges (p-adically)** to the p-adic number α if $\lim_{n\to\infty} \mathrm{ord}_p(\alpha - \alpha_n) = \infty$ (the limit is taken over the reals) and we write $\mathbf{lim_{n\to\infty}\alpha_n} = \alpha$. We call α the **p-adic limit** of the sequence $\{\alpha_i\}$. If α is a p-adic number, $\alpha = \sum_{i=r}^{\infty} a_i p^i$, $r \in \mathbf{Z}$, and $s_n = \sum_{i=r}^{n} a_i p^i$, $n \geq r$, then, since $\mathrm{ord}_p(\alpha - s_n) = \mathrm{ord}_p(\alpha - \sum_{i=r}^{n} a_i p^i) \geq n + 1$, s_n converges to α, as desired. Thus all p-sequences which represent α converge to α and every p-adic number is the p-adic limit of a sequence of rational numbers.

In order to connect with the more familiar notion of convergence where limits tend to zero, we define a norm and a metric on $\mathbf{Q}_p$. Choose a real number ρ, $0 < \rho < 1$, and let

$$f_p(\alpha) = \begin{cases} \rho^{\mathrm{ord}_p(\alpha)}, & \text{if } \alpha \neq 0, \\ 0, & \text{if } \alpha = 0 \end{cases}$$

(some authors take $\rho = \frac{1}{p}$ as a canonical choice). ρ will not be indicated in our notation unless needed.

From the definitions of f_p and ord_p, and properties (1) – (3) of ord_p, we see immediately that

(4) $f_p(\alpha) > 0$ for $\alpha \neq 0$,
$f_p(0) = 0$,

(5) $f_p(\alpha + \beta) \leq \max\{f_p(\alpha), f_p(\beta)\}$,

(6) $f_p(\alpha\beta) = f_p(\alpha) f_p(\beta)$.

From property (5), we obtain the triangle inequality

$$(7)\quad f_p(\alpha+\beta) \le f_p(\alpha) + f_p(\beta)\ .$$

A **norm** f on a field F is a function $f : F \to \mathbf{R}$ satisfying properties (4), (6) and (7). Thus f_p is a norm on $\mathbf{Q}_p$, the **p-adic norm**, and if we let

$$\mathbf{d_p}(\alpha, \beta) = f_p(\alpha - \beta)\ ,$$

then $(\mathbf{Q}_p, d_p)$ is a metric space. Clearly, from the definitions,

$$\begin{aligned} \lim_{n\to\infty} \operatorname{ord}_p(\alpha - \alpha_n) = \infty &\iff \lim_{n\to\infty} f_p(\alpha - \alpha_n) = 0 \\ &\iff \lim_{n\to\infty} d_p(\alpha, \alpha_n) = 0\ ; \end{aligned}$$

so our definition of convergence corresponds to the usual notion on an appropriate metric space.

Warning: Many authors call f_p, not ord_p, the p-adic valuation.

Note that the ordinary absolute value is a norm on the field of real numbers, What distinguishes it from the p-adic norm is that although $|r|$ satisfies the triangles inequality (7), it does not satisfy the sharper inequality (5). Norms satisfying (5) are called **non-Archimedean** and all other norms are **Archimedean**. A norm f on a field F is non-Archimedean if and only if $f(\overline{n}) \le 1$ for all positive integers n, where $\overline{n}$ is the result of adding 1 to itself n times in F (exercise).

With our definition of convergence, the standard theorems on limits over the real and complex numbers are also true over $\mathbf{Q}_p$. Thus the limits of sums and products are the sums and products of limits and sequences of p-adic numbers which are bounded above in the norm f_p contain convergent subsequences. Moreover, in $\mathbf{Q}_p$, convergent sequences are absolutely convergent, i.e., if the terms of the convergent series $s = \sum \alpha_n$ are rearranged, then the resulting series also converges to s. Convergence of series can be decided more easily than over $\mathbf{R}$:

$$\text{if } \lim_{n\to\infty} \alpha_n = 0, \text{ then } \sum \alpha_n \text{ converges.}$$

Convergent sequences in $\mathbf{Q}_p$ are Cauchy sequences:

$$\text{if } \{\alpha_n\} \text{ converges, then } \lim_{m,n\to\infty} f_p(\alpha_m - \alpha_n) = 0\ .$$

Conversely, in $\mathbf{Q}_p$, Cauchy sequences converge:

$$\text{if } \lim_{m,n\to\infty} f_p(\alpha_m - \alpha_n) = 0, \text{ then } \lim_{n\to\infty} \alpha_n = \alpha\ ,$$

for some $\alpha \in \mathbf{Q}_p$, i.e., $\mathbf{Q}_p$ is a complete metric space.

Using the idea of Cauchy convergence, we can construct $\mathbf{Q}_p$ in another way. We recall Cantor's construction of the field of real numbers from the rationals as equivalence classes of Cauchy sequences. The absolute value is a norm on $\mathbf{Q}$. We say that a sequence $\{r_n\}$ in $\mathbf{Q}$ is a Cauchy sequence if $\lim_{m,n\to\infty} |r_m - r_n| = 0$. We call two Cauchy sequences $\{r_n\}$, $\{s_n\}$ equivalent if $\lim_{n\to\infty} |r_n - s_n| = 0$. The field of real numbers consists of the equivalence classes with addition and multiplication defined as follows:

$$(8)\quad \overline{\{r_n\}} + \overline{\{s_n\}} = \overline{\{r_n + s_n\}}\ ,$$
$$(9)\quad \overline{\{r_n\}} \times \overline{\{s_n\}} = \overline{\{r_n \times s_n\}}\ ,$$

where $\overline{\{r_n\}}$denotes the equivalence class of the Cauchy sequence $\{r_n\}$. $(\mathbf{R}, |\,|\,)$ is a complete metric space, and we call $\mathbf{R}$ the completion of $\mathbf{Q}$ with respect to the absolute value. There are other norms on $\mathbf{Q}$, for example, $|r|^t$ for any real t such that $0 < t < 1$, and completion of $\mathbf{Q}$ with respect to any one of these norms yields a field isomorphic to $\mathbf{R}$. Two norms f and g on $\mathbf{Q}$ are **equivalent** if a sequence converges with respect to f if and only if it converges with respect to g. If f and g are equivalent, then their completions are isomorphic.

Similarly, if for any fixed ρ, $0 < \rho < 1$, we restrict $f_p(\alpha) = \rho^{\mathrm{ord}_p(\alpha)}$, to $\mathbf{Q}$, we have a norm on $\mathbf{Q}$. Completing $\mathbf{Q}$ with respect to this norm (i.e., taking equivalence classes of Cauchy sequences with respect to the norm and defining addition and multiplication as in (8) and (9)) yields a field isomorphic to $\mathbf{Q}_p$. Thus the norms are equivalent for all ρ.

Hence we have two classes of norms on $\mathbf{Q}$, and each norm allows us to extend $\mathbf{Q}$ to a field which is complete with respect to it (fig. 2).

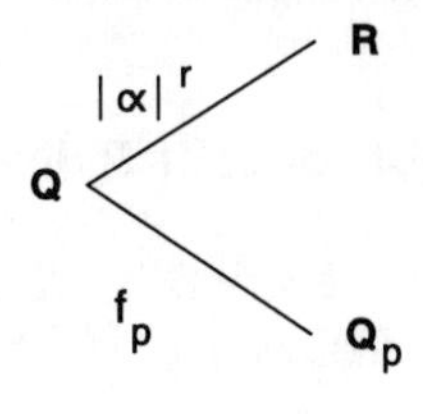

Figure 2

In fact, these are the only norms on $\mathbf{Q}$ as we know by

Ostrowski's Theorem: *Every norm f on $\mathbf{Q}$ is of the form*

$$f(r) = |r|^t \ , \ \textit{for some } t, 0 < t \leq 1 \ ,$$

or

$$f(r) = f_p(\alpha) = \rho^{\mathrm{ord}_p(r)}, \ \textit{for some prime } p \textit{ and some } \rho, 0 < \rho < 1 \ .$$

If F is a field which is an extension of $\mathbf{Q}$ and which is complete with respect to a norm f, then f, restricted to $\mathbf{Q}$, is a norm on $\mathbf{Q}$ and thus F must contain $\mathbf{R}$ or $\mathbf{Q}_p$, for some prime p.

Now that we know $\mathbf{Q}_p$ is a complete metric space, we can introduce the notion of a continuous function and more general methods of analysis, where, e.g., polynomials are continuous functions. There is a powerful theory of p-adic analytic functions [Bor -Sha], and the study of p-adic differential equations is a very active research area. These ideas are central in number theory, algebraic geometry and algebraic function fields. Thus, e.g., Thue's theorem (sec. 21.5) can be proved using p-adic analytic functions [Bor -Sha, sec. 6.3] and Dwork pioneered the development of p-adic analysis beginning with his proof of the first Weil conjecture (sec. 20.9) that the zeta function of an algebraic hypersurface is rational [Kob 1]. Our proof of the Lutz - Nagell theorem, characterizing the rational points of finite order on an elliptic curve (sec. 20.2), clearly uses p-adic ideas and could be completely recast in p-adic terms.

Techniques which involve proving theorems about $\mathbf{Q}$ by using all the embeddings of $\mathbf{Q}$ in its completions $\mathbf{Q}_p$ and $\mathbf{R}$ are often called *local methods*, and the completions are called **local fields**.

The P-adic numbers, discussed at the end of the last section, can also be constructed using valuations and norms. We pick a prime ideal P in I_K, the ring of integers of the number field K, and define the **P-adic valuation, $\mathbf{ord_P}$**, on K. For $\beta \in I_K, \beta \neq 0$, let

$$\mathrm{ord}_P(\beta) = \text{the highest power of } P \text{ dividing the ideal } (\beta) \ ,$$

and set $\mathrm{ord}_P(0) = \infty$. Since every $\alpha \in K$ can be written in the form $\alpha = \frac{\beta}{\gamma}$, where $\beta, \gamma \in I_K, \gamma \neq 0$, we let

$$\mathrm{ord}_P(\alpha) = \mathrm{ord}_P(\beta) - \mathrm{ord}_P(\gamma) \ .$$

Choose a real number $\rho, 0 < \rho < 1$, and let

$$f_P(\alpha) = \rho^{\operatorname{ord}_P(\alpha)} .$$

Then f_P is a norm on K and completing K with respect to this norm yields the field of P-adic numbers.

Now we turn to a very important application of p-adic methods in elementary number theory, viz., the theory of congruences.

5. Congruences and p-adic Numbers

As we shall see, the relation between p-adic integers and congruences over the rationals is so natural that one might expect that this would have been the principal motivation for the creation of the p-adic integers. However, new mathematical ideas often do not arise in their simplest context. Thus, e.g., Cantor's introduction of the infinite cardinals was motivated by his work on the convergence of Fourier series and not by a preconceived desire to understand the nature of the infinite, and p-adic numbers were introduced to study the factorization of ideals in number fields, not to solve congruences.

Recall our example in section 9.4, where we constructed solutions of

$$x^2 \equiv 2 (\operatorname{mod} 7^n) , \tag{1}$$

for all $n > 1$, from the solutions for $n = 1$. The solution $x \equiv 3 (\operatorname{mod} 7)$ led us to a unique sequence $a_0 = 3, a_1 = 1, a_2 = 2, a_3, \ldots, 0 \leq a_i < 7$, such that $s_n = \sum_{i=1}^{n-1} a_i 7^i$ satisfied $s_n \equiv 3 (\operatorname{mod} 7)$ and

$$s_n^2 \equiv 2 (\operatorname{mod} 7^{n+1}) . \tag{2}$$

Thus we could code the sequence as a p-adic integer $a = \sum_{i=0}^{\infty} a_i 7^i$ so that the partial sums yield the solutions to (1).

Since the polynomial $x^2 - 2$ is a continuous function in the 7-adic topology and $s_n \to \alpha$, $s_n^2 - 2$ converges 7-adically to $\alpha^2 - 2$. By (2),

$$f_7(s_n^2 - 2) \leq \rho^{n+1} \to 0 ,$$

so $s_n^2 - 2$ converges 7-adically to 0 and thus $\alpha^2 - 2 = 0$ or $\alpha^2 = 2$ in the 7-adic integers.

A careful analysis of our construction (sec. 9.4) of solutions of general congruences modulo p^n, $n > 1$, from the solutions modulo p, together with the elementary properties of congruences in $\mathbf{Z}_p$, immediately yields a proof of

Hensel's Lemma (One Variable): *Let $f(x) \in \mathbf{Z}_p[x]$. If a is a rational integer such that $f(a) \equiv 0 (\bmod p)$ and $f'(a_0) \not\equiv 0 (\bmod p)$, then there exists a unique p-adic integer α such that*

$$f(\alpha) = 0 \quad \textit{and} \quad \alpha \equiv a (\bmod p) .$$

This lemma is analogous to Newtons method for approximating real roots of equations and is sometimes called the p-adic Newton's lemma (see [Kob 1] for details). The lemma can be generalized to:

Hensel's Lemma: *Let $f(x_1, \ldots, x_n) \in \mathbf{Z}_p[x_1, \ldots, x_n]$. If $a_1, \ldots, a_n$ are rational integers such that $f(a_1, \ldots, a_n) \equiv 0 (\bmod p)$ and, for some i, the partial derivative of f with respect to x_i satisfies $f_{x_i}(a_1, \ldots, a_n) \not\equiv 0 (\bmod p)$, then there exist p-adic integers $\alpha_1, \ldots, \alpha_n$ such that*

$$f(\alpha_1, \ldots, \alpha_n) = 0 \quad \textit{and} \quad \alpha_i \equiv a_i (\bmod p), \quad \textit{for} \quad 1 \le i \le n .$$

Note: In some accounts of this lemma, our assumption that the a_i are rational integer is replaced by the assumption that the a_i are p-adic integers. This yields an equivalent form of the lemma which is useful for generalizations.

While Hensel's lemma allows us to lift solutions of congruences modulo p to p-adic solutions of equations, the more basic relation between equations with solutions in $\mathbf{Z}_p$ and congruences with solutions in $\mathbf{Z}$ is given by

Theorem: *If $f(x_1, \ldots, x_n) \in \mathbf{Z}_p[x_1, \ldots, x_n]$, then $f(x_1, \ldots, x_n) \equiv 0 (\bmod p^k)$ is solvable in rational integers, for all $k \ge 1$, if and only if $f(x_1, \ldots, x_n) = 0$ is solvable in p-adic integers.*

When f is a form, the theorem can be refined to

Theorem: *If $f(x_1, \ldots, x_n) \in \mathbf{Z}_p[x_1, \ldots, x_n]$ is a form, then $f = 0$ has a non trivial solution in $\mathbf{Z}_p$ if and only if $f \equiv 0 (\bmod p^m)$ has a solution in rational integers with not all terms divisible by p^m, for all $m \ge 1$.*

See [Bor - Sha] for proofs which are based on convergence and continuity properties of the p-adics. These results clarify the relation between congruences and p-adic integers and reinforce the idea that p-adic integer solutions of an equation are a convenient method of coding the solutions to an infinite set of congruences.

6. Hasse's Principle; The Hasse-Minkowski Theorem

Now we turn to the deeper question of the relation between solutions of polynomial congruences and rational and integer solutions of polynomial equations. A brief discussion was presented in section 9.7 where we introduced the Hasse principle. Now we introduce a more sophisticated formulation based on p - adic concepts. A class C of polynomials with, rational coefficients, satisfies the **Hasse principle** (*or the local - global principle*) if the existence of solutions of $f = 0$, $f \in C$, in all *local fields* (i.e., solutions in $\mathbf{R}$ and in $\mathbf{Q}_p$, for all p) implies the existence of a *global solution* (a solution in $\mathbf{Q}$). Of course, by the results in the last section, solutions in $\mathbf{R}$ and modulo p^m, for all p and $m \geq 1$ also imply the existence of a global solution.

Example: Let $C = \{x^2 - r | r \in Q\}$. Assume that $x^2 - r = 0$ has solutions in all local fields, where $r = \frac{c}{d}$, $(c, d) = 1$. A solution α in $\mathbf{Q}_p$ implies that $\alpha^2 = r$ or $\text{ord}_p(r) = 2\,\text{ord}_p(\alpha)$ and thus p appears with an even exponent, which can be zero, in the prime factorization of either c or d, but not both . Hence $r = \pm s^2$, for some $s \in \mathbf{Q}$. But $x^2 - r$ has a solution in $\mathbf{R}$; so r must be non negative, $r = s^2$ and $x^2 - r = 0$ has the global solution $x = s$.

For forms, there is always the trivial solution $f(0, \ldots, 0) = 0$; so, in this case, we are only interested in non trivial solutions and we must add the qualifier 'non-trivial' to all solutions in the formulation of the Hasse principle . A form f in $R[x_1, \ldots, x_n]$, R a commutative ring, **represents** r in R, if there are elements $r_1, \ldots, r_n$ in R, *not all zero*, such that $f(r_1, \ldots, r_n) = r$. One of the central theorems in number theory says that the Hasse principle holds for quadratic forms. We state it in three equivalent ways.

Hasse - Minkowski Theorem: *(I) If f is a quadratic form in $\mathbf{Z}[x_1, \ldots, x_n]$, then f represents zero in $\mathbf{Z}$ if and only if it represents zero in $\mathbf{R}$ and in $\mathbf{Z}/p^m\mathbf{Z}$ (i.e., modulo p^m), for all primes p and positive integers m.*

(II) If f is a quadratic form in $\mathbf{Z}[x_1, \ldots, x_n]$, then f represents zero in $\mathbf{Z}$ if and only if f represents zero in $\mathbf{R}$ and in $\mathbf{Z}_p$, for all primes p.

(III) If f is a quadratic form in $\mathbf{Q}[x_1, \ldots, x_n]$, then f represents zero in $\mathbf{Q}$ if and only if f represents zero in $\mathbf{R}$ and in $\mathbf{Q}_p$, for all primes p.

The equivalence of (I) and (II) follows from the last theorem in section 5. To see the equivalence of (II) and (III), first note that solving $f = 0$ with rational coefficients can be reduced to solving an equation with integer coefficients by multiplying $f = 0$ by the least common multiple of the denominators of the coefficients of f. Moreover, $f = 0$ with integer coefficients can be solved over $\mathbf{Q}$ (resp. $\mathbf{Q}_p$) if and only if it can be solved over $\mathbf{Z}$ (resp. $\mathbf{Z}_p$) since we can write the terms of the solution with a common denominator b and multiply the equation by b^2.

Recall that a quadratic form $f = \sum a_{i,j} x_i x_j$, $a_{i,j} \in \mathbf{Q}$, can be written in terms of matrices, $f = X^t AX$, where $X^t = (x_1, \ldots, x_n)$, $A = (a_{i,j})$, and t denotes the transpose of a matrix. If $\det A \neq 0$, then f is called a **non singular quadratic form** and if $\det A = 0$, then f is a **singular quadratic form**. If a non singular quadratic form with coefficients in a *field K* represents zero in K, then it represents all elements of K [Bor - Sha, suppl. 1]; hence

Corollary: *If f is a non-singular quadratic form in $\mathbf{Q}[x_1, \ldots, x_n]$, then f represents r for all r in $\mathbf{Q}$, if and only if f represents zero in $\mathbf{R}$ and in $\mathbf{Q}_p$ for all primes p.*

We shall not prove the Hasse - Minkowski theorem [see Bor - Sha], but a note on reducing the general case to a special class of equations will prove useful. By a non-singular linear transformation of the variables, $X = BY$, $B = (b_{i,j})$, $b_{i,j} \in \mathbf{Q}$, $\det(B) \neq 0$, the form $f = X^t AX$ can be transformed into a diagonal form

$$g(y_1, \ldots, y_n) = \sum_{i=1}^{n} b_i y^i ,$$

with $b_i \in \mathbf{Q}$. Since the transformation is bijective from $\mathbf{Q}^n$ to $\mathbf{Q}^n$, f represents zero in $\mathbf{Q}$ if and only if g represents zero in $\mathbf{Q}$ and thus one only needs to prove Hasse-Minkowski for diagonal forms. If f is a singular form, then some of the b_i's are zero and clearly f represents zero. Hence the general Hasse-Minkowski theorem is reduced to the case of non singular diagonal forms.

In the case of three variables, it is easy to reduce this further to those diagonal forms

$$f(x_1, x_2, x_3) = ax_1^2 + bx_2^2 + cx_3^2 , \tag{1}$$

where a, b, c are in $\mathbf{Z}$, non-zero, square free, pairwise relatively prime and not all of the same sign. This is the same class of polynomials to which Legendre reduced the problem of deciding if a rational conic has a rational point (sec. 18.3).

Legendre's Theorem: *If a, b, c are rational integers, non zero, square free, pairwise relatively prime and not all of the same sign, then*

$$f(x_1, x_2, x_3) = ax_1^2 + bx_2^2 + cx_3^2$$

represents zero in $\mathbf{Z}$ if and only if $-bc$, $-ca$, $-ab$ are quadratic residues modulo a, b, c respectively.

Clearly the criteria can be checked in a finite number of steps. We prove the equivalence of Legendre's theorem and the Hasse-Minkowski theorem for the forms given by (1), which, as noted, is equivalent to the general Hasse-Minkowski theorem for three variables.

Legendre $\Longrightarrow$ Hasse-Minkowski: Assume that f is a form of the type given in (1) which satisfies the conditions of the Hasse-Minkowski theorem. We prove that f satisfies the conditions of Legendre's theorem and thus f represents zero in $\mathbf{Z}$.

By assumption, there are integers $(x, y, z) \not\equiv (0, 0, 0) (\text{mod } p)$ such that

$$ax^2 + by^2 + cz^2 \equiv 0 (\text{mod } p) . \tag{2}$$

Let p be a prime dividing a; then

$$by^2 + cz^2 \equiv 0 (\text{mod } p) \tag{3}$$

and, multiplying by c,

$$-bcy^2 \equiv c^2z^2 (\text{mod } p) . \tag{4}$$

If $p|y$, then, since $(p, c) = 1$, we have $p|z$ by (3). But if $p|y, z$ and thus $p^2|by^2, cz^2$, then, by (2), $p^2|ax^2$ and, since a is square free, $p|x$. Thus if $p|y$, then $p|x, z$, which contradicts $(x, y, z) \not\equiv (0, 0, 0) (\text{mod } p)$. Therefore $y \not\equiv 0 (\text{mod } p)$ and there exists a y' such that $yy' \equiv 1 (\text{mod } p)$. Multiplying both sides of (4) by y'^2, we have

$$-bc \equiv (czy')^2 (\text{mod } p) ;$$

hence $-bc$ is a quadratic residue modulo p.

Since a is square free, we can use the Chinese Remainder Theorem to paste together the solutions of $-bc \equiv u^2 (\text{mod } p)$, for all p dividing a, to get a solution of $-bc \equiv u^2 (\text{mod } a)$; hence $-bc$ is a quadratic residue modulo a. A permutation of the letters yields $-ca$ and $-bc$ as quadratic residues of b and c respectively and we are done.

Note that the only part of the Hasse-Minkowski conditions that we used were the existence of solutions modulo p, for those p dividing the coefficients. However the other conditions follow for any form of type (1) from the conditions on the coefficients, as is illustrated in the next proof.

Hasse-Minkowski $\Longrightarrow$ *Legendre:* Assume that f is a form of the type given in (1) which satisfies the conditions of Legendre's theorem. We prove that f satisfies the conditions of the Hasse-Minkowski theorem and thus f represents zero in $\mathbf{Z}$. Actually, we shall only prove this when 2 does not divide abc (this case requires a special argument).

By assumption, there is a u satisfying $-bc \equiv u^2 (\bmod a)$. Then, for a prime p dividing a, we have

$$-bc \equiv u^2 (\bmod p) .$$

Since $(a, c) = 1$, $c \not\equiv 0 (\bmod p)$ and there is a z' such that $cz' \equiv u (\bmod p)$ and

$$c^2 z'^2 \equiv u^2 (\bmod p) .$$

Hence $-bc \equiv c^2 z'^2 (\bmod p)$ and

$$-b \equiv cz'^2 (\bmod p) .$$

Therefore

$$b(1)^2 + c(z')^2 \equiv 0 (\bmod p)$$

and, since $p|a$,

$$f(1, 1, z') = a(1)^2 + b(1)^2 + c(z')^2 \equiv 0 (\bmod p) .$$

To find solutions in $\mathbf{Z}_p$, we apply Hensel's lemma to $f = ax_1^2 + bx_2^2 + cx_3^2$. Since $p \not| b$ and $p \neq 2$, $f'_{x_2}(1, 1, z') = 2b \equiv 0 (\bmod p)$, and by Hensel's lemma there exist p-adic integers α, β, γ such that

$$f(\alpha, \beta, \gamma) = a\alpha^2 + b\beta^2 + c\gamma^2 = 0 ,$$

with $(\alpha, \beta, \gamma) \equiv (1, 1, z') \equiv (0, 0, 0) (\bmod p)$. Hence f represents zero in $\mathbf{Z}_p$ for all p dividing a, and by a permutation of the letters, f represents zero in $\mathbf{Z}_p$ for all p dividing a, b or c.

If $p \not| abc$, then by Chevally's theorem (sec. 9.6), $f = 0$ has a non-trivial solution modulo p, and we use Hensel's lemma again to lift this solution to a non-trivial solution in $\mathbf{Z}_p$. Therefore the Hasse-Minkowski conditions hold in all $\mathbf{Z}_p$.

To prove the conditions in **R**, recall that a, b, c do not all have the same sign; say $a, b > 0$ and $c < 0$. Then, setting

$$z = \sqrt{\frac{a+b}{-c}},$$

we have

$$a(1)^2 + b(1)^2 + c(z)^2 = 0,$$

i.e., f represents zero in **R**.

The Hasse principle does not hold for the set of all polynomials, e.g., $3x^2 + 4y^2 + 5z^2$ and $x^4 - 17 - 2y^2$ satisfy all the Hasse - Minkowski criteria but have no rational solutions [Cas 2]. The true scope of the Hasse principle remains an outstanding unsolved problem.

7. Valuations and Algebraic Number Theory

As discussed in section 1, there is another thread of development for the foundations of algebraic number theory as opposed to Dedekind's ideal theory. This development is more in the spirit of Kummer's original work and based on the notion of a valuation. One modern version of Kummer's theory is developed via an axiomatic theory of divisors in algebraic number fields. We present a brief outline of this theory and take some liberty in not stating all conditions with complete precision.

A theory of divisors consists of a commutative ring R (say the ring of rational integers or the ring of integers in any algebraic number field) and a homomorphism of R^*, the non zero elements of R, into a semigroup S, which has unique factorization of its elements into prime elements, satisfying several simple conditions. The semigroup of ideals in an algebraic number field together with the mapping that sends the non zero integers of the field to the principal ideals they generate is an example of such a divisor theory, but we will present an alternate approach using valuations.

First we can prove that if R has a theory of divisors, it is unique, i.e., any two such theories for R are isomorphic. Thus the different constructions are just two ways of looking at the same factorization theory, but each approach provides different insights into the deeper theorems.

Next we need to show when a divisor theory exists and how to construct it. Here is where valuations enter the scene. If R has a divisor theory with semigroup S, $\alpha \to (\alpha)$ is the mapping of R^* into S, and P is a prime element of S, we define the (non Archimedean) **valuation, $\mathbf{v_P}(\alpha)$**, for any α in R^*, as the highest power of P which divides α. Hence

$$P^{v_P(\alpha)} | (\alpha) \text{ and } P^{v_P(\alpha)+1} \nmid (\alpha).$$

We also set $v_P(0) = \infty$. The valuation satisfies

$$v_P(\alpha\beta) = v_P(\alpha) + v_P(\beta) \tag{1}$$

and

$$v_P(\alpha + \beta) \geq \min(v_P(\alpha), v_P(\beta)). \tag{2}$$

This valuation determines a valuation of the quotient field K of R as follows: if $k = \alpha/\beta \in K$, where $\alpha, \beta \in R$, let $v_P(k) = v_P(\alpha) - v_P(\beta)$. The conditions (1) and (2) still hold.

Since the prime factorization of any (α) is given by

$$(\alpha) = \prod_i P_i^{v_{P_i}(\alpha)},$$

where the product is over all i for which $v_{P_i}(\alpha) > 0$, it can be shown that the divisor theory is completely determined by the set of all valuations of K which are extensions of the valuations v_P for all prime divisors P.

All the (non Archimedean) valuations of the rational integers are easy to describe; for each prime p, $v_p(n)$ is the highest power of p dividing n. This set of valuations, for all p, yields a theory of divisors, where we don't have to add new elements since we have unique factorization in $\mathbf{Z}$. Now we can construct all valuations of the ring of integers I of a number field K (a finite extension of $\mathbf{Q}$) by constructing extensions of the valuations of $\mathbf{Z}$. Hence we obtain a theory of divisors for I, i.e., every non zero element of I will have unique factorization into prime elements in a corresponding semigroup.

Given a valuation v_P on a number field K, one can extend the field to its "completion" K_P. This construction generalizes our completions of the rational numbers to the fields of p-adic numbers (this was discussed a bit in sections 3 and 4). One can then piece together information about the completions to study properties of K. This is analogous to the Hasse principle (sec. 6).

Everything we have described takes some real work to prove. Our description follows the development presented by Borevich and Shafarevich [Bor - Sha], which is the clearest introduction to this theory.

Algebraic number theory developed via valuations and their completions is often called the "local theory" of algebraic numbers, while the approach via ideals is called the "global theory". Some of the deeper theorems in the theory are more easily proved by one approach and some by the other. Knowledge of both theories is requisite for a mastery of algebraic number theory. It is interesting to contrast the local theory as presented by Borevich and Shafarevich with the global theory presented by Hecke [Hec].

The study of arithmetic on curves as well as that of equations over finite fields (numbers of solutions of equations) can also be approached via valuations on functions fields. This yields very strong analogies between these two areas and algebraic number theory, and explains to some extent the central role of zeta functions in all three fields. These analogies have played a key role in the development of much of 20th century number theory.

Bibliography

[Ada-Gol] W. W. Adams and L. J. Goldstein. *Introduction to Number Theory* (Englewood Cliffs, NJ: Prentice Hall), 1976.

[Ahl] L. V. Ahlfors. *Complex Analysis*, Third Edition (New York: McGraw-Hill), 1979.

[And 1] G. Andrews. *The Theory of Partition, Encyclopedia of Mathematics and its Applications*, Vol. 2 (Reading, MA: Addison-Wesley), 1976.

[And 2] G. Andrews. *Number Theory* (Philadelphia: Saunders), 1971.

[Apo 1] T. M. Apostol. *Modular Functions and Dirichlet Series in Number Theory*, Second Edition (New York: Springer-Verlag), 1990.

[Art, M] Michael Artin. *Algebra* (Englewood Cliffs, NJ: Prentice Hall), 1991.

[Ati] M. Atiyah. "The Role of Algebraic Topology in Modern Mathematics," *Jour. London Math. Soc.* **41**(1996), 63–69.

[Bac] G. Bachman. *Introduction to p-adic Numbers and Valuation Theory* (New York: Academic Press), 1964.

[Bak] A. Baker. "Effective Methods in the Theory of Numbers," in *Actes du Congrès International des Mathématiciens 1970* (Paris: Gauthier-Villars), 1971, 19–26.

[Bak 2] A. Baker. "Linear Forms in the Logarithms of Algebraic Numbers I, II, III, IV," *Mathematika* **13**(1966), 204–216; **14**(1967), 120–127, 220–228; **15**(1968), 204–216.

[Bak 3] A. Baker. *Transcendental Number Theory* (Cambridge: Cambridge University Press), 1975.

[Bak-Coa] A. Baker and J. Coates. "Integer Points on Curves of Genus 1," *Proc. Camb. Phil. Soc.* **67**(1970), 595–602.

[Bas] I. Bashmakova. "Arithmetic of Algebraic Curves from Diophantus to Poincaré," *Historica Mathematica* **8**(1981), 417–438.

[Bax] R. J. Baxter. *Exactly Solved Models in Statistical Mechanics* (Boston: Academic Press), 1982.

[Bel 1] E. T. Bell. *Men of Mathematics* (New York: Simon & Schuster), 1937.

[Bel 2] E. T. Bell. *The Last Problem* (New York: Simon & Schuster), 1961.

[Ber] L. Bers. *Riemann Surfaces*, notes by E. Rodlitz and R. Pollack, Courant Institute of Mathematical Sciences, New York University, 1957–58.

[Bir-Mac] G. Birkhoff and S. Mac Lane. *A Survey of Modern Algebra*, Fourth Edition (New York: Macmillan), 1977; Fifth Edition (Wellesley, MA: A K Peters, Ltd.), 1997.

[Bli] H. F. Blichfeldt. "A New Principle in the Geometry of Numbers with Some Applications," *Trans. Amer. Math. Soc.* **15**(1914), 227–235.

[Bor-Sha] Z. I. Borevich and I. R. Shafarevich. *Number Theory* (New York: Academic Press), 1966.

[Bri-Kno] E. Brieskorn and H. Knörrer. *Plane Algebraic Curves,* English translation by J. Stillwell (Boston: Birkhäuser), 1986.

[Bue] D. A. Buell. *Binary Quadratic Forms: Classical Theory and Modern Computation* (New York: Springer-Verlag), 1989.

[Bus-Kel] H. Busemann and P. J. Kelly. *Projective Geometry and Projective Metrics* (New York: Academic Press), 1953.

[Cas] J. W. S. Cassels. "Mordell's Finite Basis Theorem Revisited," *Math. Proc. Camb. Phil. Soc.* **100**(1986), 31–41.

[Cas 2] J. W. S. Cassels. "Diophantine Equations with Special Reference to Elliptic Curves," *Jour. London Math. Soc.* **41**(1966), 193–291.

[Cas 3] J. W. S. Cassels. *An Introduction to Diophantine Approximation* (Cambridge: Cambridge University Press), 1957; Reprint (New York: Hafner), 1972.

[Cas 4] J. W. S. Cassels. *An Introduction to the Geometry of Numbers*, Second Corrected Reprint (New York: Springer-Verlag), 1971.

[Cas 5] J. W. S. Cassels. *Local Fields,* London Mathematical Society Student Texts, Vol. 3 (Cambridge: Cambridge University Press), 1986.

[Cha] J. S. Chahal. *Topics in Number Theory* (New York: Plenum Press), 1988.

[Che] C. Chevally. "Démonstration d'une Hypothèse de M. Artin," *Abhand. Math. Sem. Hamburg* **11**(1936), 73–75.

[Cho] S. Chowla. *The Riemann Hypothesis and Hilbert's Tenth Problem* (New York: Gordon and Breach), 1965.

[Chr] G. Chrystal. *A Textbook of Algebra*, 2 Vols. (Adam and Charles Black), 1900; Reprint (New York: Dover Publications), 1961.

[Cip] Barry Cipra. "Big Number Breakdown," *Science* **148**(1990), 1608.

[Coh] Harvey Cohn. *A Second Course in Number Theory* (New York: John Wiley and Sons), 1962.

[Cohe] H. Cohen. "Elliptic Curves," in [W-M-L-I], 212–237.

[Con] J. Conway. "An Enumeration of Knots and Links and Some of Their Related Properties," in *Computational Problems in Abstract Algebra*, edited by J. Leech (New York: Pergamon Press), 1970, 329–358.

[Coo] R. Cooke. "Abel's Theorem," in *The History of Modern Mathematics*, Vol. 1, edited by D. E. Rowe and J. McCleary (Boston: Academic Press), 1989, 388–421.

[Cox 1] D. A. Cox. *Primes of the Form x^2+ny^2: Fermat, Class Field Theory, and Complex Multiplication* (New York: Wiley-Interscience), 1989.

[Cox 2] D. Cox. "The Arithmetic-Geometric Mean of Gauss," *L'Enseign. Math.* **30**(1984), 275–330.

[Cox, H] H. S. M. Coxeter. *Introduction to Geometry*, Second Edition (New York: John Wiley and Sons), 1969.

[Cox, H 2] H. S. M. Coxeter. *The Real Projective Plane*, Second Edition (Cambridge: Cambridge University Press), 1955.

[Cre] L. Cremona. *Elements of Projective Geometry* (London: Oxford at the Clarendon Press), 1893.

[Dav] H. Davenport. *The Higher Arithmetic: An Introduction to the Theory of Numbers* (London: Hutchinson University Library), 1952; Reprint (New York: Dover Publications), 1983.

[Dav 2] H. Davenport. "L. J. Mordell," *Acta Arith.* **IX**(1964), 3–12.

[Dav 3] H. Davenport. "The Work of K. F. Roth," in *Proceedings of the International Congress of Mathematicians*, edited by J. A. Todd (Cambridge: Cambridge University Press), 1960, lvii–lx.

[Dav 4] H. Davenport. "Minkowski's Inequality for the Minima Associated with a Convex Body," *Quart. Jour. Math. Oxford* **10**(1939), 119–121.

[Dav-Mat-Rob] M. Davis, Y. Matijasevič, and J. Robinson. "Hilbert's Tenth Problem. Diophantine Equations: Positive Aspects of a Negative Solution," in *Mathematical Developments Arising from Hilbert Problems*, Proceedings of Symposia in Pure Mathematics, Vol. 28 (Providence: American Mathematical Society), 1976, 323–378.

[Den] J. Denef. "The Rationality of the Poincaré Series Associated to the p-adic Points on a Variety," *Invent. Math.* **77**(1984), 1–23.

[Dic 1] L. E. Dickson. *History of the Theory of Numbers*, 3 Vols. (Washington, DC: Carnegie Institute), 1919–1923; Reprint (New York: Chelsea), 1971.

[Dic 2] L. E. Dickson. *Introduction to the Theory of Numbers* (Chicago: University of Chicago Press), 1929.

[Dir] P. G. L. Dirichlet. *Werke*, 2 Vols., edited by L. Kronecker and L. Fuchs (Berlin: George Reimer), 1984; Reprint (New York: Chelsea).

[Dir-Ded] P. G. Lejeune-Dirichlet and R. Dedekind. *Vorlesungen über Zahlentheorie*, Fourth Edition (Braunschweig: Friedrich Vieweg und Sohn), 1894; Reprint (New York: Chelsea), 1968.

[Dun] G. W. Dunnington. *Carl Friedrich Gauss: Titan of Science* (New York: Hafner), 1955.

[Edw 1] Harold M. Edwards. *Fermat's Last Theorem: A Genetic Introduction to Algebraic Number Theory* (New York: Springer-Verlag), 1977.

[Edw 2] H. M. Edwards. *Galois Theory* (New York: Springer-Verlag), 1984.

[Edw 3] H. M. Edwards. "An Appreciation of Kronecker," *Math. Intelligencer* **9**,1(1987), 28–35.

[Eis] G. Eisenstein. "Eisenstein's Geometrical Proof of the Fundamental Theorem for Quadratic Residues," English translation by A. Cayley, *Quart. Jour. Pure Appl. Math.* **1**(1857), 186–191.

[Eis 2] G. Eisenstein. *Mathematische Werke* (New York: Chelsea), 1976.

[Ell, W-Ell, F] W. and F. Ellison. "Théorie des Nombres," in *Abrégé d'Histoire des Mathématiques, 1700–1900*, 2 Vols., edited by J. Dieudonné (Paris: Hermann), 1978, Chapter 5.

[Euc] Euclid. *The Thirteen Books of Euclid's Elements*, 3 Vols., Second Edition (Cambridge: Cambridge University Press), 1926; Reprint (New York: Dover Publications), 1956.

[Eul 1] L. Euler. Extracts from "Theoremata Circa Residis ex Divisione Potestatum Relicta," in *Opera Omnia*, Series I, Vol. 2 (Basel: Birkhäuser), 493–518; English translation by R. J. Stroeker as *Euler Power Residue* with commentaries in [Str].

[Eul 2] L. Euler. *Elements of Algebra*, Fifth Edition, English translation by J. Hewlett (Longmann, Orme and Co.), 1840; Reprint (New York: Springer-Verlag), 1984.

[Eul 3] L. Euler. *Introduction to Analysis of the Infinite*, English translation by J. D. Blanton (New York: Springer-Verlag), 1988.

[Eul 4] L. Euler. "Observationes circa Divisionen Quadratorum per Numeros Primos," in *Opera Omnia*, Series I, Vol. 3 (Basel: Birkhäuser), 497–512.

[Fal] G. Faltings. "Endlichkeitssätze für Abelsche Varietäten über Zahlkörpern," *Invent. Math.* **73**(1983), 349–366.

[Fer] P. Fermat. *Oeuvres de Fermat*, 3 Vols., edited by P. Tannery and C. Henry (Paris: Gauthier-Villars), 1891–1896.

[Fla] D. Flath. *Introduction to Number Theory* (New York: John Wiley and Sons), 1989.

[For] Otto Forster. *Lectures on Riemann Surfaces*, English translation by B. Gilligan (New York: Springer-Verlag), 1981.

[Fow] D. H. Fowler. *The Mathematics of Plato's Academy: A New Reconstruction* (Oxford: Clarendon Press), 1987.

[Ful] W. Fulton. *Algebraic Curves: An Introduction to Algebraic Geometry* (Reading, MA: W. A. Benjamin), 1969.

[Gau 1] C. F. Gauss. *Disquisitiones Arithmeticae* (Lipsia in commissis apud Gerh. Fleischer Iun), 1801; English translation by A. Clarke (New York: Springer-Verlag), 1986.

[Gau 2] C. F. Gauss. "Theoria Residuorum Biquadraticorum: Commentatio Prima," in *Werke*, Vol. 2 (Göttingen: Königliche Gesellschaft der Wissenschaft), 1876, 65–92.

[Gau 3] C. F. Gauss. "Theoria Residuorum Biquadraticorum: Commentatio Sequnda," in *Werke*, Vol. 2 (Göttingen: Königliche Gesellschaft der Wissenschaft), 1876, 93–148

[Gel] A. O. Gelfond. *Transcendental and Algebraic Numbers*, English translation from the Russian by L. Boron (New York: Dover Publications), 1960.

[Gol, D] D. Goldfeld. "Gauss' Class Number Problem for Imaginary Quadratic Fields," *Bull. Amer. Math. Soc.* **13**,1(1985), 23–37.

[Gol, J 1] J. R. Goldman. "Hurwitz Sequences, The Farey Process, and General Continued Fractions," *Adv. in Math.* **72**,2(Dec. 1988), 239–260.

[Gol, J 2] J. R. Goldman. "Numbers of Solutions of Congruences: Poincaré Series for Strongly Non-Degenerate Forms," *Proc. Amer. Math. Soc.* **87**,4(1983), 586–590.

[Gol, J 3] J. R. Goldman. "Numbers of Solutions of Congruences: Poincaré Series for Algebraic Curves," *Adv. in Math.* **2**,1(1986), 68–83.

[Gol-Kau 1] J. R. Goldman and L. H. Kauffman. "Knots, Tangles, and Electrical Networks," *Adv. in Appl. Math.* **14**(1993), 267–306.

[Gol-Kau 2] J. R. Goldman and L. H. Kauffman. "Rational Tangles," *Adv. in Appl. Math.* **18**(1997), 300–332.

[Gol, L] L. J. Goldstein. *Abstract Algebra: A First Course* (Englewood Cliffs, NJ: Prentice Hall), 1973.

[Gol, L 2] L. J. Goldstein. *Analytic Number Theory* (Englewood Cliffs, NJ: Prentice Hall), 1971.

[Gra] J. J. Gray. "A Commentary on Gauss's Mathematical Diary, 1796–1814" (with English translation), *Expos. Math.* **2**(1984), 97–130.

[Gro] E. Grosswald. *Topics from the Theory of Numbers*, Second Edition (Boston: Birkhäuser), 1984.

[Gra-Knu-Pat] R. Graham, D. E. Knuth, and O. Patashnik. *Concrete Mathematics* (Reading, MA: Addison-Wesley), 1989.

[Had] C. R. Hadlock. *Field Theory and Its Classical Problems*, Carus Mathematical Monograph of the Mathematical Association of America, Vol. 17 (Washington, DC: Mathematical Association of America), 1978.

[Hal, M] M. Hall, Jr. "On the Sum and Product of Continued Fractions," *Math. Annalen* **48**(1947), 966–993.

[Hal, T] T. Hall. *Carl Friedrich Gauss: A Biography*, English translation by A. Froderberg (Cambridge, MA: MIT Press), 1970.

[Han] H. Hancock. *Development of the Minkowski Geometry of Numbers* (New York: Macmillan), 1939; Reprint in 2 Vols. (New York: Dover Publications), 1964.

[Har] R. Hartshorne. *Algebraic Geometry* (New York: Springer-Verlag), 1977.

[Har 2] R. Hartshorne. *Foundations of Projective Geometry* (New York: W. A. Benjamin), 1967.

[Har-Lil-Pol] G. H. Hardy, J. E. Littlewood, and G. Pólya. *Inequalities* (Cambridge: Cambridge University Press), 1964.

[Har-Wri] G. H. Hardy and E. M. Wright. *An Introduction to the Theory of Numbers* (Oxford: Clarendon Press), 1938; Fifth Edition, 1979.

[Hea] Sir Thomas L. Heath. *Diophantus of Alexandria: A Study in the History of Greek Algebra* (Cambridge: Cambridge University Press), 1910; Reprint (New York: Dover Publications), 1964.

[Hec] E. Hecke. *Vorlesungen über die Theorie der algebraischen Zahlen* (Leipzig: Academische Verlagsgesellschaft), 1923; English translation by G. U. Brauer and J. R. Goldman with R. Kotzen as *Lectures on the Theory of Algebraic Numbers* (New York: Springer-Verlag), 1981.

[Her] I. N. Herstein. *Topics in Algebra*, Second Edition (Lexington, MA: Xerox College Publishers), 1975.

[Her, C] C. Hermite. "Sur la Fonction Exponentielle," *Comptes Rend.* **77**(1873), 18–24, 74–79, 226–233, 285–293; Reprinted in *Oeuvres,* Vol. III, 150–181.

[Hil] David Hilbert. "Die Theorie der algebraischen Zahlkörper," in *Gesammelte Abhandlungen*, Vol. 1 (Berlin: Springer-Verlag), 1932, 63–539; Reprint (New York: Chelsea), 1965.

[Hil 2] D. Hilbert. "Über die Transcendenz der Zahlen e und π," *Math. Annalen* **43**(1893), 216–220; Reprinted in *Gesammelte Abhandlungen*, Vol. 1 (Berlin: Springer-Verlag), 1932, 1–4.

[Hil 3] D. Hilbert. "Mathematical Problems," English translation by M. Newson, *Bull. Amer. Math. Soc.* **8**(1902), 437–479; Reprinted in *Mathematical Developments Arising from Hilbert's Problems*, Proceedings of Symposia in Pure Mathematics, Vol. 28 (Providence: American Mathematical Society) 1976, 1–34.

[Hil-Coh] D. Hilbert and S. Cohn-Vossen. *Geometry and the Imagination*, English translation by P. Nemenyi (New York: Chelsea), 1952.

[Hil-Hur] D. Hilbert and A. Hurwitz. "Über die diophantischen Gleichungen vom Geschlecht Null," *Acta Math.* **14**(1890), 217–224.

[Hor] T. Horowitz. "On Jargon: Elliptic Curves," *UMAP Jour.* **2**(Summer 1987), 161–181.

[Hou] C. Houzel. "Fonctions Elliptiques et Intégrales Abéliennes," in *Abrégé d'Histoire des Mathématiques, 1700–1900*, 2 Vols., edited by J. Dieudonné (Paris: Hermann), 1978, Chapter 7.

[Hua] L. K. Hua. *Introduction to Number Theory*, English translation by Peter Shiu (New York: Springer-Verlag), 1982.

[Hum] G. Humbert. "Sur les Fractions Continues Ordinaires et les Formes Quadratiques Binaires Indéfinies," *Jour. Math. Pures Appl.* **7**(1916), 104–157.

[Hur 1] A. Hurwitz. "Über die Reduktion der binären quadratischen Formen," *Math. Annalen* **45**(1894); Reprinted in *Werke*, Vol. 2 (Basel: Birkhäuser), 1933, 1963, 157–190.

[Hur 2] A. Hurwitz. "Über die Kettenbrüche, deren Teilnenner arithmetische Reihen bilden," in *Werke*, Vol. 2 (Basel: Birkhäuser), 1896, 276–302.

[Hur-Kri] A. Hurwitz and N. Kritikos. *Lectures on Number Theory*, English translation by W. Schulz (New York: Springer-Verlag), 1986.

[Hus] D. Husemöller. *Elliptic Curves* (New York: Springer-Verlag), 1987.

[Igu] J. I. Igusa. "Complex Powers and Asymptotic Expansions I," *Jour. Reine Angew. Math.* **268/269**(1974), 110–130; "Complex Powers and Asymptotic Expansions II," *Jour. Reine Angen. Math* **278/279**(1975), 307–321.

[Inc] E. L. Ince. *Cycles of Reduced Ideals in Quadratic Fields, Mathematical Tables IV* (London: British Association for the Advancement of Science), 1934.

[Ire-Ros] K. Ireland and M. Rosen. *A Classical Introduction to Modern Number Theory*, Second Edition (New York: Springer-Verlag), 1990.

[Ita 1] J. Itard. "Joseph Louis Lagrange," in *Dictionary of Scientific Biography,* Vol. VII (New York: Charles Scribner's Sons), 1972, 559–573.

[Ita 2] J. Itard. "Adrien-Marie Legendre," in *Dictionary of Scientific Biography,* Vol. VIII (New York: Charles Scribner's Sons), 1972, 135–143.

[Jac] C. G. J. Jacobi. "De Usu Theoriae Integralium Ellipticorum et Integralium Abelianorum in Analysi Diophantea," *Jour. Reine Angew. Math.* **13**(1835), 353–355; Reprinted in *Gesammelte Werke*, Vol. 2, 53–55.

[Jen] W. E. Jenner. *Rudiments of Algebraic Geometry* (New York: Oxford University Press), 1963.

[Jon] F. Jones, Jr. *Rudiments of Riemann Surfaces*, Lecture Notes in Mathematics, Vol. 2 (Houston: Rice University), 1971.

[Jon-Thr] W. Jones and W. J. Thron. "Continued Fractions: Analytic Theory and Applications," in *Encyclopedia of Mathematics and Its Applications*, Vol. 11 (Reading, MA: Addison-Wesley), 1980.

[Kat] N. Katz. "An Overview of Deligne's Proof of the Riemann Hypothesis for Varieties over Finite Fields," in *Mathematical Developments Arising from Hilbert Problems*, Proceedings of Symposia in Pure Mathematics, Vol. 28 (Providence: American Mathematical Society), 1976, 279–306.

[Kau] W. K. Bühler. *Gauss: A Biographical Study* (New York: Springer-Verlag), 1981.

[Ken] K. Kending. *Elementary Algebraic Geometry*, Graduate Texts in Mathematics, Vol. 44 (New York: Springer-Verlag), 1977.

[Kir] F. Kirwan. *Complex Algebraic Geometry*, London Mathematical Society Student Texts, Vol. 23 (Cambridge: Cambridge University Press), 1992.

[Kle 1] F. Klein. *Elementary Mathematics from an Advanced Viewpoint*, Third Edition, Vol. 1: *Arithmetic, Algebra, Analysis* and Vol. 2: *Geometry*, English translation by E. A. Hedrick and C. A. Noble, 1908; Reprint (New York: Dover Publications), 1945.

[Kle 2] F. Klein. *Vorlesungen über die Entwicklung der Mathematik im 19. Jahrhundert* (Berlin: Springer-Verlag), 1928; English translation by M. Ackerman as *Development of Mathematics in the 19th Century* (Brookline, MA: Mathematical Sciences Press), 1979.

[Kle 3] F. Klein. *Lectures on Mathematics*, edited by A. Ziwet (New York: American Mathematical Society), 1911.

[Kle 4] F. Klein. "Famous Problems of Elementary Geometry," in *Famous Problems and Other Monographs*, Reprint (New York: Chelsea), 1962.

[Kli] M. Kline. "Euler and Infinite Series," *Math. Mag.* **56**,5(1983), 307–315 (special issue on Euler).

[Kna] A. V. Knapp. *Elliptic Curves* (Princeton, NJ: Princeton University Press), 1992.

[Knu] D. E. Knuth. *Seminumerical Algorithms*, The Art of Computer Programming, Vol. 2: (Reading, MA: Addison-Wesley), 1969.

[Kob 1] N. Koblitz. *p-adic Numbers, p-adic Analysis, and Zeta-Functions*, Second Edition (New York: Springer-Verlag), 1984.

[Kob 2] N. Koblitz. *Introduction to Elliptic Curves and Modular Functions* (New York: Springer-Verlag), 1984.

[Kob 3] N. Koblitz. "p-adic Analysis: A Short Course on Recent Work," in *London Mathematical Society Lecture Notes*, Vol. 46 (Cambridge: Cambridge University Press), 1980.

[Kro] L. Kronecker. "Näherungsweise ganzzahlige Auflösung linearer Gleichungen," *Berliner Sitzungsberichte* (1894); Reprinted in *Werke*, Vol. 3, 47–109.

[Lag] J. L. Lagrange. *Oeuvres de Lagrange*, 14 Vols., edited by J.-A. Serret (Paris: Gauthier-Villars), 1867–1892.

[Lan 1] S. Lang. *Introduction to Diophantine Approximation* (Reading, MA: Addison-Wesley), 1966.

[Lan 2] S. Lang. *The Beauty of Doing Mathematics: Three Public Dialogues* (New York: Springer-Verlag), 1985.

[Lan 3] S. Lang. *Algebraic Number Theory*, Second Edition (New York: Springer-Verlag), 1994.

[Lan 4] S. Lang. *Fundamentals of Diophantine Geometry* (New York: Springer-Verlag), 1983.

[Lan 5] S. Lang. *Elliptic Curves: Diophantine Analysis* (New York: Springer-Verlag), 1978.

[Lan 6] S. Lang. *Introduction to Transcendental Numbers* (Reading, MA: Addison-Wesley), 1966.

[Lan-Ost] E. Landau and A. Ostrowski. "On the Diophantine Equation $ax^2 + by + c = dx^n$," *Proc. London Math. Soc.* **19**,2(1920), 276–280.

[Leg] Adrien-Marie Legendre. *Essai sur la Théorie des Nombres*, 1798; Fourth Edition as *Théorie des Nombres*, 1830; Reprint (Paris: Albert Blanchard), 1955.

[Lek] C. G. Lekkerkerker. *Geometry of Numbers*, Second Edition (Amsterdam: North-Holland), 1969.

[LeV] W. J. LeVeque. *Topics in Number Theory*, 2 Vols. (Reading, MA: Addison-Wesley), 1956.

[LeV 2] W. J. LeVeque, Editor. *Studies in Number Theory*, MAA Studies in Mathematics, Vol. 6 (Washington, DC: Mathematical Association of America), 1969.

[Lin] F. Lindemann. "Über die Zahl π," *Math. Annalen* **20**(1882), 213–225.

[Lut] E. Lutz. "Sur l'Equation $y^2 = x^3 - Ax - B$ dans les Corps p-adic," *Jour. Reine Angew. Math.* **177**(1937), 237–247.

[Mac] G. Mackey. "Harmonic Analysis as the Exploitation of Symmetry: A Historical Survey," *Bull. Amer. Math. Soc.* **3**,1(1980), 543–697.

[Mah] M. S. Mahoney. *The Mathematical Career of Pierre de Fermat (1601–1665)* (Princeton, NJ: Princeton University Press), 1973.

[Mah, K] K. Mahler. *Lectures on Transcendental Numbers*, Lecture Notes in Mathematics, Vol. 546 (New York: Springer-Verlag), 1976.

[Mah, K 2] K. Mahler. *g-adic Numbers and Roth's Theorem, Lectures on Diophantine Approximations*, Part 1 (Notre Dame, IN: University of Notre Dame), 1961.

[Mar] D. A. Marcus. *Number Fields* (New York: Springer-Verlag), 1977.

[Mas] W. S. Massey. *Algebraic Topology: An Introduction*, Fourth Corrected Printing (New York: Springer-Verlag), 1977.

[Mat] G. B. Mathews. *Theory of Numbers*, Second Edition (New York: Chelsea).

[May] K. O. May. "Carl Friedrich Gauss," in *Dictionary of Scientific Biography*, Vol. V, (New York: Charles Scribner's Sons), 1972, 298–315.

[Maz 1] B. Mazur. "Modular Curves and the Eisenstein Ideal," *Inst. Hautes Études Sci. Publ. Math.* **47**(1977), 33–186.

[Maz 2] B. Mazur. "Rational Isogenies of Prime Degree," *Invent. Math.* **44**(1978), 129–162.

[Meu] D. Meuser. "On the Poles of a Local Zeta Function for Curves," *Invent. Math.* **73**(1983), 445–465.

[Min 1] H. Minkowski. *Geometrie der Zahlen* (Leipzig: Teubner), 1896; Second Edition, 1910; Reprint (New York: Chelsea), 1953.

[Min 2] H. Minkowski. *Gesammelte Abhandlungen*, 2 Vols., edited by A. Speiser and H. Weyl (Leipzig: Teubner), 1911; Reprint (New York: Chelsea), 1967.

[Min 3] H. Minkowski. *Diophantische Approximationen* (Leipzig: Teubner), 1907; Reprint (New York: Chelsea), 1957.

[Moe] R. Moeckel. "Geodesics on Modular Surfaces and Continued Fractions," *Erg. Theory Dyn. Sys.* **2**(1982), 69–83.

[Mor 1] L. J. Mordell. *A Chapter in the Theory of Numbers* (Cambridge: Cambridge University Press), 1947.

[Mor 2] L. J. Mordell. *Diophantine Equations* (London: Academic Press), 1969.

[Mor 3] L. J. Mordell. *Reflections of a Mathematician*, Canadian Mathematical Congress (Cambridge: Cambridge University Press), 1959.

[Mor 4] L. J. Mordell. "Reminiscences of an Octogenarian Mathematician," *Amer. Math. Monthly* **78**(Nov. 1971), 952–961.

[Mor 5] L. J. Mordell. "Indeterminate Equations of the Third and Fourth Degrees," *Quart. Jour. Pure Appl. Math.* **45**(1914), 170–186.

[Mum] D. Mumford. *The Red Book of Varieties and Schemes* (New York: Springer-Verlag), 1988.

[Nag] T. Nagell. *Introduction to Number Theory*, Second Edition (New York: Chelsea), 1964.

[Nag 2] T. Nagell. "Solution de Quelques Problèmes dans la Théorie Arithmétiques des Cubiques Planes du Premier Genre," *Wid. Akad. Skrifter Oslo* **1**,1(1935).

[Nar] W. Narkiewicz. *Elementary and Analytic Theory of Numbers*, Second Edition (New York: Springer-Verlag), 1990.

[Neu] O. Neugebauer. *The Exact Sciences in Antiquity*, Second Edition (Providence: Brown University Press), 1957.

[Neu-Sac] O. Neugebauer and A. Sachs. *Mathematical Cuneiform Texts*, American Oriental Series, Vol. 29 (New Haven, CT: American Oriental Society), 1945.

[New, J] James R. Newman. "The Rhind Papyrus," in *The World of Mathematics*, Vol. 1, edited by James R. Newman (New York: Simon & Schuster), 1956.

[New] *Newton's Mathematical Papers*, edited by D. T. Whiteside (Cambridge: Cambridge University Press), 1967–1981.

[Niv] I. Niven. "Formal Power Series," *Amer. Math. Monthly* **76**(1969), 871–889.

[Niv 2] I. Niven. *Irrational Numbers*, Carus Monograph of the Mathematical Association of America, Vol. 11 (New York: John Wiley and Sons), 1956.

[Ped] D. Pedoe. *Geometry and the Visual Arts* (New York: Dover Publications), 1983.

[Per] O. Perron. *Die Lehre von den Kettenbrüchen*, 2 Vols., Third Edition (Stuttgart: Teubner), 1954.

[Poi] H. Poincaré. "Sur une Généralisation des Fractions Continues," *Comptes Rend.* **99**(1884), 1014–1016; Reprinted in *Oeuvres*, Vol. 5, 185–188.

[Poi 2] H. Poincaré. "Sur les Propertiétés Arithmétiques des Courbes Algébriques," *Jour. Math. Pures Appl.* **7**,Ser.5(1901), 161–233; Reprinted in *Oeuvres*, Vol. 5, 483–548.

[Pol] G. Polya. *Induction and Analogy in Mathematics*, Mathematics and Plausible Reasoning, Vol. 1 (Princeton, NJ: Princeton University Press), 1954.

[Poo] A. van der Poorten. "A Proof that Euler Missed ... Apéry's Proof of the Irrationality of $\zeta(3)$: An Informal Report," *Math. Intelligencer* **1**,4(1979), 195–203.

[Poo 2] A. van der Poorten. *Notes on Fermat's Last Theorem* (New York: John Wiley and Sons), 1996.

[Rad 1] H. Rademacher. *Higher Mathematics from an Elementary Point of View*, edited by D. Goldfeld (Boston: Birkhäuser), 1983.

[Rad 2] H. Rademacher. *Lectures on Elementary Number Theory* (New York: Blaisdall), 1964.

[Ran] G. Rancy. "On Continued Fractions and Finite Automata," *Math. Annalen* **206**(1973), 265–283.

[Ree] E. G. Rees. *Notes on Geometry*, Universitext (New York: Springer-Verlag), 1983.

[Rei, C] C. Reid. *Hilbert* (New York: Springer-Verlag), 1970; Reprint (New York: Springer-Verlag, Copernicus Imprint), 1996.

[Rei, C 2] C. Reid. *Courant in Göttingen and New York: The Story of an Improbable Mathematician* (New York: Springer-Verlag), 1976.

[Rei, L] L. W. Reid. *The Elements of the Theory of Algebraic Numbers* (New York: Macmillan), 1910.

[Rei, M] M. Reid. *Undergraduate Algebraic Geometry* (Cambridge: Cambridge University Press), 1988.

[Rib] P. Ribenboim. *Algebraic Numbers* (New York: John Wiley and Sons), 1972.

[Rob-Shu] Gay Robins and Charles Shute. *The Rhind Mathematical Papyrus: An Ancient Egyptian Text* (New York: Dover Publications), 1987.

[Rob-Roq] A. Robinson and P. Roquette. "On the Finiteness Theorem of Siegel and Mahler Concerning Diophantine Equations," *Jour. Num. Theory* **7**(1975), 121–176.

[Ros] M. Rosen. "Abel's Theorem on the Lemniscate," *Amer. Math. Monthly* **86**,6(1981), 387–395.

[Ros, H] H. E. Rose. *A Course in Number Theory* (Oxford: Clarendon Press), 1988.

[Rot] K. F. Roth. *Rational Approximations to Irrational Numbers* (London: H. K. Lewis & Co. Ltd.), 1962.

[Rot 2] K. F. Roth. "Rational Approximations to Algebraic Numbers," *Mathematika* **4**(1955), 1–20; corrigendum, ibid, 168.

[Rot 3] K. F. Roth. "Rational Approximations to Algebraic Numbers," in *Proceedings of the International Congress of Mathematicians*, edited by J. A. Todd (Cambridge: Cambridge University Press), 203–210.

[Row] D. Rowe. "Gauss, Dirichlet, and the Law of Biquadratic Reciprocity," *Math. Intelligencer* **10**,2(1988), 13–25.

[San-Ger] G. Sansone and J. Gerretsen. *Lectures on the Theory of Functions of a Complex Variable*, Vol. 2 (Groningen: P. Noordhoff), 1960.

[Sar] G. Sarton. "Lagrange's Personality (1736–1813)," *Proc. Amer. Phil. Soc.* **88**(1944), 457–496.

[Sch] W. M. Schmidt. "Simultaneous Approximation to Algebraic Numbers by Rationals," *Acta Math.* **125**(1970), 189–201.

[Sch-Opo] W. Scharlan and H. Opolka. *From Fermat to Minkowski: Lectures on the Theory of Numbers and Its Historical Development* (New York: Springer-Verlag), 1985.

[Sch-Spe] O. Schreier and E. Sperner. *Projective Geometry of n Dimensions, Introduction to Modern Algebra and Matrix Theory*, Vol. 1, English translation by C. A. Rogers (New York: Chelsea), 1961.

[Ser] J.-P. Serre. *A Course in Arithmetic* (New York: Springer-Verlag), 1973.

[Ser, C] C. Series. "The Geometry of Markoff Numbers," *Math. Intelligencer* **7**,3(1985), 20–29.

[Ses] Jacques Sesiano. *Books IV to VIII of Diophantus' Arithmetica: In the Arabic Translation Attributed to Qustā Ibn Lūqā*, Sources in the History of Mathematics and Physical Sciences, Vol. 3 (New York: Springer-Verlag), 1982.

[Sha] I. R. Shafarevich. *Basic Algebraic Geometry*, Second Edition (New York: Springer-Verlag), 1994.

[Shi] G. Shimura. *Automorphic Forms and Number Theory*, Lecture Notes in Mathematics, Vol. 54 (New York: Springer-Verlag), 1968.

[Sie 1] C. L. Siegel. *Topics in Complex Function Theory*, 3 Vols., English translation from the German by A. Shenitzer and D. Solitar (New York: Wiley-Interscience), 1969–1973.

[Sie 2] C. L. Siegel. "On the History of the Frankfurt Mathematics Seminar," *Math. Intelligencer* **1**,4(1979), 223–230.

[Sie 3] C. L. Siegel. "Über einige Anwendungen diophantischer Approximationen," *Abh. Preuss. Akad. Wiss.* **1**(1929); Reprinted in *Gesammelte Abhandlungen*, Vol. 1 (Berlin: Springer-Verlag), 209–266.

[Sie 4] C. L. Siegel. *Transcendental Numbers* (Princeton, NJ: Princeton University Press), 1949.

[Sie 5] C. L. Siegel. *Lectures on the Geometry of Numbers* (New York: Springer-Verlag), 1989.

[Sil] J. H. Silverman. *The Arithmetic of Elliptic Curves* (New York: Springer-Verlag), 1986.

[Sil-Tat] J. Silverman and J. Tate. *Rational Points on Elliptic Curves* (New York: Springer-Verlag), 1992.

[Smi, D] D. E. Smith. *A Source Book in Mathematics* (New York: McGraw-Hill), 1929; Reprinted in 2 Vols. (New York: Dover Publications), 1959.

[Smi, H 1] H. J. S. Smith. "Note on Continued Fractions," *Messeng. Math.* **6**,Ser.2(1876), 1–14; Reprinted in [Smi, H 2], ii, 135–147.

[Smi, H 2] H. J. S. Smith. *The Collected Mathematical Papers of H. J. S. Smith*, 2 Vols., edited by J. W. L. Glaisher; Reprint (New York: Chelsea), 1965.

[Smi, H 3] H. J. S. Smith. *Report on the Theory of Numbers*, 1894; Reprint (New York: Chelsea), 1965; Also in [Smi, H 2], Vol. 1.

[Spr] G. Springer. *Introduction to Riemann Surfaces*, Second Edition (New York: Chelsea), 1981.

[Sta] H. M. Stark. *An Introduction to Number Theory* (Chicago: Markham), 1970; Reprint (Cambridge, MA: MIT Press), 1978.

[Sta 2] H. Stark. "Galois Theory, Algebraic Number Theory, and Zeta Functions," in [W-M-L-I], 313–393.

[Sta-Whi] D. Stanton and D. White. *Constructive Combinatorics* (New York: Springer-Verlag), 1986.

[Ste-Tal] I. Stewart and D. Tall. *Algebraic Number Theory*, Second Edition (London: Chapman and Hall), 1987.

[Str] D. J. Struik. *A Sourcebook in Mathematics 1200–1800* (Cambridge, MA: Harvard University Press), 1969.

[Str, R] R. J. Stroeker. "Aspects of Elliptic Curves: An Introduction," *Nieuw Arch. Voor Wis.* **XXVI**, 3(1978), 371–412.

[Tat] J. Tate. "Problem 9: The General Reciprocity Law," in *Proceedings of Symposia in Pure Mathematics*, Vol. 28 (Providence: American Mathematical Society), 1976, 311–322.

[Tat 2] J. Tate. *Rational Points on Elliptic Curves*, Phillips Lectures, Haverford College, April–May, 1961.

[Tat 3] J. Tate. "The Arithmetic of Elliptic Curves," *Invent. Math.* **23**(1974), 179–206.

[Thu] W. Thurston. *Three-Dimensional Geometry and Topology*, Vol. 1, edited by S. Levy (Princeton, NJ: Princeton University Press), 1997.

[Thue] A. Thue. *Selected Mathematical Papers of Axel Thue*, edited by T. Nagell et al. (Oslo: Universitetaforlaget), 1977.

[Thue 2] A. Thue. "Über die Unlösbarkeit der Gleichung $ax^2 + bx + c = dy^n$ in grossen ganzen Zahlen," *Arch. Math. Naturv., Kristiania* **34**,16(1917); Reprinted in [Thue], 561–564.

[Tid] R. Tijdeman. "Hilbert's Seventh Problem: On the Gelfand-Baker Method and Its Applications," in *Mathematical Developements Arising from Hilbert Problems*, Proceedings of Symposia in Pure Mathematics, Vol. 28 (Providence: American Mathematical Society), 1976, 241–268.

[Tru] C. Truesdell. "Leonard Euler: Supreme Geometer," in *1972 American Society for 18th Century Studies* (Madison, WI: University of Wisconsin Press), 1972; Reprinted in part in [Eul 2], vii–xxxix.

[Tur] P. Turan. "On the Works of Alan Baker," in *Actes du Congrès International des Mathématicians 1970* (Paris: Gauthier-Villars), 1971, 3–5.

[van L-Wil] J. H. van Lint and R. M. Wilson. *A Course in Combinatorics* (Cambridge: Cambridge University Press), 1992.

[Veb-You] O. Veblen and J. W. Young. *Projective Geometry*, 2 Vols. (Boston: Ginn), 1910, 1918.

[Wae] B. L. Van der Waerden. *Algebra*, 2 Vols., Seventh Edition (New York: Friedrich Ungar), 1970.

[Wal] R. J. Walker. *Algebraic Curves* (Princeton, NJ: Princeton University Press), 1950.

[Wan] P. L. Wantzel. "Recherches sur les Moyens de Reconnaître si un Problème de Géométrie se Résoudre avec la Règle et le Compas," *Jour. Math. Pures Appl.* **2**(1837), 366–372.

[Wee] J. R. Weeks. *The Shape of Space* (New York: Marcel Dekker), 1985.

[Wei] A. Weil. *Number Theory: An Approach through History from Hammurapi to Legendre* (Boston: Birkhäuser), 1984.

[Wei 2] A. Weil. "Fermat et l'Équation de Pell," in [Wei 12], Vol. 3, 413–419.

[Wei 3] A. Weil. "Two Lectures on Number Theory, Past and Present," *L'Enseign. Math.* **20**(1974), 87–110; Reprinted in [Wei 12], Vol. 3, 279–302.

[Wei 4] A. Weil. "Une Lettre et un Extrait de Lettre à Simone Weil," in [Wei 12], Vol. 1, 244–255.

[Wei 5] A. Weil. "Gauss et la Composition des Formes Quadratiques Binaires," in *Aspects of Mathematics and its Applications*, edited by Frei and Imfeld (New York: North Holland), 1986.

[Wei 6] A. Weil. "La Cyclotomie Jadis et Naguère," *L'Enseign. Math.* **XX**(1974), 247–263; Reprinted in [Wei 12], Vol. 3, 311–328.

[Wei 7] A. Weil. *Elliptic Functions According to Eisenstein and Kronecker* (Berlin: Springer-Verlag), 1976.

[Wei 8] A. Weil. "Book review: *Mathematische Werke* by Gotthold Eisenstein," *Bull. Amer. Math. Soc.* **82**(976), 658–663; Reprinted in [Wei 12], Vol. 3, 398–403.

[Wei 9] A. Weil. "Number Theory and Algebraic Geometry," in *Proceedings of the International Congress of Mathematicians* (Providence: American Mathematical Society), 1952; Reprinted in [Wei 12], Vol. 1, 442–453.

[Wei 10] A. Weil. "Book review: *Mathematische Werke* by Gotthold Eisenstein," *Bull. Amer. Math. Soc.* **82**(976), 658–663; Reprinted in [Wei 12], Vol. 3, 398–403.

[Wei 11] A. Weil. "Sur un Théorème de Mordell," *Bull. Sci. Math.* **54**(1930), 182–191; Reprinted in [Wei 12], Vol. 1, 47–56.

[Wei 12] A. Weil. *Oeuvres Scientifiques: Collected Papers*, 3 Vols. (New York: Springer-Verlag), 1980.

[Wei 13] A. Weil. "L'Arithmétique sur les Courbes Algébriques," *Acta Math.* **52**(1928), 281–315; Reprinted in [Wei 12], Vol. 1, 11–45.

[Wein] S. Weinberg. *Not. Amer. Math. Soc.* **33**,5(1986), 731.

[Wey 1] H. Weyl. *Algebraic Theory of Numbers*, Annals of Mathematical Studies, Vol. 1 (Princeton, NJ: Princeton University Press), 1940.

[Wey 2] H. Weyl. "David Hilbert and His Mathematical Work," *Bull. Amer. Math. Soc.* **50**(1944), 612–654.

[Wey 3] H. Weyl. *The Concept of the Riemann Surface*, Third Edition, English translation from the German by G. R. MacLane (Reading, MA: Addison-Wesley), 1964.

[Wey 4] H. Weyl. "On Geometry of Numbers," *Proc. London Math. Soc.* **47** (1942), 268–289.

[Wil] H. Wilf. *generatingfunctionology*, Second Edition (Boston: Academic Press), 1994.

[Wile] A. Wiles. "Modular Elliptic Curves and Fermat's Last Theorem," *Annals Math.* **141**,Ser.2,1(1995), 443–551.

[Wile-Tay] R. Taylor and A. Wiles. "Ring-Theoretic Properties of Certain Hecke Algebras," *Annals Math.* **141**,Ser.2,3(1995), 553–572.

[Wym] B. Wyman. "What is a Reciprocity Law?," *Amer. Math. Monthly* **79**,6(1972), 571–586.

[W-M-L-I] M. Waldschmid, P. Moussa, J.-M. Luck, and C. Itzykson, Editors. *From Number Theory to Physics* (New York: Springer-Verlag), 1992.

[You] A. P. Youschkevitch. "Euler," in *Dictionary of Scientific Biography*, Vol. IV (New York: Charles Scribner's Sons), 1971, 467–484.

[Zag] D. B. Zagier. *Zetafunktionen und quadratische Körper* (Berlin: Springer-Verlag), 1981.

Index

SEASONS OF CAPTIVITY

AMIA LIEBLICH

SEASONS OF CAPTIVITY

THE INNER WORLD OF POWs

NEW YORK UNIVERSITY PRESS
New York and London

NEW YORK UNIVERSITY PRESS
New York and London

Library of Congress Cataloging-in-Publication Data
Lieblich, Amia, 1939–
[Ḥuts mi-tsiporim. English]
Seasons of captivity : the psychology of POWs in the Middle East / Amia Lieblich.
p. cm.
Translation of: Ḥuts mi-tsiporim.
Includes bibliographical references and index.
ISBN 0–8147–5079–6
1. Israel-Arab Border Conflicts, 1949– —Egypt—Personal narratives, Israeli. 2. Israel-Arab Border Conflicts, 1949– —Prisoners and prisons, Egyptian. 3. Israel-Arab Border Conflicts, 1949– —Psychological aspects. 4. Prisoners of war—Israel–Psychology. 5. Prisoners of war—Egypt—Psychology. 6. Prisoners' wives—Israel—Psychology. 7. Oral history. I. Title.
DS119.8.E3L5313 1994
956.04—dc20 93-20732
CIP

New York University Press books are printed on acid-free paper, and their binding materials are chosen for strength and durability.

Manufactured in the United States of America

10 9 8 7 6 5 4 3 2 1

To Yuval, Maty, and Eliav

CONTENTS

ACKNOWLEDGMENTS

My deepest gratitude goes to the ten men who were willing to share their experiences with me, to their wives, and to Rami and Menachem in particular for advising me throughout the project. It was the men's choice to appear in this book using their real first names. Minor changes have been introduced: Michal is a pseudonym for the ex-wife of Yitzhak who preferred to maintain her anonymity; Amos L. was renamed Amnon, to avoid confusion with Amos Z.; and Motti B. was renamed Benny to avoid confusion with Motti C. This book was written from their recorded narratives and it is theirs as much as mine. However, I take sole responsibility for any distortion or false interpretation of their accounts.

Since most of the book is about men, and in fact the vast majority of POWs are male, I have used the masculine form wherever I make general statements. I hope that my female readers will not be offended by this choice.

Two women helped me in transcribing the tapes—Neta Shaked and Zipi Shmaya; without their help I would not have been able to carry on. Racheli Adelman and Alex Zehavi of Schocken Publishing House in Israel gave their professional attention to the Hebrew book, titled *Only the Birds,* on which the present English version is based. Leora Sherf and Eda Flaxer helped in editing the English manuscript. Eric Schramm copyedited the manuscript. The Hebrew University of Jerusalem provided financial and administrative help for the project. My warmest appreciation to all of them.

INTRODUCTION

S*easons of Captivity* is a book about the experience of a group of ten prisoners of war (six of them military airmen) who shared a single jail cell in Egypt for more than three and a half years. It is a unique case study of survival on both the individual and the collective levels, an oral-history account that is presented from the personal perspectives of these ten men, as shared with me in 1987 in a long series of in-depth interviews.

The story unfolds in ten personal voices, starting with the men's capture in 1969–70 and continuing through the first six months of interrogation, torture, and isolation; the period when they were joined together in a common room; the establishment of an almost utopian social system and its subsequent maintenance and problems; and the release of the prisoners and their return to Israel in November 1973. The narrative chronicles the process of reentry into family and social roles and the personal impact of the experience on the men's lives years after their liberation. In addition, it presents the experience of the wives of the five prisoners who had been married before their capture, introducing their own story of separation, reunion, and aftermath of the experience.

Following are a few quotations that succinctly demonstrate the impact of the experience eighteen years after being captured (or fourteen years after liberation), as well as some of the issues concerned in recounting it:

Dan: Some things can be repeated hundreds of times, but you find out you speak in a different voice, and only someone who has shared your experience will understand. Maybe this is what concentration-camp survivors mean when they say that language cannot convey their experience. The language is the same language, but the seasons are different.

Menachem: This was the first time we got out, without our blindfolds, from the courtyard. An interesting thing happened: Suddenly I discovered the horizon. Out there, on the edge of the desert, lay the infinite horizon. I felt dizzy. All these years I had seen nothing beyond the eighteen meters of our room and courtyard; only the birds up in the sky.

Rami: When we returned, I immediately appeared in public at my kibbutz and told the story of our captivity. As I heard the members' reactions, I had the urge to say, "Hey guys, stop pitying me." We lived, we acted, and that's it. Only when you live through an experience, you know that it is not that awful. Later on, however, from the reactions of my listeners, I realized that what we had done in prison was significant, exceptional, perhaps even great.

Since then, I have become aware of a message I would like to convey, namely that when you take your present condition as a starting point, you can always achieve much more than if you sit and cry for sympathy. Every group can progress from the point it's at; each individual makes progress at his own pace. There is something to be learned from every condition: from a group gathered by sheer chance, from the experience of a solitary cell, from the interrogations. People's resources exceed others' estimates by far, even their own estimate. The moment a person says, 'That's it, I can't muster up any more strength, I can't take it anymore'—yet he does, he turns over a new leaf. There is no limit to human endurance.

Motti: I read those things Menachem told the reporter for the Air Force journal. If he did say those things, I don't know what happened. People don't remember the details and instead just say what they like. They say, "I did this, I said that," but it isn't true. We translated *The Hobbit,* all of us together.

Benny: When you talk to me about Benny the POW, I feel as if I'm standing aside and watching, as if there are two of us: the real Benny, the way I am today, and Benny the prisoner, who doesn't exist anymore. He is something totally different. I cut myself off from that Benny, as if somebody else had experienced all the suffering. In fact, I know everything that happened and I can talk about it, but I'm not touched by the story.

In 1986, I was approached by Rami for the first time with the idea of writing an oral history of the group's experiences in captivity. I asked him to contact each of the ten men and discuss the project. A year later, with the cooperation of all the men ensured, I started to interview each of them separately. The interviews took place in the men's homes, in my home, at my office at the Hebrew University in Jerusalem, or—rarely—in coffee shops near the interviewee's workplace or home. Several of the men were accompanied by their wives, who sat and listened to our conversations, rarely interfering. When I completed these interviews, I invited their wives to talk to me separately, including an ex-wife of one of the men. Basically, I attempted to get a full, chronological narrative, starting from the period before they were captured and concluding with the present. I minimized questioning and probing and intervened only when the flow of the story came to a halt, to ask for clarification or examples, or to mention associations from previous conversations. I never confronted the men with inconsistencies in their own stories, or with conflicting versions of other group or family members. I believe that I conveyed to the interviewees great respect and empathy, yet I tried to be an objective recorder of their experiences, never judging them. (For a discussion of the relationship between the researcher and the narrative, see chapter 15.) All interviews were tape-recorded, and later transcribed. These verbatim accounts provide the data for the present book.

The length and schedule of the interviews varied greatly, according to the wishes and needs of the men and their wives. Some were marathon sessions of long weekend visits, others were spread over a year. Although all the men expressed consent and willingness to participate, much resistance had to be overcome in the process of recalling and sharing the narratives. Some men did not keep scheduled meetings, which had to be rescheduled after delays of weeks or months. Often, during our meetings, we took a break when talking became too painful, went out for a drink or a walk, and eventually resumed our interviews. The number of sessions per person ranged from four meetings for two men, to more than twelve recorded sessions for two men who also met me for many informal visits and conversations in between.

The narratives of the men cover different stages, with distinct characteristics. Some relate to single events, such as being taken captive or being united in the common room. Others deal with longer, yet discrete phases, such as the stage of interrogation or the immediate period following their release. The major chapters of the book, however, on organization of the group, its social life, and the inner world of the POWs, deal with the three years the men spent together in the common room. This was a long period without clear-cut markers to organize the men's accounts. Their stories seemed to wander forward and backward in time. Some topics were grouped according to association, others by the remembered order of their occurrence. It was hard to place or order the events precisely on the time scale.

Certain men presented stages in the experience of their joint captivity, while others stressed gradual evolving changes. One of the common presentations of stages went as follows: from a lack of social organization, through the construction of an organizational system, to a life of routine, until liberation. Other stages presented included: living under the shadow of trauma, then adjusting to regular life in prison; the temporariness of one's existence in Egypt and the eventual acceptance of this life as constant; and living under close control and then with relative autonomy. Reported changes in captivity included raising the standard of living, achieving educational goals in the study program, which continued throughout their imprisonment, and the growing harmony and understanding among the individual prisoners. These stages give the story a sense of progression in time.

Various time markers were also utilized in order to convey a sense of fluid history, such as holidays celebrated in captivity. The men mentioned the first, second, and third Passovers, for example. Rosh Hashanah, the Jewish New Year, was another important date. Personal birthdays and the birthdays of family members provided similar anchors in time. However, it is clear that the first holiday or birthday was best remembered, and perhaps represented a condensed version of all three occasions.

Some important events in the life of the group could not be placed precisely or even approximately in time, because the interviewees varied in these matters: for instance, the first time the courtyard was unlocked for an entire day, the removal of the Egyptian guards from the yard, the end of the disciplinary regime based on reveille parade and lights out, the removal of the shutters from the windows, and the first radio reception of news broadcasts from Israel. All these events were important in the men's narratives, but could not be placed unequivocally in a certain period. Consequently, some obvious inconsistencies appear in their accounts, which were naturally maintained in the following chapters of the book.

While the primary source for the book was the captives' recollected accounts, additional documents were utilized to complement the narratives. One such document, the secret collective diary, grew out of a unique feature of group life in captivity: the weekly assembly meeting. On February 19, 1971, two months after all the POWs were gathered in the common cell at Abassiya prison in Cairo, they started to conduct Friday night meetings to discuss various matters pertaining to their lives. Each of the ten prisoners, following the alphabetical order of their first names, served his turn as the chairman of the meeting. They decided to keep a record of these meetings and the decisions made at them. Each of the chairmen wrote down the minutes of the meeting in his own style, and Avi made a clean copy of the proceedings in a notebook that they brought back to Israel with them. One hundred and thirty-nine meetings were recorded in the group's diary up to the men's release on November 16, 1973; the last one took place on November 2, 1973.

As I was collecting the oral histories for this book, Menachem had the diary typed by one of his secretaries, a gift to each one of the captives. In his introduction to the diary Menachem wrote:

> The minutes of the meetings recorded herein are an excerpt of the lives of ten men, who spent all their days and nights together. Each one observed the other in his grief and joy. Each one, according to his ability and sensitivity, saw it as his duty to contribute to the general welfare, to save our boat from sinking, God forbid. In fact, we managed to keep afloat most of the time, and if we erred here or there, at least we had the best intentions.

Thus, this collective diary, as well as part of the prisoners' correspondence with their families and the Red Cross reports, provide real-time documents that shed additional light on the obtained oral history. A sample of these is included in the chapter entitled "Testimony," which divides the men's accounts between stories of their captivity and those of their release.

Even though the story unfolds through trauma and pain, the

generally positive outcomes of the men's experience come to the forefront. These could be attributed to the leadership that emerged in the group and to the dominant value system governing their lives—utilizing national, military, and kibbutz values, and humanistic-existential beliefs—that gave meaning to their suffering and inspired their productive lifestyle. Several other aspects of life in captivity may have helped the men maintain sanity and well-being, including the stability of lifestyle, the variety of activities, and the norms for tension release and interpersonal support that evolved in the common room. While these were the predominant aspects of the POWs' lives in captivity, friction among individuals of very different backgrounds, competition for leadership, frustration from the loss of freedom, deep longings for home, and a sense of wasted time were some of the natural negative trends displayed in the men's accounts. Some of the men maintained feelings of loss and bitterness even fourteen years after their liberation.

Before the introduction of the narratives, a brief review of relevant previous research on POWs will be presented.

THE PSYCHOLOGY OF CAPTIVITY

The experience of a captive or hostage starts with a trauma and becomes a chronic state of stress of a complex, multifaceted nature. At the moment of capture, the POW loses his former status and identity. He faces a new reality where nothing is known and his life is in constant danger. He is separated from his family and friends, his home and country, his occupational rank and setting. All of a sudden he has a new status—a POW, without any control over his fate, schedule, or behavior. His environment changes abruptly, and he encounters new functionaries—wardens, interrogators, fellow prisoners. He has to unlearn his former habits and acquire new ones. His most salient goal is survival (Miller 1974).

The difficulties in the POW's status may be divided into three categories: (1) physical difficulties, namely injuries from the battle prior to capture and those inflicted during capture, during subsequent medical maltreatment, various forms of beating and torture, and deprivation of essentials such as sleep, food, water, and daylight; (2) emotional manifestations, such as despair, fear, helplessness, humiliation, guilt, tension concerning interrogations, worries about the future, longing for one's family, and regret about lost time and freedom; (3) social problems, namely loneliness, if in isolation, or, if in a group with other prisoners, interpersonal friction or violence and adjustment to cohabitation in crowded quarters with people of different backgrounds, habits, and tastes.

In clinical terms, capture is undoubtedly a traumatic event, which is further marked by the length of its duration. The *DSM III-R* (American Psychiatric Association 1987) characterizes trauma as "a psychologically distressing event that is outside the range of usual human experience" (247), while according to Friedman (1991, personal communication), a trauma has four components, which all exist in the case of becoming a POW: the shattering of the stimulus barrier, of the self concept, of the concept of others, and of the concept of the world. How does the trauma affect the captive during his experience and afterwards?

Various psychological aspects of soldiers during World War II and the Korean and Vietnam wars were studied extensively, yet relatively little is known about POWs or hostages, the outcomes of their experiences, or the factors that may contribute to their survival and coping. Military organizations have a particular interest in studying the subject, and some, such as the Center for POW Studies at the Naval Health Research Center in San Diego, California, specialize in it. Much of the research about American POWs and their families is summarized in a special study conducted by the Veterans Administration (1980), and recently by Hunter (1991) based on numerous projects (e.g., Hunter 1976, 1978, 1982, 1983, 1986; Cohen and Cooper 1954; Nefzger 1970; Segal 1974; Segal, Hunter, and Segal 1976). The common conclusion of these studies is that, although each POW or hostage experience can be quite different, it is always traumatic and leaves long-term effects on the personality and functioning of former prisoners.

One should remember that ex-POWs, who can be studied, are a subsample of survivors of the experience of captivity. Their chances of survival and future physical and psychological health were determined by personality and situational factors. Some factors of the situation that affect survival and subsequent health are the historical context in which the incident occurs, the culture of the captors, the duration and harshness of the captivity, and the support received from others. The captive's appraisal of the stressful situation and his ability to cope are the major dimensions to be considered from the personality perspective. These, in turn, depend on "the captives' innate predispositions or temperaments, commitment to whatever ideology or tasks placed them in jeopardy initially, their maturity, personal value systems, and satisfaction with family relationships during the pre-capture period" (Hunter 1991, 745).

Regarding the outcome of the experience, Hunter's own studies indicated that "basic personality did not change, even after extremely harsh, prolonged captivity. But it appeared to have solidified [the captive's] basic traits" (ibid., 752). On the other hand, she contends that, as studied in the U.S. and elsewhere, the stress and deprivation of captivity have lifelong effects on subsequent physical health, family adjustment, and occupational history. Many POWs suffer permanent psychological damage (Segal, Hunter, and Segal 1976). Most of the residual symptoms described in clinical literature are well defined by the syndrome of Post-Traumatic Stress Disorder (PTSD) (American Psychiatric Association 1987), which includes intrusions of the traumatic memories, nightmares, and a high level of tension, anxiety, or irritability—or the opposite: numbness of responsiveness to stimuli, low energy, depressive moods, difficulties in concentration, and a lacerating guilt, with a possible delayed onset. The prevalence of these may be inferred from the fact that, while POWs are only 1.7 percent of American veterans, their requests for compensation for psychological damage, especially anxiety neurosis (the older term for PTSD), account for 5 percent of all compensation requests (Veterans Administration 1980).

As one might expect, the duration and intensity of these aftereffects are correlated with the length and severity of the experience in captivity. Thus the experiences of torture, isolation, humiliation, and hostility, disrupted contact with the outside world, malnutrition, and medical neglect all contribute to the intensity of the psychological aftermath (ibid.). Among the most shocking results is that ex-POWs held by Asian captors were 40 percent more likely to die (of accidents, disease, suicide etc.) than other American males of the same age who had not been POWs (Nefzger 1970). From a different angle, a third of the Vietnam era POW marriages ended in divorce during the first year after release, and

50 percent had ended by the end of the fifth year. At the same point in time, there is only an 11 percent divorce rate among Vietnam veterans who were not taken captive (Hunter 1982, 1983).

Conventional epidemiological studies have rarely discovered positive residuals of the experience of captivity. Yet Hunter says that "the effects [of captivity] are not always on the debit side of life's ledger" (1991, 752), even though there is very little to support this claim. A unique study by Rutledge, Hunter, and Dahl (1979) reported a permanent shift in values as the result of the trauma and the experience of regaining one's freedom.

As a summary of her review, Hunter (1991) offers several generalizations, which can be paraphrased as follows:

1. All individuals can cope with much more stress than they believe they could.
2. All human beings can be made to behave in ways they did not think possible.
3. Older and more mature individuals with firmly ingrained values and an internal locus of control cope better with capture.
4. Commitment to a cause, such as family, country, or God, helps one to endure traumatic conditions.
5. Group support, especially from those with similar experiences, during the stress or after liberation is of utmost importance to POWs and their families.
6. Good communication with loved ones during and after captivity is key to adjustment.
7. Flexible homecoming plans and counseling are important to returning captives.
8. Preparation for capture, in the forms of information and code of behavior, may help in coping.

While none of the above is new to students of stress and coping, Hunter concludes that "we have learned a great deal from studies of former prisoners of war" (1991, 754). Academic literature fails to portray captives and their mechanisms of coping and recovery in a similar manner to autobiographical (e.g., Risner 1973) or fictional works (e.g., Clavell 1962). Works of more subjective nature depict the experience of captivity as at least partly profitable, in the sense that it empowers the individual, providing him with new values and confidence about his ability to master crises of different kinds.

The phenomenon of capture in the frequent wars between Israel and Arab countries is also incompletely documented and researched. Few autobiographical or journalistic books were published on the topic (e.g., Ha-Meiri 1966; Kfir 1974), and among those that were, most dealt with the experience of Israeli POWs in Syria, which is considered to be the cruelest captor among the Arab countries. In the field of psychology, the study of Avneri (1982) is an introspective autobiographical account of his captivity in Syria during 1973, which attempts to apply psychological analysis in terms of defense mechanisms used during the trauma and afterwards. Another outstanding work in this area was done by Barnea (1981), who formed a matched control group for a group of Israeli Air Force ex-POWs and assessed their personality via objective and projective techniques. Like Hunter (1991), Barnea concluded that, in general, the experience of captivity did not leave significant marks on the respondents' personalities. However, detailed analysis of the Rorschach test results indicated that ex-POWs were more balanced and flexible, more open to experience, more interested in deep relationships, and less defensive. Two expert clinicians, who blindly scored the Rorschach protocols, were asked to sort them into two groups differing in "oversensitivity and traces of traumatization." They systematically misplaced ex-POWs as "normals" and the controls as "traumatized." In other words, with the help of a less direct and perhaps deeper means of personality assessment, positive outcomes of captivity were detected.

LIFE IN THE EGYPTIAN PRISON AS A METAPHOR

The story to be recounted in this book does not represent a single historical event, or a typical example of the life of POWs everywhere. Rather, it can be taken as a metaphor for human society in general, demonstrating what human wisdom can create—or withhold—in the midst of adverse conditions, and how positive values can be drawn from, objectively speaking, the most negative situations. It can easily be generalized to a wide range of human conditions in which individuals are threatened by physical and psychological dangers or have to accommodate a lack of freedom to privacy or extreme personal exposure.

The story of the captives and their recovery resembles a dramatic legend, repeating basic human themes or myths about the Fall and Return. First of all, it is a story about the fulfillment of human hope for redemption, freedom, and a safe return home. Nothing can evoke our deepest identification like a plot consisting of an abrupt fall into the hands of hostile powers for an indefinite time and the subsequent struggle for survival that concludes with the hoped-for return to loved ones (Polster 1987).

Furthermore, the narrative demonstrates basic human existential dilemmas, as formulated by philosophers, as well as in cultural myths and literature from *Robinson Crusoe* to the *Lord of the Flies*. In a condition of complete helplessness, while confronting an extremely hostile environment and with an immediate threat to life, man faces the most essential choice of his existence, and herein lies the truest test of his personal freedom: whether to hope or despair, whether to create a meaningful existence from nothing or give up struggling, fall apart, murder, or commit suicide. In our case, the captives, both as individuals and as a group, created a life of value and interest with reasonable harmony between them, governed by pride and the joy of productivity, to the extent that on the day of their liberation one of the men said, "I need two weeks more to complete my projects." This creation of a world from nothing may serve as a lesson, not only for people in adverse conditions, but for all men and women in so-called normal circumstances, whose life seems to be a doomed-to-fail struggle against death.

The story of their POWs portrays their uncertainty regarding time and existence. The prisoners faced a time interval stretching from complete temporariness to the unknown infinite. At every single moment, they might have been liberated and returned home, yet in the same vein, they might have been held in jail forever, until death. Under these conditions, what were the men's choices? One alternative was "to sit with their suitcase packed," waiting for the hangman or the redeemer, maintaining loyalty to a far-off world to which they did not belong anymore. The other alternative was to create a worthwhile existence, disregarding the uncertainty of its duration. The metaphors for this dilemma were numerous: for example, should the POWs take apart the cardboard boxes and order bookshelves for their cell? Should they continue drinking their coffee out of empty jam jars, or have their families send them ceramic mugs? Should one feel resentful of his wife, who sent his favorite cardigan to Egypt against the cold? Should one ask his parents to send his record collection? Should one proceed to take university courses by correspondence while it was unknown whether they would be of any use?

The existential point of view claims that the moral choice is to remain in one's own reality, accept it for what it is ("create your own home playing field," in Rami's words), and make the best of it. This is human destiny, even in a normal life that seems to be running its natural course, without traumas or disasters. Should we spend our lives "sitting with our suitcase packed," wasting resources, waiting for the end to come? Alternatively, we can exhibit courage and creativity and use our potential to build a life of significance.

The story of the ten heroes of this book teaches us that this is actually possible. While it relates the case of a concrete universe, a group of ten men sharing a single cell in jail, the narrative clearly reflects the basic dilemmas of human life with their potentials and hazards. By this book we thus convey a message of human survival and hope.

1

CAPTURE

The ten Israeli men who later formed the group in the common room all fell into the hands of the Egyptians between December 1969 and July 1970, during what is known in the Middle East as the War of Attrition. The first four were Dan, thirty-seven, severely wounded in the attack preceding his capture; David, nineteen; and Motti and Benny, both twenty-one years old. They were each captured in combat by Egyptian commando units who had penetrated Israeli territory. The next six men were Air Force pilots or navigators whose planes were hit during missions over Egypt. These men, twenty to thirty-one years old, were taken prisoner after parachuting from their planes. One of them, Menachem, was seriously injured during his jump.

Each man's memory of the traumatic event was very clear and detailed, although cross-validation showed remarkable disagreement concerning shared experiences. Their narratives came in response to my request to tell me a little about their lives prior to captivity and to reproduce the circumstances of their capture.

DAN

It happened during the War of Attrition. The soldiers in compulsory service had a hard time coping with the extensive period of being exposed to constant shelling. Many were killed or wounded on the waterfront of the Suez canal. Moshe Dayan had therefore asked the reserve officers to volunteer for active service. I was impressed by his request, as well as by what I had seen when I visited the canal area, so toward the end of October 1969 I volunteered. I was at the time thirty-seven years old, a father of three. Nowadays my decision is, perhaps, hard to understand. But for me Zionism was never an empty word, and I felt obliged to get out of my routine and to give up some of my personal comfort at a time of crisis. I am a member of Kibbutz Ein Hashofet, where I was born and worked as a farmer. So I made up my mind to become one of "Dayan's Tigers"—a nickname I didn't like even then.

It wasn't easy to say goodbye to my wife and children up in the north, and some people considered me to be out of my mind. But I believed I would be away only for three or four months. As a lieutenant, I received a commanding position in one of the strongholds on the canal, the one nearest the Bitter Lakes. There were several compulsory soldiers and reservists in the post—I can't give you any numbers because I made myself forget those during interrogations. In addition, I had two tanks under my command and some citizens who worked in construction on the line.

The Egyptians shelled the stronghold and its surroundings several times a day. The worse thing was to be exposed to an attack while outside the bunker. Egyptian marksmen were sitting on the trees across from us and covered the area. Sometimes we aimed at them too, but the morale of our soldiers was low. They wanted to go home. Under these circumstances, it was hard to maintain discipline and alert routine. As in any small group that's closed in, there was pressure, and from time to time violence broke out. My presence, as an older man, was supposed to calm people down.

When I went home on leave, I discovered the immense difference between the quiet life in the country as opposed to the tension at the front line. We had many casualties then. I remember how, just as I had left home on Friday, we had completed digging a fine latrine outside the post. Driving home, as I reached headquarters, I received a message: "Sorry, Dan, the latrine got a direct hit, it's all gone" [laughing].

Since my arrival at the canal, the thing I feared most was going back and forth to the post. While driving on the way, I felt exposed to the enemy and it disturbed me. On my return from leave on Sunday, I stopped at our headquarters and could not find any transportation. I waited for half a day, and finally I was given the commander's jeep, with another soldier as a guard, and a driver who was supposed to return the vehicle. We hit an Egyptian commando raid on our way. It happened in an area that later, during the Yom Kippur War, was nicknamed "The Chinese Farm." The enemy unit consisted of six well-armed soldiers. They shot at us and immediately hit the jeep. The guard was killed on the spot, and the driver was severely wounded. He managed to jump out of the vehicle and died later in the hospital. I was hit in the legs. I jumped over the driver and rolled on the sands. When I wanted to get up and run, I couldn't make it. Both my legs were smashed with six or seven open fractures. I was lying there, while they continued to fire hysterically till they reached me. I raised my arm, handed my pistol over to the officer, and thus, I believe, stopped him from killing me. At that moment, I became a prisoner of war.

I recall having said to myself just then—I often talk to myself—"Dan, you're out for a long, long trip." I assumed I would remain in captivity for six or seven months.

The day was December 14, 1969, less then two months after my mobilization. This was the time when the Air Force was bombing deep in Egypt, and the Phantom was the fear of the land. Shortly, however, the Egyptians managed to conquer the Phantom. When I finally met the Air Force pilots in prison we didn't feel like great winning heroes. . . .

I was in terrible pain. They lifted me up, put me on their shoulders, and started to run. My head faced the ground and my legs shook all over. It was sheer agony, yet I didn't lose consciousness. I had always been a great fat guy, so I had to encourage the soldiers who carried me, in Arabic: "Go on, you can make it." Suddenly there were shells flying all over the place, and they threw me on the ground. I don't know whose guns were shelling us, but as I was lying there, somebody pulled one shoe off my shattered leg, and the detectives found it later and wondered what it meant. Finally I was hauled again and dragged with my leg hanging on by the skin, all the way to the waterfront. There I was rolled down the sand embankment to the commando soldiers who had been waiting near the water. They transferred me over to the other side and dragged me up, somehow, into a small tin post.

We were under shelling of the Israeli artillery, and everything shook around me. But I didn't expect any help, since the Egyptians were fast and efficient, while our forces weren't at their best at the time. Or perhaps it was the way my mind always works: I cannot change reality, therefore I have to accept my lot. Suppose I would be very angry at the Israeli army for not having rescued me, would that be of any help in my state?

In the meantime, the Egyptian soldiers poured out all their fear, excitement and frustration on poor old me. They beat the life out of me, pulled off my bars, and took away my watch and documents. I didn't lose my senses, however, and I used my poor English to say, "Doctor, what sort of thing is this, how can you hit a wounded soldier?" The Egyptian officer finally reacted; he gave a huge commanding yell and all the soldiers rushed out of the bunker. Later on they tended to my wounds, gave me a morphine shot, and waited for darkness.

AVI

I graduated from high school in a suburb of Tel Aviv in 1963. I had wanted to become an aeronautic engineer and had already passed my examinations for college, in an arrangement that enabled me to defer my service until I graduated, but at the last moment I was called to do my military service right away. So I went to the aviation course instead. I was drawn to the freedom of flying. When I was eighteen, I believed that someone who flies alone in the sky is a king.

I completed the course in 1966, and since I was one of the best pilots, I was transferred right away to the Mirage, which was the top plane of the Air Force then. I had two desires at the time: to hit Migs and to marry Yardi. Since I managed to fulfill these two wishes, I was happy. But only for a short time: I married Yardi in January 1970, and forty days after our wedding I was taken captive.

During the last year of my service I was an aviation instructor, but I also flew on missions, deep into the Egyptian territory. We were like hunters—waiting for an Egyptian plane to take off and going down on it. It was thrilling. I didn't feel any fear whatsoever. We kept telling ourselves that good pilots never crash, and I knew I was good.

I was taken captive on February 9, 1970. It was an ordinary flight. Egyptian airplanes were detected crossing the border, and we were alerted. Two Mirages were sent toward them. I managed to hit two, but I was too close to the second when I fired, so my own plane was hit also. My mirrors reflected fire at the rear. Soon the whole plane was on fire and spinning down. I knew I had to jump. I tried to eject myself by pulling the upper handle, but it didn't work. I tried the emergency handle and was ejected. Outside I heard planes firing at each other. To my utter amazement, the ejection mechanism threw me out with the chair, so I was in the ridiculous position of parachuting while sitting on a chair [laughing]. It took me about ten minutes to land, and I found myself in a marsh area, something like a big rice field.

While I was parachuting, I was planning how to get rid of some of the things I carried on me. I remembered a story of a pilot whose finger was almost chopped off because of his gold wedding ring. I threw away some things, including my watch and the maps I carried. When I finally landed on my chair, it was early evening. I looked for a shelter to hide in until I would be rescued. I was near a village, surrounded by flat, muddy marshland. I was afraid of sinking if I were to continue onward. I buried all the items I had on me in the mud, except for a radio and my pistol. My radio was supposed to transmit SOS signals to the rescue team. I saw my Number 2 passing above, still in combat; then more planes arrived, circling above me. At that moment two men approached. They were dressed in rags, as fellahin [farmers] are, carrying hunting rifles, and I realized that I didn't stand a chance against them. And I was certain that within moments I would be rescued. So I raised my hands with the pistol and threw away the radio. The fellahin were promptly joined by others, all highly excited. They formed a line and took me to the village.

They asked me who I was, and I said, "Anna Russie" [I'm Russian]. I hoped they'd believe me. They behaved politely and even offered me a hand-rolled cigarette. It disgusted me, but I didn't want to hurt their feelings, so I smoked it. Then we walked on, toward the north, with the village Mukhtar [chief] leading the way. Several men joined us from the fields as we were walking. An Israeli helicopter appeared above us and, not knowing Arabic, I tried to explain in mime that we should motion the chopper to land and take me to Cairo, where I would report to my superiors. The farmers were arguing what to do, and someone did signal the pilot, who was very low and right above us. But for some reason, the pilot hesitated, the Egyptians hesitated, and we moved on in procession, with the chopper just above us.

It was strange—this Israeli helicopter was so close, and more aircraft were above him, yet the Egyptians didn't even try to chase them away. I was so sure I'd be rescued! For some reason, however, the chopper went away. Later on, its pilot claimed that he hadn't identified me. But I know I stood out very clearly against the background; my parachute was spread on the ground, and I was wearing the grey aircrew jumpsuit among all those farmers wearing their black rags. Anyway, the chopper went away; I imagine the pilot was too afraid to land.

In the meantime, the Egyptians started to understand who I really was, and they began to flare up. Luckily, the chief of the village was a strong man, and he protected me from the mob. We were walking north on the water canal, and I heard the men arguing all the time. The young ones sounded particularly aggressive, but the chief calmed them down. From time to time, someone approached me, kicked me, or poked me. Once I was even slapped in the face. This was the first time in my life I saw stars. That's why the first photographs show me with a swollen face. I was afraid that the Mukhtar might lose control.

A police boat finally arrived. The Mukhtar explained that I was an Israeli pilot, and I realized I couldn't deny it anymore. We were put on the boat, and I sat there quietly, trying to collect my thoughts in order to plan my next steps. The transition from being a pilot in the sky to a captive is so extreme and abrupt that I suffered, obviously, from shock. Yet I kept thinking: what would be my cover story?

During the aviation course we had to undergo drills preparing us in the event that we would be captured. It concentrated on the very first stages of being a POW. We were blindfolded for three days, humiliated, and not allowed to use the toilet. But we knew it was make-believe, and I myself did not experience a crisis at all.

We were also somewhat prepared for interrogations. As I reviewed this, I told myself I wasn't a hero. I knew that telling only my name, number, and blood type was out of the question. So on the boat I created some kind of a story, with several points on which I could fall back on if I decided to change my mind under torture. It was a sort of mental rehearsal for the interrogations.

The men on the boat were polite and offered me coffee and cigarettes. We were going to the district town, where the military and police headquarters were situated. As we arrived, it was already dark and a mob awaited us at the pier. Two rows of soldiers were formed, and I passed between them, but I could see a hanging tree with a rope all prepared for me. The mob had obviously planned to have me lynched. From then on, it was as if a curtain had dropped down on my senses: I was an observer of somebody else's fate. A mass of people tore me away from the soldiers and dragged me to the pole, but the soldiers pulled me out of their grip. This continued back and forth for five minutes, which seemed like an eternity. Finally, the soldiers managed to take me into the building.

All this time, the screaming went on outside, and men were knocking on the locked metal doors. [He is silent for a long time, as if he hears the sounds.] I myself was in shock. I felt like a rare animal of sorts. All kinds of high-ranking officers came to have a peek at me. The phone was constantly ringing, and I heard excited voices. But I was treated politely. I remember watching, disgusted, as the men came and went, kissing each other on the lips. I tried to detach myself from the present, from what was going on outside, and prepare myself for the future. During these moments, I also said goodbye to my wife and parents, deciding that they would not have a part in this story. In my heart I told them, "So long, we will not see each other for three, four months." That's how much time I thought my captivity would last. I remember clearly how I established this distance, telling myself that from now on I was all alone; there was no one to help me, and I had to struggle all by myself.

BENNY

I was twenty-four years old, having completed my compulsory service in the Armored Corps, and I wanted to make money. So I found this job as a military canteen worker, but I was a civilian. Many civilians worked in the Sinai desert at the time, and I wasn't afraid of being killed or taken prisoner. I'm not the type who's scared by danger.

The shelling on the canal was intense during that week. Today, I think that had I been more aware of the danger, I might have asked for a transfer. But I didn't. On Saturday night, in town, I saw a movie on American POWs. On Monday, at Sinai headquarters, I saw a movie on POWs again. On Tuesday, a group of three Egyptian soldiers was captured and brought to the camp. I said to Motti, my partner, "They probably curse the moment they were captured." Next morning, one of the construction workers asked me out of the blue, "Suppose you are taken prisoner, do you have any orders to be carried out?" I didn't even want to give it a thought, so I sent him to Motti for an answer.

I drove the military mobile canteen, a big truck full of products for the men on the line. Our route was divided between the different drivers. On that particular day, I was given a new man to train for the job. Motti, my partner, was also new on the job, after his transfer from the north. So I sent the new man with a truck to one region of the canal, while I took Motti to show him his area. I simply volunteered to do this.

It was February 11, 1970. At noon, we arrived at a post named Cobra, and we sold our products to the guys. As we were leaving, the officer warned us not to go on the way by the canal, but to take a roundabout, longer road, away from the waterfront. However, I wanted to arrive at the next post in time. Because of the shelling, we had not been there for four days already.

I drove on the regular, shorter way, when all of a sudden my front wheel was hit by a shell. I yelled at Motti to jump off before we caught fire, and I myself jumped out and rolled on the sand. I tried to crawl to a ditch, but I heard Motti cry that he was injured. I saw him running erect, his arm bleeding, so I grabbed him and pushed him down and we took shelter behind the dune. At that moment everything was crystal clear in my head. I was planning my steps, while Motti warned me that a hand grenade was rolling towards us. I was considering grabbing it and throwing it across the canal, a funny thought. Just then, Motti was screaming, "Look, Arabs." I turned around and saw him standing with his hands up. So we didn't have a choice anymore, and we were taken.

As it turned out, a whole commando unit of forty men had infiltrated the canal line. Had they known that we were only canteen workers, with a truck full of drinks and cookies, they might have waited for something better to come up the road. But they mistook our truck for a personnel carrier; they didn't recognize the canteen sign.

With their daggers, they forced us to march to the water, where commando boats waited. Across the canal, a truck full of soldiers joined the men—they were extremely well organized. I think that they were disappointed they had caught only the two of us.

I didn't hear any response from our posts, and this puzzled me. They must have heard the commotion; the Egyptians were firing all over the place. Anyway, during the first moments I considered the whole event a joke, like a dream from which I'd soon wake up. It didn't sink in.

MOTTI

Three or four days before the event I had several signs of what was ahead. But I wasn't taking the signs seriously enough. On Saturday I saw a movie about American soldiers being taken captive by the Germans. On Sunday I went down to the Sinai, my job being a civilian canteen worker. I had been posted in the north, but after a few mishaps I was ordered to the canal as a punishment. That evening in the Sinai, I saw another POW film, about Italian soldiers. On Monday I heard from the commander of the base that a unit was out chasing Egyptian infiltrators, and indeed at noontime three dead Egyptian soldiers and one prisoner were brought into headquarters. Somebody saw me and ordered, "Keep an eye on the prisoner," who was lying there handcuffed with a wire. I remember Benny saying, "He must be cursing the day he was born."

On Tuesday, a construction overseer asked me, "What would you do if you hit an ambush with your truck?" I told him to go into my cabin and see my firearms. "I would break their neck," I responded. "And if there are thirty of them?" he insisted. This was the exact number of soldiers who attacked us three days later. Strange, isn't it?

I don't believe in fate, but the most serious sign foreshadowing something was that I switched my route that day. Had I gone the route assigned to me first, this would not have happened. Ask others, even the pilots. Many people were captured when they went out of their way to exchange schedules with their friends.

We went on our way early on Wednesday. We were accompanied by some guards up to Cobra, which was a huge post accommodating only about twenty soldiers. I had all the information at the time. . . . When I told the interrogators that Cobra was manned by twenty soldiers, they said, "Don't lie to us!" and the translator advised me, "Stop making fun of us. Give us a bigger number and you won't be beaten anymore." So I said, "Sixty-five," and still they didn't believe me. The truth is that the whole line was a huge pretense, a bluff. From one end to the other, it was manned by perhaps six hundred soldiers, no more.

Anyway, at Cobra I asked to be accompanied to Zehava, the next stop on our route. The commanding officer said that he had no spare guards, and we were late as it was. So I told Benny to come anyway. We were in the truck, and all of a sudden I felt it rising up in the air. I heard a big explosion and saw fire in the engine. I didn't understand what was happening. We were about three hundred meters from the waterfront, but we were shot at from a very close range. I told Benny to jump, and I was going to jump right after him. For years I had been telling myself that in this situation you have to jump in the direction of the vehicle, but at that moment, I was afraid of stopping or slowing the truck down. I took my Carl Gustav [rifle] and two magazines and jumped in the *opposite* direction from the car. I rolled on the road like a ball until I stopped at a sand dune. To this very day I can't believe I survived the jump.

The gun had torn my back, and I was wounded all over from the jump. I tried to crawl, but I heard several shots nearby. I still believed they were on the other side. The thing which puzzled me was that an Israeli armored truck which was present at the scene didn't come to our rescue. It's out of the question that they didn't see us! I kept crawling in the dunes. I didn't see Benny anywhere. Suddenly I saw three Egyptians running right next to me. I fired my machine gun, and five more appeared. I think I got three of them, and that's why I was so badly beaten afterwards. . . . Anyway, I saw thirty commando soldiers with all their equipment; they were model fighters. It didn't occur to me that I would be captured alive; I was sure they'd kill me on the spot. That's why I gave them a fight. Since I was wounded in my shoulder and in great pain, my vision was foggy and I saw them as if through a car window on a rainy day. I heard yelling in Arabic: "Come! Run!" Somebody stuck a commando knife in my back. I don't know how I had the energy, but I started to run. I made some turns, however, and the knife stuck deep into my flesh. And still I was thinking: It's inconceivable that I'll be taken so easily! Troops from the ground or the air will come and help me.

When we arrived at the canal, I saw four rubber boats with frogmen waiting in the water. I was pushed and rolled down into the salty water. The salt burned my wounds, although it was probably a good disinfectant. The pain made me shiver, but the soldiers didn't pay any attention and threw me into the boat.

As we made the crossing, I was cursed the whole time. On the other side, I was forced to run through prickly bushes, and suddenly I noticed lots and lots of people emerging from holes in the ground. They were some kind of cavemen, living there. They all wanted to kill me, but the soldiers protected me and pushed the mob away. A woman, about forty years old, managed to get close to me, however. She spat on me and pushed her finger into my wound.

At some point, I was joined by Benny. The soldiers drove the mob away, and we were thrown into a bunker. I was taken into the toilet and was left to sit there with a guard. Later on, I was dragged, my back bleeding severely, into a large room where Benny was also seated, confronting six big Russian guys. All I remember is one question that was asked repeatedly. It was asked in Arabic, and Benny translated for me: "Are you a soldier or a civilian?" I said "civilian," but they didn't believe me. I was dressed as a civilian, with hiking shoes, but Benny was, for some reason, in military uniform.

DAVID

I was captured on May 30, 1970, at the northern part of the Suez Canal. I was in the paratroopers, with about six months remaining before the termination of my mandatory service. I was posted in a stronghold on the canal. Every morning we had to inspect the road and open it for traffic, a daily reconnaissance duty carried out by a team of three tanks and three armored trucks. I was the commander of the first armored truck that day, and towards the end of our mission the commander informed us that an Egyptian ambush was ahead. We jumped off the vehicles and looked around. Suddenly I heard a burst of fire behind me and saw that my armored truck had been attacked. I threw two hand grenades and ran toward the truck, but it had escaped to take cover at the nearest stronghold.

Intense shelling had started in the meantime. [He is quiet for a while.] When I lifted my head between the shellings, I saw burning tanks and bodies all over the place. I was sure there was nobody alive inside the vehicles anymore. This was the "black Saturday" of the paratroopers; fifteen men were killed that day.

To this very day, I walk around with the feeling that no one should have been killed in that incident, since we had been warned about the ambush. I don't know why so many men remained inside their vehicles, instead of running to the ramp. I don't know why nobody returned the Egyptian firing from the posts. There are four or five versions of what really happened that day.

I took cover and watched what was happening. When the shooting was over, the Egyptians detected me in the sands. I fired my Uzi and ran toward the marshes. A shell pit provided me with a hiding place for a while, but the Egyptians came from the rear and picked me up. As they were dragging me toward the canal, I saw an IDF [Israel Defense Forces] tank and I was sure they realized I was being taken away.

I didn't believe my eyes when I saw about fifty men on the waterfront. They were all hiding below the ramp, a dead area for our observers. We waited there for darkness, and at about 5:30 we made the crossing.

All this time I had been tied by my hands and legs and around my neck. During the crossing, they put a life belt around my body, tied me to an Egyptian guard with a knife in his mouth, and, with a rope, pulled me to the other side, until we reached a small pit. There, I was met by an intelligence man who spoke Hebrew. I believe they were out to catch prisoners that day. He offered me water and a cigarette and started with basic questioning. When the soldiers tried to hit me, he drove them away. He even tried to cheer me up.

As it was turning dark, our Air Force started to bomb the area. I was terrified that they would hit our pit, yet at the same time I was hoping that a helicopter would come to my rescue. None of these happened, however.

RAMI

I was born in Kibbutz Mishmar Ha'emek. I served in the Air Force for five years, and a year before my release I married Nurit, from a neighboring kibbutz—Hazorea—and we moved to her kibbutz. We lived in this kibbutz for six years, had two kids, but throughout this time, I had the feeling that the Air Force needed me more than the kibbutz. In 1967, after the Six Day War, I signed up for three years of service. In 1969, when the Air Force was getting the first Phantom planes from the United States, several pilots were selected to go and train with the new planes. So Nurit, the children, and I went to the States for almost a year. We returned at the peak of the War of Attrition and formed the first Israeli Phantom squadron.

It was a strange war, you know. It was very selective, fought by a small number of soldiers, pilots among them. I was under so much work pressure that I didn't realize that the whole country was almost unaware of the war. We were still mastering the new plane, we had to develop its firing system, we were in charge of younger pilots whom we were training on the Phantom, and, on top of that, we had very frequent missions in enemy territory. Everything was new; there was no one to learn from. In the meantime, the Egyptians got Russian anti-air missiles, and we learned to evade them. Personally, I have never felt afraid or endangered. This was my job. Since I rarely had the time to see my kids, I moved my family from the kibbutz to the Air Force base. Nurit was pregnant again, and I wanted to have some time with her. These were the craziest days of my life.

On June 30, 1970, we went on practice flights in the morning, and in the afternoon we went after the Russian missile launchers deep in Egypt. My navigator was Los. We hit the southern missile battery, as planned. On our way back, we saw a missile behind us. I was on the alert and tried to evade it. The technique was to change direction abruptly, after having waited for as long as possible. Six seconds later, however, a second missile appeared. We were going rather slowly already, and couldn't escape the second one. The plane was hit at the rear, and its tail torn away. Since we were very high, we figured we had enough time to save the plane, but it started to spin like a leaf in the wind and I decided to jump. Because of the fast spinning, I needed incredible effort to pull the ejection handle before a third missile would come and finish us off. Finally I made it, and we were out, at a very high altitude.

You have to understand, the principle is the following: The pilot pulls the handle, the navigator is ejected first, and after him the front pilot, myself. It worked. Since we were at such a high altitude, we parachuted for about half an hour, a very long time indeed. The interesting thing is that Yitzhak was still on the ground when he heard on his radio that we had jumped. Nevertheless he took off, bombarded the enemy targets in Egypt, was hit on his way back, ejected from his plane after a lot of mechanical trouble, and still landed in Egypt before me. That's because he had been very low.

Being in the air for so long, I did a great deal of thinking. First, I considered the technical problems of the ejection process, which no one had previous experience with, since we were the first Israeli Phantom to fall. I was telling myself: you must remember to tell the guys that the chair becomes steady as it's ejected! Then, I took my knife out and tried to cut off the boat, but I cut off the whole survival kit by mistake. I was making contact on the radio and I informed our people that when I landed, I would make a run for the hills.

I remember seeing Los parachuting not far from me. He seemed bent down in a strange way, and I assumed he was injured during the ejection. I wasn't able to reach him on the radio. I checked myself and felt intact, except for a cut in my cheek caused by my mask. When I finally landed, I started running toward the area I believed would provide a hiding place, but I saw right away some soldiers running toward me, and I knew I would be captured.

I think that thirty minutes in the air provided ample time for me to enter the role of a POW. I told myself I would be back home in six to eight weeks, because this was the duration of captivity of previously captured Israeli pilots. This thought kept me going in the beginning. I also prepared myself for being beaten and decided that I wouldn't say anything during the first interrogation, because whatever I would do, they would beat me anyway. I would have plenty of time to tell them things later on.

I cannot recall any premonitions about this event, but years later Nurit told me a very strange story about my mother. She said that after she had received the bad news, she went to my mother, in her kibbutz, to inform her. She found my mother just awakening from a bad dream. She must have had the dream while I had been in the air, you see. In the dream she saw me in my pilot's overalls, with a number of other men in similar uniform in a circle around me. I implore them to help me get out, but they stand there and don't provide any help. I am fully convinced that the great effort I was exerting to reach the ejection handle—this was the only means for my survival then—was expressed, somehow, in my mother's dream.

I forbade myself to think about any possible errors that might have caused my fall, because it was useless. I had followed my orders. However, the idea that it might have been preferable to fly at a lower altitude, and thus avoid the missiles, has lingered beneath the surface. Only years later, during the debriefing of the return, did I permit myself to analyze the situation.

YITZHAK

I was born in the United States, and when I was fourteen we came to Israel. I was educated in Kibbutz Mizra and was in the first group that went to study the Phantom in the United States. At twenty-five, I was the youngest and had a certain advantage in the course because English was my native language. I learned to run both the back and front cabins of the Phantom.

When we came back to Israel, we worked for three or four

months to bring the Phantom into our operations. We also practiced and instructed other pilots—each of us was under a very great personal load. I myself instructed very complicated technical matters. The day after I was captured, I was supposed to give a group lesson on a certain topic on the care of the plane. I had already prepared a diagram on the blackboard, and when I came back from Egypt, the guys from the squadron took the blackboard out of the storage room. They had taken it off the wall and saved it with the diagram still drawn, for three and a half years. "You owe us a lesson," they said.

There was a nice atmosphere, a kind of pioneer spirit in the squadron, from December 1969, until I was captured in June 1970. We were a very small group of pilots who were setting up the squadron. We felt so important; we knew this plane was a special contribution to the Air Force.

After a very short time, we started flight operations with it—that is, sudden penetrations into Egypt. We were making between five and ten penetrations a week, sometimes even twice a day, and on the same day we would also be practicing and teaching. I was young, enthusiastic, and I don't know exactly how my wife and two little daughters, who were left behind at home, felt about it. But I think we became used to this way of life. No one had been captured yet, and we all had this feeling of superiority. A pilot needs to have such a feeling, that "it won't happen to me," otherwise he can't function properly. I'll say it even more strongly: Arrogance is in the blood of a fighter pilot as part of his resilience.

Everyone wanted to go out on flights. The lineup was run according to a "fairness schedule" hung up on the notice board. It was hard to convince the younger ones that sometimes things were complicated and it would be better if the senior pilots did them.

June 30 was an ordinary Friday, but I remember that we were already aware that the Egyptian missiles were heating up the atmosphere, and there was a possibility we'd be shot down. (This awareness was accompanied, of course, by the feeling that "this wouldn't happen to me!") There was going to be a party at the base that evening, and we were preparing for it. I had a practice flight in the morning and prepared the blackboard for the lesson the next day. At three in the afternoon, we got a message that batteries of missiles had been advanced toward the canal and we had to bomb them. The bombing was scheduled for dusk, about 6:30 in the evening. Two formations would go on the mission, with half an hour between them. I was to lead the second formation. Our takeoff time was exactly the moment the first formation began dropping its bombs.

I was in the plane ready for takeoff. David Ya'ir was the rear pilot, and we were listening to the radio. It was completely quiet, because during attacks no one used the radio. Suddenly, half a minute before we were to take off, we heard a strange conversation on the emergency channel. It was the voice of Ehud Hankin, who said something like, "Try to hide. I'll come right away to get you out." I remember saying to David, "Somebody jumped, somebody abandoned a Phantom." Then I said, "Let's shut off the emergency channel. Anyway we can't help him, and on the way back we'll find out what happened."

We took off and along the way we got a message to fly a course different from the one planned. This happens. We flew due west, and it was hard to see the targets because the sun was in our eyes. Suddenly a missile warning appeared on the panel. We didn't get scared; we were trained for this, but it was a very sharp warning. We turned and saw two missiles coming down on us. We reported to the plane behind us. He succeeded in evading them, but the missiles came really close to us. That's when I knew I should have stayed in bed that day. . . . It's hard to explain, but the Egyptians—with the help of their Red friends—sent their missiles that day in an unfamiliar method, and it surprised us very much. The first two missiles exploded very close nearby, and I heard a loud noise, like gravel falling into an empty barrel. There were explosions and smoke, and I knew we had been hit. The plane was already in a downward dive when I saw the target, and I said to David, "First let's bomb and then we'll try to go home." We were still in control of the plane and went down to ten thousand feet. I released the bombs and straightened out, and we flew in the direction of the canal. Both the engines were on fire; the cabin was full of smoke, and nothing worked, but the plane still reacted. To stop the fire, I turned off the least important engine and we continued to fly. In another minute or two we would be across the canal, and then we could jump. I remember saying to David that I didn't think the plane would explode; we could try to drag it out for another minute.

The second plane in the formation started to shout into the radio that our plane was full of fire and smoke, and I answered him that I was trying to reach Israeli territory. Then he told me another missile was heading my way. I remember telling David to jettison our fuel tanks, maybe they would attract the missiles. We actually did this, but were hit directly anyway and the tail of the plane was cut off. Now we really had to abandon the plane.

I remember saying to David, "OK, are you ready?" and I pulled the ejection handle. He ejected with the seat upwards and then went off into the distance. I thought to myself: Now me. Then I remembered there is three-quarters of a second between the ejection of the two seats, and the American pilots had told me that when your life is in danger, it seems very long. All this time, the plane was revolving and going down. I was waiting for ejection. Finally I realized that my ejection mechanism really wasn't working. I had to manually perform the ejection: untie my straps, open the canopy, jump out of the plane, and open the parachute, same as a parachutist in World War II. As fast as I did it all, I was in the air only about two seconds and then I hit the ground. The plane fell next to me. I had escaped at the very last moment.

While I was trying to get out of the plane, I was sure that I was going to die. And I thought, "What do people who are about to die think about?" They say that their whole life passes before their eyes—and that's what happened. I really saw pictures from my life, not like a movie but like stills, from all the periods of my life. This did not keep me from doing all the complicated actions and jumping. All this took about twenty seconds or half a minute. But time has a quality of lengthening. I identified the pictures I saw. I understood what they were saying to me and even thought, "Hey, this is what happens to a man who is about to die."

No, I don't remember today the pictures I saw, but most were from when I was younger. What I do remember is that I functioned without panic. My hands did what was necessary, as if by instinct, even though one hardly ever practices for such a situation. I only felt that everything was happening much too slowly. I was very frightened, of course, but it didn't prevent me from acting, from keeping cool, and even seeing some humor in the situation. That's what saved my life.

Years later, I took a course in the United States for test pilots. We had a lesson on the Phantom, and the officer was explaining that it might be possible to get out of the plane manually, but that no one had ever done it. A student in the course to whom I'd told my story said to him quietly, "You want to bet?" It may not be routine, but I'm sure that I'm not the only person in the world who's done it.

I remember the entire time after we were hit we were talking to each other, David and I, and we joked about the situation. I remember him asking me, "What's happening?" and I told him the cabin was full of smoke and I couldn't see anything, but maybe if I went closer to the panel I could see what it said. A minute later I remember saying, "No, it's not worth it, you don't want to know what it says." I can't say this humor was controlled, but that's how I function.

At any rate, I fell like a sack. I found myself on the ground. I didn't lose consciousness, and I knew I was intact. I had burns in the places that weren't well protected from the fire in the cabin when I jumped. The plane came down next to me, full of fuel and ammunition, and I was afraid it would explode. I didn't feel any pain. I saw where the sun was; Israel was across the canal on the other side. I took the water in my survival kit, my gun, and my two-way radio, and started running east, toward the canal.

A Skyhawk pilot in the area searching for Rami and Los, who had jumped before, intercepted me on his radio. I told him that I had jumped, that I was trying to hide, and that I didn't know where the rear pilot was. Unfortunately the plane had fallen about three hundred meters (a thousand feet) away from an Egyptian post, and the soldiers had obviously seen me—such a treasure doesn't fall out of the sky every day! The next few moments were like a Western: I looked westward into the sun and saw a truck coming up the hill, about thirty soldiers inside. Here I was, running and trying to escape in a desert that was completely flat. Of course they soon caught up to me, and about two hundred meters away, they all jumped out of the truck, stood in a row, and shot at me. I was a sniper too and said to myself, "As long as you run, you can't be hit at this distance." So I kept running. I heard the bullets all around and even reported on the two-way radio that soldiers were running after me, and someone should come to rescue me right away. I remember the idiotic pilot—poor guy, he was more agitated than me—saying over and over, "Try to hide," even though it was obvious there was just no place to hide.

I ran quite far, so the soldiers went back to the truck, stopped, and started shooting again about two hundred meters away. It happened a third time, too. I saw there was no point in running away anymore and had to give up with a last message on the radio: "You haven't come yet, and I can't fight the whole Egyptian army all by myself. I'm destroying the radio." The radio would tell the helicopter where I was. I threw it down and shot at it, and almost got myself killed, too. At this point the soldiers reached me and poured out all their anger. They hit and kicked me and tore off my watch and my clothes. They tried to kill me—shouting the whole time. I didn't understand much Arabic, but I understood enough to know that I really shouldn't have gotten out of bed that morning.

It was my bad luck that there was no officer with them. Again I saw death, but it didn't happen. In the end, they threw me onto a truck and brought me to some army camp—the only thing I remember is trying to protect my body from their blows. I pretended to faint, when I suddenly saw an officer approaching, clearing a path through the soldiers. I asked for water and hoped that if the officer saw that I was still alive, maybe he would stop them. He did. He really saved me. He got the soldiers away from me, shouting and shooting in the air. I decided to pretend that I was a Russian pilot, and I started to utter some words in Russian—a language that I don't know, of course. But I wasn't sure that he believed me.

I was hurting badly from the beating and burns, and I remember only that they put me into a jeep and we rode to some field clinic. There was a human body diagram on the wall, with muscles on half the body and the nerves on the other half. I said to myself, "Finally I'm in the hospital; now everything will be all right." Again I asked for water and said to myself, "It seems I'll stay alive. I have to think what's the best thing I can do now."

AMOS

I was drafted in 1964. I was an Air Force pilot until I was taken prisoner in 1970. I used to fly the Mister, and I was a flight instructor as well. I learned the Phantom in the second course that was given in Israel. Six months before I was captured, I married Dalia, and we planned to move north and live at the base. In the meantime we lived near Tel Aviv, because Dalia was a student, and I flew a Cessna every day back and forth to the base.

The War of Attrition was the hardest war I have participated in—mainly because of its long duration, at least from the Air Force's point of view. For three years, we were actually in operation all the time, starting from the Six Day War in 1967. The contrast between reality on the front and life in the cities was difficult to take. During the day, I'd be on a mission, fighting and bombing, and at night, I'd go out in Tel Aviv, where life kept its normal course.

Every flight involves tension, especially when you cross the border above enemy territory. From time to time excellent pilots had to abandon their planes. We learned to live with this possibility and ignore it at the same time, because once you start to think about probable disasters, you can't fly anymore. Troops pay with their lives everywhere, but during the War of Attrition the Air Force was fighting all by itself. It was only toward the very end of that war that soldiers who manned the posts along the canal also did their share. Just the same, I don't remember feeling tired or burned out. We were young and trained for that kind of work. There was even some "combat joy" in our performance. It never crossed my mind that I might end up a POW.

The months of June–July 1970 were extremely dangerous, really crazy times. I was shot down five days after Rami, on July 5. We had been bombing deep in Egypt and accomplished a lot. We surprised them where they had never expected us. Their resistance was nothing to speak of. Their missiles were ineffective and we had no casualties. In the middle of June, however, their missiles got better, and the atmosphere changed. Everyday we practiced another evasion technique. Then two Phantoms were shot down on the same day, Rami's and Yitzhak's, and five days later it was my turn. Two more, including Menachem's, were hit a few days later. You have to remember that the Phantom was the pride of the Air Force, and when we started losing them, a very bad mood prevailed. We had the feeling that the Air Force commanders didn't have any answers for the new situation.

During that time we were going after the missile batteries and their dugouts, day and night. I went on a mission with Amnon as my navigator, but we weren't a regular team. It was a routine flight for us, and nothing special happened until we were hit. We received a missile alert on our panel, but I wasn't worried, since this happened often. A missile that you see is not dangerous, because you can evade it. We released our bombs on the target, and seconds later we were hit. The timing was critical, because while you're dropping bombs, you can't maneuver and escape. We were hit directly by a missile or by heavy anti-aircraft artillery. The plane exploded right away and its tail caught fire. The engines stopped reacting, all the alarm lights went on, and the front windows were blasted. Wind blew into our cockpit, yellow smoke was infiltrating from the air conditioner, and the whole place smelled of burning plastic. We were in the middle of Egypt. It was clear we wouldn't be able to make it home.

I was looking for a deserted location in which we could hide. I pulled the plane across the highway and then abandoned it. We jumped from an extremely low altitude, at the very last moment. I touched the ground the moment my parachute opened. In other words, I was in the air for a second or two, that's all.

The whole ejection process went smoothly. We weren't wounded, probably because of our slow speed. The whole thing happened automatically: I pulled the handle, the chair separated from the cockpit, the parachute opened, and before I had the chance to think about it, I was on the ground.

I saw Amnon near me. The plane also fell nearby. I thought we had parachuted in the desert, since that's how it looked from above, but in fact we found ourselves in the middle of an army post. Before we reached each other, we were surrounded by soldiers.

I remembered stories which led me to think that the first moment is the most dangerous. I hoped that we'd live through it and be taken to the authorities. Luckily, an officer came immediately after the soldiers. He was very excited, pointing his gun as he talked. I was worried that he might shoot by mistake; in fact he did. The bullet went past my head and hit one of the Egyptian soldiers. It calmed everybody down.

We were taken to a bunker where a Russian officer and several Egyptians were waiting for us. They gave us water. The Russian got all interested in Amnon, because he saw that his face had Slavic features. The atmosphere was friendly. One of the officers kept talking excitedly on the phone. We heard the soldiers cheering outside.

In the beginning I didn't grasp that I had been taken prisoner. It was as if it wasn't happening to me, since I was supposed to return to the base in half an hour. I was watching the scene from the outside, as if in a movie. Only several days later did I realize it was me there. It is tough to admit such a reality and enter the new role of a POW. Until the last moment I had been a heroic pilot—and here I was a helpless captive, a victim for any soldier's whim. The transition is too sharp. At the same time, I kept thinking and planning my steps.

I had never been prepared for the eventuality of capture. In the years after our fall, they started to prepare pilots for captivity. I didn't know what to expect. I thought about a Hilton, or another five-star accommodation. . . .

AMNON

I grew up in the shadow of the Six Day War. My generation admired the army, its victories and heroes. I had not considered my military service as a duty or a job to do, but rather as a mission. When I had planned my service, I wanted to become a combat soldier, but since I hated to run and was afraid of water, the only option for me was flying. So in October 1967, I started my flying course. I was so deeply engaged in the course that I detached myself completely from everything going on around me. I hardly even knew that the War of Attrition was going on.

At the end of the course, I was placed as a rear pilot [navigator] in the Phantom squadron. We had a two-month course specifically for the Phantom but were sent on missions even before we completed it. There was a lot of pressure on the Phantoms because of the war. Most of the Phantom pilots were old-timers in the Air Force, who had been retrained for the plane. I, like the other new graduates, was very young; I had just turned twenty. The veterans had some experience in fighting, they had already lost friends in the wars, they had visited bereaved families. Compared to them, we were like newborn babies. We had never seen the leading plane explode right in front of us or seen somebody abandon a plane. Yet we received our rank and joined the squadron as equals.

I had a survival workshop during my flying course. It also included preparation for captivity. We were "caught" during a hike, blindfolded, handcuffed, and put in confinement for three days. Some of the people were beaten during interrogations, too. After my release, people inquired whether I had found the workshop helpful at all. My reply was that, for me, with my complete lack of previous war experience, for the first few days this workshop had been a lifesaver. Living surrounded by the smells of your own urine and feces, dizzy, in a state of hunger and humiliation, had taught me what to expect. But the missing part was the great fear I felt when I was really captured. Even when I had been interrogated nonstop for five hours, I knew it was make-believe. Physically, the workshop had been quite tough, but, as I learned later, one forgets pain really fast once it's over.

At any rate, four months after my graduation from the flying school, during my third operational mission, I became a POW. This was the time when our flight instructors, like Rami and Yitzhak, were going on missions like this everyday, and we all got into this "hunting atmosphere." I remember myself as being extremely ambitious then, and the reward for my ambition was that, among my class, I was the first one to go on a penetration operation deep in Egypt, and the first one to get screwed!

The belief that "this can never happen to me" is central for pilots' way of thinking. Missiles were flying in the sky like rain, but those of us who had graduated from the course were convinced that we were in full control of our fate, the best, supermen in the sky—while the enemy was stupid and primitive by comparison. . . . Those who didn't develop this kind of belief couldn't cope with the situation.

One episode which is relevant to my capture: at the end of our course we were divided into the northern and southern squadrons.

Since I had been a good boy, the school commander had promised me that I would have the choice. I selected the northern squadron, but at the end of the course I found my name on the other list. I called the commander and implored that he keep his promise, and I was transferred to the north. The man who had taken my place in the south was also involved, later on, in a Phantom operation that was interrupted by missiles. In his case, the veteran front pilot was wounded and lost consciousness, and my young pal, the rear pilot, returned the plane safely to the base. Both of them received citations, while I. . . . Thus I learned one shouldn't switch places in life.

My first mission outside of Israel, my virgin flight, so to speak, was to bomb PLO camps in Syria. And the second, to Egypt, took place on June 5, 1970. Our plane had to bomb a military airport near Cairo. It's a fantastic sight: six Phantoms flying in the blue sky, heavy with ammunition, like metal birds. I remember that not far from the target we descended and I saw the delta with its irrigation canals, mango and avocado trees right underneath, a fantastic sight. We were flying very low, only thirty feet above the ground, and everything was completely quiet around us, no missiles or anything. Our Number 1 dropped its bombs, and we too identified our target. Just then we received an alert, and the sky suddenly filled with missiles. We pulled upwards, and at 7,000 feet we suddenly heard an explosion. We were hit at the tail, and the plane caught fire. We tried to release the bombs, but the mechanism didn't work. We informed the leading plane that we were hit. We pulled toward the canal and looked for a spot to jump.

It was clear that we wouldn't make it across the canal. The gliding range of a plane without engines is quite limited. I wasn't afraid then; I was totally absorbed in surviving. I had to act with maximal efficiency in order to escape this trap. We were rather low and flying slowly. I remember being ejected like a missile. The chair spun once, and I was wondering—would the parachute open or not? When it did, I noticed the great contrast between the cockpit, full of noise and smoke, and the complete silence outside. It was quiet, like in paradise. That's when I realized that I still had hope, that I might survive. Suddenly they started firing at us from below. I pulled my knees up, protecting the most sensitive spot. . . . I saw Amos landing about a hundred feet from me, and I touched the ground.

Another memory: our Number 1 circling in the sky above us to see if we parachuted all right, so that he could report it. I cursed him and said: "Go away before they get you, too!" My second thought was: "You go home, while I stay in hell." I was in a state of shock. Just a moment before I had been an eagle-like knight in the sky, ignoring all the dots below me, but suddenly these dots had become people and now I was at their mercy.

The aircraft exploded nearby; the noise was awful, and it drove the soldiers away. But a moment later we were surrounded by hundreds of them. I didn't manage to report on my radio, I just released myself from the parachute, cleaned myself a bit, and saw that I was intact. I started walking toward Amos, whose neck was bleeding. The ring of soldiers around us drew tighter, and I saw Amos lifting his hands up. I was startled: how could I, a fighting machine, surrender? I wasn't able to do it, but I heard Amos saying: "Amnon, put your hands up!" and at that moment I knew it was all over. I was a POW.

The Egyptian soldiers were hysterical, but luckily three officers arrived and drove them away. At the end, we were put in a bunker of sorts. Inside they treated us decently. Two big Russians were sitting there, in Egyptian military uniforms. A small Egyptian major was sitting between the two, talking on the phone excitedly, reporting he had captured two "kites." People came and went. I asked for water again and again and drank huge quantities. We were offered cigarettes and asked some preliminary questions, like name, rank, and military number. After about twenty minutes, we were blindfolded, our hands were tied in the back with metal wire, and we were mounted on a jeep. The jeep went on its way. I was in such a state that I didn't even try to communicate with Amos, who was probably in the same vehicle.

It was a rather long drive. One of the men near me was smoking, and from time to time he gave me a puff. I fell asleep, as if intending to gather strength for the terror to come, or perhaps as an escape from reality. My head dropped on the Egyptian at my left, and I remember how grateful I felt when he moved a bit, so that I could lean on him comfortably. It was the last humane gesture I experienced for a long while.

Another thing I recall is that from the moment of our jump, the thought of my parents kept striking me like lightning: How would they take the bad news? How would my mom be told about it? In addition, I was thinking about myself: What was going to happen to me? There were about a hundred soldiers who had seen us alive, so they couldn't just kill us. So what would the next stage be? Would I be tortured? How long would it take? Also, was I to blame for our fall? Had it been inevitable? How was Amos, was he all right? And in spite of the paralyzing fear, I managed to sleep.

MENACHEM

I was twenty-four when I finished my service in the Air Force. With financial support from the army, I went to study engineering, and in return I signed on for many years of future service. During my first period of service, I had been a Voture navigator. When I returned from my studies as an aeronautics engineer, I was assigned to the first group that was sent to the United States to study the Phantom. We returned during the summer of 1969 and formed the first Phantom squadron. A month after our return, we were already fighting in the War of Attrition.

In the Phantom, the role of the navigator, or rear pilot, is to run the fighting system—the bombing, electronic fighting, interception and radar functions. It is a complicated plane and cannot be run by one pilot alone. I was the senior navigator in the squadron.

As the time went by, all the fighting duties were transferred to the Phantoms. The Air Force had, in fact, two missions: one, a "fun mission," was to shoot down the Egyptian Migs. This was a mission that ended up, almost always, with great success. The other job that was assigned to us, the Phantom squadron, since our aircraft was larger and could carry a much heavier load, was to penetrate deep into Egypt and bomb various targets there. At that time we would destroy almost daily some missile batteries and launchers in their dugouts. In addition, we bombed other targets, especially military camps, in Egypt. This was a terrible war for the Egyptians. Stories had it that the army had evacuated its camps because of fear of bombardment, or that about a thousand soldiers had been killed in an attack carried out by two of our Phantoms. Our missions were highly complicated and involved planning to the smallest detail. Going on these missions was exciting.

These were extremely hard days [sighs], involving instruction in the squadron, practice flights, and, on top of these, almost daily operational missions. That whole year we were working around the clock like slaves, never having a free day, not even on Saturday. I remember that once, after a long absence, I went to Tel Aviv, and I was amazed to see all these people sitting in the coffee shops, ignoring the war. I envied them, yet, on the other hand, I was happy that our work enabled people to live normally in the city, and that in a little while we too would join the fun.

With the help of the Russians, in June 1970 the Egyptians had managed to build a huge missile system near Cairo, and started to bring it forward to the front. A missile system is self-protective, since while you may hit one battery, you will be attacked by another.

I became a POW on July 18, 1970. I clearly remember the three preceding Saturdays, when we tested new electronic devices to avoid the missiles. At that time, we already felt the high risk of performing those penetrations. There was also growing resentment: Why don't they use the whole Air Force in a major attack against the missile system? Why are they using only the Phantoms, in a small-scale operation? I was young at the time and not interested in politics. . . . But I felt burned out by the war, and indeed a cease-fire was to be declared three weeks after my capture.

Usually I teamed up with the squadron commander, Chetz, as a front pilot. We were also good friends. On the Friday prior to our fateful flight, we had been at the squadron with all the rest of the pilots, preparing for our Saturday missions. After everybody had left to go home, I remained with Chetz, rechecking our maps, aerial photographs, and routes and preparing the briefing for the others. Later on, the wing commander also joined in. He too was a close friend of mine. I remember him asking whether we had teamed the flights for tomorrow. We showed him the assignments on the blackboard, and he asked, "Why do you two fly together? I don't like the idea, because you two have the best knowledge of the Phantom right now, and if both of you go, it will be disastrous for the squadron."

I recall Chetz's answer. He told a story about the Six-Day War, saying that its lesson had been that in the most complicated missions the best, most experienced pilots should lead the way. Our wise wing commander didn't argue, but in the end he was right. The next day Chetz was killed, I was taken prisoner, and the squadron was indeed in a great crisis.

I remember having a heavy feeling the night before. We had lost several planes already, and that week Amos and Amnon had abandoned their plane, too. We had the feeling we were failing our mission. There was also fear, naturally. I never concealed my fears; only stupid guys are unafraid. I remember clearly how Chetz and I were apprehensive as we sat in the squadron late the evening before, how we looked to each other for support. Later on, Chetz came for dinner at our house, together with his wife. It was a depressing evening. I never used to talk to my wife about my work. She is very naive and optimistic by nature, and I trained her to believe I'd always be okay. But of course she knew that pilots had been captured recently, and that some were killed. A few months before that she had found out she was pregnant, and I demanded that she abort the baby, because I didn't want to have a child while the war was going on. My decision was influenced by the death of a friend, a pilot who had left a wife and three tiny kids. So we all knew that our missions were dangerous, although we didn't say so openly.

In the morning, at seven, we were near the planes. I remember one of the pilots looking at Chetz's palm jokingly and saying, "Your palm tells me you're going to die." Then he looked at my hand and continued, "No, not you, though, you're okay." It's unbelievable, after what really happened that day.

Chetz and I were good friends, but we had never been sentimental with each other. Now, when two pilots enter their cockpits, you know, they are separated from each other for the rest of their flight. Although I could see his back, each of us was isolated in his cockpit. So before entering, we hugged each other that morning, saying, "See you later." This was something we had never done before. We must have had some premonition. And in fact, when I was lying wounded on the ground, surrounded by Egyptian soldiers, I was feeling relieved and thinking: So that's it, the nightmare is over. Now it's all right; we know what will happen from now on.

The flight itself was uneventful. At one point, we were alerted that some missiles were chasing us. We followed instructions, namely we didn't try to evade the missiles because our electronic warfare devices were supposed to take care of that. We were well disciplined and followed our orders, against our instincts, one may say, which warned us to escape. So the missile approached and was indeed exploded by our devices. However, the explosion was much too close to the aircraft, and we were hit as well. We released the bombs so that we'd be lighter in order to escape, and we turned back, toward the east.

We didn't notice fire or smoke in the cockpit, but it was clear that pieces of the missile had hit the plane. We tried to lower our altitude and speed up, and this was in fact possible, once we got rid of the extra weight of the bombs. Ten seconds before crossing the canal, the plane suddenly tilted left and downward, and Chetz was saying in the intercom, "Here it goes." I understood that we had lost control, and immediately raised my hand to pull the ejection cord. I don't know if Chetz was doing the same, but at any rate, my act should have ejected both of us, with me going first. I was out, and since we were quite low then and at high speed, I was wounded by the blast. I suppose Chetz didn't make it. The plane exploded and he was killed.

The Egyptians never returned his body. . . . Three and a half years later, after the Yom Kippur War, we returned to the location and found several human bones there. So we gave him a funeral. [His voice is very sad.]

From the moment we were hit, until I jumped, I remember a sense of control, even calm. All my senses were completely clear. I wasn't afraid of anything anymore. The moment I was ejected, however, I lost consciousness, because of the speed. I regained consciousness for a second when the parachute opened above me. With the pull of the cable, I detached myself from the chair and fainted again. I woke up on the ground.

I would like to add that the moment of jumping is a strange one. You are actually shot out of the aircraft, and you feel like you're breaking through a boundary of sorts. It's a unique experience, something you never practice for, of course. You feel something shattering inside you, like in a breakthrough, perhaps. Furthermore, it is an immense transition. A moment before you were sitting inside a powerful instrument, king of the sky, and now you are lying helpless on the ground. When pilots jump from high altitude, as Rami did, they may use the time to adapt a little to their new situation, but in my case it was all over in seconds.

I found myself on the ground and immediately realized I was badly wounded. I remember lifting my head and checking my body. My right arm was twisted in an unnatural position behind my back, like a strange, flexible part I hardly recognized. My left leg was stretched out unnaturally on my side [he laughs heartily]. It was a surprising sight. Later on I found out that my right arm had been fractured in three places, including the shoulder bone, and my left leg was severely broken, with an open fracture at the ankle. I had also been injured in the groin and had an open, bleeding wound there. I was slightly wounded in the right leg as well. But the pain was not too bad then; my surprise and fear were greater. I realized I was lying wounded in enemy territory. I searched for Chetz. I saw the plane in flames quite close and assumed Chetz must have landed between me and the aircraft, but where was he? Had he remained inside?

I didn't see any Egyptians, and I knew it was important to transmit the message that I was alive. I believed that if I'd manage to do this, the Egyptians wouldn't dare kill me and claim that I had died in the explosion. I had realized before jumping that our landing spot was a missile site, so that there was no chance that the Air Force would come for me. I succeeded in pulling the two-way radio out of the survival kit. I even started it and got a connection, but I couldn't recall the call sign, which was changed daily. So I told myself, What's the matter with you? Use your own name. And I did. Immediately I heard Ehud's voice; he was another Phantom pilot in our squadron. I had a long conversation with him, but I don't remember what about. All I recall having said was that I didn't see Chetz anywhere, and that I was badly wounded. At the end of this conversation, I informed Ehud that I was surrounded by soldiers and I said, "See you soon and give my regards to my family."

The soldiers arrived in a truck, and it was good to see an officer among them. They circled me and looked for my personal rifle. I had never carried a gun on my flights, so they couldn't find anything. I did have a commando knife, which we used for cutting the parachute strings if they tangled. It had a folding blade. The officer opened it, but couldn't make the blade go back in. This was a funny moment: I'm lying on my back with everybody pointing their rifles at me, and I gesture to the officer: Hand me back the knife and I'll show you how it works. I remember laughing inside, probably feeling superior.

I told you earlier about my sense of relief. All the tension that had accumulated in the previous weeks had sort of been released. So that's the end, the catastrophe I had imagined had materialized, and I had survived. Maybe also: I won't have to go on such missions anymore. It would have obviously been more natural to feel that way had I parachuted over Israel, where I would have been taken to a nice hospital with all the pampering and attention waiting for me there. But just the same, that's how I felt in Egypt.

The officer started to ask me questions. I speak Arabic, because my family emigrated from Iraq, but I decided to pretend I don't, to gain an advantage over them. The officer ordered the soldiers to put me on the truck. I was afraid they'd hurt me, or maybe it was just a show, but anyway I started screaming that I was in great pain. This must have impressed them. They lifted me gently, spread the parachute on the floor of the truck, and put me on top of it. One of the soldiers straightened the fractured leg and put his gun along it for support. I also remember that one of the soldiers sat close and put my head on his knees. It was wise to make them worry; they took good care of me.

2

INTERROGATIONS

Each of the men went through a period of interrogation that lasted from a few weeks to three months. The Egyptians used various techniques, including torture, threats, isolation, and deprivation of basic needs. The prisoners developed individual strategies to avoid the pressure and minimize their pain, trying to maintain standards of behavior that they considered honorable.

All the men agreed that this was the hardest time in captivity, a period they had tried to forget and about which they rarely talked. Revealing these experiences to me was simultaneously difficult and cathartic. Periods of silence were frequent during this talk, especially when thoughts of betrayal or humiliating episodes were shared. It was painful for me to listen to these details, yet the men realized how interested I was in sharing these hidden memories. When the emotional reactions became very intense, we would take a break and later resume our conversation.

DAN

Before the morphine started to have its effect, I had an insight, and I knew what my line of defense would be from then on. It might have been a result of my life experience, or the outcome of having read all those spy novels, but whatever its origin, I developed a cover story and I stuck to it for the next four years. With all due modesty, I think this was a brilliant line of defense. I decided to say that this was my first day in the Sinai, that I was a reservist on my way to the stronghold to meet my commander for my orders. Therefore, I was completely ignorant of the military situation along the canal. Only somebody who has experienced six months of intense interrogations, and then three and a half years of imprisonment, where the door can be opened any second for further interrogations, can appreciate the difficulty of holding on to this cover story, never budging from it, not breaking down. That's one of the reasons why I emerged from my captivity feeling pretty good about myself, after all [laughing].

During interrogations, one may reach the point of making the decision to talk, because it's difficult to hold on to the silence. You may allow yourself to tell about A, but once you've told them about A, you'd probably continue with B, and so forth. One cannot lie all the time, but one can lie partially. I stuck to my defense line. They didn't get anything out of me, though I paid dearly for my silence.

On one of the hardest days, my interrogator suddenly drew out of his files the Israeli newspaper *Yediot Achronot* and said, "You keep lying to us. Here it says that you had volunteered to serve on the canal front. How dare you say that you don't know anything?" What a miserable journalist must have put this item in the paper; how irresponsible could a paper be?! As a result of this I suffered a great deal.

For a while I was afraid that they might catch another prisoner and compare our stories. In that case I would be even worse off. This is what had happened with pilots who couldn't hide a thing from their interrogators, because several of them had been captured together.

The Egyptians didn't believe me, naturally, while I kept saying, "I don't know; I have never been there." They could not believe that a big guy like me could be only a lieutenant. They knew that my father had been a big shot in the IDF, and for them, family is highly influential. My family name caused lots of problems for me throughout my captivity.

I was taken out of the bunker during the night, and this was the start of a four-year blindfolded journey. I was put on a stretcher and carried to a truck or an ambulance, which drove off. On the way I heard the sound of shovels digging the ground; they must have been digging ditches, I suppose. To this day I can hear the sound of shovels hitting the stony ground. Several times we stopped so I could be shown around to the soldiers, who would cheer. I was a good catch. I was afraid I'd be lynched. Blindfolding frightened me for the rest of my four years in Egypt.

At one bunker, on our way, the rag covering my eyes was suddenly taken off, and I saw many faces around my stretcher, including blue-eyed Russians. I saw many men of rank all around. Before I realized what was happening, I was photographed with my head down; I felt deeply ashamed of my condition. This was a picture that made my family back home quite miserable. Today I realize that soldiers should be trained to look their best for the enemy cameras, so that their families can recognize them. At that time I had a big beard that I let them shave off at the hospital for reasons of hygiene. I have never let it grow back.

After the picture was taken, we drove to Al Mahdi hospital, which is the largest and most modern in Cairo. I was taken up to the famous floor of Israeli POWs, where they have a high security wing. I spent six to eight months there. At intervals, I was sent to a solitary cell in jail, because I "misbehaved." Had I "behaved," I might have spent the whole period in the hospital, like other wounded POWs.

My interrogation was difficult, probably because my father had been one of the important commanders of the War of Independence in 1948. In addition, I think that I was punished for the Phantom bombardment deep inside Egypt. . . . Sometimes they missed their targets and hit a school or a factory. The Egyptians started trembling at the sound of an airplane, and all this anger was taken out on me.

At the same time, I went through medical treatment. I was operated on during my first night. I had to have surgery on both my legs, which had been shattered, a lengthy procedure. For about a year, both my legs were in a cast up to my groin. I was lying down all the time. Psychologically, this was a peculiar condition: all the men who stood or sat around me seemed to me bigger than I was.

When I woke up from my first operation, I didn't see a nurse in white but a team of interrogators sitting at my bed. I was still drugged, yet already felt excruciating pain, while they were ready with their questions: "Who are you? From which battalion? What were you doing in the Sinai?" I gave them my story, but they refused to accept it. Disbelieving me, they put pressure on me to produce another story. It was tough. I used to tell myself that I had fallen into a deep, dark pit, but every passing day, every interrogation, was like going one step up, out of the pit. And so it went on and on. After each of my operations, I would find all these interrogators, like birds of prey surrounding a dying animal, around my bed. They never let me be at peace.

They continued with their pressure. They wouldn't let me sleep, deprived me of food for a number of days, and beat me on the head. To this very day my ears buzz from their treatment. They used to kick me and shake the bed, causing agonizing pain in my broken legs. Once they entered the room with an axe and threatened to amputate my leg. . . . I didn't believe they would be that cruel. Their questions were repeated again and again, because they couldn't report to their commanders with the kind of evasive answers I had been giving them. Usually I was interrogated by two men, with a Palestinian interpreter. The interrogations scared me to death, yet I knew I shouldn't let them realize this. Every time the door opened, I'd feel this awful fear. What was I afraid of? Naturally, I was scared of the beating and torture, but mainly I was afraid I might break down and tell everything I knew. The moving of chairs toward my bed was enough to frighten me badly.

I suppose that the interpreter had an additional role in the team. One day, after the interrogators had left the room, he told me, "Listen, you must talk, or it will be much worse for you; you will be taken away." He said that he had come from Jaffa and knew the Egyptians very well. I answered naively that he didn't understand the immense hatred the Egyptians felt toward the Palestinians, and that only with us, the Israelis, did they stand a chance of living in peace. He went excitedly out of the room and returned with the interrogators who screamed, "Snake! How dare you speak like that to your interrogators!" I had just added oil to the fire.

I found out that they knew a great deal about the IDF. More than I or any average officer did. They expected me to be acquainted with details about this camp or that camp, to identify various signs and symbols, information I really didn't have. I tried to present myself in the lowest manner, as a farmer, a fellahin, nothing more. I was willing to tell them about the kibbutz and its social structure. When I broke down, so to speak, I was willing to divulge the military defense system of my kibbutz. I was constantly trying to move the focus of the interrogation from the canal and its strongholds to my kibbutz in the north. At the end I found out that they had a map with all the kibbutz shelters and posts marked on it!

Sticking to my line of defense throughout the period was much harder than inventing it in the first place. In addition, I couldn't judge which were the details I could tell under the circumstances, things that the Egyptians knew anyway; because the Egyptians weren't blind or stupid, on the one hand, and information becomes outdated quickly, on the other. When you're a POW, all by yourself, you have no support, and you have only your own inner strength to rely on. Most people emerge from interrogation with a deep sense of guilt, because they suspect that they have given away state secrets. This guilt is so strong that we have never compared notes on our experience during interrogations. In my opinion, a POW has no way of evaluating the importance of his information vis-à-vis the price he's required to pay for withholding it.

I passed several stages of interrogation: before the solitary cell and after that terrible time, always after my operations, when I was still under the influence of the anesthetic. Finally, the interrogations gradually ceased. They came to the conclusion that I wouldn't talk. What was their choice? Kill me? I had been there for a long time already, and the Red Cross knew about me. In the meantime, they had new prisoners to deal with.

AVI

After an hour or two in the local headquarters, I was blindfolded and taken out to the car. The angry crowd outside greeted me with screams, hitting the car with sticks, whatever they could find. I assumed we were heading for Cairo. On the way, I heard my name in the news, in Arabic naturally, and I realized that they had been announcing my capture. This served me very well later on, because it was clear that people knew I was alive, in the hands of the Egyptians, in spite of the fact that the interrogators often tried to create a completely different impression.

We stopped on our way at a certain MIG-21 squadron. All the pilots gathered around me and asked me questions in English. I answered freely, even proudly. During the dogfight we were two Mirages against six of them, and we hit three.

It was nighttime when we arrived in Cairo, at some interrogation center that belonged to the Egyptian Intelligence. When the rag covering my eyes was taken off, I found myself in a tiny cabin that was totally bare except for a concrete slab, supposed to serve as a bed, and a strong light bulb above. My hands were handcuffed behind my back, and I was left alone. After a while a sergeant came in with a pita bread. I didn't want to eat, but he said: "Eat, eat, you'll need it later" [laughing]. From another cell, I heard the screaming of people being beaten up. I was sure this was not a show, but I couldn't imagine what was going on. A little later, I needed to use the toilet and was taken to a little cell, its floor covered with feces half a meter high. "That's it," my guard said. I was barefoot. This made me aware of my situation: What I had eaten at a gourmet restaurant in Tel Aviv was coming out in this shithouse in Egypt. Somehow I accepted it as a fact, and with no other choice, I started to adjust to my current situation.

After a short while the interrogator came, carrying a tiny desk and a chair that fit exactly the size of my cell. He sat down, ready to write whatever I'd tell him. I myself was seated in the corner, my hands still tied behind my back. I guess that this first interrogation lasted for about twenty-four hours. The interrogator went out for a rest a couple of times, then he returned.

In the beginning I spoke Hebrew, and an interpreter translated. This procedure made me nervous, however, so we decided to converse in English. I then had an advantage: since I had a limited knowledge of the language, being fluent mainly in the technical English of flight manuals, I could pretend to misunderstand and ask them to repeat their questions whenever I felt I needed extra time.

In the beginning I decided to divulge only basic facts, but right away I discovered that they weren't playing games with me. They threatened: Either you talk, or it will grow much worse for you.

Indeed it grew worse. I was hung up with my hands behind my back, on a hook on the wall. It was unpleasant. My hands fell asleep all the time; I think that several nerves were damaged in the process.

Two men interrogated me. One was Aziz, who was in charge, and the second, Shamel, from Intelligence, who alternated between being the good guy and the bad guy. They wanted to find out about the Air Force bombardment. It was rather easy not to answer them, because I really didn't have the information they wanted. Sometimes they asked me about airports, and when I provided false information the interrogator said, "You're wrong here," and showed me a photograph, complete with all the details. "So what do you want from me?" I'd ask. And he replied, "I'd like to know from where fuel comes into the field." I didn't have all these details, so sometimes I'd invent them.

I didn't think I was providing facts they didn't have anyway. A year later, we were taken again into some interrogations from our common room. I met my interrogator again then, and he said, "You rascal, you didn't tell me anything. Do you know what I got out of all these pilots who were captured after you?" This was a tremendous boost to my ego, although I understood that I had been in a better state since I had been alone, and they didn't have other pilots to validate my answers with, as they did with the Phantom pilots. They kept asking me where the Air Force squadrons were located, how they were organized, but I answered none of these questions. When they got sick and tired of me, they called in their "beater," who hung me up on the wall and flogged me on the soles of my feet. I'd tell myself, "Bravo, you've done it again," and fall into the sweet slumber of unconsciousness.

Before I fainted, I yelled with pain, naturally. Yet when they reached the point that they had to beat me, I felt as if he, the interrogator, was failing, while I gained some points. I don't know if it was due to my luck or my stamina that I never reached the point in which one is willing to divulge even what one considers top secret. For me, some subjects were more important than life, and I am happy to say I could maintain my silence about them.

Later on some new interrogators started to show up. They were specialists in certain subjects. One was an expert on firearms and ammunition, for example. The total time of the intensive interrogation was perhaps two or three weeks. Throughout this period my hands were tied behind my back. I even got used to sleeping like this. Most of the interrogations took place in my cell, but sometimes, when a more important interrogator got the job, they'd take me out to a fancy office in the building. They couldn't have an important officer pass by that shithouse!

My survival and sanity were all I was concerned with at the time. The moment I could think a little, I realized I wasn't cut out to be a hero. I wasn't particularly pleased with myself, because I knew that due to my error, my plane had caught fire and I had lost it, as well as four years of my life.

BENNY

When we arrived at their stronghold, the Egyptians went into hysterics. Motti was dressed in civilian clothes, while I had a military uniform and paratroopers' shoes on. That's why I was beaten by the soldiers more than he was. They removed the money I had in my pocket, too. Later on, a major arrived and conducted a preliminary interrogation. I suppose that somebody had told him about the money, because the soldiers who had taken it were promptly brought in. He slapped the thief on his face, right there in front of me, then took the money and put it in his own pocket. . . . I saw two Russians in the corner of the room. I remember telling Motti, "Look, these are Russians, with their pale skin and blue eyes." An interpreter was brought in, and they kept trying to identify us. I explained that we were civilian canteen workers, but obviously they didn't believe me.

Later on they blindfolded us, took us to a jeep, and drove us to Cairo, about two hours from where we were. We were taken to a cell in the Intelligence Center, where a doctor examined me. My clothes and shoes were removed, and I received overalls to put on my naked body. After a while they took me to be interrogated in a cabin. I kept insisting that I was a civilian working for the army, but it didn't help. They put me facing the wall and started to beat my back with a rubber club. I think they beat me for about two hours. When the doctor saw my back, he was amazed at the sight. He took care of me and then I was thrown back to the solitary cell, where my legs were chained and my hands tied behind my back. I was left there without food or water, but I fell asleep just the same.

The next morning I was awakened by a shoe kicking my head, and I was again taken for interrogation. I repeated my story and they repeated their beating. What I didn't understand then was that only an intelligence man would be captured like this: in military uniform, yet without identification tag. The canteen job would have been an excellent cover story indeed. The worse thing was that Motti said he hadn't known me before, that he had met me just the previous day. When they asked me the name of the director general of the military canteens, I couldn't remember it. They believed Motti, who was caught in civilian clothes and even had a salary slip in his pocket, and decided that he had been the truck driver. As for me, they decided I was from Intelligence, on a hike with Motti.

They kept trying to make me admit their version. I, on the other hand, decided not to say anything about my former military service. I told them that I came from a poor family and was exempt from service. I clung to the story that I knew absolutely nothing about the IDF. I pretended to be a dumbhead, easygoing and simpleminded. Later on I regretted it, because once you start playing a role, you must go on with it, even when you're asleep.

The presence of an interpreter allowed me to think for a while before answering the questions. I could tell them about things I saw as a civilian, a military base here or there. I told them about the canal posts but only what I was sure they had known already.

They tortured me severely. They sent wild dogs to bite me, they used to hang me up by my hands, and they beat me cruelly again and again. The interrogation lasted for a month. They put me in a special torture cell. Every night they sent thousands of bedbugs into the cell. I don't know how they did it, but every evening I'd see all those bugs crawling in from under the door; it was amazing. In the morning they'd slowly walk away. I used to crawl on my knees from corner to corner all night long. The moment I stopped they'd be all over me, ready to kill. During the night I'd fall asleep finally somewhere, exhausted, and wake up in the morning swollen from their stings. They gave me shots for it, which cured me for the day, until the next night and so on and on.

At a time like this you have to be as strong as you can. You mustn't become indifferent, because it's too risky. You might go out of your mind.

The interrogations usually took place in the evening, and during the day I was confined to my cell. The wardens treated me badly. I was like in a no-man's-land; everybody could come in and beat me up. As time went on, I got used to the fact that I was a POW. Gradually I was willing to tell them more about the strongholds, because I was convinced that the others had already talked. After six weeks, they probably realized there was nothing more I knew, and they transferred me elsewhere, to a solitary cell in jail.

MOTTI

At the first interrogations, still in the bunker, I tried to explain that I was a civilian and what a military canteen is, but nobody seemed to understand. No one spoke Hebrew or English; I can't speak Arabic, and they had no translator there. I felt superior to them, and believed that any moment an Israeli chopper would come and save me. Later on I discovered they weren't as stupid as I thought. When I asked for a bandage, they pretended not to understand.

I was taken out to the toilet, where somebody tried to bandage me over my sweater, which was all soaked with blood. I had to explain that first he should cut the sleeve off, and showed him how to bandage the wound. A major suddenly came in, slapped me on the face with all his might, tore the bandage off, and shoved me into the toilet. He cursed me saying, "Jewish son of a bitch," or something worse. Then I was blindfolded and thrown like a potato sack on the floor of a command car. The truck was full of soldiers, who put their feet on me. I felt I was dehydrating and asked for water, but they ignored me. I reflected on my condition: how did I become such a miserable thing! At the same time I was saying that they were petty people and wouldn't harm me.

Finally some good guy wet my lips with water from his flask. I hit him, and all the water spilled over me. I was revived. I remember passing about five or six roadblocks until we arrived at the military base, and there I was put in confinement with my eyes blindfolded.

The cell was dark, one meter high, one meter long, and about sixty centimeters in width. You could only sit on the floor in it or crouch on your knees. They removed my clothes and looked for documents on my body. I had managed to throw everything away. Then they took off the rag from my eyes, and threw me some overalls to put on. I remember seeing myself naked covered with a lot of dried blood. I tore off one of the overall sleeves and made a bandage out of it. Ten minutes later a huge dog was brought to the cell, and he started to sniff my clothes. I thought he was brought in to frighten me a little, but later I realized they just wanted the dog to recognize my smell, in case I were to run away. Later on I was taken to another place, a hut of sorts, where I was seated on a chair, blindfolded, and asked all these questions. They were simple questions: where I was from, my ID number, etc. I interrupted them and asked for an antitetanus shot. They were surprised, and asked me what for. I explained that I had been wounded all over. They said, "We'll see later."

Again they asked me where I served, and again I explained I was a civilian. I tried talking in English, but it was too complicated. I asked for a Hebrew interpreter, hoping to gain some time by the translation. The interrogator asked, "Why don't you speak English?" And without any preparation, he slapped me so hard that I flew off the chair. I think his slap had this effect because I couldn't see it coming. Later on I discovered that if I didn't resist the blow, but flew with the force of it, like a dancer, it was easier.

The first interrogation lasted for about three hours, and went on without an interpreter. Then I was put back into confinement. Every day I'd be interrogated for about eight or ten hours, sometimes with my eyes shut, sometimes with an interpreter, sometimes without. They treated me as if I were a security officer. They didn't believe me that I had worked in the canteen, and said that this was my cover story. I remember I insisted and argued that certainly the newspapers in Israel had reported two canteen workers missing. They said that the papers had nothing on the subject, and that I wouldn't come out alive from these interrogations. "If you're going to kill me anyway, why should I answer your questions?" I asked. And he replied, "Because you have no choice. I'll beat you up until you tell me everything."

The interrogations continued for two months. All this time, I was chained by my legs and hands, in confinement, with daily torture and beatings. Whenever I said, "I don't know," two men entered the room and started to beat me up. They sat outside all the time, two big black guys with extremely long arms. I had never seen such people in my entire life. As time went on, I discovered that the interrogator had a button on the desk, and when he wanted these men to enter, he pushed the button and a red light went on outside the door. Sometimes I was ordered to stand in the corner of the room without moving. One of their popular tortures was to put me on a pipe with my feet and hands cuffed, lift the pipe up, and flog me on my soles. My feet became so swollen I couldn't stand up.

The worst part in confinement was when I needed to go to the bathroom. They gave me murderous blows on my way. They wanted me to restrain myself from going. This was another way of tormenting me. They didn't give me any water, and I used to drink some of the water they spilled on the floor when they washed the cell. I got used to sleeping all bundled up on the concrete floor.

They always had a good guy and a bad guy: the interrogator was the bad one, while another man came and inquired: Do you need anything? What's going on? The name of the good guy was Shamel. I implored him to take me to the hospital, because the wound on my neck was completely infected. He promised to take care of it, and of course nothing happened. The interpreter also pretended to be good when he was left with me in the room. He used to promise me that if I would divulge all I knew, he would see that I was well treated. Once I asked him for his pencil. He cut it in two and gave me half of it. I managed to sharpen it and mark the days on the wall of my cell.

They asked me about the different posts, about plans for the war, emblems of the units. They brought some air photographs for me to explain—they were things I have never seen before in my entire life. The interrogator would be sitting with a huge pile of papers, writing down everything I said. He said that when the papers were all full, the interrogation would be over.

I hadn't seen a doctor for three months. The wound got all infected and when I finally arrived at the hospital, they couldn't believe their eyes. They needed to cut all the dried pus away until they drew blood.

After two years in jail they suddenly decided to interrogate us once more. They took us blindfolded into a room, asked all kinds of questions I can't remember, hit us with a broom—I think it went on for about twelve hours. After living in relatively comfortable conditions, going back to that was terribly difficult. It reminded me of all that I had tried to forget. Afterwards they wanted to compensate us for it and took us on a tour of Cairo.

DAVID

At nightfall I was blindfolded, put in a car, and driven to a place where I saw some Russians. The investigators lit a bright lamp, shone it into my eyes, and started to interrogate me. In my pocket I had some lottery forms, and they thought they were some codes of great importance. So for three hours they tried to break me down and decipher the code, until they realized what it was.

The interrogation was conducted in Hebrew, because I said I couldn't speak any other language. At some stage they asked me if I needed anything, and I asked for a pill for a headache. A soldier brought five pills and put them in front of me with a cup of tea. Without a second thought, I swallowed the whole lot, because I believed they might help me during the next couple of hours. A medical officer entered the room suddenly, asking for the pills. The soldier motioned to the desk, but they were all gone. I pretended not to understand. They brought an interpreter in, and I explained that I had taken the pills. The officer started to scream that those were sleeping pills, and he smacked the soldier.

The pills didn't have any effect on me as long as I was active. Late at night, I was transferred to Cairo. When we arrived, I was put in confinement. My uniform and shoes were taken away, and right away I was brought before my interrogators. The first questions were fairly easy, and I had an interpreter, which allowed me extra time for thinking. Suddenly they asked me something about the tanks, and when I said I had no idea, they beat me. I fell off the chair and went to sleep right there for twenty-four hours. To this day, I don't understand what happened. I remember that they were trying to wake me up with water and slaps on my face, but I didn't react. I fell into their hands on Saturday, and woke up on Monday night.

On Tuesday, when I was interrogated once more, I was well

rested. From then on, I was constantly interrogated, sometimes in the daytime and sometimes at night. Some of the interrogations were long and others were short. Sometimes I would be given "homework" to do in my cell. They were interested in the Armored Corps, or the military mail addresses of hospitals, for example. Of course I wasn't able to supply those details, so they beat me. I was unable to answer most of their questions.

The blows were of all kinds, but I stopped paying attention to them. When you are hit morning, noon, and evening, you get used to it. They would enter my cell and beat me. I'd say to myself: Very well, beat me up so that I can go back to sleep. I think that whoever passed my door and saw the paratrooper uniform lying there felt obliged to come in and beat me up. Just for fun, without any connection to the interrogations. The body can adjust to a great deal of pain, apparently. One cannot guess, in one's wildest imagination, how much one can take and still continue to function.

In the beginning, they used to promise me that if I'd be good, I could get my freedom and go anywhere I choose, except for Israel. Or they'd offer me a cigarette. After three days of this, I said I didn't smoke, and I really stopped smoking. Even when I was given a cigarette, I didn't smoke it.

RAMI

I was transferred to Cairo with my eyes closed, but I sensed we were traveling for three hours. During the drive I was telling myself: That's it, know that you're a prisoner of war and act accordingly. In other words, you're not a fighting hero; they can do anything they want with you. You're not an enemy, you're not their equal. You have to survive, and there's no law in this jungle. You'll face the problem of information divulgence, and you'll have to decide what to tell, and how fast.

I was thinking about my family, too. I told myself that I'd probably miss Nurit's delivery, which was expected six weeks from then. But I told myself that in two months I'd be back home.

The first interrogation was meaningless. I didn't say anything, as I had already decided while parachuting and on the drive to Cairo, and they beat me a great deal. For the next four or five days

I wasn't able to function, so they resumed their investigation only a week later. In the meantime, some more pilots were captured and they had a lot of business on their hands. When they returned to me, I was over the initial shock and had already acquired the mental state of a POW.

During the interrogations I had to struggle both with the torture and beatings and with the problem of information divulgence. It was like a regressive system of sorts: Every time you reveal some secrets, and you have to remember what you had said before. Between interrogations, you're considering what things you may tell them, what not to tell, and how to tell them. No POW knows how to behave beforehand. We know, of course, the Geneva convention, which provides silly instructions, namely to tell your name, rank, ID number, and blood type. And that's all. Nobody can behave accordingly, especially not a pilot. I never meant to be a hero who doesn't tell a thing.

I adopted another rule of thumb: You shouldn't make your interrogator angry; it's simply not worth it. When you make him angry, you endanger your life. I was certain that they needed me, because I was an important source of information. Therefore, as long as I'm in control and I don't make him mad, my life is guaranteed. In order not to upset my interrogator too much, I realized that I must supply information all the time.

During my first interrogation, I told them about the Ouragan, Myster and Supermyster—old Air Force aircraft—as well as about the previous air base I had served on. When they asked me about the Phantom, or about my present air base, I replied that I wasn't allowed to answer. I noticed they were smiling under their mustaches, and I knew that the game we were playing had already been determined. Because I was "bad," they called in their hitmen, and they finished me off for a week. After I had fainted, they dragged me to my cell and threw me down there, with my hands and legs chained and my eyes covered. I was unconscious, and I was lying on the bunk in this two-by-three meter cell. When I recovered, they took the bunk away and left only a rug of some sort. A pita with some water was brought in once a day. When I knocked on the door, they unlocked it and took me to the toilet. I remember I was semiconscious most of the time. I hallucinated that the ceiling was moving, or that Los was with me in the cabin.

I told Los, "Don't pay attention, they move the ceiling just to confuse us."

Most of the time, they used a rubber club to beat me. Sometimes they also used wooden clubs or beat me with their hands. These were "dry" blows all over my body, but especially on my back. My back was black from the blows. Apparently they knew their job, because no marks remained on my body later on. Whenever one part of my body was in extreme pain, they moved on to another part. Once when they kept hitting me on the tail bone, I thought I was going to come apart. I felt every blow from head to toe, and I said to myself: That's it, I'm going to ask them to bring Captain Aziz in, and I'll tell him whatever he wants. Just then they stopped. This was the time I was closest to my breaking point, but luckily they weren't aware of it. All in all, they treated my body with a certain respect. They never beat me in the nude, but always left my underwear on. They never beat me in the genitals or the stomach. They never screamed or cursed while they hit me. They behaved like soldiers doing their job. After all, the Egyptians are good people.

The worst torture was when they twisted my hands behind my back and tied me to a high window. I had to stand like that. After some time, my shoulders would become sort of paralyzed and I was in agony. Once I thought they had left me standing like this for twenty-four hours, but it turned out to be only six. I lost my sense of judgment or control. I thought about the possibility of letting my body drop down, but I was afraid my arms might be broken or dislocated, which would make me suffer even more. Finally I did let myself drop and felt a certain relief. When they saw this, they took me off the window and let me lie down.

They used to give me sixty or seventy floggings, which they called "falakas," until I'd faint. When I woke up on the floor, drenched in water, they had to drag me to my cell because I couldn't walk.

I suffered many blows and fewer tortures. During one of the interrogations, they brought a big kitchen knife and a chain and said that since I wouldn't talk, they'd amputate my leg and report that I had been injured during the jump. I thought it was funny, but I tried not to laugh. I kept staring at the knife and pretended to be frightened. Another time they brought dogs into the cell.

This was supposed to scare me, so although one of the dogs licked my hand, I pretended to be afraid.

There were good and bad interrogations. During the bad ones, the hitmen were called in and beat me up, and I had no control over the situation. I had a dilemma: I didn't know whether to scream or keep quiet. The natural tendency is to yell, because this way you release some of the tension, but this isn't very nice or very adult. I decided not to scream because I was ashamed to, and also because I was afraid that the screaming might turn into hysteria, and then I'd be completely out of control. As you probably know, I'm afraid of losing control. On the other hand, I was worried that the hitmen weren't getting any feedback from me, so they didn't realize I was in pain, and therefore they might go on and on. In the end, I figured I was right in restraining myself, because the Egyptian have great respect for strength. They think highly of a person who can take a beating like a hero. In the same vein, they respected me later on for exercising in my solitary cell. I used to jog in the cell for two hours a day, doing figure eights 630 times, which amounted to six kilometers. The wardens would stand there and admire me, but that was much later.

Being beaten has two components: the pain and the humiliation. How come I'm being slapped in the face or on the ears? It's terribly insulting. Very soon I convinced myself that the blows were part of the rules of the game, and my part was to receive them. The moment I overcame the sense of indignity, the blows became much easier to take.

The major lesson I learned from that stage of beating and interrogations was as follows: after reaching that point when I said to myself that I couldn't take it anymore, yet that moment had passed, and I did take it—a new leaf was turned. You see, part of the game was that they didn't know what my breaking threshold was, for I was not supposed to show them how much I hurt. Whenever I behaved like this, I felt I was easily winning the game. Later on I became totally immune to the beating.

A person under interrogation develops a unique perspective on his life, according to his momentary condition. For a while I was "bad," so I was transferred to a smaller solitary cell, a room that was two-by-two meters with a concrete slab serving as a bed. It had no window, but it did have a tin roof that was hot as hell that summer. I was ordered to stand erect, facing the wall, all day long, and whoever passed by would hit me. My hands were chained behind my back, and they became all swollen and hurt a great deal. I lost all sensation in two fingers, probably due to some nerve damage. But one of the wardens was nice, and at night would remove the handcuffs and clean my wounds. All that week they didn't give me any food, so he had pity on me and late one night brought me some baked goods with halvah from the marketplace. This food, however, made me terribly thirsty, and then it gave me diarrhea. I felt I was dehydrating. I remember standing there and feeling how my tongue was swelling in my mouth. I was hallucinating: I saw a pile of soft drink cans in front of me. I knew that they were all empty except for two, but I had only one chance of making a choice. . . . I was dressed in extra big pants, which I had received from the Egyptians, and in that condition, while I was defecating in my pants, with my ankles black from infection, I managed to pull one of the pants legs aside, so that I dirtied just one of my legs, while the other remained clean. It's hard to describe the sense of mastery I felt from this achievement, the feeling that I can take good care of myself!

Since I couldn't stand all night long, I would sit down on the concrete slab, standing up the moment I heard the door open. This was a game, in which I was winning every time I wasn't caught sitting.

In the morning they cleaned the floor. How? By pouring two buckets of water on the floor and sweeping it away. Since my cell was lower than the corridor, water flowed into the room, and I'd kneel down and drink it. When I started to taste dirt, I knew I had enough. I didn't feel it was terrible at the time, because this was my reality then and there: not to be caught sitting down and to find water when thirst was my punishment. Such deeds made me feel in control of my situation.

Sometimes they would punish me with starvation. This was some time later, when I was already in a solitary room at Abassiya jail and I was "bad." Every time I was taken to the toilet, I'd find something to eat in the garbage pail. I think these were the dogs' leftovers. Finally, after a long time of starvation, they brought me food—some beans on a plate. According to my calendar, this was on Yom Kippur, so I told the guard I wouldn't eat today, since it was a holiday and I was fasting. You can't imagine their surprise. Needless to say, I never before used to fast on the holiday, but the food was so meager that I preferred to impress them, to show I was in control.

I am familiar with three different attitudes for coping with interrogation. The first is: I'm a brave Israeli pilot, a hero, a defender of my fatherland, and even when captured I will not say a word. I could perhaps behave like this, up to a breaking point when I'd start talking. This happens to every POW, and when he talks, he feels like a traitor to his country. Furthermore, he feels frustrated, because he couldn't keep to his principles. Or even worse—he breaks down and has absolutely nothing to hold on to.

The second attitude is: I'm a clever Jew; they are dumb Arabs; I can trick them. I can mislead them, make up "stories," and get out of the mess. However, when one invents stories, one finds oneself divulging details that should remain secret. And this stupid Arab, he may trick me somehow. (Captain Aziz was a superior investigator; he tricked me several times!) And again, when you realize you've been had, you may break down and spill everything.

I developed a third view, according to which we are playing football. The interrogators are on the attack, and I can only defend. So we start with a given situation, when he runs with the ball, and I can either stop him or withdraw. If I manage to withdraw very slowly, I gain some points. When I withdraw too fast, I lose some points. That's the game, and actually the sides are equal. You win some and lose some. When my interrogation was over, I could relax in my cell and consider my conduct today: Did I gain or lose? Was I all right or not so good? I had a way to fight back, and perhaps tomorrow I'd gain more points. There were moments when I really experienced a partial victory.

Captain Aziz conducted the interrogations of all the pilots, and he was under a terrible work load. I remember at two or three o'clock in the morning, after four days of interrogation, he would wait with the interpreter for the car. The interpreter would ask, "When shall we start tomorrow?" And Captain Aziz would reply, "I'll come and get you at seven o'clock." So I was saying to myself—you have a lot of work, while I can sleep leisurely until eleven o'clock. They used to give me something to read or write in preparation for the next interrogation, and except for this I'd be free until late afternoon. And they, poor things, worked all day long! With great joy I overheard Captain Aziz send the interpreter to the canteen to get him some aspirin for his headache. It gave me the sense that I was winning. Sometimes I even felt that in this game I had a better chance than he did. I am in the powerful position, while he, the interrogator, is in the weaker one, since I am the one who dictates the pace of the process. It's true that I'm beaten and tortured, but it's a struggle. Sometimes he offers me tea, and sometimes he throws the ashtray at me, but just the same, the tempo of this process is up to me. For three whole days I explained the workings of one technical detail in the Phantom. In his eyes I saw the gleam of satisfaction when he finally got it. It was most important that he didn't get frustrated. This was as long as he had something to write on these huge white pages, and as long as he felt he was making progress.

Most of the time I was feeding him "stories," obviously. But true facts popped up between the lies. When he caught me lying, he'd throw the ashtray at my face and call the hitmen. The interpreter would whisper, "Listen, it's not worth your while." So I'd withdraw some more and admit, "Really, I thought I could lead you on. Now I see you know a lot, and from now on I promise to tell you only the truth." The captain would throw the previous page in the wastebasket and take out a clean one. That was the game. Later on I used to think that the interrogator, too, could be insulted sometimes. Suppose I'd tell him the whole truth right away; how would he have the satisfaction of getting it out of me with his great efforts and skills? Once I felt I was in charge of the interrogation, I could cope with the situation much better than the others.

In fact, all I know about the behavior of the others are hypotheses. When we were finally joined in the common room, we never discussed the interrogation. It was better that way, because I don't think we could have managed this topic.

Three years later, during the Yom Kippur War, some new Israeli POWs arrived in prison. At night, when we were lying in our beds, we could hear the men screaming. They were being beaten and interrogated. The men said, "We must do something about this!" while I said, "So what if we're here? What should we tell the Egyptians—that we can't stand the screaming? We've gone through this too; now it's their turn. That's it, in a short time we'll be going home."

YITZHAK

When I was on the doctor's table at the military clinic, I told myself: From now on, it won't be easy. The easiest thing to do was to faint or to get some pain relievers. So I started screaming that I had burns all over, that I was in great pain. I screamed and yelled, until the medic gave me morphine, which dulled my consciousness for the next two days. I think this was a great help.

I remember waking up naked at the hospital. The doctors were standing around me; they looked rather terrified by the burns and injuries caused by the blows, and the broken ribs. Right after that, some intelligence men came and started their investigation, with the usual questions one asks a pilot. I was still under the influence of the drugs, and I pretended to be unconscious, but I was alert enough to decide that I'd tell them that the plane I abandoned was a Skyhawk, so that they wouldn't search for the second pilot. This was the interrogation where I was severely tortured and my toes were broken, but the morphine still helped. I was aware of what they were doing to me, but I hardly felt the pain. When they saw that I wasn't reacting, they left me alone, and the doctors returned to treat me. Since I had said that I was a Skyhawk pilot, they informed Israel that I was dead. But a week later they brought in a piece of the tail of my plane and while hitting me, they said: "Is this a Skyhawk, or what?" They asked me why I didn't tell them I had abandoned a Phantom, and I said that I didn't want them to go looking for the second pilot. So they said that they had caught him anyway and were holding him, and that he had told them everything. In fact, David had been rescued on the first evening and had been safely returned to Israel.

For the first time, the Egyptian doctors put some ointment on my burns, put bandages all over my body, and gave me some Egyptian pajamas. This was my only garment for the next eight months. I never received any medical care after that first time. My internal injuries and broken toes healed themselves.

I woke up in an interrogation cell of the Intelligence. The intensive interrogations went on for three months, and were accompanied by physical and psychological torture. Often I was deprived of food and water; sometimes I was confined to a tiny solitary cell in inhuman conditions. I was telling myself this was like a card game, in which my partner had all the aces, and my part was to survive. The game had several characteristics: I was the owner of some important information they wanted to get, while I had to prevent them from getting it. I made some rules and committed myself to keeping them. But as much as they were really good, excellent rules, iron rules I wasn't ever supposed to break, I knew I would have to change them when I was forced to, maybe in half an hour.

On a conscious level I was telling myself that once I permit myself to divulge things that are okay to tell, it would be an endless process. So I started with the rule that I'd say nothing at all. I insisted that as long as I didn't see the Red Cross people, I wouldn't tell them anything besides my name and number, as declared by the Geneva convention.

It is immensely difficult to make the rules about what to divulge and what to hide. During my interrogation I told them various things, and with every additional detail I felt that I was betraying my country. "I'd rather die than betray Israel," I told my investigators. And they replied, "Very well, so you will die."

At an early stage I told them that I had been born in the United States, and I spoke English to them. My interrogators were proud to prove that they were fluent enough in English to conduct the investigation without an interpreter. This had both an advantage and a disadvantage. The disadvantage was that I had no time to prepare my answers, while the advantage was that my English was definitely better than theirs. Often I'd make words up. The interrogator would write down a word on his pad, and a few hours later, or the next day, he'd come and declare, "There's no such word. I looked it up in the dictionary." Naturally I was beaten for it, but I insisted and asked what kind of a dictionary he used, to try to divert the investigation to other matters. When they found out that I had been born an American, their response was, "You are an American, a mercenary: we'll hang you in public in Cairo to show the whole world that American pilots are flying for Israel." They used this threat quite often, and I'm sure that had they believed it would serve their needs, they wouldn't have hesitated to carry it out. This threat, however, didn't break me down, because I wasn't afraid of dying.

Several men investigated me, but one was in charge, and in addition there were men whose job was to beat me up. The interrogator used to watch how I was tortured. Frequently I tried to faint and lose consciousness. Sometimes I thought I'd like to commit suicide—not because I had lost hope, but because I couldn't stand the physical torture. Besides blows all over my body, they used electrical shocks; they pulled out my toenails, like you read about in books. Part of the painful reaction was, of course, the insult of all this. I sensed that my life was worthless in their eyes. I had been brought up in a different culture. After a while, when I realized that this was their mentality, I felt some relief.

The torture worked, naturally, and I started to remove some of the taboos I had made previously. I realized quickly that maintaining my sanity was my major goal, so that I'd still be able to channel the interrogation and say things of least importance. Once you start talking, however, it is terribly difficult to accept the sense of failure. When I was returned to my cell, lying there in the dirt, I'd try to reproduce what I had just divulged, and I would be so angry with myself, because I could always discover some links between what I had said, which was perhaps insignificant, and some other things which I considered important to conceal. Thus, for example, I permitted myself to tell them all I knew about my former kibbutz, including the names of men, where they served in the IDF, etc. What I was trying to hide were the secret details concerning the Phantom. It was very difficult to prepare myself for the next interrogation.

They were pretty clever in using what other pilots had already told them, or what they said they were told. Naturally it is much easier to tell things that your interrogators probably know anyway. For example, they asked me the number of planes in a squadron. I said: "Twelve." A few days later they returned and said, "Liar, S. told us there were thirty!" Possibly S. had said nothing of the kind, but I was in a rough spot anyway. By the way, S. was a pilot who had been severely injured when he abandoned a Phantom, and they returned him to Israel quite soon. His navigator, a very young pilot of about twenty, was interrogated in the cell across from mine. One night I woke up to the sound of terrible screaming in Hebrew. They were beating and torturing him, and I heard him cry out in Hebrew, but definitely not answering any of their questions. I couldn't take it, and I shouted, "Goldwasser, tell them what they want to know or they'll kill you!" But he must have already been beyond hearing. And indeed he was killed. [He is silent for a long time.]

This went on for about three months, day and night, at this intelligence interrogation center, which consisted of several solitary cells and all these torture devices. I lived in the midst of urine and feces, flies, and rats. I was often starved, and the water I had was what they had used for washing the floor. After about six weeks I said to my investigator: "Look, we discuss different matters, but if you keep starving me I'll stop talking to you. I'll simply die." He asked: "What do you want?" And I replied that, being born in the U.S., I was used to drinking milk. Three days later, all of a sudden I'm getting a small bottle of milk every day. Later on I found out that so did the other POWs on our wing, who didn't understand why. It was about the only bright spot in the darkness of that time.

I believe that I was interrogated by about twenty men, and of all those I had met only one who was really smart. Some of the investigators were Russian. I remember seeing one with Russian letters on his watch. Some were Air Force technicians, who were interested in the technical features of the Phantom. The worst part was when I was asked about certain individuals, when I had to admit, "Yeah, so-and-so serves in my squadron." I was sure I was killing him. I was afraid that the Egyptians might send over a unit to murder his wife and kids. This information divulgence tormented me for years to come. I was afraid that I had betrayed the country and provided the enemy with facts that might harm others. That's why I wanted to die. But it isn't easy to die. I didn't manage it.

In bad times, one has the terrible urge to reach the point of telling everything, if only one could be left alone. That situation is worse than death. Luckily I didn't reach it. Had the interrogations continued, though, I might have. For a time I thought that had they kept me in confinement for two weeks more, in the same inhuman conditions, I might have broken down and told everything. But obviously one can never know. I feel that we were all ashamed of our behavior during the interrogations, therefore we never discussed it among ourselves. When you're out of it alive, you start doubting: perhaps I could have been stronger, could have said nothing, and survived it just the same.

They used all kinds of psychological pressures, although not very sophisticated. During one interrogation, they put an envelope on the desk and said that if I'd be "good," I'd get this letter from my wife. Psychological pressure has a way of accumulating, and it leads to mental attrition. Luckily they weren't clever enough in their use of it.

After a while, there was a gradual improvement in my conditions. Perhaps they had already gotten all they wanted from me. They were less violent and started to ask me about politics, strategy, even about Judaism and Zionism, topics that didn't have immediate relevance to the army. I gathered they were trying to compare the Egyptian and Israeli ways of life, trying to understand Israel better from a political, social, or economic point of view. When we gathered in the common room, we found out that we had all gone through this stage at the end of our interrogations, and we called the man who had conducted this part the "sociologist." I told him about the kibbutz and asked, "Why don't you have kibbutzim?" This way I would divert him from putting pressure on me. At that time, food also got better; there was a weekly shower, and sometimes whole days passed without any interrogations.

Some improvement occurred in the interrogations as well, especially after the first visit of the Red Cross agents. At that stage I had developed my game against my interrogators to a high degree indeed. I learned how not to divulge anything, so that I'd be at peace with myself, yet keep talking just the same so that I'd not be beaten. For example, on the subject of the aircraft—some things you can read in any manual, but when I spoke about them, it was like I was giving in to their pressure. I continued to confuse them with my English. I learned how to blackmail them, promising my cooperation for a shower or some better food. I remember how they said to me one day, "Why don't you tell us anything? All your friends speak freely and therefore receive no blows." I replied that

I wasn't speaking because I was filthy, and if I was promised a weekly shower, I might be more civilized toward them as well. Such negotiations helped.

There were also some funny moments. Once, when I was interrogated by an expert on the radar system of the plane, I managed to divert his attention and he ended up telling me about his life and family in Alexandria. Among other things he told me how he had been in a camp bombed by Phantoms, and a big shell fragment entered his bunker and stopped miraculously just a little above his head. He was keeping this shell fragment for a souvenir. I asked him for the date of that attack, and figured out it was me in that Phantom.... Naturally I didn't admit it, but imagine, what a small world!

AMNON

The jeep stopped at some field hospital. Everyone was shouting and gesturing hysterically. I became an animal. They dragged me; they beat me. A doctor looked me over but found no injuries. At one point they took off my blindfold and my handcuffs. From the shouting in the background, someone shot out from behind with the question, "What number squadron are you from?" I gave him the cover story and was beaten for it. They broke some of my toes—but I didn't feel much. Apparently I was still in shock from being captured.

Afterwards they blindfolded me again and we rode to a place that I later discovered was intelligence headquarters in Cairo, near the Abassiya prison. I remember the car arriving, the barking of dogs, voices in Arabic, and being pulled from the car like a sack of potatoes. They brought me inside. Someone gave me a kick in the belly, and I doubled up. Then they started to beat me. This went on and on. The pain was awful, and I wanted to faint—but I'm apparently the sort that doesn't faint. I was in a small cell, on the floor, and everyone who passed by hit me. Suddenly I smelled urine and feces on the floor, and this reminded me of the survival training I had in the Air Force, and I said to myself: "OK, these are things I know."

Later someone important arrived, and it became quiet. They removed my blindfold, undressed me—I was still in my flight suit—

and photographed me from all angles. They gave me Egyptian army clothes and then chained me from behind. I was lying in some corner and the beating started again. If there's such thing as 'seeing stars,' I really saw them then. I remember worrying about screaming. I was Rambo then, so I held it in. They finally left me alone. I was aching and suffering, and scared. What next? I was lying in the middle of the shit and urine and I told myself to cope, not to cry about my fate, and then the door opened and they took me for my first interrogation.

I was used to walking with my back straight, chest out, bars on my shoulder, and here I was with my back bent, head down, and chained like an animal being led to slaughter. I got a lot of blows and kicks along the way, and then the interrogation began. I remember it well because it was the cruelest of all.

I told them my name and my ID number, but I lied about the squadron number. They screamed that I was lying, blindfolded me again, tied me up, and began the torture. They beat me on the head with wooden and rubber clubs, threw me on the floor, stepped on me, and brought in dogs. I wasn't scared of the dogs, but all in all it was real hell. I realized I couldn't play the macho role any longer because they would kill me. I couldn't faint, and I finally screamed out. It was fairly intentional—I would play the bewildered child, maybe that would stop them. But actually the screams made it easier, and they eventually stopped, put me in a chair, and continued with the questioning.

This was repeated over and over. The interrogator asks the name of our squadron commander; I lie; he knows, and then starts another round of beating and torture. When I finally told them that the commander was Avihu Ben-Nun, they said "Good boy." And I was the bewildered child, the stupid child, for the entire interrogation. I developed childish concepts to stress my stupidity, thinking I was showing them I wasn't a serious source of information, and that I really didn't know very much. This wasn't far from the truth, as I had actually been a pilot for only four months. But each time I lied, the beating continued even worse than before.

My first interrogation lasted almost ten hours, and then they threw me into a dilapidated cell. It happened over and over. I had a strategy that I thought would help me survive, thinking that the pilots captured before me were all heroes who hadn't blabbed. You tell them your rank, your blood type, and your ID number, that's all. Obviously I couldn't hold out. The Egyptian interrogator wasn't dumb at all, and he wouldn't take any lies. I felt worse when I gave him names of people who had been with me in the squadron. I tried to change names, but had to remember all the information I had given in order to be consistent. Each time I broke, I had the feeling I was the first to hand over state secrets to the enemy. I even thought of suicide. Afterwards I found out that I had told the least of anyone.

They would give me homework, questionnaires to fill out in the cell. I kept on acting the stupid child, making awful spelling mistakes, in unclear handwriting. I reported lots of nonsense and almost nothing of importance. I remember that when they put me together with everyone, I still had two of these pages with my answers in writing, and Rami read them and laughed hysterically.

My physical condition deteriorated and I lost a lot of weight. I excreted nothing but urine for about a month. I was sure that they were putting something in my food and that I would die of blocked intestines. I'd pick at the rice, but I couldn't eat anything because I had cramps all the time from the stress of the interrogation. This was for the entire first month.

Most distressing during the interrogation were the handcuffs. I had what are called royal handcuffs—they have teeth that can be adjusted. . . . For a long time they handcuffed me behind my back. They cut me so badly that I almost got gangrene on one arm. It was so swollen that you couldn't see the arm. I remember the pain even now.

I had hallucinations at first, and I remember I hallucinated that a helicopter had come to rescue me, and I begged the pilot, "Take off my handcuffs;" he would answer, "I don't have the key." I begged him to bomb them, shoot at them, and he said no, it would injure me. Another of my hallucinations was of big glass bottles of Coca-Cola. I suffered a lot from thirst.

After about two weeks I lost any sense of time. There was one nice guard who would sometimes come into my cell and take my handcuffs off, or bring me a cup of water. He took me to the toilet. We spoke a little, even though I didn't know Arabic. I saw him as an angel sent to save me. He came at night and would talk to me for an hour or so, draw something and ask how to say it in Hebrew. But at the slightest sound he panicked, blindfolded me, closed the handcuffs and locked me back in the cell. During the day, interrogation and torture continued. I spent the whole time preparing for the interrogations, trying to remember what I'd already said in order not to get caught contradicting myself, and get more beatings.

I grew up in a nonreligious home. My mother was from a socialist kibbutz, and my father was far from being religious, too. When they took me for interrogation, I remember trying to find something that would give me strength. I would lift up my eyes and pray: "God, give me the strength to withstand this," and it would help. And when I went to sleep, I said a prayer I had made up, thanking God for helping me get through this awful day. Like the Jews who were burned at the stake. . . . I remember envying religious people, because they had the inner strength to believe, but I was also helped by it.

The Egyptians used all sorts of punishments and rewards. I would sometimes be offered a cigarette as a reward during interrogation. Years later, during a trip to the United States, I smoked marijuana, and it reminded me of something I had already experienced. I realized then that the cigarettes the Egyptians had rewarded me with sometimes contained hashish. Those cigarettes completely dazed me, and a terrible tiredness would come over me. Apparently, they thought this would make me talk.

MENACHEM

I remember being lifted from the truck at some place, probably a field hospital. I was asked several questions concerning my identity, and then they sedated me and I went to sleep. After a long time I woke up in a nice, big hospital room with four beds. I remained there for many months to come. Later I found out that this was Al Mahdi hospital in Cairo. The first time, I woke up from the flash of the camera when they took pictures of me. The horrible picture they took appeared in all the newspapers afterwards. I remember seeing five or six men around me, and one of them spoke good Hebrew and asked, "What's your name?" I don't remember what name I used in my reply. Then he asked me who was flying with me, and I answered.

Without any preparation, I received a blow on my shoulder, and he yelled that I was lying, that my name was Menachem, and my copilot had been Chetz. I didn't wait for what he had to say, and all bandaged and in casts as I was, I sat up in my bed and yelled at him: "You will not hit me! I'm an Israeli officer, and don't you dare beat me!" I don't know whether this was the cause, but from then on they never touched me, except once. I don't know if any prisoner can influence the course of his interrogations that much, but in my case, my reaction gave them a shock, and for a moment it became unclear who was the master in the room. He started apologizing saying, "But you lied to me," while I repeated: "I'm an Israeli officer, you never touch me again!"

I discovered that the POW is not helpless. He has his own power and authority, even if it's limited. I also discovered that they knew who I was, and actually it was quite evident, since I was the seventh pilot captured at the time. (Actually the sixth, because one of the pilots was rescued and brought back to Israel.) I realized I couldn't lie to them. They had already investigated the others, they had photographs; there was a lot that wasn't worth concealing.

I realized that I had been injured all over my body. I knew I was in their hands, but hoped to be returned home in a short while. There was no one who had been held prisoner for a long time in an Arab country before us. As it turned out, Nasser refused to hand Dan back, because he was the son of a famous commander from the War of Independence. He also wanted revenge for the deep penetrations of the airplanes, and he punished the pilots. He declared that he'd never exchange us. Luckily we didn't know this for a long time.

During the first days my consciousness was blurred most of the time, because of the drugs I was getting, and I believe I was asleep for about five days. I didn't eat anything and lost a lot weight. When I woke up I felt terribly thirsty, but they wouldn't give me any water. I don't know if this was a punishment or some medical precaution. Anyway, during the coming weeks, food or water deprivation was indeed used as a way to punish me from time to time. I remember that I managed to get out of my bed and walk to the sink, trying to drink there, but the guard caught me and pushed me roughly back to my bed. This was a very primitive guard; he was mean and I hated his guts. The guards used to change, and one of them was a good man. When he saw that I couldn't eat by myself yet, he washed his hands in my sink and fed me with his hands.

The guards stayed outside my door. There were four rooms along that corridor. A veranda, from which a beautiful view of the Nile could be seen, ran all along the rooms. Later on, I found out that in the next room they kept Yair, and on the other side, Los, both of them Israeli wounded POWs. The toilet and baths were across the corridor, and I'd be taken out there in a wheelchair, with a blindfold on. It was a pretty fancy place. The whole wing was locked and guarded by two men. They used to walk around in pajamas all day long.

I remember some sort of male servants who came in to bring food or to change the bed. Nurses would come in, too, and Doctor Sami, a very nice man. He was an orthopedic surgeon, and had operated on me several times. His treatment of me was really dedicated, and he was sincere with me whenever he reported on my condition. I felt that I was in good hands.

Once I had a visitor who didn't look like the professional interrogators, but who must have been a military pilot. He was well-mannered, and questioned me politely and in good English. Another man who visited called himself Shamel, and acted the "good guy." I noticed that all the guards were trembling when he appeared, though. He was probably a high officer in the Intelligence, not directly in charge of our interrogation, but only of general matters pertaining to our lives. He would ask in good English, "What can I do for you?" I remember that at his first visit I asked for a toothbrush and the Bible. He said: "Very well," but I got nothing. And why should I? But he was a good guy, just the same, perhaps because he wasn't one of the interrogators.

I was very occupied with my health, I was worried about my crooked leg, and so on. I regarded myself with self-pity and thought I was a really miserable creature. One or two weeks later the interrogations began. I remember very little of the first interrogations. I think they asked me only about my identity and background.

Actually, I was the last pilot to fall into their hands, and they had to finish interrogating the others, who were held in the intelligence center. They didn't ask me general questions about the army or the Air Force at first, but concentrated on the electronic warfare of the Phantom. They considered me an expert on that, but in fact I didn't know much and had to make up my answers. The interrogator was using a spiral method: he'd go from one pilot to the next and compare their responses. He sat with me daily for two hours. He was a very good investigator and understood his subject. Once he asked me about a certain technical matter, and I said, "I don't know." So he replied, "Let me show you," and took out a piece of paper, started to draw a diagram for me, and gave me a lecture in electronics.

I should tell you about the interrogator who had a heart attack in my room. He was older than me (about my age today, but at that time he seemed ancient in my eyes) and once he felt faint and hurried to take some pills from his pocket. The next day he visited me in hospital pajamas and said, "You see, I, too, am now a patient here." Interrogation is hard on both sides. You also have some advantages; you can wear your interrogator out.

It goes without saying that the interrogations were highly demanding mentally. There were many questions I didn't know how to answer; but the interrogator cannot be sure you're telling him the truth, so he puts more pressure on you. I imagine these are the situations in which the others were beaten, but they never beat me. At that point, I'd lean back and tell myself: You really don't have the answer, so take it easy. What do you care? He can exert all the pressure from now to tomorrow. It was much more difficult to conceal information I did have; it demanded a great deal of preparation and effort. The interrogators were really alert. We spoke English, because I got irritated by the interpreter. Once they started the interrogation promptly after an operation, as I was waking up from the anesthesia. They planned it that way, for sure. I woke up in terrible pain, after my fractured leg was put together with the aid of a metal screw, and the cast was full of blood. But the moment I saw the interrogator, the pain went away. There must be a physiological explanation for it; mental concentration probably drives pain away. The moment the interrogation was terminated, I was in agony once more.

Since there were healthy pilots to interrogate, they'd use me mainly for verification of their replies. Once I saw the interrogator holding a typed page of answers concerning electronic warfare, based, probably, on Rami's answers. I peeked at the page and said exactly what Rami had. Afterwards they let me be for a while. Often I regretted that I didn't have more information, because I would have given it gladly. All they asked me was, anyway, published in the American manual on the Phantom. I didn't know anything else but they kept pumping me for more and more.

The most serious questioning, on general matters of the army, started at a later stage. It turned out that they interrogated me after they had finished with the others. A new man came to question me on operational matters concerning the army. He was brilliant, and it was tough for me, but I passed this stage as well. I remember the day the interrogation was over, it was October 24, more than three months after my capture.

These last interrogations were more difficult, because I had the answers and decided I must not divulge them. He put lots of pressure on me, and I'd be edgy because of the need to decide what I may or may not say. I was glad whenever I really didn't have the information. Some of these interrogations went on for a whole day, but at night they usually let me have my rest. It was hard, just the same. At this stage, they punished me physically for the only time: they took my mattress away, and I had to lie on the wooden boards. The bad guard wanted to "improve" the punishment, so he took off one of the boards from under me as well. This took place when I was still wounded, and it was very painful. When he decided to replace the board, he managed to break one of my ribs. . . . The interrogator himself wasn't in the room at the time. He didn't touch me, but he had instructed the guard to do what he did. After that single incident, they never touched me again.

At the same time I underwent all kinds of treatments and operations, bone resections, etc. One of the bones healed itself, without any intervention. I remember they wanted to break it again, so that it would join better, but they couldn't do it [laughing]. This was the only time when I felt sick and dizzy during the local anesthesia and asked them to stop; I didn't mind if the bone would be a little crooked, and they left me alone.

I don't know if I could tell myself: Well done, you cope well with your situation. I couldn't grade my behavior at that time, since I had no experience to compare it with, mine or another's.

3

ISOLATION

Throughout the period of interrogation, the prisoners were kept in solitary confinement, handcuffed, sometimes in extremely small, cold, or dark cells. After the termination of the intense stage of interrogation, solitary confinement went on for about three more months. Loneliness, lack of time-space orientation, and insecurity characterize this stage, together with continuous physical suffering and rare hallucinatory or suicidal states. Gradually, however, most of the men recovered from the initial trauma and found ways to structure the empty time. The initial visits of the Red Cross agents and letters from family members relieved the darkness of this period.

Most of the interviewees described these months as an intermediate stage between the rough time of interrogations and the positive time of living together. Others depicted isolation as an important facet of their experience during interrogations. Some expressed pride at the methods they devised to overcome the ongoing strain of isolation, as well as the pain they felt in narrating this part of their stories.

DAN

I didn't try to keep track of time, so I can't tell you when the interrogations were terminated, or when the first visit of the Red

Cross took place. I knew that when I would be freed, I would be freed. I was not in control. I could not contribute anything to make my release come sooner. There were others taking care of this. It could be that this attitude made things easier for me. I had more patience than the others.

One time when I was "bad," they came into my hospital room, rolled me off the bed onto a stretcher, and blindfolded me. I felt they were taking me downstairs on the elevator, after that, through the halls [he is quiet]. These were the most frightening moments, when they took me from place to place, with the sheet over my face. I was always afraid of being lynched if people found out I was an Israeli soldier. I put my hand on my heart while I was on the stretcher, as though I could protect it if someone suddenly stabbed me [laughs].

They brought me to a solitary cell, cast and all. The room was about as long as the concrete bed that was in it, without windows and with a steel door. There was a small mattress on the bed, full of fleas, and they rolled me onto the bed, tied my arms with chains, and closed the door. A blinding light burned day and night on the ceiling. I began to feel cold. I was wearing an open cotton dressing gown. After a few hours I needed to go to the bathroom and started to scream. There was no response. Finally a guard came in, gave me a little tin can, and said, "Do it here." Of course it was impossible. He went away, and I went on screaming—nothing happened. They apparently wanted to turn me into an animal. So I did it on the mattress. After a while the guards came in, with great joy and happiness; they were so pleased! See! All to humiliate me. If not by beating, then perhaps this way.

This was in January, I think. Evenings were really cold. I screamed to the guard: "I'm cold, freezing." He comes in and asks, "What do you want?" I say, "A blanket." He laughs and goes away. This went on for about a month or two. In chains, relieving myself on the bed. After a while I hardly did it, because I didn't eat anything. But I didn't notice the dirt anymore. When the guard brought me a pita with cheese once a day, he had to cover his nose with a handkerchief because of the stench. Twice a day they brought me a cup of water, always in a cracked and leaking cup. I never knew if it was day or night; sometimes I had delusions. Once in a while I would hear them beating prisoners in the cell next to mine, with cries of "Mother" and "Father" in Hebrew. To this day I don't know if it was a recording, or if they were really torturing them there.

They didn't try to interrogate me then. Once some investigators did come with a cart full of papers, and stood with handkerchiefs over their noses. They asked me the same questions they had already asked before, and I was still smart enough, or stubborn enough, to keep up my part: "I don't know. I don't know."

They didn't let me sleep. Whoever passed by would kick the door. I didn't sleep for weeks. I got awfully thin and could count my ribs. I got so thin that the cast became loose and I could feel my shriveled legs, full of blood and pus. And I had one nightmare, that the bugs and lice and fleas that were crawling on my little mattress would get into the cast and start living on my pus. It goes without saying that no doctor saw me the whole time.

One day, without thinking, I told the guard who brought me the pita, "Listen, buddy, if you don't bring me water in an uncracked cup, I won't eat." You could say I started a hunger strike. Within five minutes a whole delegation arrived: "If you don't do as you're told it will be even worse. Eat! Drink!" I repeated my intention. They saw I had reached a point that I really didn't need food. So I won, and got my water in an uncracked cup.

A few days later they tried to break me using dogs, but I love dogs, and I remember being actually happy when the guard came in with a round-headed white dog. I patted it and asked the guard its name; "What a nice dog," I said. I heard one of the interrogators outside yelling at the guard—well, what can you do?

Near the end of this period I was half crazy, but I also knew this was a crisis point. The knowledge made me strong. At the end two intelligence men came in and said, "Dan, you're going back to the hospital." They brought an old razor and started to shave me with cold water, and tried to wash me a little. I went into a state of physical shock. Apparently my body temperature went down dramatically. I was cold, my whole body trembled, a really bad fit of shivering. For weeks afterwards in the hospital they didn't let me look at myself in a mirror.

We passed another stage. This was an extremely difficult experience, both for me and for Egypt. Because after all, even though you are an enemy, a certain relationship is created, and I think they respected me. I didn't give myself away; I saved face. Perhaps it was my age, or my big mustache. In the battle with the solitary cell I felt I had triumphed.

After a few days I was a little better and they took me for a complicated grafting operation. It was clear that if it didn't succeed they'd have to amputate my leg. The doctor explained that I would be completely covered by a cast, but if my loss of blood was only below the knee, we were saved. If there was bleeding higher up, they would amputate. As I woke up from the anesthesia, my interrogators were standing around, thinking that this was the best time for questioning me. I kept silent, but everyone was happy when it was clear that the operation had succeeded. When the bleeding stopped there were cheers, not only mine, but also of my wardens.

I don't remember exactly when I first saw the Red Cross, but it was always in the hospital. If I was in solitary, they would take me to a hospital room for the visit. The first time I didn't really believe it was the Red Cross representative. But they gave me a postcard from home and later brought me a Bible and the Passover Haggada of the kibbutz—it was apparently around the time of Passover—and a book of poems by Natan Yonatan, who had been my teacher in the kibbutz when I was a child. This was my first reading material. They also told me I could write letters home.

After some time in the hospital, they blindfolded me and took me outside on a stretcher, into a car. Again I had a bad feeling. In such situations it was especially clear I wasn't master of my fate. A person behind bars is apparently very sensitive to changes, afraid of changes. When you're in a fixed situation, you know what to expect, the bad things become familiar, too. A change can always be for the worse.

We traveled for half an hour. At the end they threw me onto a bed and took off my blindfold. I was on a bed in a huge room, totally bare, about six by nine meters [twenty by thirty feet], with one small barred window high up. I don't know what happened, but I broke then. I started to cry. I cried and cried. It was the first time I cried, and also the last. It was terrible. Apparently the fear during the ride was so great that I felt I had no more strength to stand the situation. It was total helplessness, hopelessness. The whole burden of the tragedy of the war between Israel and Egypt seemed to be on my shoulders, and I couldn't carry it any longer. No one saw me like this. If they had interrogated me just then, I would have told them everything. They say that crying is tough, but for me it was good. It emptied me completely. After half an hour I felt calmer.

Later on I was told that I was in the Abassiya prison, where I stayed till the end. It was the room in which the Israeli Navy prisoners had once stayed, and also our friend Farkash, a madman thought to be a spy—a room with a brilliant past. Long days of isolation began there, but I had a few small victories, like when I turned over on my side and could see the eucalyptus branches from the high window, and the sunlight. Or when I gave a cigarette to one of the peasant prisoner-servants, who was in a more shameful condition than me, and overcame the fear of what might happen if they caught me. I smoked then, and after I left the hospital they gave me cigarettes, although no matches. At some point they brought me a phonograph with two or three records, but it was far from the bed and I couldn't turn it on alone. Later I got some books. But most of the time was a kind of twilight, punctuated by trips to the hospital.

They didn't interrogate me any more. One day the interrogator came into my room, sat down on the bed, asked me how I felt, asked about my children, and said, "I hope everything will be all right." He even sort of patted my head. Only later did I realize that he had come to say goodbye to me. The fear of interrogation never left me until I was released. I still wasn't sure that they would leave me in peace. Whenever they slammed a door or took me for treatment, I was afraid.

I also had some positive experiences during this time. One day the nurse, who disinfected my wounds occasionally, brought her baby to visit. I think that this was when I was taken to the hospital again for an operation. The baby was called Amalia, about a year old. I remember I touched her hand. A little humanity in the sea of cruelty that surrounded me. Another time, a nurse lifted the sheet from my face as I was wheeled in for an operation, and said, "God bless you" in Arabic. They may have risked their necks with these little gestures, but this was the only humanity around me. I must add a story from much later, when we were all together. On

Passover Eve, after they had already locked the door for the night, a guard came in and wished us that the next year we could celebrate at home with our families. No one told him to say that—he would have been punished—but he took the chance.

I was still badly wounded, and my release was requested on all sorts of levels, right up to the United Nations, but the Egyptians refused to return me to Israel. I knew about these attempts from the Red Cross agents. General Sharif from Intelligence once came in, looked at me, shrugged his shoulders, and said, "I did what I could." And so I remained in prison.

Avi, Motti, and Benny were in another wing of the prison. I sometimes heard sighs from the other side of the wall, and once I saw traces of blood on the chamber pot they brought me. I understood there were other Israeli prisoners. The Red Cross people confirmed it. I was happy that maybe we would get together, but I was sad, too, for them and for their situation.

One day they brought in chairs, candy boxes, records, and books to my room, took away my blankets, and put white sheets, like in a hotel, in their place—a real party. They sprayed against flies, and a team of photographers came in and took pictures of the prisoner in his wonderful conditions. I was lying down, smoking, and marveling at what was going on around me. As soon as the photographers left, the guards returned. One of them spilled a cup of coffee on the white sheets; then they were removed and the old blanket came back. The other luxuries disappeared the same way, like with Cinderella at midnight. The coach turned back into a pumpkin. Later on I found out that they took pictures of the other three prisoners, in the yard, sitting on green lounge chairs. Apparently they were making a propaganda film for the world to see.

AVI

My first solitary cell was at the interrogation center, where I was confined for about two months. It was a room without a window, with a bright lamp on all the time. I never saw sunlight. I would try to guess the time from the events and sounds that penetrated my cell, such as the muezzin calling for prayers.

During the first stage of interrogation I was totally absorbed in surviving. I was completely helpless, and I knew they could do anything to me. I kept asking to see the Red Cross agents, but they retorted that in fact nobody knew I was there. They claimed that, in Israel, I was reported dead, therefore they could kill me any time they chose to. I kept reminding myself that I myself had heard the news broadcast in which my capture was announced; therefore they must have known in Israel that I was alive. But as time went on, I became so feeble that it took all my strength to cling to that memory. Sometimes I thought it was nothing but a fantasy, and in fact nobody knew I was alive.

I saw the Red Cross agent after about two months. He arrived after I had been transferred to another cell in the prison. This was a dark cell on the fourth floor, containing a bed without a mattress. The whole prison staff used to come by and peek at me as if I were a strange animal. The cell was close to the minaret. I recall the first time the muezzin came out on the small veranda to call for prayers. His "Allah Akbar" startled me terribly. It made me realize the magnitude of my solitude—a man alone, in a Moslem state, in jail, so far away from home! Another thing I remember about that cell are the rats, which were very active at night. But the muezzin calls nearly drove me crazy.

When the Red Cross agent came for the first time, a candle was brought into this dark cell. I remember asking him to make them open the shutters so that I could have some light. He promised to take care of that, but nothing happened. Two weeks later he returned and told me that the Egyptians said they didn't have a ladder high enough to open the shutters. . . . These Red Cross agents had a way of reporting the silliest things in a most serious manner. That was their act.

Slowly the interrogations were cut down, and I had more time for myself. I had almost stopped eating. The Egyptians tried to feed me, especially before the Red Cross visits, but their food disgusted me and I couldn't take it in. It was dark and I couldn't see what I was eating, so I had the impression I was getting leftovers and garbage. After the Red Cross visit, they made an arrangement for a pita with falafel to be delivered to me from an outside store every day.

The Red Cross visits were the only light in the darkness. Mr. Beausart had suddenly become an important figure in my life. Between his visits, I would plan what to tell him next time he came. He became like a father to me, someone who brought in the smells of the outside world. He used to tell me jokes about the Egyptians, which raised my morale. He comforted me, saying that things weren't so bad, that the Egyptians were idiots. It was surprising, the things he used to tell me, and how much he trusted me. After each visit, for two or three days, I used to reconstruct our conversation in my mind, considering every single word for its significance.

I received the first letter from my kid sister. The letters from adults weren't allowed through. The Red Cross agent brought me the Bible. I obtained some candles and could read. They allowed me to take a shower, but I was still alone. In order to occupy myself, I would sing folk songs, the songs of the Beatles, different operas—music was always an important part of my life. I started to jog in the cell. But I still had a lot of time for thinking: What will happen to me? What will my life be like after my return? When will I go back to flying, to studying? I was fantasizing my future life, always focusing on myself. These fantasies filled my world. Then I started to write letters to my family, which was another important occupation.

One frightening episode from that stage of my confinement comes to mind: the Red Cross agent had arranged for a barber to come and give me a haircut. He came with a razor blade; it was very scary.

On the last week of my stay in solitary, the Red Cross agent managed to have an electric lamp installed in my cell. I started to hear voices of other Israelis around. By that time I had already succeeded in communicating with my guards in Arabic, although I could lead only very low-level conversations. I didn't try harder because I was sure that in a short while I would be returned home. That's what I was told, but plans changed as new POWs were captured.

MOTTI

My solitary cell was totally dark, but after a few days I noticed that a single ray of light was coming in, moving slowly as the day moved on. One day, the good guy Shamel came in, bringing me a piece of candy. What a great day—up to then all I got was pita and salted cheese. I took the candy and used it to mark lines on the wall. I compared the position of the light ray on my wall to the time I saw on my guards' watches, and this way made a clock all for myself. After this, I tried to figure out if they had a fixed daily schedule—do they take me for interrogation at a certain time, how long does it last, etc. The time organization occupied my mind.

I was in solitary confinement for three months. I heard screams in Hebrew now and then, so I figured I wasn't alone in the wing. Sometimes I heard footsteps, men walking with chains on their legs. The Red Cross agent arrived during the third month. I was taken to the hospital and my back was bandaged in his honor. . . . Mr. Beausart told me that he knew I was being beaten during interrogations and promised to do all he could to stop it. He gave me some postcards to write to my family. He also told me that they knew I was alive, and he promised to have me moved to a better cell.

Some time later I was indeed moved to the prison. The room I had was larger than the first one, with light coming in from a high window. It had a bed, but its walls were all covered with blood stains, and graffiti in Hebrew and Arabic. At first I didn't understand what it was. At nightfall, however, thousands of bugs started to fall from the ceiling. I squashed them and smeared the blood on the walls, like former prisoners had done before me. I couldn't sleep at all. I covered myself in a blanket and sat cramped in the corner figuring what to do. The war against the bugs went on throughout my captivity.

I stayed in that cell for a couple of months. Every third day I was taken for interrogation. They asked me whether I knew the other prisoners they were holding at the time. One day I started to sing in Hebrew and Benny heard me from his cell and joined in, until the guard started yelling. I decided to leave notes for Benny at the toilet, but I got no response. Later I found out that he was taken to another toilet, but the notes disappeared, just the same. I wonder where to. When I had more free time, I made up word puzzles and also gave myself complicated arithmetical problems to solve. I did all this with the pencil stub I had obtained from my interpreter, using old newspapers that I snatched from the toilet to write on. These things kept me busy for hours.

At that time I developed some friendships with the guards. Sometimes they'd take me out for a walk in the yard. Some of them sat down sometimes in the cell for a chat. I had a large written inscription on my wall, so one day I asked the guard to tell me what it said. "But this is in French," he protested, and I understood he was illiterate. Another guard couldn't tell the time on his own watch and used to tell me he wasn't permitted to tell me the time.

One day I was given a piece of coarse soap, probably for my shower. I removed a spring from the bed and started to carve the soap. I did whole sculptures from soap cakes, and this occupied me for weeks. Suddenly I discovered I had artistic talents! I brought some of these sculptures back home when I was released. Finally they brought me a book, too. It was a five-hundred-page novel about a doctor and his wife; I read it four times at least. It's a pity I didn't ask for the Bible; this could have kept me even busier.

I received only two brief notes from my family at that stage, but just the same they changed my whole world. Only then I realized they knew I was alive. I received also a small parcel of food and discovered a mouse living with me in the cell. . . . I planned a trap for him, another occupation. He was smart; he ate the peanuts I had placed for him yet avoided the trap. I never managed to catch him, but I was occupied with the hunt. I also had two birds. I left some pita crumbs for them on the window, until they learned to come into the cell. So that was my universe—the bugs, a mouse, two birds, and a primitive guard.

DAVID

I was kept in confinement for six and a half months, three of those in a tiny solitary cell. The interrogations continued all the while. The Air Force men, who had been taken captive after me, were moved to the common room before me. I don't know why I was so long in solitary; maybe it was their hatred for the paratroopers. My interrogator hated paratroopers so much because he had been one of the Egyptians who escaped them during the Sinai campaign.

I didn't delude myself. I kept telling myself that when I volunteered for the paratroopers, I knew what lay ahead. Friends of mine had been killed or wounded, and nobody complained. This awareness made it somewhat easier to bear. I never considered killing myself. Another idea that had kept me going was that our tank crew had seen me being captured, so people in Israel knew I was alive, despite what my interrogator was saying.

They used to bring me a pita with salted cheese. I refused to take the cheese, which made me thirsty. I cut the pita in two halves and used to hide a piece for later, for after my interrogation. The thirst was terribly difficult. Three times a day a glass of water was brought in, but it was almost or entirely empty, according to the guard's whims. They often left a leaking faucet nearby to make me feel even thirstier. I got used to drinking the water left after washing the floors, with dirt, soap, and everything. Things like that can break down a person who hadn't known what he was heading for.

I was left in my underwear and wasn't allowed to wash. My body was still covered with salt from the water of the canal, which formed a crust all over me. I smelled awful. When I finally was permitted to take a shower, my underwear crumbled as I took it off.

One of the intelligence agents who visited me frequently was pretending to be a good guy. I called him the watchman because he had a wrist watch with the date; I tried to peek at it. He told me that if I'd be good, he'd bring me letters from my family. But I heard how he was scolded by the interrogators: "For two months we keep telling him he's believed to be dead, and now you come in and ruin everything!"

I knew about other prisoners who had been captured before me—Dan, the canteen workers, the pilot, too. One night I clearly heard how Goldwasser was interrogated, and he refused to say anything. They came into his cell, which was next to mine, every ten minutes and beat him up. I couldn't do anything for him. In the end they killed him.

One day I was brought to a new interrogator. He confronted me with drawings of different aircraft and started to ask me about their antennas. I told him I didn't know anything about planes, and I realized they had confused me with Amnon. But I was beaten a lot while they were figuring out their mistake.

Suddenly, when I had already been transferred to the prison, they brought me a pile of clothes and some cigarettes, although I had stopped smoking. I was given soap and sent to take a shower. It was due to the visit of the Red Cross, which took place the next day. A man came in like a storm and introduced himself as Mr. Beausart. I didn't believe him until he showed me his card. He was eyeing me suspiciously, too, and I noticed he was looking for something in the cell. He said that he had expected to find me in pieces, because they refused to let him see me for three months. He figured they were trying to fix me first. . . . We started to chat. When he heard that I had never been to the yard, he demanded that I'd go out right then. We walked around the courtyard for half an hour, and he told me about the other prisoners. I heard about the other wounded paratrooper POW, Yair, who had been from my unit. Mr. Beausart couldn't believe I had no books, and he went to one of the other POWs and returned with a Bible. It was from Dan. The nicest thing was that he told me to climb to the window and then indicated the common room of the other Israelis. I could only see the building.

From then on I had Red Cross visits every week or two. They brought me books and letters. I started to receive small parcels from my family. They were packed in shoe boxes, and obviously some of the stuff had been stolen. It was a hard time: the interrogations were less frequent, I had more free time, and I suffered from loneliness, especially since I knew that the others were all together already.

RAMI

Coping with loneliness and the struggle of the interrogations were part of the same reality. Once the interrogations were not that intensive, how to pass the time became the major problem. I spent seven weeks in the interrogation center, and then moved to a solitary cell in Abassiya prison, where I was isolated four months more. I had daily interrogations only during the first three weeks. Afterwards, they were getting less frequent. So what can a person do with all this time on his hands? I used to imagine the world outside my walls, constructing it in my fantasy. I tried to figure out, for example, the direction of my cell, how many guards were posted on the wing, and when they changed, who the Israeli prisoners in the adjacent cells were. I rarely thought about the world farther away. It helped me to concentrate my thoughts on my small universe, and to cut myself off from the rest.

Numbers provided my major occupation. My bed was a shabby carpet, and I used to count its fringes. I could do this only by focusing my eyes downward through the crack of my blindfold. Later I started to calculate how many prime numbers there were between one and one thousand. I still remember that I found 170 such numbers. This was a project that lasted two days! Later on, when my hands and legs were untied, and when I wasn't blindfolded any longer, they gave me paper and pencil for writing "homework" for the interrogation. I saved part of the paper and, using formulas that I remembered, computed the tables of logarithms and sines. It took me the entire seven weeks of my stay at the interrogation center to complete these two tables. I made a complete slide rule for myself, and it worked precisely.

In addition, I used the time to design machines for our kibbutz plastic industry, where I used to work before. The wall of my prison cell was painted in white up to a height of 1.8 meters, and since I already owned a pencil, I drew my plans on the wall. When the Red Cross agents visited, they brought me some parcels, and I used the packing paper and cardboard for my work. I made rulers and even a compass. I designed a checkerboard, and made little checkers from bread. In my mattress I discovered some cotton seeds and grew them near my bed, so I had a small garden, too.

The cell was full of bugs and fleas. On the first nights, I couldn't sleep for a moment. Gradually I found a way to get rid of them. I sat in the corner, covered in a blanket, and let the odor of my body attract them to that place. When they all came, I hurried to the bed and tried to fall asleep before they followed me. Every day I hunted a whole hour for the bugs and looked for them in all their hiding places. In two weeks I almost exterminated them all.

In the evening I used to jog in my cell, making figure eights, so that my head would not turn from going round and round. This took two hours of my daily time, in which I covered the distance of six kilometers. I developed a method of counting the "eights," and when this became automatic, I used to direct my thoughts to different topics. I decided to reconstruct my life history as I was jogging and tried to go from age to age, following my school years, extracting the most detailed recollections from my mind. I arrived at a state in which I saw myself quite objectively: Rami the lonely kid, the rebellious teenager, the overactive young man never giving a thought to himself, and so on. I didn't hide anything or deny my defects. Since then, I know myself and can't use excuses anymore. I am aware of all my faults and I function as best I can with these limitations. I am unable to evade myself and say, "I just didn't feel like doing it." Today I know that there is a reason for everything, including emotional responses, which seem to be so impulsive and uncontrollable. This is the most important achievement of my experience in solitary, a lesson that is with me to this very day.

I was very proud when I finally learned to stand on my hands. This was an old wish of mine, from the time I was a chubby little fellow. A bell was sounded in prison every two hours, and this was my sign to stand on my hands. At first I had to lean on the wall, and my hands would get sore and tired. Gradually I improved a great deal and managed to stand for a long time without any support.

All in all, I was completely isolated from the world outside for eight weeks, until the first visit of the Red Cross people, who came on August 31. My conditions improved considerably from then on. I received books to read, among them some designing manuals I had requested. I used these for designing industrial machines, as I explained before. It's a shame, but at one of the searches in the common room, all these designs were taken away and never returned to me. The Red Cross also brought me some news about my home. On the first visit, I was informed about the birth of my twins, although they couldn't tell me the sex of the babies. Only ten days later, when the letter arrived, did I find out that I had two baby girls.

I managed to close my mind to thoughts about matters that were out of my control. I told myself not to worry about my release, because I assumed everything was being done for that in Israel, and those who are concerned with that worry enough. I managed to occupy myself quite well. It surprises me that I didn't even try to establish any contacts with the other Israeli POWs, who I realized were not far from me. We were all actually in adjacent cells. The Red Cross agents told me that they put pressure on the Egyptians to join us all in a common room. I told myself that when we would be joined, we'd be joined. The others tried to communicate by notes left in the toilet, or by knocking in code on the walls. When I found this out, it astounded me; Why didn't I make such efforts? I did feel the need for company, but obviously this was not strong enough to push me to make an effort. On the other hand, I concentrated all my efforts on organizing my own time, in my own field. I felt that this was under my control, and if I'd try to extend myself, I'd probably fail.

Producing a creative daily schedule, consisting of a variety of activities and covering the whole day, was my goal. I used to leave some unfinished business for the next day, so that I'd be sure to have something to do right from the morning. As time went on, I worked out a regular daily program for myself, including such matters as how many cookies I could eat a day (when I already received parcels from home), and at what time, so that I'd have enough for two weeks. In the afternoon, I used to sit at my window and watch the tree, which gave me a sense of feeling at home. That was the hour I allowed myself to think about my home a little. I never indulged in self-pity. I felt I was using the time well.

YITZHAK

When the interrogations became less intensive, I was moved to a completely dark cell, but it was more spacious. I believe I was already at Abassiya prison then. Later I was moved again to a cell which was three by four meters, with an iron bed, although without a mattress. The room had a high, narrow window, through which I could watch the sky. That's where the first visit of the Red Cross took place. I remember it vaguely. When they told me I'd meet the Red Cross people, I didn't believe them. Next day, however, a handsome, tall man came smiling into my cell. This was Beausart. I believe it was the end of August. He brought me a small parcel, with chocolate, nuts and the Bible in Hebrew. He also gave me a special postcard and told me to write home a few words. I remember talking to him for about twenty minutes. I asked for a doctor. I asked him how many Israelis had been captured, and he winked but refrained from answering. I was afraid to talk to him freely, because his visit took place when I was still under interrogation, but I was extremely happy to see him. It was proof that people outside knew I was alive. I assumed that after this visit, the Egyptians wouldn't dare continue their threats and torture. The truth is, however, that my tough interrogations continued just as before. I think that it was, in fact, harder to sustain the interrogations after being reassured that they wouldn't kill me. At that stage I didn't want to die as much as I did beforehand.

The most difficult loneliness was in the first solitary cell. That's where I also suffered from terrible uncertainty. I was lying in filth, and when I heard footsteps in the corridor, I never knew whether they were bringing me a pita or taking me to be executed. Mostly they came to take me to my interrogation. I was dizzy from the beatings, but I never allowed myself to relax.

Later on I had another cell, with a window, where I could differentiate between day and night. I remember the day Nasser died. It was at the end of September. I heard very loud sirens, and the muezzins were yelling from all the minarets from morning to night. It was an extremely hot day, but no food or water was brought to the cell. I didn't see anyone and didn't know what all the commotion was about, and why there was this sudden change. Three days later I was again taken to be interrogated, and asked what had happened. They refused to tell me. They all seemed to be out of balance somehow.

Routine set in when the interrogations were over. I was taken out daily for a walk in the yard, and once a week to the shower. The food was somewhat better as well. At this stage, I arrived at the conclusion that I should avoid getting sick in captivity, therefore I had to take really good care of my health. As someone who grew up in the U.S., I asked for vitamins, but I didn't get any. I worked out a lot. All this time I didn't see a doctor, but the wounds and burns healed themselves, especially after I started to receive good soap from my family. I got some parcels with clothes and books. Until then I was in rags all the time. I remember that Beausart had asked me what sort of book I wanted, and I mentioned an astronomy book, because I could see about twenty stars from my window. He brought me a tiny children's book, a nice booklet translated from Russian to English. What I read most of the time was the Bible. I had a good edition, with interpretations, and I read it through three times. It's an excellent book for captivity; it is rather difficult so it demanded effort on my part, and it gave me material for thought. I could identify with many of the biblical stories. Later on I received three or four English books to read, so this provided me with something to do.

One of the first books I received was Shakespeare's *Comedy of Errors*. I was lying in my cell and laughing my head off. I realized that I needed humor for encouragement. It was difficult to bring myself to laugh when I was all alone, but when I joined the others, I often made us all laugh.

I made a daily schedule for myself. I didn't want to sleep late and waste my time. I got up at eight o'clock, exercised for about half an hour, jogged a kilometer inside my cell, and then dedicated my time to reading and thinking. Every day I made myself concentrate on abstract thoughts for an hour. I used to lie on my bed, watch the ceiling, which was awfully dirty, and let my mind go. In the beginning, I used to think a lot about the Air Force and my family. But this gave me a headache, and made me feel self-pity. Later I taught myself to visualize colors and clean my head of wandering thoughts, and started to consider philosophical issues instead. I thought about war and peace and such matters. I forbade myself to think about the future, and concentrated only on my past and present. I tried to recall memories from my early childhood and school days. I remembered my first loves, my first experience of flying. I didn't dwell on flying too long, however, because due to my interrogations there were some things I'd rather forget. I tried to refresh my days with new activities, so that I'd feel in control. Of course I was totally dependent on decisions out of my control. It was their part to provide me with food and water, or to determine when I'd go home. But within the small world of my cell I tried to make my own decisions according to my will. Thus, for example, I used to change the subject in the middle of an interesting chain of thought, when the time allocated for "thinking" was over. I'd tell myself: Tomorrow you'll continue from here.

I kept track of the time with the aid of my biological clock, the calls of the muezzin, and especially by the movement of light and shadow in my cell. I knew that the sun in that season moved fifteen degrees per hour, and I calculated the projection of the shadow on the floor. It wasn't important whether my time estimate was precise or not, as long as I felt in control over my time, and I knew I could allocate the time for my different activities.

My strangest activity was killing ten bugs a day. These bugs were killing me. They bit me and I was full of sores. I couldn't smash them with my fingers, as Arabs do, so I used to step on them instead. Finally I asked the Red Cross people for some bug spray and I sprayed the room so enthusiastically that I was sick for two days. I almost poisoned myself together with the fleas.

I looked for creative outlets. I used to wet a pita and form some tiny balls that I played with, but I wasn't very sophisticated in comparison to the others. My only goal was to occupy myself enough so that I wouldn't dwell in self-pity. I wasn't so miserable, but just the same it was tempting to occupy myself totally in my own suffering.

Some time later I managed to communicate with the other POWs. I realized they were nearby. I identified Rami by the heavy steps of his giant feet. He has a unique way of walking, and there was no mistake about it. We were all barefoot then, but for a long time we had chains on our legs, and this helped me keep track of the movements in the corridor. Once I heard somebody being interrogated. He was screaming in the next cell. It was apparently David, the paratrooper, whom I didn't know yet. I was afraid to call out to the people in the corridor, because I believed that this might lead the Egyptians to put us at a distance from each other. But I remembered reading about POWs who established communication between cells by Morse code. I didn't remember the code, and my next-door neighbor didn't provide any help, so I made up a system, which was rather primitive: one knock for *A*, two for *B*, three for *C* etc. Naturally, this is difficult with twenty-two [Hebrew] letters. I tried to improve the code, and this occupied a lot of my time. Using a small rock, I wrote the message I wanted to transmit on the wall, and then coded it and knocked on the wall. After many attempts, Amos, my neighbor, finally understood. It drove me crazy, waiting a whole week until he realized what was going on. Then one day he was responding: Amos. We were afraid that the guards would discover the secret, so we limited ourselves to ten sentences per day. In fact, I didn't know Amos or Amnon before; they abandoned their plane two weeks after us, and I found out about their captivity only through this communication.

The realization that I was surrounded by other Israelis had a positive effect. I remember asking the Red Cross people about their identity, and added that I wanted to join them, because according to the Geneva convention prisoners of war should be put together. The chance that we would be put together, at least all the officers, made my return home seem more probable, too. On our first conversation about this, Beausart gave me a mysterious smile and asked, "What makes you think there are other Israelis here?" Then he added that he was optimistic about that. He was always optimistic. We used to joke about that a lot later on, he in his French accent, I in my American. But I was willing to bet that if one kept demanding something, it would finally be given.

I was happy to sense that our reunion was close. On the other hand I wondered who all the other POWs were. Suddenly I worried that if all the Air Force was in captivity, who would take on our jobs? I was afraid to find out that the squadron commander was also in jail, and worried that the pilots who remained might not know how to use the Phantom, because it was so new in Israel.

AMOS

After about a month of interrogation, I was moved to Abassiya prison, and there the number of interrogations gradually diminished. We were all put in one wing, which was built around a courtyard, three cells on each side. We were taken out to the courtyard, one at a time, and didn't meet. As it turned out, Yitzhak and I were on the same side, and Rami, David, and Amnon on the other. After about half a year we were put together.

The Red Cross agent visited me a short while after I was moved to the prison. On the first visit, only Beausart came, and he brought me the first letter from home. A few days before that, Shamel, the "good" officer, came for a visit and brought me two Russian books translated into English. They were horrible books; one of them was about the stars.

A wall separated my cell from Yitzhak's, and we communicated by knocking on it. We made up a code, which was rather difficult, and more suitable for English than for Hebrew words. The first time we succeeded in communicating, I said to him that I'd leave him a note in the toilet [laughing]. Later on we exchanged notes frequently, because transmitting messages by knocking was quite awkward. I don't know why we didn't include others in our note communication, because we were all using the same toilet. Perhaps it was logistically too complicated. This contact was important for me, because it somewhat relieved my loneliness, which was very hard for me to take.

I occupied myself mainly in reading, especially after they brought me a Bible. I wrote letters, even though for three months I didn't get any from Israel. Some of the time I was sick. I had the shivers and felt bad, so I asked for a doctor. No doctor came, naturally, but I got better just the same.

Two sparrows were living in my cell. They had their nest in it, and they used to come and go through the tiny window. I watched them a lot, and they provided me with some company. I hoped they'd lay eggs, but it was the end of the summer and fall, and these are not the seasons for that. [He is quiet for a long time.] There were many fleas there; I was bitten all over. I used to fight them ferociously, and this was another occupation of mine. I had to locate them in their hiding places and exterminate them. This problem kept us busy in the common room, too.

Loneliness was the hardest thing at this stage, especially before we started our note communication. I was thinking a lot, whether I wanted to or not. I thought about home, about my wife who was pregnant at the time. I think I was told about the birth of my daughter on the first visit of the Red Cross. [Here Amos tries to reorder his memories, thinking about the various dates, and ends up discovering that he had been detained three months, rather than one, in the interrogation center. At this point he looks somewhat disoriented and says, "Never mind the dates."]

I was a happy father in jail: my first child, a girl, was born! My head was spinning with thoughts about this. I didn't make up problems to solve but just remembered and imagined different things. At first I was concerned with the fact that I had become a captive. I was obsessed with the idea that had I acted differently, perhaps I could have saved the plane. These ideas depressed me, and I had to force myself to abandon them. I couldn't let myself become deeply depressed in jail.

Toward the end of this period we were questioned by a sociologist. These were friendly conversations, without any pressure. He also hinted that soon we would be joined in a common room. Once he said to me, "Go on, finish your story, and you'll be moved in with the others."

AMNON

On August 12, I was moved out of the solitary cell and blindfolded and put in a car. They must have been driving me round and round in the camp, but I was certain that since they now had all the information out of me, they were going to execute me. Finally they moved me to another wing, where I gained a small improvement—a new and larger cell with a bed. They took off my handcuffs and blindfold and left me alone. Apparently the cell had been painted recently, and with the great heat the walls were radiating a strong, suffocating odor. I felt I was being gassed and could hardly breathe. I lay on the bed, with its thin mattress, and waited for them to come and get me for the next interrogation. Five hours of terrible anxiety passed by, yet nothing happened. It was night already. I was lying on the bed, trembling, with silence all around.

It was a large room. It had a big window high up, which was kept open, and a heavy door, like a vault. As I was lying down, I noticed a big nail sticking out of the door, about the height of my forehead. All of a sudden I felt an urge to kill myself, because I considered myself a traitor. I approached the door, measured the height of the nail, and examined the possibility of running from the end of the cell so that my head would be crushed on that nail. Would the nail penetrate deep enough to kill me? I was walking round and round considering this possibility, until suddenly the opposite idea popped up in my mind: The hell with it! They had put this nail here on the door on purpose, to make me think about suicide. I am going to win by not doing it! But you can guess my state of mind from these types of ideas.

Later on the door opened, and a man whose head looked like a donkey, with a distorted face, frightening, like in horror movies, popped in. I was startled, but he smiled, and handed me some bread with halvah. I couldn't believe it. I ate it trembling; nothing had ever tasted better than this in my entire life. Suddenly I was flooded with indescribable bliss. Here was a human gesture in the midst of hell. A few moments later he peeked again, this time with a lighted cigarette for me. Right then we heard footsteps, and I panicked. I put the cigarette out and swallowed it, so that the guard wouldn't suffer on my account. When the steps had passed he came in once more, and with his gestures asked me where the cigarette was. I indicated that I had swallowed it, and we both laughed. From then on we were friends. A few days later he came back, gave me soap, and took me to the shower. For six weeks I hadn't been allowed to wash.

So that was the beginning of a somewhat different stage. The first shock of captivity was over, and I had more free time for myself. The physical conditions were less threatening, or perhaps I got used to them. My fantasies were highly active during that stage. I was constantly visualizing my rescue or escape from prison. These were uncontrollable visions. I was sure that, in Israel, all efforts were being made to release me, that the whole nation backed me up, but I didn't know when this liberation would take place.

I remember clearly my first contact with the civilized world. I heard footsteps and the dragging of boxes outside, and suddenly a man in a suit came in, introduced himself as Shamel, and informed me about the Red Cross visit on the following day. I think this happened about two and a half months after the beginning of my captivity. He made some gestures to the effect that if I uttered one unnecessary word, it would be the end of me. At the same time, two parcels from my family were brought in. It took me some time to realize that I was finally in contact with my former world. He gave me two white envelopes. I immediately identified my mother's handwriting on one. I have no words to describe my excitement. I remember having the feeling that all the fluids in my body flushed toward my eyes. The whole Nile was in my throat—I just wanted him to get out of my cell. But I couldn't hold it in; I sat on the bed and burst out crying. It was an uncontrollable outpour of everything. . . . These letters were more precious to me than any diamond. Apparently Shamel was moved by my behavior and he approached me and patted me on the head, saying, "Don't cry, Amnon, everything will be okay." At least that's what I remember.

Later on I opened the letter, which I remember by heart to this very day. "Amnon my darling baby, how are you, my precious. . . ." I couldn't have prayed for more.

The two first parcels contained some clothes, shorts, and two books—the Bible and a novel titled *Salambo*. I remember allowing myself to read only a couple of pages every time, so that I wouldn't finish it too soon. It was a book about the Carthaginian wars, full of descriptions of battles with thousands of casualties. I'd read it and tell myself: So what are you complaining about! At least you're alive and well! The Bible filled my existence. Finally I had time to read it to the smallest detail, to explore its depths. It's a great book. Later on Rami knitted a cover for my Bible.

The second letter was from my girlfriend. This is another story. Several months later I asked her, in a letter sent from the common room, to terminate our relationship. I couldn't live with the feeling that she was waiting for me, while I wasn't sure I still loved her.

My life was less stressful from then on. Instead of the interrogations, I had questionnaires to fill out, which they collected and read. It worried me that I didn't get any response. I made things up or I would tell them the truth—just the same, no reaction. From time to time, I'd be taken for an interrogation. In one of those they got mad at me for pretending to be so dumb. They humiliated me in different ways, just for fun. But all the same this was a better period, from August to December, until I joined the common room. The food was somewhat better, and I received three meals a day.

I had a window in my cell. The sun would make shadows on my bed, and I used their movement to mark lines on the wall until I had a pretty accurate clock. I was given ten cigarettes per day, so I scheduled my smoking times by the lines on the wall. I reserved some cigarettes in case they stopped providing them. Smoking was my only pleasure besides reading the letters I received. Reading was also important. I didn't exercise, because I was young and in good shape and didn't need to invest energy in that. I thought a lot. Most of all, I felt sorry for myself. I was thinking that had somebody else from my family experienced what I was going through, and had I known about it, I wouldn't have been able to take it.

Sometimes I thought about other matters. I tried to plan some projects, like a big, legal brothel or a casino. I tried to think about all the details involved. Sometimes there were funny episodes as well. I knew that other Israelis were detained in the same wing, but they were very careful to keep us apart. One day, one of the prisoners was being taken for interrogation, another asked to go to the toilet, and with all the movements my door was unlocked by mistake and all of a sudden I saw Yitzhak standing in front of me. A Jew, an Israeli—there before my eyes. It was five months since I had seen an Israeli. Seeing him moved me so much, as if he were one of my forefathers coming alive. The moment was interrupted with the appearance of the sergeant, screaming that he'd kill the guards, and all the doors slammed shut.

I tried to communicate with my next-door neighbor in Morse code, but didn't get any response. When we were joined, I found out it had been David. He said that he did hear some knocks but assumed this was a trap set up by the Egyptians and he was careful not to respond. We were all terribly suspicious, naturally.

One day, toward the end of the period, they took me out to a nice room, with armchairs, where I met a nice looking man who said he wanted to talk to me. Of course I contracted my body, expecting the familiar interrogation. I was especially startled to see the interpreter, whom I remembered from my very first interrogations. But the new man wanted me to tell him all kinds of things about Israeli society, and he didn't ask anything about military matters. He told me that when this stage would be over, I'd be transferred to a room with all the other POWs, but naturally I didn't believe him.

At some stage in the meetings with that psychologist or sociologist he asked me to make a drawing. When I finished drawing, he asked me to sing a song in Hebrew. I searched my memory for a children's song, but for some reason I felt like singing the national anthem. Perhaps just because that meeting was more humane, I felt even more humiliated and helpless, having to obey any command. So I stood up, and with tears in my eyes, and with my entire soul, sang "Hatikva." I felt like a person facing his execution, saying his last credo.

I regretted this later, however. I heard that the Egyptians used the recording of my singing for their propaganda broadcast at the canal posts, like they used to do with various so-called recorded messages from the POWs. But perhaps hearing my singing there gave our soldiers a good feeling, who knows.

That was the end of my interrogation. With my "Hatikva" I touched even the investigator's heart. I was returned to my cell, and two hours later they told me, "Get dressed, you are moving!" My eyes were blindfolded and I was taken to join the gang.

MENACHEM

There were four beds in my room at the hospital, but for more than six months I stayed there alone. When I had been in my aviation course, I went through preparation for captivity. We were taken on a hike, kidnapped, and put in jail. I was captured on the fourth day of the hike, and when I was brought in there were some pilots who had already been there for four days. As I passed by in the corridor, I remember one of them saying, "Please tell them I'm willing to talk now, because I can't take being alone any longer." And this was only a game, we all knew that. But until then I wasn't aware how difficult solitude can be.

And there I was, half a year alone in the room. A doctor came in from time to time, as did the attendants. I wasn't confined in solitary cells like the others. I discovered that I didn't mind being alone. It was much worse when I shared the hospital room with Los, but much better when I joined all the others in jail.

I always knew the date and the time of day. When the attendant came in, I looked at his watch. At the same time I noticed the location of the sun rays in the room, so next time I remembered the hour by the place of the sun. I became very precise in my estimates. One day I heard the nurse tell the attendant to take me for physiotherapy at 1:00. When he came at 1:15, I asked him why he was late. He said, "I'm not late." I answered, "But it's fifteen minutes past the hour now!" He remarked, "Good God, how do you know the time without a watch?" This was a very primitive guy. I used to chat with him in Arabic, and at the same time I told him that I didn't know the language [laughs].

They weren't aware that I was fluent in Arabic, although I used to talk to the guards from time to time. They were all orthodox and used to say their prayers in my room. I enjoyed talking to them about God, also about peace. Once I tried to explain to the attendant that if we'd have peace, Israel would advise Egypt about agricultural matters. He retorted, "What for? We have had our agriculture for the past five thousand years!" Some of the guards used to come in and play dominoes with me in the evenings.

When you're locked alone in a room, you become quite sensitive to noise. You don't know what has happened, and you're helpless and alone. Once there was a fire at the hospital, which frightened me. But it was worse on the day Nasser died. I woke up to terrible sounds, as if a huge demonstration were taking place outside. I thought they had a revolution and expected the worse. Later on I found out the reason for all that commotion, from the guards.

My world was confined to that room for a whole year. For some of the time I was in a cast up to my loins. I was taken out to the toilet in a wheelchair, or used crutches to get to the sink in my room. I was in bed all the time. After about eight months, the cast was removed. I got up and started to exercise, but I still spent most of the day in bed. Naturally I wasn't allowed out of my room, but I could watch the wonderful view of the Nile, or follow the rebuilding of other hospital wings, from my window.

Even before the termination of the interrogations, I received the first visit of the Red Cross men. Beausart told me, whispering, that Los was hospitalized in a room next to mine, and he told me about the POWs in jail. He was a wonderful man, assisted by young Olivier, both of them from Switzerland. I am not sure, but I think they brought me the first letter from home, and also a parcel for the Jewish New Year, and the Bible. Many tears were shed on that first letter. [He is quiet for a while.]

The books completely transformed my world. I remember that I received *The Godfather*—it was an excellent book to read in captivity. It has all the juicy elements—murder and sex and whatnot. Esther, my wife, used to select some of the books for me, but most of them were selected by the Air Force.

My only occupation was reading and thinking about myself. Reading in captivity is kind of strange, and I felt it in the common room, too. I had never experienced reading so intensely before. One becomes so sensitive and vulnerable, that identification with heroes, as well as rejection of them, is extremely strong. Any text with some emotional color was greatly amplified. It was wonderful to read that way; I used to cry a lot while reading. Perhaps we blocked our feelings about ourselves and permitted the expression of them indirectly, through the heroes of the novels we read.

Los was in the next room, and I used to hear him. Once, when I returned after an operation, I had a strong urge to talk to him. Since my hand was supposed to be in great pain, I was expected to scream. So I yelled to Los, "Listen to me, I only pretend it's the pain, but I just want to talk to you!" I screamed so much that I ended up throwing up [laughing], but Los didn't respond. Four months later, when they brought him into my room, I asked him whether he had heard me then, and he said that he hadn't.

Los was severely injured. He was wounded in his spine and suffered various degrees of paralysis. Since no wound could be seen from the outside, they thought at first that he was pretending. For the first three or four days they had beaten and kicked him, and he had almost died. After six months alone in my room, they brought him to join me. We stayed together for six months more, until I was well enough to join the others in jail. Los remained a few months alone at the hospital, and then he was returned to Israel. He was very heroic in his efforts to learn to walk again, and did it all by himself. It was a great experience watching him.

I was very happy when Los was moved into my room, but my happiness was brief. After two or three days I found out that it was suffocating to be with another man in the room, twenty-four hours a day. For me it was worse than being alone. But nine other men were all right. . . . Maybe Los was a difficult person to be with. We could hardly talk to each other; we didn't do anything together; I can't tell you why.

My transfer to the prison was delayed a great deal, because the Red Cross had hoped to be able to define me as "severely wounded" so that I'd be returned to Israel, like what finally happened with Los. Perhaps they had Los's welfare in mind, too, and didn't want to leave him all alone. But eventually I was completely cured and my hospitalization couldn't go on. One day they came for me saying, "We're taking you to stay with your friends in Abassiya, so pack your belongings and let's go."

4

GETTING TOGETHER

The moment of meeting was extremely moving for the men and resulted in a definite change in their lives in captivity. It was, however, a single event of short duration. As the reader will soon discover, the interviewees retained the sense of the emotional impact of the event, yet had different recollections about the people they met, or the order of coming together in the common room. The following versions are somewhat contradictory, probably due to the great excitement of the moment.

MOTTI

One day they transferred Benny and me into a common cell. I have hardly any recollections of that period. I think we were brought into the big room, and soon afterwards Dan was moved in as well, so that we would take care of him. I don't remember anything else; you'll have to ask the others.

BENNY

Motti was located at another wing of the jail. One day I was moved into his cell. That's where his wound had been treated. At that time he was already given three meals a day; in other words, he was better off than I had been in solitary. They had probably believed that he was a civilian, not an enlisted man. Shortly after that, I believe it was about a week before Passover, we were both moved to a much larger room. I immediately sensed that there had been other Israelis in that room. Indeed, several hours earlier Dan had been taken out of that room for his medical treatment at the hospital. That was the room in which we stayed until our liberation.

We felt the closeness of other Israelis. Sometimes we heard the tunes of Israeli songs whistled in the courtyard. We knew that an Israeli Mirage pilot had been captured two days before us. One morning, the door opened and Avi was brought into the room. A few days later, Dan arrived. The four of us stayed together for about three months.

AVI

For me, being brought together with the others was simply part of going home. The Red Cross was telling me that they were going to return us. They had four Israeli captives, Dan, Motti, Benny, and me. Everything was agreed on, and now they would return us. In preparation for our return they would put us together to raise our morale and give us better food to fill out our bodies. It would all be a matter of days.

At the end of May or the beginning of June they finally brought me into a big room where Motti and Benny were already staying. Dan was brought in to join us a day or two later, and then there were four of us in the room. It was a very great relief to be with the three others after so many months alone. First of all, there was a great need to talk about what happened. Second, there was now someone to lean on, to share my pain. There was also a very clear feeling that we were going home. The Red Cross gave me the feeling that this was the end of the episode. And while waiting, it would be good to be together with other people in the same situation, who could understand me.

Motti and Benny had already been together for some time. They had already talked enough, so now it was my opportunity to share my experience. A few days later they brought Dan in from the hospital. It made me so happy. With him was a man from the Intelligence whom I recognized as Shamel. Others knew him as

Hyman—both fictitious names meaning left and right in Arabic. I slapped him on the back and said, "You're really okay." Two of the guards grabbed me, covered my eyes, and threw me into solitary [laughs]. A punishment. The guys started to ask about me, and the Red Cross came. After a day or two they brought me back.

I remember each one as he joined us in the room. Yair came in totally crippled but with a healthy sense of humor. A few months later, they started to bring in the pilots. It was in the winter, raining all the time. We had only a small light bulb, and it was very dark in the room. Rami came in barefoot. . . . Amnon came in shock. He recognized me, but I didn't know him at first. I remembered him as such a handsome guy, but now I saw a broken man. I remember thinking, "Poor guy." He was the only one who hadn't performed any activity in solitary, he even gave up fighting against bugs. He was full of bites and still has scars. He just wasn't yet an adult. He had finished the aviation course two months earlier, and he was already captured. Later he shaped up—even his handwriting changed completely—and he came out a new man.

David also seemed to be in a very bad shape when they brought him in. His face was swollen—I don't know from what. But I was most moved to see Yitzhak, since I knew him well. We were once in the same squadron. The fact that they had brought us all together was a breath of hope: soon we were going home.

DAN

Shamel came to my hospital room one day and said, "Today we're taking you to the other Israeli prisoners." Naturally, I pretended to be greatly surprised, as if I hadn't realized they had captured other Israelis. I was blindfolded and taken back to jail. I was brought into the room in which I had already been living for six months. They put me on the bed, and when the rag was pulled off my eyes I saw three young guys. They approached me and said, "I'm Benny," "I'm Motti," "I'm Avi." I noticed that beds had been added to the room, and understood that they had been living there for several weeks. I remember having mixed feelings of sadness and joy. "Who are you?" "I'm a Mirage pilot, I was captured about six months ago, two months after you." "We're canteen workers, we were delivering our goods to the canal posts and kidnapped there." I was the veteran among them. My two legs were still in a cast. We had absolutely no medical assistance, and I wasn't able to walk to the toilet or take a shower. From then on the Israelis would help me, I felt, because I needed all the help I could get.

RAMI

I knew about the other POWs from the Red Cross people. They told me about four old-timers who had already been put in a common room in jail. The idea that I might join them was very exciting, almost like the idea of being returned home. For six months I hadn't had anyone to talk to in Hebrew, except the interrogators, with whom I can't say I "talked." I communicated with the Red Cross people in English and with my guards in poor Arabic. The guards were okay, but they were simple and I couldn't really relate to them.

I remember being moved to the common room on December 12. I had the feeling that I should restart counting my days in captivity. I looked back on the previous period and realized how lonely I had been.

I was the first pilot to join the old-timers. Yitzhak was brought in on the next day, and in the following three or four days all the others arrived, till we were ten. Several months later, Yair, who had been severely injured, was returned to Israel. But later on Menachem joined us, so we were ten again.

I remember the old-timers had all kinds of rituals; for example, concerning what one might or might not tell the guards. They were sure that the walls were bugged and were careful all the time. My impression was that they had completely adopted the mentality of a prisoner in enemy territory. Furthermore, they had developed some kind of interdependence with the guards. This was their major concern. Another issue was how to prepare breakfast. They were like a primitive society in a way. At that time I wasn't able to express my impression, but today it's clear to me that they lived without an ideological pivot or focus, which might have organized their lives. Things totally changed after our arrival.

AMOS

I kept demanding that they put us all together. Loneliness was terrible for me. On the day I was moved, finally, I experienced tremendous joy. I found out that the common room was actually very near my solitary cell. When I was brought in, they were all there already, except David. Yair was with them still, and Menachem hadn't come yet.

I remember that my first impression was of a weird group of men. They had strange civilian clothes on, with big beards and mustaches. Men tend to grow long hair when they're alone or in captivity.

It took me some time to realize the significance of being together with others in the room. Then the stories started to come. I hadn't seen Amnon since we jumped, and I hadn't seen Rami. I was captured only five days after Rami, but during these days I had seen Nurit. I flew her out to another base one day and could give Rami regards from her. In the meantime my daughter was born, and so were Rami's twins—there was a lot to share.

By some sort of an implicit agreement, we didn't talk about the interrogations, but we shared more about being taken in captivity. I remember Yitzhak's story about his trouble in being ejected, and how he was almost lynched on the ground. We, on the other hand, could tell him about the rescue of his copilot. I didn't know that Los was at the hospital. I didn't know Dan and the canteen workers.

YITZHAK

One day I heard somebody being taken out of the next door cell but didn't hear him being returned. On the same afternoon there was another one. I could hear quite well what was happening in the corridor. So I was saying to myself, either they're being moved together or executed. I decided to be on the optimistic side. The next day they came for me, too. This was the day they used to take me out for a shower. They had always blindfolded me before going out of the cell. But on that day I heard the guards talking: "Take this and that, too." All I owned was a pair of extra shorts, one pair of underpants, and some books. I realized I was being taken to the common room. And indeed, when the blindfold was taken off, I could see Dan, Avi, Amos, the canteen workers, and Yair. I was the seventh in the room, where I stayed for the next two and a half years.

The move was so exciting that I was in a state of shock. I couldn't utter a word after "Shalom." I hardly knew the men. Amos was vaguely familiar from the Air Force. I knew about Dan and Avi, who had been captured before me, but was surprised to find the canteen workers. I found a bed and sat down among them. At that moment we didn't know how to behave. It was very hard to start a conversation, after being isolated from your own people for so long. We were groping for words. We said very little about our experience in captivity. I was interested in what had happened in the Air Force during the two weeks between my and Amos's capture. This was extremely important for me, even though six months had already passed. Rami arrived on the same day, or later in the evening, and then Amnon, the kid. Finally David was brought to the room, and we were ten.

When I arrived the room was completely bare, and it looked very big. It had only military beds. On one of the walls there were two or three high, barred windows, closed with shutters all the time, so that we couldn't take a glimpse of the outside world. A single light bulb hung from the ceiling.

You're saying I didn't mention happiness at joining the others. It's true. I had taught myself not to be glad or to expect anything. Because after you're glad, you get disappointed and this is even harder. For six months I hadn't been joyful. Whatever I wished for had never arrived. So I learned to expect nothing. I lived with what I had, so that I wouldn't be disappointed. One develops such a numbness, which is a defense, and then you feel neither sadness nor joy, love nor hatred. You stop laughing and crying as well. I was aware of this process, numbing myself from feeling what normal people usually feel. That's why my initial reaction to the others in the room was confusion. Moreover, they seemed to be scared. No one said, "Well, man, it's good to be together finally," or "How have you been?"

I should add that for half a year we believed that they were taping our conversations in the room. We ruled out the possibility that they were watching us or taking pictures, but we felt that they might be listening. At the end I told Rami that had they really listened, they would have punished us for what we said. Just the same I remained on guard, a process that went on until three years ago. In my pictures from that time as well one can notice the frozen, numb expression, even in that picture with the cake, two days before our liberation. I looked as if any moment all hell might break loose.

AMNON

After six months of interrogation and isolation I arrived at the common room a broken man. I had been too young and unprepared for this hardship. I had lacked self-confidence. I was especially apprehensive about what I had divulged when I was questioned. I was afraid I had betrayed the country. It was very important for me to discuss my interrogation with Rami, and he made me see that I had behaved more or less okay. My conversations with Rami relieved some of the mental tension I had suffered since my "betrayal." Rami also saw the questionnaires I had completed and they made him laugh uproariously. He enjoyed both the style and the contents. This feedback put me back on my legs, and allowed me to put my pieces together. From then on I regarded the world more optimistically.

When they came for me saying, "Let's go, you're being taken to your pals," I was extremely happy. At the same time I was keeping some margin of defense in case this was a bluff. I was blindfolded and led through a low gate, then a door was opened, the rag was pulled off and I saw . . . Rami. Everybody. Yitzhak, Avi, Dan. Several seconds had to pass before I could focus my eyes. What I saw was like . . . a birthday party with candles. They were all standing there and talking to me, but I didn't respond. I was too excited. I was the ninth in the room, the last one to join before David.

I had been thinking a lot about that moment: how it would be, what my pals would look like, how we should behave. Whenever I have a great wish fulfilled, I have a sense of emptiness, some kind of sadness. I remember that I had experienced a similar feeling at the graduation from the aviation course. Here, too, at one single moment all my previous hopes and expectations were realized—and gone. The moment was not such a peak experience as I had imagined it would be. It was one of the moments of this world, after all.

There was something else: throughout the interrogations I claimed not to know any of the other POWs. And here they were right in front of me, and what if the Egyptians were watching or listening in? Avi was standing somewhat removed from the others, near his bed, asking, "Don't you remember me?" I answered, "No," but nodded "Yes" with my head. I did remember him from the air base. I knew all the other pilots superficially as well, but I couldn't contradict in my behavior what I had claimed for half a year. Suddenly I noticed the way they were glancing at each other and realized they were taking me for a nut. They were probably thinking that I was distraught from all that beating on my head. I wanted to check all the walls and openings for hidden microphones. I sat down on one of the beds, and gradually I could listen and answer. I was euphoric.

The two boxes with my stuff were full of fleas, and the guys took them right out of the room to the courtyard. I hadn't fought the fleas; I didn't even know what they were. They took out the bed I had sat on, too, and started to disinfect everything.

The common room was rather dull. It had coarse, unpainted walls, and the windows were all shut. We had iron beds and very primitive mattresses, a wretched table. During our stay, however, we turned that room into a jewel.

DAVID

Every day I was promised that I'd be taken to the common room. But it didn't happen. That was the time I was meeting with the sociologist. A couple of days before Hanukkah (and I was captured in May, mind you!), two guards came in and started to roll my mattress and collect all my things. By that time I had accumulated about fifty boxes of cigarettes, because I had stopped smoking. They told me, "Come on, we'll take you to the gang." It turned out that the common room was about thirty or forty meters from my solitary cell.

They were all there before me, and as I came in they examined me, searching me for signs of torture. They didn't find a thing. Then they gave me a bed, and I gave them my cigarette treasure.

"I'm coming from the canteen," I teased. I approached Yair, and asked him how he ended up there. After all, he had been in my armored truck, which had kept going for shelter at the post. He told me that they had a second attack, that all the soldiers of our unit had been killed, that only he had been wounded and captured.

The guys questioned me about my time with the Egyptians. It was difficult to get used to the drastic change—from more than half a year in solitary—to a room with people speaking my language, to the sense of support of nine men. It took me two days to get to know everybody and realize where I was.

MENACHEM

The Red Cross men used to give me news from the common room when I was still in the hospital. So I knew, more or less, what was going on there. Furthermore, during the first Passover, in 1971, Rami and Yitzhak came to visit us at the hospital. Their stories gave me an idea about the size of the room and the courtyard and the people who were living there. I knew all the Israeli POWs who were being held at that time in Egypt, since I was the last one among them. There was just one more Phantom that fell after mine. Its navigator was killed during interrogation, apparently, and its pilot was severely wounded and promptly returned to Israel after his leg amputation.

I remember being blindfolded in a truck on my way to the prison. I couldn't understand why they had to blindfold me, when all I could see were the streets of Cairo. Were they ashamed of their city? I was happy to join the others, but also apprehensive, as always when a change is introduced into my life. I recall thinking, wait a minute, here I am well organized, I adjusted to my environment—and now they're upsetting me all over again. Furthermore, I was quite disappointed at the failure of the Red Cross to have me returned to Israel as a severely wounded POW, and I was feeling sorry for Los, who had remained all alone in the hospital. But the moment I entered the room with all the guys I felt a great joy. One might say that this was the only really happy moment I experienced in all the three and a half years of my captivity. I realized that I wasn't alone anymore, that there were other souls with whom to share my fate, other people to talk to.

In captivity, people tend to damp their feelings. You don't allow yourself the luxury of great joy, because you never know when you will be hit again, where the next disappointment will come from. But I was happy to meet Rami and Yitzhak, who had been my friends before, during our studies in the U.S. I didn't know the others, not even the pilots, but I was glad to meet them.

At first, they jumped all over my boxes, to see what I owned, especially the books. They were taking turns reading *Love Story* in the room at the time, so right after all the hugs and kisses they wanted to know if I had read this novel, and they said I had to start reading it right away. It's so funny. I immediately sat down to read it, in English, and it is indeed a book you consume breathlessly, crying the whole time. . . .

I was shocked at their standard of living. They had already been in that room together about half a year, and Dan had lived there for a whole year. They were gathered together during the winter, at about the same time Los and I were put together in the hospital room. They had already developed a daily schedule; even the Friday night meetings had been institutionalized and worked well. But the conditions in the room were awful.

What struck me more than anything was the darkness and the overcrowding. The place was highly unattractive. I could probably see that because of the contrast with my hospital room, which had been spacious, full of air and light, with windows along its entire length. In comparison, this was a pit. The yard looked terrible, too, because of the three-meter wall surrounding it. I was also disturbed by the Egyptian guards who used to sit in the yard. They were some kind of prisoners whose job was to keep our courtyard clean. They were crouching in the corner, and whenever a eucalyptus leaf dropped on the ground, they hurried to pick it up. It was a depressing room. The windows were small and very high up, and at that time, nailed shut by wooden boards so that the prisoners couldn't peek outside. An electric bulb was on all day long. The concrete floor was rather clean, but the walls were filthy. All the men's belongings were kept in the cardboard boxes in which the parcels had been mailed from Israel. The room was crowded with beds, with a dilapidated dining table at the center.

It irritated me. I wouldn't have minded if the room had been arranged to stress the temporariness of this state. I didn't object to that. But I felt it was simply neglect.

So that was my impression. On Thursday I arrived, and read *Love Story*. On Friday night, during the assembly, I brought up the things I saw, and we started changing them. That Friday's minutes started, "Here he comes!"

5

ORGANIZATION

This chapter, as well as the next two, is based on an open-ended request to describe life in the common room, which provoked long, detailed, and varied narratives. As each of the men presented his version of the collective experience, three general topics emerged: organizing the group as a solution to the men's individual needs, the social aspects of life in the group, and their private experiences while living so closely together. The stories about the joint captivity often wandered back and forth in time, and many inconsistencies are obvious in the following accounts. Selections from the group diary, which appear in a later chapter, shed additional light on some of the men's recollections.

As the group grew from five to ten men, the need to organize their common reality became apparent. The strain of ten men from different backgrounds, with various needs and moods, living in a crowded environment without respite, might have produced a catastrophe. For several reasons, structure and routine overcame the potential chaos. In their accounts, the men tried to capture this transition, epitomized by establishing the Friday night assembly meetings, the egalitarian rotation of chores in the room, and the study program—all these initiated within the first six months of the group's history. The emerging social system resembled a kibbutz, and was often compared to a utopia.

RAMI

The most important aspect of the first stage of our lives together was the new grouping. We were five additional men who joined the five already in the room. Gradually we changed our whole way of life. But in the beginning they were the old-timers, and by virtue of their seniority they had a relative advantage over us.

At first everything revolved around food and eating. They used to play canasta, and in the evening they told plots of films to one another. There were a few books there that everybody read. They wrote letters. After we arrived we began to define the day: when to eat, when to prepare food, when to wake up, when it was permitted to make noise. After about two months we started the Friday meetings, and after that lessons—all this in an attempt to build a system with more value.

At first, we were unorganized. Just being together was the experience. Getting up in the morning stretched from eight to ten o'clock and sometimes until noon, each according to his own time; it was forbidden to make noise. An order was formed naturally of those who got up early and those who got up late; this prevented lines at the toilet. But the morning was unorganized; everyone did what he liked, read, exercised. The first focus was the noon meal—what would we prepare for today? What kind of soup should we make and who should prepare it? Conditions were still very bad. There were no cooking utensils; there was one knife that the old-timers had found. It was a German pocket knife, just a blade, really, for which we later constructed a handle. There was one electric hot plate, on which we did everything, including the laundry. We didn't do any complicated cooking—we got most of our food from the Egyptians, mixed with the packages from home. We didn't yet buy from the canteen. The food was poor, but we made an enormous business out of preparing it. I remember, for example, baking our first cake. We got flour from home, but needed to improvise baking utensils. We finally put the mixture into one pot and a larger pot on top of it—and out came a cake. This was the first of a magnificent series of cakes. We baked cakes for every Friday—we had a festive Friday night meal—and afterwards another cake for Sabbath morning. They brought us an oven and we achieved magnificent results with whipped cream and fruit.

There was a heater for the outside shower. Twice a week the heater was lit. We calculated how much time each person had for the shower so that we had enough water. It was winter when we were put together, and although it was possible to sun oneself in shorts in the yard during the day, an evening shower in cold water was unpleasant, so we had to organize a fair distribution. There was also the problem of Dan's shower, getting him there, not wetting his crutches. We busied ourselves with this for an hour in the beginning.

Shaving was also organized. Like other prisoners, we tended to allow ourselves to be basically lazy in shaving, until we developed a "formula" obligating everyone to shave three times a week. On Sunday, when the Red Cross never visited, it was okay not to shave. Therefore we shaved on Monday, Wednesday, and Friday. All these little things were our major occupation in the beginning, the center of our lives.

In the beginning they allowed us out into the yard for an hour every day. Later, they left the entrance to the yard open from seven in the morning to eight in the evening, and we were able to go in and out freely. In the beginning there was also lights out around eight in the evening. We finished supper in the dark, with the door already locked. And then we developed a culture of candles. Later on, they didn't turn our lights off any more.

According to the diary we began our assembly meetings in February 1971. I don't remember exactly how we came to the idea. My background as a kibbutznik certainly contributed to it, and there were other kibbutz members there. It seemed obvious that one of the simplest solutions for running a complex social system is through the creation of some sort of democracy.

At the first meetings, we dealt with our basic daily routine: getting up, eating, putting things in order, keeping clean, consideration for one another. There was a problem about common living arrangements, and we agreed on majority rule. Thus we solved the problems of who decides. Until we agreed on democratic decision making, the old group had tended to let Dan decide, as the senior member of the team. He was like everyone's father. But when we arrived it became obvious that Dan couldn't take responsibility for every decision that came up. I remember our efforts to give him the deciding voice: "Dan, as the senior member of the team, you decide." But although he had been able to play this role in the old group of five, he couldn't continue after we arrived, and he crumpled up in his corner.

I think there was another factor in the change of the decision-making structure. After all, Dan was wounded and crippled. Until we arrived, the guys moved him everywhere—to the shower, to the toilet—and they didn't pressure him to get out of bed. Three or four days after we arrived, I said to him, "Nu, get up, we're walking." He had crutches, but he didn't dare use them. There wasn't anyone to confront him. To me it seemed that his legs were intact. The wounds, which because of his diabetes developed complications later, weren't large yet, and I was determined to get him up so he would walk. At first he got to a sitting position, and afterwards we helped him walk. It's true one foot couldn't bend, but he only had to learn how to put it down. At first, when he started to walk with crutches, he didn't know how to turn around, so we turned him around. It all took a few days, but in the process he gradually lost his authority with the group. I remember it made me feel uneasy with him.

From then on, the deciding forum was our common conversation. It didn't matter what we talked about at the meetings every week, the important thing was that we sat together and talked about any topic that was bothering any of us. We became a self-sufficient society. I immediately suggested that there should be a rotation of chairmen and that an orderly diary of the meetings should be kept.

Our first meetings were held during the period when there was still an early lights out, and they took place in the light of an improvised candle in a sardine can. We felt we were putting one over on the guards. Someone would climb up and cover the windows. In this, we created a tradition of discussions by candlelight, and even when the guards no longer turned off our lights, we continued to hold our meetings by the light of candles that we got in our packages from home.

After we had better organized our daily routine, we tried to discipline ourselves to get up more quickly at seven o'clock, the hour when the door to the yard was unlocked, to have morning exercise together. But we never succeeded. Each person exercised according to his own private idiosyncrasies. Some of us jogged;

some joined me in handstands. We never tried to create an obligatory regime; even the morning exercise was only obligatory for those who wanted it. In general, we didn't want to make our daily routine the main focus. We still got up in our natural order, at our ease, and the process took about two hours, from seven to nine o'clock. I was usually one of the early risers. We listened for what the others were doing in the process getting up—when the previous person finished with the shower and the toilet, it was worth getting up. We also listened to one another in the matter of preferences, what each liked best—without discussing it directly. I remember that sometimes I would let someone else get up first because I felt that he wanted to.

We decided to study in a systematic manner. This led to an organization of the morning hours. Each of us prepared his own breakfast, and lessons began at nine o'clock. The person on duty would cook, and after lunch and cleanup we would play cards—at first canasta, but later we learned bridge and had two bridge tables every afternoon for two hours. Only eight people could play, and so the question would be raised—who would not play today?

We set up the rules in our discussions. I think it was the person on duty who would not play that day. After that there was a natural weeding-out process, and in the end we were one table.

What was the attraction of bridge for us? When I was playing—card games, or later with the kittens we managed to raise—it was as though I was not in captivity. I was doing these things of my own free will. Each such activity helped to get me out of prison.

Nevertheless there were long empty stretches of time. We had a little homework and individual study. And sometimes someone wanted to be alone, and that was also accepted. But we were successful in creating a way of life. We didn't just sit and stare at the ceiling and ask: When will the time pass? What do you do now? In our free time we could keep busy with all kinds of work and reading and games.

I knitted a lot. I made dresses for Nurit and our daughters, and for Independence Day I knitted a blue and white flag that I later presented as a gift to Golda Meir. I remember trying to involve the others in knitting, in planning and counting stitches. Sometimes Amos or David joined me. It was actually Nurit's idea. She sent us wool and wooden needles because obviously we were not allowed to receive sharp instruments, but they didn't give us the wooden needles either. I peeled off wooden strips from the baskets in which they brought us our vegetables and knitted with those. Nurit wrote to ask me to make a dress for Netta's doll; she was having trouble convincing the children that I was still alive. And that's how it started. Some of the others wove rugs, or made batik, but for me knitting was very relaxing and time-consuming. Benny found some matches and built nice things. In the evening we ate a light supper, and then it was culture time. We read, translated, wrote letters. We listened to the radio or records or watched television until eleven o'clock. After this hour it was accepted that we preserve the quiet.

Each of us found a way to express his own creativity. It is interesting that these activities, even though they were individual, were never done alone. When Benny planned to build the Eiffel Tower with matches, we all occupied ourselves with calculating the measures and proportions. All we actually had was a picture! From this we had to prepare a three-dimensional projection. We didn't say we have to teach Benny because he doesn't know how to do it himself. But Benny asked for advice, and immediately three of us jumped on him, each one with his own opinion.

Even our reading wasn't completely solitary. While reading someone would say, "Listen to something extraordinary"—and he would start reading aloud, even if it was out of context. Good books passed from one to the other with recommendations, and we would hold discussions. I read a lot and kept a list of all the names of the books I read. I read more than three hundred books in captivity—that's about one book every three days, and at any rate more then I read in the 10 years before or after. We collected a library of three thousand volumes.

Reading in captivity was a unique experience. You have time to sink into the book, to enter into and experience the feelings. There was an intense identification with heroes; we often shed tears. Perhaps it's good to share someone else's loss, and sometimes it's comforting to discover that there are people more unfortunate than you.

Friday was different from other days. This was the day for showers, and the whole place would be cleaned. There were those who baked the cakes and made ice cream. We prepared a more elegant meal, and immediately afterward went on with our meeting. On the Sabbath, it was permitted to sleep late, and there was cake for breakfast and no lessons. We also celebrated the holidays, especially Passover, which we did in a very elaborate way. For Purim we put on long performances, for Israel Independence Day I knitted a flag, for Rosh Hashanah I prepared a greeting card. It broke up the daily routine.

Another thing that broke the routine was our only trip in Cairo. The Egyptians took us for this tour about a year after our stay in the common room. I don't know what they had in mind. We were taken in pairs. Some of us were taken to a nightclub as well. Amnon was offered a prostitute. We wanted to go on another trip, but it never happened again.

DAN

In the beginning, there were four of us—Benny, Motti, Avi, and me. First of all, there are no words of praise high enough for the way Avi took care of me and helped me throughout the three and a half years we spent in captivity together. For about a year I was in bed. Day and night he was there for me, to help me with the toilet, bathing, laundry, feeding—everything. Afterwards he would sit at my bed for a chat. From him I received all the help I could dream of. He was younger than me, one of our best pilots, a sensitive guy with a generous soul.

One day they took Motti and me to the hospital. He needed treatment for his wounded shoulder, and it was time to take my cast off. After the cast was removed, every touch hurt my legs like hell. Both of my legs were as thin as matches, covered with layers of dead skin, blood, and pus. I remember they put me on a high stretcher, which suddenly broke, and I fell down, in awful pain. I was returned to the common room like that, without even a bandage on my legs. With my last drop of energy, I pulled myself to my bed. Motti, Benny, and Avi were staring at me helplessly. We were so miserable. Avi tried to peel some of the dried blood off with his nail. We didn't even have warm water then. I was lying in bed thinking, will I ever be able to walk again? Will I remain an invalid for the rest of my life? I made a vow not to let them break me. The next day, lying in bed, I tried to move my big toe, then the others. Later on I started to move the whole foot, repeating back-and-forth movements thousands of times. Then I started to move my knee and lift my legs up, and so on until I could stand.

At that time all we were eating was the Egyptian food: three times a day we received rice and beans. Twice a week a piece of meat was added; it was fatty and swarmed with worms. A tomato or an egg was a rare addition to our menu. I ate everything, because I knew I had to grow strong.

As the summer came, we were attacked by mosquitoes and I suffered from the incredible heat. Avi used to pour water on me and my bed to cool me off. We tried to keep the room clean. Once a week we took everything out to the yard, burned the ends of the beds against the bugs, and disinfected and washed everything. I believe that our cell was the only place clean of bugs and lice in all Egypt.

We started some activities. We stole the tiny saws that were used by the nurse to cut off the bottles of my injections, and we kept demanding soap. We then used the little saws as a carving tool and started to make soap sculptures. The Egyptians wondered what the Jews used so much soap for. . . . When I got alcohol for disinfecting my wounds, we used some of it to make a lamp, which we put on after lights out. This, too, was a minor victory over the Egyptians.

After several months of this kind of life, new prisoners began to arrive. The first one added to our group was Yair. He had lost an eye and an arm and arrived in terrible shape. Indeed he was released shortly afterwards. His body was full of splinters that hurt him terribly. We operated on him in the cell, taking out some of these splinters. The first one was huge. We gave him a towel to clench his teeth on; we cleaned the spot with alcohol, held him down, and opened the area with the saw, until the splinter came out. Avi was the surgeon. After the first place healed, we repeated the operations frequently, and Yair kept all these splinters in a little jar for a souvenir.

During that period, I was very sick too. My left leg became infected in one of the operations, and the infection led to gangrene. My leg was swollen to the size of an elephant's, and I fainted. The guys screamed until a doctor arrived, but he said, "I'm afraid to treat you, in all seriousness, because I'll be blamed for curing the

Zionist enemy." We all heard him saying this. He gave me a penicillin shot. The gangrene would go down a bit and then return all over again. When one of the Red Cross agents filed a complaint for medical neglect in my case, he was sent back to his country. We never saw him again. Later the doctors diagnosed me as having diabetes too. I believe it was an outcome of all the stress and tension. I was under constant stress because of all the military secrets I kept to myself, and the fear that one day another soldier from the Canal might be captured and my interrogations would start all over again.

Having all ten of us in the room contributed to my recovery. With a lot of effort, I taught myself to move my legs, but I didn't dare stand up—that is, not till all the guys came and said, "That's it, Dan. Today you're going to stand on your feet." Then eight or nine men surrounded me and lifted me up. The whole world turned around me, and I sat down. The next day, they lifted me up again, until I made a step, leaning on the guys, walking between them. Eventually they helped me walk up the step, and I was out in the courtyard. Like a baby, but without the flexibility of an infant's body, I started to walk. The support of the group was extremely important, but the work was, naturally, all my own. The pain went on, hard as hell, but I progressed until I could even jump rope! I received special shoes from Israel and a pair of crutches, too. When I was walking relatively freely I demanded to share the chores and be on duty like the others. Gradually I became the chief lunch cook. I like to eat and I like cooking, so I enjoyed my role.

When there were ten of us, it was essential to fix a daily schedule. If everything around you is chaotic, you must provide order and stability. I thought it was very important to eat our meals together and at a fixed time. I demanded that we complete the chess competition we had started, even though people got tired of it in the middle. It is easy to fall apart under such circumstances, and that was something I wanted to prevent at any cost. This is actually our greatest achievement in captivity: that in spite of all the tensions and crowdedness, in a situation where an opening door could bring either a letter or interrogation, either the Red Cross agents or the wardens, we maintained our sanity. It's much easier to give in to the circumstances, to the dirt and to bickering, to get up in the morning and curse your luck. I think we had fewer fights among us than in a normal family, and we managed to preserve our humanity in a cohesive community, which we maintained to the very last day in jail. That's why we returned as healthy people.

Later on we established a daily school, and we formulated rules for contact with the guards. We had some confrontations about norms of behavior regarding the guards, some of whom were real troublemakers for Israelis. In the beginning, I thought that only by strict military discipline would we be able to maintain some order in the room. Others objected to discipline. They argued that they had coped well with their hardships so far, and would do fine without being ordered around. It turned out that it was possible to maintain order in a different manner—by using common sense and understanding, through our group discussions.

We tried to celebrate the Sabbath and make it different than the rest of the week. We put on clean shirts and cooked a special meal. For some time I tried to add something for the spirit, also, by reading selections from the Bible or poetry. But for some reason this habit faded away, maybe because of the lack of intellectual stimulation.

In the beginning, the Egyptians interfered with our lives a great deal, but as time went on, more and more autonomy was granted. For example, during the first stage, one of the prison "generals" would come in every evening, count us, put the light out, and lock the door from the outside. This regular visit raised arguments among us: Should we stand up for the high-ranking Egyptian officer? What if his rank is lower than that of some of our men? In a well-organized POW camp, everybody salutes an officer, even those of lower rank. But we maintained our Israeli arrogance, and I can't remember what was decided. After a while, they stopped counting us and didn't enter the room in the evenings. We only heard the lock being turned from the outside.

Our physical conditions constantly improved. Look, at that time, the Israelis were holding sixty-seven Egyptian POWs, who had been captured on Shaduan Island. For every improvement that we got in our conditions, they received five. But the Egyptians turned down the offered exchange. "Let them sit and rot," they said. "These Israelis will not be returned."

Our studies introduced an additional dimension of order into our lives. The fact that we had to get up in the morning in order to be in class on time was more important than the studies themselves. Learning together and the Sabbath and holiday celebrations provided a framework for the group. We got tired of the studies after a while, but as long as we kept going, they were terribly important. We had a bridge addiction, too. It is funny to recall our arguments about who would play with the pretty or the old cards—as if it had been a question of life and death. You see how minor daily details were blown out of proportion.

One of the major events was the day we received the first broadcast from Israel on the radio. One of our guards used to buy small radios in Cairo for sale in his village. At night he wanted to leave the merchandise in a safe place, so he put them in a closet in our cell. We saw him doing this, so during the night the pilots broke the lock and we had a radio in our hands. They all gathered around my bed, and we heard Israeli programs half the night. From then on it became a habit. One night we heard on the news the Red Cross reporting our conditions. It was proof that we had not been forgotten in Israel, and it made us all very glad. We also had a neighbor who wanted to keep us updated on the news. This was an Egyptian gentleman who had been the Minister of Defense and was sentenced to prison because of the failure of the 1967 war. He had luxury conditions in his cell, and he owned a big radio which he used to leave playing full volume in the corridor, broadcasting either Israeli or BBC programs. The guards were unaware that the radio wasn't speaking Arabic. . . . These are some of the episodes that gave us strength to go on.

When Passover arrived, we prepared for the holiday as we used to do in the kibbutz. Each one received a part to recite, and we cleaned the room better than ever. The Seder celebrated with the rabbi was the peak experience. Actually we celebrated two Seders. The first one took place in the officers' club, according to the Jewish tradition, with lots of officers who came to observe us. Afterwards we returned to our room and had a second kibbutz-style Seder only for ourselves. Meeting the local rabbi for the first time moved me a great deal. He seemed to be a fearful Jew, a symbol of the diaspora for me, the Sabra. I felt that we, the prisoners, were encouraging him! In his humiliated demeanor I could see all the anti-semitism I had never experienced myself. Since I was the oldest of the group, I was seated next to the rabbi. He was happy to discover the similarity between the traditional and the kibbutz versions of the ritual. In the middle of the Seder he leaned toward me and said, "All of them are troublemakers for Israel." I think I knew what he meant. We did not see the rabbi other than on Passover, once a year.

There is no doubt that the threat I felt constantly, even when we were living peacefully in jail, had some justification. All this insecurity resulted from the surprise factor. Out of the blue, in the middle of this routine period, they decided to interrogate us once more. They took us out in pairs to the club, asked questions, tied our hands in back, hit us on the head, and kicked us. A few days after the interrogations, we were taken out for a trip to Cairo! We went to see the pyramids and the Sphinx. We ate at the Hilton and went up the revolving tower, where we could see Cairo in full view. Some of us were taken for a second tour of Cairo at night. I was happy to be spared this one, because I didn't feel safe outside the prison, and I didn't trust the Egyptians to defend us if that would be necessary. The trips provided us with a new topic for conversation: What did you see? What did you do? How were the girls, the cars? Did you see the Nile? After all, it was a day out of prison. But when Shamel approached me after the trip and wanted to shake my hand, I withdrew it and said, "Not after your behavior toward us a couple of days ago!" I have never trusted them.

Such changes completely shook the stability we had apparently obtained in our lives. I think that to break a prisoner down, it is enough to move him to a different cell every day. I remember the shock we experienced after one night when the guards suddenly came in and without any explanation separated us, each one in a solitary cell. Later we found out it was some drill for a war, but when it happened I felt totally helpless. A similar event was the death of Nasser. At first we heard the shouts "Nasser, Nasser," like the waves in the ocean. Then we heard the rush of heavy boots in the corridors. We were lying in our beds at night and didn't know what to expect. Nobody opened our door to the courtyard the next morning, and the guards didn't bring in any food. We were very afraid. Only the following day the doctor of the jail came in and told us that the leader had died, but we shouldn't worry since nothing would change in our conditions. I remember a funny episode from that event. On the following day one of the guards was talking to Benny and Motti, and Motti told him that we had heard that Nasser had died. So he said, "Well, yes, you know because you understand Arabic, so you heard it on the radio. But how do all the others know?"

One day an Egyptian plane was shot down by the Israelis, for some reason. Since a highly popular Egyptian movie star was on the plane, we didn't get our fresh food and everybody was mad at us. Our neighbor silenced his radio from then on as if we were responsible for the disaster. Such small changes undermined our sense of security.

BENNY

For about three months we were four in the room, and we felt we were about to be returned home. We figured out we would be returned in May. Dan was seriously wounded, and it was clear he would be returned. They started to feed us better, bringing us restaurant food three times a day, so that we'd gain a little and look healthier. Then they captured two paratroopers, and we understood that our release would be delayed until they finished interrogating them. Then the pilots were captured, and Nasser declared that he would never return them to Israel, so we were really stuck. For a long time, however, Motti and I continued to believe that our release was near, because we were civilians.

One day they brought Yair to join us in the room. He was terribly injured and I volunteered to be his caretaker. Avi and I became surgeons in jail, too; every night we took splinters out of Yair's body. We were so busy with the wounded that we had no time to think about ourselves. I was happy to be healthy and capable of helping others.

The most difficult thing for me was getting used to living behind a locked door. I was a young guy and was used to going out a lot. I had never stayed home in the evening. And here—this feeling that you washed and got dressed but the door stopped you from going out almost drove me nuts. After many years of this you get used to it, though. I feel lucky that we were joined by fellows like

Rami and the others, with whom it was possible to share feelings. Somebody who is unrealistic would certainly break down.

When the pilots arrived, we organized the study program, physical exercises, bridge, and balanced meals. It was actually Menachem, who arrived later on, who took responsibility for the school. He believed this would keep us busy, so that we would stop getting on each other's nerves. I was busy with my writing, too. Every night I sat at the kitchen writing. I made up small plays for the holidays, also simple novels—it helped me a great deal. Afterwards I started building with matches. I built more and more complicated designs, until I made an Eiffel Tower, which took me about a year to complete.

MOTTI

At first there were just three of us—Dan, Benny, and myself—with nothing to do. We started to make some games. All we had was a nail, a pencil, and some paper or cardboard. Out of these we improvised some guessing games that kept us very busy. I started to carve soap, and all the others followed my example. Then I made a slide rule. It was a monumental job, but the result was very accurate. They started to give each of us ten cigarettes a day, and we smoked a lot. One day the doctor came to give Dan a shot; we all started talking to him, and in the confusion I managed to steal his tiny saw. This became our surgical knife when we treated Yair later, removing splinters from his body. From the wires tying our parcels, I improvised pincers for the same purpose. When Yair was returned home, he took a whole bottle of splinters we had removed from his body.

When all of us were put in the room, a completely different life began. We started to study in the mornings, which was excellent. I improved a lot in mathematics and English. Then we all worked to make a radio that would enable us to listen to news from Israel. As a child I had built a radio, so I had some background for that. The first night we picked up Israel on the radio, nobody went to sleep, we were so excited. People were jumping all over the place with joy. This was in secret, obviously. After a while, we persuaded the Red Cross agents to tell our people in Israel that we had a radio, and from then on we used to receive regards on the radio. They were often transmitted a couple of minutes before the midnight news.

I got hooked on bridge. Chess also occupied me for hours. I could play from morning to night, but when we got more books, I started to read a lot. My reading became very fast in jail. I also enjoyed talking to the pilots and learning all I could about airplanes. I was very curious.

We made some nice things together. Amos learned to cook really well, and for Friday night and the holidays we baked cakes, especially after we received the cookbooks. We conducted a Bible quiz. We performed different plays, which Benny wrote. We shifted roles—some acted and the others were the audience, and then we switched.

The most difficult thing was finding something to do. I tried about a thousand different things. I knitted and made sculptures, collected animals, studied, played cards—everything. Not everybody was that busy; it is a matter of personality, I guess. Every day I thought about what I would do on the next day. I always looked for new occupations. I started to copy good passages from books and had a notebook of quotations that I tried to memorize. Every morning I got up early and did my homework, because I discovered that early in the day was my best time for studying. My life was full of interest.

AVI

As long as there were four of us in the room we believed we were about to be returned. All of a sudden Beausart came and told us about two paratroopers who had been captured, so our return would be delayed. But right away, in six weeks time, when their interrogation would be completed, we would be returned [laughing]. That was his style. It was clear that we would be returned together, that we wouldn't leave anybody behind. That's how we heard about the capture of David and Yair. So we had the feeling we were waiting for them. Then in June, the pilots started to arrive. At first, we heard about it from the wardens, but we didn't believe them. One day Beausart came, took me aside—I was the group's representative then because I spoke English better than the others—and gave me the names of all the pilots. He used to say, "Listen, things are not so good, they captured two more of your pilots." I would ask, "Who? Who?" being afraid that they might have somebody else from my squadron, who would contradict the stories I had told them in my interrogation. Besides that, the Air Force was so small, everybody was friends, and indeed I knew all the pilots who were captured after me, except Menachem. Beausart was never hesitant about sharing the news with me, and he updated me on the condition of the wounded in the hospital, too. This was very bad news for us. Each capture was like being hit on the head with a hammer.

That was the time we tried to establish some daily organized routine. It was tremendously important for me to be able to listen to music, and indeed we received a player and some records. Music was my rescue for the whole duration of my captivity. I used to take the two tiny loudspeakers, put them close to my ears, and fly away with the music. But when we asked for a radio they laughed at us, and the Red Cross agent provided strange excuses: "You'll listen to the Israeli broadcasts and this would not be good for you as long as you're here."

We were occupied a great deal with Dan's bad health. In the beginning he couldn't get off the bed, not even to the toilet. He was too fat and wouldn't maintain a diet responsibly. For that reason he couldn't use crutches. Every time we had to take him to the toilet, it was a big collective effort. I undertook his treatment. Motti was also wounded; his shoulder was full of bullets and he couldn't move his arm. It seemed natural that I would become the caregiver. It must be part of my nature.

Yair arrived a little later, a real invalid. He, too, needed a great deal of attention. He was all shriveled, missing an arm and an eye, his body perforated with lead. Since his right arm had been amputated, he had to train himself to use the left one and had difficulty eating without help. He came with terrible stories about the way the Egyptians had treated him. Once, after an operation, they took the urine from his pot and poured it over his wounds, and it felt like fire. They tortured and humiliated him even when it was clear there was nothing he could tell them. He was new in Israel and had scarcely any information about the country or the military. We helped him by taking the splinters out of his body—an episode he dramatized too much when he was interviewed after his return.

So our day evolved around the chores of taking care of the wounded and reading. There was no organized schedule, but Yair and I started to study together. I taught him Hebrew and he taught me reading and writing in Spanish. I used to jog every day in the room and later on in the courtyard. We played some chess and card games, especially canasta. At a certain stage Yair started to teach us how to play bridge, and Olivier from the Red Cross helped in the instruction, so we all started to play.

After all we had been through, we were simply five broken men thrown together. We needed to tend our wounds and to adjust to each other. In spite of that, some semblance of order emerged. A spontaneous structure came into being in the room, based on our contacts in conversations and games, but it was not the kibbutz, organized way of life. This developed afterwards.

When there were only five of us, we had an early lights out, and we used to go to sleep early. The room was quite dark all the time, and there was nothing driving us to get up in the morning. It was a passive lifestyle, and one of the reasons for it was the feeling we shared that we were going home soon. We didn't want to plan anything for the coming time, because our release was supposed to happen any day. I remember that a cease-fire was declared in August 1970, and Kissinger started to make his rounds in the area. We told ourselves, that's it, we're going back. How much will they interrogate the new prisoners, what do they need it for, since the war is over? We calculated that by October all of us would be freed.

Although this was the atmosphere, I was busy all day long. I didn't lie in bed. I got up, read a lot, took care of Dan and Yair, jogged, played cards in the afternoon. I received and sent letters. The day passed somehow, and I never sat down and asked myself: When will this day be over? Benny, too, knew how to occupy himself. He started to write adventure stories, and I used to correct and edit them a bit.

My crisis occurred in March 1971, when Yair was returned to Israel, and again in the summer of that year, after the release of Los from the hospital. That was when I realized that we were doomed to a long stay. There were now ten of us already. I remember asking Yardi to send my records over, as if I had transferred my home to the jail. At the end, after three years, when they came to inform us that in three days we'd go home, I said in my heart, how can I leave now? There are so many things I'm in the middle of doing! I had plans for two weeks more. It shows how we rooted ourselves in the place [laughing].

I remember that during the first stage Motti asked for a guitar and wanted to learn how to play, but they didn't allow it. At that time they also didn't allow us to get textbooks for our studies. Also, for the whole time they never gave us permission to study Arabic in a systematic manner, with books and all, because they claimed that if we could speak Arabic, we'd be able to escape. Surprisingly, however, when Motti asked for a miniature football table, we received it, and later on a ping-pong table as well. It was hard to understand the principles guiding the Egyptians in their decisions. Afterwards we found out that our improvements were tied to what the Egyptian POWs received in Israel. The Israeli authorities would often grant the Egyptian POWs things we had demanded, hoping that they would be reciprocated. This sometimes happened in six months' time, after a lot of persuasion by the Red Cross people. The Red Cross and their visits played a major role in our lives. Everything revolved around their visits. On top of that, they were very nice people.

When all ten of us lived together, it became quite crowded. But then the Egyptians opened the courtyard for the whole day, so you could find a quiet corner outside. We had the ping-pong table in the yard and a big tree in the center. Privacy is relative, you see. If you sat behind the tree, or at the corner, you could almost feel alone. I slept on the high bunk. I had about three meters above me to the ceiling, and when the windows high up were finally opened this was another place I could go for privacy. I got a good place when they brought the bunk beds. I did want to remain close to Dan, who had a bed of his own. Yitzhak slept in the bunk under me, and Motti was near me on the next bunk bed. Benny preferred to sleep below, because he wanted to be close to the record player.

I wasn't disturbed by the noise in the room. When we had arguments about our lifestyle, music was often brought up. The group assembly was a good forum, because we reached decisions that couldn't have been obtained in any other way. The assembly helped in maintaining order, and provided the men with a sense of democracy in which everybody had their say. When a decision was made by the majority, everybody respected it. I don't think there was any other way to organize our lives there. We couldn't have made it by military discipline. There was a time that the Egyptians tried to force us to put on military uniforms, but we refused. We grew long hair and wore shorts all the time.

We did create an orderly life, but a terrible uncertainty bothered me. For example, one night in October 1970, terrible hysterical screams were heard from the direction of Cairo, which was about fifty kilometers away. It was the night of Nasser's death. Being closed in a place at a time like this is terribly frightening. You hear the screaming mob and ask yourself: Who could they blame, after all, but me? Or one day a guard fired his gun by mistake in our yard. Immediately we realized how fragile our existence was. When the new POWs arrived during the Yom Kippur War, I felt very anxious. They were interrogated next to our room and we could hear how they were beating them, from the screams. I remember how I tried to stick cotton and newspapers into my ears so that I wouldn't hear it, because I felt I was going nuts. The sense of temporariness never went away. I didn't think about it consciously, and I functioned all right, but this temporariness was always in the background.

YITZHAK

The two and a half years in the common room could be divided into two periods. The first was a transitional stage. We had to adjust to living in a less hostile environment, and also to get to know each other. There were many personal struggles among us. Each one had to regain his balance and overcome the first harsh six months. Everyone did it separately, for himself, because we had never discussed our interrogations and torture. I think it was due to the sense of guilt and the wish to forget. We were still afraid that the hostile attitude and even the interrogation might be renewed, although when we analyzed that possibility, we gave it a slim chance. During this transition period we got used to seeing life in prison as not completely transient. While in solitary or during the interrogation I had known that this would be my lot for a short time. When I arrived in the common room I realized that now I might be in that situation for longer than I'd prefer to think. This is really strange. In solitary, when I heard footsteps outside, I could make up a fantasy of a good world, to delude myself that they were coming to free me, and I was going home. Everything was interpreted as a sign of the coming liberation.

I tried to preserve this sense of transience in the common room. I felt that as long as I didn't form close relationships with the others, the loss in case of separation or some disaster that might happen to one of us would not be too great. My contacts with the others were superficial. We exchanged some jokes, that's all. After all, it was a time for recovery, and each one recovered alone inside his shell. Furthermore, life in the common room was not an improvement in all respects. In my solitary cell, for example, I had already gotten rid of all the bugs, while here I had to fight them all over, and this made me really mad.

I think that this period lasted for about a month, until we were all joined together. When there were ten of us I felt the noise level had become unbearable, and something must be organized. In fact, noise remained one of the major problems of living together throughout this time. We were ten men, sharing a very small space, without any corner for privacy. It was difficult even to write a true letter home, something from the heart, when all the guys were around. I think that only after four or five months were we allowed to keep the door to the courtyard open at all hours of the day, and the shutters of the windows were removed even later. I always felt suffocated in that room. In spite of the big temptation to peek outside through the windows, once they were opened we were determined not to do this, not to make our wardens angry, so that they wouldn't close them again. It was extremely important that the windows remain open.

It occurred to us to have classes for different subjects. We believed that if we started studying, the Egyptians would permit us to obtain more serious books, because up until then all we got in our parcels, with the exception of the Bible, was very lousy literature. We believed that we should not ask for things that might be refused, and since it is common for POWs to study, we expected the Egyptians to agree. We changed our philosophy afterwards, though. We arrived at the conclusion that it was preferable to ask for as much as possible, because this increased our probability of getting at least some of our requests granted.

The Egyptians were ridiculous. They refused us a blackboard and chalk, claiming that we'd use this to plan our escape. They considered us supermen who might design a helicopter or something like that!

Participation in class was not obligatory, but somehow a norm of attendance came into being and people felt they had to study. Rami and I believed that it would be much harder for those who wouldn't study to cope with the reality of captivity, that they might feel like outsiders in the group, and we, too, would find it harder to include them in our life. At one of our assembly meetings we arrived at a decision to oblige everybody to attend the classes. It was somewhat egotistical, basically, I believe. In the same manner we obliged everybody to attend the morning exercises.

The time for the assembly meeting was fixed for Friday evenings and this helped to create a different atmosphere for Sabbath eve. We didn't look for a religious framework, but the meetings provided a break in the routine of the weekdays. We also had a school break on Saturday, and the guys used to sleep late.

The schedule of our classes and breaks was not so rigid. Even with the study program and all the creative projects, we had a lot of empty time on our hands. I think we did everything slowly.

The quality of our food improved gradually. Half a year after coming together in the common room we were in the position of having all the supplies we needed. In fact, some of us even had to start watching our weight. We were allowed to shop at the canteen, like other political prisoners in Egypt. We gave the guard a shopping list every morning. The canteen staff went to the market for fresh produce for us. We didn't lack a thing. We also received very nice packages from home. We arrived at the point of arguing what was preferable to order from home—this or that brand of cake mix, for example. Since we were cut off from real life, we overrated all the minor details of our life.

The man on duty was in charge of the daily cleaning. Cooking, however, was done on a voluntary basis. Several of us loved to cook, like Amos and Dan. Others were inspired from time to time and cooked something special. I remember that one day Rami and

I decided to produce our own catsup. Some people used to put catsup on anything they ate, so the quantities we received in our parcels were never enough. Rami said he knew how to make catsup, but the first attempts didn't amount to much. We kept trying, like in a chemistry experiment, until we got an impressive result. Others specialized in cakes or ice cream.

I remember when we decided to celebrate Passover as our major holiday in jail. This was an occasion to inaugurate our new table, and it was a very pretty holiday, indeed.

After a while my brother sent me Tolkien's books and got me very excited. This is some sort of escapist literature, about the adventures of fictional beings who conquer evil. Since only four of us could read the books in English—and we couldn't stop discussing what we read—we tried to share the experience with the others. I don't remember who was the first one to initiate the idea of translating *The Hobbit*. I am sure that at least four of us could claim the idea for their own. But it doesn't matter whether it was Rami, Avi, Menachem, or myself. We said that since we were translating the book orally anyway, why shouldn't we write it down, maintaining a high standard, like serious people? We started the written translation as part of our English lessons, but it didn't work out so well. So then we got organized in the Air-Force style: I read the book in English and dictated a verbatim translation to Avi. Rami went over the draft and made style corrections, and finally Menachem took care of the grammatical form. But in fact it became a group project. Everybody was discussing the atmosphere of the text and its meaning, and how it could be transmitted in Hebrew. We argued a lot about that. Rami specialized in the translation of the riddles and the rhymes, and he did a beautiful job. Every Friday night we read aloud the outcome of the week's work. This occupation became a highly positive experience. For me, it was important to bring the product of our efforts home, so that we'd feel that we hadn't wasted our time. Actually we brought the manuscript back to Israel and it was published.

I think that we didn't continue with the translations because we got tired of it. It took us four months, and we didn't feel we had the energy to start another such project. But perhaps the translation also divided us into those who were more active in it as opposed to the rest, and this was something that disturbed the group as a whole.

AMNON

As I arrived in the common room it occurred to me that this would be a good place to fill the gaps in my education. At school I used to be a wild boy who didn't study. I completed the minimal requirements for a high school diploma and then for the aviation course, and that was it. I knew how much studying I had missed. On my second day, I told Yitzhak that I intended to make him my English teacher. I don't remember any time in between. I started to study on my second day.

Learning English from Yitzhak was a catastrophe at first. Every day we sat with our legs crossed for hours on my bed, and I memorized about fifty new words a day. But Yitzhak was a lousy teacher; he taught me Shakespearean English and had absolutely no sense of grammar. Six months later I tried to speak English to the Red Cross men and they couldn't understand a word [laughing]. He knew English well, for sure—but that's not enough to be a teacher. We tried using the record player, translating the lyrics of the songs, so that I'd learn some common expressions. Finally we received some instruction books and from then on I used to explain the grammar rules to him .

After several weeks of private studies, Yair joined me, and Motti, Benny, and David also wanted to study. So Yitzhak formed two groups: Yair and me in one group, where we progressed quickly, and the others who learned more slowly in the second. Learning English was a great achievement for me. When I came out of prison I knew English very well.

I also wanted to learn trigonometry. This was a subject they taught us in the aviation course, but I had memorized the formulas without understanding them. I overcame this lack of understanding in jail and went on to different subjects. As the time passed, we got good textbooks from home. When Menachem joined us, the whole study program became more organized. We sat around the table, as in a class. Rami taught from his own knowledge whereas Menachem used the texts, in a more formal manner.

Afterwards we obtained language instruction books, the Berlitz series for studying French, Spanish, Italian, German, and Russian. The books came with records. Avi and Amos started to learn French, I started Spanish, Menachem, German, and Rami, who always had to climb the highest mountain, tried to learn Russian. In fact we didn't keep it going, but we tried for a while.

I was a passive listener in the translation of *The Hobbit.* I had my share of translation before, when I used to translate songs from English to Hebrew. I wasn't too good in arts and crafts—all I accomplished was one small carpet, which I sent to my brother for his wedding, although it arrived for the birth of his son. . . . We played a lot. At first we played dominoes every night after lights out, with candles. Then we started card games, which saved our afternoons. Canasta was a nice game but bridge was much better since it required concentration, and you could improve your performance in the game. Bridge occupied us until our liberation.

I remember contracting jaundice and being isolated for three weeks. It wasn't a good period. When I returned to our room I noticed how ugly it was; it was simply disgusting. We argued endlessly about changing our standard of living and finally we decided to renovate the room. We started by chiseling the walls with a hammer and a chisel. Then we received wallpaper through the Red Cross and covered all the walls. The place changed its appearance. When we found out that the Red Cross could help us in ordering furniture, we did that. At about this time we were given autonomy in all matters of cooking our meals, and we started celebrations in high style. Rami used to give the guards our shopping list for the day, and at noon we cooked a meal for everyone. We had plenty of food. At the age of twenty-two, I learned to cook a little.

When the first Passover arrived, we had a big debate. Rami, Dan, Yitzhak, and Amos wanted a kibbutz-like Seder. Coming from the city, I resented this idea. Not only was I far from home, and in Egypt of all places, but I wouldn't be able to celebrate the holiday of freedom as it should be celebrated. We made a compromise and decided on having two Seders—one traditional, and the other kibbutz-style. I remember that the Jewish community in Egypt sent us some homemade cheese as a gift for Passover. It was the best cheese I'd ever eaten in my entire life. The Jews of Switzerland sent us a pile of chocolates.

AMOS

I remember how at first nothing was organized, and it was each one for himself. We read a lot but had no common activities. As the time went on, we occupied ourselves with the improvement of our conditions. Due to our initiative, and the things we received from Israel, we obtained a fairly high standard of living.

In the beginning our room and courtyard used to be cleaned by prisoner-servants who were nicknamed "duffas." Gradually we persuaded them that we would rather do everything by ourselves, and they stopped coming into our area. The Egyptians' attitude toward us was rather good. We lived in the officers' wing, and from the corridor I frequently heard the wardens speaking respectfully to the prisoners: "Yes, Effendi; no, Effendi." It didn't seem like they hated us, or that they wanted to punish us. The period of interrogation was over, and when they had beaten us, it was for the purpose of obtaining information. I realized how fairly we had been treated when during the Yom Kippur War I was placed for several days in a wing with new Egyptian prisoners and witnessed the awful treatment they got.

I know that the famous psychologist Maslow claimed that only after obtaining your basic material needs can you free your energy to obtain higher goals. That was the case with us, too. When our material needs seemed to be satisfied, we started searching in other directions.

I had always wanted to study engineering at the Technion, so I started to move toward this end. The study program determined my routine in jail. At that time we already got up together for exercises. We had fixed hours for our meals, the chores, and the different activities in the afternoons and evenings. I think it took us about six months to establish this orderly life.

I think I was the first to initiate the study program. I asked for academic material from the Technion, and it arrived, so we all started to study together. We started in study groups before Menachem's arrival, but he had an important contribution because he could be an academic instructor for me, and our work was accredited by the Technion when we came back. I accomplished quite a lot. I covered the material of a whole semester in a year. We studied mathematics out of a book translated from Russian, and Yitzhak taught us English.

In the afternoons I played bridge and did different craft work. I made a leather handbag, I embroidered, and I loved to cook, especially cakes and ice cream. One of my embroideries is exhibited in Geneva in the Red Cross museum. I gave it as a gift.

When you're confined in such a tiny world, all your proportions tend to change so that minor things become highly important. Thus, for example, finding a name for the kitten we adopted was much more important than all kinds of events happening in Israel or outside, sometimes even more important than what was happening with my daughter. We dedicated two meetings to naming the kitten. When you have no contacts with the outside world, it fades away, and what remains are the tiny details of your own small universe.

I think that life in the common room had three stages: chaos in the beginning, like living with boxes; then organization with the daily schedule, the studies, our rotation of chores and the assembly; and the third, a two-year period of near-routine life in prison.

MENACHEM

In the assembly meeting the day after my arrival I told the men that it was impossible to go on living in such conditions. I argued that the room had to be painted and that small shelves should be ordered for each person. I explained that if we painted the room white, there would be more light in the room. I also asked to order a cover for the ugly table. I proposed starting a regular study program.

The proposals led to arguments. Everybody agreed that the walls should be whitewashed. But regarding the shelves, some said that it was preferable to leave the Red Cross alone until we had more important requests. Their outlook was that the less we asked for, the higher the chances that we would really get what was important. I offered a pragmatic approach: let us ask for anything we have in mind, and if worst comes to worst, we won't get it. All my propositions were approved. More important was the resolution concerning our lifestyle—about order and cleanliness, studies and conduct. My ideas were intended to prevent deterioration of the men, something I dreaded. Perhaps I became aware of this threat somewhat later. It was easy not to get up in the morning and to stay in bed instead. So I proposed to make the morning physical exercises obligatory for all. I didn't support the kibbutz-like approach, which let each one do his own thing.

I didn't need to be with the group for long to realize what the essential changes were. I had some experience with the Red Cross and I knew there was no harm in submitting requests. Moreover, I felt that the men needed occupation, so what was better than asking for paint and brushes and offering to do the paint job by ourselves? The same ideas guided me for the next two and a half years. I was trying not to let the group die out.

One phenomenon repeated itself in many forms: people claimed to behave this way or the other in order to simulate a sense of home in jail. But we were ten men in the room, and the question was, whose home are we referring to? When I arrived in the common room, they used empty jam jars to drink coffee, and this disturbed me a great deal. Every now and then a jar broke and was replaced by another. I suggested that the Red Cross order ceramic mugs for us, the same as I had at home. A typical argument followed: Rami claimed that the jam jars gave him a feeling of home. Some admitted that they didn't care what they drank out of. Others believed that if we demanded too many, we wouldn't be able to get the essentials. There was also a childish kind of opposition: I object because it's your idea, not mine. But I was very forceful on the subject of raising our standard of living; it was almost an obsession with me. As I had demanded, we received nice mugs from Israel. They all liked them and enjoyed drinking out of them. Other improvements concerning the aesthetic aspect of our life were introduced in a similar fashion. We designed, for example, a new table that would be both pretty and clever. Imagine: six engineers designing a single table! It was actually very convenient. Once we had a table, it was time for chairs, which would replace the former stools. The chairs had an additional advantage—they could be moved outside for reading in the sun; why not?

The food we bought and received in our parcels enabled us to prepare a varied menu. Afterwards we got an electric refrigerator, too. All these benefits, including the food from the canteen, were funded by the State of Israel, through the Red Cross. I know that all in all it didn't cost more than $500 a month, not too much for the maintenance of ten men.

We used to eat supper around the table in an orderly fashion. Once the men sat down to eat while I was conducting some business with an Egyptian officer in the yard. I was very angry and protested: "Why, in my home we wait for my three-year-old daughter to sit down, if she's playing, and we don't start eating until we're all there. And I wasn't just playing!" Again we faced the same problem—feeling like we were at home, but whose home?

The group had strange eating habits. In the morning, five milk bottles were opened and everybody was permitted to drink a half. Olives and sardines were equally divided and put in one's plate. I resented this dividing business, when we had all the food we could eat. It's true that I had never suffered hunger in captivity, while the others had. But what has passed is over with. I asked, "At home, do you also count olives and sardines and divide them among family members?" I started to sabotage this perfect order. One morning I declared that I would like to drink a whole bottle of milk all by myself. David started to yell, but Rami saw my intention right away. I asked David, "What's your problem? Do you want a whole bottle, too?" Suddenly it was discovered that when food is not rationed, less is consumed. But it took us some time before all this dividing was abandoned, and of course, when Swiss chocolate arrived, even I understood the need for equal division.

The Egyptians used to heat our shower water only once a week. I have to take a shower every night before I go to bed, so I used to take a cold shower, which was quite unpleasant in the wintertime. However, we gave in to the rule of the Egyptians. After two years I proposed asking the Egyptians for warm water three times a week, since we were the ones to pay the bills for the fuel anyway. Lo and behold—we got it right away! These were small achievements but they were important.

I didn't like the way of getting up in the morning either. When I arrived, a sergeant used to come in and open the door to the courtyard at seven o'clock every morning. After this, one of the guys would get up and put on a record of Blood, Sweat and Tears very loud, the same record every day, which would wake us up. I resented having to start my day like this. It was certainly nice music, but who said I had to listen to the same song every morning? So I stopped this habit. Somebody would get up and walk around, and naturally, this would wake up all the others. If somebody kept sleeping, we would approach him: "Hey, it's eight o'clock already." So a new routine developed. We got up, drank something quickly, cleaned up the room, and the first class would begin.

The organized study program started after I joined the group, and it continued almost to the end. It was fantastic. Each lesson was well prepared, a daily routine emerged, and the mornings were very interesting. We used the same table for studying and eating. Those who didn't have a class, read. Perhaps not everybody was a natural student, but a group norm of studying evolved and no one said, "Well, I don't feel like studying." It's not easy to be the odd one out in such an atmosphere. Some of our studies were later accredited by the Technion, and in fact I received a formal title of lecturer for that purpose, and Amos was considered to be my student working toward his degree in engineering. We didn't study on Saturday and the holidays, so we felt the difference on these days. That had some importance as well.

Outside of studying we read a lot. I listed all the books I read in jail and the number reached four hundred. We had very serious books, which I studied, like the works of Plato or the history of the crusaders. I remember a funny episode concerning that history book. I told Rami that he had to read that book, it was great, and Rami answered, "But I have no time." This shows how busy we were in jail.

The translation of *The Hobbit* was the climax of our activities. We dedicated about three months to the project. Four of us worked together originally, and the others joined in later. Some made copies of the original manuscript, so that we would be able to hide it in different spots. In the end we sent home one copy via the Red Cross people, but it never arrived at its destination. Probably the Egyptians kept it for themselves.

Another activity was our game of bridge every afternoon. I was about the only one who didn't get hooked on the game. Bridge is a great game, and I don't know why it never grabbed me, but I enjoyed sitting behind one of the lousy players watching his mistakes.

In addition, we took good care of ourselves physically. We exercised regularly, later on with the help of equipment that was sent to us. We had a bike we connected to the water pump, thus solving the water shortage in our room. It was only one of our many inventions, and most of them are still secret.

6

SOCIAL LIFE

This chapter is based on collected quotations that focus on relatively informal interactions among the men. It reveals how, on the one hand, the prisoners experienced a great deal of togetherness, even when doing something individually, such as reading or writing letters; on the other hand, respect for individual privacy was highly regarded and exposure of feelings or intimate material was rare. The chapter describes the tensions among the men and the means used to reduce them. The emerging picture indicates the different roles of some of the individual members: Menachem, who insisted on organized activities and improvement in standards of living; Rami, who established norms of interpersonal behavior and became the father figure for others; Yitzhak, the joker; Amnon, the growing adolescent; and David, who was always difficult to please.

RAMI

Our reality in jail was an endless experience of togetherness. Even when someone was busy with an individual project, such as making a sculpture or studying, he knew he was being watched by the others and had to accept this. It was crowded, and that overcrowding produced stress. Yet rarely did someone declare that he wanted to be alone, away from the rest. Our willingness to live and act together dominated. I never wanted to be alone. My first six months in solitary were certainly enough. But perhaps it was relatively easy for me to live in the group because of my status in it. Our life was organized more or less according to my pace, so I wasn't under pressure.

I remember an exceptional episode; I believe it happened after Amnon's return to the common room from his medical quarantine. Benny announced that he would request to be moved to a solitary cell. He explained his request as a means for getting out of the tense atmosphere in the room, but I think it was a test for us, to see our responses. Since we all reacted negatively, he never repeated it.

The group discussions continued regularly until our liberation, and they provide the strongest expression of our collective life. We acted as a democratic society, yet at the same time I was appointed the group's spokesman vis-à-vis the authorities, namely the Red Cross and the prison administration. It was clear that we needed a representative, but it wasn't clear it should have been me. Actually, I was the highest ranking officer in the group until Menachem's arrival. I was then a major (Menachem and I were both promoted to lieutenant colonels in prison on Independence Day, 1972), but Benny had a significant advantage over me in that he was fluent in Arabic. I think I got the job because the guys noticed that I was cool in my contacts with the authorities and could speak for us calmly.

Menachem's arrival, about six months after we had all been there, had a great effect on the group. He contributed much to the study program, but from my perspective that wasn't his major contribution. He and I had the same rank, and we had many arguments about our lifestyle. Menachem probably competed with me for the leadership position, because he's that type of guy; he'd compete with the entire world, wherever he is. However, this pleased me very much, because with him around I had a worthy opponent. We talked openly, sometimes argued to the point of screaming, yet we respected one another. We didn't manage to convince each other frequently, but a high quality argument doesn't have to end in persuasion or a compromise; it's enough to be exposed to an opposing view. I was glad when he confronted me and yelled: "That's not the right way!" To this day there are very few people who can confront me like that.

I wouldn't call myself the leader of the group, not even the "strong man," but rather the one holding responsibility. I felt committed to the system, to keep it functioning in the most effective way, so that we wouldn't shame ourselves, and that Israel would not be ashamed of us.

I accepted responsibility for the quality of our life. I was among those who drove each one to find his own creative outlet, whether it was building structures with matches or knitting. When Amnon, for example, expressed interest in studying the geography of South America, we ordered him books and maps. I was also responsible for our approach to personal crises or moodiness. We adopted the habit of not rushing to the depressed man with consolation, but rather providing him with a framework that would help him find his own way out of the crisis. We avoided the development of intimate relationships, which might lead to carping at one another, but we created a sympathetic, supportive system. When an argument or a fight broke out between the men, I was careful not to apply my own standards to determine who was right or wrong; I always preferred to have a matter-of-fact discussion of the problem at hand, ignoring the underlying hatred, envy, competition, or frustration that had evolved due to our circumstances. For example, I remember a piercing, bitter argument between Menachem and Benny about the nightly use of the fan. One was too hot, while the other couldn't sleep with the fan on, and neither of them was willing to exchange their beds. We finally found a technical solution by turning the fan at a certain angle. I never confronted them with "You dumbheads, what are you fighting about?!" This way I prevented the possibility of slipping into emotional territory, and we remained on the technical level of problem solving. When a conflict erupted about the hours for playing music on our record player, I never hinted that it was really a cover for a power struggle. At first I did that intuitively, but as I gained more experience I did it intentionally. I decided to avoid getting into men's intimate lives and preferred to lead the group on the superficial level of everyday life arrangements. This was perhaps the outcome of my failure to conduct group dynamic sessions. It was a sort of defense: as long as we don't deal with emotions, they won't destroy the fabric of our togetherness. Furthermore, it was evident that I couldn't force the men to act in accordance with my moral level, or persuade them to do so, so it was better to have matters decided by simple majority rule.

If I was the leader of the group, it was because I was better able than the rest of us to integrate all the facts and select the optimal course of action. It wasn't I who always pushed the group forward—that was Menachem's role, while I often slowed him down for the sake of the others. With their sense of humor, men like Yitzhak, Amos, and Avi helped in releasing the tension. I took what everyone had to offer, and out of that built our common system.

I was the brake when it came to our standard of living. My attitude was that we shouldn't endlessly improve our standards. I was wrong about that. Had the group accepted my attitude, we'd have kept on living at a very low level. But I felt it wasn't respectable to make more and more material requests. I believed that facing our existence with all its deficiencies was an honorable choice, demonstrating self-restraint and self-control, which I highly value. We had a lot of arguments on this, most of them between Menachem and myself. Under Menachem's leadership, the majority voted against me, and I think it was all for the best. Only later did I realize how physical conditions contribute to the life of a group like ours. It turned out that the men really cared about the appearance of our room. When the walls were covered nicely and everything had its place, I, too, enjoyed the change.

Menachem and I had other subjects to argue about. He was for the separation of officers from soldiers, as had been the practice in many POW camps. He claimed that even in prison we should see ourselves as a military unit, in which the status of the officers provides them with authority vis-à-vis the simple soldiers. I objected to this approach and struggled for absolute commitment to our democratic system. On this subject, I was the winner.

We could have established the lifestyle of a military unit in jail. In that case, the highest officer would issue orders to everyone that would cover all aspects of life, like when to get up, how to dress, what to do. Such a regime could have stopped the constant clashes between the men and helped them cope with the circumstances. This was Menachem's minority stand, while I fought for the establishment of a more natural framework, which would be active and productive for its own sake and would enable each man to start from his current level and from there make progress at his natural pace. As a principle we didn't force anybody to do anything, and I think we succeeded, because each one felt accepted as a person of value with the justification to be himself. Because of the legitimacy of arguing and the democratic decision-making procedure, we avoided deep conflicts within the group.

The way I see our history, we joined an already existing group of veteran POWs who had their own culture, and the two groups merged gradually. The old group preserved its identity by means of their own childish sense of humor and a lot of noise, dominated by Yair's personality. With our arrival, and when he was gone, a productive culture emerged with reading and craftwork at its center. This became possible also because we were getting many more parcels from home. I don't think it was significantly related to the division into Air Force versus non-Air Force POWs. Avi was a pilot, yet he clearly belonged to the old group, while David, who was a paratrooper, arrived with the new one. As time went by and we occupied ourselves more and more with studying and reading—especially when we translated *The Hobbit*—a division of the more educated versus the less educated men emerged. Motti, Benny, David, and Dan were among the less "advanced," but so was Amnon, in spite of belonging to the Air Force. It was a matter of busyness, not of status; a distinction between those who got up earlier in the morning to attend the first class, and all the rest. It was true that one mathematics class dealt with multiplication problems, while the other studied calculus, but the multiplication problems weren't that easy and we respected the mental effort required of the students. From the social point of view, Avi belonged with Dan and David; they shared jokes together. I was putting in a lot of effort to keep the gaps from dividing us, and I always looked for common ground.

Avi and Dan were close friends, but no subgroups were formed except for the close relationship between Yitzhak and me. From time to time we separated ourselves from the rest, sharing our personal life with each other. He became like a brother to me. Never before had I had such a close friendship with a man, and I think that only due to our isolation in captivity could I have allowed this to happen. Generally, it is dangerous to form that kind of relationship because it may damage the delicate fabric of the collective.

During the first stage, we often heard complaints like, "Listen, that's not how you do this or that"—as if people of a higher culture had arrived. This pride disappeared later on, because the men changed and learned to relate to each person's strengths, rather than to their weaknesses and outward behavior. I remember that Menachem used to say that we were wrong to create for some people the illusion that they were worth more than they really were, because they were bound to be disappointed when they'd return to Israel. This indeed happened, in a way, after our liberation.

The matter of equality became salient with the packages. At first, we were spoiled more than the others by the Air Force: they used to send six items of everything. It took them about a year to realize the implications of this policy, and from then on the Air Force adopted the whole group of ten. In one of the assembly meetings, it was decided that when parcels arrived from our families, the recipient should identify the personal items, and then all other stuff that was not private would be the property of the whole group and divided either equally or randomly among us all.

The Hobbit was initially translated for David, since he couldn't read the book in English. He was the group's kid, in a way, and we wanted to educate him. Coping with Tolkien's English was a challenge. We tried to convey the atmosphere and the spirit of the story, and this raised our creativity to its climax. Four of us worked very hard to enable the others to read the book. The four months dedicated to this project was a beautiful period, full of elation; it gave us a sense that we were winning against the whole world.

But when it occurred to us to translate the whole trilogy of *The Lord of the Rings* I decided against it, so that we wouldn't increase the gap already existing between us. I had another argument: this was a project far greater and more serious than the translation of *The Hobbit,* and I was afraid we wouldn't succeed.

Naturally there was tension among the men. It couldn't be avoided, since each one of us lived so close to nine other men he didn't choose to live with, and in a situation he didn't like. There were arguments, but it is important to stress that only twice did they involve physical violence. Menachem and Benny used to argue a lot. I believe that being able to bring up controversial issues for group discussion in our Friday assembly was the main tool we used to release tension. But the size of the group was an even more important factor. We were ten—neither three nor thirty. Our group was small enough to enable personal contacts among all of us, and big enough to prevent the formation of intimate relationships. That's why we failed in our group dynamics attempts. We formed a network of relationships that made our lives easier. After our return, I heard about a group of three Israeli POWs who were together in Syria and formed much deeper friendships than we did. But ours was a supportive system that alleviated the group tension and personal depression and kept them at bay, without dealing with them directly. At the same time, we got to know one another so well that you knew what someone was going to say before he uttered a word. People didn't have to hide or be ashamed of anything; we were very open to each other. Only Yitzhak and I created a deeper bond involving the two of us alone.

MENACHEM

There is no doubt in my mind that the group was lucky that Rami was among its members, because he is a very special person indeed. He is solid and mature, unusually wise and humane. Luckily he was also senior in rank, until my arrival, and had the longest record of service. God knows what might have happened had I been of higher rank.

I had my own place in the group, but I never reached the position of saying, "Hey guys, listen to me and don't follow Rami." I was deeply disturbed by the democracy he had established prior to my arrival, but after the fact, I am not so sure that had we behaved otherwise, our life would have been better. Deep in my heart, I sometimes admit that it was the best method of all. What was the alternative? That I would dictate to a poor captive how to live, because I was older, wiser, and of higher rank than he was? We could have established a military framework and conducted our business by discipline, but this was never tried.

Today it is hard to explain what it was about our democracy

that disturbed me so much. I think that the democratic process might have oppressed the minority in a cruel manner; for if nine men agree on a certain question, let's say on drinking coffee out of old jam jars, it's easier on them because they're together on this. The tenth man, however, feels trapped, really beaten. Not only is he miserable because he's alone, but due to the external circumstances there's no way for him to get out of this group. That was why I argued that we should take into consideration the single person, rather than the nine. But my argument sounded like Greek to the others. The guys couldn't comprehend it. Yet I felt that had my principle been accepted, it would have solved some of the hardships of the weaker members of the group. Imagine that by the majority rule they had decided to read only the kibbutz Haggada on Passover, because that was Rami's wish, and he was the unquestioned leader of the company. Had this been the case, I would have rebelled against the majority. I would have sat outside and not participated in the Seder. But all the same, the democratic solution was the best in most cases.

Once Rami and myself agreed on a matter, there was no question it would be adopted by the group, since together we were completely dominant, each in his own way. By nature, I am not a leader. I do love power and influence, but I don't enjoy a situation where people hang on to me.

After my arrival, you could say generally that the group developed an elliptical structure, with Rami and me as its foci. I think we should all be grateful for that, because had they stayed with Rami alone, they would have all eventually fallen asleep. When I joined the group, I formed a new focus of power, and thus changed the former equilibrium. I pushed forward, whereas Rami supported the people and formed them into a kibbutz-like group based on equality and cooperation. I don't think there was an individual in our group that I didn't respect. I had arguments with Rami about that, too. He used to scold me: "Why do you keep arguing with Benny, leave him alone!" I said that by arguing with him, I was demonstrating my respect for him, because it proved that I considered him a worthy opponent, and that I was not lowering myself to his level. I claimed that Rami, the democrat, was a greater snob that I was. I was. . . . like a spoon that stirs the tea in the cup. I didn't let the group sink. Just the same, had I been there alone with the men, without Rami, it would have been a disaster. I wouldn't have left the people alone for a moment, and this might have driven them nuts. So neither an excess of rest nor restlessness is healthy. Together we balanced out one another, although if one of us had been there alone, this essential balancing process might have taken place anyway.

I assume that all these arguments between Rami and me were good for the group. This way, some of the weaker members discovered that they could confront Rami: "That's what you say—but I don't want it that way. You say it's stupid to ask the Red Cross for more chocolate, but it's my right to have another opinion."

Rami didn't provide direct emotional support to any of the men except for Yitzhak. From time to time they would withdraw from the group and sit together in the corner of the courtyard. It became clear that they had closed themselves in a bubble and didn't want anyone to join them. There was no other couple like that among us.

I don't think we formed subgroups either, certainly not a subgroup of pilots versus non-pilots. Rami may tell you that he stopped the translation project because he was afraid that this occupation might have broken up the group. Perhaps he was extra sensitive to such matters, but I didn't experience this danger. The group consisted of individuals of varying power and quality, but an elite didn't emerge. It would have been a disaster if three of the pilots would have formed such an elite. People must have sensed that somehow and avoided the formation of small stable cliques, and thus saved us from extra tension. I didn't experience a special intimacy with any of the others, and I didn't miss it, either. Living in such conditions, it was preferable not to become too close to one another. Each of us had his high and low periods, naturally, yet nobody tried to understand why his companion felt that way. No one tried to expose himself, or to uncover the secrets of the other.

We established this special social equilibrium: we did a lot together, yet each one preserved his privacy. It was typical of the group that we developed a very high moral standard, maybe due to the background of some of us in the youth movement. No lies were accepted; nobody cheated the others. We never gossiped maliciously about members of the group, not even in our private letters.

YITZHAK

The fact that several of us had a kibbutz background probably affected the structure of our group. In the kibbutz one lives under the influence of public opinion, which obligates everyone. I reached the conclusion—which I think Rami reached independently—that since we share a room we have to live as a group. In other words, it is unacceptable that each of us will do his own thing. Had we been given a larger space, where people could be alone some of the time, this might have been possible. But not in our crowded quarters. You couldn't decide not to get up at a certain hour because in that case I wouldn't be able to put the light on and start reading or studying. The meaning of our democracy was that an order would be decided upon, in such a manner that it would satisfy the needs of the majority, so that no one was allowed to disturb the others.

In the system we formed, one would never say: "You'll have to get up, exercise, and study because I say so, and I'm stronger." The vote of the assembly obliged us all, and we thought that this would diminish the arguing among us. We voted on everything, even though sometimes it was funny. Some things were decided upon with two against, one in favor, and seven abstaining. We did have a leader, naturally, and we were lucky that it was Rami. Thanks to him we endured this period like we did and not much worse. Rami is a great man. Menachem and I also had authority of sorts.

The leadership of the group decided to prevent the formation of subgroups. We were apprehensive that a sense of discrimination and bitterness might have resulted had it been otherwise. Fully aware of the matter, we took care to practice a complete equality with no group differences. I think this was a wise step, so that life for the weak members would not be even harder. This behavior was for everyone's benefit.

The order of the bridge games, for example, was clearly related to this social system. We all wanted to play, and the formation of the foursomes playing together became an important matter. This game demands concentration and understanding, even some education, so naturally there were better and worse players. But we didn't want the good players to play their own game, so that a separate table would be left for the poor players. We also didn't want to shuffle the cards in such a manner that we could know beforehand who would be the winner. After many debates, a rotation of the players and tables was agreed upon. Rami was the only one who didn't care who he played against. He played equally enthusiastically with any player, and he was also the only one who maintained his enthusiasm to the bitter end, until our release.

The presence of Menachem contributed a great deal to our studies, especially to the discipline of us all. Menachem is an orderly person. He sees things either in black or white and rejects compromises. He doesn't use the word "maybe." It was not easy to live with him. He was less sensitive to the needs and problems of others. Rami was exactly the opposite. Menachem could function only in a structured system, while Rami could be spontaneous and use any opportunity to initiate something. Menachem objected to the democracy because he felt he knew better than the others, therefore his ideas should be accepted. Rami was a proponent of direct democracy, like in a kibbutz. But the contrast between him and Rami was very important. The arguments between them clarified the different perspectives and led us to find the best compromise.

There were two groups—leaders and followers. I am a quiet type, yet I was together with Rami and Menachem in the first group. Dan was difficult to classify. He was the sort of kibbutz member who had never been exposed to another kind of environment. He had very strict principles, and we respected him for that, as much as for his age and seniority and the fact that he was a wounded man. It was, however, hard to get him to see the limits of his principles. He was rather slow to comprehend and to adjust to new situations.

Rami and I became very close and we also spoke about intimate, personal matters, which we wouldn't share with anybody else. This friendship was highly important for both of us, and it helped us cope with being in captivity. Menachem also talked to me from time to time, but it was a one-sided communication. This was the case with Amos, too. I am apparently a person who is capable of listening. Dan loved Avi, who took care of all his needs in a most devoted way, but I don't know if they talked. The Egyptians treated Dan awfully. He had wounds in his leg, which turned gangrenous, and we had no means to help him. We used to bang on the door: "Doctor!" and when a doctor finally came, he would put some sulfa on the wounds, bandage them, and go away. We were sure that he would lose his leg in the end.

There were also a couple of nay-sayers, who used to make a lot of noise in order to get attention. We tried to control them, especially when they were fighting. We were not willing to have the Egyptians witness violence among us.

There was tension among us. I suffered a lot from the smokers, especially as long as the courtyard was closed most of the day. The smoke in the room suffocated me. Motti, Benny, Amnon, Dan, and David all smoked heavily. I understood that they were addicted to it, and it was easier for me to accept that than for them to stop. Afterwards it was decided that smoking was allowed only in the yard, but I hated the full ashtrays left around the room. Once I flared up when I found a full ashtray on my bed, and I spilled its contents on Amnon's bed.

Sometimes I was disturbed by the others' neglect, concerning the cleanliness of the toilet, for example. I didn't try to change them, though, and only when I was on duty did I clean as I thought it should be done. Usually, however, even in this area everybody's conduct was acceptable.

People were emotionally restrained, generally. I never saw anyone cry. I know I cried, and possibly each one did under his blanket. Perhaps they thought that expressing one's feelings was unmanly. I don't remember anything more expressive than staying in bed in the afternoon instead of going to join the bridge game. Nobody suffered from a long spell of depression. On the contrary—we laughed all the time. Everything was turned into a joke. You could never expect a serious answer to your question the first time you asked it. Sometimes it was too much.

Much of the humor was focused on our contacts with the guards or the prison authorities. Some of the jokes expressed how superior we felt in comparison to them. There was, for example, the story of the pump. Water pressure was low in Egypt; we used to say that the water was tired. When our water container had to be filled, several prisoner-servants in blue overalls, the duffas, were called for the job. One climbed on our roof, the other filled a bucket, and they hauled it up with a rope. Our first idea was to order a pump from Israel, financed by our government, obviously. We explained that to the Egyptians. One day an old plumber, who knew all the pipes by heart, came for an inspection. Two months later he returned, smiling with his toothless mouth, bringing a manual pump. From then on we could fill the water tank without the help of the duffas. At the same time, we ordered an exercise bike from Israel. The first one sent to us was confiscated by the Egyptians, because the Medical Corps wanted to examine it. . . . When we found that out, we ordered a second bike, and asked the Egyptians if one would be enough for their study. They answered, "Inshalla," God willing. Finally we received the bike and it was placed in the yard for the men to exercise on. One day Rami was operating the pump while I was riding the bike. We both noticed the similarity of our rhythms. That was how the idea of connecting the two was born: we used a broom and a rope to tie the pump to the bike, and from then on, whoever exercised was also pumping water to our tank.

This was a clever idea, and it got the Egyptians very enthusiastic. The commander of the jail arrived, gave it a look, and said, "The Jewish genius at work." The funniest episode happened when the mayor of Cairo came for a visit. He asked, "What's that?" I told him it's an original invention of ours. He asked how it worked, and I invited him to try it for himself. And so it happened that ten Israelis were tanning in shorts in the sun while the mayor of Cairo was pumping water for their shower in his best suit.

A similar thing happened when Rami was demonstrating something in his physics class. It was a simple experiment, a demonstration of a vacuum, and the guards who interrupted him suddenly saw water rising in a glass turned upside down, as if by magic. One of them yelled, "Electronics!" and rushed out. We knew it was trouble, and indeed, several minutes later the commander of the prison entered, finding us all sitting innocently around the table. He asked, "Where are you hiding this electronics? Electronics is strictly forbidden." Rami repeated his demonstration, using a bowl of water, a glass, and a match. The shocked general walked out of the cell keeping a dignified face. For the life of him he couldn't have comprehended the experiment.

The Egyptians placed a great value on education. Some of the guards wanted to learn from us. I remember teaching some of them reading and writing in Arabic! One of them brought a third-grade reader, and I recall how shocked I was to see the caricatures of Jews in the book. It was anti-Semitic literature like during the worst times in Europe. It made me think that there would never be peace between us and them, if that's the education they got. I think that Menachem managed to teach one of the guards to read Arabic, in spite of the fact that he kept insisting to them that he didn't know the language. We all ended up knowing some Arabic, though.

Most of our guards were good guys, who didn't know us during the torture period. Several of the old-timers, who had been with us since the beginning, sometimes maintained their mean attitude. After some time, only one of those remained, and we named him "the dwarf." We discussed our plans about him at several of our Friday meetings. We debated whether it was advisable for us to try and get rid of him, and whether it was moral. Finally the majority decision was that since he was so bad, we had the right to plan our revenge. When we received gift packages for the Egyptian Revolution Day, we had an idea. Each of our gifts included a bottle of rosewater, an Egyptian perfume we found repulsive. When the "dwarf" informed us that he was going home for the holiday vacation, we gave him the ten perfume bottles as our gift. The moment he left, we called for the prison commander and told him that our perfume bottles had been stolen. They checked at the gates, and naturally caught the "thief." We never saw him again. I don't know if he was jailed or transferred to another location. I'm sure he yelled that he had been given the perfume as a gift, but nobody believed him. . . .

The other guards were good, but we didn't form any close relationships with any of them. Some were posted outside our door, others brought us the food we purchased from the canteen. I think they were afraid to talk to us; perhaps these were their orders. When we needed the doctor or the commander to come, we used to bang on the door.

We created a kind of an island, isolated both from Egypt and from Israel. We developed our own culture and humor and accumulated daily common experiences. We formed a tiny universe, fed from the outside, but within it we felt as if it were autonomous. From a certain point of view, it became our haven. We conducted an orderly life in it, and we didn't miss a thing. Even during the Yom Kippur War, being the citizens of an enemy country, we were reassured of our safety. I think that some of us started to feel that nothing wrong might happen to us there. Perhaps we were even afraid to go home, where we didn't know what to expect.

AMOS

Having no choice leads to adjustment. When you have no choice, you learn to share a small room with ten men, whose background and needs are quite different from yours. It was hard at the beginning. We needed to reach a balance and it took us some time to learn how to be together. But after we learned, we lived well. Let me give you an example. We brought from the Air Force very cynical attitudes and style of speech, and all of a sudden we found out that others didn't understand what we meant. I had to learn how to converse with Benny, for example, in a style entirely new for me. Gradually I learned how to talk to each of the men so that he would understand. I started to take into consideration each one's idiosyncrasies. All in all I learned to listen better and be much more open to others.

I remember that after my return, the style of speech in the Air Force sounded strange to my ears. We dropped all cynicism and became very open toward each other; there was simply no other choice.

Rami was an excellent model for us, because he knew how to create wonderful human relations. I wasn't too bad myself. Rami was always willing to support whoever was in need of support, was able to solve any problem at all, and enjoyed doing it. If he was ever in a bad mood himself, we weren't aware of it, because he has a very special personality. Most of the men learned to be sensitive to the others and acquired a much more tolerant attitude. We lived in a closed space, and had we not changed accordingly, a terrible explosion would have occurred. No one had another place to retire to, there was not even a small cell at the side where someone could hide for a moment away from the crowd. Afterwards, when the yard was opened for us, it became somewhat easier.

Everyone had his days of being down, when he required an outlet. We had to look out for these moods and allow that person some rest when he needed it. A very delicate situation emerged in our closed-in, crowded reality. It allowed privacy within the crowd. At the beginning we used to upset each other, but afterwards we stopped.

There was no division into Air Force and non-Air Force groups. There was no basis for such a formation; we just felt we were people living together, all equal. However, only those of us who were fluent in English could negotiate with the Red Cross agents. Rami's leadership was not determined by his rank, but by his qualities as a human being. Menachem's rank was identical to Rami's, and I think he wanted to lead the group, but we didn't accept him as a leader. Menachem created tension in the group. He made people angry when he didn't take them into consideration. But in spite of the tension I was much more pleased with the ten of us together than I was alone. When I remembered my loneliness, I found it easier to cope with the crowd. I loved the bridge game each afternoon, for example.

I supported the democratic system that we had established, because I myself was born in a kibbutz and lived there until I was fourteen. We wouldn't have been able to live by another system, such as a military unit. They tried to make us wear uniforms but didn't succeed. We were willing to wear only Israeli military uniforms and refused to put on the prisoners' clothes. For the duration of our captivity, we continued to wear shorts all the time. I am sure that a group of men living closed in for such a long time can achieve something only by agreement. That's why the democratic system was appropriate.

AMNON

There were great age differences among us. I was three and a half years younger than Amos, and he was three and a half years younger than Yitzhak. Rami and Dan were much older and so was

Menachem. The age gap was very significant for me. I had not yet left my family at the time, and the only reality I knew outside of my family was the aviation school. I had not matured yet. I had only had superficial acquaintances and had not yet managed to form a deep relationship with a woman. The others were married and had children, so I tried to absorb their experience. I don't think I was hurt by feeling the least important of all the pilots, because I was indeed younger and inexperienced. So I sat for hours and talked to Yitzhak in English, to the others in Hebrew, and asked them to tell me about their past experiences, about marriage and family life, about raising children and sexual relationships. Every evening, before we went to sleep, we used to have long, quiet conversations on almost anything in the world.

My position as the child of the group changed, however, as time went on. Yitzhak, who had problems in his family, withdrew from me and turned to Rami for support. When Menachem arrived, we formed a good relationship. He can be a nice man. With him, too, I was the sponge, trying to absorb his wisdom and experience in the world. We were good friends for about a year, until a blow-up between us. Menachem was obsessed with cleanliness. We all used to soak our underwear in a bowl, and then wash it and hang it out to dry in the yard. Menachem needed the bowl every day. One day, as my laundry was soaking, Menachem approached me and said, "Take your washing out already, because I need the bowl." "In a little while," I answered. "No," he insisted, "right now." You can see in that instance the underlying dynamic of a closed group. Because of the existing tension and the crowdedness, some daily problems would frequently acquire disproportional dimensions. We argued, and Menachem was so worked up that he said, "If you don't take your things out right away, I will dump it all on the floor." I disregarded his threat, and he carried it out.

I restrained myself, out of my respect for Menachem, his age, and his rank, and also because of the friendship we had up to then. I just dumped his laundry out and returned mine to the bowl. In the evening, as we were jogging in circles in the yard, he tried to stop me to apologize, like a kid. He tried to talk to me, but I refused to answer. For two months we didn't exchange a word. Then we gradually resumed our relationship, but it was never as good as before that incident.

Yitzhak and Menachem were the two men with whom I was most intimate. I never felt good vibes with Rami. I cannot befriend a man who lacks emotions, sensitivity, or delicacy. I am highly sensitive to aesthetic matters, which Rami didn't care about. He was a rough kibbutznik with the skin of an elephant.

There was, for example, the story of the radio. For hours I worked on it, trying to get broadcasts from the Israeli stations. Rami used to come over, peek at it from above, and say, "It's nice Amnon, but it won't work." After three days, I connected the receiver to the radio, and we picked up Israel. Rami had not helped at all; he took me for nothing. That's why I'm quite angry with him. Because after all, I had been just an unformed child, while he was a man of thirty-one. He had the security valves provided by a loving wife who supported him from a distance, while I had to swallow everything all by myself. He couldn't see that. Anyway, I managed to build a beautiful radio, but it was very fragile. Rami became ambitious and built another one, which was stronger, but ugly and clumsy. It's true he was the leader and contributed a lot to the group, but I didn't love or admire him.

I had a strange relationship with Amos. He is an introverted guy with poor communication skills, but he opened up a bit during captivity. From time to time we used to fight jokingly. Since he was the pilot with whom I had been taken captive, he was sort of overprotective of me, and I resented it: "Who are you to tell me?" He was not mature enough for me to accept him as an authority, and I reciprocated with a kick for every attempt he made to boss me around. But we had fun pulling pranks together. One day, for example, we both shaved our heads, promising ourselves that by the time we had hair, we'd be going back home.

With all the others it was simpler, just good relationships. Of course there was tension among us from time to time. Once Menachem tried to remove the kitten from the room, while Benny wanted to play with it on his bed. This argument turned into a violent fistfight, and we had to force them to stop. Taking into consideration the level of tension, however, there were only a few such outbursts. This was due to Rami's presence. He knew how to calm everybody down. He also took care of keeping us all busy most of the day. Studying, too, was therapeutic in that sense.

Some of us were able to release tension by listening to music.

But different men had different tastes; some wanted to listen to their records at high volume. It was easier to settle this after we received earphones. You could see people being drawn into the sounds. Our pets also helped us let some steam out. I myself tended to an injured young falcon who had dropped off our tree. Sports helped to take some energy out. Reading was a good escape from reality, you could just fly away with the story. We all read a lot. Reading the books of Dreikurs helped me in building my self-confidence. And in spite of all these activities, sometimes I would lie in bed, under the blanket, and cry. Nobody noticed it.

A group of ten men in jail could be organized as a military unit; that had been the case in Vietnam. But the Israeli mentality is different. We are less square, less obedient. Furthermore, the majority had a kibbutz background, and only a few, like me, were city-raised. Rami struggled to establish an egalitarian framework, and this obliged us all to disregard the different origin, background, and rank of each. It was natural, though, that age and life experience had their effects. Some of us, namely Dan and all the pilots except myself, were already in the standing army. They were the elite group, while the others were on a lower level. At the beginning, when we were joined together, there was a distinction between the Air Force group and the others, because we got better treatment from our corps. Some sense of discrimination developed, but we tried to diminish it by having the Air Force adopt the whole group. I remember that whenever I received personal parcels, I would share them with everybody. The kibbutz background influenced the formation of a rotation of chores, which included cleaning, cooking, and other duties.

As long as I was in captivity, I didn't take the Friday meetings very seriously, because I was too young to understand. But when I think about them now, I find that providing everybody with formal equality was an outstanding way to reduce the pressure. In the assembly, we were usually divided into two parties—a small one supporting Menachem, and the larger one with Rami. But belonging to one of the parties was not fixed. In addition, it was clear that any decision endorsed by the majority would be followed by both Rami and Menachem. Rami was objective and fair in his conduct, in spite of the fact that he had clear opinions of his own on every topic. Menachem competed with him for the leadership position, but the results of this competition were clear before it ever started. For me, the Friday night common dinner was very important, because it reminded me of my home, and made our group a little like a family. In fact, the program of our Friday nights consisted of one event—starting from the festive dinner and moving on to the assembly meeting. Rami determined the atmosphere of our meetings, and made us sit quietly and deal seriously with the agenda. When we started to take down the minutes, we were foreseeing the need to document the period many years later. But perhaps we glorified reality a bit. All on all, it was clear that we shared our bad fortune and the awareness of this helped us in overcoming it.

DAN

Slowly they all came to our room. At that stage we faced a new challenge: how to build a communal social life. Some of the men were more suited for it, others less so. The majority understood that the only way for survival and sanity was democracy. Perhaps it was our luck that a large number of us came from kibbutzim. Even though they were not members anymore, several men grew up in that system and were educated in its values.

Pilots have at least one advantage [laughing]: they are taught to be together. It's true that each pilot is trained to be self-sufficient, to be alone, to attack by himself, yet at the same time he is instructed to follow the leader closely, not to separate from his partner, and to take him into consideration, even if it's against his instincts.

The facts were clear: a six-by-nine meter room and an eight-by-ten meter courtyard, which was not always unlocked. Each person lives together with nine men, opens his eyes with them in the morning, and goes to bed with them in the evening, unable to say in protest, "I don't want to play this game anymore." It was an unusual human density, much harder than family life, because you didn't go out to work, you didn't meet other friends for entertainment. You are sick of their odors, the smell of cigarettes, the full ashtrays, the volume of their music on the radio or the record player, their moods—a thousand and one scenarios that could have turned our life into hell. It was quite probable that some would return home insane, perhaps all of us. The secret of our success was that we realized we had to take each other into consideration as much as possible; but it wasn't always a bed of roses.

The confrontations continued even after some preventive measures were developed, and explosions might have burst the bubble for any sort of trifle. We lived in a pressure cooker, where surplus energy accumulated all the time.

The transition from a company of five to that of ten was complicated. After we had reached a balance, it was interrupted when Menachem arrived from the hospital. His arrival was like adding pepper to the pot, but it increased the tension and disturbed the group. Menachem was extremely ambitious and too individualistic. He was of the opinion that each one should decide for himself, yet he was very demanding toward the others. He was certain that he was the wisest and had all the answers, but in our condition wisdom meant being quiet and giving in. Right after his arrival, he suggested new ideas, which were pretty good actually, but caused a lot of friction among us. Looking back after all these years, I think that this agitation was good for us. Expecting improvement in our conditions created a positive tension and raised our morale. On the other hand, Egyptian logic demanded ten improvements for their POWs in Israel for each one we got.

It was nice to see how people changed their conduct under the influence of our group norms. Some were not used to reading, but they became readers in jail. Benny was like a sponge, and he went through an intellectual transformation. As time went on, each of the young men became more restrained in behavior and demonstrated more readiness to understand others. It was not moderation resulting from deterioration, but the adoption of a new norm of conduct.

I was somewhat outside the group, because I was in bed all the time. I tried to alleviate the tension of others, and did as much as I could. Nine guys were running about, working, exercising—while I was in bed, asking them to bring me this, bring me that. Sometimes it was hard to keep asking; sometimes it was hard for them to supply what I needed. I used a bedpan in the beginning, then the guys started to take me to the toilet and the shower. I was heavy even for all of them together. Sometimes I'd crawl on my bottom and sit on the floor, and they'd turn on the water to wash me.

Yair was soon returned to Israel because of his condition. I was taken to the hospital from time to time to be treated. One day the wardens came in and announced, "Today Dan is going home." I didn't believe them. They took me, with my eyes blindfolded, without any packages, and it turned out to be another treatment for my legs. In the afternoon I was returned to the room, and the guys were shocked: they were convinced that I was at home already. They had already given my bed to somebody else and divided my things among them. . . . They didn't save any lunch for me, naturally. This was also an experience.

You have to remember that our life was uneventful. How long can you talk about your interrogations and tell others how they hung you up by your feet and beat your soles with rubber clubs? We were afraid to talk about our military service because we suspected that they might be listening in. Once I sat with Avi in the yard and he drew an airplane on the sand and whispered, so that the Egyptians wouldn't hear.

One of the causes for tension was the inequality of the parcels. I wasn't disturbed, because my wife sent me plenty of things, with help from the whole kibbutz. But the Air Force had discriminated in the beginning by sending stuff only to their men. This was a mistake, and the IDF shouldn't have allowed it. There was a time that we had nothing to wear, and suddenly six sweatsuits arrived for the Air Force members. I know that some of us are still bothered by that to this very day. I am sorry that people in Israel didn't realize that we should be treated equally. I hope they learned a lesson from our experience. However, some packages made us all happy, like the chocolate sent by Amos's parents.

I remember that some time after the issue of the parcels came up, Benny told me he would like to go to a solitary cell. This was the climax of the group's tensions. We managed to convince him to drop the idea. Afterwards the packages sent were for ten men. We had plenty of things; it was absurd. The Egyptians around us lived in utter poverty, while we were flooded with goods in prison.

The democratic system was not only a mechanism to solve problems and alleviate friction, but it proved to be therapeutic as well. We mainly discussed technical matters, actually, because we were afraid to open up on emotional topics; but just talking in public was a relief. Getting together, whether your mood is good or not, and eating ice cream together was excellent group therapy. At some assembly meetings we did discuss love, family or human relations directly. As time went on we got to know each other better than a husband knows his wife, because we were constantly together, with nowhere else to go.

BENNY

I was very active in the group life, in the organization of holiday celebrations, in writing plays, in making fun. On the one hand you learn to live with people, to restrain yourself, not to have a fit on every matter. On the other hand, you need some privacy. You are constantly in the room with nine guys, so what do you do when you want to be alone? I used to talk to myself, saying how much I needed to be alone with myself. I felt this especially when letters were distributed, and I felt like retiring from the crowd, maybe even crying. But there was no place to go. One could go out to the yard, but it didn't help. Even when I sat in the corner and wrote in my notebook, someone would peek over my shoulder. When one of us put a record on, there was always one who disliked that particular music. Dan wanted to listen to concerts, while I wanted to hear a rock band, like the Beatles, at a volume that I liked. It was never possible to listen alone. People didn't respect privacy, not even a little bit.

I wanted to make the men aware of the problem, so finally I brought it up at the assembly. I said that I wanted to go to solitary. In fact I did not intend to, because of all the implications of such a step. But I had to express my suffering. Naturally they objected to it and I was voted down, but I had made my point, and from then on there was more consideration for privacy.

I didn't feel that there were subgroups among us. The Air Force people didn't act superior. Maybe I felt this way because I tended to be cooperative with others. I was able to express what was bothering me, and got everybody to pay attention to me. Motti and I received many packages from the military canteen service. They sent mostly food. We used to joke, "See what we got compared to the Air Force men!" I never felt discriminated against. On the other hand, I'm grateful to the Air Force people for teaching us in jail. I wasn't envious of any of the men who shared my captivity. I think David felt envious, because he didn't have a military unit supporting him.

There were very few outbursts in the room. I had an outburst once, two days before our return to Israel, when we were all under pressure. Menachem said that we had to throw the cats away. These were cats we had raised in the room; they were born there and lived with us all the time. It is difficult to explain how attached we got to them, because of the prison conditions. Anyway, Menachem and I had an argument, until I slapped him. After that, we all calmed down. I regret it to this day.

MOTTI

Gradually more people joined us in the group—Yair first, and later on the pilots. Each one had his caprices and we had to get organized to take everyone into consideration.

I was indifferent, and took things easily, not paying attention. I was almost the only one from the city, while all the others were kibbutz members, who had been used to living in a collective society. As time went on, I saw that I had no choice and I learned to live with them. I noticed that the married men were under more pressure than the single men, probably because they missed their wives and kids. I used to yield to them in many ways, like in the line to the shower.

Rami had the superior role among us in jail. He is an exceptional person. A real genius. He is an expert in many areas, both practical and academic. He reads a book and remembers it all. He can study everything by himself. We never caught him making an error, not even when he argued with Menachem. But the best thing about him was that he was so friendly and modest in his behavior. He read all kinds of psychology books in prison and studied how to behave in a situation like ours. He was also physically very strong. When we came off the airplane I heard people from the Air Force saying, "This is Rami, the genius of the Air Force!" Often I think that it was worth being in captivity just to get to know him. . .

Rami taught us to give in to one another. At first I didn't understand it, but as time went on I realized that this was the only way for us to continue our life together. I gave in even when I was right. In some cases I brought matters up in the meetings. The meetings solved all our problems.

Most of the men learned to give in, and there were no outbursts. We sometimes pushed each other jokingly, or fought for fun. Even when we had serious conflicts, they ended up all right. I think that the pilots had learned to live as a group during their training; it was more difficult for us in the beginning.

When Benny asked to go to the solitary cell, we didn't agree. We were afraid that if we'd agree, we'd be helping the Egyptians. Earlier they had wanted to separate the officers from the enlisted soldiers, while we had wanted to stay together. I had seen many films so I had a suggestion: since the highest ranking officer in captivity had the right to promote others, Rami should give David, Benny, and myself the rank of officers, so that we wouldn't be separated from the others. Rami said he had to consider the idea. I think he wrote a letter about it and received an answer confirming what I had proposed. Rami said that if a separation were proposed again, he would promote us all. When the prison commander heard about this plan, he dropped the separation idea.

I never felt as if the officers felt superior to us, or that the Air Force people behaved in another manner toward their own members than toward the rest of us. If there was such a tendency, Rami would have stopped it right away. He was the smartest man in the room, and everyone felt they could learn from him. He did not declare himself a genius, but everyone knew he was above us all. He was a modest man and had the patience to hear anyone out.

DAVID

Menachem and Rami were of the same rank, but Rami had the authority among us, because of his personality. He used to say, "If someone needs to let out steam—do it on me." He could absorb anything. He was willing to be cursed and kicked without responding.

I was mad at the way they sent us parcels. Some were general parcels that came from the IDF, so to speak, but there were only six of each item. For every holiday these packages arrived—with six shirts, six towels, six sweatsuits. It drove me nuts. I told Rami that at some stage this unfair treatment would break the group apart. Rami said that he hadn't noticed that the packages were made for six. But once I made him realize the fact, he contacted Israel several times through the Red Cross and asked them to stop doing that. They didn't, though.

As the second Passover approached, there was a big outburst about it. Everything was prepared already for the holiday, and we were all dressed up in honor of the rabbi. Suddenly an argument erupted between Menachem and Motti, and in the heat of the argument Menachem said, "You should be thankful for wearing my clothes!" I decided that after the holiday I was going to have an answer to this.

For some reason, I never received clothes in the packages sent by my family. Perhaps they were taken out by the Egyptians, maybe out of hatred for the paratroopers. So I used to wear the extra clothes that belonged to the others. After the holiday I returned everything to the guys, and I put on the Egyptian garment that had been given to me in solitary, so that nobody would tell me to be thankful for wearing his clothes. When Rami saw this, he asked what had happened, and I explained that I wasn't like Motti and couldn't take the insults. This bad feeling stayed with me to the end. For the whole time, I never deluded myself that when we'd return, we'd keep being friends. And in fact, I saw that after our return, too, the Air Force people were treated better. That's when I found out that the families were treated differently as well. The Air Force families were invited to meet with the government people and were taken to meetings with the Red Cross near the Suez canal. They were briefed about what was done, what was going on. But nobody came to give such explanations to my family in a poor neighborhood in Jerusalem. A year passed before my commander visited my parents. It was clearly the fault of the authorities, the total social-military system, and not the men we sat with in prison. This feeling, that we were not treated alike, was the worst part of my experience in captivity—that, and not the stress or length of time.

I didn't have close friends in the group, but neither did the others. I slept near Amnon, and we used to chat a lot at night. I'm not the type who makes confessions easily, nor do people come to me to confess. It was especially hard for me to be together with Motti; he kept pestering me, and once we even had a fistfight. Afterwards I learned to keep at a distance from him. But a fistfight is not the worst. Daily friction among the men, in words and curses, caused much more pain than the physical fighting. We didn't want the Egyptians to see us fighting; that's why we restrained ourselves and only fought rarely.

I wasn't very impressed by the Friday night conversations, because I was used to it. I was educated in a kibbutz, where I had been for seven years prior to my military service. The major decisions of our life were not reached in the assembly. Many of us liked the meetings only because of the ice cream we got at the end. During the week, also, one could approach Rami and say, "Look Rami, I want this or that," and things would be arranged, somehow.

7

THE INNER WORLD

The prisoners' emotions of longing, fear, or despair, their fantasies and dreams concerning escape or release, were rarely displayed in public. Various defense mechanisms were utilized to fend off these feelings, which constituted a major part of the inner world, and found virtually their only expression in the men's letters to their loved ones. These interviews, which occurred so many years later, in a safe situation, enabled the men to face their inner world at that time. Some of the men brought these up spontaneously during our conversations, as in the long detailed narratives about the pets in the common room. For others, my introductive question was, "Do you remember feelings, moods, fantasies, or dreams from that time?"

RAMI

In some unplanned, unconscious manner, we never pried into each other's intimate world. When someone got a letter, he would climb up one of the high beds to read it alone. Nobody asked him, what's in your letter? In a group of ten very different men, we didn't allow ourselves to build closer relationships among us; somehow we felt that we would not be able to cope with the commitments and problems resulting from such intimacy. After getting letters, we felt a deep longing for home. Someone might say that such-

and-such happened at home. We never discussed real things, no matter how homesick we felt. We were careful to avoid a collective homesickness in the group. Only through humor could we express part of our feelings.

I had a famous saying: "In two months we will be out of here." Why in two months? There was always an answer: "In two months it will be Passover," for example. I used to explain to the men that "two months" was a fictitious period of time. It wasn't too long, so that we wouldn't despair, and yet it wasn't too short, so that we couldn't just sit packed and ready to leave. If it was "in two months," we had enough time to go about our business.

I never dealt with the problem of how much longer I would be held by the Egyptians. I always tried to plan ahead for the next two months. When the interrogations were over, I had no ways to occupy myself for a while, and time didn't pass. I was bored, and it was difficult. When they moved me to solitary, however, I found a lot to do—clean the floor, kill bugs, practice handstands, draw on the walls. Time didn't bother me any longer. By nature, I must be a "here and now" person. Things that I can't control don't bother me. After my return, I felt guilty for never having worried about Nurit, and how she managed alone with four kids at home. It was out of my control anyway, you see.

I never considered the idea that I might grow old in prison, and I was never bothered by what I was missing back home, in my career or otherwise, except for raising my kids. I missed home, naturally, but especially—after the passing of two, two and a half years—I regretted that I was not there to watch my children grow, particularly the little twins. Whenever I received letters, I would miss them painfully. It was a physical sensation of longing. During holidays, too, I would wish so much to be with my family, at home. But all these were passing feelings.

When ten men live together, whenever one loses heart, the others support him. Frequently one of us would enter a period of intense longing, for his family or for freedom; once or twice we all felt it together. Once we were caught peeking out the windows; the Egyptian nailed boards over them, and we could not see outside anymore. From then on it was always dark in the room, and it made us all feel awful. We had the lamp on all day long, and it was depressing. I told the others: "What did you expect? We were caught doing what we weren't supposed to do, and were punished for it." It was only much later that the boards were removed and we enjoyed daylight in the room again.

Later on, about three months before the Yom Kippur War, we were all depressed again because Beausart, the Red Cross agent whom we had all liked so much, was replaced by another man with whom we had nothing in common. Somehow we came out of it; I think it was during the New Year. I gave the men a peptalk, and it worked.

Missing a woman is something else. After three and a half years without women, I felt no sexual desire. We never had anything that could be interpreted as homosexual behavior or attraction. Our only references to sex were in jokes that seemed to express our fears about potency after such a long deprivation. I had wet dreams, completely normal dreams, and I think they saved me the need to masturbate. I remember masturbating only two or three times throughout the whole time. We received *Playboy* and also some porno books. We enjoyed reading and looking at the pictures, and that's all. When I got stimulated, I was pleased that my sex drive seemed to be in order, but I never felt the desire for a woman at the time.

I remember a few days before our liberation, a team of American TV reporters came to interview us, with a young female photographer. I shook hands with her and was amazed at the size of her hand in mine. At that moment I realized that for more than three years I hadn't been with a woman.

Letters were our only emotional outlet, yet we knew that the letters were censored and read by many along the way. Because of the censorship, our letters were somewhat removed from actual events and took a long time to arrive. I wrote some technical letters, about innovations for our kibbutz factory, and lists of things I needed. But I also wrote intimate letters to Nurit, and these letters constituted the only world I had all to myself. I never shared this world with anyone. I told Nurit about my thoughts, about books I had been reading—I tried to share with her all that was inside me. They weren't letters about captivity or prison, rather about life and philosophy. All the psychological truths that came to me in jail are documented in my letters to Nurit.

Never before had I had the opportunity to develop my thoughts like this.

I also received photographs of my family, as did the others. The pictures were like public property. We hung them on the walls and later on put them in albums. Taking care of the photographs and writing brief comments in the album was the only manner in which I too could participate in raising my children.

We raised pets in our room, which gave us a sense that we were in a place like any other. It brought us out of prison, in a way. At the moment you play with a cat, you're not in jail. Our first cat was Dina, and it belonged to us all. Later on we had many more. We discussed how to take care of them, where they should spend the night, and things like that. We gave a lot of thought to naming them; it's documented in our diary. We had a bird, too, for a while. It had an injured wing and it fell off the tree in the yard. We treated it, and named it Lucy, and it lived in our yard. One day an Egyptian officer came for a visit and wanted to show us how brave he was, so he took a brick and smashed the bird's head.

We had dreams, of course. In one of my recurring dreams, I was home on leave, but I knew I had to return to prison. In the dream, I kept arguing with myself: should I return or could I stay out of jail? This dilemma was never resolved, and I think it manifested my sense of responsibility for the group. Even in my dreams, I didn't allow myself to escape all by myself.

We did consider escape, but whenever we applied a logical analysis to the situation, we realized it was useless to try. Even if we managed to change our clothes and get out of jail—we would be so conspicuous in Cairo that we would be caught immediately. Moreover, since it was clear that not all of us would be able to make it, I was certain that the Egyptians would make life unbearable for the ones left behind. Therefore, even if I had had an opportunity to run away, I would have resisted the temptation, and I think that's what the dream was all about.

We received various car magazines and each of us selected a car he would get when he returned. I chose a family van that would allow me to put everybody in for an overnight stay, so that we could plan different trips together. It was fun to argue about the different cars. We used to share fantasies about our return, and by the way, our fantasies were very close to what happened in reality later. But I never planned my life for after the return. The others did have plans, like Motti with his fashion boutique, for example. I didn't participate in that kind of conversation.

MENACHEM

Coping with time dragging along was the worst problem in captivity. I entered prison after a year in the hospital. When I looked back then, I had the sense that a year had been an awfully long time. Had somebody told me just then that I would be imprisoned for two and a half years longer, I don't think I would have been able to take it. Even now, when I recall that moment, I am overwhelmed by the meaning of an additional two and a half years in jail. How did the time pass, after all? A week after another week after another week. There are records of over 130 meetings we held in jail. When I joined the group, about twenty such meetings had already taken place. In other words, we lived together for 110 weeks. All this time we didn't see anything beyond the yard and its surrounding walls. Even when they took me out to see the prison dentist, my eyes were blindfolded. The first day I saw the horizon was on the day of our liberation. When we were taken without our blindfolds to the bus outside, I felt dizzy: for the first time after such a long time the eye was focusing on the infinite distance. Yes, sometimes we saw a plane in the sky high above us, or kites fighting ravens. But we didn't see anything else, only the birds.

I was in my worst moods when I thought about the time I was losing in jail and the unknown longer period still ahead. I felt I was missing the experience of being with my daughters as they grew up, and this was very painful. When I was captured, the oldest one had just completed the first grade, and here the second one had already reached about the same age, and was starting to write letters to me! During my absence, my wife graduated from college; she renovated our private home, even though she stayed at the air base; she bought a car. All these things were happening without me.

The sense of waste was terrible. I am an engineer; I had already contributed to the department of firearms development, and I knew what else I could contribute. I was about to be significantly promoted in my career just before I was captured, and all of a sudden I was stuck in jail. I used to walk around in the courtyard and think, only peace or war could lead to my return home. Peace was out of the question, so only war remained. . . . I remember they had blackout maneuvers in Cairo at the time; every now and then another neighborhood would have to keep dark, but no one saw this as preparation for a war. We lived with the sense of an infinite future in captivity. I remember one of the radio talk shows just a couple of days before the outbreak of the Yom Kippur War, which said that war was not expected for at least ten years. I had the feeling of being in a real trap.

One of the things I hated most about Rami's behavior in captivity was his announcement that in two months we'd be going home. The two month index was advanced, naturally, with every passing day. This was sort of a game for him and the others, and it deeply upset me.

At the beginning of captivity, the POW is in trauma: he has just become a prisoner, he meets the enemy face to face, he is interrogated and tortured. When captivity goes on, however, conditions improve, you are left alone, and the trauma seems to be over. I believe, however, that other traumas replace the first one. These have to do, for example, with the prolonged period of captivity. Coping with this is much harder than with the first shock. Everything gets minimized compared with the awareness that another year has gone by, and you've missed your daughter's birthday again. We used to drown all these longings in daily activities, or, for example, in the preparation for the Seder in jail. We took such care with all the details, what we would read in the ceremony, what we would buy and cook, how to make it as pretty as possible. When it was all over, I suddenly realized that another Passover had gone, and I was still in prison in Egypt. It was extremely difficult. [He is silent.]

I find it hard today to recall emotional expressions in captivity, other than in reading books and writing letters. I remember that sometimes I would cry in the courtyard. I didn't want anyone to notice this, so I used to walk alone in circles in the dark. I think that Dan was the only one to notice. He would come out with his crutches, at his slow pace, and make me walk side by side with him. We walked, and he would talk to me softly about various matters, as if he knew and didn't know that I had been crying. If there ever was anything like a feeling of being caressed in jail, it was in these slow walks with Dan in the courtyard.

My correspondence with Esther was at the heart of my emotional world. After some time, we adopted a certain pattern for our letters. We used to hand letters to the Red Cross agents once every two weeks. I wrote daily during this interval, and before the visit of the Red Cross men I copied what I had written into neat handwriting. My letters to Esther reached two thousand, even three thousand words. I told her about the books I had read, about my experiences, my thoughts; I opened up to her. Her letters to me were much shorter, naturally. She wrote them in one stroke, not over the two-week interval. She didn't expose herself in the letters the way I did, but it was an emotionally rich correspondence just the same. Actual events were rarely referred to, because of censorship.

We store these letters at home. We have three daughters and today they're twenty-three, nineteen, and twelve years old. The older ones, who were born before I was captured, have asked to see these letters several times. We respond that when the time comes, we'll take the letters out of the box and let them read them. Time has made us more inhibited about this, however, and we still haven't shared them with our daughters.

I used to refer to sex in my letters, but they weren't erotic. Our attitude toward sex was an aspect of the inner equilibrium that we created unintentionally, in order to avoid extra pressure. When there is nothing to arouse you, you're not bothered by your sex drive, so our deprivation didn't bother us as one might imagine. We received *Playboy* magazine, but without the centerfold, which was removed by the Egyptians. When the *Newsweek* photographer—a pretty young woman—arrived to take our pictures, I exclaimed, "How nice to see a beautiful woman—the first woman I've seen in two and a half years!" It was pleasant to see her.

Since news magazines were forbidden, our families sent us other kinds, like architecture, gardening, or car magazines. We sat in that cell surrounded by these magnificent pictures. We started to fantasize what we'd get for ourselves when we returned. The family photographs reinforced our fantasies. I dreamt of the beautiful family, beautiful wife, a sports car—everything. This was quite natural. We had had a fairly high standard of living, and realizing our fantasies seemed quite probable. At the same time, I was worried that we were doing some harm by raising the expectations of some of the men, who were bound to be disappointed when we'd return. Perhaps getting married right away after our release, or being bitter about not having found a wife, or becoming resentful toward the government for not being compensated enough were all reactions to those expectations formed in jail. On the other hand, these fantasies helped us to overcome hard times in jail. If we did some harm in the process, there is nothing to do but take care of the damage.

YITZHAK

At first, letters were instruments to get me to cooperate. In my first interrogation, I was told that they could easily kill me, because nobody knew that I was alive. On the other hand, if I'd tell them what they wanted me to, they promised to let me write a letter home. Clearly, once my family knew I was alive, it would be harder for the Egyptians to kill me. In addition, I was terribly worried about my family; what they might be suffering, thinking that I was dead. I wrote several letters right away, even in solitary, but later on I discovered that the Egyptians hadn't mailed them. All I wrote was that they shouldn't worry, that I was okay, eating and sleeping. . . . In fact, this is all I could say during the first six months. The thoughts about my family's whereabouts and the hope that they didn't suffer because of my absence were immense, and the interrogators knew how to take advantage of this. Luckily, they were not sophisticated enough to use this tool more effectively.

In some of the letters I did receive meaningful news. For example, in one of her first letters, Michal wrote to me about David and his wife, who had a new baby. David was my copilot, with whom I had parachuted, so when Michal told me that he was helping his wife by getting up to feed the baby at night, I gathered that he had been rescued and was safely back home, and not, as I was told by my interrogators, that he was right there in the next cell, and collaborating better than I was. . . .

I received only two letters by the beginning of November, one from my mother and one from Michal. Afterwards, when the Red Cross visits became regular, we got many more letters. The guard would come in and throw a package of letters on the floor. Our own letters to Israel were transferred only through the Red Cross to the Egyptian censor. Arriving letters always followed the same pattern: everything is okay, life goes on, the children eat well, start school, go for a trip. They tried not to offer too much sympathy, not to be too philosophical, for fear that such letters would be kept longer by the censor. Evidently, they couldn't write about attempts made to liberate us, although this was hinted by the Red Cross men. We never felt abandoned or forgotten.

Our hopes for a quick release went up and down. I remember a moment of crisis when Los went home, while we stayed. He wasn't part of our group, actually, and only Rami and I went to visit him in the hospital, but somehow we believed we'd all be returned together. When only he was liberated, it drove home the idea that we wouldn't be going back so soon. Our hopes went up again after the six Syrian officers had been taken prisoner by the IDF, and we thought that an exchange might take place. When we managed to intercept Israeli broadcasts on our radio, our morale increased for a while. But our prevailing opinion was that only peace or war would lead to our liberation, and we did not believe that either of these events were likely to occur right then.

Our captivity lasted a long time, without any meaningful events to break the routine. When I was in solitary confinement, going through the interrogations, I figured that I couldn't take it more than another two weeks. I was afraid of reaching a state when I'd be willing to answer anything they'd ask. I don't know what made me think about two weeks—perhaps I could have taken the situation much longer. This same experience reoccurred in the common room. Although we realized that we might remain in Egypt for a very long time, we couldn't imagine not seeing our family and friends for, say, ten years. As a defense mechanism, we developed the notion that we'd be released for the next holiday. We all shared this pretense, not having planned it. And as the Jewish holidays are spread nicely over the year, there is one every two or three months. We didn't need a big holiday to construct our fantasy. We said, we'll be back by Hanukkah, and if not, certainly by

Purim. The holiday to wait for was naturally always moving forward, but the intervals till the next one allowed us to live without despair in the meantime.

We put in a lot of effort to prepare our first Passover in captivity. We had a strong feeling that we'd be liberated right after it. The same feeling accompanied every holiday, as if we told ourselves that this was our last holiday in prison. Birthdays also served the same function; they divided the time for us. When we became adjusted to our life in jail and improved our standard of living, we used to bake a birthday cake, and add cards with our wishes. We celebrated the birthdays of our children back home, too. We felt that as long as we didn't forget those family events far away, we wouldn't be forgotten by our families either.

The realization that my girls were growing up and changing while I was absent was my most painful experience. I had the continuous sense of loss, as if life ran along and I was only an onlooker, as if I had died. Somebody who has died never ages, his picture remains the same, and that was me. I was wondering: How will I close the gap?

The only way to cope with our longing was by being active all the time. I badly wanted not to feel that my time was wasted, not to have spare time. I was worried that I wouldn't be able to fly when I got back. At the same time I was telling myself that flying is like bicycle riding: once you learn it, you never forget. I was afraid I wouldn't be able to get adjusted to working, and I worried about what might be salvageable from my interrupted career. I didn't want to make a career as an ex-POW; I was determined to get back into my profession. One thought was forbidden: that I'd never be home again. I kept building a colorful picture of my future return. I would see how we'd be welcomed by everybody at the airport, how I'd meet my grown daughters, how we'd all be happy—an idyll of sorts. But I didn't dare revive this fantasy too often.

Usually I don't remember my dreams, not even in jail. I didn't have elaborate escape or life-out-of jail fantasies either. But we amused ourselves from time to time in constructing fancy escape or rescue plots, like what really happened later on at Entebbe. We were wise in planning, but when we calculated the odds, we concluded that our rescue would cost Israel a helicopter and ten casualties. Dan couldn't have made it, because he was badly wounded. When all was taken into consideration, we decided not to recommend the project, because we didn't want to be liberated at the cost of anybody being killed.

We examined the idea of taking Egyptian hostages for bargaining, too. I think others also had the same idea and gave it up. Based on the assumption that war broke out every ten years, I predicted that in 1977 we'd have one, and this might lead to our release. I was saddened, however, by the thought that only war, with all the killing involved, could liberate us. This was indeed what happened, but sooner than I figured out.

The letters I sent home were mainly of encouragement. My mother says that I never showed her such warmth before my captivity. I wasn't disturbed by the fact that half the world was reading the letter before it was delivered. I expressed in my letters strong feelings of love and longing; but once I had put the letter in its envelope, I didn't dwell on those feelings. Sometimes I talked to Rami about my feelings, and it was good. For the first time in my life I had a close friend.

The letters I received from home were tremendously important. I was very depressed if others received letters while I didn't. Even when I knew what were the objective reasons (the Egyptians took a terribly long time to censor our mail), I would blame my family. I believe that Michal indeed wrote less than the other wives, and also sent me fewer pictures of the girls. We never shared our letters. The contents of our letters were the most personal matter in the system we established.

I remember a period of about two months when letters didn't arrive. The Red Cross people explained that the Egyptian censor office was overworked, and they wouldn't transfer letters to us before they had been translated to Arabic and checked. I asked Beausart to find out whether these interpreters were soldiers or civilians, and when he reported they were civilians, I suggested that we pay them for the extra hours it would take them to process our letters faster. It cost Israel an Egyptian piaster per word—a lot of money—but as a result letters started to arrive more regularly.

My thoughts about my wife and girls and the family, about what awaited me at home, preserved me for a while. But as time went on, my memories faded. I wasn't able to build a picture of what was really happening at home from the letters of Michal and my mother. I don't think any of us asked himself questions about his wife's behavior and fidelity. Anyway, we didn't talk about that. I suppose nobody imagined they might have a problem at home. Avi had been married a month before he was taken captive, and he didn't receive mail for a long time. He looked concerned, but we didn't discuss it. The question never surfaced.

During the first year I was never sexually aroused. My thoughts about home never entered the bedroom. We didn't discuss sex, either. Once we tried to raise the issue of love and sex in our Friday meeting, but it wasn't a success. When I talked privately to Rami, we sometimes shared our experiences with women. A need to masturbate appeared gradually, and I did that quietly under the blanket, in the dark. I think that the others did the same. It's amazing that we never talked about this. Perhaps we were afraid of too much closeness, or of homosexuality. I recall that once they tried to bribe Amnon with a prostitute, but he never gave us a clear account of what happened.

We had pets in the room. Our attitude toward them, I think, was an expression of our humanness. Even when our conditions were bad, we found the warmth to care for another creature in need. Possibly the warmth expressed for the cats was a replacement for all the feelings we blocked in relation to each other. We needed to cuddle someone. Furthermore, the kittens were not part of Egypt; they sort of belonged to another world, or to that part of Egypt that wasn't mean. Not all of us cared for the pets to the same degree. Dan was very attached to his cat, while some of the others were pretty wild. For example, one day we decided to test the maximum height from which a cat would fall on its paws. . . . Today it may seem cruel, but it was, perhaps, the only case in which we had total control over another creature. The end was ridiculous: when we were about to go back home, some of us refused to move without the cats. Maybe some of us didn't want to be completely separated from jail, since it was such a secure haven.

AMOS

I wrote the first letters in solitary confinement on my knees. When I was in the common room I used to write a draft first, then copy it. Writing letters home was a serious business.

The letters from Israel arrived in packs. When I didn't receive any I was angry and concerned; when I did receive them, I was depressed and full of longing. When a letter arrived there was a great need for solitude, but there was nowhere to hide.

I used to dream about the world outside, but I can't remember those dreams now. I didn't fantasize about material things like a home and a car, but more about my wife and about the children I didn't have yet. Some of my dreams were about escape also, but they always ended up in failure. I think that had there been a war in Israel, we might have considered escaping. When I asked myself how come so many POWs escaped German camps in Europe, I concluded that they were motivated to rejoin their units and contribute to the ongoing war effort. In our case, there was no war, so we didn't feel the drive to escape. We discussed the possibility and analyzed the odds. We believed that it was possible to get out of prison, but we estimated that we might get stuck in Cairo, and even if we reached the Canal, we wouldn't have any means to cross it. Whenever the idea of escape was brought up, we reached the conclusion that it wasn't feasible.

One cannot live with complete uncertainty about the time. Rami's solution was to behave as if we'd soon be released, say in two months. This was the approach encouraged by the Red Cross people as well, especially in the beginning. They used to ask me, "When is your daughter due to be born? In October? Well, by then you'll be home for sure." Or: "By your birthday, in December, you'll be back." Such an expectation leads to a sense of living with packed suitcases ready to move, but I think it's impossible to live that way. Had I known the time of my liberation, I could have seen the interval moving toward the end, and it would have been much easier. We knew about POWs who had been returned quickly; but all the previous dates that fit those precedents went by, and we still weren't released.

In the end, I treated captivity as if it were my whole world: this is where I am, and I don't expect to be out. I made up my mind to live with what I had and tried to cut off my ties to anything else. It was difficult, and I had to work hard to obtain this state of mind. We had to see our conditions as static and permanent, and this approach is apparent in my letters to Dalia. I did express longings for her and for our small daughter whom I didn't know, but I also expressed an acceptance of our separation. We decided to order good mattresses from Israel, because we realized that we shouldn't see our state as temporary. Our behavior reflected this reconciliation with our fate to remain in captivity for a long time. When we reached this understanding, we started to study seriously and established a stable routine for our daily life. Once I had plans for my future in jail, I began to feel much better. Even the others, who claimed, like Rami, that in two months we'd be home, behaved as if they were in prison for good.

I did continue to cling to the hope of liberation, naturally, but this belief was somehow pushed to the side.

AMNON

I started to think about the time in my solitary cell. Based on my information about previous POWs, I estimated that I might be there for a year. It seemed as long as eternity to me. I was worried about how I'd be able to pass the time, namely the next day, the next hour. Sometimes I told myself that I was imprisoned for life until further orders.

After six months, at the end of my isolation, I joined the others and started to hear different bits of information. One of these said that Nasser had announced that as long as the war situation continued, the POWs would not be released. A living proof of this policy was the fact that Dan wasn't sent home, in spite of his bad health. That was when I changed my estimate of the duration of our captivity to two years.

The third stage has to do with the story about my twin brother and the fortune teller. When I was in the aviation course, I was taken by my girlfriend to a fortune teller. She was a very impressive lady and surprisingly accurate in what she said. Among other things she said that she saw me having a difficult training, which I would complete successfully, but after that—she saw all black. And indeed, I was taken captive. This same girlfriend had maintained a relationship with my family and told the story to my twin brother. So one day he went to the same fortune teller. What she said is a miracle to me. She told my brother that he had a relative in prison far away, and not because he was a criminal. She promised my brother that he'd see his relative in three years, three months, and three weeks. The amazing thing is that this fits exactly with the date of the outbreak of the Yom Kippur War, which made our liberation possible.

I think that we didn't permit ourselves to look forward to a certain liberation date in order not to be disappointed. We all read in prison Victor Frankl's book about the Jews in the camps who had set their hearts on being liberated on a certain date, and when it didn't happen, they had no resources for coping and they died or committed suicide. It is naturally unfair to compare our situation to that of the concentration camps. We had a whole country behind us, and we had hope. But after two years, when all the predictions proved to be wrong, and when I had just convinced myself to behave as if I were imprisoned for life until notified otherwise, I suddenly received my brother's letter about the fortune teller and I hung onto it for dear life. Believing in something is undoubtedly important in such conditions.

The letters from my family provided me with lots of encouragement. At first, I had a girlfriend in Israel, and she also wrote to me beautifully, but her letters bothered me. I thought that I might be held for a long time in captivity and I didn't want her to wait for me. I knew I wasn't going to marry her or make her the mother of my children, and I didn't want to delude her. I wrote her about that and also asked my brother to delicately persuade her to let go, but the letters took so long in each direction that by the time she understood the message she had managed to send me several moving love letters, which I couldn't stand.

I wrote many letters, but it was difficult not having one person before whom I could open up completely. I wasn't married, and my letters to my parents and brothers were circled around. I felt that I couldn't reveal my inner feelings, and I even held back from writing the things that we were allowed to write. I tried to encourage my family and tell them that it wasn't so bad in Egypt. I had every intention of showing them that I remained sane, both physically and mentally, so that they'd stop worrying about me. But I didn't manage to write anything besides that, in spite of the fact that I put a lot of time and effort into the letters.

At later stages, the Egyptians tried to deliver the letters before the Red Cross visit, so that we'd have time to prepare our replies. They visited us every two weeks, at best, but sometimes it was once a month or even once in two months.

The moments when we got the letters were astonishing. Each person went to his corner, like a puppet, and tried to soak in this drop of family or country. When mail was delivered by the Red Cross agents, we waited impatiently for them to go, so that we could be alone with our letters and photographs. We tended to withdraw from each other and attach ourselves to this bit of home, ignoring the jail. We didn't share the contents of our letters, but since we knew each other so well we could sense what was in them. Afterwards, we were overcome with nostalgia and sadness and the whole room would be quiet. People walked along the walls, avoiding each other. We looked like cats who had taken a beating.

Still later, we shared some selected items of news about the country, or about our families. Only one or two days later things went back to normal. If someone hadn't received any mail, or had received less than the others, he was in a rough spot. But it usually didn't happen.

Throughout the period I tried to imagine the day of our liberation. I attempted to construct it to the minutest detail. These visions cheered me up a bit, and I used them when I was especially depressed.

We lived together, ten men, without any reference to sex. In spite of what you may have heard about men in jail, there was no hint of homosexuality among us. Perhaps it was due to the fact that, above all, we felt like delegates of the State of Israel, who had to represent it honorably. During the first year we were under the effects of the trauma, and the subject of women never came up. Moreover, there were absolutely no stimuli to arouse us. Subsequently, when our conditions improved, we received a TV, which showed women often, but Egyptian ladies were not my cup of tea. The individual solution of masturbating was the only way, and it was done without anybody noticing. There was absolutely no manifestation of lust among us, neither in words nor in conduct.

I tried to bring up the subject of marriage and love in several of our Friday night meetings. Sex was also mentioned in these conversations, but in an abstract manner. I believe that the men pretended to be holier than the pope on the topic of marital fidelity. The married men displayed an ideal picture of their life, although I know several of them had had extramarital affairs. It was natural, though, to idealize the things we missed. On the one hand, we lived together and felt we knew each other like an open book. On the other hand, each of us left many hidden parts out of public view. We are all complicated beings; with several layers of personality; what we shared was just one of these layers.

We never saw a baby, a child, or a dog during our captivity. Our neighbor was an important Egyptian who was permitted to keep two dogs with him in jail. One day I heard barking, and peeking through a crack in the gate I saw two strange creatures on four legs. It was astounding to see their funny shape after two years. It is hard to explain, but one loses the correct proportions of things which are long unseen. The sight of the dogs excited Menachem and me, and we asked the guards to bring them into our cell to play with. Since our neighbor had been nice to us, we decided to pay him back by giving a bath to these dogs. We filled a pail of water and washed the dogs. To our amazement, they turned from yellow—probably the color of the desert dust—to a glimmering black.

The proportions of a woman were just as strange in my eyes when I saw the first one after three and a half years. I remember shaking the hand of the lady photographer and being overwhelmed by her tiny fingers in my hand. When we went to the airport on the bus, I saw two little girls. Their voices were so funny, and they looked like storks with their thin legs and huge knee caps. . . . You see, we were used to male bodies, with or without clothes, and the eye and brain got adjusted to those proportions.

AVI

For a long time I clung to the sense of transience in captivity, as if my release was close, and I avoided organizing myself for a long stay. Perhaps it would have been easier had I known then what

Yardi knew. After I was captured, she went to a palm reader, who told her that we'd be released in May, as was, indeed, the original plan. We didn't return in May, though, so he told her that something went wrong, and now she had to wait much longer. He invited her to come in a year, and so for the next three years she saw him once a year. On her last visit he gave her the exact date of my return, but she didn't share it with me.

In the second stage of our confinement, we used to imagine a month or two months of future captivity. We'd get up in the morning and say, in two month's we'll be back home. This was one of the ways of overcoming the great pain and difficulty of being confined there. Each one of us tried to somehow define a time frame in his life, but the sense of waste became harder to take. Even in the last days, when I declared that I hadn't completed all that I had planned to do, my four wasted years were foremost in my mind. I was aware of the fact that during this time I could have graduated from technical college or done any number of things, like investigate apes in Africa or join a mission to Alaska. I could have been productive instead of being stuck in a spot. We lived with this awareness that captivity was forced upon us, and that we had no control of our fate.

The sense of lost time was the hardest thing to take. I couldn't free myself for a moment from the uncertainty about my future. This soon developed into a fear of dying. I was constantly afraid that I might die the next day. The fact that I had already been imprisoned for three years, and there seemed to be no authority strong enough to release me, aroused in me a deep feeling of anxiety. As a result I believed, for example, that I'd never be a father (all the other married men had children already)—in other words, that I'd leave no trace in the world. Deep down it was the same as dying.

At the last stage I lost all hope of return. I still repeated the slogan, "In two or three months we'll be back home," but deep down I didn't believe it. We were already getting newspapers and listening to the radio, and we knew what was going on in Israel. It seemed like life was going on beautifully for everyone else, and only we were in jail. I used to ask myself: If all is so well, where am I? And the Red Cross people kept saying: There's nothing new, we don't know. What then? We'd be buried there for life?

Before, in the interrogation center, whenever I'd bring up the image of my family it would weaken me. My longing for them was too strong. In addition, I felt that had they seen me in my present shape, their heart would be torn. So I taught myself not to think about them. The only person I did call to mind was my father. I saw him as a strong man, someone who might help me in my situation. As long as the interrogation and isolation went on, I had the sense that my father was with me, as if we were in contact.

I broke down for the first time when I received the first letter, which was from my eleven-year-old sister. It was at one of the first visits of the Red Cross agents. By that time I had taught myself to feel all alone, and suddenly this letter brought back all my attachments. I was reattached to my loving family which was sitting far away and waiting for me, the family I had willingly put out of my mind. This breakdown was not an expression of weakness, but a state of letting go a little.

When there were only four or five of us in the room, our expectation for the mail was accompanied by a lot of tension. When will letters arrive? If the Red Cross men came without mail, or didn't have letters for everyone, the disappointment was terrible. Waiting for the letters was our major occupation then. I recall similar tension later on, when Yitzhak didn't get any letters from his wife for a long time. Rami used to read Nurit's letters to Yitzhak, although I can't comprehend how this was supposed to console him.

I didn't stop missing Yardi and my family for a single moment. My longings never abated. Later on, we formed a good contact via letters, and it helped to maintain our relationship. I think it took us about a year to learn to disregard all the stations these letters were going through—the censorship, my commanders, intelligence in Egypt, and so forth. We wrote freely about everything, completely open. We shared all our dreams. This correspondence supported me. My parents and sister wrote mostly informative letters. My world was full of Yardi's letters and my letters to her.

There are no words to express what it meant to be unable to touch someone physically in any way for four years. It is such a basic need! I consider this deprivation to be much harder than the absence of freedom, the inability to open the door and walk out.

Our sex drive found its only expression in masturbation, each one in his private realm. Talking about sex among us was taboo. Another unmentioned topic was the conduct of our wives during our absence. I never asked myself whether Yardi had met somebody else while I was away. Perhaps I was too afraid to face such a question. I never doubted that she was waiting for me, and I was certain that our relationship would be as good as before when I'd return. I think no one knew what had happened to Michal at the time. I heard that she had wanted to tell it all frankly to Yitzhak but was forbidden to do so. Perhaps Rami was aware of what was going on, I'm not quite sure. We saw ourselves as a family in captivity, and we imagined that our wives and children in Israel had the same feeling between them, and that they supported each other. As it turned out, this wasn't the case.

During the first two years, I often had dreams about my escape. But it was always a failed escape: I would stumble into a market in a strange city, then people would discover right away that I was a foreigner and catch me. Pictures similar to my near-lynching experience came up repeatedly in my dreams. After two years these dreams went away.

We didn't plan to escape, actually. Some of us, like Motti, did try to. Motti was studying ants, he saw how they were digging tunnels, and this gave him an idea. But I had flown over the land and I knew that we couldn't cross the Nile and the delta. It was a dense country, where all people were of one kind, and men like us would stand out immediately as strangers. Only Menachem could perhaps pass unnoticed. In my wildest fantasies I could see myself crossing to West Cairo, and hijacking an international plane—but it was far from reality. Had we escaped, we would have had twelve hours before the morning inspection, and this wasn't enough. We became a fact of life in Egypt, part of the prison landscape. Once, at the stage when I believed we'd be captive forever, I wrote Yardi an extremely nihilistic and picturesque letter. The people in the Air Force who read it believed it was a cover for an escape plan. They thought it was describing an escape westward, and tried to decipher it. . . . This letter arrived at the beginning of 1973, when the Air Force commander promised our wives that if by the end of the year nothing had changed in our situation, they would rescue us by means of a military operation. Somebody thought that my letters referred to the same thing.

MOTTI

I managed to occupy myself so well in jail that I didn't feel I was missing home anymore. I told the guys that it didn't interest me to hear "What will become of us?" over and over again. You feel homesick only if you're idle and you complain. Of course when I received a letter from a girl, telling me about the parties she went to, I was aware of what I was missing. But when no such letters arrived, I didn't pay attention to anything outside our world in prison.

There were no sexual stimuli in jail. We did read that such conditions may produce homosexuality, and we discussed it in one of our meetings, but none of this happened to us. We received some pornographic magazines in our packages, but since I can't read English I didn't get aroused by them.

I don't remember homesickness as much as escape plans. The Egyptians noticed this, and that's why they never took us out of the cell, not even within the prison—like to the dentist—without blindfolding us first. But we realized that had we made it to the canal, the Israelis would have killed us. I am convinced that this was the only problem. I had a complete plan. There was one guard who used to leave his clothes—rags, really—in our cell. I planned to put them on and walk out. I had asked the pilots about airfields nearby, and they had even instructed me how to fly a plane. I could hijack a plane. It was possible. I would have managed to escape and was willing to be the leader, but the guys put all kinds of obstacles in my way. They knew that if one of us got away, they'd all be punished for it. In spite of that, some of the men told me, "Make a good plan, and I'll join you in running away."

One plan was to dig a tunnel under our shower. The floor tiles were loose, and I saw sand underneath. Again and again I was interrupted in carrying my plans out. I know that the others also dreamed about escape; Avi told me he had a dream of becoming so tiny that he'd escape in the Red Cross man's pocket. I heard later that in Israel they had some plans to come to our rescue, especially after the return of Yair, who could provide some details about our situation. But they didn't act either.

We were afraid that they wouldn't allow us to take back the diary of our meetings. This gave me an idea to conduct a diary through my letters. I wrote what I did every day, copied it into a letter, and mailed it. This way I created a complete record of my captivity. I had a notebook where I kept all my drafts. Writing occupied me for hours, and by this means I could feel as if I were outside.

I am not a dreamer; I like to act. Just the same, there was one repetitive dream I recall, in which I stood in front of an Arab audience and gave them a speech explaining why they were wrong in the Middle East conflict. We had many thoughts about our return. Dan and I planned to open a restaurant as partners, which would also have a little pool for fish and ducks. Last year, when we all met for a celebration in a restaurant in Jerusalem, we said, "Couldn't we have opened a better one ourselves?"

I had lived on a farm until I was five, so raising animals was in my blood, and the urge became quite pronounced in jail. One day I saw a cat walking on our wall. I wanted to take her down, but the guard was sitting there in his tower and he wouldn't let me do it. Finally I seduced the cat with food and all kinds of motions, and she came to me. She must have been hungry. I took her into the room and named her Dina. She became our pet. She used to go out from time to time, looking for a mate, and then come back and have kittens. All in all she gave birth to twenty or thirty kittens. Most of them were killed though, by one of us who has a killer nature—I don't want to mention his name. But in spite of that Dina managed to raise several kittens in the room. It was fascinating to watch how she'd get in heat, and sometimes perform the act right on our wall. Right away she would become pregnant. I watched her in labor several times, too. Later on we adopted some more cats—one was almost blind, another so thin that we named it Dry Bone. But Dina was our favorite. She was beautiful, like a Persian cat, and she had a big, thick tail. We fed her the best things we cooked for ourselves, and in the end, we took her back home with us.

I was interested in ants, too. Once I captured a whole troop of ants and put them in a jar. I put some sand and food in, I captured a queen and added her to the workers, and I conducted daily observations, recording it all in my notebook. Later I found out that they don't nest in a jar.

One day, while we were playing bridge, I heard Dina squeaking in the yard. I rushed out and saw her with a tiny parrot in her mouth. I saved the bird and put it in a box. When the Red Cross people came, I asked for a cage and a mate for the parrot, and this was the beginning of my parrot colony. Again I used to feed them, observe them, and record my findings. When they were fighting, I pulled them apart. Afterwards, the female had eggs, and we had tiny new parrots, and so on—until I had a huge cage with about sixty birds in it. They were pretty and colorful. I gave our neighbor a pair as a gift, to amuse his grandchildren when they came to visit him.

This was my own hobby, by myself. Not everybody liked the pets. The parrots were noisy, and at noontime the guys wanted to take a nap. I used to cover the cage with a black piece of cloth and they'd quiet down. But one of them would always shriek all of a sudden. At the end I agreed to take the cage into the yard, and I hung it on the wall. I had planned that when I would be released, I would release the birds as well, but in Israel, not in Egypt. In fact, I brought back the big cage with me, with fifteen parrots in it. I gave some of them to my sister and friends, and the rest I set free. The only problem was that they couldn't fly so well, once they were out in the air. . . .

DAVID

The Egyptians used my mail to put pressure on me. For three months they didn't give me any of my letters from my family, only from my adopted family on the kibbutz. These letters also cheered me up, but I wanted to know that my old parents were okay, too. The truth is, however, that during the first stage I was so overwhelmed by my own experience that I hardly gave any thought to what was happening at home.

We had many cats. It all started when one of our wardens brought us a little kitten, which could hardly stand on its feet. We made a pacifier for it, and we fed it from a bottle. That was our beloved cat, Dina, that grew to be so pretty because she got so much attention. She had very thick, unusual fur. She used to get out through the walls, but she always came back. When she grew up she had kittens, and other cats also arrived. We used to argue about the names of the pets. Five or six of them returned with us to Israel.

The cats were with us for two and a half years, and in some way, they made us feel as if we were out of jail. You pet such a delicate creature, and you forget where you are. They gave us a sense of warmth, of release. At night they used to lie in our beds, warming our legs in the winter. All day we played rough games with them, which helped take some of our aggression out. Our hands would be all scratched from these "battles," and then we'd pet them, and it would be all over.

We had parrots as well, but I didn't care for them. We had a young kite in the yard, which was wounded and couldn't fly yet. One day a sergeant came in and smashed its head with a brick. He wanted to test the Jews, see their reactions. All these Egyptians were standing there and staring at us. They could have taken the kite away, if it was forbidden to raise it, but why did they have to kill it right before our eyes?

8

TESTIMONY

Following are some translated protocols and excerpts from the diary, and a sample of the Red Cross reports to the families. Rather than one prisoner's impressions, this diary records the decisions made and the guidelines set up by the group as a whole. As the reader will note, the diary is both cryptic and full of humor, revealing daily aspects that were not emphasized in the men's narratives. The Red Cross reports add a new perspective, from observers of the captives in "real time." All these documents are arranged chronologically, so that the reader can follow some of the developments in the group's life.

FROM THE DIARY

In a preliminary meeting which took place today, (February 19, 1971), it was decided to set up a council of ten members to discuss problems of the room and make decisions accordingly.

The decisions that were taken at the preliminary meeting:

1. The council meets once a week on Friday after lights out.
2. The chairman serves for one week.
3. The chairmanship is assigned in alphabetical order of first names.
4. A majority for decisions is six people.
5. In case of a tie vote, the chairman has the power to decide.

6. If all of the suggestions brought up at a given meeting have not been discussed, they will be postponed until the following week.
7. A memorandum of the meetings will be recorded in a notebook, preferably brief and to the point.
8. Suggestions for the agenda will be submitted on Friday unless the chairman is willing to accept them during the week.
9. The symbols for votes in the notebook will be: + for, − against, and x abstention.

Meeting No. 1. February 19, 1971. Chairman: Avi

The following problems were discussed and decided upon.

1. *Attitudes toward the wardens:* An attitude of restrained superiority, one must never raise one's voice, joke within the limits of good taste without going into personal details, cigarettes to a reasonable degree. Official contacts: Benny—language, Rami—rank.
2. *Turn-taking:* There will be a permanent rotation of duties on the following matters:
   (a) Cleaning the stove once a day. (b) Sweeping the floors once a day. (c) Keeping the toilet in decent condition. (d) Emptying the ashtrays in the room. (e) General cleaning of ashtrays. (f) Cleaning the table. (g) Washing the dishes in hot water once a day. (h) The person on duty on Friday washes the cleaning cloths. Everyone will help the person on duty by stacking the dirty dishes on the table after meals.
3. *Procedure after lights out and in the morning:* After lights out there will be a reasonable degree of quiet. In the morning the lights will be turned on not later than 9:50. If there are still two people sleeping at that time, it will continue to be quiet. On the days of Red Cross visits everyone will get up by 8:00.
4. *Food distribution:* Sugar, and spreads not in cans—without limit; cans and chocolate rationed; tomatoes—to be counted; one bottle of milk per person per day.
5. *Smoke sticks for blues* [prison cigarettes for the wardens]: With permission from the guard.
6. *Phonograph and records:* The phonograph should be kept on the table, records should be cleaned before use.

7. *Carrying out decisions:* What is written in the notebook is obligatory. It would be better not to make personal comments.

Meeting No. 2. February 26, 1971. Chairman: David

Manners: This situation should be understood, each one comes from a different background with different education, so tolerance is essential, e.g., the way we refer to each other, cleaning the table after the meal.

Soccer: It is forbidden to play during dinner.

Representation: We are within a military framework, therefore we have to shave twice a week, wear clean, neat clothing, behave well. Important: finish all interpersonal business inside the room. Above all, represent the country well.

Laundry: Time for laundry will not be allotted, but everyone has to try and finish as fast as possible.

Meeting No. 3. March 5, 1971. Chairman: Dan

Requests from the Red Cross: It was agreed that all requests to the Red Cross will be transmitted through Rami. Good manners during the agents' visit. Divide the gifts after they leave.

Cleaning the room: Once a month, on the first, we will do complete cleaning, so that other life [insects] in the room will be prevented. The member-in-charge will wash the rags in boiling water every Friday.

Cleaning the toilet: The request to keep the toilet cleaner was repeated. The member-in-charge will take the papers out.

Behavior: Two members were severely reprimanded for misbehaving. We have to behave moderately, in a way which becomes Israelis of our level in a place like this, since all eyes are watching us!!! Behave seriously and quietly with the wardens. They are still suspect!

Soccer and chess leagues: Accepted by all. We will allow a week for the formulation of the rules, and start the games next week.

Meeting No. 9. April 23, 1971. Chairman: Amnon

Getting up: After we were caught red-handed by the commander (fast asleep at 11:00), it was decided that we will get up earlier: no later than 9:50 in the morning.

Chess: The chess manager, Rami, announced publicly the failure of the competition, and asked that it be cancelled. His suggestion was accepted: 6 in favor, 1 against, and 2 abstaining.

Meeting No. 10. April 29, 1971. Chairman: Rami

Preparing reading for Friday night: Although it is hard to find appropriate texts for reading aloud, Rami will help Dan, and we will keep trying, since people want spiritual uplift for the Sabbath.

Lights out procedure: People who want to sleep are disturbed by the noise and yelling in the room. It was decided to keep the music at a minimal volume, try to yell less during the domino games, and adjust the light so that it does not disturb the men who sleep.

Reveille: It was proposed to put lights on earlier, i.e., 9:00. The majority voted to keep it as is, i.e., 9:50. (Update: In the meantime we all get up at 7:00! Added by Avi when copying the diary in 1972.)

Meeting No. 12. May 14, 1971. Chairman: David

Inspection: Because of a cleanliness competition in jail, it was decided to get up early and clean up the room, in case we get a visit from the commanding officer.

The dwarf [*a nickname for one of the guards*]*:* Following several conflicts with the dwarf—once involving Benny and once involving Motti—it was decided to report him to the Red Cross and not to the commander of the prison. Minimize contacts with him and do not make fun of him. (Is it possible?)

Communication: If we want to talk to an officer, we ask for an officer—whoever comes, it doesn't matter. See you all in Rami's house with Netta—his daughter!

Meeting No. 17. June 18, 1971. Chairman: Amos

A week of tours in Cairo: The Hilton, the pyramids, and Bourj.

Bridge: Rami started with a speech on competitive attitudes. It was decided to stop fighting. Each one will agree to play with any member of the group. Later, Benny suggested that the better table will play with the worse card deck. David proposed that both the table and the deck will be randomly drawn. Results of the vote: 1 for Benny, 4 for David, the others abstaining. It was decided to go on playing bridge by drawing lots for the foursomes.

Getting up: The subject was raised once more. Again it was decided to get up quickly, etc. The popular philanthropist, who owns two sets of underwear, donated one of his towels for general use in the room.

Meeting No. 21. July 16, 1971. Chairman: David

Menachem arrived yesterday from the hospital, after a year there. Yair's bed is finally back in use. Our new member, master of wonderful ideas, who, during his convalescence, has had all the time in the world to think, is the originator of all the proposals tonight. The others must be burned out already. [In English in the original:] Here he comes!

Shelves: Yitzhak claims that we should forget about them, and ask the Red Cross for more important things. Rami says that painting the room is important, but shelves are unnecessary, and would take too much space.

Menachem: There is no space problem and the room will be more pleasant. As for other requests: if we are permitted to get them, we will, and if it's forbidden, we will simply be refused.

Rami: Shelves are a luxury. The less we ask for, the better the chances that we get what is really essential.

After a discussion of these matters, we voted. Painting: 7 for, 2 abstaining, 1 against. Shelves: 5 for, 5 abstaining. It was decided to ask for shelves.

Movies: It is rumored that the POWs from the other side get to see movies. We want to see movies too. It was decided by the majority to ask the Red Cross for movies.

Tablecover: It was decided to ask the Red Cross to purchase a tablecloth for the dining table.

Other matters that will be discussed next week: a study program, including subjects, hours, discipline. It was decided that nobody will be forced to study, although it is good to push people a bit, so that they get in the right frame of mind. It seems that the room is being neglected: beds are left unmade, with clothing lying on them. We should correct ourselves in this matter. This finishes my third meeting as a chairperson. Hopefully the next one will take place in Jerusalem.

Meeting No. 22. July 23, 1971. Chairman: Dan

It is proposed to have breakfast earlier so that more hours will be free for studying.

Studies: Following the discussion of last week, Rami suggests the following subjects: Mathematics, Physics, English, Bible.

For most of the subjects, the group will be divided into two levels:

Mathematics: Level a: Trigonometry. Teacher: Rami.
Level b: Calculus. Teacher: Menachem.
Physics: Level a: Mechanics. Teacher: Rami.
Level b: Electricity. Teacher: Menachem.

English:	Level a: Advanced. Teacher: Yitzhak.
	Level b: Very advanced. Teacher: Yitzhak.
Bible:	Teaching this subject will be the responsibility of Avi.

Menachem: A matter of principle: Is the study program obligatory or voluntary? Given our long stay here, and the probability of some time more, do all men have to study all subjects or only two of the four? Is someone who started a course obligated to complete it until the bitter end?

Amnon: I don't think we can force people to study. It is a right rather than a duty. I object to making studies obligatory.

David: Whoever starts a subject must stay.

Yitzhak believes that everyone has the right to choose his program. Menachem thinks that studying is obligatory, and we have the right to force it, in order to create a positive social atmosphere.

David: I wasn't asked if I wanted to come to Egypt in the first place. So I cannot be told how to pass my time here. In fact I want to study, so why do we keep arguing?

Amos: The argument is unnecessary. Probably each of us would prefer to study two subjects. Let's vote and see.

Rami: Classes will not take place at the same time, so that each one will be able to participate in as many classes as he wishes.

It was decided that:

(a) Studying is voluntary.
(b) Starting a class is a commitment to continue to the end of the term.

Practically all want to study, even more than two subjects.

Gymnastics: It was decided to have daily exercises in the mornings, organized by David. Studies and exercises will be conducted daily, except Friday, for the sake of the Moslems, and Saturday, for the sake of the Jews. Someone proposed canceling work on Sunday, for the sake of the Christians. Another suggested Monday for the Chinese and Tuesday for the Japanese.

It was decided to keep Friday free for cleaning, and Saturday for rest.

FROM THE RED CROSS REPORTS TO THE FAMILIES

Visit on July 24, 1971

The place: The men asked to paint their room. After some arguments with the authorities, it was decided that the agents will supply white paint and brushes. The guys asked for small dressers or shelves for each of them, to keep their personal belongings. This request was passed on to the authorities.

Hygiene: Satisfactory.

Food: The guys are generally content. They asked for more meat. The agent delivered this request to the authorities, and they promised to oblige.

Canteen: On July 8, 1971, a sum of 150 Egyptian pounds was deposited by the agents in the group's account at the prison canteen. This will also suffice for their excursions. The cooler is working, and the guys reported that ice is delivered daily, as are the cold drinks.

Clothing: The men complained that they still did not receive the caps that were purchased by the agency. They asked also to purchase sandals, shorts, and underwear. The agency will make a list and see how to deliver these items.

Studies: The men have started to study. They asked for a blackboard and chalk, pens, and five notebooks for each. They still did not get the English dictionary, which had been purchased by the agency.

Visits: The guys expressed their wish that one of them would be sent to visit their wounded comrade every week.

Correspondence: Each of them got two parcels. No letters were delivered. Each prisoner wrote two letters to be sent home.

Care: There are no complaints. Since our last visit the guys were taken on tours of Cairo at night. Every evening two of them were taken on the tour, four evenings out together. They had dinner out, and enjoyed the tour very much. They asked to be given permission to go to the movies. They hope that these outings will continue.

General comments: The visit lasted from 10:30 to 14:30. The agents were allowed to see the men immediately, without any restrictions.

FROM THE DIARY

Excerpt from Meeting No. 28. September 3, 1971. Chairman: Amnon

Studies: Menachem suggests that the more advanced classes be held before breakfast. Reasons: (a) Classes are ending too late. (b) With the approaching winter, days will be even shorter.

The proposal was adopted 3 for, 2 against.

Avi will update the class schedule.

Meeting No. 29. September 11,1971. Chairman: Rami

Due to lack of topics for discussion, it was proposed to discuss happiness: what, why, when, and how. We did not reach an agreement, but the common idea was that happiness is a fleeting sensation, following the fulfillment of a need (to be differentiated from a mere feeling of satisfaction). Happiness is never static, and has to be repeatedly refueled. Consequently, a person has to be sure to have unfulfilled needs in stock. . . . Menachem expresses reservation about wasting the word "happiness" on such a simple matter as the need for satisfaction. He believes (rightly so) that for a man to say "I'm happy," he really has to be happy.

Excerpt from Meeting No. 33. October 8, 1971. Chairman: Yitzhak

Parcels from home: Are the personal parcels to be divided?

Rami: Every parcel is personal, but if there is excess, or items that one does not like, they will be divided. If personal items arrive, everyone will decide how to deal with it. Everything is personal until one declares something else.

Motti: Toilet items should be public, even if sent from home, such as shampoo, after-shave, etc.

Menachem: Personal items that are sent from home should not be divided. Dividing them makes them lose their personal value.

Avi: Since all the things delivered by the Red Cross—except specific items requested by individuals—are equally divided between us, what is the problem?

Motion: Every item which is identified as personally sent belongs to the individual unless he decides to share it with others. For: 8, Against: 2.

Meeting No. 39. November 19, 1971. Chairman: Rami

Requests to be made to the Red Cross: bookshelves, sponges, playing cards, glue, tape, a knife, flour, red paint, black thread, pens, a ruler, Hanukkah menorah—if they can bring it. It was decided not to ask for a mixer.

Aquarium: An argument. 3 for, 5 against. Gone forever.

There were many more jokes, but no room to write them.

Meeting No. 42. December 10, 1971. Chairman: Dan

Friday before the 104th Sabbath here. Not for everyone. A discussion about letters that are delayed and/or lost. Rami thinks it has to do with censorship.

Menachem: We should number the letters and keep track to see how many are lost.

Benny: We should demand all the letters from the Red Cross.

Additional requests to be made to the Red Cross: A book to study French, discuss matriculation examinations.

Miscellaneous: Rami tells a joke, Menachem provides an interpretation. In the meantime we eat Jello. An eternal problem is debated: the cake, the Jew, and the preserved milk. Benny volunteers to bake a cake. David seconds. Rami plans to make gefilte fish for Friday night.

All our problems are solved!

Excerpt from Meeting No. 47. January 14, 1972. Chairman: Amos

A name for the cat: Rami is worried about the determination of the sex of our cat, and the fact that she has been given a masculine name. He brings evidence from *Alice in Wonderland.*

Excerpt from Meeting No. 48. January 21, 1972. Chairman: Amnon

Bilbo: It was decided to keep Bilbo the cat. He will be fed with leftovers and sardines.

Dina: Dina the cat is too fat, and Yitzhak is shocked.

Excerpt from Meeting No. 49. January 28, 1972. Chairman: Rami

Pets: David does not want kittens in bed at night. Amnon does. After a big argument, Amnon promises to keep the cats in bed only as long as he is awake. David declares that any cat which gets into his bed at night will be kicked in the ass.

Meeting No. 52. February 18, 1972. Chairman: Dan

I opened with greetings to everyone, since it is the first anniversary of our assembly meetings.

Rami summarizes a year of academic and social activities, and especially commends the Friday night sessions. He thinks that as much as being in captivity is unfortunate, the period has been positively used. We all hope to return soon to our homes and families. But in order to improve our feelings as long as we are here, he suggests that we elevate the level of our Friday night meetings, so that not only daily matters will be discussed, but some more spiritual issues. We were taken captive together; that's a fact. But by engaging in certain types of conversations, we can develop our group further, to the benefit of all. General subjects can be discussed, in areas of human relations, love, and anger, for example, only if we agree not to use the material brought up to make fun of each other during the week between our meetings. It may be difficult in the beginning, but with time we will gain experience. Rami is willing to bring up topics and lead these meetings.

Yitzhak supports the idea, and thinks it will be very interesting, as long as it does not develop into arguments.

Menachem: Each one will express his opinion in the order of sitting.

Avi: I don't think that everyone wants to express his thoughts on each one of the subjects. It may create too much tension for some of us.

Menachem is willing to talk about and listen to any subject. Following Benny's question, Rami explains that what he proposed is termed "group dynamics," which is a psychological method for therapy by self-expression in a group. He hopes that this kind of conversation will lead to tension reduction. This part of the Friday meeting will come first, before discussing the other daily matters, such as requests to the Red Cross. Eating pudding will close the evening.

Three rules are proposed by Rami:

(a) Talk only after being acknowledged.
(b) Do not interrupt each other.
(c) Take each other seriously.

Meeting No. 54. March 3, 1972. Chairman: unmentioned

[Comic account of a meeting with the Red Cross. In English in the original:]

B.: Okay, shall we sit? I have some good news for you, gentlemen. From the next visit there will be no letters. The parcels will arrive from the cannal *[sic]* generally. Yes, every three months the door will be closed during daytime to prevent some little shebabs [guys] that are mobbing around oozing their charm in this time of the year, and be patient. Well, gentlemen, I have an appointment now and I have to leave. Oliver *[sic]* will stay with you. Rami, will you please come over here for a moment? There are some problems to hire the ten trucks to bring the materials you ordered. But it will arrive, inshalla [God willing], through the cannal and will reach you within 48 hours, that I can insure *[sic]* you.

[In Hebrew:]

"Well, Rami, what's up? What did you talk about for 20 minutes?"
Rami: "Guys, everything is okay. We're not going home tomorrow."

Excerpt from Meeting No. 60. April 21, 1972. Chairman: Avi

We celebrated Independence Day with Amnon in isolation. Is it jaundice? From Sunday to the last moment, we saw Rami working with his wooden needles, counting the stitches, knitting furiously with two colors. When the holiday arrived, at the festive meal, he presented us with the Israeli blue and white flag. Israel's twenty-fourth anniversary celebrated with a flag at Abassiya prison in Egypt.

Excerpt from Meeting No. 64. May 19, 1972. Chairman: Benny

Problems of pets (dog): According to David, we should not take a dog in, because the pets cause too much crowding. Menachem says that if he has a choice between cats and dogs, he is for cats. Yet no one talked about a choice. Amnon believes we need not vote on that, just hear the men. He says that when the first two cats were adopted, nobody voted on it. David repeats that it will be very difficult to take care of a dog in the room, and clean up after it.

Vote: 2 for a dog, 4 against, 4 abstaining.

Excerpt from Meeting No. 66. June 2, 1972. Chairman: Menachem

Rami informed us that the "group dynamics" meetings will be discontinued, because members did not show enough maturity for honest participation. Since Rami initiated the activity, his proposal to discontinue was accepted by all.

Studies: Benny raised the subject of the attitude toward our studies in general, and the English course in particular. In the subsequent discussion people expressed their opinions about the place of the study program and its influence on our lives here.

Excerpt from Meeting No. 68. June 16, 1972. Chairman: Amnon

Bridge: The subject exploded because of differences in skill. It was agreed that some skills needed refining. Therefore:

(a) Bridge classes by Amos will be given to all those interested.
(b) We all agree to use the "blue book" method for calling. It is necessary to master the system, i.e., Goren's and Scheinwood's books.

Meeting No. 71. July 7, 1972. Chairman: David

Bridge: Well. . . . Amos refuses to continue organizing the bridge lessons, because of too many personal arguments. Menachem supports Amos, and he proposes to stop playing bridge altogether.

Radio: Since it is disturbing that people sit on the lower beds while listening to the radio, it was suggested that, after the door is locked, the radio be placed on the table. This is Dan's suggestion. Rami says that it is technically impossible and too risky. I think Dan's idea is crazy—extending wires with 220 volts from the wall to the table. . . . The only solution is to go back home.

A good ending—everything would be good.

Meeting No. 76. August 11, 1972. Chairman: Menachem

Social Problems: It was decided to bring to the attention of the Red Cross the difficulties that we encounter here. The matter will be presented in such a manner to express our need for help from home and the Red Cross. Furthermore, the Red Cross will be asked to deliver this information to the Israeli authorities, but not to our families.

The Diary: The fate of the notebook with the diary was discussed. It was decided to consult with Beausart, and then to decide.

Meeting No. 83. September 29, 1972. Chairman: Yitzhak

Dan raised the issue of the ventilator. The last days have been very hot and dry. Dan wants ventilation. An argument started, because Menachem catches a cold if any air touches him at night. He sleeps exactly between Dan (who needs air) and Benny (also needs air). There were some hints regarding tolerance. . . . Finally, Dan gave in for the sake of everybody, and prayed for an improvement in the weather. Amos seconds.

Benny suggests that Dan be relieved of the chores, but Dan refuses. Recently he has been "resting" because of back pains and an inflammation of his leg. But as soon as he is better, he will join the rotation and wash dishes, too.

Many requests to be made to the Red Cross: we want a refrigerator, ice cream, chocolate.

Meeting No. 91. November 24, 1972. Chairman: Dan

A program for improvement of the room and the kitchen: Menachem proposes a plan to improve the room and its style of furnishing. It was agreed that we will make the requests in the following order, hoping that the Red Cross will provide the first items, if not all.

(a) Bookshelves with a special place for the TV.
(b) Kitchen cabinets.
(c) Small personal dressers for clothes.

It was agreed unanimously. In his visit, Beausart took all the blueprints to carry it out.

Excerpt from Meeting No. 97. January 5.1973. Chairman: Amos

Wallpaper: Menachem proposes to cover the walls with wallpaper. Rami thinks we have lost our minds, and he objects. Yitzhak says that we determine our standards, everything is relative. The things that we get from home will not change our image. It was voted: 4 for, 4 against. Benny asked to vote again: 5 for, 4 against.

It was decided to ask for two sets of wallpaper of the same pattern and different colors. Our wives will pick out the pattern.

Excerpt from Meeting No. 103. February 16, 1973
Chairman: Yitzhak

Yitzhak raised the subject of studying bridge with three men of his choice, outside the regular time for the game. He feels that he is not advancing in the game as much as he could. Yet making a foursome outside the regular hours may cause some friction in the group. Moreover—with selected members! He is afraid that others may feel rejection and discrimination. When he had consulted with Rami, Avi, and Amos, with whom he would like to study, they agreed that it might cause tension in the group. So the damage of this tension weighs against the benefit to Yitzhak. Therefore he brought this up at the group meeting for discussion, asking the whole group to decide what is right.

Benny said that it should be allowed, since he views this foursome as the team of the best players, who have the right to play without the weak players, so that they may perfect their skill. No tension will result from this, and because he knows Yitzhak for two years now, he believes that he can become a better bridge player.

Menachem objects, since this is a foursome of Yitzhak's choice, and he [Yitzhak] cannot do what he wants, and furthermore the group will be hurt by this act of discrimination and exclusion. David says that he feels like Menachem, and that it is a criminal step that should not be taken. When Menachem asked what Yitzhak expected to get from raising the matter in the assembly, Yitzhak responded that he had hoped to be given permission to carry out his idea. When a person decides to take a step that might hurt others, he should inform them, so that they may be aware of it, and then try to talk him out of it, if not by speaking, then by the atmosphere in the room. Yitzhak withdrew his proposition.

FROM THE RED CROSS REPORTS TO THE FAMILIES

Visit on March 5, 1973

The room: During the last visit, the Red Cross agent received permission to supply a table to the captives. This is a table that can be enlarged, and it will replace the former one, which was old and unaesthetic. A new lamp and mirror were installed in the bathroom. With the additional furnishing, the room looks more cheerful.

Hygiene: The Red Cross officers found all the POWs clean, as was their clothing. The room is clean. A new cabinet for food keeps the flies away.

Food: The captives have no complaints about the food. They receive enough food in parcels from their families. All food is checked by a military physician before its delivery.

Canteen: There are no complaints about food purchased in the canteen.

Exercise: As in the past, most exercise involves football and ping-pong games. A new ping-pong table was delivered during our last visit, but the prisoners still wait for permission to use the general prison courtyard for their football game. So far they are able to play only in their own quarters. They hope to get an exercise bike soon.

Requests: The POWs made their requests for books, a TV, a radio, and materials for hobbies. They want to get a film from Israel every Thursday. The Red Cross agents asked the Egyptian authorities to hasten their censorship procedure in that regard. They keep studying mathematics. Since language books from a self-study series were supplied, they plan to study Russian, German, Spanish, and Italian.

Medical conditions: All the captives feel well. They have had general checkups and the Red Cross will get a report on each of them. They received immunizations for typhoid and cholera. In the psychological sphere, no changes could be observed. Their morale is satisfactory.

Contacts with the external world: During the last two months, the POWs received eighty-five letters, pictures, and drawings from their family members. In addition, forty parcels were delivered. Two letters home per person were collected during our visit. Due to censorship, letters take about three weeks to be delivered. The agents demanded that the Egyptians respect the agreement to deliver letters in both directions within two weeks.

Relationships: There were no complaints about the guards, soldiers, or officers.

General comments: The men are physically and psychologically well, and they are treated in a satisfactory manner.

Subjects for further improvements: In future visits, the following subjects will be brought up again with the authorities:

(a) Using the big courtyard for a football game played by the captives.
(b) Taking the POWs out on occasional tours in Cairo.
(c) Taking pictures of the captives to send to their families, especially pictures of the Passover celebration.

FROM THE DIARY

Excerpt from Meeting No. 113. May 4, 1973. Chairman: Yitzhak

Dan opens with a long and convincing explanation of his need to participate in the rotating chores and dishwashing. He insists that we allow him to do his share. Some reservations were expressed from the corners, in the sense that "it's not out of pity but because we love you so." At the end everybody agreed to his wish. We hope that it is not too difficult for him, and that he learns to wash dishes, after being the cook.

Excerpt from Meeting No. 127. September 10, 1973
Chairman: Amos

Cats: It upsets David when they get dirty from being outside, as they spread disease, etc. He suggests that they be kept inside at night.

Vote: 1 for, 5 against.

Now we know what David means: to keep only Dina indoors, and put the others outside. In spite of Benny's reservations, it was agreed to keep Dina indoors, because she is about to have kittens, and she tends to run away.

Meeting No. 133. September 21, 1973. Chairman: Yitzhak

The problem of "stereophonic noise," which has come about in the room of late, has been raised once more. Who is right? The first one to turn on the noisy device? We have a conflict between the radio and TV, for example. Two men argue: one wants to listen to the hit parade, the other wants to watch an old Gregory Peck movie. Is the winner always the first one? Or perhaps fairer way of deciding should be adopted? How do you determine fairness? Is majority rule to govern individual choices?

The discussion of the matter was free and honest (for a change). The key word was the magic "awareness"—sensing the other's reaction to the act you perform or plan to undertake. Some say that the will of the majority is nothing but the summation of all the individual wills, and there is no way to compare them. The will of the individual is as important as the will of the total group. When no solution is satisfactory for both sides, it is usually because of the refusal to find a solution. This, too, can be tolerated. For after all, one cannot deny a person, who is locked up with people he did not select, the right to burst out in anger occasionally, thus relieving some of his inner pressure. Naturally, this is undesirable (yet it cannot be avoided), since, by this chain reaction, the pressure builds up in others, who are quite tense as it is! To sum up, take all these matters with a grain of salt.

By popular demand I record that the wallpaper project was finally carried out. We will do some additional painting toward Rosh Hashanah. The room looks like home. Poor me.

Meeting No. 136. October 12, 1973. Chairman: Menachem

On Yom Kippur, the sixth of October, the war started. We were divided in pairs in different cells, except for Dan and Avi who remained in the common room, and prepared food for the rest. On Monday night, after two days and seven hours, we were reunited. It was a great relief.

On Friday the TV and radio were returned.

In the meeting it was decided to be very cautious with information.

Meeting No. 137. October 9, 1973. Chairman: Amos

Happy birthday to Amnon. He is twenty-four today.
The study program will restart on Sunday.
We are having a party. Join us if you want.

Meeting No. 139. [Last one, undated.] Chairman: Rami

One moment of silence "in memory" of Dan, who is already on his way home.
Benny wants us to do something so that we'd be released, too.

The cats:

(a) Since we are in the midst of a political struggle, we cannot find time to name the new cats.
(b) Also, some of them will be given away, so why name them?

Cats: what will happen during the winter? They probably don't mind sleeping outside.

The serious part is over. Now let's have something instead of ice cream.

9

THE RETURN

The release after close to four years in prison, the return to Israel, and the reunion with family and friends was, naturally, a peak experience for the men. It took place in November 1973, within the framework of a massive POW exchange, following the termination of the Yom Kippur War between Israel and Egypt (as well as other Arab states). Dan, who was still suffering from his injuries, was returned a few days earlier, while the remaining nine prisoners were released together and received in public with great honor.

The circumstances of the return, however, were far from simple. Not all the men liked being in the limelight. In addition to the personal transition and adjustment, Israeli society as a whole had been traumatized by the recent war. Upon their return, the liberated men discovered that the casualties included some of their relatives and friends and experienced the general crisis atmosphere among the population. The initial moments of exuberance were followed by confusion and a sense of being flooded by information. The narratives also reflect joy mingled with pain.

DAN

Yom Kippur was on Saturday that year. By accident, Yitzhak put on the BBC, and all of a sudden we heard that Egyptian and

Syrian forces had penetrated Israel, and that war had broken out. There was a very heavy feeling. I still didn't connect the outbreak of the war with the approach of our redemption. The Sudani commander of the prison arrived, all pale, and exclaimed, "Your brothers are fighting my brothers, but if you remain quiet, no harm will come to you." Several armed guards came in afterwards, took away our radio and TV (we had just managed to take apart the receiver that we had built to intercept Israeli broadcasts), and directed the men, in pairs, into different cells in jail. Avi and I were ordered to stay in our room, and we were in charge of supplying food for all the others. I remember that when I went out of the room to wash some grapes in the courtyard, the Egyptian guard on the wall suddenly fired in my direction. I stood close to the wall, watching how he was grazing the eucalyptus branches above me, until he had used up all his ammunition. He didn't hit me, but as I returned to the room I saw that Avi was white as a ghost; I guess I wasn't looking any better myself.

Thirty-six hours later, all the men were returned to our room. A day later, the TV and radio were back. We then started to hear the sound of screaming and beating from the new POWs. It was terribly hard to listen to, and I was afraid I might be interrogated again. On the TV we saw armed Egyptian soldiers walking inside our canal posts, stepping on the Israeli dead. On another program, I recognized some of the Egyptian doctors who were taking care of the new wounded prisoners. The news was an ongoing victory report. But one night, when the guards forgot to lock our yard door, we saw a missile flying above us, and we knew that it was a step toward our liberation. The anticipation of freedom was coupled with a lot of pain, though.

On the morning of October 30, I was cooking eggs, when General Sharif came in and announced, "Dan, in ten minutes you're going home." I thanked him and sat down to eat my breakfast. I knew I had a long day in front of me. I packed all my belongings in two boxes. The soldiers came, blindfolded me, took the boxes, and out we went.

I took my leave of the guys with mixed feelings. I was happy to be going home, but I was concerned for them. In addition, I wasn't sure I was really being released. After passing the prison yard, Sharif ordered my blindfold to be taken off and said: "Come, Dan, let's go to Cairo and shop for gifts for the wife and children. After that you'll meet the Red Cross men and go home."

What could I tell him? That I didn't want to buy gifts for "the wife and children"? I went into the car, and for the first time in four years I rode like a king, with my eyes open. He took me to a supermarket, where I bought some stamps for my children, a purse and some jewelery for my wife, all very touristy items, ornamented with camels and the pyramids. Sharif paid for the gifts, probably from the budget of the translation of our letters. . . . Later on he took me to a fancy restaurant that hosted many high-ranking officers from the army and the government. They all kissed each other on the mouth, a repulsive custom, I think. For a while I was left alone while the general went to take care of some business. I don't know if somebody else was watching me from the side, but I sat there alone for half an hour. We then went to a building in Cairo to meet the Red Cross agents. I was moved to another car and introduced to General Gamassi (who was, at the time, in charge of the negotiations on the exchange of POWs), and i said goodbye to General Sharif. He called me "Captain Dan," while I was careful to correct him: "Lieutenant Dan." Even then, I couldn't be caught off guard. I knew that any problem might hinder the whole process.

We took a long ride out of Cairo. Suddenly, in the middle of the desert, we came to a stop. Only one officer, a soldier, and I were left in the car, while the whole convoy moved on. Anybody could have killed us; we waited for a long time. The Egyptians offered me a pita, which I refused. After my liberation I found out that new problems had come up at this stage of the negotiations for the POWs, and that I was returned all by myself as a token of the Egyptians' good will. We waited there for a couple of hours, until another vehicle arrived and the officer called me: "Come, we'll take you to meet the Israelis."

One kilometer further, I suddenly saw an Israeli bus, an agricultural produce van, water carriers, and a whole camp of tents—the typical Israeli mess in the middle of the desert. General Yariv came to greet me, with some other Israeli officers. It was the famous 101-kilometer point, at which negotiations and POW exchanges were taking place.

I took my leave of the Egyptians. General Gamassi saluted me and I returned his salute. A commotion started right away on the Israeli side: "Don't take any pictures! Nobody is allowed to talk to Dan!" Just the same, the photographers were taking their pictures—I knew I was back home [laughing]. I was very quiet and reserved. We boarded a helicopter and were taken to a military air base, where we transferred to a light plane and flew to Tel Aviv.

My father was waiting for me at the airport with the Chief of Staff and all the top generals. I was asked about the new POWs and how they were. They realized I was sound and well and talking coherently, so they let me go home. [He is quiet for a while.] I was home with the wife and children for a day, and on the next day I was taken to the hospital for an examination.

MOTTI

During Yom Kippur we were watching TV and suddenly heard an announcement that Israel had declared war on Egypt. We were glad on the one hand, because war might lead to our exchange. On the other hand we worried about how the Egyptians would treat us during the war. A couple of hours later armed guards came and divided us into groups of two. Each pair was taken to a different location in the prison. They told us it was for our protection, so that if the prison was bombed we would not be hurt. We knew that no Israeli force was going to bomb this jail as long as we were inside. All the same, the experience was unpleasant. I was put in a cell with Amos, and our room was just across from where new prisoners were admitted. We had a window and we could watch some "humiliation scenes." Egyptian prisoners were physically abused and searched for drugs in a terrible way—it was awful. We were suddenly out of contact with the other POWs and had no news of the war. We heard shelling in the distance. I remember also getting prison food again and having to eat it with my hands. It was unpleasant.

But after a few days we were returned to our common room. We knew that our liberation was near. One afternoon, a few weeks later, we were ordered to clean our room because we were getting visitors. Some reporters arrived, with photographers, and told us that the next day we would be going home. Later I heard that my picture was on Jordanian television that night, and my parents had seen me. We were very excited and didn't sleep the whole night. We talked about what would happen upon our return. We promised to meet at a restaurant, as we had imagined all the years in prison.

Next morning we were taken out and led through the prison. For the first time we were not blindfolded; it was strange. Several ambulances were parked outside the prison to take us to the airport. We passed between two lines of Egyptians. They didn't touch us, but they cursed. . . . I had my bird cage with me, and the soap sculptures, and that's how I boarded the plane. It was an old plane full of wounded Israeli prisoners from the recent war. I drifted into the pilot's cabin and spent most of the flight there.

At Lod [now Ben Gurion] Airport we were welcomed like kings. Golda Meir was there, Moshe Dayan, all the government and the Air Force. They gave us a huge reception. It was exciting. Somebody interviewed me for the radio, and I heard myself on the midnight news. I was in a dream, walking on air. It was unreal.

I remember meeting my family. My little sister had been fourteen when I was captured, and there she was, a young lady, a woman. I almost didn't recognize her. My little sister!

BENNY

We were taken out of prison on a minibus. I was the first one to board the bus, dressed as a civilian in my best clothes, with long hair. Three wounded soldiers had been inside; one had an amputated arm, the other was a Bedouin tracer. I offered them some cigarettes, but they refused to take any. They didn't believe I was an Israeli. When I told them I was one of the War of Attrition prisoners, they nearly fainted. After so many years in Egypt, how could we look so well? We got the same reaction when we landed.

One of the Yom Kippur POWs asked me whether we were not excited to be back. He saw us sitting on the plane, discussing what we'd buy, what we'd do, and couldn't understand. I told him, "Listen, buddy, when you sit for four years, you live through the moment of your return so many times that when it finally happens, you just experience what you imagined all those years. You wear the clothes you've prepared, you board the bus. It's like having the fantasy once more, only this time it's for real. So you're not excited. You accept it for what it is, that's all." Even when I saw the land below it didn't move me too much; I wasn't swept away with all the enthusiasm of the people waiting for us.

DAVID

Once we heard about the war on the radio, each one of us withdrew into his own thoughts, making his own speculations regarding our situation. We knew the trouble caused by wars, yet we understood that for us, the end was approaching. We would be the only ones to profit from that war. The prison commander explained that for our own good, it was better to separate us for a while. I was put in a cell with Rami, in a wing in which Egyptian prisoners were detained. Once more it was a solitary cell full of bugs. But after two days we were returned to the common room, where we tried to resume our routine life. The excitement of the war, however, penetrated our world. We heard the Egyptian stories, such as the one about a whole Israeli tank battalion being defeated by a unit of seven Egyptian tanks. New POWs arrived, and their presence was felt even though we didn't see them. Gradually we sensed that our framework was dissolving. We didn't have any organized activities anymore, and each one was waiting separately for the news from Israel, the news about our liberation.

One evening the guards came in quite late and told us to get ready for an interview with some TV people. A large group arrived, together with the Red Cross men and the Egyptian Intelligence. The reporters probably thought that after four years in captivity, they'd find a group of prisoners climbing the walls, and discovered human beings instead. The Red Cross agent told a couple of us that we might be returned tomorrow. Once they all left, we started to pack.

But we weren't returned the next day. We were waiting with our boxes until 2:30, and nothing happened. On our radio, however, we heard on the Israeli station that the plane with the POWs of the Attrition War had taken off from Cairo. I remember that moment clearly. Avi was washing the dishes in the sink, and when this item was announced, he dropped some dishes and they broke. He said, "If I'm on that plane, how come I'm here?"

The Red Cross agents came once more in the evening, to promise that tomorrow would really be the day. We left early in the morning. It was Friday. I remember looking at the prison building from the outside, saying a nice goodbye to the place where I had spent three and half years of my life.

I had built some large constructions from matches, among them a battleship that took me about half a year to complete. The Red Cross agent said that the Egyptians wouldn't let me through with the boat, so he offered to bring it to me once I was on the plane. I gave it to him a week prior to our release, but he never showed up at the airport and I lost track of the boat. Motti, however, had constructed an Eiffel Tower from matches and had no problem carrying it with him.

I insisted that we take the cats with us, but we weren't sure the Egyptians would permit it. We took an apple cart and put the cats inside. Dina had just delivered some kittens, so we took her with three or four of her babies. The Egyptians didn't examine our baggage. The warden put everything on the bus, and from there we went straight to the airplane. After takeoff, however, we had some problems. Apparently cats have difficulty in adjusting to air pressure changes, and Dina didn't feel so well. She tore the box apart, went out, and came to me. She sat on my lap for the duration of the flight. At the beginning I brought the cats to my sister in Jerusalem, but it didn't work out. So I moved them to my brother in the kibbutz and just kept one kitten, who resembled Dina very much, for myself. It was run over by a car later on.

It was strange to meet the POWs of the Yom Kippur War. Some of them were wounded. When we greeted them on the bus, they didn't respond, because they didn't recognize us. I sat next to a Bedouin tracer, and I said to him, "I am a POW myself." He turned his head from me and looked through the window. It took him some time to comprehend that we were Israeli too. When we boarded the plane he asked, "Are you really the prisoners from the War of Attrition? You were detained for four years, and now you comfort someone who has been in captivity just a few weeks!" The truth is, the transition was harder for them; they were still in a state of shock.

I think that we sat on the floor of that plane. I was sitting there reflecting, now I'll arrive in Israel and meet all those people who caused me so much anger, because they didn't write letters, didn't send parcels; yet at the same time I was longing for them. Who had changed? What had happened to them in the meantime? My father was seventy-five years old, and I tried to visualize him. Did they tell me the truth in their letters, when they said he was feeling well, or were they hiding anything from me?

Finally we arrived at Lod Airport. We noticed a big crowd below. I remember people grabbing us, searching for their relatives. The wounded men were taken off first, while I remained sitting in the plane, until I said, well, I should be getting off, too. I took my two boxes and was among the last ones to descend. It turned out that my family was waiting elsewhere, and I almost missed the bus with all the War of Attrition prisoners. Somebody was yelling inside the bus: "Just a moment, there's one more!"

I met my family with hugs and kisses. They searched for signs of the torture I went through but couldn't find any. Two of my brothers had been married while I was away. From what I could see, everybody was looking well indeed. After I saw the family, a strange officer approached me and invited me to try on a new military uniform. I followed him, then returned to the family, and I remember Golda was speaking. Finally each went his own way.

I was asked whether I wanted to go to Jerusalem with my family or to the kibbutz. I preferred Jerusalem, and so that's what we did. Our neighborhood received me with "Welcome Home" posters; people grabbed me out of the taxi and lifted me onto their shoulders. I can hardly remember how I got home after all. The apartment was full of flowers and pictures. All my brothers, even those serving in the army, came home to see me. They wanted to hear all my stories right away. I didn't even notice that a TV crew was following me, until they asked me for an interview. I told them to wait, and that was it.

RAMI

When war broke out, we immediately understood that we were going home soon. We were the only ones who profited from that war. The Egyptians told us, "We have plenty of new prisoners, but it doesn't concern you. You are our guests and we'll treat you the same as before." Dan was returned on October 31, and we knew that our day was coming too. But we maintained our regular routine to the last day; we even renovated our bathroom just then, having received a new sink, tiles, and a closet.

We did start to pack. Everyone was allowed to take two boxes. People began complaining right away: "I have to take this and that; I'm not going home without it; if we're not allowed to take this I'm staying here." I packed a whole box with the craftwork I had made and many notebooks. Avi took the only copy of our diary. Somebody else took *The Hobbit* translation. We had a meeting to discuss what to do with our common property. We had by then about three thousand volumes and a hundred records. Our decision was to donate the books to the Air Force library, and to divide the records among us. We decided to leave all the food behind for the wardens. So the packed boxes were stored under our beds while we waited to be summoned.

On Wednesday we were told that we'd leave on Thursday. On the next day, however, they said they had had some difficulties in organizing our transportation to the airport [smiling]. Another day passed. On the following day, a truck arrived and took us [he says it very slowly].

I remember that we were joined by the wounded prisoners of the recent war. I sat next to a pilot, who was still in a state of shock from the war and from being captured. I made him talk, and I listened to his account. I had only a vague notion about the Yom Kippur War. I was still stuck three years earlier.

We landed after a while. All the Air Force officers whom we had known were on a bus in the field, and we were all taken to a location where the families were waiting, and. . . . our captivity was over.

All that time I was worried about the Air Force casualties of the war. Nurit's brother was a pilot, and so were my sister's husband and a cousin. I was certain we wouldn't get through this war without paying a price. When I hugged Nurit I asked, "Which one of them?" and I found out that both her brother and my cousin had been killed. It was very traumatic, even though I had been prepared for it.

We were given a reception at the airport, and from there taken with a helicopter to our base for another reception, and then we went to the kibbutz for a third one. All the kibbutz members formed two rows from the gate to the dining hall, and I passed between the lines, shaking hands. We reached the club, where I gave a talk, and so did some other people, and only after all this could I go home with Nurit. She brought the twins from the children's house, and I saw them for the first time. [Silence.] After that we started to get organized at home.

It was a pile of things, all happening together: the happiness of the return and the mourning over those who were killed, all over the country, in the Air Force, and in the family. I couldn't separate those emotions. And the world seemed strange to me—whether because three and a half years had elapsed, or because society had really changed due to the trauma of the war. It wasn't just me who had an adjustment to make; everybody seemed to be in a process of accommodation to the new reality.

MENACHEM

There were several stages in our return. First we heard about the war on TV. (It was the month of Ramadan, and they had programs the whole day long.) We were watching a program we all liked about Flipper the dolphin, when it was suddenly interrupted and the Egyptian attack on Israel was announced. It wasn't very clear. I was the only one who understood Arabic well enough to follow the news, and what I heard was worrisome. We switched to the Israeli station and to our surprise, they were broadcasting [in spite of the High Holiday]. Thus we realized that something serious was really happening. We were very anxious.

In the afternoon we were divided into pairs and sent to small cells—for our own good, they said. Two days later, however, General Sharif arrived and exclaimed, "What is this? These are our old guests," and we were sent back to our room. The radio and TV were also returned, and we got the daily papers, so we were able to keep up to date.

It was unpleasant. On the one hand, this war gave us hope for our release, yet on the other hand, we knew it was a war that endangered us. We saw on TV lines of new Israeli POWs being taken out of the canal posts, and we felt a rush of activity in the prison. All this produced mixed feelings.

I remember that this war seemed to go on forever. I told myself that we had to be out by October 24, since this was my oldest daughter's birthday. In fact this was the day of the cease-fire. From then on we knew that the end of our captivity was near. We stayed on another three weeks after the end of the war, though. Negotiations were taking place at the 101-kilometer line. A week prior to our release Dan was taken away. Our turn arrived on November 16.

When we gathered in the common room we tried to return to our previous routine, but it wasn't easy. I don't think we resumed our studies. One day Beausart came, after a long absence, and this was a good omen. He had returned to Egypt for the POW exchange, and his presence made us feel more secure. He arrived for the second time with a TV crew, including a woman photographer, and said, "So, are you ready to go home?"

They were waiting for our arrival in Israel on the next day, but we were delayed once more. They told the families that we weren't returned because we hadn't finished packing, but naturally this was a lie. We had all been packed for weeks. I would have sold my soul to the devil that day to be able to go home twenty-four hours earlier.

On Friday we got up early, made coffee, and put our best clothes on. There were nine of us then. It was the first time we went out without our blindfold, and it made me dizzy to see the horizon and all the things far away.

We boarded a bus and saw people in pajamas. We realized they were the new prisoners, but they couldn't comprehend who we were. We didn't look like prisoners. Some of them were eighteen or nineteen years old, and had never heard about the POWs of the War of Attrition. They took us to be Egyptians in disguise and ignored us. But later on, when they finally realized who we were, we had a very emotional encounter.

I met a pilot about my age on the airplane, and he talked to me the whole way. This was very important for me, because he updated me on the news. He told me about some of my friends from the Air Force who had been killed in battle. I remember he mentioned one who had been about to get an important position in the development department, and I had a feeling this would be my job. And that's just what happened.

I went to talk to another pilot, a wounded young man lying on a stretcher. He told me that he had been wounded after his ejection, when Egyptian farmers nearly killed him with a pitchfork. He looked so young, and I asked him what kind of plane he was flying. "A Phantom," he said. He was a member of my squadron! When I introduced myself he was so amazed, as if I were some kind of national hero.

Throughout the period of captivity we had dreams about the return. I never shared those dreams, but I had a clear picture: our group crossed the canal, and all the celebrities of the country were waiting for us on the bank. In reality, as we finally landed, all the celebrities were in fact there to greet us. We didn't see our families, but Golda and all the ministers were there; only the president was missing. The former commander of the Air Force, a man we all admired like a father, came aboard the plane, and I have a picture showing me coming down the steps with him arm in arm.

The next stage, which I had so often visualized, was meeting my family. After descending from the airplane, I had a feeling I was getting numb. I was dizzy. We were told that our families were waiting for us at a military base nearby. Two buses were parked to take us there, and near the buses—a surprise, a line of all the Air Force officers above colonel. It was a breathtaking moment! Why? Because when we had shouldered the whole burden of the War of Attrition with our Phantoms, we had had the feeling that we were the Air Force. And here indeed, the whole Air Force was giving us a reception, hugging us as if we were babies. Once more we felt part of the Air Force family. We were wild with joy.

We boarded the bus and went to meet our families. It's a good thing my mother and brothers weren't there but only Esther and our daughters. It was a meeting. . . . [laughing] I have no words to describe it. For a moment you don't believe it's happening. You touch a person, but you can't believe you really do.

From there we were taken to the air base, where fire trucks sprinkled us at the gate. My mother and brothers were waiting there. We all went home, and the place was full of flowers.

That was the beginning of the next stage: it was the house in which we lived for about a year prior to my captivity. I remembered it as a very nice house, but all of a sudden it seemed so small and crowded. It was perhaps because of all the flowers in the living room. Right away I noticed things that needed repair—a broken door on the refrigerator, for example—as if there had been three and a half years of neglect. I realized that Esther wasn't cut out to live without me, and didn't make an effort to. These tiny things gave me my first shock.

AMOS

The moment we understood that a real war was going on, we knew we'd soon be back home. The Egyptians isolated us in cells, because they were afraid we'd be rescued by Israeli forces. After two days we were returned to our room. I recall how we watched TV and saw the capitulation of the canal posts. We didn't know the real facts of the battle; normally we took the Egyptian reports as exaggerated, but it was obvious that some people were killed. We experienced great pain and happiness together.

I remember that Dan went back, and we received greetings from him on the radio, in code, naturally, as we had previously agreed. This made us even more confident that our time was near. The waiting during the last couple of days was quite pleasant. We were treated really well, because the wardens hoped that we might leave them the stuff we had accumulated, and in fact we left them all our food. For men who were used to eating only pita and beans, this was a cause for great celebration. I think there was a deal that we'd leave the large items in prison in exchange for similar items purchased by the Egyptian POWs in Israel.

On the last evening a group of reporters came and formally announced that we'd be going out the next morning. They came to see our reactions to the announcement, so we pretended to be really surprised. . . . The next day we were taken to the airport. We flew in an old piece of junk.

The return was extremely exciting. It's hard to describe it. I was in a kind of euphoria, as if I were walking a meter above the ground, floating. On the one hand we were flooded with happiness . . . which one cannot depict. On the other hand, we were overloaded with information. This had already started on the plane, when we met some Israeli POWs, and they told us what was happening in Israel.

Dalia came to the airport, but she didn't bring our daughter.

The whole government was waiting for us, and so was the Air Force. This was obvious; I expected it. Golda came too, and cried a little. There was some kind of reception, then we were taken in a helicopter to our base. I remember a big commotion around me, but it didn't disturb me because I withdrew into myself.

My daughter was waiting at home. Our meeting was funny and strange for both of us. She didn't really know who I was. She had always heard "Daddy, Daddy"—and here Daddy really appeared. I think kids take things naturally and only parents complicate matters for them. She seemed smaller than I had visualized her, such a three-and-a-half-year-old miniature. They all looked smaller than I had remembered, Dalia too. These were my first impressions. I had no problem in orienting myself, but I was flying high in my euphoria.

AVI

I got the news about our release when I was in the middle of a painting. One of the newspapers showed a photograph of me and my work. Right away I thought about the study program that I hadn't yet completed, and I said that I needed two more weeks here before I could leave. I think this was a kind of defense mechanism against too much emotion and the fear of disappointment. I had been thinking about this moment so often, when Yair was returned, when Los was—and it was always followed by disappointment. But when Dan was taken back, I started to hope again.

Anyway, I behaved as if we really were going home. We started to pack and we discussed what to take and what to leave behind. The nonsense we took back! Cats, for example. This is understandable, after all, because a pet that I had fondled for the last four years had satisfied a great need, and it was normal to get attached to it. More ridiculous were the conflicts about some pants, for example. My entire world was packed in two boxes—the books I received, my records, and a few clothes. With great regret we decided to leave our pots and pans behind.

I had been afraid that we wouldn't get sent back, but I wasn't afraid of what I'd find there. To return meant to go back to the woman I loved, to a family, to a warmth that I had missed so much all those four years. To this very day I still ask myself, what did I do with that warmth all that time?

The moment I arrived, I wanted to be alone with Yardi, but we were surrounded by too many people. It even started on the plane. The Chief of Staff came up to us with Golda and said quietly, "Hey guys, give her a kiss, you don't know how much she worked for this!" Then we were put on a bus with all the Air Force commanders, men I had always disliked. They kept saying, "Hey guys, what's going on?" Our reception was so typical of the army—they didn't really think about our needs. They were happy to see us, sure, but that had nothing to do with our own needs at the moment.

I remember the moment we were reunited when we stepped off the bus. You can still see it in the movie they filmed. Strangers really fell all over us—the soldiers on the base watched in amazement—and our families had no chance of getting close. When family members could finally approach, they started to search for their relatives like in a market place. It was such a mess. I saw Esther looking for Menachem. . . . and I remember hugging Yardi, then hugging Esther, and hugging Yardi again. Suddenly a young soldier came to me: "Do you know who I am?" he asked, and he started to explain that he was married to a remote cousin of mine.

And that wasn't all. A new commotion started at the club of the base near the airport, with Golda again, and all the big shots. True, they had worried about us all these years, and now they were happy to see us—but this was their problem! Later we drove to our own base in a car together with the commander, and we didn't have a moment of privacy. On the base my whole family had gathered; my parents and sister; my father nearly collapsed and my mother fainted in front of five hundred soldiers—what a celebration! We had a big lunch and were driven to the squadron. They said, "Come, get into an airplane," and I did [laughing bitterly].

Finally I saw Yair, who was there too in our honor, and I asked him for his car; I had to get away. Yardi and I drove home, but there we found another group of friends. They were sitting in our living room and they wanted to hear my stories the whole evening. When the nine o'clock news came on, they said, "Hey guys, let's watch the whole story on the TV!" This was too much.

I think only after nine o'clock were we finally alone. Yardi and I started to collect ourselves and see where we were. It was a very strong experience, yet we had no time to experience it.

The next morning my family came for a visit, and the squadron commander called to ask whether I wanted to fly, and all the crowdedness and excitement repeated itself. And the war had just finished—and where was I?

10

BACK TO LIFE

The ten former POWs painted very different pictures of the sensitive first weeks back home. Some tried to return to their former routine as soon as possible, while others felt changed by their experience and looked for a place that would be adequate for their new needs. Their career paths went in different directions—some returned to military careers, others chose to explore the civilian marketplace. Some of the men had difficulty negotiating with the Israeli authorities for compensation for their captivity. Their return to their families was even more complex. Readjusting to a marital relationship and the parenting of grown children was a task of varying difficulty for the men, culminating in a painful divorce for Yitzhak. Except for debriefing and a medical check-up, the men remember getting little guidance in their reentry, and several of them still regret it.

AMNON

I was so eager to be back home; I had imagined the moment of returning so often and had created an idealization of what would happen to me. Actually we returned at a very difficult time, right after the Yom Kippur War, when the country was full of pain and bereavement, and this atmosphere destroyed much of the joy of our release. I suddenly missed some people who had been killed.

On top of that there was this whole business of coming back to reality, after the fulfillment of the dream. No wonder I was disappointed; I had always been sad following the realization of my wishes. It was the same when I had completed the aviation course. There was an episode that exemplified the feeling of the first day: in their excitement, my parents had forgotten to put on the hot water boiler, and when we arrived there was no hot water for a shower. It sounds so stupid, but this bothered me a lot. I think that had I been married, my wife I could have helped me through these first steps, but I was single. When you develop such high expectations, it's obvious you'll be disappointed. So during the first period of my return, I was a sad man.

On top of my own expectations, there were those of others. They wanted me to be the happiest man on earth, and this made me feel even worse. I pretended for them, but deep down I was depressed. I used to argue with myself: What are you crying about? You're back home, after all.

For some reason, the Jewish mourning custom comes to mind. You sit at home for seven days and all the people come and comfort you, while you have no time to cope with your loss. Once they're gone, you become aware of your tragedy. When I returned, I was welcomed with a great celebration. Parties and receptions were given every evening; I was never left alone. This is the similarity: when the celebration was over, I had to return to normal, and coping was tough.

One of the minor problems had to do with the overload of information—too much stimulation. Soon I was satiated and tired. I remember myself for hours and days, going, traveling, visiting—while already saturated. It was as if I had a mask on, and people couldn't realize what was going on underneath.

One event is symbolic of the whole period. While in prison, I had a dream that when I'd be released, I'd go to Jerusalem to visit the Old City and the Western Wall. I arrived at the Jaffa Gate and started to walk. A small Arab boy put a bomb right in front of me. It exploded right there, and an innocent woman blew up, with her inner organs flying onto us. For me, too, it was a near miss. My life story could have ended right there. From then on I always carry a gun with me. In a way, this is the dream of the return and the reality one encounters in a nutshell.

For a long time after my release, I was disconnected from reality, really out of it. I had returned a different man, and couldn't go back to being a child in my family. I didn't want to share these difficult problems with my parents and brother, so I kept it all inside. I do the same to this very day. Only when I started to write my memoirs was I able to release some of the inner pressure from my captivity. I had no framework to return to. I was cut off from my parents and had no wife to return to. The guys were sure I'd get married right away, but I did it only a year and a half later, and to a wife who tried forcefully to get me back to earth.

During the first stage I tried to avoid evaluating the loss caused by my captivity. I referred only to the profit. I often repeated how I returned from jail much more educated. All the other consequences were revealed only gradually.

I asked for two months of leave from the Air Force, and when this was over, I returned to flying in my squadron as if nothing had happened. I still had several months of mandatory service to complete. The familiar routine of the army helped me regain my normal self, but outside this framework I was very lonely.

As it turned out, we had developed a special kind of communication in jail. We knew each other very well, and therefore we could be completely open and frank with each other. When all the festivities of the return were taking place, we used to get together in a corner, like chicks around the hen, and resume our habitual conversations. People who were listening in were amazed. I think we all had to relearn how to talk to ordinary people.

RAMI

Nurit and I decided to get out of the center of activity so that we might build a new life for ourselves. I requested a transfer to a small air base in the Sinai, where we could be removed from others and get to know each other again. I felt this was our best chance, and it turned out to be a wise step.

During that time, I flew a little and got back to my level of performance. At the Sinai base I felt like I was on leave. I was in charge of instructing Skyhawk pilots. I was overqualified for the job but requested to be assigned to it just the same. For a year we lived on that base, with about twenty families, and went on many trips in the desert. We had a lot of time for ourselves. When the year was over, I was appointed commander of a large base near my kibbutz, where I served for five years.

My time in the Sinai desert was magnificent. Nurit and I felt right away that we could resume our life together, but we still wanted time to test ourselves and the kids as a family. With the children it was more complicated. I was a new dad for the twins, who had had only a photograph of a father before. For their entire childhood they never turned to me for advice in any substantial matter, only for technical help. Only at adolescence did they realize I was a person with whom they could share some of their problems. The older kids also used to turn to Nurit with everything. I had to rehabilitate my relationship with them, and the long trips we took in the desert contributed a lot toward this goal. We used to go out every weekend; it was terrific.

I have never tried to fill in the gap of time I missed while in captivity. The Yom Kippur War also had its effect in making the previous period insignificant. As life went on, the sense of my lost time diminished. My absence didn't damage my career in the Air Force, and I didn't become a base commander any later than my peers who hadn't been in captivity.

I tried to do everything slowly after the return. I discovered that I had difficulty in functioning in big, loud groups. I allowed myself not to socialize too much. For a while I had no patience for books and movies. I used to read or watch for ten minutes, get the idea of it, and lose interest. Today it's less severe, but I still have no time for nonsense.

MENACHEM

During the first week there was not a single free evening when our house was not flooded with visitors. I enjoyed it, but it turned out that Esther did not. She wanted us to get in the house and lock the door behind us; but the door remained open. Lots of people came in, and I wanted to go out as well. I needed to see people and places. I remember that even on the first evening I wanted to go and visit Chetz's widow, and we did. There were plenty of things I wanted to do.

I don't remember the very first period after my liberation. We had briefings during the first week or so; the Air Force wanted to know what we had said in our interrogations. It was tremendously important for me to get an evaluation of my behavior, and I was happy with the feedback I received. A week after my return I was called to the Air Force commander for a conference about my future job. Since I had been bothered by my lack of productivity, I was keen to resume my work right away. I was offered a chance to go to graduate school for an advanced degree in engineering, but I preferred work to studying. The Air Force commander offered me several positions, and suggested that I give it thought. I made my choice right there, however, and selected the position in the arms development department which I had heard about in the airplane coming from Cairo. I felt I was getting back to the best position in the corps, and a few days after the appointment I came to the office and started work. As it turned out, I hadn't lost anything in my career by being absent for four years.

I returned to regular work very fast. I felt like jumping into deep water, and it was a real experience. I felt good at work. I adjusted to the staff and I didn't realize I was burning myself out in the effort. I really had limited abilities then, and I needed to put in much more energy to do my job adequately. The results of my efforts were good, but it would have been better if I had let myself rest a little. Only later on did I realize that it was an exhausting time in my life.

Today I'm sure that I returned to work and assumed responsibilities much too soon. I needed to relax and probably get psychological help—which I didn't. It was an extremely hard year at home. I wanted to be as free as a bird, to come and go as I desired, while Esther wanted to keep me all for herself and the family; that's what she had wished for four years! So there was a contradiction between our needs. I didn't feel I was rejecting Esther by going out so much and committing myself to all sorts of activities. I felt that our relationship was superb. Apparently, however, Esther wasn't satisfied with what I was willing to give, and justly so. There were a lot of small episodes in which I "vanished" when she wanted me around. Esther had difficulty accepting me for what I was at that time, namely a man not really ready to be domesticated. I was entirely sure of Esther's and our daughters' love for me. But evidently I needed to put more effort into these relationships. Esther had changed while I was gone; I think for the better. She coped excellently. But the distancing and the changes that had occurred in each of us demanded readjustment. As I said, it was a hard time.

When I had been taken captive, our oldest daughter was six and a half. She is a strong girl, and she coped well with my absence. The little one, who had been three, suffered much more. She was prone to anxiety and feared death, and she refused to eat, so that she wouldn't grow old and die. When I returned, she was in first grade, and immediately she started to get stronger, like a wilted plant after it is watered. It was as if a huge block of ice that had been inhibiting her started to melt down. Gradually she got rid of her fears. I am very happy that it was still a reversible process. Both are wonderful girls today, and so is the third one, who was born a year after my return.

I don't remember a period of feeling estranged from my daughters. I didn't have any difficulty in resuming my role as their father. But we had our problems, and often I didn't pay enough attention to the girls because I was so engulfed in myself.

AMOS

I was under the pressure of information overload right away. For a long time after my liberation I made many errors in talking to people, as if I had a huge black hole in my knowledge. I had to close the gap.

Time was another source of pressure. I was afraid of being late. When I had an appointment, I couldn't decide how much ahead of time I had to leave the house. I hate to be late, and I was anxious about that.

You can't say that they knew how to treat us here. We didn't get any professional treatment, except for the military debriefing. Somebody was supposed to teach us how to return gradually to normal life, but instead we were thrown right away into deep water and expected to swim by ourselves. A wife and children are a great help in returning back to normal but some of us were single and had no one for support in the process. It took them a long time to find themselves. I think nobody paid attention to our needs, perhaps because the whole country was under the trauma of the war. I believe that both we and our families should have been prepared systematically for the return, the single men in particular.

I was at peace with myself and had no doubts considering my conduct in captivity. I think I behaved honorably throughout the experience.

The Air Force didn't put any kind of pressure on us to return to duty. There was another war of attrition going on, however, with Syria, and I felt that I had to go back to flying as soon as I could. I had to struggle for it. I went to the Air Force commander to convince him to put me back on duty. He didn't make it easy for me, since he wanted to make sure I really wanted to rejoin the ranks and wasn't doing it out of social conformism. I demanded to go back as a combat pilot. I wasn't interested in any other job, or in flying for my own pleasure. In the end, I was the only one who returned to the Phantom squadron to which I had belonged. Less than a month after my return I was already flying regularly, and I continued for the following two years. Today I am still a combat pilot on reserve.

But it wasn't easy to return to the Air Force. Those who had graduated with me from the aviation course had, in the meantime, been promoted to various commanding positions, and I couldn't just go on being a junior pilot. Finally I was given a more professional task.

We moved to live on the base—which I hadn't done before. I combined studying with flying and in three years graduated from the Technion. I don't notice any major change in my life after captivity. I resumed my former friendships, which were mostly with other men in the Air Force. There was a certain shell I needed to put back on, because during my captivity I had become very open. Had I continued to live according to the ways we had adopted in jail, I'd be too vulnerable. The men in captivity regarded each other more positively than others do outside jail, otherwise we couldn't have survived the experience. In normal life, at least in the Air Force, it was all different.

Dalia and I had been married for seven months when I was captured, so our separation was much longer than our living together. Since I was in isolation, I had the sense that the whole world had stopped in its tracks, and I went back to where our life had been interrupted. Our daughter was an addition to the family, of course, but I behaved as if nothing had changed. From a certain perspective this period of three and a half years in jail had no effect on my life.

Anyway, my family put me quickly back on the ground. For better or worse—I really don't know. I suppose it had a positive effect since I returned to normal, both at work and also as a husband and father who goes out in the morning and comes home every night.

YITZHAK

I remember that when I arrived at the base, I asked Michal for the car keys, because I wanted to drive home. She looked at me and asked, "Are you sure you can?"

The first days are blurry in my memory. The only thing I remember is how important it was for me to return to the Air Force. On the second day I went for a medical examination, and two days later I was flying with the squadron leader right behind me. As we landed he confirmed that I hadn't forgotten anything. Apparently I had done a lot of unconscious flying during captivity. . . .

On top of the excitement of the return, and the sense of the lost time with its future consequences, there was the story with Michal. I think that for four days or a week we acted like a normal couple. Then, one night, as we were in bed, Michal told me about her affair with Gideon. She told me that this relationship had been going on for a long time, and that while she really didn't want to hurt me, she wanted to get a divorce.

I was stupefied. I went outside and sat there for a couple of hours. Then I took the car we had just received from the Air Force and went to Rami's house. It was five o'clock in the morning when I arrived there, and we had a long talk. That's when I discovered that everybody knew the story. To this day I'm not sure whether Rami found out about the affair in Egypt, and frankly, I'd rather not know.

It was hard to accept, not only because of the personal significance, but because of the public aspect. It's so unfair to do this to a prisoner held abroad, while he's not even around to defend himself! The trauma was so great that I couldn't accept the idea of a divorce. I had no mental strength left for coping with a new situation. I was particularly hurt by the feeling that everybody had known it all along. This upset me even more than the affair itself, because I'm the kind of person who cares a lot about other people's opinions of me. This left me with a scar that hasn't healed to this very day. Even now I find it hard to reconcile myself to the facts. Now I don't blame Michal anymore, and I'm perhaps out of the mental anguish, but I still can't take the shame.

When all this happened, I tried every possible way to mend the situation. I appealed to Michal's common sense and asked her for a second chance. I don't know whether I convinced her or just made her surrender to my pressure; in any case, because I was so surprised, I wasn't ready to take the situation for what it was. I regarded our marriage prior to my captivity as a good one, but today I know that we were not in the process of getting closer, and we didn't reach a deep level of understanding. At that time, however, I didn't feel anything was missing.

Right after the crisis, I was sent on a mission to the U.S. I regarded this as a real opportunity to save our marriage, and I demanded that Michal sever all her contacts with Gideon while we attempted to salvage it. Things were happening very fast, while I was sort of out of balance. In January we arrived in Washington. Right away I had a car accident, and I don't think I was fit to fly, but nobody else seemed to notice. My debriefing was cut short because of the trip, and I didn't even have a chance to see the mental health professionals. As it turned out, I think it was a mistake to leave things alone like that. My commanders said: "Let them go to America and everything will work out. . . ."

But it didn't work out. A month after our arrival I found a letter from Gideon in our mailbox, and Michal admitted that she had maintained contact with him, that she missed him and wanted to go back. I think it happened on my birthday. I agreed that she return immediately, but first I wanted a divorce. In the Israeli embassy, where everyone was aware of our situation, we were helped right away, and our divorce was obtained through the military rabbinate. In two weeks it was all over.

Michal and our daughters returned to Israel, while I stayed for the year in Washington. Today it seems to me that my life at that time was sort of unreal. I remember the pain of separation from the girls. I was flooded with longing for our brief moments of happiness in our bungalow, when I came to kiss them goodnight. It was painful to remember their sweet smell. . . . But I couldn't take any contact with Michal, so my relationship with my daughters was eventually severed, too.

I went through a second childhood, and then went for a highly demanding study program to become a test pilot. I remarried, had two daughters again, and divorced my second wife. Today I am married for the third time, and I work as a test pilot for a firm in the States. I would have liked to have gotten such a job in Israel, but it hasn't been available. One doesn't always get what one desires.

AVI

I used to dream about two things in prison: to be with Yardi and to fly. These were the pivots of my entire life. I wanted with all my heart to be sure I could fly again, so three days after my return, I was given a ten-minute course and took off by myself. I did two or three flights that same day and was satisfied.

The parties went on. A big one was given by the Air Force at the base. I didn't even have a uniform yet and borrowed one from a friend. While we were in captivity, the country had experienced a war, people had died, many things we didn't know had happened, and getting to know about them was pretty traumatic. I began to hear about men who had grounded themselves or parachuted from their planes out of fear, and about squadrons who had lost half their pilots. Slowly I realized what had happened in the country and the Air Force while I was away.

The big excitement we were received with disrupted our privacy. Being alone with Yardi was my top priority. It was disappointing to my parents, too. It was difficult to be open to all the people and their demands. I felt the need to run away.

Several of my meetings with the army were unpleasant. As a group, we never agreed on what we should get as returning POWs. Rami, for example, thought that we weren't entitled to anything. The authorities hadn't made their minds up. I requested to be supported through college, but the army demanded that I should sign up for years of service in return. When we went to the personnel commander with our list of requests, we found him under a great deal of pressure. His own son had just been wounded, and he said, "What do you want? People have been killed and injured recently, while you're sound and well. There are others to rehabilitate before you." We came out of the meeting feeling quite bad.

Nobody tried to reach out to us and take care of our needs. We had been imprisoned for four years, after all, and were somewhat traumatized! I was sent to see the Air Force psychologist. After thirty minutes, I felt it was no use; I got up and left. It was my second disappointment.

Later on I was invited to give a lecture at one of the air bases on my captivity. I prepared a short presentation, but when I came to talk to the commander he started to make restrictions: "Don't mention this; don't tell about your torture," etc. I said, "Goodbye, I won't speak at all." I realized nobody wanted to hear me out or use my information.

What disturbed me more than anything was the pressure put on me to go back to regular flying. I was interested in becoming a test pilot, but the job was given to Yitzhak, because he was "miserable." Yitzhak is an excellent pilot, but in the competition between us the decision was made in his favor even before anyone saw me. I was considering a college education, but I didn't want to live in Haifa, so I finally decided to become an El Al pilot. When they heard about it, one of my officers proposed that I return to the squadron with the promise that I'd be promoted to squadron commander pretty soon. I explained that I refused to fly over enemy territory, while he warned that flying El Al would ruin my family life. As if I didn't know the Peyton Place of the Air Force bases. . . . When they found out that I was considering leaving the army altogether, they warned me that I couldn't afford the home Yardi had built during my absence in a neighborhood where many pilots were living then—despite the fact that we had paid for it all already. I was about to go to court, when at the last moment the matter was settled.

So these were some of my encounters during the first months. I felt that I had no place in that system and went with Yardi for a trip abroad. Throughout my captivity, I used to correspond with Yardi about our future trip. We had in mind going to Rio for the carnival to rediscover each other. After my return, it didn't matter to me where we'd go, as long as we'd be together, share the excitement of new places, and recover some of what we had lost during our separation. We went away to Europe and the U.S. for six months. People abroad hosted us and helped us finance our vacation. I was alone with Yardi, and I overcame my crisis. It was magnificent.

Since I had some time left till the beginning of the training course for new El Al pilots, I went back to the Air Force for that period, but again they demanded that I resume regular duties. Most of the men who had been POWs did this. One of the squadron leaders, a man who had been a POW during the 1967 war, came to persuade me: "What are you saying, of course you can recover, look at me!" I told them: "I love flying. Find me an appropriate position and I'll stay." I knew I had something to contribute even if I stayed within Israeli borders, and I wasn't as yet ready to go on dangerous operations. My memory was too fresh, and Yardi also refused to let me risk my life again.

I knew a pilot who had been captured during the Yom Kippur War, and was held in captivity for only six weeks, but he came back much more shaken than I did. Probably since he had two traumas to overcome simultaneously—being taken captive and returning. He started to fly right away and resumed all duties. He told me that he was under severe pressure, and I encouraged him to go and ask for leave, to relax a bit. He was killed when we were abroad; he was so intent on hitting the target that he aimed his nose right into the sea.

I had my struggles with the corps. I tried to construct a role for myself, but nobody understood my situation. I even agreed to photograph enemy territory, because I knew it wasn't as dangerous as other operations. But I couldn't stand the constant criticism I was getting due to the fact that I wasn't ready to resume all of a pilot's duties. When they distributed albums of the Yom Kippur War to all the participating pilots, I didn't get one. I left the Air Force with a bad taste in my mouth.

I realized that had I been willing to conform, the military system would have been good to me, as it was to its other members. But I couldn't do it just because others did, so I left. I have been flying El Al for fourteen years now and have no misgivings about my retirement. What puzzles me is how the others resumed their duties in the Air Force, especially Amnon, who had been more afraid than any of us.

BENNY

I returned from captivity well prepared for life. I was all organized. I had time to reflect on everything and I knew what I wanted. I even visualized the woman I'd marry. When I returned, I went right back to normal life. I got married in four months, had three kids, and built a house. I wasn't surprised at reality.

We received a salary for the whole period of our imprisonment, as if I had been a sergeant in the regular army all that time. The money was saved in my bank account, and it provided me with what I needed for the wedding and for starting a family. In addition, we received a certain sum as compensation; it was no big deal. I was declared an invalid, because of permanent damage to my legs and teeth—probably the result of my torture and confinement in solitary—and was entitled to a small pension. At first I was ashamed of asking for anything, but Dan's father convinced me it was just and fair. The Ministry of Defense arranged a cabdriver's license for me, and that's my occupation. I don't compare myself to Amnon or Menachem; I don't ask whether they are doing better than I am. I am not the IDF's only child. I believe I received everything due me based on the criteria of what POWs are entitled to, and I'm satisfied. Whatever I have now is the result of my own efforts.

During the first period after my release, I used to dream that the Egyptians had kidnapped my son and would keep him if I didn't return to jail. Sometimes I dreamt that I was just on leave and had escaped my obligation to return to prison. To this very day I sometimes dream about captivity, with clear pictures from our life in jail. It's strange—when I was in prison, I used to dream that I was home on leave but had to come back to jail.

When you talk to me about Benny the POW, I feel as if I'm standing on the side and watching, as if there are two of us: the real Benny, the way I am today, and Benny the prisoner, who doesn't exist anymore. He is something totally different. I cut myself off from that Benny, as if somebody else had experienced all the suffering. In fact, I know everything that happened and I can talk about it, but I'm not touched by the story. There was a third Benny, actually—the young man who wanted only entertainment and nice clothes. Today I believe it's good I had a good time in my youth, because when I returned from captivity I was prepared to get married right away.

I think that my captivity didn't affect me in any negative way, perhaps because I'm strong and don't take things too seriously. I am realistic, and I know that life goes on even if I make a mistake, and when I die nothing is going to stop either. It's clear to me that a man must be realistic and never complain about his fate.

I should add that a month after my return, I demanded to go back to military reserve duty. I'm still thrilled by danger. In the Lebanon War I was called for combat duty, probably by mistake. I went to war, and only five days later, when they discovered who I was, I was kicked out. There is no cure for this thrill.

DAVID

There were several debriefings and many questions were asked in the beginning. I still felt they didn't dare ask everything, as if they were careful not to open any old wounds that had healed in the meantime. I think that I've never told the whole story to anyone up to now.

I tried to cope with my new reality and adjust to it, but it was hard. I remember taking a bus to the end of town, and finding out it wasn't the right line, because the numbers had been changed. I was confused. I had money again, I walked in the street. . . . People recognized me and followed me. One day I passed near a restaurant and three men approached me, saying, "Come in, eat whatever you want on our account." I realized people were really happy about our return, and it couldn't be avoided. This was especially true in the family, where I couldn't come and tell them I wasn't interested in all these celebrations. They had waited so long for that moment! On the other hand, perhaps a POW could be allowed to resist all this exposure in the media. Some may argue that it wasn't just a private matter, the whole society rejoiced at our return; there's something to that. During the first days the house was constantly packed with people, and I can't recall who they were. I had to be present; I had no choice; I couldn't run away. Nobody gave us any help in finding ourselves. There was no hand to direct us, no authority to turn to.

That was when the second period of the return started. . . . As a matter of fact, I had no family to take me in. My father was an old man, my mother had died when I was little, and I never got on with my stepmother. My brothers were spread all over the country. I didn't want to return to the kibbutz which had raised me, because I felt they hadn't treated me fairly. I was twenty-eight years old, and I didn't know what to do. My military commanders offered me a leave of two months, for adjustment to my new life. I got formally released from the army four months later.

There were a series of meetings at the military headquarters concerning our rehabilitation. We were interested in finding out what our rights were. For me it was a very long story. The Air Force men had conducted their own negotiations, and I never saw them again. The rest of the group didn't carry much weight. Dan was wounded and he was a kibbutz member, so he was out of our negotiations, too. Benny, Motti, and I remained alone in trying to solve our economic problems, and frequently I was alone in my meetings with the authorities. I got into a mess when I was promised a loan for the purchase of an apartment by one officer who had resigned, but nobody was willing to back up his promise. For several years I didn't let go of the matter, until I was forbidden to enter the headquarters base. Instead of using violence, I started a hunger strike near the Knesset. Throughout my struggle, I had appointments with many important men, including Knesset members, but nobody could help me. At the end, I wasn't interested in buying an apartment or getting a loan any longer; all I wanted was for my just struggle to be recognized. I was deeply insulted. I'm not cut out for this kind of struggle. I did receive a loan after my hunger strike, from a civilian, not a military source, but it aggravates me to hear people say that we got a fortune in compensation for our captivity. Looking back on the whole matter, I have nothing but praise for the Air Force. This is a corps that knows how to take care of its men.

DAN

One day after my return, I was admitted to the hospital for examinations. They had a whole wing vacated for me. The TV had made my return a public event, and everybody recognized me. But I think it was good, since it made the public realize that people can cope with a lot of hardship and maintain their sanity. I was especially anxious to give reassurance to the families of the other POWs who were still detained in enemy countries. The night I returned I was given a lovely party in the kibbutz. I talked a little there, careful not to divulge anything that might harm those who were still in Egypt.

What does it mean to return? One returns to a wife, children, the kibbutz, houses that were built, trees that grew tall, a new street. It isn't easy. I left in the "miniskirt" period, and returned to see the "maxi." My adjustment to the community was interwoven with my medical treatment. I had examinations at the hospital for ten days. I remember one day I was referred to an office to take care of some documents, and I realized I couldn't handle it. It was even more difficult to get from my kibbutz to the hospital. I felt sort of helpless, a very peculiar feeling.

People accompanied me for the first few weeks, but gradually I took longer trips on my own. I remember, however, one evening when friends came to take me out of the hospital to go to their house for a visit, and on the way I asked to go back. All of a sudden I felt like a beaten dog looking for a hiding place. You obtain freedom too fast, in too large a portion, and you need time to digest it.

Right away I was able to resume a good relationship with Chaya. Four years had elapsed, the children had grown, and she had become an independent woman. We shared our attempts to rebuild a good family life and we succeeded, because we were open with each other. She felt protective of me in the beginning, censoring things she believed I wouldn't be able to handle. I wasn't aware of it at the time. Things had happened, obviously, during my absence, both in the family and in the kibbutz. Some of us tried to fill in the missing time, but I realized it wasn't possible anyway. I learned to reconcile myself to the loss, and I left a wide open hole in my history.

I think it was at least a year until I got somewhat organized. I didn't work for a whole year, because I was going back and forth to the hospital. I had ongoing infections in my legs, problems with my teeth and ears, and the doctors didn't know how to cure me. Various doctors offered different treatments, going all the way to leg amputation. At the same time some other professionals started to inquire why I didn't return to work or start some kind of a study program. I wanted to go through the process at my own pace. I think I was wise to do this. As long as I had the medical problems, they provided me with a good excuse. But the truth is that I lacked the mental resources necessary for taking on some kind of a routine. Prison life protected me from the need to make any decisions, and here I was called on to decide. It was a sharp transition.

I wandered between the specialists, with their different ideas about my legs and my condition. Each had his own advice: Live with the pain, amputate, operate, what have you. One day I told myself, stop chasing doctors, you're becoming a hypochondriac! I selected one physician at a hospital nearby and followed his suggestions. Several years and numerous operations later, my bones are clean, there are no discharges, my walking ability is limited and the pain is always with me. But I can live like this.

After a while I went for vocational counseling and decided to study social work. I worked in this profession for five years. I had good relationships with my clients, mentally retarded young adults, but had problems in dealing with the establishment. Today I'm a simple factory worker in the kibbutz. About a year ago my wife died of cancer. What can I add? The way back home has not been easy. Some things can be repeated hundreds of times, but you find out you speak in a different voice, and only someone who has shared your experience will understand. Maybe this is what concentration-camp survivors mean when they say that language cannot convey their experience. The language is the same language, but the seasons are different.

11

PERSONAL CONCLUSIONS

Toward the end of our interviews, I asked each of the men to reflect on the lessons he had learned from captivity or on what remained of the experience. It was surprising for me, again, to find great diversity in the men's responses and the variety of moods expressed at this request to summarize the experience in its totality.

MENACHEM

I won't say that I was left with the feeling of being a hero. Throughout my captivity I felt like a defeated Israeli soldier. I have seen many times how our country gives every returned POW the sense that he is a national hero, how we are ready to pay almost anything for the exchange of a prisoner, but I disagree with this approach. A pilot who had to abandon his plane probably had no other choice; he couldn't fight to the bitter end, even if that's what he was taught in the IDF. A pilot is attached to his machine and when it's hit he may either escape the machine, or die with it. Not one of us had been taught to die rather than be caught alive. Despite these rational assumptions, I ask myself again and again: What is the meaning of captivity? You went out to fight the Egyptians, and lost. There are several degrees of defeat: one can return the plane without accomplishing the mission; one can lose the plane but return to Israeli territory; one can abandon the plane and fall into enemy hands. So in what sense is a captured pilot a hero? After all, he went on a mission and failed.

In the second stage, after the return home, a POW is entitled to feedback about his conduct in prison, especially during interrogations. There are two poles to this dimension—one is either a hero or a traitor. I believe that only rarely does one reach one of these poles, although I know a pilot, formerly a POW in Syria, who exhibited impeccable behavior and perhaps deserves to be called a hero. When he stepped up to the podium at a recent Air Force event, the whole audience gave him a standing ovation. On the other end, I know of some men who behaved in an unforgivable manner, yet nobody placed charges against them; they were just quietly expelled from the Air Force.

Between these two poles, one finds the middle way, where the norms are not well defined. I couldn't evaluate my own behavior, but when my debriefing was done, I knew that I had done what was expected of me. I knew that I had tried to be as wise as possible, always on the alert, and didn't let myself divulge any important state secrets easily. I'm rather glad this was also the opinion of the authorities about my conduct three and a half years later.

I saw a TV program in which Rami and another POW from Syria made it look as if being taken captive was great and made them better people. I'm sure they didn't mean it, and I'd explain the matter a little differently. Once you're doomed to captivity, you can take advantage of the situation if you create the right conditions within the framework. I felt I was immunized by the experience, but I can't tell you exactly against what. It was a remarkable experience, from which I gained some wisdom that others may not have gained. In addition—and it may sound somewhat cynical to you—society compensates the POW in several ways. But I repeat: all these outcomes are not worth the price I paid; I had a wife and children and a fascinating job, and the fact that they were out of my reach was horrible.

During my captivity I developed a new attitude toward war. I discovered what a colossal waste it is. You can acknowledge this after paying a personal price like we did. I remember the night in Egypt when I heard that the Yom Kippur War had ended with

3,400 casualties. I couldn't restrain myself. The courtyard was locked for the night and I had nowhere to hide. I went to my bed and cried like a baby. It was awful.

The Lebanon War was even worse, because it wasn't forced on us. There, too, we wasted almost a thousand lives, isn't it terrible? Now that I have paid the price of three and a half years of my life, I have an entirely different attitude toward these facts.

I got to know myself better in captivity. I'm resentful of people who are dependent on me, so I didn't want to become a source of support in our group. I dislike being in the company of men, and that's exactly what I had to do for such a long period. Since my captivity, I find it hard to stand noise; even music sounds noisy to me. Moreover, since my captivity I'm a restless man. I notice that some people have an easier time with themselves. . . . Who knows, I might have been this way anyway, even if I hadn't been captured.

YITZHAK

Today, when I look at my life, it seems to me that prior to my captivity I was a sensitive man, while during that time I had to close the lid over my feelings. This went on until about three years ago, when I met Lea, my third wife. I had never cried since my captivity, even when I experienced great disasters. I could sit in a corner, read a book, and weep a little, but never for real people or events. I could realize what was happening to people around me, but the moment I let it affect me, I felt as if I were weak. The night I heard a man murdered during interrogation next to my cell, I experienced a state of shock. I had only been in captivity ten days then. Afterwards I closed myself off to all feelings and stayed like that for a long time after my liberation.

I think that mentally I grew in captivity. I got to know myself. When much later on people helped me open up, I discovered that I had known myself all along from my captivity, but that for years I didn't allow myself to see anything positive in that period of my life. I profited in terms of my education as well. I learned subjects I would have never been exposed to otherwise, and I read a lot. But I still view this period as a waste.

My captivity entirely changed the course of my life. Had I not been captured, I'd still be married to Michal, we would have had another child, and I'd be living as an Air Force pilot in Israel, and not as a professional pilot in the U.S. I have no reason to assume that my life would have been any different. Instead, I got married twice more, I had two more daughters, and I acquired a place among the top professional pilots in the entire world.

Without my captivity I wouldn't have gained the emotional maturity I now have. I would have stayed in the same place. The Air Force was like a hot-house, in which I was exempt from making any decisions and assured of automatic promotion and companionship. Suddenly this whole route was interrupted, and I had to exert effort. At the test pilots' school, I had to apply myself academically for three years like I had never done before. Now I'm learning how to put effort into my relationships as well. I am building an entirely different kind of marriage for myself, and restarting my relationships with my daughters.

One of the most positive things I gained in captivity was my friendship with Rami. A few years ago people staged Rami's life story as a surprise for him, and I attended too. Rami said at the time that some brothers are from your family, and some brothers come as a gift, like us. It's really true.

AMNON

For years after my captivity I refused to talk about it. When asked, I'd say that I was bored of repeating my story. But it was an excuse. I understood people who had survived the Holocaust and couldn't describe what they had lived through. I repressed the memories inside me, locked them in, and threw the key away. A few years ago I started to write a little, and my experiences suddenly surfaced. I also began to talk more about that period, and I discovered that by saying it bored me I had been cheating myself for many years. I still say that when I grow up, I'll write my story from a unique perspective.

Today I think that during captivity I lived another incarnation, so to speak, and I have to make an effort to remember what had happened to me. Clearly it was a big breakdown in my life. In my education, my captivity provided me with the opportunity to close some gaps. I read a lot, and it had a good influence on me. Had I not been captured, I would never have become a reader. The prison reality made a student out of me, and later on I almost completed graduate school. I acquired tools for learning that serve me to this day.

I emerged from my captivity somewhat frail, but also stronger. There are some events that upset me easily—like an abrupt noise, or sudden footsteps when everything is quiet. It probably reminds me of the wardens coming to get me for an interrogation. On the other hand, I have acquired some toughness, and I may be less scared than others in certain circumstances, like when I'm stopped for a traffic violation. I say—so what, this is a Jewish policeman, after all! As long as I'm alive, that's the main thing. Being alive keeps me high all the time, and I don't need drugs to reach that effect.

I think that we were all thrown into jail like into a cauldron. We entered it dirty and emerged clean and purified. Now I'm more sensitive to the needs of others. Often it's problematic; I tend to give in since I see more easily the other's point of view. I take others into consideration too much. It's the outcome of the deep family-like relationship we developed in jail, where I learned to live in a group and to be able to read the cues of other people.

Among the main things that came out of my captivity, there's my immense need for freedom. I can live in a framework, but it's not easy. I have many conflicts about this matter, because it's hard to maintain one's freedom in a family. During the first years after my release I suffered from restlessness; I tried my luck in several places, here and abroad. I was floating and couldn't settle into any kind of routine. I think that people didn't notice my turmoil and therefore I wasn't offered any help, which might have been good at the time.

AMOS

Today, more than fourteen years after my release, I'd say that captivity was no more than three and a half years of my life, and it doesn't amount to much; when I reach eighty, its weight will be even less. The interrogations lasted much less time, and in spite of their intense effect, they're nothing but a small point in my life. I can't evaluate the effect of captivity on my life, but I guess the more I live, the slighter its effect.

It's possible that captivity changed my order of priorities, and I gained a somewhat new perspective on life. I realized one lives only once and has to enjoy life to the utmost. Before my captivity I had always been busy flying and fighting wars, and I disregarded my need for amusement and relaxation. My whole world consisted of the Air Force and Dalia. During my captivity I comprehended that I have to see the world and not miss any experience, since a passing year would never return. I have no interest in material things; I don't mind the old furniture in our bedroom, but I want to travel and enjoy my life.

One of the things I retained from my captivity was a sort of attachment to Amnon. He's become like a relative, and as much as you don't choose your relatives, I can't say I selected him because of his traits. We simply have this relationship. Presently we also own a business together. As to the other ex-POWs, we see each other occasionally, but our deep attachment is a matter of the past.

DAVID

I think that each one of us still has a load on his mind. I can't explain my own load, whether it's the result of the long time or the beatings, or perhaps the bitterness I accumulated because of the way I was received back here. I was left with mistrust toward people. I haven't found a proper occupation and have changed jobs many times. I can't be locked in an office, because I have to work in the open, and I can't find the right job for me. I haven't married either, despite the fact that I once even mailed out invitations to my wedding. Others have been married and divorced; perhaps it's the result of captivity. I look at myself in the mirror and say, "You're not actually injured," but when I have to make decisions something is blocking me. Perhaps I needed counseling right after our return.

My reserve unit was changed a couple of times, too. It's hard for me to get used to a new group of people every time. My adjustment to new people is really difficult. When I serve in the reserves in the occupied territories and come upon a demonstration, I don't wait for a rock to hit me but fire to hit them first, in spite of the strict orders against it. I escaped death once, and I don't want to risk it again.

Outwardly I look okay, but there's something people don't notice. Sometimes I'm unable to be by myself and I do anything I can to be with others constantly. I often spend the night with one of my brothers, and not in my own home. I don't know how to define my condition, but I blame the establishment and the discrimination I experienced for it. The discrimination I began to feel in prison cannot be forgiven or erased. The ones who acted against me were perhaps unaware of their deeds, but I suffered from the experience just the same. Since my captivity I have often been hurt again, and I have lost my strength to go on struggling. I'm sure others also carry scars from their captivity. Perhaps this book is going to spread my message around and this will make me feel better.

BENNY

I got a great deal out of my period of captivity. I gained an education for myself. Before, I had had only two years of evening high school. In prison I improved my English and math, things that I doubt whether I would have done otherwise. It may sound horrible, but my time in captivity helped me a lot in life. I learned a lot and gained life experiences. Today I can cope with any problem that may come up, as difficult as it may be. After overcoming what happened to me, I know that I can overcome anything at all, and I'll never give in to difficulties. Furthermore, I have learned how to live in a group. It's not easy to live with nine men in a closed room. It could be a recipe for a successful marriage. I know how to calm down my wife and children and behave tolerantly even when they are pretty upset. Before my captivity, I had been a violent, stubborn character, and now I'm different.

AVI

When I reflect on how much I could have accomplished during those four years, I see that I could have done much more; but if I was doomed to be detained, and I evaluate what I did gain from it, I see that I have accumulated information and improved my social skills considerably. I acquired a good knowledge of English, which I couldn't have done otherwise. My level of English was higher than Yardi's, who had studied in the university all those years! I also made progress in my studies toward a degree in engineering, which I had planned to obtain. The experience of living daily with nine men, the organization of our time and space, became a significant lesson for life. Just recently I heard about three couples who had returned from a week-long yacht cruise, and they told me how awful it was to be stuck with each other for so long. . . .

Each one of us absorbed a lot from the others, and we all matured as a result. It was illuminating to be exposed to a wide variety of opinions on every subject and to listen to the life stories of each of the men. Although we were careful not to expose our intimate life, many barriers came down in time.

In spite of this, I see my four years in captivity as wasted, unnecessary, and unheroic. I would have been able to learn the same lesson elsewhere or to live without it. I feel no nostalgia about the group or the period, and I rarely give it a thought. I never leaf through my letters or our diary. I don't feel driven to share my experience and its conclusions, as some of my friends do.

On the other hand, I'm upset by the fact that the army didn't make use of our experience. They debriefed us, for sure, but I don't know whether they put any of the information to use. I was under the impression that nobody wanted to draw the proper conclusions from my experience, despite the fact that it was up-to-date and quite important.

I have never recovered from the unprofessional reception we were given in Israel, even though a lot of studies exist in the world on this subject. They thought that once they had thrown us into deep water, we'd have to swim. This was a mistake, however.

It may sound absurd, but our long captivity cured us of the shock of being captured. We had time to process the event and we returned home more grounded and balanced. This is mainly because we had enough time to recover from the trauma of the failure of falling into enemy hands. By building ourselves up during the long captivity, our sense of failure was abated somehow.

We had enough leisure time to analyze and work through the trauma. Each of us found a personal direction for development—whether building with matches or a high school diploma. One painted, another raised birds—and thus we found ourselves anew and got back some of what had been lost with the trauma of being captured. This was the result of our lifestyle in captivity.

This interval provided me with the resources for my struggle with the Air Force later on, and led me to make the right decisions. I couldn't be influenced easily, like my neighbor who got killed right after his return from captivity. I didn't give in to pressure, not even to blackmail. I knew very well what was good or bad for me, and I didn't need permission from the Air Force for that. I used to be different before my captivity: I needed Air Force approval for everything, I was dependent on it, as if it were my family. I wanted to be loved by everyone. After captivity, I liberated myself from this Air Force environment, which can be so suffocating in its warmth.

I rarely paint nowadays, less than I'd like to. And I still haven't completed my college education, in spite of the fact that the Air Force promised me a fellowship. I will use it perhaps in the future, when Yardi completes her Ph.D.

DAN

I regret to this very day that the lesson from our experience was not taken seriously by the system. Nobody wanted to learn from it. I know a lot about captivity, how to prepare yourself for a certain defense line, how not to talk too much at home while some soldiers are in prison; but when I offered my conclusions to the military authorities, they advised me to write a book. I didn't want to write a book no one would read. I regret to say that nobody has learned the lesson that could be learned from our experience.

On the personal level, I keep asking myself whether these four years in prison had made a better or worse person out of me. It's hard to say. I'm a square, in my kids' jargon. I stick to the old norms. I dislike changes. A man with an earring still makes me angry. I regret many of the changes introduced in the kibbutz way of life, such as having the children sleep in their family's apartment. Perhaps I find it harder to change than others do. Could this be the outcome of my captivity? When I'm asked to tell the story of my captivity I stress the friendship and mutual support, which guaranteed our sanity. I feel that I myself have become more tolerant and accepting of others, but I don't know if this is related to my experience in Egypt.

What I do know is that my experience in captivity contributed to my sense of self-respect. I don't talk about it in public, but I tell myself that I passed my test honorably, and as such I have also honored the country. I did not retain any guilty feelings; not that I think anyone has to reward me for that, but the kibbutz could have made my life somewhat easier after what I had been through. It's not that heroic to be captured, but if the choice is between that and dying, the first is sometimes more difficult to take. We had all fought to the last moment before we allowed ourselves to be captured, and to return sane like we did is also quite an achievement.

I believe that had I not been captured I wouldn't have gone into social work. My captivity made me a more tolerant and understanding man.

Rehabilitation is an endless process. I get up every morning with sharp pains in my ankle and go to work. This struggle is going to be with me for life. The main thing is that I emerged in peace with myself and with my environment.

MOTTI

My captivity taught me how to live with others; I used to be selfish before. But basically I have tried to erase this period out of my mind. I have no nightmares, and I have no dreams about captivity. It was a hard time, especially the torture, and I wonder how I managed to tell you all I did. I also find it hard to remember anything that happened prior to my captivity. Ten months after my return I got married, and this started an entirely new chapter in my life.

I believe that in prison I became more aware of my limitations and began to cope with them in a healthy manner. Today I have many more friends, I know how to get along with people, and they enjoy coming and talking to me. I think that I took after Rami. He is a perfect human being, and he was our teacher. He built us all in the proper way. I have never said it to him, but living with a man like him in one room for three years proved to be an experience for a lifetime. It's perhaps worthwhile to have been taken captive just for that.

RAMI

I think we took advantage of our conditions in captivity as much as possible. It's easy for me to say, since I was one of the two foci of our existence. The biggest thing we've learned from captivity is that from every starting point, one has the possibility of climbing up or falling down, and it's a matter of choice which it will be. One may draw something good from any condition, and once you discover that, you can be happy or unhappy with what you've got. I give credit to the guys for discovering this truth when they were in such a difficult situation as imprisonment, at the very bottom.

For the first time in my life, I had enough time in jail for thinking and for reading important books, which taught me a great deal. I learned that a man determines how he feels under different circumstances. You have no control of the facts, naturally, but you have control over your attitude toward them. This principle worked for me in jail and seems to be working for people everywhere.

Before I was taken prisoner, I never had time to reflect on these things, but apparently they somehow had been part of my understanding, and that's why I could apply them so fast once I found them expounded clearly in the books I read. This sense, that a man is the master of his feelings, that he's the one who produces them, has been with me since I was fifteen and could understand the world. My father, who I have always remembered as the person who brought light to our home, was killed in a car accident when I was twelve. Mother stayed with seven children in the kibbutz. I was her sixth child and grew up quite alone, and at fifteen I was able to see things I could formulate only at thirty. Up until my captivity, I had run away from thinking toward doing. I never had the leisure, security and environment to sit still and find out what was I thinking. If I hadn't been captured, it might never have happened.

I think that a person who grows up in a warm, normal family doesn't discover this truth so easily. Since I was hurt as a child, and again in captivity, I could do it. I told myself, this is your field, go play your game. That was the difference between me and the others. That's why I found it easier to cope with our conditions. Feelings like rage, frustration, or helplessness, questions like, "Why me?" don't exist for me. That's why I coped well with my life in captivity, and my behavior helped the others too. The more I read and reflected, the better formulated my philosophy. There is no doubt in my mind that this was the most important lesson I drew from the experience. But I made some additional gains—like learning to stand on my hands, for example.

I discovered that becoming a leader was completely natural for me. What's the criterion for leadership? The ability to cope with conditions that others find hard. Anyone can make a salad, but the one who can negotiate fairly with the prison commander under stress, when everyone else is scared stiff, gets others to listen to him. It became clear very fast that I function better than others under stress, perhaps because I'm less open to feelings. I had similar experiences frequently afterwards, on trips or in the army. When the system is stuck with a problem, I come out as someone to be followed. I can say confidently what's to be done, and once I start acting, the others follow along. This is because of the complete confidence I convey in whatever I'm doing—until it proves to be wrong. When that happens, it's not too difficult for me to admit I'm wrong and offer a different course of action with the same confidence. It may be this opportunity for leadership that compensated me for my longing and suffering during my imprisonment.

On the other hand, I'm aware of my limitations. The horizon of my emotional life is pretty narrow. When I need to cope with strong emotions, I tend to block them and pass on to action. These repressed emotions have not disturbed me so far. Possibly every leader has to isolate his inner personality and protect it from vulnerability. He does it by building an inner cave that's inaccessible. This way he can function, while frustration, disappointment, or despair stay out of his experience. I myself can withdraw into my inner cave, and when I do I make fun of myself there, but I'm still pleased with it.

My behavior has been formed over time, but my captivity intensified the learning process. It's true that my traits prevent me from becoming really close to others (and I don't regret it), with the exception of Nurit, who is part of my inner world. However, I emerged from captivity well equipped with tools to communicate with others, especially on a one-on-one level. Today I can form a direct, open contact with anyone. I feel comfortable among strangers, because I can open all my channels and discover very quickly the wires that may provide a connection. Before my captivity I was introverted and inhibited; today I'm not. But I dislike big groups, I dislike starting things that I know I won't be able to finish. I'm not prepared to throw myself away when I'm not sure of the consequences.

I don't think that I developed any new traits in jail; old characteristics that had lain dormant became more pronounced and legitimate. The main thing is that I know myself much better and I don't need excuses anymore. What's no less important is that I was opened up to warm human contacts. I can get more excited now, I can even cry sometimes. Within my limits—which I dictate—I can be happier and sadder than before. After spending time as a family at the Sinai base, I developed a more profound relationship with Nurit and the children, and I'm thankful for it.

12

FROM THE WOMEN'S PERSPECTIVE: CAPTURE

■

The next three chapters are based on interviews with the five wives of the POWs who had been married at the time of their captivity. Dan's wife, the mother of his three children, died of cancer two years ago, and was the only wife whose account could not be obtained. I also interviewed Michal, Yitzhak's ex-wife, who divorced him six months after his liberation. Generally, my interviews with the women revealed a depth of emotion sometimes subdued in their husbands' accounts. Their stories present an unusual display of courage, which is frequently omitted in war-related accounts. This chapter explores the period prior to the event and the exact circumstances of receiving the shocking news of their husbands' capture.

NURIT (RAMI'S WIFE)

We returned from the U.S., where Rami had been training to fly the new Phantoms, and landed right in the middle of the War of Attrition. We went back to the kibbutz at first, and every morning Rami left for the Air Force base. Although his base was nearby, we hardly saw him. It was a very tense period. We had a two-way radio at home, and when Rami finally came home for the night he was often summoned back on an emergency. The children hated that radio.

In the meantime I became pregnant, and we decided to move to the base so that Rami could be with us a little more. As soon as we started to unpack, the alarm sounded and Rami left for his plane. A neighbor came to help me arrange the kids' bedroom. Rami came back the next morning, but we didn't manage to unpack our books before he was captured.

Two months later, on June 30, while Rami was going on the night shift, I planned to go with the children to the school party in the kibbutz. The children were seven and five at the time. Rami came home for lunch. I remember telling him that the washing machine was out of order, and he promised to check it the next day. He asked me to call him later, to give him a certain phone number he needed.

Before leaving for the kibbutz, I made the phone call. The receptionist behaved strangely. Instead of taking a message, she transferred my call to one of Rami's friends. He too behaved strangely. I told him that I was going to the kibbutz and just wanted to leave a phone number for Rami. He said, "Don't go. Wait another five minutes." I didn't like that at all.

Later on I found out that just as I called, they were communicating with Rami, who was parachuting out of his plane. They didn't want to tell me anything until he landed. But I felt something was wrong. In five minutes I called again, and this time my call was transferred to Rami's squadron commander. I asked him, "What's the matter? I'm mysteriously being transferred from one person to another, and all I want is to leave a number for Rami." He said that nothing was the matter and took the number from me. I told him I was leaving for the kibbutz.

I drove the car to the gate, and there we were stopped by the guard. "Are you Nurit?" he asked. "If you are, the base commander, Yallo, asked for you to stop because you forgot something at home." At that moment it became crystal-clear that something had really gone wrong, and in a minute they would come with the news that Rami had been killed. This was the only possibility that entered my mind, because almost no one was taken captive at the time. The kids stood at the gate next to me, all upset; they wanted to get to the party. Meanwhile I saw the car of the base commander approaching the gate, with several men in it, and I knew the worst was coming.

Yallo told me right away that Rami had ejected from the plane, that he was okay, and that the Egyptians had captured him. First of all, it was a relief. . . . It was such a relief, I can still feel it today. Because for me he had already been dead.

They gave me all the details they had, and explained that they had waited to tell me until they knew that Rami was alive in the hands of the Egyptians. Yallo told me he'd been Number 2 in the formation, following Rami and Los, and he had seen how their plane had been hit and that they had ejected. We had very little experience with pilots who were captured, and all I could think of was that Rami was alive. This relief sustained me for the first few days and I didn't realize that I had so little to be happy about. . . .

Some of the men went with me to the kibbutz. While the children went to their party, I went to tell Rami's mother the news. This was a strange experience. We found Rami's mother with her sister-in-law, sitting in a dark room. She had had a bad dream. When she saw me with the other pilots, she immediately understood. I told her that Rami was alive in the Egyptians' hands, and then she told me about her nightmare: she had seen Rami in his pilot's outfit, stuck in his cockpit, and he couldn't get out. Several men in similar outfits stood around him and she screamed at them: Why don't you help him get out? That's when she woke up. She had this dream at the same time that Rami was struggling to eject his chair from the burning cockpit.

The news spread, and people kept arriving. I was numb. Only when Shulamit, the wife of another pilot who had been captured by the Syrians, came to visit, did I cry for a moment. She told me that it had been Rami who had kept her spirits up for the past three months, promising her that her husband would soon be back, and "now he, too, is a prisoner."

ESTHER (MENACHEM'S WIFE)

I don't remember any particular fears in the days prior to Menachem's capture, but I was restless. Menachem could hardly ever get out of the air base, so I used to take the girls and go visit my friends. It was clear we couldn't make any future plans then and that all decisions should be postponed. At the same time I can't say I was anxious or worried for Menachem. Perhaps these feelings were repressed, because otherwise we couldn't have taken the daily tension.

I relied completely on Menachem. During the Six Day War I was pregnant with our second daughter, and Menachem used to call me twice a day and say, "Why should you worry? The hit rate of airplanes is only 0.001%!" I believed him, because he had always told me the truth.

It was the same during the War of Attrition. He kept telling me about all the new devices they were using to avoid the danger of missiles, and I believed him. Despite this, when I saw the base commander standing at my door and was informed that Menachem's plane had been hit, I responded right away: "So there, it happened to me, too." As if I expected this to happen all along, in the same manner that you expect to give birth after nine months of pregnancy. They told me that Menachem had been in touch with our pilots until he was taken by the Egyptians, and that he had been injured. After twenty-four hours, his picture appeared in the press, proof that he was really alive. I wasn't bothered by the fact that he was wounded. I realized that I had to prepare myself for my new circumstances and keep waiting.

We heard right away that his copilot had been killed. When I thought of Menachem's survival in contrast to the death of Chetz, I didn't feel my grief was in any way legitimate. Compared to Chetz's wife, I had hope, and this made all the difference. I thought we'd wait for a month, and he'd be back home.

DALIA (AMOS'S WIFE)

At that time I was a student in a hospital, I lived at my mother's, and I was pregnant for the first time. We planned to move to the base when I graduated from school. The day he was captured, I called Amos at the base to ask when to expect him, since I used to pick him up in our car from the airfield every evening. Another pilot said he couldn't call him to the telephone and gave some lame excuse. I told him I was going out shopping and would call again a little later. He said, "Don't go out now," which was strange. I felt confused and asked myself what was happening. It turned out that he knew that several officers were on their way to tell me, and he wanted to keep me at home. We were still on the phone when the doorbell rang, and the moment I saw the men I said, "Oh, I understand," and hung up. The two officers spoke together: "He's okay. He abandoned his airplane, and he's all right." This was good, because I was expecting worse, naturally. They also told me that he had been on the radio and had talked to the squadron commander before the Egyptians got him.

We had been under pressure the whole time up to this. But I was convinced that nothing bad would happen to me, and that Amos was responsible for that. I was busy with my studies and ignored what was going on around me, and anyway—you can't worry all the time! Now, when Amos is back flying, I keep telling myself: This won't happen again to us, it simply can't. I guess it's because of these convictions that life can go on.

Late in the evening, after Amnon's twin brother was located, it was announced on the news. In the meantime people started to arrive: my friends from college, where the officers had looked for me earlier, and people from the Air Force. One of my good friends came and said to me, "I have to be really cruel, but you must prepare yourself for an absence of about three months." I calculated the time left until my baby would be due and said, "That's okay. He'll be here for the delivery." On the following day I calmed down and went back to school. I accepted the idea of being three months on my own.

YARDI (AVI'S WIFE)

We had been a couple since we were ten, off and on until we were married on January 1, 1970. I was twenty-one and Avi was two and a half years older than me. We rented an apartment in Beersheba where I studied at the university and Avi commuted to his air base. Forty days later, Avi was taken captive. It happened on Monday, February 2, 1970.

On the previous Friday and Saturday he was on duty, and I went to my parents in Tel Aviv. These were the heroic days when all the young pilots competed to see who would hit more Migs. As Avi's girlfriend, I had already lived for a while on the base and I had breathed the atmosphere of the Air Force—the joy of hitting, and the losses when our own men were hit. But when Avi was on duty, I never worried; it was this familiar sense of "it won't happen to me." When I discussed the risk with Avi, he used to say, "Only an ass falls out of the sky, not a professional. Every single case of being hit is due to the pilot's error." Since I counted on him to be highly professional, I felt immune to disaster.

On Monday I was on my way back to Beersheba when I met a friend, also a pilot's wife. She invited me to come and spend the night in her apartment on the base, because there was some program for the evening. I hated to stay overnight in our flat alone, so I agreed. When we arrived at the base, she saw her husband's car parked near a house, and went in to see what he was doing there. I waited outside in the car. A few moments later she joined me, giving some explanation, but she seemed sort of strange. At home too she continued to behave in a restless, unfamiliar manner, but it didn't occur to me that it had anything to do with me. As I found out later, she had heard from her husband that Avi had abandoned his plane, that he was being searched for, and that a rescue mission would be attempted. They didn't want to tell me anything as long as his fate wasn't clear.

A little later, the commander of the base and some other officers came in, as if for a visit, and started talking. I didn't realize they were talking to me. They were saying something about a guy who had abandoned his plane, but only after some moments did I comprehend they were telling me about Avi. Once they completed telling me the facts, their wives started to give me advice. There was one woman especially, whose husband had been a prisoner before, who said, "Don't worry, it's a matter of three months, not more. It would be good for you to take on a project for that time. I renovated our house in the meantime." I couldn't digest it; all I remember is that it was terribly strange.

I left the crowd, to be with myself in the bathroom, to try and digest what had happened. I knew by then that he hadn't been injured, because they reported that his helmet was found clean. Apparently they tried to rescue him with a helicopter and took an Egyptian prisoner instead. During the night we heard the Red Cross report about Avi, quoting the Egyptian authorities. This was a great relief, because I assumed that from now on they would be responsible for his life and well-being.

Immediately after I heard what had happened, I was taken by plane to Tel Aviv to talk to our parents. I went first to my parents. My father was deeply disturbed and started looking for my mother, who was visiting her sister. We found out from my aunt that on hearing the news, Mom had immediately felt that it had to do with Avi and was on her way home. We were still on the phone when she arrived, gave a single look at me, and said, "I have known all along." Then we left to tell Avi's parents. Their reaction was even stronger. So a new period began in my life.

MICHAL (YITZHAK'S EX-WIFE)

We lived on the base. Being a young mother and a pilot's wife is a full-time job, because he doesn't have regular hours, and when he's home—he can be alerted any moment. I didn't work or study, although I intended to. I enjoyed the warm, protective environment. In my life I felt that I had exchanged one protective environment, namely the kibbutz on which I grew up, and where I had met Yitz, for another—the air base. It's a hot-house where everything is taken care of by a single telephone call. Everything is known, there are no surprises, and the families live in a tight network. On the other hand, we all lived with an immense, often unbearable, tension. You hear every takeoff and unconsciously wait to hear the safe landing. When you see smoke, you immediately know a disaster has happened. High tension is part of life on a base, and you never get used to that.

When we were sent for the Phantom training in the U.S., I was pregnant with my second daughter. I had to return earlier for the delivery, and Yitz barely made it before she was born. I had a complication while giving birth and spent several weeks at the hospital. Later on, I was feeble, and all I could do was care for the girls and myself. I hardly noticed the War of Attrition, although I realized that Yitz was very busy flying and hardly had time to see me. It took me three or four months to recuperate, and then it was only a short while before Yitz was captured.

I remember that a ball was scheduled for that night, and I went to the hairdresser in town. On my way back I dropped in for a visit at my friend Etti's, whose husband, Yair, was Yitz's navigator on the flight that day. She was expecting a baby at the time.

Their house was near ours, and I could see our entrance from her window. Suddenly I noticed the car of the base commander coming to a stop at our gate. My heart simply stood still. I knew he wasn't just visiting, not at that time of day. I watched the scene, paralyzed, and saw the doctor following the officer to our door. I touched Etti and said, "Etti, they're coming to me."

We were watching together as they approached Etti's house, and then heard the knock on the door. I felt as if my legs were cemented to the ground and couldn't move. I will never forget these moments as long as I live. The commander of the base finally came in and said, "You and you." I couldn't even hear the rest of it. Naturally we had no idea that they had gone to Egypt, or that they had been together on the plane.

Etti was about to faint, and the doctor seated her and tended to her. I was, as I said, paralyzed, standing in the midst of all the commotion with the baby in the carriage and my older daughter clinging to me. Gradually it sunk in that they had both parachuted from their plane, that Yair would be rescued tonight, while they were waiting to hear the Egyptians' report about Yitz's capture. I understood that the moment we'd hear that report, it would be a good sign.

I remember that they were quite hopeful about the situation, while I can't tell you what I felt. They asked me to attend the ball, to show high morale—this is the mentality of the Air Force. You belong to the collective, not just to yourself. Yitz's mother arrived, and she cried. I asked her to babysit for the girls, and I went to the ball. Everybody looked stupefied that night, and I felt as if I was activated by some forces outside of me; I really didn't comprehend what I was doing there.

All that time I was waiting for the news. They were careful to update me during that night, and I followed the process of Yair's rescue until four o'clock in the morning, when he was out of Egypt. On the one hand I was happy. On the other, I was thinking, what about me? What about Yitz? Why can't they take him out too? It was certainly a limited happiness.

My parents also arrived from our kibbutz that night, and in the morning we got up and went to buy a high chair for the baby, exactly as we had planned to do with Yitz. I had the feeling that I must stick to our routines, the only way to show I was still normal.

Furthermore, I didn't want the girls to miss anything because of this "nonsense." But again, underneath these activities I was asking, "What am I doing here? What's really going on?"

I think that several days had gone by before the Egyptians released the information about Yitz's capture. I felt an immense relief. His photograph appeared in a Lebanese paper, and as much as it was an awful picture, it was our evidence that he had survived.

13

FROM THE WOMEN'S PERSPECTIVE: LIVING ALONE

"What happened later?" I asked. The women provided their individual reports of the next three and a half years of their lives as wives of missing husbands and as single mothers. The outstanding features of this existence were, of course, knowledge that their husbands were alive, the inability to directly communicate, and the uncertainty about the time of reunion. It seems that the emotional responses to separation were more heterogeneous for the women than for the men. In addition to the normal variability in their personalities and marital relationships (as in the case of the men), the women's lives differed in circumstances. Although they formed an informal group that met from time to time, they lived their individual lives independently as well as they could. In some cases, I felt that episodes which might threaten the couple's relationship were omitted from the narrative, or vaguely hinted at; perhaps some secrets concerning this separation will never be disclosed.

Interestingly, the women's stories describe the major role of the Air Force in the lives of the POWs' families: the Air Force helped them solve daily problems and maintain contacts with the prisoners (directly or through the Red Cross), and also supervised the wives in a variety of other ways, almost taking the role of the missing husband.

NURIT

I spent the first night at my mother's in the kibbutz. Some friends tried to convince me to stay there, but in the morning I went home to the base. First of all, because I believed it would be a short separation. It was certain that something would be done, since four pilots were taken in captivity that same day! I was sure Rami would be back soon and resume his service, so why should I move away from our home? I was reluctant to go back alone to the kibbutz, when the baby was almost due, and then I'd have to go every evening and put three children to sleep, each in a separate children's home. Furthermore, I felt the need for the company of the Air Force people. I felt that right there I'd feel closest to Rami and would get all the news concerning him right away, which proved to be true.

Four days later, I went to see my gynecologist to check what all this excitement was doing to my pregnancy. She started inquiring about twins in the family, but I convinced myself it was simply a big baby. I refused to take an x-ray and waited patiently for the delivery.

We went back to normal. The house was full of visitors, and it wasn't easy. The kids kept arguing whether or not their father had died. We didn't hear anything from Rami except for an item about his capture on the news and his picture in the papers. In the meantime there were many more Air Force disasters. Planes were hit and more men were captured, all of them very good friends of ours—it was terrible. I was filled with fear and despair; every day brought bad news, as if the Air Force was gradually establishing a wing in Egypt.

That's when it occurred to me for the first time that this captivity might be a long one, like the war itself. It was an awful blow after the relief I had experienced in hearing the news that he had survived. Two months had gone by, with no news whatsoever. I was terribly depressed and nothing could comfort me. My pregnancy dragged on and on; I was very big and heavy.

I was still pregnant when my little brother graduated from his aviation course. This was my kid brother, whom we raised after the death of my father, and I knew he was following in Rami's footsteps. So I decided to attend the ceremony and was given a ride on a light plane to get there. Four days later, on August 4, I gave birth to twin girls.

I was taken by a friend of ours to the maternity hospital and when they took me in, the nurse said, "Just a second, let me go out and tell your husband that he can go home and leave you here, because it's going to take some time." That's when I suddenly broke down. I was weeping hysterically. The nurse started to inquire what was the matter, and the story came out. Till then I had had to function with the kids and everybody, and only there, at the hospital, did I let myself go.

The delivery went okay—after seven hours the twins were born. It was like a dream. The room was full of flowers from the Air Force, from Golda Meir, and others, and I was the saddest and the happiest woman in Israel. I felt great joy mixed with great worries—how would I get along with two infants all by myself?

Talks about cease-fire between Israel and Egypt started immediately after the delivery. On the night of August 7 I had a dream I can still see clearly: In my dream Rami comes into my hospital room, where I am with our twins. He is wearing his uniform, with his bars on, and he says smiling, "Nurit, it's all over. The war is over, and I'm here." [She is very moved.] An agreement was indeed signed, but without Rami.

I came back home with the two babies, and an extremely difficult time started. I had to feed two infants, one of whom was sickly. The two older kids also needed my attention, especially since we hadn't heard anything from Rami. The boy used to cling to me all the time, saying, "Don't you leave me too." The girl was sure her father was dead. She said, "Don't believe the Air Force. He won't be back. You better get married to another man."

I wanted the children to understand more about captivity, so I invited over one of the pilots who was an ex-POW. He told them some stories about what prisoners get to eat and drink, what clothes they wear; that's what interested them. I knew something about the nature of the interrogation process, but I was so overloaded with my daily chores with four kids that I hardly gave it a thought.

I do remember a feeling of intense expectation to hear something from him, though. If somebody would only tell me they had seen him, that he was really alive. . . . Just then Nasser made his declaration that he'd never return these POWs to Israel. Finally, on August 30, I got the news from the Red Cross that they had seen the prisoners. They reported that Rami had a slight wound on his face, which had already healed. The first Red Cross postcard from Rami arrived two days later.

News and mail were transmitted to me via a liaison officer of the Air Force, whose job was to keep in touch with families of the POWs. At first I received the letters via the commander of the air base in person. Afterwards they began sending the mail or Red Cross reports with their drivers, who were exceptionally kind to me. I remember one in particular who said, "Your husband is a hero. You'll see that he'll come back. We'll return him to you. The whole country waits for him." It was very touching.

My contact with Rami was quite problematic at first. I wasn't sure what I might or mightn't write in my letters to him. The instructions of the censor were ambiguous. We didn't know what to send in our packages, either. We just tried our luck, to see what would be transferred. When the first letters from Rami arrived, it was a cause for a lot of excitement. But other than that, I wasn't occupied with the matter. I was totally mobilized to the emergency condition of being a single parent of four kids. The load was such that I had nothing left for anything else.

One additional burden was the instruction we received not to talk about the prisoners, because there were people who would attempt to listen in on our conversations, and the outcome would be more difficult interrogations for our men. We obviously didn't know what they disclosed and what they managed to hold in. It was our task to educate the Israeli public to shut up. I accepted this as my personal duty. I thought that I could save Rami from additional interrogations. That's why I was mad at the media, who is always looking for a story. I was also mad at the kibbutz members for talking too much. For example, they repeated the fact that Rami had been to the USA to train on the Phantom, while Rami claimed not to speak any English . . . And how can you silence a whole kibbutz? I felt, however, that by silencing people I was contributing something to Rami's efforts. It was so difficult to feel that helpless, while he was suffering far away. When we improved our communication, it helped to relieve some of these feelings.

Gradually we formed a network of the POWs' wives. First I became friends with Dalia, who lived on the same base, and then with Esther and the others. After a year had gone by, we started to realize that a long separation lay before us. It was the kind of situation that can be terminated in a minute—or continue forever, both equally likely. I remember the Israeli Ambassador to the U.S., with whom we went to discuss possible plans for the men's release, saying, "You must prepare for a separation of ten, even twenty years." But nobody can prepare for such an absence! It led to questions of loyalty and fidelity, naturally. Each of the wives saw the situation her own way, just like the nature of her relationship with her husband before his capture. There were many parts to this, a lot of misery involved, and I can't blame anyone for what she did. I remember one of the young wives saying, "And what if he returns an old man, and we'll never have children?"

I had many male acquaintances, but my brother Yoram was the most important. He was a young pilot and served on our base. He and his girlfriend supported me all the time. The children needed a male figure, and they adopted Yoram as a father. We took lots of trips together. I remember promising myself that I'd repeat each of those trips together with Rami, and indeed we did.

My friends from the Air Force were extremely helpful. The base commander visited us daily to see what I needed. I experienced the Air Force as a huge, warm family enfolding us. I have no words to express my appreciation. Sometimes I told these men, "That's enough. I'm fine now." But they wouldn't let go. They explained that in case they were captured, they'd like to believe that others would provide similar help to their families.

In spite of all this, when half a year went by and I realized that I had to adjust to a long separation, I made up my mind to return to the kibbutz. At that stage I saw that my place in the Air Force was assured, and they wouldn't forget me even if I lived outside the base. I waited for the end of the school year, and then I moved. Despite having made this choice myself, the move was a big crisis. It represented my reconciliation with living alone for a long time.

It's a strange feeling. You have a husband and you haven't. Since I was young I have known death, but this is a different situation, not any easier. For all practical purposes, your partner is gone, as if he's dead; at the same time he's highly present, and you can't ignore him. Whenever I made a decision, which happened hundreds of times, I felt a tremendous need to have Rami's consent. I used to try to figure out what he'd do. It's not that I was afraid of his negative reactions, but I didn't want to harm the chances of our rebuilding our relationship later. As time passed, his image became more distant, yet I never abandoned hope of his immediate return. It's a difficult situation; you cannot finish any of your business.

Parallel to this, I developed a new kind of existence for myself. I formed new friendships, some of them pretty deep. The group of the POWs' wives grew in significance for me. I kept myself away from intimate relationships with men, but we often discussed the subject among us.

Returning to the kibbutz was difficult for a number of reasons. Living in a collective was stressful. People used to say, "But you get everything, you get a car when you ask for it, you travel a lot." It's true, I had to go to Tel Aviv quite often. I was helping in the selection of items to be sent to our men, buying good books and materials for their various activities and hobbies. I was visiting the other families, I was meeting people of authority to consult about possible ways to obtain release for our men. I was highly involved and was called often to meetings. This was important to me, because it made me feel I was helping Rami a little, or sharing some of the experience with him.

I met Golda Meir a couple of times, and Moshe Dayan and others. I met the Red Cross agents frequently. I knew we had to keep reminding the government of our husbands' predicament. Obviously, the ministers were busy with millions of things, and they saw tragedies much worse than ours, but we couldn't let go of our efforts. Since our men had been sent on a state mission, the state had to find a way to bring them back. We weren't popular with some of the ministers, and when our husbands returned, some of them said we shouldn't have been so insistent. I would do the same today, however, if I had to. If somebody had told me that the only way to help was silence, I would have kept my mouth shut. But the advice I got was the opposite—to keep the fire burning all the time. I insisted that it was our familial duty to apply pressure, though I objected to the style of some of the wives who got really disrespectful and violent on certain occasions.

Apparently these activities provided me with a certain freedom of movement that most kibbutz members are denied. Some people were envious of that, but most of the time I felt that the love and appreciation for Rami was far greater than those negative reactions.

The relationship of a husband and wife is not the same as with parents or brothers. This leads to another complication, concerning Rami's letters. In the beginning he used to write to me thinking that I'd share his letters with the entire family, which is an extremely large one. His letters used to pass through all his brothers and sisters, some of the older nephews and, of course, his mother and mine. The Air Force authorities also read the letters, for their own purposes. Usually I received a photocopy of the letter, while the original was kept for the commanders. Every one of Rami's letters received a tremendous amount of exposure. All our kibbutz members used to share the news of the letters. I used to hear conversations in the dining hall: "Have you read Rami's last letter yet? It's something!" When I walked down the sidewalks in Rami's mother's kibbutz, I'd be stopped by people who were complete strangers to me, saying, "Why didn't you tell us that another letter arrived?"

I hated this situation. I was ready to give up the letters altogether, if I could only get some personal regards from him. After a year I felt I had to do something about that. After consulting with our liaison officers, I asked Rami to write separately to me and to the family. He agreed and understood me, but it was hard to make this big family realize that my letters were private from then on. I still had to provide a brief report about the contents of my private letters, and I often showed them to Rami's mother, but they didn't leave my room. This personal correspondence was very good; it opened up a new form of communication between us.

During the second year I went back to college, made new friends with whom I studied, and finally received a good position as a biology teacher in the kibbutz regional high school. The children grew up and life went on—until the outbreak of the Yom Kippur War.

ESTHER

Although I was familiar with both Michal and Nurit, I hadn't visited them in the interval between their husband's capture and Menachem's. I'm not the kind of person who hurries to visit after a disaster.

I knew about former POWs of the Air Force, and that they had returned in three to four months. That was the length of absence I expected in our case, too. For seven months I was expecting him daily, and only after Nasser's declaration did I realize that ours was a different story. That's when I experienced my crisis.

I remember that I turned to one of the Air Force ex-prisoners I knew. This was a man I liked a great deal, and I had hoped that he'd help me figure out how to plan my life with Menachem's long absence in mind. He told me a very wise thing: that it would be a mistake to evaluate what was happening to Menachem on the basis of my own experience. He explained that Menachem was living in a world of his own for the time being, with his own joy and pain, which are determined by an entirely unique set of factors. He told me that I couldn't fathom this world of captivity, but he could reassure me that it was a full world, with a complete spectrum of emotions. These words comforted me a great deal. I had always relied on Menachem, and I was convinced that with his wisdom, enthusiasm, and stamina he'd build a meaningful life in captivity. From then on, I realized that I had my own life to live separately, and I felt a great relief.

Throughout this period I followed the reports concerning Menachem's health, and his progress with the treatment he received. I was worried about possible change in his physical condition and found it difficult to accept the possibility of permanent damage. But gradually I understood that he was recuperating. The Red Cross provided me with medical reports, and Menachem wrote about this in detail—he was walking a little, dragging his leg. . . . Now he was starting to write again with his right hand. I had always believed what he said. After about fifteen months we received photographs of the guys together in the common room, which moved me a great deal. I could see that outwardly he was the same man and seemed to be in control.

The efforts to obtain their release went on all the time, but especially for Menachem and the other wounded men. They made him stay at the hospital far longer then medically necessary, to make his release on humanitarian grounds more feasible. I was involved in that a little. I was sent to the Red Cross headquarters in Geneva to apply some personal pressure, as was Dan's late wife. When Los was returned alone, we realized that our attempts had failed. We kept seeing the government officials from time to time, and I met with Golda Meir too. The meetings were good, but in several of them the wives talked too bluntly, which is not my style.

Once I was sent to see Golda by myself, and I visited her privately in her apartment in the evening. I told her that the wives were in bad shape, but in fact she seemed to me more tired and exhausted than we were. She showed me photographs of various women pioneers and said, "Since when couldn't the Hebrew woman bear her pain?" I felt tiny and ashamed in front of this lady. I remember feeling that although we were six lonely women, was this a national problem? The truth is, however, that Golda gave our story a lot of personal consideration. I was convinced that everybody was doing the utmost to help our men, but we kept asking for appointments with different functionaries and offering our own ideas. Once I proposed that we exchange our men for Moslem art treasures from Israeli museums, for example.

The realization that we were being given all this national backing was highly important and also, in a certain way, obliging. I remember, for example, the moving moment when I was in a taxi with the girls on the eve of Rosh Hashanah and heard on the radio how each one of our POWs was mentioned by name, with wishes for a happy new year. I felt I wasn't alone.

Five of the pilots were married, and the Air Force made the wives into a group. Meetings of the wives' group were initiated frequently, and we became quite close to each other. At first we took responsibility for sending packages, but later on this was given to more professional people. It was hard to decide what to send, and I think that often we were childish in our choices. I remember that Menachem had asked for clothing, and I shopped all over the city for the fanciest items, while all he needed were comfortable simple clothes that he'd be able to wear all day long. When the men were waiting for certain things, and after a long interval something arrived that did not exactly fulfill their expectations, they reacted in great pain and anger. It's hard to describe the effort that we put into this whole matter. I think that had I been required to have the sole responsibility for the packages all along, I wouldn't have been able to take it. So it was good that the Air Force took it upon itself. They could remember that one wanted this kind of coffee, while the other required cocoa, etc. They supplied the study materials and books, too, in accordance with the men's requests. We just helped with personal items such as homemade cookies or our kids' drawings.

When after a while Menachem asked to be registered as a graduate student at the Technion, I made all the arrangements for him. I also obtained for him the position of instructor so that his students in the common room would receive credit for their course with him.

Learning to write letters was a long process. I changed my style entirely after the first seven months, with the realization that our separation would not be terminated soon. During the first months I briefly reported about the girls and myself. I discovered that Menachem wasn't satisfied with these letters, and I didn't know how to fulfill his expectations. I used to sit every Saturday night and try to sum up our week for him, telling him also about our friends and relatives, but it still wasn't good enough. Finally I learned to provide very vivid and detailed descriptions, so that he'd feel as if he had shared our experiences, rather than summarizing events for him. I learned to write about my feelings more openly; it got better gradually.

It was obvious that our correspondence was passing through many hands. I delivered the letters to the commander of the base, although I don't think he personally read them. But somebody from security certainly did, and these people knew me and met me later on. It took me a long time to convince myself that I wasn't that interesting for them; reading letters was part of their job. Gradually I got rid of the feeling of being spied on, and I was able to write more intimately. This happened when I realized that our separation might last as long as ten years, and I had to do everything to maintain our relationship.

Menachem's letters were very interesting and meaningful, and each of them reinforced our bond. They described his feelings and thoughts and reactions to the books he read. When he wrote about feeling bad, he never exaggerated. I greatly appreciated this. He kept complaining that my letters were too brief. He teased me that I used big handwriting to fill the page, while I actually said very little. . . .

At the same time I built a life of my own, well organized, full, and interesting. I enjoyed my role as a mother, and thought that my girls were adorable. My challenge was to move forward all the time, as I had before while living with Menachem. This meant studying, living fully, having experiences. I continued to live on the base, because that's where my closest friends were living, and I thought that was what Menachem would have wanted me to do. I was financially well off so I kept a maid at home and drove daily to Tel Aviv University, where I studied counseling. I must admit that often I enjoyed my life as a woman of some importance. I also tried to give my daughters the best of everything. Our family life wasn't sad or in mourning. The girls knew that their father was a POW, and it added some dramatic aspect to our life. I lived much more spontaneously than when Menachem was at home. I would get up in the morning and often decide on a trip or a visit without having to plan for it beforehand.

My little girl suffered, apparently, from her father's absence. She didn't comprehend the situation of a missing father who is nevertheless alive. As a result, her existence was kind of foggy. That's perhaps the reason for her declaration that she didn't want to grow up, since grown-ups eventually die, she said. Every evening at bedtime I had to promise her that her heart wouldn't stop working while she was asleep, or that in case it did happen, I'd have a machine revive it so that she wouldn't die. She was very attached to me, and wanted to be carried in my arms even when she was already five or six. I was sure that she needed her time to grow and gain some trust in the world, and in fact all her problems disappeared later on, especially after Menachem's return.

The status of a POW's wife, living alone, is fertile ground for all kinds of involvements. There is a romantic, heroic flavor to her very being. I established very clear boundaries around me. I led my social life mostly with married couples, where I felt protected. The story of what happened to Yitz and Michal could have happened to anyone under these circumstances, especially when the couple's relationship had had its flaws prior to captivity. Menachem and I, however, were separated during a good period, and we felt deeply for each other. I was ready to do everything to have this kind of relationship continue and hoped that Menachem would return to me at exactly the same point in which our life had been interrupted. If anything had to change, I wanted it to be a result of our life together, and not of his absence. (This reminds me of a story one of the pilots' wives told me once—that for a long time she restrained herself from fighting with her husband in the morning, because she used to think: What if he goes out on an operation and gets killed?) I think that the quality of the couple's relationship has a lot to say about their ability to cope with separation. Actually, sometimes I envied the wives who had affairs during their husbands' captivity; they were having fun. But I'm not like that.

All in all it wasn't a terrible period, perhaps because right from the beginning I understood the message: "How dare you complain, since he's alive!" So from the initial panic response of "My God, how will I manage without him?" I switched irreversibly to the other end, saying, "Everything is okay. Actually nothing has happened to me." Underneath it, however, I accumulated a lot of anger toward Menachem, as if he had abandoned us, and I suffered from a great deal of loneliness. I was surrounded by nice men, but none of them could replace Menachem. Most painful were the days of celebrations with the girls, in the absence of their father. I remember the pain of sitting by myself in the audience of a school show, for example, watching my daughter perform. I was acutely aware of Menachem's empty seat and felt very lonely.

Often I felt that I lived my life as I chose to and therefore had no complaints or demands from others. I didn't feel entitled to an award for my coping. I tried to convey this feeling in my letters to Menachem, describing the wonderful life I was living, showing that I wasn't in need of any compensation. At the same time I did sometimes feel like the heroine nobody recognizes, and that Menachem owes me something for it. It's not all clear to me. To this day I'm sometimes angry and disappointed that he doesn't realize what I have been through and doesn't treat me especially well for it, but I can't formulate exactly what it's all about.

DALIA

At the beginning I didn't form any relationships with the other families. I lived in the city and went to school. I remember one of the first meetings, where all the families were gathered to hear former prisoners at Abassiya, who were even able to describe the common room to us. Gradually the contact with the other POWs' wives became more significant; it was less for the exchange of information than for mutual support. Later on I moved to the base, where I felt part of a family. Nurit lived on the same base, and so did another woman whose husband had been captured by the Syrians. The three of us and our children formed a kind of family.

My pregnancy was very difficult. I was aware of the fact that Amos was being interrogated at the time, but I didn't want to think about it. After our baby was born, I lived for a while at my mother's; she took care of the baby until I graduated from school. During that year our contact with Amos became firmer; we got more letters and knew what was going on. When our daughter was nine months old I moved to the base, and the long period of waiting continued for almost three years.

I preferred living on the base because that's where we had planned to live after the delivery, and I was attracted to the company of the other wives and the Air Force people. They were better company than my parents. I wasn't pampered on the base, but it was a kind of hot-house, which helped me cope with my life alone. I found a job at the hospital nearby and got a babysitter for the baby. Thanks to our daughter, we had a family framework. I had to get up in the morning, take her to the babysitter (and later to nursery school), go to work, and be back on time to pick her up; it provided structure to my life, which was missing in the life of those who had no children.

I lost my normal anonymity and wore the halo of a captured pilot's wife. I didn't enjoy this. When I gave birth to our daughter it was a news item in all the papers, and everybody was moved. On the base, at least, people knew how to behave sensibly toward me. They didn't bother me with constant questioning about my private business.

At first, none of us knew how to send parcels and what to put in them. There were duplicate attempts, by us privately and by the Air Force liaison department. Many things simply disappeared on the way. I remember how satisfied I felt when Amos finally asked for some specific things, which I could provide. Letters were also awkward at first. I felt I couldn't express myself freely with all the people between us reading these letters. I knew some of the young female soldiers whose job it was to censor our mail and, in a way, they shared our experience. It was difficult to be so exposed. The best thing to send were photographs, and this was confirmed by Amos' reactions to them. I was deeply moved to receive some of the artwork he had made for us in jail, too.

I remember several of the meetings with the Red Cross agents, which were arranged for us at the Suez Canal. That's when I felt closest to Amos. The most difficult thing was being alone. When I'm asked how I overcame the experience, I really don't know how to answer. Life simply goes on, and it's stronger than anything. I continued to study and work in my profession as a physiotherapist. I used to ask myself what Amos would have wanted me to do, and that's what I chose. In some cases he would write to me what to do, but most of the time I could guess even before the arrival of his letter.

YARDI

I thought that Avi's captivity would last for three months. When this period had gone by, a friend from the Air Force told me, "Yardi, it will still take a long time." I asked how long. He said, "Half a year." I was so mad at him that I wouldn't talk to him for several weeks. I was certain that this couldn't go on for such a long time. But in fact nobody knew how long it might take.

I am known for my belief in the occult, so I went to one of the astrologists I knew. On my first appointment he told me that Avi would be back by May. There was, indeed, some excitement in the air that month. Apparently my mother heard similar rumors from somebody in the government, and one day I caught her baking Avi's favorite cake. She said she had a feeling that he was about to return.

The truth is that in Egypt, too, they were preparing the release, but just then new prisoners were captured, and a whole new process began. When I returned to the astrologist he told me that now it would be a long time. He told me to return after my next birthday, namely in fifteen months. When I came then, he said, "Too bad, it's still far in the future." My third visit to this man was a few days before the Yom Kippur War, and at that time he said that Avi would be back after my birthday, which was exactly what happened.

Look, I was a very young woman, without kids, when Avi was captured. I felt that my whole world had fallen apart, not just because of the worry and the lack of love and contact, but because my entire life order had collapsed, as if I lost my grounding. I was in a state of deep depression for many weeks. I didn't do anything and I stopped studying. I stood in front of the calendar all day long and counted the days, doing all kinds of calculations and predictions about his return.

From the present perspective I can say that during that first period all my previous personality defects surfaced. I had never become independent before. I left my parents' protection only to join Avi, and had never lived successfully by myself. I married Avi as a dependent woman, not as one who has an identity of her own. This condition became apparent when Avi was gone, and for the first time I was forced to construct an adult personality.

I resisted the temptation of going back to my parents' home, and moved to a southern Air Force base, not far from the university where I had started to study. With the encouragement of my professor, I resumed my studies. I did it half-heartedly, and I didn't have any brilliant achievements, but it was the only framework to which I could hold on. I completed my studies in English literature during Avi's absence.

It was an awfully difficult time for me. Today I can divide it into stages. The first two and a half years were my mourning period. I lived constantly with the hope that Avi would be back tomorrow, and I didn't accept the idea that his return might be delayed for a very long time. I maintained my relationships with three families that lived on the base, where I felt at home. All I talked about was my misery and longing, my dreams and expectations. My female friends shared this reality, supported me, and tried to read the future with me.

After that period I changed, somehow. I started to live as a student. I stopped being as self-centered as I had been, developed several new relationships in the university, and participated in life again. I think that only then I began to believe that I might stay alone for a very long time, and I had to organize my life in a different manner. I went out more and repressed my grief.

Some people treated me as a single woman. Naturally I was under all kinds of pressure, and I was lonely. What can you do when you're told: "Look, he may be gone for another ten years"? Normal excitements and needs didn't leave me untouched. Yet I struggled to maintain my identity as a married woman and keep the memory of my love for Avi alive. Often I thought that it was more difficult for me than for the other wives, since I didn't have children yet. A child is a reminder of the relationship, a focus for daily activity, and a channel for giving and receiving emotions. I lacked all that. At the same time I was pleased that I didn't have a kid. I was aware of the pain of the prisoners' children, the result of the long absence of their fathers. I had always planned to have a child only when I'd be totally ready for giving.

I hadn't known the wives of the other POWs before, but I became their friend. I was the youngest in the group. They all had children, mostly two or more, and had some experience in their marriage, for better or worse. In comparison, I was fresh and green, yet I felt that I had more intense feelings toward Avi and greater confidence in his love for me. I found out—what I should have realized anyway—that years of relationship do not provide a guarantee for its quality. I felt that there was something unique in our love. When I look back, I feel that we have retained that depth throughout the years.

Our love was expressed in our correspondence. For Avi this was the only intimacy outside the room. He wrote such beautiful letters! They were the kind of letters you had to read again and again to get all they meant. I discovered the depth of Avi's personality through his letters. I saw how with the aid of painting, music, writing, and reading he preserved his inner being in spite of the hostility and crowdedness all around him. He experienced the captivity as a separate man, though never distant or in conflict with the others.

Once I received a very poetic letter, written like a myth about captivity and return. It was a beautiful, touching story, nothing more, nothing less. Two days later, however, men from the censor's office came to ask for my help in deciphering the myth, because they had the notion that it was a code for a proposed escape plan. . . . I was sure they were wrong, since I was already familiar with Avi's literary style of writing.

Our group of prisoners' wives had its own dynamics, which, as I discovered later, were somehow parallel to that of the common room. Some of the wives felt more important than the rest of us, perhaps since they realized that their husbands had dominant positions in the group in Abassiya, or because they were older and more experienced in the Air Force. Suddenly I discovered that some things were happening without all of us participating. Perhaps they needed to feel powerful, or perhaps they had plans to release their husbands separately, by themselves, once the attempts to liberate the entire group had failed. I remember, for example, a plan that two of the "senior" wives would go to Geneva to talk to President Sadat's wife. It was some sort of grandiosity, which disturbed the harmony among us, and I resented it.

Michal was the closest to me. She shared with me what happened in her life more than she did with the rest of them. I think that she treated me like a single woman, while in comparison to the other wives she felt very deviant. They were totally dedicated to their husbands and to waiting for them, while she needed something else. I realized she was looking for someone to love. She was open to that possibility, so that what happened to her wasn't a chance process. I think that her relationship with Yitz wasn't strong enough to withstand the separation. All their friends saw this; only they themselves did not. Perhaps it's always most difficult to become aware of your own shortcomings.

In comparison to Michal, I felt that I wasn't searching for another man in my life. I did have my conflicts about how to live in Avi's absence in the most satisfying manner, without hurting our relationship. I had no doubt in my mind that I loved Avi more than anyone else, and that he was my choice of a partner to build my life with. Today it's hard for me to understand how I could wait for him for so long. Probably it's because I didn't know beforehand how long it would take and I expected him to come back soon. One cannot guess what would have happened if the captivity would have continued a year or two longer.

MICHAL

I remember that I was told very early on to prepare myself for a long separation from Yitz. People said that only another war, or a peace treaty, could lead to the liberation of our POWs. Nasser declared that the pilots would never be returned to Israel, not before the final victory. . . .

I was very scared. I didn't know how to go on living without Yitz. [She is silent for a while.] We had just started our lives, I was only twenty-six, with two tiny daughters. It was not that I had been so dependent on Yitz in daily matters, but emotionally and physically I needed him desperately. How can you live your life alone? It was our daughter's birthday. Then the older one began first grade. How can you go through such events without sharing them with someone close? No one could replace Yitz, not parents, not friends.

The only one who understood my feelings was Yitz's mother, Hanna. She had been very close to Yitz, very proud of him. I hadn't gotten along with her too well before—she was a pushy lady. But after Yitz was taken captive, we drew closer because we both missed him so much. We spent weekends together, sometimes at her home in Tel Aviv, sometimes at the base. She helped me with the girls, and she was always willing to babysit for me.

The girls seemed to grow up normally. They were sweet and I enjoyed them. The little one suffered from allergies and a slight asthma. I didn't realize that these might have been the outcome of her emotional problem, of being left without her father.

I grew close to the wives of the other POWs, especially Esther and Yardi. The Air Force took good care of us; they appointed a special liaison officer for our needs. I saw him often, visited him and his family in their home. He encouraged me to write letters all the time, even though it was a long time before we got answers. In writing letters, he said, I could relieve some of my longings, and pressure the Red Cross as well to bring messages to our men.

I remember the day the first letter finally arrived. It was a celebration. All the family assembled to read it, Yitz's brother phoned from the States, and I got flowers from my friends as if a baby had been born. Gradually, his letters became regular. They were beautiful letters and always a major event.

The Air Force had people read all our correspondence, however. Once I wrote to Yitz that I felt so lonesome that I would have given anything for a phone call from him. A woman officer censored the letter and deleted the passage, yet she felt so sorry for me that she called me that evening herself to ask how I was doing. . . . Actually I had a big argument with her about that passage. Why wasn't I allowed to express freely my feelings for Yitz? Her argument was that I'd expressed what I'd felt at the moment, and by the time my letter reached Yitz, I would have felt many different things. She did not convince me. I had the sense that Big Brother Air Force was controlling me all the time. It made us comfortable and secure, but at a price in terms of our individual freedom. As time passed I saw many examples of this patronizing attitude, which was, for me, often irritating.

When both of the girls started kindergarten and school, I looked for work to occupy myself. I considered the university, but I felt that I didn't have enough energy for study. I got a job at the local post office. I worked every morning, I saw everybody and I enjoyed making some money on my own. Actually I had no financial problem. On the contrary, I saved money, bought a new car, and joined a group of pilots building private homes near Tel Aviv. When I drove my new car in the base, some people seemed to criticize me: how dare she, spending her husband's salary while he is in jail in Egypt! But I couldn't have cared less. Buying the house was my bravest decision at that time. I managed to get loans all by myself, I combined all our resources, and I went for it. Building our house gave me an additional goal and interest in life. I often took the girls and Grandma Hanna to visit the house, to see what progress was being made, to plan for the future.

I had the girls, the house, and my job. On the other hand, time went on, summer followed winter, birthdays and holidays, year in, year out—with no hope for a change. It was horrible. Once a friend came to visit and I told him that it might have been easier to be a widow. At least that's a final state, and you begin to adjust. When I look back and know that it was just three and a half years, I am, of course, wiser. But when I was experiencing the longing, the loneliness, and thought it might be for ten years or forever—there was no end in sight.

There was pressure of all kinds. My inner needs were the result of my loneliness. I needed someone to love me, to care for me, to share my load. I had become everything, Mom and Dad for the girls. I had to always be strong just when I felt so weak. Male friends reacted in different ways. Several of them propositioned me. They tried to take advantage but I didn't let them. I felt that affairs would not satisfy me; I was ripe for something much deeper.

I met Gideon two years after Yitz was captured. He was on reserve duty at our base, and I saw him at the post office, in the swimming pool, in the supermarket—I seemed to be following him everywhere. I asked one of my friends about this stranger and I discovered that he, too, was married, a father of two, and lived in Dimona (a southern development town). In the evening I met him at a party, and I noticed that he was looking at me as intensely as I was looking at him. I started a conversation, and we have never separated since. I don't know how things like that happen; I know it sounds crazy. On the first night he told me, "We'll be married, you'll see."

It was incredible love. I remember that one of the pilots I talked to right in the beginning told me, "Michal, if that's how you feel, don't give it up. Let it develop, it may be the true love of your life." That's exactly what I felt. I felt no conflict whatsoever, and I didn't worry about our future. I simply let myself go. The pressure was all from the outside, but I didn't want to give in, and I struggled for my happiness.

It was a difficult time. Public opinion was unequivocally against me. Everyone's movements are public knowledge at the base, and people knew all about me and Gideon even before anything had happened. My friends and neighbors stopped talking to me; they ignored me on the street. I had nobody to turn to. When I told my own parents, my mother, too, blamed me for my infidelity.

Grandma Hanna was the only person from whom I tried to hide the affair. I knew that once she found out, our relationship was over. With all that we had built between us, Yitz was her son, after all. Later on, I started to reveal the story to her gradually, in parts. I discovered that she refused to know. She considered Gideon just another male friend that I dated now and then. When people told her the whole truth, she denied it and defended me.

As my relationship with Gideon grew in depth and commitment, I became dismayed about my letters to Yitz. I wanted to be fair with him and write honestly about what had happened. The liaison people from the Air Force strongly objected to it, however, and advised me to continue writing as if nothing unusual had happened. The officer in charge of us told me that I had no choice—I must continue my normal correspondence with Yitz. In fact, the liaison officer knew Gideon too, and told me he was a wonderful guy. . . . It is such a small country that nobody can stay anonymous. My story became the hottest news item all over the place. I was afraid that someone would write to Yitz, or to one of the other men and he would find out indirectly, which I considered much worse. It was remarkable that this did not happen. I guess people understood that revealing my affair to the prisoners was too much of a responsibility; they were afraid it would break the morale of the whole group.

You can imagine how the story became a scandal on the base. Pilots would tell me, "If you behave like this, what would my wife do if I'm captured? What do you expect me to think? That she, too, would take a lover and betray me?" It was so unfair, since some of the same men had propositioned me before, when they came to console me in my loneliness. . . . Suddenly I was taken as an example for everybody, and I was treated like trash. To this day I can't understand why nobody tried to understand me. I became an outcast. The women were horrible, too. Only the other prisoners' wives were a little more sympathetic and willing to accept me. One of them even told me that she envied me for my courage and my ability to build such a relationship under the circumstances.

In this atmosphere we remained in our home on the base. I had nowhere to go. Gideon was still married, the house I was building was not yet ready, and my parents had rejected me. I also didn't want to take the girls out of their school; but had the situation lasted a little longer, I would have left and looked for a place of my own. Anyway, I knew that I had no way out. Whatever happened, as much as I was punished for it, I felt that my relationship with Gideon was unbreakable, notwithstanding all that I had to pay for it.

14

FROM THE WOMEN'S PERSPECTIVE: THE RETURN

The return of the men was, naturally, the happy conclusion of the long waiting period of their wives and children. At the same time, they viewed this return with a certain apprehension about the need to adjust to living together again. Most of the wives felt that they had acquired strength and independence and were ambivalent about giving that up. Since they had to adjust to the changes that had occurred in their husbands as well, all of which had occurred during and after the recent war, the transition period was not easy.

From a long-range perspective, the couples seemed to have gained in depth and mutual understanding as a result of the separation and near loss. The case of Yitzhak and Michal stands out as an exception, an example of a marriage that could not take the strain of the long separation and the uncertainty regarding the future. Their traumatic story provides another dimension to the life histories of POWs and their families.

NURIT

My brother was ordered to his squadron the morning of the Yom Kippur holiday, and so were other pilots. In spite of this, when the war was announced in the early afternoon, I was deeply surprised and bewildered. My major concern was for the safety of our men in Cairo. Two days later, we got a report that they were okay, and then a new period began. . . . with disasters all around me. So many pilots were killed, all of them wonderful young men, it was incredible. First it was Rami's cousin from his home kibbutz, and on the following day, my brother and my best friend. On the third day—the commander of our base, also a close friend. It was impossible to bear.

I remember that on the second day Yardi came to see me and exclaimed, "That's it, it's all over, they're coming back!" while I couldn't rejoice, not even listen to her. It took me some time to realize that something good could also result from this terrible war. I had had this in mind on the first day, and I had talked about this to the children. But all the casualties and bad news had driven the joy out of my mind. It was so hard to be glad about Rami's probable return with the great grief over my brother Yoram. He was so close to me and the children. When I told the twins about Yoram, they said: "No, it isn't true. Daddy is killed and Yoram is just a prisoner." After Yoram's funeral I talked to the older children about the approaching return of their father. They asked what he would do now, and said that they didn't want him to fly airplanes anymore. They inquired if he'd remember them, and I told them that he would, since we kept sending him pictures. But the little ones found all this too hard to grasp.

On November 15 we were informed that they were coming. We flew to the airport in a light plane to wait for them, and after several hours they said, "Not today" [laughing], so we flew back home.

It was awful. The kids were mad at me: "We won't get him back, no matter what they tell us." I was aggravated too, and I said to the liaison officers, "Don't take me to the airport again. When Rami comes—bring him to the kibbutz. I can't take it again."

In the evening they called and promised that tomorrow they'd surely be back. I felt I had no choice but to go and wait for him, and we flew down again.

I remember my entrance into the hall where all the families were assembled. It was so quiet, and I felt terrible. It turned out that a few moments earlier the names of the POWs reported on the plane had been read, and Rami's name was not among them. Fifteen minutes later, however, a new list arrived, including Rami's name. Luckily I hadn't been aware of it.

I remember the immense excitement when the plane landed. A few moments later, the bus arrived to where we were waiting. The door opened and Rami appeared, very pale, thinner than I remembered him, with short hair. He looked very well. I stood right there, with our older children, each one holding a hand, and. . . . It's a moment one cannot describe. It was all over in a second, that's what I felt. At that very moment the whole thing was over.

Rami asked right away what was going on with the war, and it was so hard. I was scared of his return to the kibbutz, having to face my mother and his uncles, in their bereavement. In our kibbutz alone, there were five additional families who had lost their sons in that war, and so many new widows on the base. All this combination of sorrow and happiness is something I have no words to express.

If you ask me about the first two days, or even the whole first week, I see it like in a fog. Like in the movies—I try to push away the fog with my hands, and it comes back. There was an endless string of parties and celebrations. People were crying and laughing with us all the while, and the children were in the middle of all this, of course.

When we reached the kibbutz, our daughters were in the children's home, and I went to get them. I dressed them up in the knitted dresses Rami had sent them from jail, one green and one orange. They were jumping on the sidewalk singing, "Our Daddy is back, our Daddy is back." People were standing and watching us with tears streaming down their faces. When we came home they stood still and looked at their father. He seated them on his knees and read a book to them about a puppy. When he completed the reading they said, "Okay, Mom, let's take a walk."

Only weeks, perhaps months, later were they able to accept him. It took them a long time to realize that he was the man about whom we had been telling them all the stories, and whom the older children remembered all along. Only in the Sinai did the ice break between them. The older children accepted him faster and shared all the ceremonies with us.

We were rushing from one reception to the next, and my mother was always near us. Luckily she's a woman who can control her emotions, and with all her pain she was indeed glad that Rami had returned. She wasn't a burden, yet our tragedy was with us all the time. Rami started to give public accounts of his captivity right away, first in our kibbutz, then in his mother's. He went to his debriefings, and when he was home the place was packed with people all the time, a big commotion. I especially remember one reception given by Golda Meir, where Rami gave her the flag he had knitted in jail and announced the liquidation of the Egyptian squadron. . . .

We were invited to go abroad, but we declined and went to Eilat instead. In the evening, with the full moon on the desert, it suddenly dawned on me that that's where our family could rehabilitate itself. I told Rami that I couldn't go on with the stormy emotional climate in our extended family and kibbutz and had to get some peace for myself. Rami agreed enthusiastically and it was arranged for him to serve a year at the air base in the Sinai.

It was a wonderful decision, and it liberated us. It wasn't easy to explain to my mother and to members of the kibbutz that I had to leave for a while. How could I tell her that I couldn't experience my happiness in the midst of such tragic pain? But we did it and in three or four months we moved the family south.

In the meantime we had shared Yitzhak's story and grief. Rami was the person he could lean on, and our house a place he could always come with his sorrow.

On the outside, Rami was keen to show that he was okay. I realized, however, that the company of people irritated him. Although he could speak in public very well, he tended to sit quietly when people came to visit or when emotional response was called for. Often I had the impression that he wasn't present with us, but elsewhere. Several times he tried to tell me about his difficult experiences: the torture, his diarrhea, how they hung him up on a hook, how he licked water from the floor, how he had been beaten until he fainted. It was too hard for me; I felt the pain in my own body, and I asked him to wait with his stories for a later time, if he could. Gradually I learned to separate his experiences from mine, and I was able to listen.

The year in the Sinai was great. The children went to school and I got a job at a marine biology lab in Eilat, while Rami conducted flying instruction during the day. We took wonderful trips during the weekends, with or without the children, and when I spent nights with Rami outdoors under the starry sky, I felt like I was in a dream come true.

For the kids it was a more gradual process. For years they tended to use only me as a parent. Perhaps it was not only the outcome of his captivity, but also his long years of absence due to his military service. The twins didn't know what a father was, and they didn't actually see Rami as part of the family. For the older children, he was the returning father they had waited for, but with his return, especially when the first year in Sinai was over and we went back home, they were right back in the whirlwind of disasters, and tension was typical of our home atmosphere on the base.

I was terribly afraid that we'd have problems between us—but we didn't. For years I had lived on the hope of seeing him again, and this hope had sustained me. It's a hope unlike anything else, since I had no idea when it might be realized. At the same time I had developed as an individual and discovered that I could enjoy my independence. I had had time for myself, and I had opened up to new experiences, such as studying biology or going to the theater. From Rami's letters, I comprehended that he, too, had changed. So naturally I was anxious about our adjustment to each other, and I wasn't sure whether our different expectations would match the new reality.

The realization of my hope was divine, however. Our encounter was stunning, as if we had never parted from each other. It was hard to grasp at first; I have seen so many people depart forever—my father, my brother, lots of friends. In a certain sense, it was as if Rami had also died, since he was completely out of our lives. And then he was back; and his return into the fabric of my existence gave me the feeling that I had been incredibly blessed! What else could I ask for? I experienced something that people very rarely experience—that my man returned to me from nowhere, the darkness, the void. The immensity of this good fortune gave us the power to do the right thing and go to the Sinai, the very best thing we could do.

Another thing: for years I had seen myself as small in comparison to Rami. I had given no importance to my own achievements.

Life in the Air Force dwarfs women, you know. Men are the heroes, they act and experience, while their wives keep the home and family going. But when Rami was in captivity, I had managed all by myself, and it was a positive experience. I could hold on to my independence as long as we were at the Sinai base, but with the return to our normal environment, with its extremely tense reality, the old pattern was resumed. I discovered that Rami could accept my decisions if I would be more assertive, but I'm still fighting for my place in our family, and we haven't resolved this problem. Only recently, when Rami left the Air Force, did I decide to take a serious job for myself, and I told Rami that he'd be responsible for the house and family from then on. So far it seems to be working.

At the same time, some good things had indeed happened to Rami as a result of his captivity. He became more sensitive to others and learned to talk to people much more than before.

Fifteen years are over now, and the period of captivity is losing its grip on our present. But it happens now and then that I realize suddenly how lucky I was. There were so many misfortunes around us, and I reflect on what would have happened had the missile that hit Rami's plane been aimed just a little sideways. This awareness of barely surviving makes my life worthwhile. One should be grateful for what is. I have lost my tolerance for pettiness, for routine, for complaints about the food in the dining hall or the quality of the laundry. I don't pay attention to gossip and trifles; I try to live fully, with my entire soul, heart, and brain.

There is one problem—our only son is also a combat pilot now. I didn't feel I was entitled to shape his choices, all I wanted to be sure of was that he wasn't just following in his father's footsteps, or proving something to him. He convinced me he was doing it for himself. I shared my fears with him, but we respect each other's emotions. Recently I've adjusted, and I don't think that I worry more than any other mother of a soldier.

ESTHER

When the Yom Kippur War broke out, I had these terrible mixed feelings: fear of the war, and with it, my personal celebration. I knew that Menachem would come back, and mentally, I started to prepare myself.

We were told that our men would arrive with the first group of released POWs. On November 14 we were informed they would fly in the next day. It was a Thursday. I didn't send the girls to school and we all went to the airport. After hours of waiting we were told that the return had been delayed. We went home empty-handed. It was awful. The girls went to sleep and I locked my door—something I never do—and tried to sleep. I didn't answer the phone or any knocks on the door. I had no more energy. I didn't want to live anymore.

Later on I heard one of my best friends yelling at the door: "If you don't open up, we'll break in." Finally I let him and his wife in and told them how I felt. But he made me dress up and go out for dinner with them. We didn't mention the return or my disappointment; it was self-evident.

Next morning I decided to send the girls to school, so that they would not see my despair. I thought they would be too upset. I picked them up from school only when I was assured that the plane with the POWs was already on its way. So we went to the airport for the second time. [She is quiet for a while.]

I saw him coming off the bus. He was wearing the nice outfit that I had sent him years before, and he looked well. He seemed excited, and completely self-absorbed. He almost didn't see us. I could understand him, accept him, but it was impossible to connect with him. I felt disappointed and asked myself: Do I deserve this, after all the waiting and suffering? For the whole first year this was my dominant feeling, that I deserved better.

I had arranged a leave of absence from the school where I was teaching to stay home with Menachem, but he encouraged me to go back to work right away. This, of course, was another big disappointment. Every night the house was full of visitors, or we were invited to parties and receptions. It was difficult to get up in the morning, but mostly I didn't want to see people all the time. It irritated me. I wanted to close the door and have Menachem all for myself and the girls, but Menachem couldn't have enough of people and events. He was questioning everybody about all the things he had missed. Once, after midnight, at some friends' house, I wanted to go home; I was completely exhausted. He said, "Go home by yourself." I went home and cried my heart out. This was about a week after his return. When I later asked him how he could have done that to me, he answered that he had three and a half years to catch up on.

This was our situation during the first year. He couldn't see my needs, while his own were quite different. As parents we seemed to function well. We always maintained a strict daily schedule. The girls went to school, each one had her afternoon activities or friends, and at seven we all sat down for dinner, each of us telling about our day. We gave them a bath and put them to bed. Menachem would tell a bedtime story to one of them, and I to the other, and that was it. Our personal struggle would start when their lights were off. I believe that the girls were absolutely unaware of it. We provided a secure framework for them. Menachem did all that was expected of him as a father, as he had in the past.

I was thirty-four years old and keen on getting pregnant. When I did get pregnant, however, I miscarried. I was emotionally hurting all the time. I asked Menachem repeatedly if he cared about our relationship, and he always answered positively. I believed him. Indeed he wanted me to be happy and was causing me all this pain unintentionally. At the same time, I constantly felt that he was angry at me, although he kept denying it. I used to tell him, "Let's sit down and think about it. Do we really want to pull out of this mess? Do we want to keep this family intact?" And he always answered, "Yes, it's very important to me." Then he went alone to the U.S., on some Air Force business. . . . I understood that he had to experience his freedom, but emotionally it was terribly hard for me. It was hard to accept that his offensive behavior toward me and his insensitivity to my needs were outside his awareness, intention, or control.

I remember telling myself that this was just a crisis situation and we would soon be out of it. I got pregnant again and gradually calmed down, and so did Menachem. Somehow we managed to get back on the right track, but it was awfully difficult. Menachem is an individualist all the way. He can be considerate of others, but he has to feel free, to be in control. I think that these traits were reinforced by the experience of his captivity.

I keep returning to this period in my mind. Later on I thought that Menachem had been angry at me, in part unconsciously, for being so strong and competent in his absence. In my letters I went on about my accomplishments. I renovated our house, I bought a new car, I graduated from the university, and I had a nice savings account that he would be able to use when he returned. Obviously it pleased him to have such a wife and family back home. On the other hand, it wasn't easy for him. I remember that after his return, Golda Meir's secretary called for some reason and mentioned to Menachem how wonderful I had been during his absence. He reacted by saying, "Why is everybody telling me how wonderful she was? What about me?" I realized that my accomplishments were hard for him to take. He saw me as the young girl he had rescued from a miserable childhood. I have always let him feel like the boss of our family, because I liked it that way. But all of a sudden life had turned me into my own boss, and I had managed well, almost effortlessly.

I had, for example, a savings account. I opened it in his name and gave it to him when he returned, telling him that he would not need any of the loans offered to returning POWs. He wasn't happy about it at all. He took the whole amount and bought equipment for our new kitchen, and that was it—as if telling me: "Take this fancy kitchen I organized for you, just so you know who is the boss around here." The truth is I didn't have any intention of taking his authority from him. I have always been willing to allow him all the initiative and authority he was willing to take; I dislike being in such a position anyway.

What has remained with me from the time of Menachem's captivity is the fear for my family. I am happiest when we are all together. When someone is missing, I feel anxious. I think that if I ever had to cope again with a similar situation, or with any significant crisis in one of our lives, I would not be able to stand it. What remained with Menachem is the need to stand out, to excel. He had it before, but it became more intense. All I want in life is to be a normal wife and mother, in a normal family. For Menachem it has never been enough, but we have learned to live with our differences.

DALIA

Right after his return, it seemed as if Amos was a different man. He was very open and talked nonstop from morning to night, until I was completely exhausted. He surprised me in several ways. It was hard to believe he could speak so much and share his experiences. I had warned my friends from the hospital not to rush him, because he's very shy. I was afraid he wouldn't even want to meet them. But in the end, it was amazing: he accepted everybody with so much affection, even total strangers. I was thinking: how wonderful, he benefited from being in captivity! Two weeks later, however, Amos went back to flying and regained his normal self. He hadn't changed at all.

I wanted another child but decided to wait for a while and see how we got along together after such a long separation; I wasn't sure we'd be happy together [laughing]. We have four kids now.

What are the results of that period that remain with me? I hate to be alone. Even though I was socially active in Amos's absence, and people invited us over, especially on holidays and weekends, the negative associations of being alone are part of me still. In spite of Amos's scorn, I'll never go to the movies alone. For some time I couldn't watch people fight with each other; when this happened, I used to get up and leave. Today I understand that fighting is part of every family's life. I can't let the phone ring, or disconnect it when I'm home; I must answer it. I know that this is from the time I used to wait so desperately for a message from Amos. I can't live without a daily paper—I'm still attentive and alert to news. It shows how much we appreciated every hint or piece of information when our men were in captivity.

YARDI

[The interview took place at her parents' apartment, where a huge photograph of Avi and Yardi in each others' arms near the plane that brought the POWs back is displayed on the piano.]

The story of the return wasn't an easy one. We knew that only peace or war would bring about the prisoners' return, but none of these appeared to be on the horizon. In the summer prior to the war, I went to South Africa with friends, and that's where I received the news about the mounting tension in the area. I knew that war was coming, although even Israeli Intelligence didn't, and I managed to get back to Israel on the eve of Yom Kippur, twenty-four hours before the outbreak of the war. I had these premonitions—they still give me the shivers.

I called my friends on the base right away, and they told me that women and children were being evacuated to a safer location, and I could come with them. However, I couldn't imagine being quarantined with all the wives and kids, waiting for the evening to get the casualty lists. . . . I invited one of my friends and her three kids to stay with me in my parents' house in Tel Aviv.

We lived together for ten days, while I helped her take care of her kids. It was an awful period, and we were hysterical. After ten days we felt the need to be where everybody else was, to get the news firsthand. We went to the family camp and got an apartment to share. As I had imagined, it was indeed terrible. Every evening the officers came and announced the names of the dead men and the captives. We used to avoid being outside so that we wouldn't see this trio of men with their horrible announcements.

At the same time my friend kept saying to me, "But for you, Yardi, it will all end in happiness, because when all this is over, Avi will come home." I felt it was my chance, but I was worried that our victory over the Egyptians might lead to revenge on the prisoners. It could be an unhappy ending, and I was haunted by anxiety and the fear that my dream would never come true.

The war ended after all, and the month of the exchange negotiations too, and one night I got a call telling me to be ready for the POWs' return the next morning. Naturally I didn't sleep the entire night, and in the morning, the whole country was there at the airport waiting. Big confusion, little kids and everything and at 2:30 only a group of wounded prisoners arrived. It was. . . . the last straw after all this long waiting.

I went home engulfed in grief. I asked not to be informed about their return until they were about to descend. And so it was. They called me the next day only after the list of POWs on the approaching plane was confirmed. I had an hour to get to the airport to receive Avi.

Such geniuses! The families were put in a distant hall, while the big shots from the Air Force and government could see them descend and be right there near the airplane. I don't know whose idea it was. Every additional minute of waiting was like an eternity.

The moment of our encounter was very strange. The bus with the men arrived, and I was standing right in front of it. I saw Yitzhak, Menachem, Rami, everyone—but Avi wasn't there. Meanwhile Avi was standing on the steps, searching, but couldn't see me. We were facing each other but we didn't see! I think it was our reaction to the huge excitement. We couldn't take it. This blackout lasted for seconds, than we saw each other, and we met.

We went to the base, where our parents were waiting. They had to wait one hour extra. Everything was so vulgar, and it took hours before we could be alone and quiet together. When we ran away from the crowd in the evening and drove to our flat, we found another surprise party there [laughing].

Our love stood the test of the separation well. Avi told me quite a lot about his captivity; he's very perceptive and intuitive. I was glad to discover how in the midst of all the power struggles and suffering he maintained his true self, without damaging his inner harmony and peace of mind. I grew up in his absence, and when he returned I was ready to become a mother. It seemed appropriate that at the peak of our reunion we'd produce a child, and in fact I became pregnant. We went on a long trip and returned for the delivery. We had a fabulous time.

What has remained from all this? For years I used to look back and say to myself: "It wasn't my own story, it couldn't have happened to me." Perhaps it's typical of me to deny and keep a distance, but still it's not like a childhood trauma one can repress. Avi was prone to anxiety; whenever he left me, he was afraid to go. But I, too, was always thinking, what if that's the last time I see him? What if it happens again? This kind of irrational fear is still with me.

MICHAL

When we received the news of the prisoners' return, Gideon slipped away on to reserve duty in the Sinai, leaving me to figure out what to do with my relationship with Yitz. I decided to tell Yitz in my own time.

We had some briefings with a psychologist prior to the return. He was especially annoying in his attitude to me. Before leaving him, I asked: "Where have you been for three and a half years? Now you remember me? I know what I have to do."

At this stage no one dared to interfere with my life anymore. I simply didn't let it happen. They threatened that Yitz might commit suicide, and I answered that I'd be very considerate. Actually I wanted to see what I would feel in his presence.

At the grand reception on the base, one of my male friends complimented me: "The whole base could be illuminated with your sparkling eyes." It's hard to explain, maybe it's hard to believe, but I was happy—because I knew that my torment was over. Now I'd be able to make my decision and go on with my life. I realized that even in the presence of Yitz I was feeling completely committed to Gideon, and I had no doubts whatsoever that I had made the right choice. The problem remained when and how to reveal the matter to Yitz.

After several days, when we were in bed, I turned to him and asked, "How come you don't ask me?" He said, "Ask about what?" And I said, "Whether I had someone else." He gave me a shocking response: "I know there couldn't have been, so I don't have to ask." I said, "And what if I tell you it's not true?" And he answered, "Never mind, I'll live with it."

He didn't provide any opening for my confession, and I felt terrible. For a whole week I didn't talk about it, but I was truly apprehensive that he might hear the story from another source, so I decided to do it. It was Saturday night, eight days after his return, when I couldn't hold it in anymore. We were in bed, and I started to cry and told him everything. His first reaction was amazing: "I don't blame you at all, but these rascals from the Air Force, our good friends, where were they? Instead of helping you and understanding you, they victimized you and only pushed you into Gideon's arms. How could they throw you away like that, when you had no choice? You had to hold on to somebody. I'm not angry at you."

We both cried a lot, and then he got up and left the house. He didn't return that night and I didn't sleep a wink, because I was worried stiff about him. In the morning I called Rami in the kibbutz and he told me that Yitz was in a safe place, and that he'd return home when he could. Rami came to see me later, and we discussed the matter frankly. I think that Rami wanted to demonstrate that we were all human and had our weaknesses. I appreciated him a lot for what he said. Finally he explained that right now Yitz was attached to me in an inseparable way, and that for the time being I'd have to maintain this relationship even if I loved Gideon.

A deep conflict developed, and for a while I really lost my way. Gideon called saying:,"Look, he's your husband and he's back. He needs you. Go back to him and I'll vanish." This reaction made it even worse for me. At the same time, Gideon received some phone calls from Air Force men threatening his life if he set foot on the base. And the whole country was talking about our scandal. . . .

The Air Force, with its ingenuity, found a solution. We would be sent as a family to work in the Israeli consulate in Washington. By the end of December we were already on our way. Yitz went to Dimona prior to our departure and had a talk with Gideon and his wife. They agreed that all contacts between us would be terminated for the duration of our stay in the U.S. There would be no telephone calls nor letters nor anything. The departure was supposed to heal the wound, but actually I gave it no chance at all. I wanted it as a gesture to Yitz and as an opportunity for the girls to be with their father. I hoped that we'd have a chance to talk and understand each other better. One thing was clear—that I couldn't forget Gideon. He was on my mind twenty-four hours a day.

The agreement didn't work; Gideon started sending daily express letters, and calling on the phone. I told Yitz: "It's no use, I eat, sleep, and live with Gideon even when I'm here." At that time Yitz asked for an immediate divorce, and I agreed. The next period was tough. We slept in separate bedrooms, and we shared the matter with our daughters. I remember that the older one started to weep, but the younger said: "Okay, so I'll have two fathers from now on." They both knew Gideon, but for the younger one he was already a father figure then.

We lived together in Washington for three months, until our divorce came through. Again it was a matter of public interest, and I felt isolated while everybody sided with Yitz. The Air Force dealt with all the formalities. I didn't even follow the arrangements and was willing to give up my financial benefits in order to free myself and get custody of the girls. I returned to Israel a poor woman: the girls and I had no part in the lovely house I had built, and all that remained in my possession was my car. But I was naive at that time, and all I wanted was to go back home.

I remember how Yitz took us to the airport and departed right away. I remained waiting for four hours with the girls, and I cried the whole way to Israel. I got what I wanted, yet I felt miserable. I felt I had been kicked out, although I had clearly been responsible for this development. Perhaps I was worried about my future, knowing that I had no way back. Gideon was still not divorced. I wasn't indifferent to Yitz's pain either. I knew I had left him in bad shape.

Gideon was waiting for me at the airport. We sent the girls to my parents in the kibbutz, and we went to be on our own for a week. With all the happiness of our open reunion, I had lots of unfinished business to take care of. For three and a half years I had fought like a tigress for my place, my conscience, my kids, and my right to love, while everybody had been against me. In that first week with Gideon I allowed myself to break down; I let myself go, and was willing to be led by somebody else.

There were many additional struggles. When I came for our belongings to the air base, nobody greeted me. When Passover arrived, my mother announced that she wouldn't invite Gideon and me home for the holiday. In Gideon's town, Dimona, where we moved, I had another crisis to bear. I was the second wife, an outsider to the community, and they didn't accept me at all.

Gideon received his divorce finally, and we got married in September. Immediately afterwards we were sent on a mission to the U.S., and lived there for two years. In the meantime I recovered, and people had other matters to gossip about. My girls decided, of their own accord, to call Gideon "Dad". Gideon and I had a son, and when we came back to Dimona, a new page was turned.

That's my whole story. It's clear to me that whatever happened to Yitz and me was not because we hadn't lived well before his imprisonment. I keep telling the girls that we had had an excellent relationship. I'm often told that I tend to idealize Yitz. It makes me sad that he has suffered so much. We used to love each other intensely, but perhaps for every age there's a different sort of love.

I'm not one of those women who look for brief affairs, for momentary satisfactions; I'm a person of great loves. For better or worse, I live things in their totality. I have never regretted what has happened, although my life as a pilot's wife would have been much more bright and comfortable. Possibly Yitz and I are victims of the war situation; I find no better description than that.

15

SURVIVAL AND COPING: ON NARRATIVE, TIME, AND CONTEXT

■

The evidence presented in this book provides ten different perspectives on the experience of a long captivity. The physical and psychological difficulties of the various stages of confinement, and the mechanisms for coping with them, both on the individual and group levels, are manifested in the personal narratives and the shared diary. We have learned about hidden feelings that reflect the inner reality of fear and uncertainty and also about external behaviors that display the interrelationships of a small group of men incarcerated together for almost three years. We have felt the men's longing for their former world of attachments and have followed the process of utilizing various techniques to alleviate this pain, as well as the eventual return and restoration of these attachments after their release. Parallel to this, the stories of the families who were severed from their loved ones for an indefinite period complement the picture to provide a whole, multifaceted narrative, which demonstrates the resilience, courage, and ingenuity of people undergoing trauma and prolonged stress.

This narrative may serve as a prototype for many other conditions in men's and women's lives, in which a host of physical, emotional, and interpersonal elements combine to create stress and thus demand unusual coping ability. Furthermore, it may provide a model for more ordinary conditions we all encounter, as we negotiate our life course among various human problems of living.

The aim of the book was to portray this narrative as it was shared with me, accompanied by the emotions that emerged as an aspect of the storytelling process, thirteen to seventeen years after the experience itself. In other words, as accurate as it may be, the book presents a psychological or narrative truth (Spence 1982; Gergen and Gergen 1986), rather than a historical one. Before going into an analysis and interpretation of coping with adverse conditions in light of the men's stories and my perspective, I would like to deal with the matters of time and audience, which are essential to understanding and evaluating this narrative.

The passage of such a long time is, of course, a factor affecting the story as told. Would we get a different narrative if the study had been conducted immediately after the men's liberation? The answer is obviously yes. As time goes by, different processes of forgetting, selection, and elaboration take place in people's minds (Ross 1991). However, such processes may operate, as we know, even while experiencing or perceiving events, and not only in their later recall or memory (Loftus 1980; Carr 1986). Thus, variations in the men's accounts of their shared experiences, as for example in their reports about being joined in the room after solitary confinement, or about the outbreak of the Yom Kippur War, may have been prevalent in their stories immediately after their release. While there would have been less time to forget the experiences, the fear, pain, and disorientation of a new environment would have been greater, perhaps coloring or distorting the narrative in their own way. The story told more than a decade after the fact is unique in its way, as any story would be.

While there are great individual differences in this respect, psychological as well as legal and literary work indicate that people often avoid talking about their traumas close to their occurrence, as the pain of the re-lived experience is too hard to bear. In other cases, victims refrain from sharing their experiences in order to protect friends and relatives from knowing about extremely cruel realities. Shame haunts survivors of humiliating traumas, as if they share, somehow, the responsibility for their misfortune (Lifton 1973; American Psychiatric Association 1987).

A critical period of silence may be necessary before victims mobilize enough courage to deal with their past suffering by recounting it. The evidence of the Holocaust in Europe is a clear example: for many survivors it was buried deep down their memories for several decades until surfacing recently (Davidson 1980; Danieli 1983). This cultural phenomenon is probably due to the safe distance in time and place, as well as the emerging shared climate of establishing historical records of the Holocaust for the sake of younger generations (Bar-On and Gilad, in press).

Although the story of the POWs presented in the book does not simply belong to the categories of trauma or victimization, since it has as much to teach us about resilience, healing, and coping, the heroes of the present narrative did not look for a writer before most of them had gone through a long process of recuperation and reconciliation with their fate. At the time of our interviews, many had never told their closest relatives and friends in detail about the worst hours of their captivity. For these relatives, listening to the stories told during the study (as was the case for Amos, Amnon, and Yitzhak, for example, who brought their wives along to the interviews about their torture), or reading the transcripts of the recorded sessions or the book in its final form (as with Dan's children, for example), was their first encounter with the harsh experience of their loved ones' captivity. Many mechanisms may explain why this could not have happened earlier.

As people grow older, and especially as they pass their midlife marker of forty, they tend to become more introspective and reflect on their lives (Neugarten 1964, 1968; Brandes 1985). They also feel the "generative" need (Erikson 1963) to transmit the legacy of their life experience and wisdom to the younger generation and to leave their personal traces on history by making them public. Men, in particular, are not inclined toward self-observation or similarly inner-oriented mental activities in their youth, but tend to engage in these exercises more in the second half of their lives (Jung 1933; Gutmann 1975). It was therefore not accidental that, as the heroes of the present drama passed their midlife transition, they became more willing to tell their stories and better equipped to do so in great depth. Furthermore, their ability to review their lives in perspective, in the context of later developments, adds a rich texture to their narratives.

Another factor to take into account is that stories are always told within a context of an interpersonal relationship, whether real or imagined; they are told for someone to hear. In the present case, the accounts that provided the basis for the book were given to me alone, either in the privacy of my office or at the narrator's home, or in the presence of a relative—a wife in three cases, and a grown daughter in another. As the ex-POWs were sharing their narratives with me, I had the impression that they often had their families in mind, too—elderly parents or growing children—with whom they wished to share the complete story of this important stage of their lives. All sessions were tape-recorded, and several of the men asked for a copy of the full records of our meetings, which I know will become family mementos. All the men realized that a book based on their narratives would be published, for Israeli and foreign readers to witness and evaluate their experiences. Although it is hard to estimate the cumulative effect of these underlying sensibilities, one should not disregard them. Deception and self-deception cannot be ruled out. Social desirability is a well-known human propensity, leading people to try to make the best impression on others. Heroism is a major aspect of Israeli culture and Israeli male identity (Gal 1986; Lieblich 1978, 1983, 1989), so that one may claim that each of my narrators wanted to be a hero for himself. The experience of humiliation, so typical of captivity, is perhaps the most shameful, antiheroic aspect of the stories, and may explain why it is so difficult and rare for Israeli men to make such disclosures; the interviewees may have "doctored" their narratives in this respect. I cannot discount such possibilities, but my personal experience as a listener to the men and their wives gave me the strong sense that the narrators were honestly trying to provide a truthful account as they believed the events had occurred, and that relating them in itself provided a cathartic relief of long pent-up secrets and pains.

Every interviewer participates in the creation of the narrative by her or his explicit and implicit interaction with the storyteller. In addition to my identity as a writer and psychologist, I am a female listener, and this perhaps allowed the men to trust me with their secrets rather than relate to my presence competitively as their witness and record taker. Women have less expertise or experience than men in the military realm, and even less so in being taken captive (an exception, during the Persian Gulf War, is Corum 1992). Thus I conveyed the message that my interviewees could teach me things I had never realized, and that I deeply respected them for it. At the same time I was a recent widow when this study took place, and I believe that at least for some of the men and women, my loss fostered a tendency to "teach" me some other lessons, too, in particular that people can recover from traumas and cope with their fate, if they have the right attitude. On the other hand, I probably listened to their narratives with a search for comfort and courage, too. I believe that the narrative truth shared with me was colored by all these additional hues. Rather than damaging or biasing the story, I see these factors as the rich sound of an orchestra accompanying a melody performed by the soloist, which together produce the final musical synthesis.

DIFFERENT PERSPECTIVES ON SURVIVAL AND COPING

The ex-POWs and their relatives were able to cope effectively with their captivity, and most of them drew positive conclusions from their misfortune. In other words, this is not a tale of trauma and its irreversible psychological damage, but rather of survival, resilience, and coping. In the remainder of this chapter I will describe the POWs' successful coping from my point of view and emphasize the major factors that probably contributed to this state of affairs.

Before embarking on this psychological analysis, one simple explanation for the POWs' survival should be discarded, namely, that the present study depicts cases of easy or luxurious captivity. Although some POWs' accounts, both in Israel (e.g., Ha-Meiri 1966) and internationally (e.g., Risner 1973), are indeed more cruel, the present case can be unequivocally defined as traumatic and extremely stressful. "Psychologically distressing event that is outside the range of usual human experience" is the formal definition of trauma used by the *DSM III-R* (American Psychiatric Association 1987, 247). While the first stages of captivity, namely interrogation and isolation, meet any definition of stress or trauma, the subsequent isolation, crowdedness, and deprivation of freedom that characterized the POWs' long incarceration constituted of a severe, on-going strain. That physical conditions in the common room gradually improved and that items such as learning materials became available are not merely positive environmental factors, but at least partly an outcome of the men's successful struggle.

In studying the presented narratives, we can focus on two phases: the isolated, individual experience that includes being taken captive and the initial stages of interrogations, torture, and solitary confinement; and the experiences of the group of ten men, almost totally separated from the external world, living together in the common room for three years. In addition, we can distinguish between short-term adjustment during the captivity, and the long-term aftermath in the more than ten years following the termination of the traumatic period itself. Finally, one can draw conclusions from the men's personal perspectives, some of their comments about each other, their wives' points of view, and my own impressions regarding survival and coping.

Each of the ten former POWs evaluated his own adjustment during the first, isolated phase as good, yet they varied in the degree of their self-satisfaction. While all of them described their overall coping during their interrogations as honorable, some retained a sense of guilt regarding their conduct during specific times of questioning and torture, and admitted there were moments when they felt they were betraying their country, so that they wished to die. When their long period of captivity ended, however, most of the men recuperated from these feelings and regained their self-respect, probably as a result of the curative group factors which will be discussed in the following pages. All of the men took care to recount that during their debriefing by military authorities after their return to Israel, they were given positive feedback on their behavior as individual POWs. Records of some of the men's public appearances in Israel right after their release indicate that even then, their presentation of their experience emphasized resilience and coping rather than trauma and suffering, partly because humiliation and pain are considered to be unheroic. A further process of "smoothing their narrative" (Spence 1986) with time and maturation may have taken place as they advanced in years. As an end result, according to the accounts of most of the men, the positive outcomes of their captivity outweigh the negative ones. The majority considered their time in captivity as "good" or "not bad," and were able to formulate personal lessons from their misfortune.

The survival of the men as a group is, however, much more unique in this case. In the long narratives about the group lifestyle and routines, both the absence of extremely negative phenomena and the abundance of positive influences indicate a remarkable adjustment of the ten men as a group. The idea of confining a number of men together for an indeterminate time provokes associations of extreme violence, either as a result of the formation of subgroups that compete over limited resources or as a result of the stronger members' scapegoating the weaker ones. Several factions of prisoners may evolve, or a corrupt leadership may take over, protecting only those whose loyalty is guaranteed. The frustration and personal tension may result in aggressive behavior within the group. Such interpersonal developments may lead to self-destruction or exploitation of the weak. In addition to these social phenomena, other deviant behaviors may develop, such as forced homosexuality or various addictions.

None of these violent phenomena could be detected in the men's personal accounts and their references to each other. In addition, their reports include no evidence of insanity, mental breakdown, long periods of depression, or symptoms of self-destructive behavior. On the other hand, given the forced isolation in enemy territory, limited resources, and the extremely crowded living conditions, the group accomplished unusual achievements in study, leisure, and social activities. When they landed in Israel more than three years after their traumatic capture, the ex-POWs seemed stronger, healthier, and more composed than one might have objectively expected.

Another perspective on the men's survival considers longer-range effects. According to the psychological diagnostic manual (American Psychiatric Association 1987, 247–50), maladjustment to trauma includes reexperiencing the traumatic event, avoiding stimuli associated with it, a numbing of general responsiveness, and increased arousal. Without therapy, these symptoms may last for very long periods of time. In the present case, almost none of the men complained of such symptoms. On the contrary, rather than the typical avoidance of thoughts and feelings associated with the trauma, the men demonstrated their willingness to share in detail the narratives of their captivity.

When no well-defined symptoms can be diagnosed, the professional literature does not agree about criteria for well-being and resilience. We can ask such questions as: Is the man mentally healthy? Is he satisfied with his life? Does he lead a relatively stable life, or does he change, shift, and wander in his choices of relationships, residence, and vocation? Did he obtain significant achievements in his profession? Are his accomplishments suitable to his potential? Is he a productive citizen in his community?

From an observer's point of view, I suggest that all the men are much closer to the positive than the negative side in the survival and adjustment continuum. Only one—Dan—remained a physical invalid as a result of his captivity, yet socially and professionally he functions very well. In the vocational sphere, the men are quite accomplished at present; only one was not satisfied with his achievements in this area. This man—David—is the only bachelor among the ten, which is a normal proportion for this age group. Divorce rate in the group—two out of nine—is also not unusual and fits statistical data for Jewish couples in Israel. Surprisingly, however, the two divorced men were divorced twice following their captivity and are presently married for the third time. On the other hand, five of the six couples that predated the men's captivity survived the long separation and still live together. (Dan's wife died of cancer recently, so she was not included in the research.) All in all, it seems that the ex-prisoners live normally, and notwithstanding their differing levels of adjustment and happiness, they survived their captivity trauma.

INDIVIDUAL COPING

Several processes and mechanisms interacted to produce the unusual outcome of the incarceration. Some occurred within individuals, while others functioned predominantly on the group level.

The captured men were, to begin with, people of high individual quality who volunteered for their jobs within the military system. In Israel, military elite units draw the best men out of all the eighteen-year-olds, who are drafted into military service for three years (Gal 1986; Lieblich 1989). The Israeli Air Force, as well as the paratrooper corps, is highly selective and rejects a high percentage of the young enlisted men who choose to serve in these prestigious military occupations. Air Force personnel are carefully selected according to the volunteers' talents and coping ability; they are then trained further to be able to function, individually and in teams, in a large variety of stressful and unusual circumstances. Dan, a mature officer who volunteered for a long reserve duty on the Suez canal posts, can also be considered as an individual of strong moral and personal background. Whether we characterize the group of ten men by the proportion of officers or by the proportion of members of elite units, we arrive at a similar conclusion, namely that it consisted of a critical mass of highly skilled and resilient individuals. These personal resources helped the men especially during the first months of their captivity, when no social support was available.

Each of the former prisoners described the moment of "falling captive" as traumatic, stressing many components: among the most prevalent factors are fear of immediate death, injury, or physical pain, a severe threat to one's ego, anxiety regarding one's responsibility for failure in the mission, and the dramatic shift ("fall" is the Hebrew term for this type of event) from a state of strength to total helplessness. The trauma lasted for several months during the phases of interrogation and isolation, during which every prisoner experienced a continuous fear of death, great physical and psychological pain, loss of control over his fate and daily routine, immense mental effort to protect state and military secrets, an uncertainty regarding the future, loneliness, extreme conditions of cold or heat, sleeplessness, hunger and thirst, and later, boredom. While each of the POWs described these first stages in somewhat different terms, according to his personality and background as well as the different objective circumstances of his capture, the elements of the traumatic experience consistently appear in all accounts. All the POWs provided details about their extreme suffering and their subsequent efforts to mobilize themselves for a struggle against their hardship.

The first defensive process appears right at the beginning of the POW's tale; it is manifested in the emphasis that their capture had been absolutely unavoidable. The pilots and navigators told me about numerous efforts to avoid abandoning their planes and falling into enemy territory, stressing the damage they caused to the enemy during their last mission. Yet their own doubts about the inevitability of this event are apparent throughout their stories. Some of the pilots and navigators told me that they willfully decided not to think about their doubts, or it would have driven them insane. In fact, talking about the subject was tacitly prohibited when the men gathered in the common room, in the same manner that behavior during torture and interrogation was a taboo subject in the POWs' conversations. Only many years later, during our interviews, did the men reveal these pervasive self-doubts, and admit that being captured is not a great honor for a soldier. It seems, therefore, that the first defensive strategy is to convince oneself that the capture was inevitable and to suppress or repress any memories that would indicate the opposite.

Another early mechanism, associated with the above, has to do with the isolation (or mental compartmentalization) of images, thoughts, and memories that might weaken the individual in his psychological struggle for survival. Some of the men felt that thinking, worrying about, or longing for their parents, wives, and children under these circumstances would disempower them, and they taught themselves to stop these ruminations. (Later on they complained that indeed they had forgotten how their relatives looked!) Related to this is a repression or willful attempt to forget details and facts that one did not want to divulge during the interrogations. On the other hand, some of the men described the opposite process—of evoking specific empowering memories and images (or internal objects), such as one's father or God, which provided comfort and support.

Processes of interference with perception, memory, and interpretation under traumatic conditions all come under the title of "depersonalization." These, too, abound in the men's early defensive attempts. During the first days, or the most difficult moments of torture, the men often experienced a split reality in which they became the observer rather than the subject of the trauma, in the sense that "this is not happening to me." This is why rare moments of human contact or kindness between the prisoner and his captors stood out in their accounts—since they broke the consistency of the dehumanization/depersonalization experience.

The men repeatedly mentioned the mechanism of autosuggestion, in the form of drawing comfort from beliefs that the time of captivity would be brief and that attempts to liberate them were already under way. Selectively recalling examples about former Israeli POWs, most of the men could cling to the hope that their time in captivity would terminate very soon. This early mechanism was later activated by the group for the entire duration of captivity, in the form of: "In two months we will go home."

As the condition of captivity continued, the need for an emotional outlet grew in intensity, as a way to discharge the continuous pain and existential uncertainty. Some of the men reported that following reflection, they allowed themselves to cry and scream, surrendering their masculine military manners (Lieblich 1983). While this may be defined as regressive behavior, it alleviated some of the POW's psychological stress.

A popular mechanism utilized by the men during the next few weeks was mental distraction, which also helped to reaffirm a sense of control over their environment. Most of the men told me about attempts to establish a daily routine, with different activities to mark the passage of time and draw attention away from fear and pain. Some stressed physical activities, while others created intellectual challenges demanding mental concentration. Interestingly, several men occupied themselves in recall activity, trying to reconstruct their life stories in as much detail as they could remember. This is obviously a process of self-therapy of utmost curative potential, and for some it changed their identity for years to come. As conditions slightly improved, all the men managed to fill the time free from interrogations with activities such as sports, drawing (on the walls), planning business projects, reading (the Bible was among the first books delivered to the solitary cells), sculpting (from soap or bread crumbs), growing shoots from seeds, and bird watching through the tiny windows. All these helped to divert the men's attention from their oppressive present conditions, and contributed to the creation of a sane reality within which they could reclaim their healthy, mature selves.

A central aspect of this massive defensive attempt consists of a struggle to regain control of the situation, even though the control may be partial or illusory. Using different cognitive mechanisms, most of the prisoners managed to devise strategies for their struggle against their interrogators and tormentors, and in this framework they could evaluate their performance and grade themselves for success. Frequently they managed to feel superior to the Egyptians who questioned or beat them, as when they coped honorably with torture or deprivation, when they managed to distract their interrogators from the main line of questioning on forbidden topics, or when they succeeded in planting a lie to cover the truth. Many of the POWs were able to feel as if they had dictated the tempo or contents of the interrogation process, and thus regain some sense of self-confidence they had lost during the first period of utter helplessness. Only by arriving at this point of feeling in control—each according to his own methods and criteria—could the men survive the ongoing trauma and maintain their sanity in the midst of chaos.

Contacts with benevolent reality agents eventually contributed to the healing process. When the interrogations were over, the prisoners were visited by Red Cross officials and received the first letters from their families. Both of these events retain extreme significance in the POWs' accounts, and prove the traumatic impact of isolation from one's support networks. Only a small proportion of the men discovered the existence of other Israeli prisoners in the vicinity, but when this happened, further relief was reported. The hope of uniting with the others, kept alive by the promises of the Red Cross officials, was almost as consoling as the hope for release.

For eight out of the ten men (all but Dan and Amnon), a noticeable healing process took place before they were brought together in the common room. Even though none of them received adequate medical aid, they recovered from their wounds. Due to the above mechanisms, they regained some confidence. As a result of the harmony and constructive atmosphere of life in the common room, the improvement in the standards of living, and the time/reality organization, these physical and psychological recovery processes continued in the next three years of the POWs' incarceration. Only Dan describes a continuous awareness of the dangers and threats to his life in prison until his liberation. This is probably due to the facts that his wounds were maltreated, serious medical complications continuously threatened his life, and he was the only man who had presented a false identity to the Egyptians and therefore was constantly afraid of being discovered. As for Amnon, who had been quite immature when he was captured and subsequently experienced the deepest sense of betrayal, shame, and suicidal desires, his recovery started to take place mainly with the help and support of the other prisoners in the common room.

As time went on and the interrogations terminated, all the men except Dan reported an evolving sense of security within this enemy territory. The common room became their haven, in which they felt they were the masters of their fate. They were not optimistic about their prompt release, because they understood that this might happen only in the context of either peace or war between Israel and Egypt, with a large prisoners' exchange in the aftermath. Naturally, they were ambivalent about war, and peace seemed a remote prospect. In spite of these dire circumstances, individually they developed coping and defense mechanisms that helped in their adjustment to the sense of lost time and the longings for their families. In later stages, the major intrapsychic process was reference to life in the common room in jail in Egypt as the only universe, so that they withdrew most of their energy from interests in the world outside, which was beyond their reach and control. While reading the diary one might regard such issues as naming the kittens or deciding on the shape of the dining table as blown out of normal proportion; however, this here-and-now orientation preserved the men's sanity. That is why letters and Red Cross visits were very welcome, yet at the same time deeply unsettling; not only were they sometimes disappointing, but they disrupted the delicate mental balance established as a result of those defensive efforts. Only through the impersonal activities of reading and studying were the men willing and able to break through the barriers of their confining walls. A number of the men remarked that reading was an extremely powerful experience during captivity, since the immense identification with heroes of fictional and nonfictional stories provided some sense of freedom of the spirit.

In their concluding remarks, many of the POWs claim that a long captivity may be less traumatic than a brief one, since it allows a full unfolding of the physical and psychological healing process. When they compared their state to that of the Yom Kippur War POWs, who were exchanged after only a few weeks, the prisoners of the War of Attrition considered themselves clearly more sane and sound, in spite of the basic fact that they had lost a much longer portion of their lives as free members of society.

COPING AS A GROUP

While these individual mechanisms, which continued to be operative throughout the period, were certainly significant, the main riddle of this narrative is what made this POW group so successful in its collective survival. I propose a number of aspects to solve this riddle.

1. The Ideological Aspect of the Situation

All the men's accounts related the fact that the group became a cohesive community as a result of its isolation from the external world and its place vis-à-vis their common enemy (Lewin 1948; Charters and Newcomb 1958; Tajfel 1982). Recent research in social psychology has confirmed the idea that people evaluate their group in contrast to other groups, and their self-esteem is enhanced if the comparison is favorable (Messick and Mackie 1989). POWs and political prisoners—in contrast to criminal prisoners—naturally see themselves as representatives of a "better" society, class, or minority in a hostile territory. This was clearly the case for the Israelis, who saw themselves primarily as representatives of Israel. They experienced national pride and patriotism and hoped to prove their superiority to the Egyptians by adopting certain moral standards and an exemplary lifestyle. As prisoners, they had a very well defined audience for their best behavior, a negative reference group (Newcomb 1943; Kelley 1952) they wanted to impress. As documented in the diary, formal decisions had the rationale of "not shaming Israel." Moreover, according to their recollections, this provided a motive that affected each of the prisoners even before they formed the group. Their basic patriotic loyalty and pride manifested itself also in the fact that at no time throughout their three and a half years of imprisonment did the men doubt the need for the missions that had led to their captivity. They were convinced that they were not forgotten, and that Israel was making every possible effort to liberate them.

Besides reinforcing group cohesiveness, being "good" or "virtuous" in the midst of an "evil" and hostile world might have had other effects. According to Jung (1959), man's nature has good and evil sides: the good part is often conscious, while the bad side or "shadow," which is responsible for hate, destruction, and aggression, is primarily unconscious. The state of captivity produced a natural polarization or split of good (me) and bad (the enemy), thus saving the individual from facing his own evil nature, and returning him to a blessed state where he was purely good and could rest from his inner struggles. The "bad guys" outside become the target for all violent wishes, while the "good guys" protect each other. This externalization of evil may contribute to our understanding of how a group of ten heterogenous men achieved such harmony in spite of the difficult conditions of overcrowding and frustration.

The ideological system that sustained the men had several additional components. Although none of the men was actually religious, their Jewish identity (against the Moslem background) provided comfort in difficult moments, and the Jewish holidays and the Sabbath became central for marking the passage of time, creating rituals to consolidate the group and structure its daily reality. Furthermore, all saw themselves as IDF men, Air Force personnel, or officers, which produced certain standards of conduct originating from the military system. The military culture, with its respect for order and discipline, provided the initial norms for the men's lifestyle. The Air Force is perhaps even stronger than other IDF branches in educating its inductees for interpersonal cooperation and mutual help.

Finally, about half the members of the group had some previous kibbutz background. (This is just a coincidence; the kibbutz comprises less than 3 percent of the Israeli population, although its representation in elite military units is much higher.) Gradually, values from the kibbutz culture took precedence over those from the military institutions. Instead of a unit functioning according to hierarchical military rules, the group became a commune governed by the decisions made in its weekly assembly meeting. Maintaining life in the common room in a totally egalitarian, democratic fashion became one of the most salient values of the group. The fact that two of the prisoners were kibbutz members (among them Rami, the leader) and four others had spent long periods of their life in kibbutzim made the collective egalitarian system of values easily accessible.

A common element within these two value systems is the utmost respect for individual life. Both the Israeli military system and the kibbutz ideology emphasize the responsibility of the collective for the survival and well-being of each of its members, as weak or useless as he or she may be to the whole unit or group. This is also a basic Jewish value, asserting that any Jew is accountable to all others ("kol Yisrael arevim ze la-ze"). The basic trust of each of the men in his comrades, and the underlying support network among them, can be directly traced to these ideological sources.

For men who have access to such articulate value systems—here, from the Air Force and the kibbutz; in other cases, from religious backgrounds (as in Risner 1973)—the process of applying values to establish meaning and order in a strange environment is greatly facilitated. Thus, the early chaotic state of life without rules and regulations, which is very dangerous for the weak members of a newly formed group, was relatively brief. However, the transitions from laissez-faire to military, and then to a kibbutz-like group, were not entirely smooth. As the laissez-faire atmosphere was changed, Dan, and to some extent Avi, lost status in the group, and maintained some resentments about it for a while. The struggle between the military and democratic systems went on for a long time, taking the form of tensions between Menachem (who represented the first system) and Rami (who represented the second) for leadership. However, luckily for the group, this struggle, in which the democratic values won the favor of the majority, was not hostile, and Menachem, too, obeyed the collective decisions. In any case, the situation can be described as a conflict between different possible systems of values, rather than a lack of values, which might have been devastating for the group.

As time went on, a new system of values was introduced, originating from the reading and reflections of the group's leading members. This can be termed an existential-humanistic ideology, whose essence was formulated by Rami based on the lessons he had learned from his experience in captivity. As bad as circumstances may be, according to this ideology, human beings are responsible and accountable for their own attitudes and feelings, and they can turn their experience into a meaningful one. This belief, which Rami tried to convey to the others, challenged them to make a positive, constructive experience out of their misfortune.

Many existential psychologists, among them Viktor Frankl in his famous book *Man's Search for Meaning* (1959), claim that a person with a clear sense of meaning or value system will be more resilient in adverse conditions. One has to find meaning in and justification of the suffering itself, as some Holocaust survivors managed to do. Recent research in the psychology of health supports the claim that a sense of coherence, hope, or control increases the chances of overcoming various medical conditions (Steptoe and Appels 1989; Friedman 1990) and indicates the behavioral and biological mechanisms underlying this effect (Steptoe 1989). In the case of our POWs, their emerging value system, consisting of patriotic, democratic, and existential components, sustained them and contributed to their survival and health. It provided guidelines for behavior both among the men and in relation to the Egyptian authorities and helped to form their productive routines.

2. Leadership in the Group

The evolution of leadership, and its significance for the group's survival and well-being, are among the major points of general application from the present case. Undoubtedly, strong, benevolent, and consistent leadership is an essential ingredient for successful groups.

The narrative, with its sequence of leadership styles, is unique to the history of this group and may remind the reader of the classic social-psychological studies on leadership. As the ten men first gathered in the common room, they encountered the amorphic, laissez-faire system (Lippitt and White 1943) of the four POWs who were already living in the room. Dan was this group's senior member, yet his power was severely limited by his injuries. Avi, who spoke English well, was the foursome's representative to the authorities. There was little structure of time or activity in their loose leadership. Soon after his arrival, Rami proved that he was capable of solving the problems which arose and that did not lose his temper under stress. For a while he combined all the leadership functions in his peculiar style, which minimized disciplinary, coercive means and relied on persuasion and group decision. He can be characterized as the classic democratic leader (Lippitt and White 1943), or, in another terminology, as a "great man" high in the dimensions of activity, task ability, and likability (Bales 1958).

Upon the arrival of Menachem from the hospital six months later, an alternative, authoritarian model (Lippitt and White 1943) of leadership was attempted but never fully implemented. However, through the democratic system, Menachem contributed to the acceleration and formalization of various productive processes, especially the study program and the improvement of the standard of living in the common room. At the same time, he augmented tensions and frictions in the group, including constant questioning of Rami's authority, which on the one hand actually reinforced it, and on the other hand made the group members somewhat more independent.

With both men present in the group, there was an unusual demonstration of the classic distinction of task-specialist versus social-emotional specialist as leaders of small groups (Bales 1953; Bales and Slater 1955; Parsons and Bales 1955). According to this model, there are two major functions that need to be fulfilled by the group leadership. The instrumental or task-oriented function stresses the efficient employment of all members in working toward the group's goals, and an "instrumental leader" who emerges to take care of this function, typically focusing more on tasks and their accomplishments than on the well-being of the group members. Groups oriented solely in this direction inevitably create stress and interpersonal competition among their members. "Expressive leadership," which complements and moderates the above, focuses on the person. It is designed to produce a social climate in which people feel well, each according to his or her personal needs. It takes care of emotional discharge and tension release and supports the weaker members of the group.

According to this classic sociological theory, a "good" group that accomplishes its objectives—such as a harmonious family—

has its two basic functions fulfilled smoothly. Frequently, two different people take charge of each of these leadership functions, since they require incompatible behavior patterns.

In the present case, once the group stabilized, Rami and Menachem emerged naturally as its two leaders. Each of the men represented a different ideology—the kibbutz versus the military—but they cooperated well in carrying out the group's decisions. Menachem drove the group to improve their standard of living, to study in a serious and consistent manner, and to be productive in a variety of ways. Rami took care of the complex social-psychological well-being of the POWs. He was responsible for improving the men's morale, establishing sane standards for their interactions, encouraging personal expression, and devising a democratic system. He was the spokesman vis-à-vis the jail authorities and was exemplary in his morale and respect for all. When arguments started among the men in the common room, a frequent retort was "Rami says so" or "Rami does it," as an anchor and model for the men's behavior. Moreover, members of the group sometimes requested individual conversations with Rami, in which they shared their moods and concerns with him while he provided counseling and support.

In the POWs' lives, when no concrete task demanded their attention (besides their mere survival!), Rami, the expressive leader, was also the leader (the "great man") in the eyes of all. His regulatory function was extremely important for each man's sanity, since the common room was their inescapable universe, and tensions due to frustration and the crowded conditions could easily have torn the delicate fabric of the group. Indeed, Menachem, the instrumental leader, conceptualized the group tasks—such as studying, sports, and the room's improvement—as means to reduce tension among the men. Thus, while most researchers of small groups claim that the instrumental leader is usually the most powerful, the reality of captivity produced an interesting variation in which the expressive leader was clearly dominant. As in a healthy family, conflicts between the two served the group by making it reexamine its priorities and served the leaders by producing new ways to fulfill their two functions in spite of the essential contradictions between them. Undoubtedly, the emergence of these two leaders and their cooperation are among the most important contributions to the group's coping ability. The complete availability of the leaders and the stability of this structure throughout the period provided a solid feature in the men's uncertain reality in an enemy jail, helping them maintain trust, sanity, and hope.

3. The Social Structure of the Group

Except for these two leaders, none of the men seemed to be inclined to gain power or influence in the group, even though there were others with remarkable capabilities. The fact that every member participated in the assembly, both to propose his own ideas and to support or reject those of others, proved to be a sufficient channel for the men's need for control. The emerging atmosphere allowed every individual to contribute in his own way, without fierce competition for resources, attention, or influence.

Other members of the group did not have a consistent, stable role throughout the period. For a while, some were troublemakers who tended to argue and aggravate the others, some were peacemakers, and others dissipated tension via humor and pranks, but the men shifted their roles. Furthermore, there were no fixed subgroups, coalitions, or deeper friendships among the prisoners. When I asked the men, "Who did you use to talk to, or to confide in?" they produced varied responses. They talked with everybody, they had different closer friends at different periods, or they had contacts with different men for different purposes—for joking, playing bridge, smoking a cigarette, baking a cake, jogging in the yard, or working on *The Hobbit* translation. Even men who shared bunk beds or were close to each other every night did not report that they had formed more intimate relationships with their neighbors than with others in the room. More remarkably, the pilots did not tend to socialize more with each other, in spite of their common background. The better educated did not distance themselves from the less educated men, although differences in previous schooling became salient and concrete in the context of the study program. Many of the class differences that belonged to external reality, however, became meaningless and disappeared in the community of the common room.

A development of subgrouping threatened to occur on two reported occasions, which may be representative of other forgotten episodes. In one case, an initiative of Yitzhak's to form a team of the better bridge players, who would play daily together to improve their skills, was withdrawn after long debates in several assembly meetings. The allotment of men to the bridge tables returned to its former rotational assignment. In a similar vein, while translating Tolkien's book from English to Hebrew, copying and other simple jobs were assigned to the men whose linguistic skills were inadequate, so that all could participate. However, when I inquired why the captives discontinued their translation project after one book, I was told that the translation endeavor emphasized the educational gap between two groups of prisoners, and therefore threatened the harmony of the community as a whole. Thus, in spite of their heterogeneity, no subgroups were allowed to form among the men. Again, one may detect kibbutz influence in this instance. For example, older kibbutzim are reported to have concealed all former educational degrees of their members, so that individuals with doctorates would never use their titles or even reveal the extent of their education (Lieblich 1981).

Only one feature of the group structure deviated from the moderately affectionate and egalitarian network, namely the deeper friendship that gradually developed between Yitzhak and Rami. From Rami's perspective, since he provided the support for most of the men, and often struggled with Menachem, he experienced the need to release his own feelings, so that he would be able to continue in his daily fatherly role. From Yitzhak's point of view, this intense contact may have supplied the necessary warmth missing from his correspondence with his wife. Whatever the reported reasons, these two men developed an unusually deep bond that was respected by all the others. In one of his letters to his wife (April 7, 1972), Rami writes: "My friendship with Yitz protects me against all feeling of frustration and loneliness, and I may say that this, at least, is one thing that I have gained from my captivity—direct, intimate, and warm friendship." Similar expressions abound both in the letters of Rami and Yitzhak and their accounts during the interviews. For their respective birthdays, for example, the men wrote poetry in praise of their bond.

Rami claimed that he has never had such a close relationship with a man before or after his years in Egypt. Possibly a leader such as Rami, under ceaseless obligations to others, cannot function for such a long time unless he is able to mobilize private support, similar to the need of many psychotherapists for ongoing supervision and ventilating channels with other, usually senior, members of their profession.

One of the outstanding features of the group's life was its emerging wisdom in interpersonal relationships, characterized by a general lack of great intimacy and depth in their contacts with each other. As mentioned above, the members formed deep loyalty to the group as a whole, which can be conceptualized by the recent term "intergroup bias" (Messick and Mackie 1989). On a personal level, all the men were equally friendly with each other. They learned to be tolerant of each other's traits and manners. Yet as a natural outcome of the need for privacy in the extremely crowded conditions, the men refrained from forming intimate friendships (except in the case mentioned above) and did not share their letters or deeper feelings such as longings for home, for example. People did not probe into each other's past or emotions. After the failure of several weekly assemblies to deal with "group dynamics" in the common room, the men comprehended that deeply emotional levels of experience should best be kept private. On the surface, the group was characterized by a moderately friendly network of relationships, in which the individual boundaries of each member were respected. Bad moods were almost ignored, while providing acceptance and space for the individual in pain. "Had we revealed everything to each other, had we talked freely about our despair and longing, we would all be crazy by now," was a sentence repeated by several of the ex-prisoners. Yet it is incorrect to judge these social relations as superficial, since the men developed a keen sense of reading moods and emotional states. "Without speaking about ourselves," explained one of the POWs, "we became extremely open and almost transparent to each other."

This mixture of directness, mutual respect, and denial is another ingredient of the men's successful survival. In particular, the role of denial in sustaining hope under stress is of great significance. While former studies emphasized the role of denial of intrapsychic material or inputs from the environment (Simonton, Simonton, and Creighton 1980; Breznitz 1983), the present work demonstrates the function of a similar mechanism in interpersonal relations.

4. Mechanisms for Tension Release

The predominant responses to the stress of living in a small contained space are frustration and aggression, which can produce considerable friction and interpersonal violence. Devising various mechanisms for the management of tension and its release distinguishes a constructive group from a destructive one. Due to their values and leadership, the heroes of the present narrative succeeded in almost totally preventing violence among themselves. Their wisdom may serve as a model for similar group situations, even for such cases such as people sailing together or a family taking a long trip.

Tension may accumulate from frustration at a lack of personal actualization, physical activity, and emotional expression. A partial solution to the natural need for actualization was found by encouraging each of the prisoners to find his avenue for personal expression and accomplishment within the constraints of the Egyptian jail. Some of the men accomplished educational goals that served them after their release. Others were able to develop artistic skills that had been dormant. But the main mechanism for tension release for the entire group was its complete schedule of daily activities, which involved study, work, play, and hobbies. Furthermore, it created a routine that structured the day and the week in a constructive manner and minimized free time, which might have been dangerously utilized. Studying elevated the men from their present concerns and refined their manners. Some of the competitive games, especially bridge, which was played almost daily throughout the period, became channels for a sublimated expression of aggression and the need for dominance. The pent-up physical energy found an outlet through excercise, which several of the men started to practice in solitary. The lack of emotional outlet was alleviated by the care of pets, as well as by reading, art, and drama. Arguments were turned into issues to be discussed at the weekly meetings and were settled by the group decision processes.

According to the men, the estimated climax of the group's productivity involved two projects: the celebration of the first Passover together, and the translation of *The Hobbit*. The celebration of Passover, the holiday of freedom marking the exodus of the ancient Jews from Egypt, obtained tremendous ritualistic significance in jail. The men had to overcome their differences of background and education to find a common way to celebrate, face the challenge of preparing the holiday meal and ritual under these difficult conditions, and then convince the prison authorities to invite guests, among them a rabbi from Cairo and several military officers. The concrete and symbolic achievements involved in this celebration empowered the men for a long period, and were recounted and reinterpreted with great enthusiasm during our interviews many years later.

No less significant was their joint translation of Tolkien's *Hobbit,* a book which appeared in press in Hebrew after the men's return (Tolkien 1977) and contributed to the notoriety of the group. The initial motive for the translation was to share the pleasure of reading it with those prisoners who could not read it in English. The four pilots fluent in English concentrated on the effort for several months, and the others participated by giving feedback and copying the manuscript. It is probably not a coincidence that the climax of the men's productive work focused on this specific book. *The Hobbit* presents an imaginary plot about a brave, peace-loving creature who lives in an underground world and fights the dragon who had taken away the hobbits' kingdom. At the end, the victorious hobbit returns to his land and home to write his memoirs. Obviously this tale was loaded with meaningful associations for the captives; it provided a mythical narrative account, with a happy ending, of their life in captivity. The issues of good against evil, and the personality of the antihero who stands for friendship, loyalty, love, tolerance, and nobility, could not but charm those who had become victims of war and violence and were fighting for their own freedom.

5. Attitudes towards Time and Space in Captivity

Time in captivity is endless, and space is very limited. Unique attitudes (which not all of the individual members could articulate) toward these basic facts evolved, and these, too, preserved the men's sanity.

The experience of captivity is very different from the experience of being in a civil prison; since the length of incarceration is unknown, the POW has no idea how much longer he must wait before he will regain his freedom. This state emphasizes the element of man's mortality or impermanence in the world. The POWs in Egypt faced daily both the chance of being released, a hope sustained by naive optimism and the messages of the Red Cross, and the prospect of staying in captivity forever, as was often threatened by their captors. In such a state, one option is passivity—waiting for either the savior or the executioner, with one's suitcases packed. This came up, for example, in the argument on whether to take the books out of their cartons or to order bookshelves for the room, and many similar issues. This option may breed a state of boredom, interpersonal violence, despair, or madness. Although not all of its members could articulate their guiding principles, the group chose a different option, as manifested in two of its slogans: "We will be liberated in two months, but in the meantime let's make the most of it" and "This is my universe; there is nothing beyond its boundaries; I have to create a complete life right here."

"Two months" is not so long that one may despair of waiting, yet it is not so short that nothing can be planned and accomplished in the meantime. If life in prison is all there is, creating the best, most productive lifestyle and community—instead of longing for one's family—becomes a clear, moral imperative. While understanding the unrealistic aspects of these attitudes, they were more than empty slogans for the men; they dictated actual guidelines for their decisions and helped them in their daily functioning. Apparently, the prisoners adjusted well to the sense of lost time and to the feeling of uncertainty and homesickness, by referring to their reality in the common room in jail as their whole life and limiting their future perspective to this two-month stretch. However, this cognitive step entailed a voluntary limitation of interest in anything outside and beyond their control, except for the imaginative leap outward, which was produced by correspondence, reading, and study. This here-and-now orientation preserved the sanity of the POWs during their long imprisonment, a wise and moral choice evidently affected by the personality of the group leaders and the members' dominant values, thus returning us full circle to the beginning of this section.

THE PRICE OF CAPTIVITY

The analysis presented so far has emphasized the positive aspects of this particular traumatic experience and attempted to specify the different aspects that contributed to this outcome. However, it would be incorrect to assume that the ex-POWs did not pay a price for their long incarceration. First of all, they lost time that could have been utilized for their personal growth and progress in their careers or family lives. Some of the men mentioned that they could have completed formal higher education degrees instead of wasting time in captivity. Later on, several men gave up on higher education (even though the Israeli government would have paid for it), deciding it was too late to start. From a different perspective, most of the men reported some PTSD symptoms right after their liberation, especially nightmares and sleeplessness, although the degree was mild enough so as not to require any treatment. If today they suffer more anxiety and malaise than the average person, I was unable to detect it. This supports the finding of a larger and more systematic research project by Barnea (1981), which did not find any psychological aftermath in ex-POWs from Israeli Air Force as compared to a control group of pilots who did not live through a similar experience.

The transition from the prison world to free civilian status was not entirely smooth, however. While such a transition may be quite traumatic and requires time for personal adjustment as well as acceptance from family and friends, this group returned to Israel at a particularly hard time. The country had suffered many casualties in the Yom Kippur War and was preoccupied with concerns that seemed more pressing than those of the homecoming POWs. After their grand reception, several of the men felt they were being forgotten. The bachelors had a particularly difficult transition since they lacked the continuous support of an understanding wife. In fact, two of them married soon after, perhaps to correct what they thought was missing in their lives.

A unique feature of their transition was noted by Menachem. He remarked that the utopian community created in the Egyptian jail might have "spoiled" the men when faced with the normal hardships of a competitive society. In the common room, all social differences between the men were obliterated. The fact that they received exactly the same treatment and had equal power in the democratic system, gave rise to an illusion of equality that did not match realistic circumstances in normal society. The men from the lower strata of Israeli society may have forgotten the barriers they were about to encounter due to lack of money, education, and former status. Moreover, they expected their friendships with the other ex-POWs to continue, although such friendships would normally be quite rare. Returning to free society, they found out that some of their compensations were determined by their former standing in the IDF, but there were other more subtle manifestations that indicated the loss of the illusion of equality.

A related difficulty reported by the men concerns openness of expression in their interpersonal communication, which they felt was inadequate and misunderstood in the "real world." The atmosphere of continuous togetherness had led to a way of relating to others that could not be recaptured among men elsewhere and was even very hard to reproduce within their families. At the same time, some of the men were able to retain this openness as an optional way of communicating, which they used occasionally when conditions were appropriate.

Of the ten men, one—David—seems to be suffering from moderate, longer range post-traumatic effects, and another—Amnon—from a milder case. David, however, confessed having had many personal problems throughout his childhood and adolescence. He tends to feel discriminated against, and this was manifested even in the utopian prison conditions. First he believed he was tortured and interrogated more than anyone else, which he attributed to a variety of reasons the other men considered unbelievable. In the common room, he bitterly complained about unfair treatment by the Israeli Air Force, which—according to his account—sent certain items only to its own personnel. Other non-

Air Force prisoners did not express such resentment. Finally, upon his return, David entered into a struggle with the authorities concerning financial compensation for his years in captivity. He ended up demonstrating in a hunger strike at the entrance of the Israeli parliament, and to this day feels deeply insulted by the attitude of the various authorities toward him.

In the years following these stormy events, David has been unable to establish himself professionally; he has not formed a family and has lived most of the time at his sisters' homes. However, I believe that the respectful listening he received in our many evening meetings in my apartment in Jerusalem alleviated some of the lonely pain he carried in his heart. Following our conversations he found a suitable occupation, which gave him much satisfaction, and he moved into his own apartment. This may demonstrate the significance of the cathartic process of sharing one's painful narrative with a sympathetic other.

Amnon has led a lifestyle marked by changes and instability. He had many professional and academic beginnings, and although he succeeded amazingly in everything he did, he has always felt compelled to move on to a new beginning. He has been married twice and currently is having problems maintaining his third marriage. Apparently, Amnon cannot be contained in a stable environment and needs to exercise his freedom and repeatedly manifest his control over his life. It is tempting to speculate that these two men, so different in their personalities and backgrounds, were the least prepared by previous life experience for the trauma of captivity. They were, indeed, the youngest of all the prisoners in our group, and according to previous research, age seems to be correlated with better adjustment to captivity (Hunter 1991).

A MODEL FOR RESILIENCE AND RECOVERY

It is beyond the scope of this volume to discuss the various theories and research about survival, resilience, and coping in adverse conditions. However, the present narrative may provide some guidelines for a model that integrates the individual and group levels of experience.

Much has been written and researched about individual coping and defense mechanisms. A common assumption in the literature is that the mentally stronger, healthier individual stands a better chance of surviving and even being able to take advantage of traumatic events (Friedman 1990; Steptoe and Appels 1989). Notable among personality factors that have been proposed in this context are "hardiness" and a "sense of coherence." Both concepts emphasize cognitive processes, since each of them is defined by a composite of beliefs about self and world, which lead to a certain appraisal of the stress conditions. Hardiness involves a sense of control, commitment, and challenge (Kobasa 1979). Similarly, a sense of coherence refers to the conviction that the events encountered in life are explicable, that resources are available to meet the demands, and that difficulties provide challenges, worthy of investment and engagement (Antonovsky 1990).

I believe that recovery from severe trauma and hardship is more probable in a group than among individuals. The concept of social support is cardinal to any theory about coping with stress (Maddi 1990), while isolation and loneliness add strain to an already stressful situation. When group support is conceptualized, a distinction arises between support from others who do not share the same stress, such as family members of a sick individual, and groups of people who undergo together the same stressful life events, as in the present case. Not all groups are equally healing under common stress conditions, and some may be destructive. To be helpful to the individual, at the minimal level, a group should serve both as witness to one's suffering and as a comparative reference, and it may become a source of physical or psychological support and distraction from the difficult present. To be helpful at the maximal level, namely to bring about recovery and turn the trauma into a challenge and lesson for life, two ingredients are necessary: a common system of values that provides meaning and coherence to the stressful situation, and firm, wise leadership that inhibits violence and channels the members' energy toward constructive ends. As an outcome of these two components, a healing social atmosphere will be established, in which positive mechanisms for tension release will be provided, and adequate attitudes toward the stressful situation will be developed. Thus, an atmosphere of hope and spiritual growth may rule even in enduring painful situations.

REFERENCES

American Psychiatric Association. 1987. *Diagnostic and statistical manual III-R (DSM III-R).* Washington, D.C.

Antonovsky, A. 1990. Personality and health: Testing the sense of coherence model. In *Personality and disease,* ed. H. S. Friedman, 155–77. New York: Wiley.

Avneri, A. 1982. *Patterns of adjustment to captivity: Description and analysis* (in Hebrew). Jerusalem: Hebrew University.

Bales, R. F. 1953. The equilibrium problem in small groups. In *Working papers in the theory of action,* ed. T. Parsons and R. F. Bales, 111–61. Glencoe, Ill.: Free Press.

———. 1958. Task roles and social roles in problem-solving groups. In *Readings in social psychology.* 3d ed., ed. E. E. Maccoby, T. M. Newcomb, and E. L. Hartley, 437–46. New York: Holt.

Bales, R. F. and P. E. Slater. 1955. Role differentiation in small decision-making groups. In *Family, socialization, and interaction process,* ed. T. Parsons, R. F. Bales et al., 259–306. Glencoe, Ill.: Free Press.

Barnea, I. 1981. *Long-term effects of captivity on the personality of Israeli airmen* (in Hebrew). Jerusalem: Hebrew University.

Bar-On, D. and N. Gilad. In press. "To rebuild a life": A narrative analysis of three generations of an Israeli Holocaust survivor's family. In *The narrative study of lives.* Vol. 2.

Brandes, S. 1985. *Forty: The age and the symbol.* Knoxville: University of Tennessee Press.

Breznitz, S., ed. 1983. *The denial of stress.* New York: International Universities Press.

Carr, D. 1986. *Time, narrative, and history*. Bloomington: Indiana University Press.

Charters, W. W., Jr. and T. M. Newcomb. 1958. Some attitudinal effects of experimentally increased salience of a membership group. In *Readings in social psychology*, 3d ed., ed. E. E. Maccoby, T. M. Newcomb, and E. L. Hartley, 276–80. New York: Holt.

Clavell, J. 1962. *King Rat*. Washington, D. C.: Coronet.

Cohen, B. and M. Cooper. 1954. *A follow-up study of World War II prisoners of war*. Veterans Administration Medical Monograph. Washington, D.C.: U.S. Government Printing Ofice.

Corum, R. 1992. *She went to war: The Rhonda Corum story*. New York: Presidio Press.

Danieli, Y. 1983. Families of survivors of the Nazi holocaust: Some long and short-term effects. In *Stress and anxiety*, vol. 8, ed. C. D. Spielberger, I. G. Sarasona and N. A. Milgram, 405–21. Washington, D. C.: Hemisphere Publication Corporation.

Davidson, S. 1980. The clinical effect of massive psychic trauma in families of holocaust survivors. *Journal of Marital and Family Therapy* 1: 11–21.

Erikson, E. 1963. *Childhood and society*, 2d ed. New York: Norton.

Frankl, V. E. 1959. *Man's search for meaning*. New York: Beacon Press.

Friedman, H. S., ed. 1990. *Personality and disease*. New York: Wiley.

Friedman, M. 1991. *Aetiological or causal factors in symptoms of PTSD*. Personal communication, Jerusalem.

Gal, R. 1986. *A portrait of the Israeli soldier*. Westport, Conn.: Greenwood Press.

Gergen, K. J., and M. M. Gergen. 1986. Narrative form and the construction of psychological science. In *Narrative psychology: The storied nature of human conduct*, ed. T. R. Sarbin, 22–44. New York: Praeger.

Gutmann, D. L. 1975. Parenthood: a key to the comparative study of the life cycle. In ed. N. Datan and H. L. Ginsberg. *Life-span developmental psychology: Normative life crises*, 167–84. New York: Academic Press.

Ha-Meiri, Y. 1966. *Israeli captives in Syria* (in Hebrew). Tel Aviv: Am Oved.

Hunter, E. J. 1976. The prisoner of war: Coping with the stress of isolation. In *Human adaptation: Coping with life crises*, ed. R. Moos, 322–31. Lexington, Mass: D. C. Heath.

———. 1978. The Vietnam POW veteran: Immediate and long-term effects of captivity. In *Stress disorders among Vietnam veterans: Theory, research, and treatment implications*, ed. C. R. Figley, 188–206. New York: Brunner/Mazel.

———. 1982. *Families under the flag: A review of military family literature*. New York: Praeger.

———. 1983. Captivity: The family in waiting. In *Stress and the family: Coping with catastrophe*, ed. C. R. Figley and H. McCubbin, vol. 2, 166–84. New York: Brunner/Mazel.

———. 1986. Families of prisoners held in Vietnam: A seven-year study. *Evaluation and Program Planning* 9: 243–51.

———. 1991. Prisoners of war: Readjustment and rehabilitation. In *Handbook of Military Psychology,* ed. R. Gal and A. D. Mangelsdorff, 741–57. New York: Wiley.

Jung, C. G. 1933. *Modern man in search of a soul.* New York: Harcourt, Brace and World.

———. 1959. *The archetypes and the collective unconscious.* 2d ed. London: Routledge and Kegan Paul.

Keenan, B. 1992. *An evil cradling.* London: Hutchinson.

Kelley, H. H. 1952. The two functions of reference groups. In *Readings in social psychology,* 2d ed., ed. T. M. Newcomb, and E. L. Hartley, 410–14. New York: Holt.

Kfir, I. 1974. *Captive in Damascus* (in Hebrew). Tel Aviv: E.L. Special Edition.

Kobasa, S. C. 1979. Stressful life events, personality, and health: An inquiry into hardiness. *Journal of Personality and Social Psychology* 37: 1–11.

Lewin, K. 1948. *Resolving social conflicts.* New York: Harper.

Lieblich, A. 1978. *Tin soldiers on Jerusalem Beach.* New York: Pantheon.

———. 1981. *Kibbutz Makom.* New York: Pantheon.

———. 1983. Between strength and toughness. In *Stress in Israel,* ed. S. Breznitz, 39–64. New York: Van Nostrand Reinhold.

———. 1989. *Transition to adulthood during military service: The Israeli case.* Albany: State University of New York Press.

Lifton, R. J. 1973. *Home from the war.* New York: Simon and Schuster.

Lippitt, R., and R. K. White. 1943. The social climate of children's groups. In *Child behavior and development,* ed. R. G. Barker, J. S. Kounin, and H. F. Wright, 485–508. New York: McGraw-Hill.

Loftus, E. 1980. *Memory.* New York: Addison-Wesley.

Maddi, S. R. 1990. Issues and interventions in stress mastery. In *Personality and disease,* ed. H. S. Friedman, 121–54. New York: Wiley.

Messick, D. M., and D. M. Mackie. 1989. Intergroup relations. *Annual Review of Psychology* 40: 45–81.

Miller, W. N. 1974. *Captive reactions initially and through the first 12–24 hours.* Unpublished report. San Diego: Center for POW Studies, Naval Health Research Center.

Nefzger, M. 1970. Follow-up studies of World War II and Korean prisoners of war. *American Journal of Epidemiology* 91: 123–38.

Neugarten, B. L., ed. 1964. *Personality in middle and late life.* New York: Atherton Press.

———. 1968. *Middle age and aging: A reader in social psychology.* Chicago: University of Chicago Press.

Newcomb, T. M. 1943. *Personality and social change.* New York: Dryden.

Parsons, T., and R. F. Bales, eds. 1955. *Family, socialization, and interaction process.* Glencoe, Ill.: Free Press.

Polster, I. 1987. *Every person's life is worth a novel.* New York: W. W. Norton.

Risner, R. 1973. *The passing of the night: My seven years as a prisoner of the North Vietnamese.* New York: Random House.

Ross, B. M. 1991. *Remembering the personal past.* New York: Oxford University Press.

Rutledge, H., E. J. Hunter, and B. Dahl. 1979. Human values and the prisoner of war. *Environment and Behavior* 11: 227–44.

Segal, J. 1974. *Long-term psychological and physical effects of the POW experience: A review of the literature* (Technical Report No. 74-2). San Diego: Center for POW Studies, Naval Health Research Center.

Segal, J., E. J. Hunter, and Z. Segal. 1976. Universal consequences of captivity: Stress reactions among divergent populations of prisoners of war and their families. *International Social Science Journal* 28: 593–609.

Simonton, O. C., S. M. Simonton, and J. L. Creighton. 1980. *Getting well again.* New York: Bantam Books.

Spence, D. P. 1982. *Narrative truth and historical truth: Meaning and interpretation in psychoanalysis.* New York: Norton.

———. 1986. Narrative smoothing and clinical wisdom. In *Narrative psychology: The storied nature of human conduct,* ed. T. R. Sarbin, 211–32. New York: Praeger.

Steptoe, A. 1989. The significance of personal control in health and disease. In *Stress, personal control, and health,* ed. A. Steptoe and A. Appels, 309–18. New York: Wiley.

Steptoe, A., and A. Appels, eds. 1989. *Stress, personal control, and health.* New York: Wiley.

Tajfel, H. 1982. Social psychology of intergroup relations. *Annual Review of Psychology* 33: 1–39.

Tolkien, J. R. R. 1977. *The Hobbit* (in Hebrew). Jerusalem: Zmora, Bitan, Modan.

Veterans Administration. 1980. *POW: Study of former prisoners of war.* Washington, D.C.: Office of Planning and Program Evaluation.